T0314477

Signals, Systems, and Signal Processing

This innovative introduction to the foundations of signals, systems, and transforms emphasizes discrete-time concepts, smoothing the transition towards more advanced study in Digital Signal Processing (DSP). A digital-first approach, introducing discrete-time concepts from the beginning, equips students with a firm theoretical foundation in signals and systems, while emphasising topics fundamental to understanding DSP. Continuous-time approaches are introduced in later chapters, providing students with a well-rounded understanding that maintains a strong digital emphasis. Real-world applications, including music signals, signal denoising systems, and digital communication systems, are introduced to encourage student motivation. Early introduction of core concepts in digital filtering, and discrete-time Fourier transforms, provides a smooth transition through to more advanced study. Over 325 end-of-chapter problems, including over 50 computational problems using MATLAB. Accompanied online by solutions and code for instructors, this rigorous textbook is ideal for undergraduate students in electrical engineering studying an introductory course in signals, systems, and signal processing.

P. P. Vaidyanathan is the Kiyo and Eiko Tomiyasu Professor of Electrical Engineering at the California Institute of Technology. A Life Fellow of IEEE, he has been awarded the IEEE Gustav Robert Kirchhoff Award, the IEEE Signal Processing Society's Shannon–Nyquist Technical Achievement Award, the IEEE Signal Processing Society's Carl Friedrich Gauss Education Award, the Norbert Wiener Society Award, and the EURASIP Athanasios Papoulis Award. He is a recipient of the IEEE Jack Kilby Signal Processing Medal, a Foreign Fellow of the Indian National Academy of Engineering, and a member of the U.S. National Academy of Engineering. He is a recipient of multiple awards for teaching excellence at the California Institute of Technology.

"This introductory textbook on discrete-time signals and systems, written in a lucid and engaging style by a subject-matter expert with in-depth understanding of the topic's foundations, is an invaluable resource for students from a variety of backgrounds."

Ali Sayed, EPFL

"This book introduces the world of digital signal processing via the shortest path through the fundamentals of signals and systems. Without sacrificing mathematical rigor or completeness, Vaidyanathan prudently builds a firm theoretical foundation while paving the way for advanced study. Students from a variety of disciplines will benefit from this entry-point to the tools used to process the ever growing amount of data and information."

Bryan Van Scoy, Miami University

"As a reference for an important foundation course, Professor Vaidyanathan has seamlessly integrated theoretical treatise with important examples drawn from practical discipline. Concepts and topics relevant to R&D are a welcome addition. This book is a valuable accompaniment for a first course in signals and systems and yields excellent prerequisite material for more advanced subjects such as control systems, communications, and DSP."

Charan Litchfield, West Virginia University Institute of Technology

"I like the fact that the book first briefly touches upon many important concepts before delving into their theoretical details. Combined with the provided real-world applications, this allows an engineer to get quickly acquainted with some important signal processing tools. The multitude of illustrative examples, hands-on practice problems, and computer exercises make it also a very valuable textbook for undergraduate students."

Geert Leus, Delft University

"In *Signals, Systems, and Signal Processing*, Professor Vaidyanathan has masterfully crafted a self-contained contemporary textbook. The book covers the fundamentals of signals and systems, complemented with essential elements and techniques of digital signal processing. With the ubiquity of digital signal processing in all areas of modern science and engineering, this up-to-date material is suited for a one- or two-term course in multiple disciplines."

Mostafa Kaveh, University of Minnesota

"I am thrilled to endorse *Signals, Systems, and Signal Processing*, 1st edition, by Professor Vaidyanathan. This book excels in its comprehensive coverage of topics like the Fourier series, digital filters, and filter structures, offering clear explanations and practical applications that bridge theory to real-world contexts, including music and machine learning. It masterfully simplifies complex concepts, making it an exceptional resource for comprehensively understanding the subject."

Peter Taiwo, Morgan State University

"An excellent textbook for those who seek to enter the world of signal processing and beyond, eloquently written by a world-renowned scholar."

Ray Liu, University of Maryland

Signals, Systems, and Signal Processing

P. P. Vaidyanathan
California Institute of Technology

Shaftesbury Road, Cambridge CB2 8EA, United Kingdom

One Liberty Plaza, 20th Floor, New York, NY 10006, USA

477 Williamstown Road, Port Melbourne, VIC 3207, Australia

314–321, 3rd Floor, Plot 3, Splendor Forum, Jasola District Centre,
New Delhi – 110025, India

103 Penang Road, #05–06/07, Visioncrest Commercial, Singapore 238467

Cambridge University Press is part of Cambridge University Press & Assessment,
a department of the University of Cambridge.

We share the University's mission to contribute to society through the pursuit of
education, learning and research at the highest international levels of excellence.

www.cambridge.org
Information on this title: www.cambridge.org/highereducation/isbn/9781009412292

DOI: 10.1017/9781009412285

© P. P. Vaidyanathan 2024

This publication is in copyright. Subject to statutory exception and to the provisions
of relevant collective licensing agreements, no reproduction of any part may take
place without the written permission of Cambridge University Press & Assessment.

When citing this work, please include a reference to the DOI 10.1017/9781009412285

First published 2024

Printed in the United Kingdom by CPI Group Ltd, Croydon CR0 4YY

A catalogue record for this publication is available from the British Library

A Cataloging-in-Publication data record for this book is available from the Library of Congress

ISBN 978-1-009-41229-2 Hardback

Additional resources for this publication at www.cambridge.org/Vaidyanathan

Cambridge University Press & Assessment has no responsibility for the persistence
or accuracy of URLs for external or third-party internet websites referred to in this
publication and does not guarantee that any content on such websites is, or will
remain, accurate or appropriate.

Dedicated to

To my family and to the great minds that made the impossible possible

Contents

Preface

In this age of data and information there is a need to teach signal processing to students with more diverse backgrounds than electrical engineering (EE). In a typical EE curriculum, there is an introductory class on signals and systems (SS), followed by a class introducing digital signal processing (DSP), after which more advanced classes and electives follow. These introductory SS classes themselves are typically two quarters long, or even longer. But for students with diverse backgrounds it is desirable to move on to DSP quickly after getting a solid background on SS. Furthermore, having a focus on discrete time right from the beginning, with continuous-time ideas introduced as and when needed, is desirable.

This book will serve such a purpose. The focus of the book is discrete time, although continuous time is included for completeness, and to the extent necessary. First the fundamentals of SS are introduced. Next, DSP topics such as digital filtering come quickly at a basic level. Besides sampling theorems (typically covered in SS classes), this book also introduces digital filtering, the discrete Fourier transform (DFT), and the fast Fourier transform (FFT) (typically covered in DSP classes). While the discrete-time Fourier transform comes early on for the study of digital filters, the continuous-time Fourier transform is introduced much later, just before the study of sampling. Fourier series (usually seen in physics classes) are discussed for the sake of completeness, but much later, as this is not really a prerequisite for most SS and DSP topics.

This book will also serve students on a traditional EE track because depth, detail, and thoroughness are not compromised. In addition, there are plenty of optional or supplementary sections and chapters (marked with an asterisk) which the course leader can choose either to teach or to assign for reading, thereby achieving greater coverage and depth. Examples include deeper discussions on bandlimited functions, eigenfunctions of the Fourier transform, and decay rates of the Fourier transform. The student will even find brief introductions to compressive sensing, sparse recovery, and wavelets in these optional portions, although these are not usually mentioned in introductory classes on SS. This will help the curious student to get a flavor for what there is beyond the traditional classroom curriculum. The connection between Fourier series and deep neural networks is also pointed out in an optional section.

The connection between real-world situations and theoretical developments is emphasized whenever possible with the help of examples.

Homework Problems and Computing Assignments

Every chapter has many worked examples to assist in the understanding of basic as well as advanced concepts. In addition, there are about 340 homework problems in the book. These range from simple practice problems to those that require more advanced thinking. Computing assignments are often included towards the end of the problems – these can be done with the help of MATLAB or other platforms, according to the student's choice. Of the 340 homework problems, more than 50 are computing assignments. Computing assignments, while straightforward, add insight and get the students actively involved in the learning process.

Solutions Manual and MATLAB Codes for Instructors

A Solutions Manual is available from the publisher, for instructors. For the computing assignments, MATLAB codes are also made available by the publisher for instructors, which they can play with and modify to generate more examples.

0.1 Scope and Outline

This is an introductory book, suitable for a first exposure to signals, systems, and signal processing. The emphasis is on fundamentals, so that the student can get a solid theoretical foundation. The emphasis is not on any of the myriad applications which have been made possible because of DSP, but the book prepares the student for this by giving a solid foundation on fundamentals. To help the student make connections to the practical side, many real-world applications and examples are mentioned throughout.

There are many wonderful books on SS and DSP at the introductory and advanced levels. As mentioned above, this book takes a fresh approach to SS, so that DSP can be quickly introduced. Students in disciplines other than EE often prefer this and it is similar in spirit to McClellan et al. [2003], although the approach here is quite different.

This book covers all the standard introductory topics, at a comfortable pace in the first ten chapters. The book also provides rigor and depth through advanced optional sections and supplementary chapters (usually marked by an asterisk) for the benefit of students and readers who seek more depth and breadth. The last four chapters are supplementary chapters. They serve as additional material when more time is available in the classroom, or as reading assignments for students who wish to learn more.

Thus, while we do not overwhelm the student with mathematical details in the main chapters, these separate chapters and additional sections benefit the curious student who wants to go deeper.

Chapter 1 gives an overview of signals, systems, and their real-life applications. Chapter 2 introduces different types of signals, and studies the properties of many kinds of systems that are encountered in signal processing. This includes linear systems, stable systems, time-invariant systems, and so on. Both continuous and discrete-time cases are discussed. Examples are presented throughout, to demonstrate the concepts.

In Chapter 3 we look into discrete-time linear time-invariant (LTI) systems in greater detail. The discrete-time Fourier transform (DTFT) arises naturally in this context. The properties of the DTFT are therefore discussed in detail along with several illuminating examples. The z-transform is also briefly introduced in this chapter. We then discuss the continuous-time counterparts of these ideas. The chapter also introduces the idea of filtering and defines several types of filters (lowpass, highpass, and so forth). Many examples of these concepts are presented throughout. Connections to the real world are pointed out, such as the filtering effect of the human ear and the human visual system. A simple image filtering example is also presented. Chapters 2 and 3 lay the foundation for many of the remaining chapters.

Chapter 4 introduces digital filters, which are LTI systems with selective frequency responses. These filters can be represented in terms of time-domain difference equations, or by transfer functions and frequency responses in the transform domain. They can also be represented by computational graphs or structures. The chapter begins by introducing structures and structural interconnections for LTI systems. Several simple examples of digital filters are then presented. It is also shown how to convert lowpass filters into other types such as highpass, bandpass, and so on, by use of simple transformations. We then introduce linear-phase digital filters, which are important in image processing applications. Image filtering is explained in a number of sections (e.g., Sec. 3.7 and Sec. 4.6.3), and applications in noise removal (denoising) are demonstrated as well (Sec. 4.7).

In Chapter 5 we introduce *recursive* difference equations, which are fundamental to digital filtering, and more generally to the theory of realizable LTI systems. Recursive difference equations have some beautiful properties which make them very useful in digital signal processing. For example, the so-called IIR digital filters (Chapter 7) are based on such difference equations, and are widely used.

Chapter 6 discusses in greater detail the z-transform introduced in Chapter 3. The inverse z-transformation is discussed in detail. Well-known topics such as partial fractions and their role in z-transform inversion are included, and stability of rational LTI systems is discussed in depth. In an optional section at the end of this chapter (marked by an asterisk), we make a connection to the theory of analytic functions from complex variable theory, which will be of interest to students who seek greater depth on this topic.

In Chapter 7, more sophisticated types of digital filters such as notch and antinotch filters, and sharp-cutoff lowpass filters such as *Butterworth* filters are presented. We explain how continuous-time filters can be transformed into discrete time by using appropriate mappings such as the *bilinear transformation*. We give a comparative discussion of finite and infinite impulse response filters (FIR and IIR filters). A simple method to design FIR filters, called the window method, is discussed. We also study *allpass* filters and discuss some of their amazing applications. A detailed discussion of steady-state and transient behavior of filters is included towards the end. Detailed books on DSP usually contain multiple chapters on digital filters, and there are many books dedicated to digital filters alone. Since this is an introductory book on signals, systems, and signal processing, our purpose in this chapter is to give an introduction to a few methods and examples, rather than a detailed exposure to all well-known methods. Sections marked with an asterisk can be used as references for additional reading and deeper understanding.

Another family of applications of recursive difference equations is introduced in Chapter 8. Difference equations where the initial conditions are nonzero are considered. The input itself may or may not be zero, and applications of both types are demonstrated. This chapter also studies homogeneous difference equations (i.e., equations with zero input and possibly nonzero initial conditions), and the types of nonzero outputs that such systems can produce.

Chapter 9 introduces the continuous-time Fourier transform (CTFT). There are some similarities to the DTFT discussed in Chapter 3, but there are also differences. The first four sections of this chapter will be crucial when discussing *sampling theory*, which deals with the conversion of continuous-time signals to discrete time (Chapter 10). The remaining sections of this chapter, which go deeper into CTFT, are important in their own right, although not "required" for later parts of the book. The student interested in a thorough understanding of the CTFT should certainly read the entire chapter. In Sec. 9.12 we explain how Mother Nature "computes" the Fourier transform automatically in some situations, without any fancy equipment whatsoever. One may or may not love the Fourier transform, but Mother Nature certainly does!

Chapter 10 discusses one of the most important topics of signal processing, namely *sampling*. In order to process a continuous-time signal $x(t)$ digitally, it is first sampled uniformly in time and these samples $x(nT)$ are then digitized, that is, converted into binary form with a finite number of bits. At first it appears that the sampling process results in loss of information, because the values of $x(t)$ between samples are lost. But if the signal $x(t)$ has a property called the *bandlimited* property, then it is actually possible to reconstruct it perfectly for all t from the samples $x(nT)$, as long as T is smaller than a certain threshold. This chapter discusses bandlimited signals, sampling theory, and the method of reconstruction from samples. A detailed discussion of the aliasing phenomenon, which arises due to sampling, is also given. The chapter concludes with some remarks on generalizations of the sampling theorem to non-bandlimited signals.

Chapter 11 discusses yet another type of Fourier representation, called the Fourier series. This is for continuous-time signals which are either periodic or have a finite duration. The purpose of including the chapter here is for reference value and completeness, although the rest of the book does not build on this chapter. The chapter does help to make a connection to many other important ideas. For example, we include a whole section (Sec. 11.8) on music signals, musical scales, and so on. The connection between Fourier series, music, and the science of hearing is fascinating indeed. A brief introduction to wavelets is given at the end of this chapter, for the interested reader. The main purpose of this is to highlight the fundamental differences between the Fourier basis and the wavelet basis in the representation of signals. In Sec. 11.9 we briefly mention the connection between Fourier series approximations and the more general function approximation results in machine learning, which use deep neural networks.

Chapter 12 introduces a fourth type of Fourier representation, namely the DFT, of great importance in DSP and data sciences. It is the only type of Fourier transform which is "computable" on a digital machine. The chapter also discusses the FFT, which is a very famous algorithm introduced in the early days of signal processing by Cooley and Tukey [1965]. Section 12.7 summarizes the connection between the four types of Fourier representations introduced in the book. Understanding these connections is important and insightful, as it helps to obtain the big picture.

Chapter 13 gives a brief introduction to the Laplace transform, which is a generalization of the CTFT of Chapter 9. The chapter is brief because of its conceptual similarity to z-transforms, discussed in detail in earlier chapters.

Chapter 14 introduces *state-space descriptions* for computational graphs (or structures) representing discrete-time LTI systems. State-space descriptions are fundamental. They are not only useful in theoretical analysis, but also can be used to derive many structures for a transfer function $H(z)$, starting from a known structure. In addition, state-space descriptions give a different perspective on stability, and on minimality of the implementation, as we shall see. Although not every introductory curriculum includes state-space descriptions, we have included them here, giving an option to the student for self-study.

Supplementary Chapters

Chapter 15 shows that bandlimited signals are not only very important for the engineer and the signal processing person, but also appealing from a theoretical viewpoint. Bandlimited subspaces and bandlimited L_2 signals are discussed. We show that the integral and the energy of a bandlimited signal can be obtained exactly from samples if the sampling rate is high enough. A number of less obvious consequences of these results are also discussed. We discuss the fact that bandlimited signals have a fascinating connection to analytic functions that arise in the theory of complex variables, and the practical relevance of this is pointed out. The student intent upon advanced study should certainly study this chapter.

Chapter 16 provides a brief introduction to sampling based on sparsity. In the history of sampling, this is a recent development, and has led to major advances – such as compressive sensing and sparse recovery. This chapter is introductory, and gives a brief overview.

Chapter 17 presents some mathematical details relating to the Fourier transform, Fourier series, and their inverses. These details were omitted in the preceding chapters in order to enable the reader to focus on the engineering aspects. However, students with a deeper mathematical inclination often ask questions about these details. The purpose of this chapter is to satisfy the curiosity of such students. For example, the chapter discusses the difference between the ℓ_1-FT and the ℓ_2-FT. A discussion on the pointwise convergence of the Fourier series representation is also given, followed by illuminating examples. Precise mathematical conditions for the existence of the inverse Fourier transform are also reviewed.

Finally, Chapter 18 is a brief review of matrices and vectors. Even though the book makes only light use of matrices, there are some sections in some chapters which benefit from such a review.

On the Ordering of the Topics

Even though our focus on DSP led us to introduce various Fourier representations in a certain order, it is more common in books to introduce the Fourier series first. One reason for this is that many students already have an exposure to Fourier series from physics courses, and therefore feel comfortable with it. Another is that, historically, the Fourier series was introduced first, and then the other representations evolved. In this book, however, the discrete-time case is introduced first, as is natural when taking a direct path to DSP. We couldn't help it!

0.2 Possible Selection of Topics for Courses

The instructor who uses this book has plenty of choice for selecting topics for a class or a sequence of classes, depending on the total time available. The sections and chapters marked with an asterisk give optional material for deeper discussions or reading assignment. Here are some possibilities in the order of increasing availability of time.

1. Chapters 1 to 6, and Chapters 9 (first four sections) and 10. Remaining material can be chosen at the discretion of the instructor, depending on the time available. This is appropriate for a one-quarter class on signals and systems.
2. Chapters 1 to 6, Chapters 9 and 10, and portions of Chapters 7 and 12. Remaining material can be chosen at the discretion of the instructor, depending on the time available. This is appropriate for a one-semester class on signals, systems, and signal Processing.
3. Chapters 1 to 10 and portions of Chapters 12 to 14. Remaining material can be chosen at the discretion of the instructor, depending on the time available. This is appropriate for a two-quarter class on signals, systems, and signal processing.
4. Chapters 1 to 10 and portions of Chapters 11 to 15. Remaining material can be chosen at the discretion of the instructor, depending on the time available. This is appropriate for a two-semester or a three-quarter class on signals, systems, and signal processing.

These are only rough guidelines because the pace of the class can vary depending on the teacher and the students. We strongly suggest that the instructor also think of other combinations that would work out best under the constraints that are unique to their school. For the convenience of the instructor, we show the approximate chapter dependencies in Fig. 1.

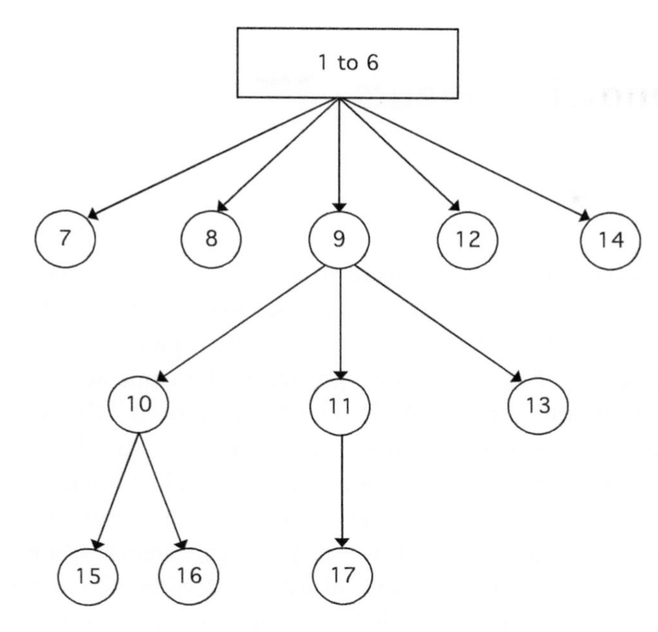

Figure 1 An approximate depiction of chapter dependencies. Note that sections and chapters marked with asterisks in the table of contents are not required for sections without asterisks.

0.3 A Few Words in Conclusion

Every author has a different way of looking at things. The discerning reader will find novelty in the writing style, in the ordering of the topics, and in the daring inclusion of some topics not commonly seen in introductory books. Much has changed since the beginning days of signal processing. Books have come and gone, teaching methods have changed, and students' tastes and needs have changed. However, the fundamentals have held on. While applications are ubiquitous and signal processing has touched every walk of our lives, the fundamentals will always remain important. This author's love is also for these, as this book will show.

Enjoy the journey!

Acknowledgments

As a student, my first exposure to signal processing was the book by Gold and Rader [1969]. Then came Rabiner and Gold [1975], which was an amazing book for its time, with in-depth coverage of many topics. I started to appreciate the academic beauty of signal processing when I later came across the legendary book by Oppenhiem and Schafer [1975]. This "orange book" (book with an orange cover) has been my favorite even after my student days, and I still have a precious copy on my shelf. This book introduced many topics in a fundamental way and included in-depth discussions. It was like reading circuit theory classics like Guillemin or Van Valkenburg, except that it was not circuit theory! My initiation to research in signal processing happened four and a half decades ago under the guidance of that other legend, Professor Sanjit Mitra, who always encouraged creativity and academic beauty more than anything else, and has remained an inspiration and support throughout my career. In my student days, some of the teachers who inspired me deeply were Professors Arun Choudhury, Brian D. O. Anderson, and A. H. Gray Jr. I hope I have been able to impart at least some of their philosophy and wisdom during my four decades of teaching at Caltech. Many other great researchers and educators have influenced my thinking process throughout the decades. For example, the grace and elegance of Professor Tom Kailath's book on linear systems, and his overview of three decades of linear filtering, will ever remain etched in my mind.

It is to all these great minds that I owe this book.

I am deeply thankful to Caltech, where all my teaching took place. Many generations of students and teaching assistants have listened to the lectures and given valuable feedback. The great intellectual atmosphere of the campus, and the presence of wonderful colleagues, has always been an inspiration.

Many thanks to those who have read the manuscript and provided valuable comments. Special thanks to Professor Yuan-Pei Lin (National Chiao Tung University, Taiwan) for very detailed comments, Professor Borching Su (National Taiwan University) and Professor Byung-Jun Yoon (Texas A&M University) for valuable remarks, and Po-Chih Chen and Pranav Kulkarni (Caltech) for help with the Solutions Manual.

Thanks also to Helen Shannon and Elizabeth Horne at Cambridge University Press, who provided crucial help during various stages. Helen provided many important guidelines for revision of the manuscript.

At a personal level, words cannot express my indebtedness to Usha for her constant love, support, dedication, and sacrifices, and to Vikram and Sagar for their love and support. Many were the evenings and weekends that were spent on the book, instead of being with my family. I remain forever thankful for their kind understanding.

Summaries and Tables

There are some summaries and tables in the book, serving as useful references:

1 Introduction and Overview

1.1 Introduction

Signals and systems (SS) and digital signal processing (DSP) have been at the heart of electrical engineering (EE) and data sciences for decades. Today we are surrounded by technology, much of which is influenced by signal processing, from cell phones to laptops, from local area networks to deep space communications.

The idea that signals can be processed using digital computers is many decades old, dating back to the works of Nyquist [1928] and Shannon [1948, 1949], who developed the theory of sampling in addition to other groundbreaking work. In order to do signal processing on a digital device, a signal in continuous time is first sampled and then digitized (i.e., converted into a finite number of bits in binary format), before being processed in some way. The digitally processed signal is then converted back into a continuous-time version, for example, a piece of music which is to be fed into an earphone. Even though its fundamentals can be traced back to 1928, digital processing of signals was not practical until the 1960s, when the **fast Fourier transform** or **FFT** was invented. In those days, when computers were large, slow, and clumsy, the FFT gave new hope. It showed that it is not so impractical after all to compute transforms and spectra using machines, and even perform what is called **digital filtering**, that is, filtering a signal using a digital device instead of electrical circuit elements such as capacitors, resistors, and inductor coils.

The world has indeed come a long way since then, and a great deal of signal processing is going on all around us because we are surrounded by hi-tech devices on all sides. Thus, a lot of signal processing goes on when you are talking to someone on the cell phone, or listening to music on your device, or watching a movie on the internet. In deep space communications, such as between Earth and Mars, or in the detection of gravity waves as in LIGO (Laser Interferometer Gravitational-wave Observatory), there is a significant amount of DSP going on, among other things. When a medical practitioner tries to separate the ECG (electrocardiogram) signal of a pregnant mother from that of the fetus (the yet-to-be-born baby), a process called *interference cancellation* is involved, which is an advanced signal processing technique.

Broadly speaking, DSP includes a wide range of things such as filtering, spectrum computation, image processing, data compression, removal of noise from signals, estimation of hidden parameters from signals (e.g., estimating the direction of arrival of a signal from a base station to a mobile device (cell phone)), estimation of pitch of voice, automatic recognition of speech and speaker from a recording, automatic generation of subtitles for audio recordings, and so on.

This book offers an introductory exposure to signals, systems, and signal processing. Our emphasis is not on any of the myriad applications which have been possible because of DSP, but the book prepares you for that by giving a solid foundation on fundamentals. It also mentions many real-world applications and examples throughout.

1.2 Overview

This section gives a synopsis of the book. The overview given here is not a chapter-wise outline (which can be found in the Preface). Here we only try to give a flavor for the many interesting topics you will learn in this book.

Signals are typically represented using plots of their functional dependence on time, as demonstrated in Fig. 1.1. A **continuous-time** signal (Fig. 1.1(a)) is represented by the notation $x(t)$, where t is a real-valued continuous variable representing time. A **discrete-time** signal (Fig. 1.1(b)) is written as $x[n]$, where the "time" argument n is integer valued (i.e., $n = -1, 0, 1, 2, 3, \ldots$, and so on). Figure 1.2 shows an example of a two-dimensional signal or an "image." Here the signal is written as $x[n_1, n_2]$, where n_1 and n_2 denote the integer coordinate location (n_1, n_2) of the "picture element," also referred to as a **pixel**. In this example the signal value $x[n_1, n_2]$ is the intensity (or gray level) of the pixel. Notations for signals will be described in greater detail in Sec. 2.2 and Sec. 2.3.3.

An important application of signal processing is to filter a signal. Broadly speaking, a **filter** or a **system** processes a signal (the input signal) to produce a new signal (the output signal). There are many types of systems, such as linear systems, nonlinear systems, time-varying systems, time-invariant systems, and so on. These will be

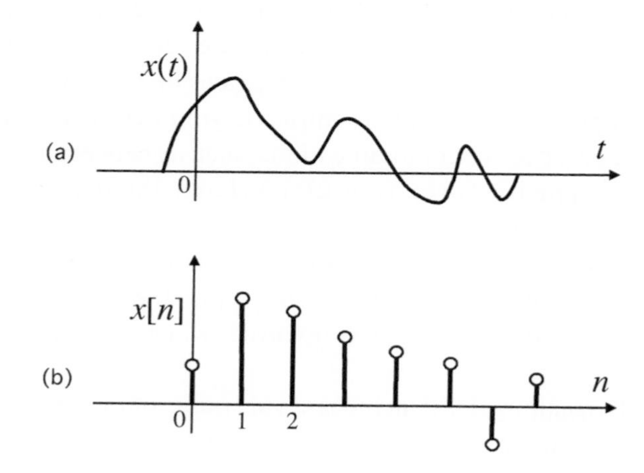

Figure 1.1 (a) A continuous-time signal $x(t)$ and (b) a discrete-time signal $x[n]$.

Figure 1.2 A two-dimensional signal $x[n_1, n_2]$, also called an "image signal."

introduced in Secs. 2.7–2.11, which also present many examples and properties. One famous class of filters is called the **linear time-invariant** (LTI) family of filters. These will be introduced formally in Sec. 2.9 and studied in more depth in Chapter 3.

One application of filtering is noise removal or **denoising**. Figure 1.3 shows the example of a signal $x[n]$, its noisy version $x[n] + e[n]$, and a filtered version $y[n]$, which is quite close to the clean signal $x[n]$. The details of this filtering operation and the detailed labelings in the figure will be explained in Sec. 4.7. Because of the great importance of filtering in science and engineering, two chapters are dedicated to this topic: Chapter 4, which is introductory, and Chapter 7, which presents some detailed **filter design** methods. These chapters introduce different types of digital filters, such as lowpass and highpass filters, allpass filters, notch filters, narrowband filters, and so on. Computational flowgraphs or structures for the implementation of these filters are also presented.

A digital filter is nothing but a filter implemented digitally, that is, on a computer or other digital device. The implementation of filters on a computer is typically done by implementing certain equations called **difference equations**. Digital filters and difference equations will be introduced in Chapter 4, and further generalizations called **recursive** difference equations will be introduced in Chapter 5. Recursive difference equations allow us to implement filters with great computational and storage efficiency, as these chapters will show. Such equations arise in many applications besides filtering. For example, the way a bank calculates your monthly payment on a loan can be described by using a recursive difference equation (Sec. 8.6). Another application is in the design of **waveform generators** (Sec. 8.5.2), which are commonly used in digital communication systems and in radar. A great deal of additional insight can be gained about recursive equations by studying their **state-space descriptions**. Although state-space descriptions are not "essential" for the rest

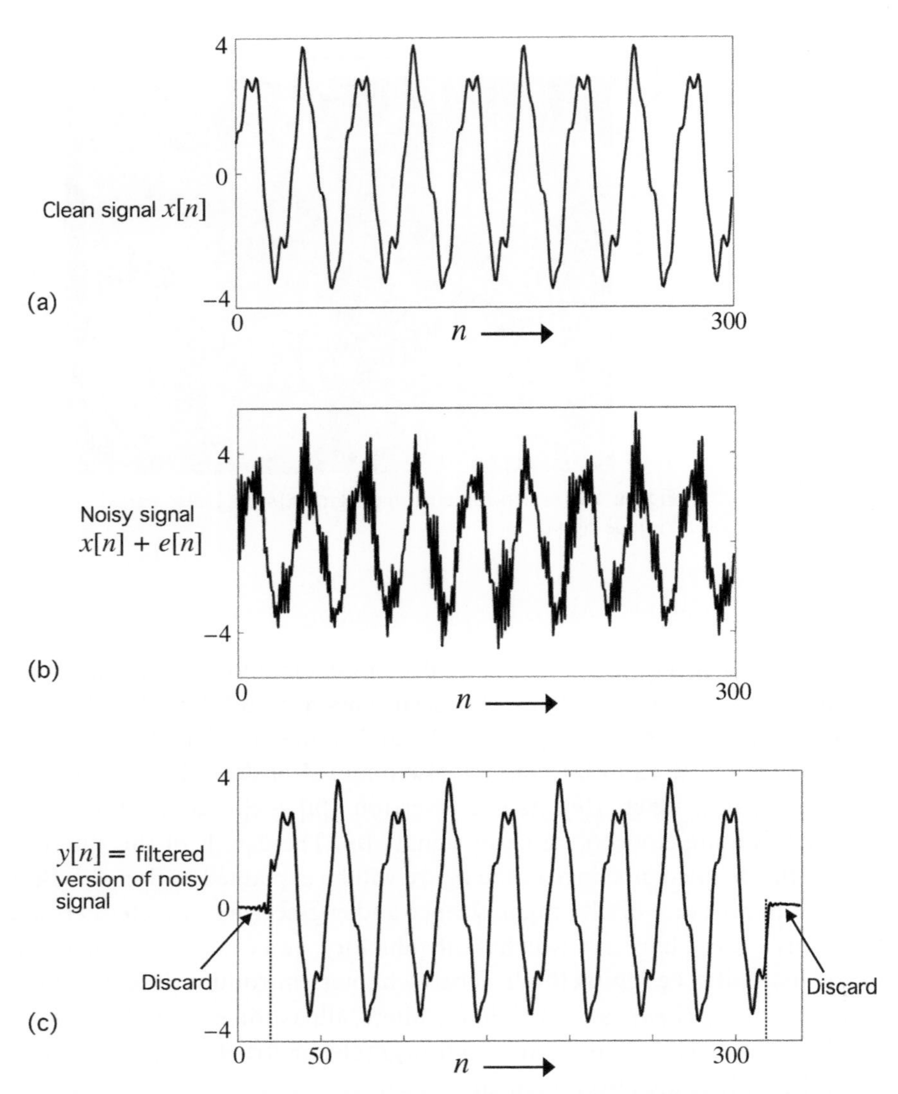

Figure 1.3 (a) A clean signal $x[n]$, (b) the noisy version $x[n] + e[n]$ where $e[n]$ is noise, and (c) the clean version recovered from the noisy version by filtering. The portion marked "discard" will be explained in detail in Sec. 4.7.

of this book, they are discussed in some detail in Chapter 14, to offer additional insights for the interested student.

As mentioned at the beginning of this chapter, in order to do signal processing on a digital device, a signal $x(t)$ in continuous time is first sampled to obtain a discrete-time signal $x[n]$. This is shown in Fig. 1.4, where the samples are indicated as $x[n] = x(nT)$. Sampling will be described in greater detail at the beginning of Chapter 10. In general, sampling a signal can lead to loss of information. But if the signal $x(t)$ satisfies a property called the **bandlimited** property, then there is no loss of information, if the sampling rate is larger than a threshold. This result is called

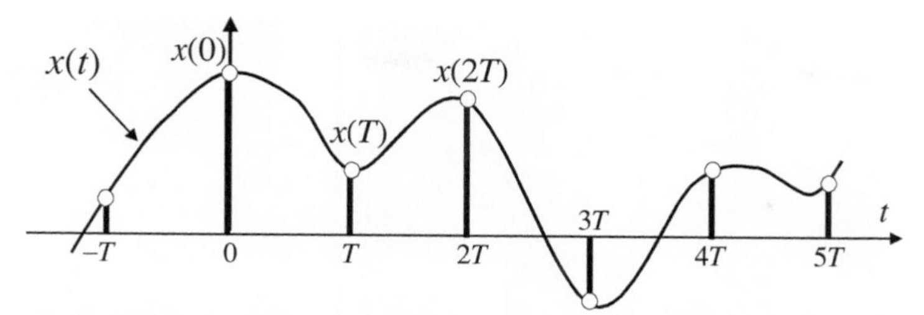

Figure 1.4 A continuous-time signal $x(t)$ and its uniformly spaced samples $x[0] = x(0)$, $x[1] = x(T)$, $x[2] = x(2T)$, and so on.

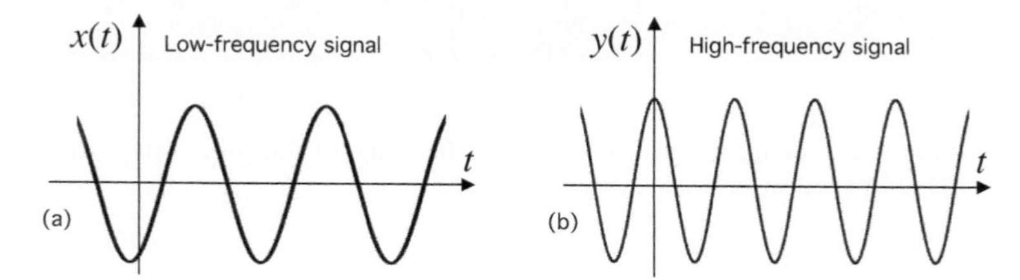

Figure 1.5 (a) A low-frequency signal and (b) a high-frequency signal.

the **Sampling Theorem** (Sec. 10.3.1). The theory of sampling is presented in detail in Chapter 10. The chapter also describes how a sampled signal $x[n]$ can be used to reconstruct the original bandlimited signal $x(t)$ for all t. Sampling, digitization, and reconstruction are at the heart of all digital signal processing systems (Sec. 10.7). More advanced results relating to sampling are discussed in Chapters 15 and 16 for optional reading. In particular, methods for sampling non-bandlimited signals without loss of information are mentioned in Chapter 16; these have created a revolution in signal processing in the last two decades, leading to major topics like sparse reconstruction and compressive sensing.

Central to many signal processing systems is the notion of **frequency**. Qualitatively speaking, a high-frequency signal has rapid variations whereas a low-frequency signal has slow variations. This is demonstrated in Fig. 1.5. As another example, the noisy signal in Fig. 1.3(b) has more rapid variations than the clean signal in Fig. 1.3(a). We say that this noisy signal has higher-frequency components relative to the clean signal. For a real-world example, think of your favorite piece of music. Here, the **bass** components are low frequencies and the **treble** components are high frequencies. The notion of frequency will be introduced more formally in Secs. 2.3.2 and 2.4. The more general term "scale" is also used instead of "frequency" in some contexts (Sec. 11.10).

It turns out that many commonly encountered signals can be decomposed into a sum of component signals such that each component has a single frequency:

$$x(t) = \sum_k \Big(\text{signal with frequency } f_k\Big). \tag{1.1}$$

More precise equations for this will be presented in Sec. 3.5 and other sections such as Sec. 9.4, depending on the type of signals under consideration. This

(a) Original (b) Lowpass filtered

Figure 1.6 Demonstration of lowpass filtering. (a) Original image and (b) lowpass filtered version.

decomposition is called the **Fourier representation**. It leads to a very important topic called the **Fourier transform**. Fourier representations and Fourier transforms will be introduced in Secs. 3.4 and 3.5. These have their origin in the pioneering work of Jean-Baptiste Joseph Fourier (1768–1830), and have had a profound impact in mathematics, engineering, and the sciences. Today we continue to see the impact in signal processing, machine learning, physics, biology, and many other areas. Several sections and chapters in this book are dedicated to the study of different types of Fourier transforms (Sec. 3.8, Chapters 9, 11, and 12). Fourier transforms are especially useful to describe the behaviors of linear time-invariant systems. Such a description is called the **frequency response** description, and it will be introduced in Sec. 3.4.1. A closely related concept, called the **transfer function**, will also be introduced in that section and forms the backbone of all our discussions on LTI systems.

A **lowpass** filter retains the low-frequency components of a signal unchanged, while attenuating high-frequency components. A **highpass** filter does the opposite. Here "low" and "high" are relative terms, and depend on the application. Figure 1.6(a) shows an image signal, and Fig. 1.6(b) shows a lowpass filtered version. Since the high frequencies are attenuated, the image looks very smooth, and sharp edges and boundaries have become vague. The details of this filtering process will be explained in Secs. 3.7 and 4.6.3. You might wonder what the purpose of such filtering is, considering that filtering only removes some information. But don't worry, all this will be explained in Sec. 4.6.3. To put it briefly, we often use filters to decompose signals into lowpass and highpass components so that the frequency components can be processed appropriately and then recombined, to produce a better signal. One example is graphic equalizers in audio systems, where the relative proportions of bass (low frequencies) and treble (high frequencies) can be readjusted according to the listener's preference. Another example is the data compression method called subband coding, which is explained in Sec. 4.6.3.

An important generalization of the discrete-time Fourier transform, called the z-transform, is introduced in Sec. 3.4.1 and discussed in Chapter 3. Its continuous-time counterpart, called the Laplace transform, is introduced in Sec. 3.9 and discussed in Chapter 13. While the z-transform is the backbone for transform-domain methods for discrete-time systems, its origin itself can be traced back to the mathematics of complex variables (analytic functions, Laurent's series, and so on), established a few centuries ago by noted mathematicians. This connection is explained in Sec. 6.9 and adds significant insight.

One specific type of Fourier transform, called the **discrete Fourier transform** or **DFT**, is especially important in digital signal processing. This is the only type of Fourier transform that can be computed on a computer. Oftentimes it can be computed using the FFT. Without the help of the FFT, computation of the Fourier transform of large sets of data would be impractical. As mentioned at the beginning, the digital signal processing revolution of the 1960s was possible mainly because of the introduction of the FFT. Chapter 12 is dedicated to the study of the DFT and the FFT.

In an engineering text like ours, the focus is on the engineering aspects and on real-life impact. We have therefore presented the material in a readily accessible manner for you, with lots of examples, insights, figures, and real-world connections. However, there is deep mathematics behind Fourier representations. After Fourier introduced his ideas, generations of mathematicians have worked on formal justifications of many of the results. There are many books in mathematics dedicated to this topic. Some of you may be curious as to what this mathematics is all about. It concerns formally proving the convergence of infinite sums and integrals in Fourier representations, establishing the conditions for the validity of inverse transforms, and so forth. These subtle points are often taken for granted in engineering texts. In order to give you a flavor for the type of mathematics that underlies Fourier theory, we have included an entire chapter on it at the end of the book (Chapter 17). This can be regarded as a reference for optional reading. You will not be lost if you do not read it, but you may gain insights by reading it. So, please do read it!

The fundamentals that we learn in classes on signals, systems, and signal processing not only allow us to understand many engineering applications, but also enhance our understanding of the world in other ways. In this book we make an effort to point out some of these amazing connections, by including optional sections for reading. One such case was mentioned in the preceding paragraph. Three more such instances will be mentioned here. Firstly, while Fourier transforms have their origin in mathematics, it is a fascinating fact that *Fourier transforms arise in Nature*, and this can even be demonstrated in a laboratory with optical devices! This has to do with the way waves (e.g., radio waves) propagate in free space. This amazing aspect of Mother Nature is explained in Sec. 9.12. Secondly, an understanding of the Fourier representation also creates a profound impact on our understanding and appreciation of **music**. For example, the reason why the notes C, E, and G played together (or in succession) create a pleasant sensation can be understood based on this. Musical instruments often generate several "harmonics" (Secs. 11.2, 11.8.1) of the frequency of a musical note. It is well known in music circles that certain harmonics (e.g., the seventh harmonic) generate an unpleasant hearing sensation. This can also be explained by using Fourier theory (Sec. 11.8.4.5). We have dedicated an entire section (Sec. 11.8) to the understanding of musical scales in the light of Fourier theory. Thirdly, Sec. 11.9 provides insight on the connection between Fourier series approximations and the more general function approximation results in **machine learning**, which use deep neural networks.

It is true that technology changes fast and often reorganizes our lives completely. But the engineering and mathematical principles and tools you gain from classes on signals, systems and signal processing have long-lasting value. They will stay with you for a long time.

1.3 Notations

In this section we summarize notations commonly used in this book.

1. $j = \sqrt{-1}$. (The letter i is more commonly used for this in math courses.)
2. For a complex number a, we have $a = a_r + ja_i$, where a_r and a_i are real quantities, called the real and imaginary parts. In polar form, $a = |a|e^{j\phi}$, where $|a| \geq 0$ is the magnitude or absolute value and ϕ is the phase. We have $|a|^2 = a_r^2 + a_i^2$ and $\phi = \arg(a) = \tan^{-1}(a_i/a_r)$.
3. $a^* = a_r - ja_i$ is the complex conjugate of a. Thus, $a^* = |a|e^{-j\phi}$.
4. $\forall$ stands for "for all"; $\exists$ stands for "there exists."
5. $t \in \mathcal{T}$ means t is in the set $\mathcal{T}$.
6. If and only if is abbreviated as iff.
7. $\mathcal{A} \Longrightarrow \mathcal{B}$ means "$\mathcal{A}$ implies $\mathcal{B}$."
8. $\mathcal{A} \Longleftrightarrow \mathcal{B}$ means "$\mathcal{A}$ is true if and only if $\mathcal{B}$ is true".
9. (a, b): open interval $(a < t < b)$; $[a, b]$: closed interval $(a \leq t \leq b)$.
10. $\omega \in [-\pi, \pi)$ means $-\pi \leq \omega < \pi$ (semi-open interval).
11. Matrices and vectors are denoted by bold-faced letters (**A**, **b**, etc). See Chapter 18 for details about matrices.
12. $\mathbf{A}^T$ = transpose of **A**; $\mathbf{A}^*$ = conjugate of **A**.
13. $\mathbf{A}^H = (\mathbf{A}^T)^*$ = transpose-conjugate of **A**; read as **A**-Hermitian. Also called Hermitian transpose.
14. For a function of z, $H^*(z)$ means $[H(z)]^*$.
15. The **tilde** notation is defined by $\widetilde{H}(z) = [H(1/z^*)]^*$, or more simply, $H^*(1/z^*)$. $\widetilde{H}(z)$ is called the **paraconjugate** of $H(z)$. For example,

$$H(z) = \frac{a_0 + a_1 z^{-1}}{1 + b_1 z^{-1}} \quad \Longrightarrow \quad \widetilde{H}(z) = \frac{a_0^* + a_1^* z}{1 + b_1^* z}. \tag{1.2}$$

That is, replace the coefficients a_k, b_k by conjugates and replace z by $1/z$. Note that $\widetilde{H}(z) = H^*(z)$ when $z = e^{j\omega}$ for some real ω.

16. Acronyms commonly used:
    - DTFT, discrete-time Fourier transform; CTFT, continuous-time Fourier transform.
    - DFT, discrete Fourier transform; FFT, fast Fourier transform.
    - FIR, finite impulse response; IIR, infinite impulse response.
    - BL, bandlimited; BW, bandwidth.
    - Hz, Hertz; sec, seconds; rad, radians.

1.4 Summary of Chapter 1

This chapter gave a brief introduction to signals, systems, and signal processing. Conventions for signal representation were introduced, and the various topics to be discussed in the book were briefly mentioned, along with some examples and figures. A list of notations to be used in the book was also provided. We are now ready for this exciting journey, which will begin in the next chapter.

2 Fundamentals of Signals and Systems

2.1 Introduction

This chapter introduces important classes of signals and systems that arise in many of our discussions. First we introduce the standard notations used to represent signals in figures and equations. We then discuss commonly encountered signal types such as exponentials, sinusoids, pulse signals, and so on, and also mention some real-world signals such as music signals and heartbeat signals. Different types of systems that are encountered in signal processing are then introduced, such as linear systems, stable systems, time-invariant systems and so on. Examples are presented throughout, to demonstrate the concepts.

2.2 Graphical Representation of Signals

Signals are functions that vary with time or space or with some other independent variable. A **continuous-time** signal is represented by the notation $x(t)$, where t is a continuous variable representing time. Figure 2.1(a) shows an example. Note that t is not restricted to be positive, and in general we have

$$-\infty < t < \infty, \tag{2.1}$$

with t measured in seconds (or milliseconds, microseconds, etc.). A **discrete-time** signal is written as $x[n]$, where the "time" argument n is an integer in the range $-\infty < n < \infty$. See Fig. 2.1(b). A discrete-time signal is also known as a **sequence**. Note the use of round brackets for $x(t)$ and square brackets for $x[n]$. This convention will be followed throughout. The discrete-time index n has no dimension (like seconds). It is a dimensionless integer. Note also that $x[n]$ is *undefined* for non-integer n. Note that $x(t)$ represents the *entire signal* as a function of t, whereas for a fixed real number t_0, the notation $x(t_0)$ represents *a number* which equals $x(t)$ when $t = t_0$. Similar remarks hold for the discrete-time case.

In some cases, $x[n]$ is a uniformly sampled version of a continuous-time signal, as in

$$x[n] = x(n\Delta). \tag{2.2}$$

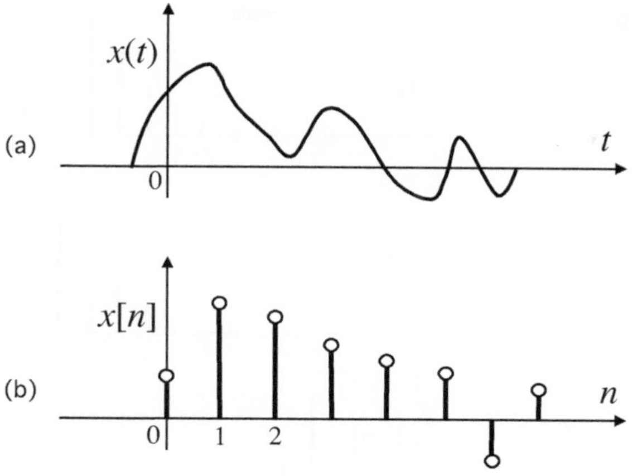

Figure 2.1 (a) A continuous-time signal $x(t)$ and (b) a discrete-time signal $x[n]$. The continuous-time variable t is a real number in $-\infty < t < \infty$ and the discrete-time variable n is an integer in $-\infty < n < \infty$. The circles on the vertical lines representing $x[n]$ are just decorative, and are sometimes left out for convenience. The vertical axis (ordinate) is also sometimes not shown.

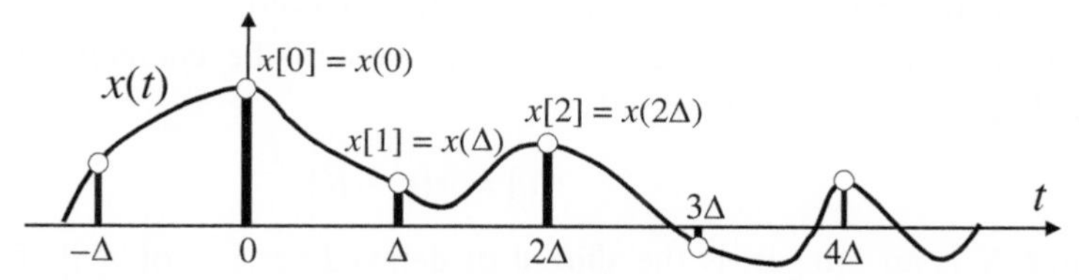

Figure 2.2 A continuous-time signal $x(t)$ and its uniformly sampled version $x[n] = x(n\Delta)$. Note the use of round brackets for $x(t)$ and square brackets for $x[n]$.

This is demonstrated in Fig. 2.2. The constant $\Delta > 0$ is said to be the sample spacing, and $1/\Delta$ the sampling rate (as there are $1/\Delta$ samples per second). The examples given above are one-dimensional (1D) signals. Signals can also be two-dimensional (2D) or higher-dimensional; examples include images and video.

2.2.1 Simple Operations on Signals

Given two signals $x_1[n]$ and $x_2[n]$, we can generate new signals using simple arithmetic operations like addition, subtraction, and multiplication:

$$x[n] = x_1[n] + x_2[n], \quad x[n] = x_1[n] - x_2[n], \quad x[n] = x_1[n]x_2[n], \tag{2.3}$$

and possibly division. In each case, the operation is carried out for each time instant n. The operation $\alpha x[n]$, where α is a constant, means that each sample $x[n]$ is scaled

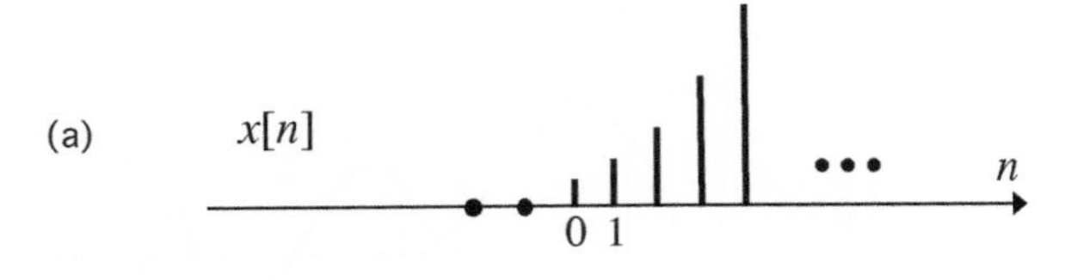

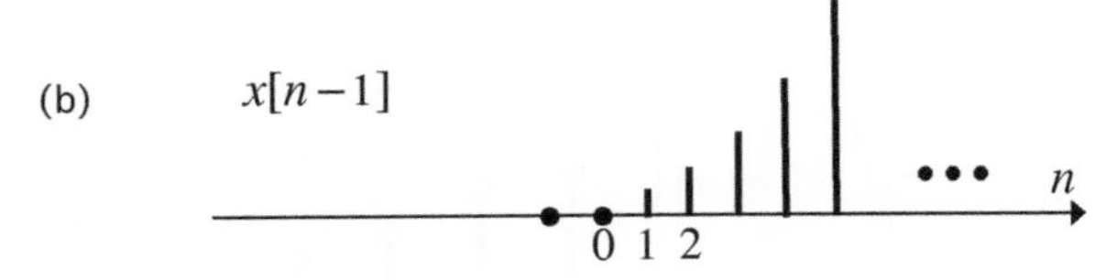

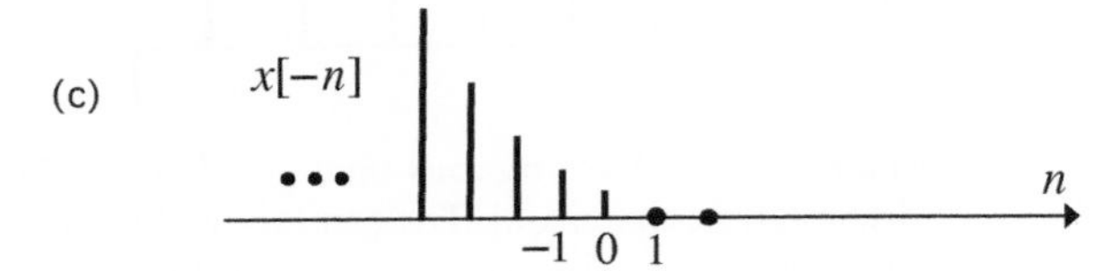

Figure 2.3 (a) A discrete-time signal $x[n]$, (b) its right-shifted version $x[n-1]$, and (c) its time-reversed or flipped version $x[-n]$.

by the same number α. The term **constant** just means that α does not depend on time. We say that

$$y[n] = \alpha x[n] \tag{2.4}$$

is a **scaled** version of $x[n]$. Similar comments hold for the continuous-time case. Note that signals and constants can in general be **complex** quantites, and not restricted to be real. The signal

$$y[n] = x[n-K], \tag{2.5}$$

where K is an integer, is the **shifted** or **delayed** version of $x[n]$. Thus, $x[n]$ is time-shifted to the right by K samples to obtain $y[n]$. Figure 2.3 shows an example for $K = 1$. Note that for negative K, the shift is actually to the left. Finally,

$$y[n] = x[-n] \tag{2.6}$$

is the **time-reversed** or **flipped** version of $x[n]$ as demonstrated in Fig. 2.3. For continuous-time signals, the shifted version is $x(t-\tau)$ and the time-reversed version is $x(-t)$, although the amount of shift τ need not be an integer. More complicated operations with signals, such as convolution and modulation, will be introduced later.

2.3 Examples of Signals

Figure 2.4 shows examples of discrete-time signals that arise frequently in our discussions. Figure 2.4(a) shows the **impulse** or **delta** function:

$$\delta[n] = \begin{cases} 1 & n = 0, \\ 0 & \text{otherwise.} \end{cases} \tag{2.7}$$

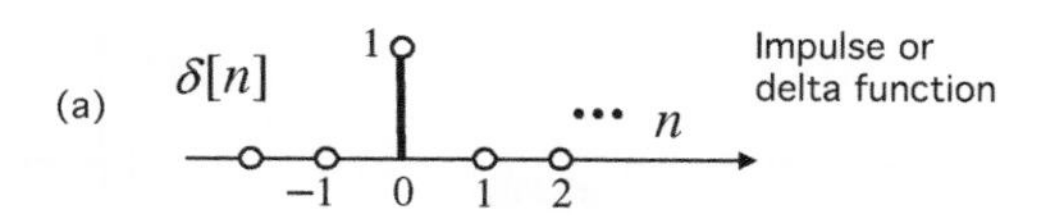

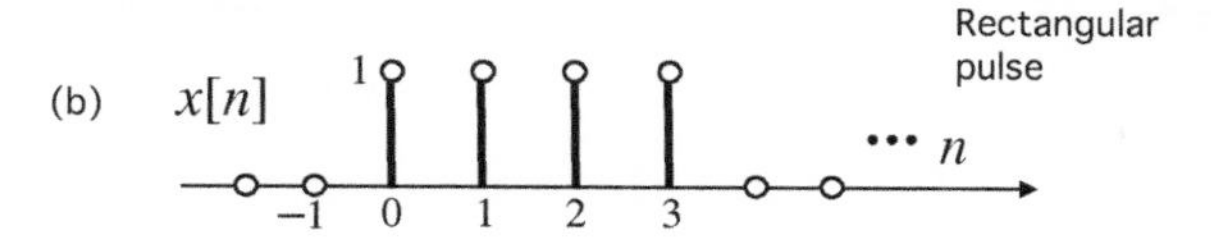

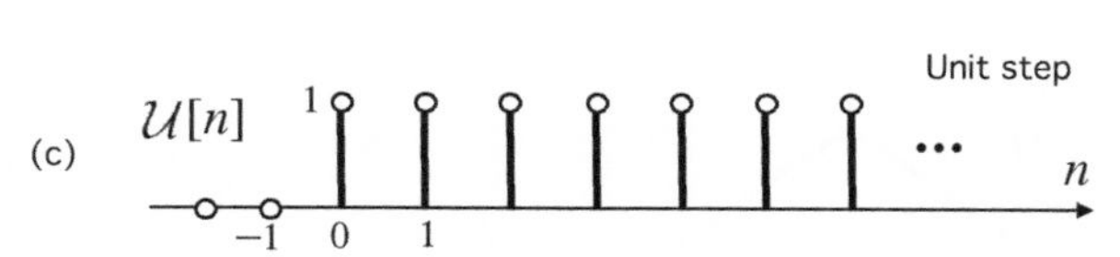

Figure 2.4 Examples of discrete-time signals. (a) The impulse or delta function $\delta[n]$, (b) a rectangular pulse $x[n]$ of duration four samples, and (c) the unit step $\mathcal{U}[n]$.

This is also called the **unit pulse**. Figure 2.4(b) shows a **pulse** $x[n]$ of duration four samples, which is unity for $0 \leq n \leq 3$ and zero otherwise. This is also called a **rectangular** pulse. Figure 2.4(c) shows the **unit-step** function defined by

$$\mathcal{U}[n] = \begin{cases} 1 & n \geq 0, \\ 0 & \text{otherwise.} \end{cases} \tag{2.8}$$

Using the definition of signal shift (Eq. (2.5)), we can express the pulse $x[n]$ as

$$x[n] = \delta[n] + \delta[n-1] + \delta[n-2] + \delta[n-3]. \tag{2.9}$$

Similarly, the unit step and the impulse are clearly related as

$$\mathcal{U}[n] = \delta[n] + \delta[n-1] + \delta[n-2] + \cdots = \sum_{k=0}^{\infty} \delta[n-k], \tag{2.10}$$

or equivalently

$$\mathcal{U}[n] = \sum_{k=-\infty}^{n} \delta[k]. \tag{2.11}$$

It is also clear that we can write

$$\delta[n] = \mathcal{U}[n] - \mathcal{U}[n-1] \tag{2.12}$$

and

$$p_N[n] = \mathcal{U}[n] - \mathcal{U}[n-N] = \begin{cases} 1 & \text{for } 0 \leq n \leq N-1, \\ 0 & \text{otherwise,} \end{cases} \tag{2.13}$$

where $p_N[n]$ is the pulse with duration N samples (e.g., $p_4[n]$ is as in Fig. 2.4(b)). Figure 2.5(a) shows the continuous-time unit-step function. This is defined as

$$\mathcal{U}(t) = \begin{cases} 1 & t > 0, \\ 0 & t < 0. \end{cases} \tag{2.14}$$

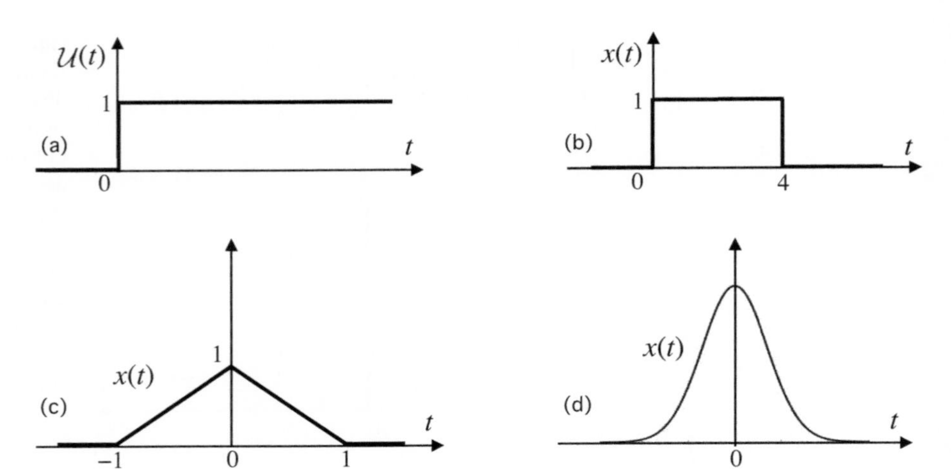

Figure 2.5 Examples of continuous-time signals. (a) The unit step $\mathcal{U}(t)$, (b) a rectangular pulse of duration 4 secs (assuming t is in seconds), (c) a triangular pulse, and (d) a Gaussian signal.

There is a discontinuity at $t = 0$, and we can take it to be undefined at $t = 0$. Figure 2.5(b) shows a continuous-time rectangular pulse of duration 4 secs. This is discontinuous at $t = 0$ and $t = 4$. Figure 2.5(c) shows a **triangular** pulse, and Figure 2.5(d) shows a **Gaussian** signal or pulse which is defined by

$$x(t) = \frac{e^{-t^2/2\sigma^2}}{\sqrt{2\pi\sigma^2}}. \tag{2.15}$$

This is nonzero for all finite t. The reader will recognize this as the famous density function which arises in probability theory. It has the property that $x(t) \geq 0$ and

$$\int_{-\infty}^{\infty} x(t)dt = 1. \tag{2.16}$$

Note that unlike in continuous time, there is no notion of discontinuity in discrete time. Thus, the discrete-time unit step is defined even at $n = 0$.

2.3.1 Even and Odd Signals

A signal is said to be **symmetric** or **even** if $x(t) = x(-t)$ and **antisymmetric** or **odd** if $x(t) = -x(-t)$:

$$x(t) = \begin{cases} x(-t) & \text{(even or symmetric signal)}, \\ -x(-t) & \text{(odd or antisymmetric signal)}. \end{cases} \tag{2.17}$$

For example, the signals in Figs. 2.5(c) and (d) are even, but the other two signals are neither even nor odd. The sine signal $\sin(\omega_0 t)$, shown later in Fig. 2.7(b), is odd. Defining the signals

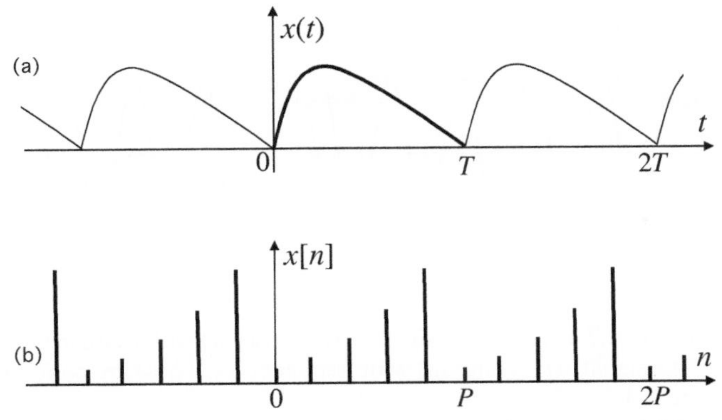

Figure 2.6 Periodic signals. (a) Continuous-time case and (b) discrete-time case.

$$x_e(t) = \frac{x(t) + x(-t)}{2}, \quad x_o(t) = \frac{x(t) - x(-t)}{2}, \tag{2.18}$$

we see that $x_e(t)$ is even and $x_o(t)$ is odd. Since $x(t) = x_e(t) + x_o(t)$, we say that $x_e(t)$ is the **even part** of $x(t)$ and $x_o(t)$ is the **odd part** of $x(t)$. Similar definitions apply in the discrete-time case. The definitions (2.17) and (2.18) are valid whether $x(t)$ is real or complex, although a variation called Hermitian symmetry is more useful for complex functions (Sec. 3.6).

A signal satisfying $x[n] = 0$ for all n is called the **zero signal**. Sometimes we abbreviate it as $x[n] = \mathbf{0}$, where boldface **0** means that all samples are zero. Similarly, the zero signal in continuous time satisfies $x(t) = 0$ for all t.

2.3.2 Periodic Signals

There is an important class of signals called **periodic** signals. These repeat periodically in time, as demonstrated in Fig. 2.6. For the continuous-time case, we can describe this property mathematically as

$$x(t) = x(t + T), \quad \forall t, \tag{2.19}$$

where the notation $\forall t$ means "for all t." Note that such a signal also satisfies $x(t) = x(t + kT)$ for any integer k. The smallest positive T such that Eq. (2.19) holds is called the **period** of $x(t)$, and we say "$x(t)$ is periodic-T." The reciprocal of the period, $1/T$, is sometimes called the **pitch**, especially in the context of speech and music signals. If T is in seconds, then $1/T$ is in **Hertz** (a unit of frequency, see Sec. 2.4.2).

A discrete-time signal $x[n]$ is said to be periodic if

$$x[n] = x[n + P], \quad \forall n, \tag{2.20}$$

for some **integer** P. Notice that in Eq. (2.20) the notation $\forall n$ means "for all *integer* n." The smallest positive integer P for which Eq. (2.20) holds is called the period of $x[n]$. A signal that is not periodic is said to be nonperiodic or *aperiodic*.

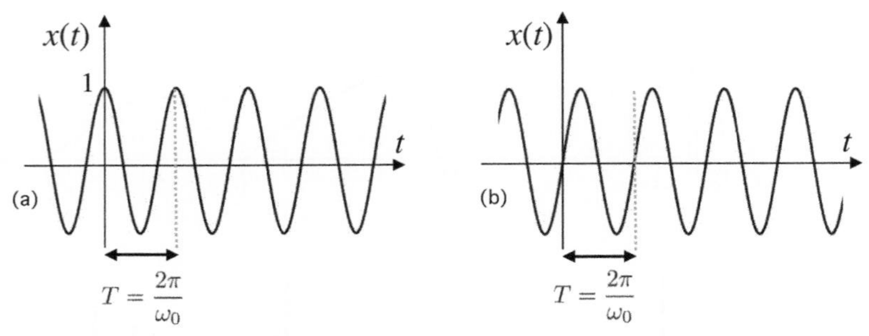

Figure 2.7 (a) A cosine signal $x(t) = \cos(\omega_0 t)$ with frequency ω_0 and (b) a sine signal $x(t) = \sin(\omega_0 t)$ with frequency ω_0. These are periodic signals with period $T = 2\pi/\omega_0$.

Sinusoids. Here is a specific example of a periodic signal:

$$x(t) = \cos(\omega_0 t), \tag{2.21}$$

where ω_0 is a real constant. This signal is shown in Fig. 2.7(a). Since $\cos(\phi + 2\pi) = \cos\phi$, it follows that

$$\cos\left(\omega_0\left(t + \frac{2\pi}{\omega_0}\right)\right) = \cos(\omega_0 t + 2\pi) = \cos(\omega_0 t), \tag{2.22}$$

which shows that $\cos(\omega_0 t)$ repeats periodically with period

$$T = \frac{2\pi}{\omega_0}. \tag{2.23}$$

The period is also indicated in Fig. 2.7(a). The quantity ω_0 is often said to be the frequency of $\cos(\omega_0 t)$, but we shall come to a more precise definition of frequency later. Similarly, the signal $\sin(\omega_0 t)$, also shown in the figure, has period T. More generally, the signal

$$x(t) = A\cos(\omega_0 t - \theta), \tag{2.24}$$

where A and θ are real, is also periodic with period T, and reduces to the cosine or sine when $\theta = 0$ or $\pi/2$, respectively. In engineering practice, Eq. (2.24) is often referred to as a **sinusoid**, regardless of what A and θ are. ▽ ▽ ▽

Real-World Periodic Signals

Periodic signals arise in many scientific and engineering disciplines. In fact, many real-life signals familiar to us are periodic in limited regions.

1. *ECG signals.* Electrocardiogram (ECG or EKG) or heartbeat signals are nearly periodic for normal hearts. Figure 2.8 shows a typical sketch of an ECG signal. The period reflects the periodic beating of the heart. For example, if the heart rate is 60 per minute (one per second), then $T = 1$ sec, and the period is 1 Hz. This is typically four to five times faster than our respiratory (breathing) rate.

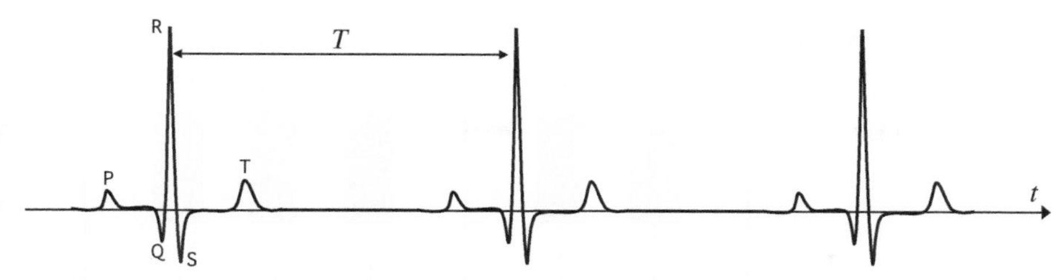

Figure 2.8 A sketch of an electrocardiogram (ECG) signal. This is approximately periodic. The letters P, Q, R, S, T mark the various components of the signal, such as the P-wave, QRS complex, and the T-wave.

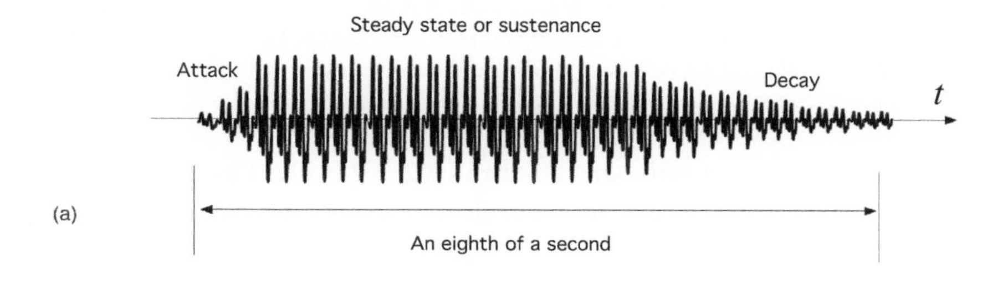

Figure 2.9 (a) Example of a signal representing a musical note and (b) example of a musical phrase (succession of notes).

2. *Music signals.* The notes generated by musical instruments are periodic [Halliday and Resnick, 1978]. Thus, if we record a musical note and plot it, the result $x(t)$ looks very periodic in the middle, as demonstrated in Fig. 2.9(a). The periodic signal grows in amplitude rapidly during the "attack" stage, and remains a steady note for a while, before going through the "decay" phase. Here $x(t)$ can be regarded as the displacement of the ear drum from normal position, as a function of time. Equivalently, it could be the electrical voltage at the output of a microphone picking up the musical signal. Figure 2.9(b) shows a succession of musical notes; here the period or pitch changes from one value to another as the note changes.
3. *Musical notes.* Continuing the example of music, Fig. 2.10 shows one octave of a musical keyboard. There are **twelve** notes in the octave (with $\dot{C}$ denoting C of the next octave), and each note has a specific pitch representing its periodicity. For example, in the middle (or fourth) octave, the pitch of A is nearly 440 Hz and the pitch of C is 264 Hz. Furthermore, the pitches of all the notes in an octave are two times higher than the corresponding pitches in the preceding octave. For

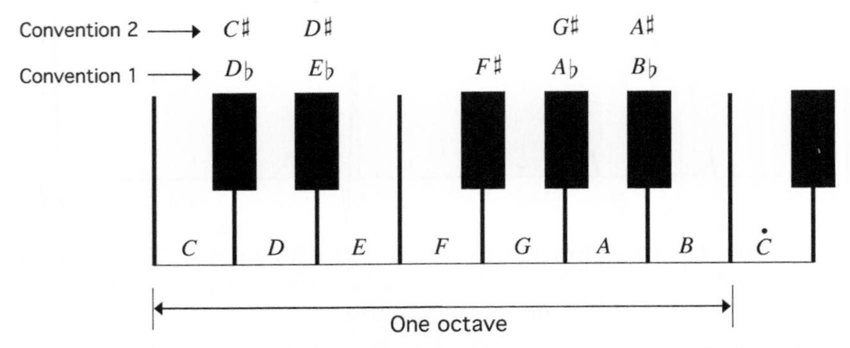

Figure 2.10 One octave of a musical keyboard. The seven notes are indicated on the white keys and the finer distinctions represented by the black keys are also indicated. Notice the two naming conventions for the black keys. For example, $D\flat$ (D-flat) and $C\sharp$ (C-sharp) indicate the same key.

example, C in the fifth octave has pitch 2×264 Hz. Finally, the ratio of pitches of any pair of notes is fixed, regardless of the octave. For example

$$\frac{G}{C} = \frac{3}{2}. \tag{2.25}$$

This is exactly so for the so-called *just-intonation system*, and approximately so for the *equitempered system*. More insight is provided in Sec. 11.8 for the interested reader.

4. *Brain waves.* Brain-wave signals measured by EEG (electroencephalography), such as alpha waves (α waves), beta waves (β waves), and so on, can sometimes look approximately periodic. Although not periodic, they belong to narrow ranges of frequencies. For example, α waves are typically in the 8–12 Hz range, β waves are in the 12–38 Hz range, and so on. Different types of brain waves dominate during different states of consciousness. For example, δ waves (0.5–3 Hz) are thought to dominate in the deep-sleep state.
5. *Signals in electronics.* The AC voltage that comes into our homes is a periodic signal with frequency 60 Hz or 50 Hz (depending on the geographical region). Some signals arising in electronic circuits, such as rectifier signals in power supplies, are also therefore periodic (Sec. 11.5.2). Some signals used in the design of radar systems, such as pulsed radar signals, are periodic as well.

2.3.3 Two-Dimensional Signals

A 2D signal has the form $s(t_1, t_2)$, where t_1 and t_2 are continuous arguments, or $s[n_1, n_2]$, where n_1, n_2 are integer arguments. A 2D image $s[n_1, n_2]$ is an example where n_1 and n_2 represent vertical and horizontal coordinates of picture elements or **pixels**, and s is the intensity of the image.

We can perform operations on 2D signals similar to 1D signals. For example, $s[n_1, -n_2]$ represents the image flipped in one dimension and not in the other. The

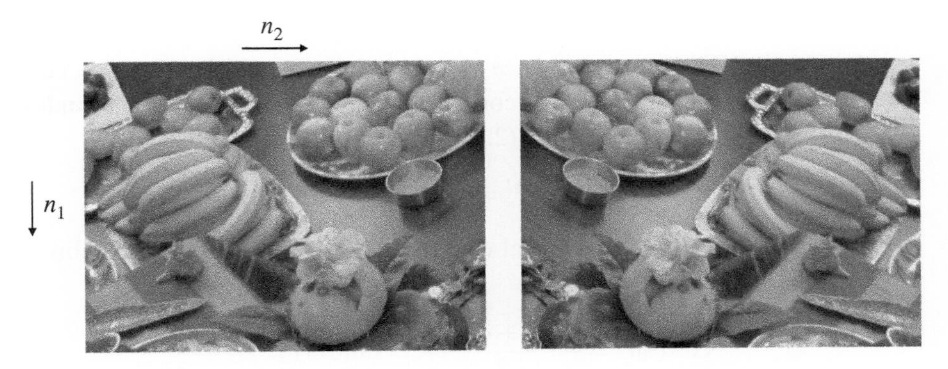

Figure 2.11 A 2D signal (left) and its horizontally flipped version (right).

image on the right in Fig. 2.11 is such a flipped version of the image on the left. As you can imagine, many more operations are possible with 2D signals (rotation, shear, and so forth) than with 1D signals [Castleman, 1996; Gonzalez and Woods, 1993]. A famous example of a 2D signal $s(t_1, t_2)$ is

$$s(x, t) = \cos(\omega t - kx), \tag{2.26}$$

where $t_1 = t$ represents time and $t_2 = x$ is a spatial coordinate (distance). These are called space-time signals. You may have encountered Eq. (2.26) in basic physics. It represents a **propagating wave**, that is, a wave propagating in the x direction, as time t advances.

2.4 The Single-Frequency Signal

The signal

$$x(t) = e^{j\omega_0 t}, \tag{2.27}$$

where ω_0 is real, is said to be a single-frequency signal with frequency ω_0. Here the notation $e^z = \exp(z)$ represents the exponential function. More generally, $x(t) = Ae^{j\omega_0 t}$ is a single-frequency signal with amplitude A and frequency ω_0. Here ω_0 is in the range

$$-\infty < \omega < \infty, \tag{2.28}$$

and the amplitude A can be complex. Recall at this point that $e^{j\omega_0 t}$ can be written as

$$e^{j\omega_0 t} = \cos(\omega_0 t) + j\sin(\omega_0 t), \tag{2.29}$$

where[1]

[1] Note that engineers, especially electrical engineers, use j instead of i to represent $\sqrt{-1}$. This is because i is usually used to represent electric current (as in $i(t)$).

$$j = \sqrt{-1}. \tag{2.30}$$

Equation (2.29) is a consequence of **Euler's formula**, which says $e^{j\theta} = \cos\theta + j\sin\theta$ for any real θ. Thus, (2.27) is a complex signal, and is also called a **complex sinusoid**. Conjugating both sides of Eq. (2.29), we get

$$e^{-j\omega_0 t} = \cos(\omega_0 t) - j\sin(\omega_0 t). \tag{2.31}$$

This can be regarded as a signal with frequency $-\omega_0$. Adding and subtracting Eqs. (2.29) and (2.31), we get

$$\cos(\omega_0 t) = \frac{e^{j\omega_0 t} + e^{-j\omega_0 t}}{2}, \quad \sin(\omega_0 t) = \frac{e^{j\omega_0 t} - e^{-j\omega_0 t}}{2j}. \tag{2.32}$$

A consequence of Eq. (2.32) is that the cosine signal $\cos(\omega_0 t)$ is a superposition of two signals, with frequencies ω_0 and $-\omega_0$. The same is true of $\sin(\omega_0 t)$. Thus, with single-frequency signals defined as in (2.27), sinusoidal signals of the form $A\cos(\omega_0 t - \theta)$ have two frequency components, with frequencies ω_0 and $-\omega_0$.

2.4.1 The Rotating Vector Concept

From Eq. (2.29) we see that a single-frequency signal has a real part $\cos(\omega_0 t)$ and an imaginary part $\sin(\omega_0 t)$. The complex signal (2.29) can be represented as a vector in the complex plane, as shown in Fig. 2.12. In this convention, the vector $e^{j\omega_0 t}$ makes an angle of $\omega_0 t$ with the real axis. So the projection of this vector on the real axis is $\cos(\omega_0 t)$. Similarly, the projection of this vector on the imaginary axis is $\sin(\omega_0 t)$. Since the angle $\omega_0 t$ changes with time t, the vector rotates in the counterclockwise direction as t increases, assuming $\omega_0 > 0$. If $\omega_0 < 0$ then the vector rotates clockwise. So, a single-frequency signal can be regarded as a rotating vector,

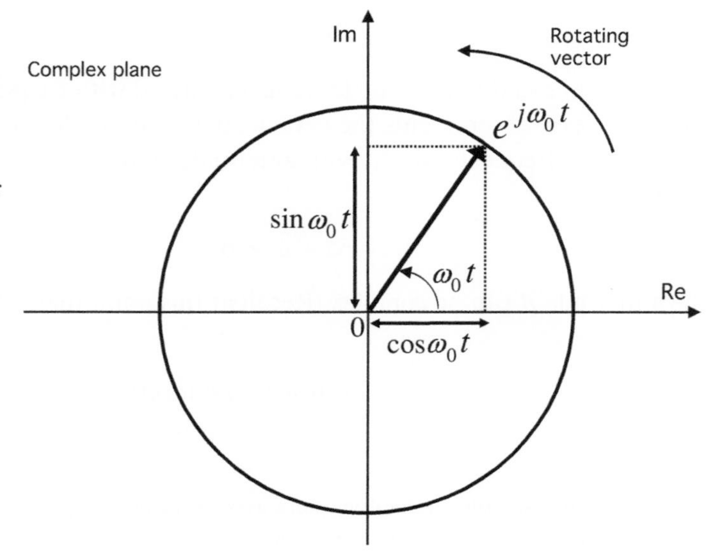

Figure 2.12 The single-frequency signal $e^{j\omega_0 t}$ regarded as a rotating vector. As time progresses, the vector rotates counterclockwise for $\omega_0 > 0$.

and the **frequency can be positive or negative**. The vector rotates counterclockwise for positive frequencies and clockwise for negative frequencies.

2.4.2 Period and Frequency

Note that

$$e^{j\omega_0(t+\frac{2\pi}{\omega_0})} = e^{j(\omega_0 t+2\pi)} = e^{j\omega_0 t}, \tag{2.33}$$

because

$$e^{j2\pi} = 1. \tag{2.34}$$

Equation (2.33) shows that the single-frequency signal $e^{j\omega_0 t}$ is periodic. It can also be shown that there is no smaller repetition interval. That is, if

$$e^{j\omega_0(t+\tau)} = e^{j\omega_0 t} \tag{2.35}$$

for all t for some $\tau > 0$, then τ cannot be smaller than $2\pi/|\omega_0|$. So we conclude that $e^{j\omega_0 t}$ is periodic with period

$$T = \frac{2\pi}{|\omega_0|}. \tag{2.36}$$

One sometimes ignores the fact that ω_0 can be negative, and uses $T = 2\pi/\omega_0$. But strictly speaking (2.36) is the correct expression, since the period by definition has to be positive. Next, consider the quantity

$$f_0 = \frac{1}{T}, \tag{2.37}$$

which is the reciprocal of the period. Thus

$$\omega_0 = 2\pi f_0. \tag{2.38}$$

Since there is one period in a duration of T secs, it follows that there are $1/T$ $(=f_0)$ periods in one second. We say that f_0 is the frequency in Hertz, which means "the signal repeats f_0 times per second."[2] For example, if $T = 0.1$ secs, then

$$f_0 = \frac{1}{T} = 10 \text{ Hz}, \tag{2.39}$$

so that the signal repeats 10 times in a second. In this example the frequency expressed in terms of ω_0 is

$$\omega_0 = \frac{2\pi}{T} = 2\pi f_0 = 2\pi * 10 \quad \text{rads/sec}, \tag{2.40}$$

where rads/sec stands for **radians per second**. Both f_0 and ω_0 are said to be the "frequency" of the signal $x(t) = e^{j\omega_0 t}$. The quantity f_0 represents the number of repetitions per second, and is measured in Hertz. The quantity ω_0 represents the amount of rotation of the vector $e^{j\omega_0 t}$ in the complex plane, in one second. In the

[2] If $T = 2$, we still say that "the signal repeats $1/2$ times in a second." This just means that it takes two seconds to complete one period.

above example, it rotates $2\pi * 10$ radians in one second, that is, the vector makes ten full 360° rotations in one second. This is the same as saying that $x(t)$ has some basic shape which repeats f_0 (= 10) times per second.

For the discrete-time case, the single-frequency signal is defined similarly, that is,

$$x[n] = e^{j\omega_0 n}, \tag{2.41}$$

where ω_0 is the frequency in radians (and *not* rads/sec, since n is dimensionless). The rotating vector concept still holds, but since n is an integer, the vector rotates in steps of ω_0 radians for every increment of n. For this reason, the unit of ω_0 is taken as rads/sample in some books (e.g., [Proakis and Manolakis, 2007]) but we will use "radians" as done in a number of books [Jackson, 1991; Oppenheim and Schafer, 2010; Oppenheim and Willsky with Nawab, 1997]. Since the rotating vector may or may not return to zero radians after an integer number of jumps, the period of $e^{j\omega_0 n}$ requires more careful discussion. We return to this in Sec. 2.6.2.

2.5 The Exponential Signal

A continuous-time signal of the form

$$x(t) = e^{at}, \quad -\infty < t < \infty, \tag{2.42}$$

is said to be an exponential signal. Here a can in general be complex, that is

$$a = \alpha + j\beta, \tag{2.43}$$

where α and β are real; α is the real part of a and β the imaginary part. More generally, $x(t) = ce^{at}$ is also an exponential for any (possibly complex) $c \neq 0$. Exponential signals have great importance in the study of linear time-invariant systems, and arise frequently in science and engineering (see Chapter 3 for more on this). Note that Eq. (2.42) can be rewritten as

$$x(t) = e^{(\alpha+j\beta)t} = e^{\alpha t}e^{j\beta t} = e^{\alpha t}\cos(\beta t) + je^{\alpha t}\sin(\beta t), \tag{2.44}$$

by using Euler's formula. Since $|e^{j\beta t}| = 1$, it is the factor $e^{\alpha t}$ which determines the growth or decay of $|x(t)|$ as $t \to \pm\infty$. For example, if $\alpha > 0$, then $|x(t)| \to 0$ as $t \to -\infty$ and $|x(t)|$ grows without limit as $t \to \infty$. The quantities $\cos(\beta t)$ and $\sin(\beta t)$ are the oscillatory parts which are bounded by unity, but keep oscillating as t changes.

When a has a positive real part as in $x(t) = e^t, e^{2t}, e^{(1+j)t}$, and so on, this signal grows unboundedly as $t \to \infty$, and decays to zero as $t \to -\infty$. When a has a negative real part, the opposite happens: the signal decays to zero as $t \to \infty$, but grows unboundedly as $t \to -\infty$. Some examples are shown in Fig. 2.13. The magnitude of the real part of a (i.e., $|\operatorname{Re} a|$) determines how fast the exponential grows or decays. When the real part of a is zero, that is, when a has the form

$$a = j\omega_0 \tag{2.45}$$

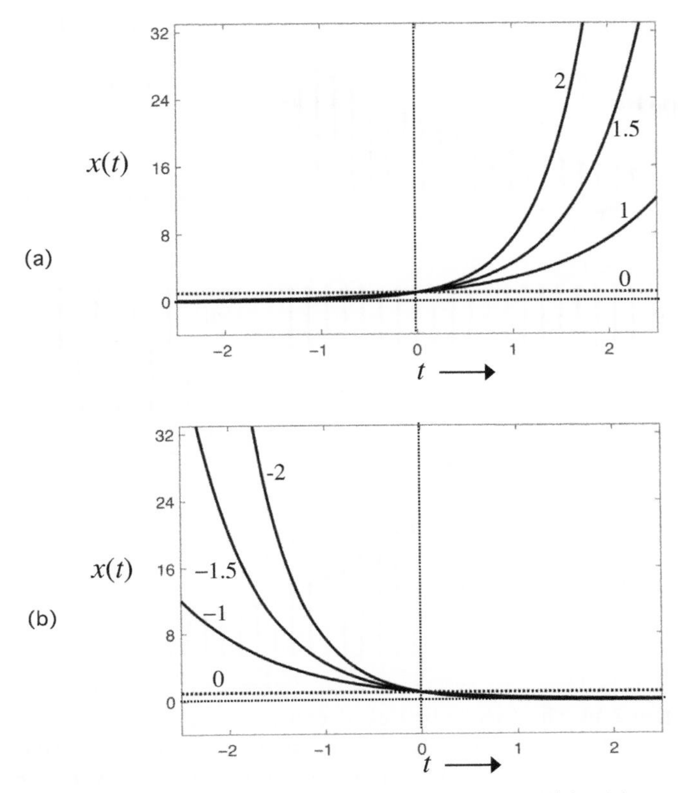

Figure 2.13 Continuous-time exponential signals e^{at} for (a) $a > 0$ and (b) $a < 0$. The values of a (namely, $0, \pm 1$, $\pm 1.5, \pm 2$) are shown next to the curves.

for some real ω_0, the signal becomes $x(t) = e^{j\omega_0 t}$, which is the single frequency signal (2.27) discussed earlier. In this case $|x(t)| = 1$ for all t, that is, the signal does not grow or decay.

2.5.1 The Discrete-Time Exponential

Equation (2.42) is a continuous-time exponential. The discrete-time exponential signal is defined as[3]

$$x[n] = a^n. \tag{2.46}$$

More generally, $x[n] = ca^n$ is an exponential for any (possibly complex) $c \neq 0$. Examples are shown in Fig. 2.14. For $|a| > 1$, the exponential grows unboundedly for $n \to \infty$ and decays to zero for $n \to -\infty$. For $|a| < 1$ the opposite happens. Notice how much faster the exponential grows for $a = 1.9$ compared to $a = 1.1$.

[3] We could have defined $x[n] = e^{bn}$ for some constant b, as we did in the case of continuous time. But since $e^{bn} = (e^b)^n = a^n$, where $a = e^b$, Eq. (2.46) is an equivalent definition, and is more convenient.

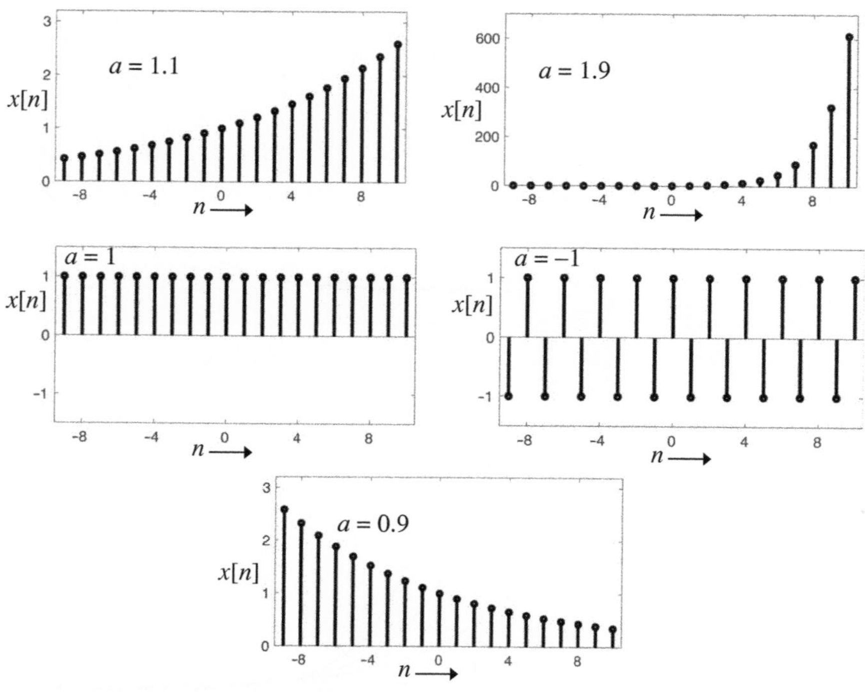

Figure 2.14 The discrete-time exponential $x[n] = a^n$ plotted for various values of a. Unless $|a| = 1$, the exponential always grows in one direction and decays in the other. Notice how much faster the exponential grows for $a = 1.9$ compared to $a = 1.1$. The cases $a = 1$ and $a = -1$ correspond to the single-frequency signal $e^{j\omega_0 n}$ with $\omega_0 = 0$ and $\omega_0 = \pi$, respectively.

When $|a| = 1$ we have $|x[n]| = 1$, so the exponential does not grow or decay. In this case we have $a = e^{j\omega_0}$ for some real ω_0, so that

$$x[n] = e^{j\omega_0 n}. \tag{2.47}$$

This is called the single-frequency signal in discrete time. It has the same form as the continuous-time version (2.27). For $\omega_0 = 0$ and $\omega_0 = \pi$, this becomes

$$x[n] = \begin{cases} 1, & \forall n \quad (\omega_0 = 0), \\ (-1)^n & \quad (\omega_0 = \pi). \end{cases} \tag{2.48}$$

In Fig. 2.14 these correspond to the cases $a = \pm 1$, respectively.

2.5.2 How Exponentials Grow

We now consider an interesting property of discrete-time exponentials. Consider the signal

$$x_1[n] = x[n] - x[n-1]. \tag{2.49}$$

This is called the **first difference** of $x[n]$, analogous to the first derivative in the case of continuous-time signals. For the exponential $x[n] = a^n$, we have

$$x_1[n] = x[n] - x[n-1] = a^n - a^{n-1} = a^n\left(1 - \frac{1}{a}\right) = ca^n, \tag{2.50}$$

where $c = (a-1)/a$. Thus the first difference is the same exponential a^n, but scaled by a constant c. This is analogous to the property that a continuous-time exponential e^{at} has a derivative ae^{at} which is also an exponential. So, if an exponential a^n grows fast as $n \to \infty$, the first difference also grows at the same exponential rate, except that it is scaled by a constant c.

Next, the first difference of the first difference $x_1[n]$ is said to be the **second difference**, which in our example is

$$x_2[n] = x_1[n] - x_1[n-1] = ca^n - ca^{n-1} = ca^n\left(1 - \frac{1}{a}\right) = c^2a^n. \tag{2.51}$$

More generally, the N**th difference** of the exponential $x[n] = a^n$ is given by

$$x_N[n] = c^N a^n, \tag{2.52}$$

which is the exact same exponential! Only the scale factor c^N is different. Summarizing, the first difference and higher-order differences of the exponential signal a^n are also the same exponential, except for scale factors.

For example, if a^n is the number of victims of a pandemic on the nth day of its onset, then the number of new cases that are reported every day *also* grows exponentially! Furthermore, the increase in the "number of new cases reported each day" itself grows exponentially, and so on.

Exponential versus polynomial growth. As a contrast to the exponential a^n, suppose $x[n] = n$, which is a *linearly growing* function. Then the first difference

$$x_1[n] = n - (n-1) = 1, \tag{2.53}$$

which is a constant! There is no growth or decay. So the second difference $x_2[n] = 0$ for all n. Similarly, if $x[n] = n^2$ (a polynomial) then the successive differences are

$$\begin{aligned} x_1[n] &= n^2 - (n-1)^2 = 2n - 1, \\ x_2[n] &= 2n - 1 - 2(n-1) + 1 = 2, \\ x_3[n] &= 2 - 2 = 0. \end{aligned} \tag{2.54}$$

All the higher-order differences beyond this are zero. Thus, **polynomial** growth is much less formidable compared to exponential growth. ▽ ▽ ▽

2.6 The Discrete-Time Single-Frequency Signal

Consider again the discrete-time single-frequency signal (2.47) reproduced below:

$$x[n] = e^{j\omega_0 n}. \tag{2.55}$$

Here, ω_0 is the frequency and n takes only integer values. There are some subtle differences when we compare this signal with the continuous-time counterpart $x(t) = e^{j\omega_0 t}$. We now point out some of these.

2.6.1 The Frequency Range in the Discrete-Time Case

For the continuous-time case, the frequency[4] could take any real value as in Eq. (2.28):

$$-\infty < \Omega < \infty \qquad \text{(continuous-time case)}. \tag{2.56}$$

This is shown in Fig. 2.15(a). But in discrete time, it is different. We have to restrict ω_0 to a finite range. To understand this, note that

$$e^{j(\omega_0+2\pi)n} = e^{j\omega_0 n}, \tag{2.57}$$

because $e^{j2\pi n} = 1$ for all integer n. This shows that any two frequencies separated by 2π (hence, by any integer multiple of 2π) are perceived as the same frequency, that is, $x[n]$ is the same signal for these two frequencies. This is unlike in continuous time, where

$$e^{j(\Omega_0+2\pi)t} \neq e^{j\Omega_0 t}, \tag{2.58}$$

because $e^{j2\pi t} \neq 1$ for arbitrary real t. But in the discrete-time case we have to restrict ω_0 to a region of length 2π, because of Eq. (2.57). Two conventions are popular, namely

$$-\pi \leq \omega < \pi \quad \text{and} \quad 0 \leq \omega < 2\pi \qquad \text{(discrete-time case)}, \tag{2.59}$$

as shown by the thick lines in Figs. 2.15(b) and (c). These two conventions are both prevalent and often used interchangeably.

Given any discrete-time frequency ω_0, one can always bring it to the range $0 \leq \omega < 2\pi$ by adding or subtracting an appropriate integer multiple of 2π. This unique value in $0 \leq \omega < 2\pi$ is called ω_0 **modulo** 2π. Thus, in discrete time, only the value of ω_0 modulo 2π matters. Here are some examples and points to observe, in connection with the discrete-time frequency variable.

1. For every negative frequency in $-\pi \leq \omega < 0$ we can find an equivalent frequency in $\pi \leq \omega < 2\pi$ by adding 2π. For example, $\omega = -0.5\pi$ is equivalent to $\omega = 1.5\pi$ (white circles in the figure).

[4] In discussions where continuous and discrete-time signals occur together, we often use Ω for continuous-time frequencies and ω for discrete-time frequencies, to avoid confusion.

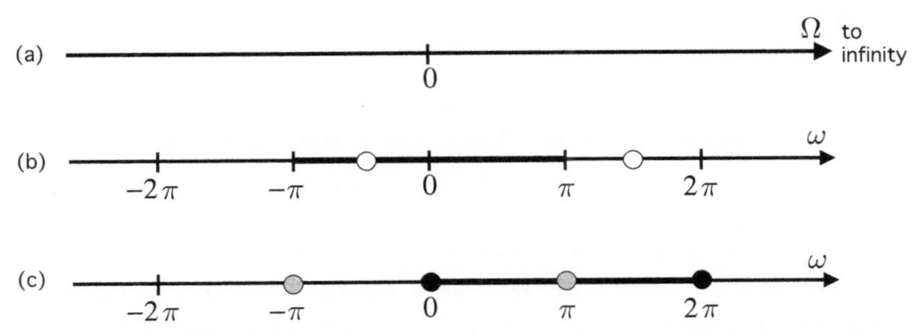

Figure 2.15 (a) The frequency range $-\infty < \Omega < \infty$ in the continuous-time case, (b) and (c) the frequency range in the discrete-time case. Two conventions are shown for the discrete-time case: (b) $-\pi \leq \omega < \pi$ and (c) $0 \leq \omega < 2\pi$. The circles with identical shades represent the same frequency. For example, the two white circles ($\omega_0 = -0.5\pi$ and 1.5π) represent the same signal $e^{j\omega_0 n}$.

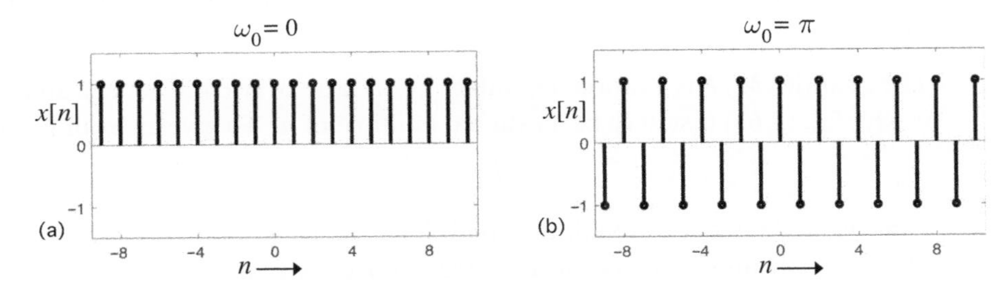

Figure 2.16 The discrete-time signal $x[n] = e^{j\omega_0 n}$ for the lowest frequency $\omega_0 = 0$ and the highest frequency $\omega_0 = \pi$.

2. $e^{j2\pi n} = e^{j0n} = 1$ for all n, so the frequency 2π is equivalent to zero frequency (black circles in the figure).
3. $e^{j\pi n} = e^{-j\pi n} = (-1)^n$, so that π and $-\pi$ are equivalent frequencies (gray circles in the figure).
4. $\omega = 0$ is the lowest frequency because $x[n] = e^{j0n} = 1$ is a constant, and shows no variations, see Fig. 2.16(a). And $\omega = \pi$ (or equivalently, $\omega = -\pi$) is the highest frequency. At this frequency $x[n] = e^{j\pi n} = (-1)^n$, and exhibits maximum variation from sample to sample for fixed magnitude; see Fig. 2.16(b). This gives a physical meaning as to why this is the highest frequency.

2.6.2 Periodicity in Discrete Time

There is another subtle difference between continuous-time and discrete-time frequencies. For $x(t) = e^{j\Omega_0 t}$ it was found earlier that the signal is periodic with period $T = 2\pi/\Omega_0$. (For convenience we assume $\Omega_0 > 0$ so that we don't have to keep using absolute values.) But in discrete time, we cannot define the period of

$$x[n] = e^{j\omega_0 n} \tag{2.60}$$

to be

$$T = \frac{2\pi}{\omega_0}, \tag{2.61}$$

unless this happens to be an integer. This is because, if T is the period, then

$$x[n] = x[n + T], \tag{2.62}$$

but this is meaningful only when T is an integer. We now show that it is possible for $x[n] = e^{j\omega_0 n}$ to be periodic even if Eq. (2.61) is not an integer, and derive the correct expression for the period (which in general is not Eq. (2.61)).

Theorem 2.1 *Periodicity of* $e^{j\omega_0 n}$. The discrete-time signal $x[n] = e^{j\omega_0 n}$ is periodic with some integer period P if and only if ω_0 has the form

$$\omega_0 = \frac{2\pi M}{P}, \tag{2.63}$$

for some integers M and $P \neq 0$, that is, if and only if ω_0 is a rational multiple of π. ◇

The quantity M/P is a rational number (i.e., a ratio of two integers M and P). That is why Eq. (2.63) is said to be a rational multiple of π. To give an example, if

$$\omega_0 = \sqrt{2}\pi, \tag{2.64}$$

then since $\sqrt{2}$ is not rational, the condition (2.63) cannot be satisfied for any pair of integers M and P. So, according to the above theorem,

$$x[n] = e^{j(\sqrt{2}\pi)n} \tag{2.65}$$

is not periodic at all

Proof of Theorem 2.1. First assume $x[n] = e^{j\omega_0 n}$ is periodic with integer period P. Then

$$e^{j\omega_0(n+P)} = e^{j\omega_0 n}, \quad \forall n. \tag{2.66}$$

This implies $e^{j\omega_0 P} = 1$, so that $\omega_0 P = 2\pi M$ for some integer M, which proves condition (2.63). Conversely, if ω_0 has the form (2.63), then

$$x[n + P] = e^{j\omega_0(n+P)} = e^{j(\omega_0 n+\omega_0 P)} = e^{j(\omega_0 n+2\pi M)} = e^{j\omega_0 n} = x[n], \tag{2.67}$$

where, (2.63) has been used in the third equality. This shows that $x[n]$ is indeed periodic. ▽ ▽ ▽

Since we proved $x[n + P] = x[n]$ when condition (2.63) holds, it is clear that P is itself the period, as long as it is the smallest positive integer satisfying (2.63). To find the period we therefore cancel off any common factor between P and M, so that they are **coprime**, and declare the period to be

$$P = \frac{2\pi M}{\omega_0}. \tag{2.68}$$

Summarizing, $x[n] = e^{j\omega_0 n}$ is periodic (i.e., has an integer period) if and only if condition (2.63) holds for a coprime pair of integers P and M, and in this case the

(a)
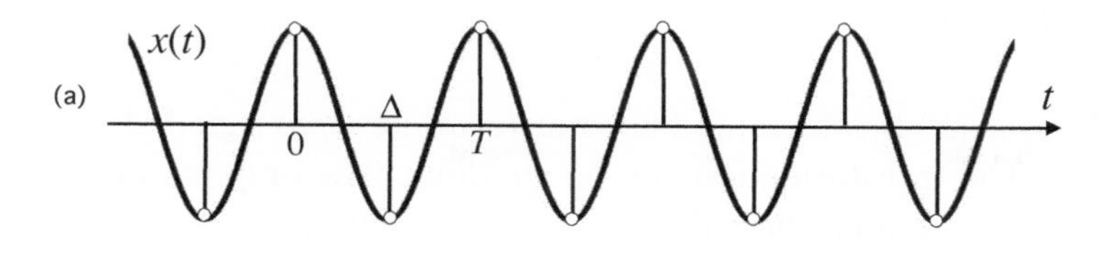

(b)
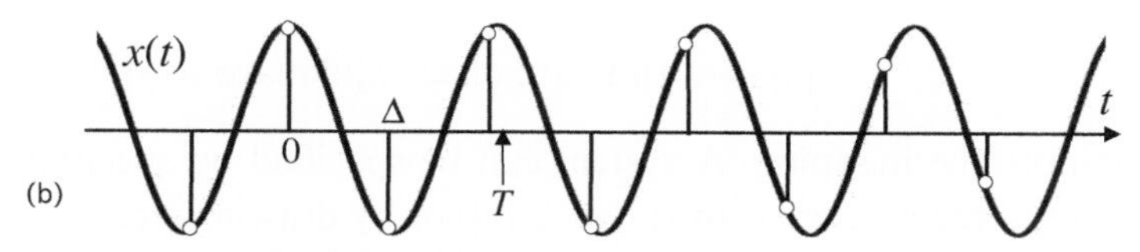

Figure 2.17 Samples of $x(t) = \cos(\omega_0 t)$, with (a) sample spacing $\Delta = T/2$, where $T = 2\pi/\omega_0$ is the period of $x(t)$, and (b) sample spacing Δ slightly smaller than $T/2$.

period is precisely P, and can be expressed as in Eq. (2.68). Notice that the integer period (2.68) can be much larger than $2\pi/\omega_0$.

Pictorial Interpretation

It is not hard to develop some insight for Theorem 2.1. Thus, consider $x[n] = \cos(\omega_0 n)$ (which is easier to plot than $e^{j\omega_0 n}$). This can be regarded as a sampled version of $x(t) = \cos(\omega_0 t)$ with sample spacing $\Delta = 1$. We know $\cos(\omega_0 t)$ is periodic with period $T = 2\pi/\omega_0$, as indicated in Fig. 2.17(a). In this figure we happen to have $\Delta = 1$ and $T = 2$, so that one of the samples $x[2] = x(T)$ falls exactly at $t = T$ (where $x(t)$ begins to repeat). After this, the sample values therefore repeat periodically. Next consider Fig. 2.17(b), where $1 = \Delta < T/2$. The sample spacing is slightly smaller than $T/2$. So, there is no sample falling at the location $t = T$ now; the samples do not repeat after $x[2]$. Now suppose T/Δ is a rational number, that is,

$$\frac{T}{\Delta} = \frac{P}{k} \tag{2.69}$$

for some integers P and k with no common factors, then the sample $x[P] = x(P\Delta)$ eventually falls at the integer multiple kT of the period T. That is,

$$\begin{aligned} x[P] = x(P\Delta) &= x(kT) && \text{(by (2.69))} \\ &= x(0) && \text{(because } x(t + kT) = x(t)\text{)} \\ &= x[0] && \text{(because } x[n] = x(nT)\text{).} \end{aligned} \tag{2.70}$$

Thus, starting from $n = P$, the samples do repeat, so that $x[n]$ is periodic with period P, although one period of $x[n]$ corresponds to k periods of $x(t)$. If, on the other hand, T/Δ is irrational then none of the samples of $x[n] = x(n\Delta)$ will ever fall at an integer multiple $t = kT$ of T (i.e., $n\Delta$ will never equal kT), so that $x[n]$ is *not* periodic. The reader should be able to see that Theorem 2.1 essentially follows from this!

2.7 Systems

A "system" is something that transforms an *input* signal $x[n]$ into an *output* signal $y[n]$. The transformation can be anything. The only requirement is that if $x[n]$ is specified for all n, then the system should uniquely specify $y[n]$ for all n. A system is also sometimes called an **operator**, a transformation, or a mapping. We use the notation

$$x[n] \longmapsto y[n] \quad \text{(read as "}x[n]\text{ maps to }y[n]\text{")}, \tag{2.71}$$

to denote the mapping. A system can be specified by specifying its input–output relation directly (e.g., as in $y[n] = x^2[n]$) or by drawing a computational graph that describes how the output is computed from the input (e.g., as in Fig. 2.18 later). There are other ways to uniquely specify the input–output relation, as we shall see. Electrical engineers often specify a system by drawing an electrical circuit and marking the input and output terminals. Some general remarks are now in order.

1. When we say that the system produces $y[n]$ in response to $x[n]$, we mean that the entire signal $x[n]$, for $-\infty < n < \infty$, is given. From this the system produces $y[n]$ for $-\infty < n < \infty$. If $x[n]$ is specified only for some values of time, say $n_1 \leq n \leq n_2$, then this is in general not sufficient for the system to determine $y[n]$ for all n.
2. A system can map a continuous-time signal to a discrete-time signal, and vice versa. That is, one can define systems of the form $x(t) \longmapsto y[n]$ and $x[n] \longmapsto y(t)$. Examples include sampling devices and systems which reconstruct signals from sampled versions (see Chapter 10).

2.8 Basic Classifications of Systems

In this section we will define a number of types of systems, and study some examples. Even though there is more focus on discrete-time systems $x[n] \longmapsto y[n]$, most definitions also hold for continuous-time systems $x(t) \longmapsto y(t)$.

2.8.1 Homogeneous Systems and Additive Systems

We begin by introducing two simple definitions. The definitions are stated for discrete-time systems, but they also hold for continuous-time systems with a mere change of notation from $x[n]$ to $x(t)$, and so on. Notice that all signals and constants are in general complex.

Definition 2.1 *Homogeneous systems.* Suppose a system produces the output $y[n]$ in response to the input $x[n]$. The system is said to be homogeneous if the scaled input $cx[n]$ produces the scaled output $cy[n]$ for any choice of the constant c, and for every possible choice of the input sequence $x[n]$. ◇

Definition 2.2 *Additive systems.* Let us use the notation $y_k[n]$ to denote the output of a system in response to the input $x_k[n]$. The system is said to be additive if the input $x_1[n]+x_2[n]$ produces the output $y_1[n]+y_2[n]$, for every possible choice of the input signals $x_1[n]$ and $x_2[n]$. ◇

Using the "maps to" notation (2.71), these definitions can be conveniently abbreviated as follows.

$$\text{Homogeneous:}\quad x[n] \longmapsto y[n] \quad \Longrightarrow \quad cx[n] \longmapsto cy[n]. \tag{2.72}$$

$$\text{Additive:}\quad x_k[n] \longmapsto y_k[n] \quad \Longrightarrow \quad x_1[n]+x_2[n] \longmapsto y_1[n]+y_2[n]. \tag{2.73}$$

For additive systems, we can repeatedly use the mapping (2.73) to show that if $x_k[n] \longmapsto y_k[n]$ for $1 \leq k \leq N$, then

$$\sum_{k=1}^{N} x_k[n] \longmapsto \sum_{k=1}^{N} y_k[n]. \tag{2.74}$$

2.8.1.1 Basic Examples

We now consider some examples. The system

$$y[n] = x^2[n] \tag{2.75}$$

is called a squaring device. This is not homogeneous because the input $cx[n]$ produces the output $c^2y[n]$, which violates (2.72). Similarly, if $x_k[n] \longmapsto y_k[n]$ then the input $x_1[n]+x_2[n]$ produces the output

$$\begin{aligned}\Big(x_1[n]+x_2[n]\Big)^2 &= x_1^2[n]+x_2^2[n]+2x_1[n]x_2[n]\\ &= y_1[n]+y_2[n]+2x_1[n]x_2[n],\end{aligned} \tag{2.76}$$

which is not equal to $y_1[n]+y_2[n]$. So the system is not additive either. Next consider the system

$$y[n] = x[n]+x[n-1], \tag{2.77}$$

which simply produces the sum of the present and past input samples. This is homogeneous because the input $cx[n]$ produces the output

$$cx[n]+cx[n-1] = c\Big(x[n]+x[n-1]\Big) = cy[n]. \tag{2.78}$$

This system is also additive because if $x_k[n] \longmapsto y_k[n]$ then the input $x_1[n]+x_2[n]$ produces the output

$$\begin{aligned}&\Big(x_1[n]+x_2[n]\Big)+\Big(x_1[n-1]+x_2[n-1]\Big)\\ &= \Big(x_1[n]+x_1[n-1]\Big)+\Big(x_2[n]+x_2[n-1]\Big) = y_1[n]+y_2[n].\end{aligned} \tag{2.79}$$

Figure 2.18 shows a schematic way to describe the system (2.77). This is called a **signal flowgraph** or **structure** or **computational graph** for the system. Here the box labeled "unit delay" produces the delayed version $x[n-1]$ of its input $x[n]$ (delayed

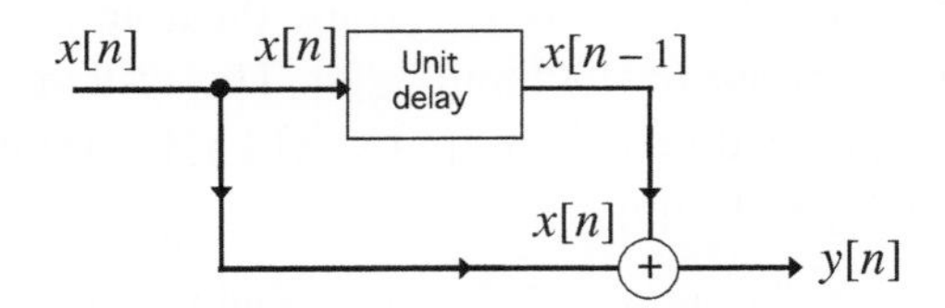

Figure 2.18 The computational graph or structure for the system (2.77).

by one sample). The circle labeled "+" is an adder unit. The black dot on the left is just a "take-off" point where the signal $x[n]$ branches off and flows in different directions.

2.8.1.2 A System that is Additive but Not Homogeneous

Consider the system described by

$$y[n] = \text{Re}\Big(x[n]\Big), \tag{2.80}$$

which simply computes the real part of the input $x[n]$ for every n. Writing

$$x[n] = x_R[n] + jx_I[n], \tag{2.81}$$

where $x_R[n]$ and $x_I[n]$ are real quantities representing the real and imaginary parts, the system can also be described by

$$y[n] = x_R[n]. \tag{2.82}$$

This system is additive. To verify this, let $y_k[n] = \text{Re}(x_k[n])$ for $k = 1, 2$ and consider the new input $x_1[n] + x_2[n]$. The corresponding output is

$$\text{Re}\Big(x_1[n] + x_2[n]\Big) = \text{Re}\Big(x_1[n]\Big) + \text{Re}\Big(x_2[n]\Big), \tag{2.83}$$

which is indeed $y_1[n] + y_2[n]$, proving additivity. We now claim that this system is *not* homogeneous. Thus, given $y[n] = \text{Re}(x[n]) = x_R[n]$, define the scaled input

$$x_s[n] = jx[n], \tag{2.84}$$

where $j = \sqrt{-1}$. If the system were homogeneous, the output would be $jy[n] = jx_R[n]$. The actual output, however, is

$$y_s[n] = \text{Re}\Big(x_s[n]\Big) = \text{Re}\Big(jx[n]\Big) = \text{Re}\Big(jx_R[n] - x_I[n]\Big) = -x_I[n], \tag{2.85}$$

which shows $y_s[n] \neq jy[n]$. This proves that the system is not homogeneous.

To prove or to disprove? Notice that to prove that the system is not homogeneous, we only have to exhibit one counterexample (as above) where the definition of homogeneity is violated. On the other hand, to prove that a system *is homogeneous*, we have to explicitly prove that the definition is satisfied no matter what the input is, as in the example (2.77). The same remark holds for all other properties one tries to prove or disprove for various examples of systems. In this sense, proving that a property is *not* true is often easier than proving that a property *is* true! ▽ ▽ ▽

The above example works because of the clever use of the complex number j as the scaling constant. This raises the following question: suppose we consider only the class of "real systems," which only take real-valued inputs $x[n]$, and produce real-valued outputs. Does additivity imply homogeneity in this case? This is rather subtle, and is discussed in Sec. 2.14.

2.8.1.3 A System that is Homogeneous but Not Additive

Consider the system described by

$$y[n] = \begin{cases} \dfrac{x^2[n]}{x[n-1]} & x[n-1] \neq 0, \\ 0 & \text{otherwise.} \end{cases} \tag{2.86}$$

If we apply the new input $x_s[n] = cx[n]$, then the output is

$$y_s[n] = \begin{cases} \dfrac{c^2x^2[n]}{cx[n-1]} & cx[n-1] \neq 0, \\ 0 & \text{otherwise.} \end{cases} \tag{2.87}$$

That is,

$$y_s[n] = \begin{cases} \dfrac{cx^2[n]}{x[n-1]} & x[n-1] \neq 0, \\ 0 & \text{otherwise.} \end{cases} \tag{2.88}$$

This shows that $y_s[n] = cy[n]$, proving that the system is homogeneous. But this system is not additive. To prove this, let us consider two inputs:

$$x_1[n] = \delta[n], \quad x_2[n] = \delta[n-1]. \tag{2.89}$$

Then the outputs are

$$y_1[n] = 0, \quad y_2[n] = 0, \quad \forall n, \tag{2.90}$$

as you can readily verify from Eq. (2.86). If we now apply the input

$$x[n] = x_1[n] + x_2[n] = \delta[n] + \delta[n-1], \tag{2.91}$$

then the output is

$$y[n] = \delta[n-1] \neq y_1[n] + y_2[n], \tag{2.92}$$

proving that the system is not additive. Figure 2.19 explains this pictorially.

2.8.1.4 Additive Systems are Also Subtractive

Given $x_i[n] \longmapsto y_i[n]$, a subtractive system satisfies the property

$$x_1[n] - x_2[n] \longmapsto y_1[n] - y_2[n]. \tag{2.93}$$

It turns out that if the system $x[n] \longmapsto y[n]$ is additive, then it is subtractive as well. The proof would be trivial if we could assume that $-x[n] \longmapsto -y[n]$, because we

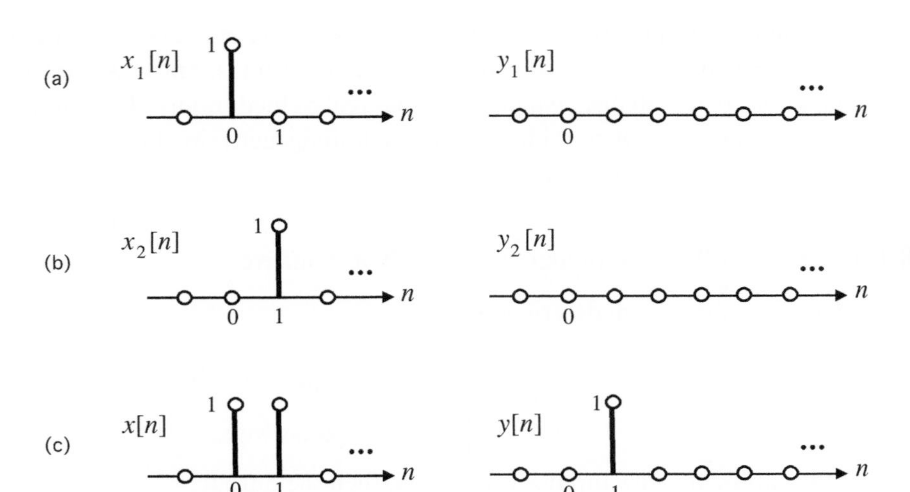

Figure 2.19 Proof that the system (2.86) is not additive. (a), (b) Impulse inputs produce the zero output and (c) sum of the two impulses produces nonzero output, proving that the system is not additive.

could then apply additivity to $x_1[n]$ and $-x_2[n]$. But since we have not yet proved $-x[n] \longmapsto -y[n]$ for additive systems (even though it is true), we are not allowed to use it. The correct proof that additivity implies subtractivity is left as an exercise (Problem 2.18), but this fact will be freely used in the rest of the chapter.

Additive systems respect negation. Since the additive system $x[n] \longmapsto y[n]$ satisfies $x[n] + x[n] \longmapsto y[n] + y[n]$, it follows that

$$2x[n] \longmapsto 2y[n] \tag{2.94}$$

as well. So the input $x[n] - 2x[n]$ produces the output $y[n] - 2y[n]$ (by subtractivity of additive systems). That is, for an additive system $x[n] \longmapsto y[n]$, we also have

$$-x[n] \longmapsto -y[n]. \tag{2.95}$$

So we say that an additive system obeys negation, that is, satisfies (2.95). ▽ ▽ ▽

2.8.2 Linear Systems

A system is said to be **linear** if it is *both* homogeneous and additive. Thus, for a linear system, if $x_k[n] \longmapsto y_k[n]$ then $cx_k[n] \longmapsto cy_k[n]$ for any constant c by homogeneity. Furthermore, by using additivity we can conclude that

$$c_1x_1[n] + c_2x_2[n] \longmapsto c_1y_1[n] + c_2y_2[n], \tag{2.96}$$

for a linear system. By repeated application of this property it follows that for a linear system,

$$\sum_{k=1}^{N} c_k x_k[n] \longmapsto \sum_{k=1}^{N} c_k y_k[n] \tag{2.97}$$

for any choice of the constants c_k, for any choice of the inputs $x_k[n]$, and for any integer $N \geq 1$. Now, a summation of the form

$$\sum_{k=1}^{N} c_k x_k[n] \tag{2.98}$$

is called a superposition or **linear combination** of the signals $x_k[n]$. So, Eq. (2.97) is called the **superposition** property. This shows that a linear system satisfies the superposition property. Conversely, if a system satisfies Eq. (2.97) then it is linear, that is, it is homogeneous and additive, as proved next.

Proof. If we set $c_k = 0$ for $k > 1$, then Eq. (2.97) reduces to $c_1 x_1[n] \longmapsto c_1 y_1[n]$, which proves the homogeneous property. If we set $c_1 = c_2 = 1$ and $c_k = 0$ for other k, then Eq. (2.97) reduces to $x_1[n] + x_2[n] \longmapsto y_1[n] + y_2[n]$, which is additivity. ▽ ▽ ▽

Summarizing, we have shown that the system is linear (i.e., homogeneous and additive) if and only if it satisfies the superposition property (2.97). In many books you will therefore find the following equivalent definition of linearity.

Definition 2.3 *Linear systems.* With $y_k[n]$ denoting the output of the system in response to the input $x_k[n]$, we say that the system is linear if $c_1 x_1[n] + c_2 x_2[n]$ produces the output $c_1 y_1[n] + c_2 y_2[n]$, for every possible choice of the constants c_1, c_2 and the input signals $x_1[n]$ and $x_2[n]$. ◇

Repeated application of this property shows that linearity is equivalent to Eq. (2.97) as well, for any $N > 1$. A system that is not linear is said to be **nonlinear**.

2.8.2.1 Examples

It was shown earlier that the squaring system (2.75) is not homogeneous nor additive. So it is nonlinear. The real-part system (2.80) is also nonlinear because it is not homogeneous. Next, the system (2.86) is nonlinear because it is not additive. Finally, the system (2.77), reproduced below,

$$y[n] = x[n] + x[n-1], \tag{2.99}$$

was shown to be both homogeneous and additive, so it is linear.

Does it help to plot $y[n]$ versus $x[n]$? Consider the system

$$y[n] = \alpha x[n], \tag{2.100}$$

where α is a constant. It is readily verified that this is linear. For this particular example, the output sample $y[n]$ can be plotted as a function of $x[n]$, and the plot is independent of n. The plot is a straight line with slope c, as one would expect based on the "linearity" property. But we have to be careful. The system defined by

$$y[n] = 1 + x[n] \tag{2.101}$$

also has a straight-line plot, although it does not pass through the origin. It can be verified that this is not a linear system; for example, it is not homogeneous (Problem 2.7). Next consider the linear system of Eq. (2.99). For this, we cannot even plot $y[n]$ as a function of $x[n]$ because $y[n]$ also depends on the past input sample $x[n-1]$. Thus, one cannot in general judge linearity by trying to plot $y[n]$ as a function of $x[n]$. ▽ ▽ ▽

2.8.3 Time-Invariant Systems

A system $x[n] \longmapsto y[n]$ is said to be **time invariant** if

$$x[n-N] \longmapsto y[n-N], \tag{2.102}$$

for any integer N. Note that $x[n-N]$ is the shifted version of $x[n]$, the amount of shift being N samples (right shift if $N > 0$, left shift if $N < 0$). The above property says that if the input $x[n]$ is shifted by N, then the new output is the old output *shifted by the same amount* N. In other words, the system behavior does not depend on *when* you use it. The system behaves the same today and tomorrow – it is not fickle! A time-invariant system is also called a **shift-invariant** system. A system that is not time invariant is said to be **time varying**. Let us now consider some examples.

2.8.3.1 The Amplifier Example

Consider again the system

$$y[n] = \alpha x[n], \tag{2.103}$$

which simply multiplies every input sample by a constant multiplier α. This is shown schematically in Fig. 2.20(a) and is called an "amplifier" with amplification factor α. If we define a new input $x_s[n] = x[n-N]$ to this system, then the output is

$$y_s[n] = \alpha x_s[n] = \alpha x[n-N] = y[n-N], \tag{2.104}$$

where the first and last equalities follow from the system definition (2.103). This shows that the system is time invariant. This is hardly surprising, as the system is just a scalar multiplier α which does not depend on time. Next consider the system

$$y[n] = nx[n]. \tag{2.105}$$

This can also be regarded as an amplifier, but the amplification factor n depends on time n (Fig. 2.20(b)). So we expect this to be a time-varying system. This is easy to

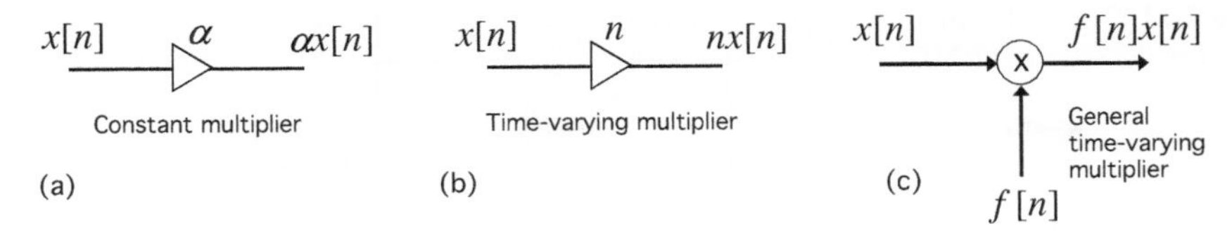

Figure 2.20 Notations used for multipliers. (a) Constant multiplier α, (b) example of a time-varying multiplier n, and (c) alternative notation for a general time-varying multiplier $f[n]$.

verify more formally. Thus, consider the shifted input $x_s[n] = x[n-1]$. Then the output is

$$y_s[n] = nx_s[n] = nx[n-1]. \tag{2.106}$$

On the other hand, from Eq. (2.105) we have

$$y[n-1] = (n-1)x[n-1]. \tag{2.107}$$

Since $y[n-1] \neq y_s[n]$, the system is time varying. A more general time-varying example is $y[n] = f[n]x[n]$, where $f[n]$ is arbitrary (but not the same for all n). In Figs. 2.20(a) and (b) the triangle represents a multiplier. When the multiplier is not a constant, it is more common to use the notation as in Fig. 2.20(c).

2.8.3.2 The Sum of Adjacent Samples

Consider again the system (2.77),

$$y[n] = x[n] + x[n-1]. \tag{2.108}$$

To check whether it is time invariant, let us apply the input $x_s[n] = x[n-K]$ and see what happens. By system definition, the output is

$$y_s[n] = x_s[n] + x_s[n-1] = x[n-K] + x[n-1-K] = y[n-K], \tag{2.109}$$

proving time invariance. On the other hand, consider the system

$$y[n] = x[n] + (-1)^n x[n-1]. \tag{2.110}$$

The output is the sum of adjacent samples when n is even, otherwise it is the difference of output samples. So the behavior of the system is different depending on whether the time index n is even or odd. This ought, therefore, to be a time-varying system. You can verify this more formally as follows: if $x_s[n] = x[n-1]$ then the output is, by system definition,

$$y_s[n] = x_s[n] + (-1)^n x_s[n-1] = x[n-1] + (-1)^n x[n-2]. \tag{2.111}$$

On the other hand, from Eq. (2.110) we have

$$y[n-1] = x[n-1] + (-1)^{(n-1)} x[n-2] = x[n-1] - (-1)^n x[n-2]. \tag{2.112}$$

So $y_s[n] \neq y[n-1]$, proving time variance. For completeness, Fig. 2.21 shows the computational graph or structure for this system.

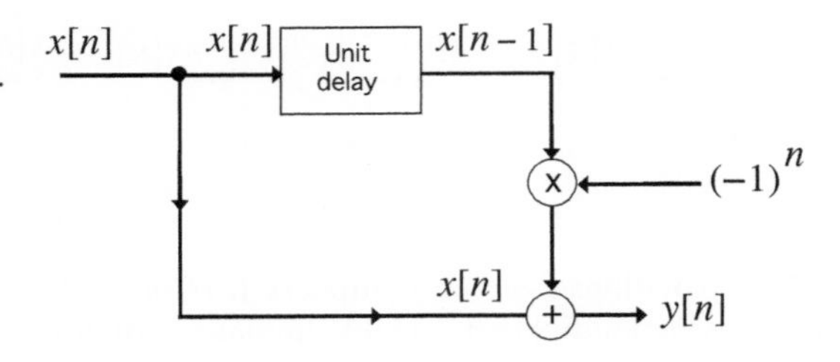

Figure 2.21 Computational graph for the system (2.110).

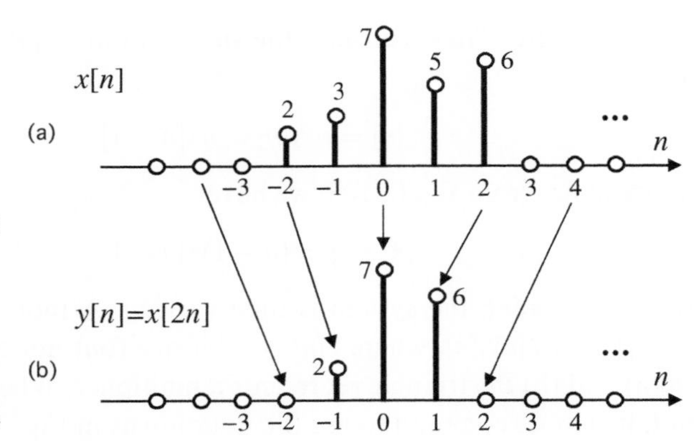

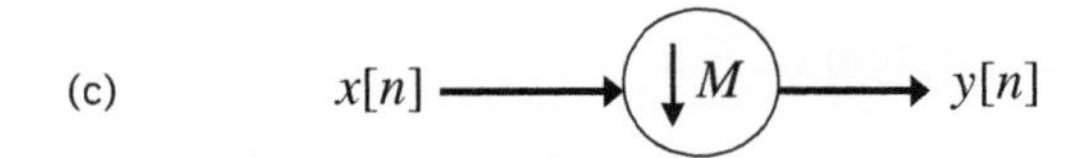

Figure 2.22 (a), (b) Illustration of how the decimator $y[n] = x[2n]$ works. The odd-numbered samples ($x[-1], x[1], x[3]$, and so forth) are discarded, and the even-numbered samples are retained: $y[0] = x[0], y[1] = x[2], y[2] = x[4]$, and so on. (c) The schematic diagram for a decimator.

2.8.3.3 The Decimator

Now consider the system

$$y[n] = x[Mn], \tag{2.113}$$

where M is an integer. Thus, only the samples

$$\cdots x[-M], x[0], x[M], x[2M], \cdots \tag{2.114}$$

are retained, and the other samples are discarded. For example, if $M = 2$, the system retains the even-numbered samples of the input and discards the odd-numbered ones. See Figs. 2.22(a) and (b). The system (2.113) is called an M-fold decimator or a **downsampler**. The decimator is schematically denoted using the notation shown in Fig. 2.22(c). It turns out that this is a time-varying system (unless $M = 1$). To prove this, consider the two inputs

$$x_1[n] = \delta[n], \qquad x_2[n] = \delta[n-1]. \tag{2.115}$$

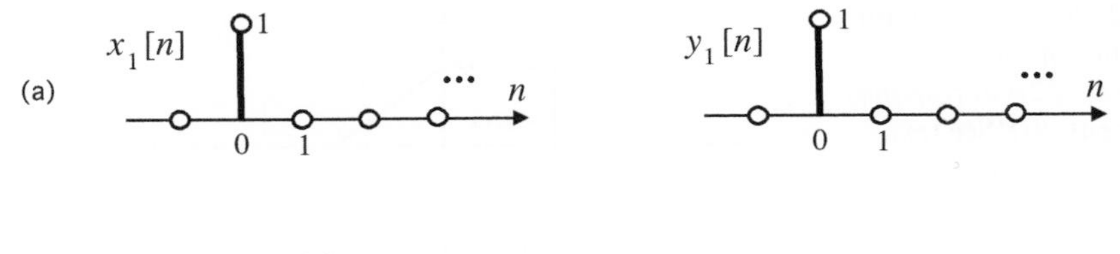

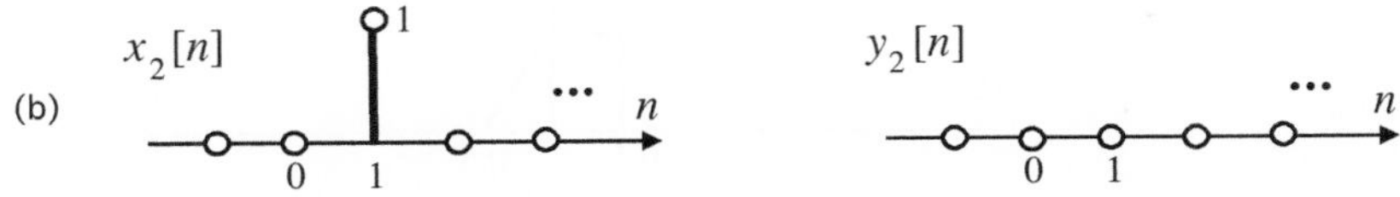

Figure 2.23 Proof that the decimator is time varying. (a) Input $x_1[n] = \delta[n]$ produces output $\delta[n]$ and (b) shifted input $x_2[n] = \delta[n-1]$ produces zero output.

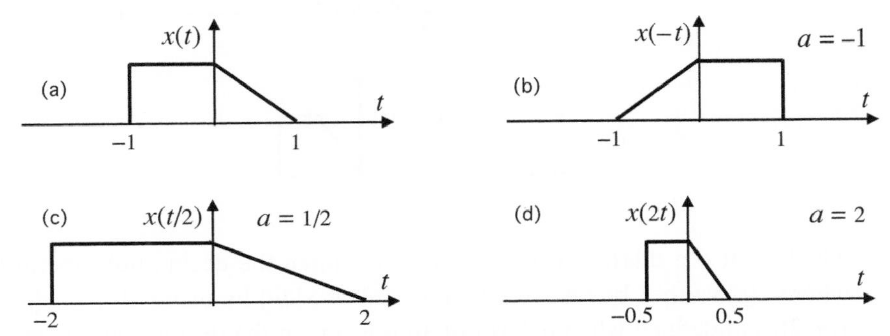

Figure 2.24 (a) A signal $x(t)$ and (b)–(d) dilated versions $y(t) = x(at)$. (b) Flipped version ($a = -1$), (c) a stretched version ($a = 1/2$), and (d) a squeezed version ($a = 2$).

The corresponding outputs are

$$y_1[n] = \delta[n], \qquad y_2[n] = 0, \ \forall n, \tag{2.116}$$

as demonstrated in Fig. 2.23. Thus, a shifted input does not produce the corresponding shifted output, proving that the system is time varying.

2.8.3.4 The Dilation Operator

In the continuous-time world the system

$$y(t) = x(at), \quad a \text{ real}, \tag{2.117}$$

is called a dilation operator. Figure 2.24 demonstrates how this works. When $a = -1$, dilation is simply a **time reversal** or **flip**. When $0 < a < 1$ this is a "stretching operation," as demonstrated in the figure for $a = 1/2$. When $a > 1$ this is a "squeezing operation," as demonstrated for $a = 2$. Note that we can recover $x(t)$ from the dilated version $y(t)$ by using

$$x(t) = y(t/a), \tag{2.118}$$

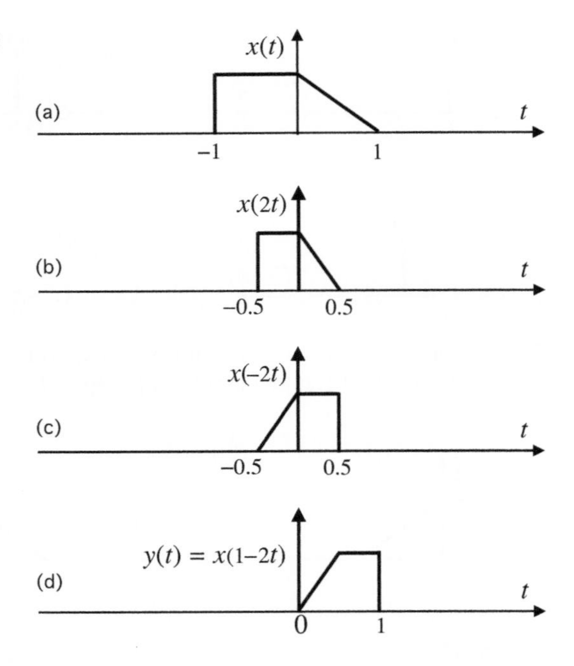

Figure 2.25 (a)–(d) Finding the output of the system $y(t) = x(1 - 2t)$ in response to an input $x(t)$. See text.

which is also a dilation operator. This is unlike the decimation operation (2.113) where $x[n]$ cannot be recovered from $y[n] = x[2n]$ by using $x[n] = y[n/2]$ because $y[n/2]$ is undefined when $n/2$ is not an integer. In the decimation operation (2.113) with $M = 2$, the odd samples are simply lost. Even though there is no loss of information with the dilation operator, it is *time varying*, just like the decimator. To see this, define $x_1(t) = x(t - t_0)$. Then

$$y_1(t) = x_1(at) = x(at - t_0) = x(a(t - t_0/a)) = y(t - t_0/a), \tag{2.119}$$

from Eq. (2.117). Thus, although $y_1(t)$ is a shifted version of $y(t)$, the shift is by a *different* amount t_0/a. This shows that the system is time varying.

Combining dilation, shift, and flip. We now consider an example where dilation, shift, and flip are all combined. Thus, let the system description be

$$y(t) = x(1 - 2t). \tag{2.120}$$

Given a plot of $x(t)$, how do we figure out what $y(t)$ looks like? We can think of Eq. (2.120) as a sequence of operations.

1. Dilate $x(t)$ by 2 to get $x(2t)$.
2. Flip the result to get $x(-2t)$.
3. Shift the result to the right by 0.5 to get $x(-2(t - 0.5)) = x(1 - 2t)$.

Figure 2.25 shows an example of $x(t)$, the dilated version $x(2t)$, the flipped version $x(-2t)$, and the shifted version $x(-2(t-0.5)) = x(1-2t)$. One can also do the above operations in a slightly different order to get the same result (see Problem 2.9).

2.9 Linear Time-Invariant Systems

As the name implies, a linear time-invariant (LTI) system is both linear and time invariant. The importance and impact of such systems in science and engineering is enormous. Even though many systems arising in practice are not LTI, they can be approximated as LTI systems over reasonable ranges of input amplitudes and time intervals. LTI systems will be studied in considerable detail in Chapter 3. They are used in a wide variety of applications such as analog and digital filtering, feedback controls, digital communications, and so on. LTI systems are also sometimes called linear shift-invariant (LSI) systems.

Among the many examples given in the preceding sections, very few are LTI systems. One of these is (2.99), reproduced below:

$$y[n] = x[n] + x[n-1]. \tag{2.121}$$

It was shown in Sec. 2.8.1 that this is both homogeneous and additive. So it is linear. It was also shown in Sec. 2.8.3 that this is time invariant. So the system (2.121) is LTI indeed. We will see in Sec. 3.2 that a discrete-time system is LTI **if and only if** its input–output description can be written in the form

$$y[n] = \sum_{k=-\infty}^{\infty} c_k x[n-k]. \tag{2.122}$$

Clearly, Eq. (2.121) is a special case with $c_0 = c_1 = 1$ and $c_k = 0$ otherwise. We will see that LTI systems are the only class of systems for which one can meaningfully define notions such as "transfer functions" and "frequency responses." These are very important ideas and are useful in characterizing many detailed properties of LTI systems (Chapter 3).

2.9.1 Zero-Input Zero-Output Behavior

If a system is homogeneous, it is easy to see that it produces zero output in response to zero input. That is,

$$\text{“}x[n] = 0 \text{ for all } n\text{”} \quad \text{implies} \quad \text{“}y[n] = 0 \text{ for all } n.\text{”} \tag{2.123}$$

To see this, let $x_1[n]$ be some input with finite sample values producing finite sample values $y_1[n], -\infty < n < \infty$.[5] Then, by homogeneity, the input $0 \times x_1[n]$ produces $0 \times y_1[n]$, which proves (2.123). Next consider an additive system $x[n] \longmapsto y[n]$. Since it is also subtractive (see Sec. 2.8.1), it follows that

$$x[n] - x[n] \longmapsto y[n] - y[n]. \tag{2.124}$$

[5] A subtle point is that if the system does not produce a finite output for any finite input, then this argument does not work. An example of such a hypothetical system is "$y[n] = x[n]$ for $n \neq 0$ and $y[0] = \infty$."

This shows that an additive system also satisfies (2.123). Since a linear system is both additive and homogeneous, it also satisfies (2.123). Summarizing, the zero-input zero-output property (2.123) is satisfied by all of these: homogeneous systems (assuming they produce a finite output sequence for at least one finite input sequence), additive systems, linear systems, and LTI systems.

2.9.2 A Summary of Earlier Examples

We conclude this section by summarizing some of our earlier examples which are not LTI systems. The reasons, indicated in brackets, were justified earlier.

1. $y[n] = x^2[n]$ (not homogeneous, nor additive, clearly not linear; Sec. 2.8.1).
2. $y[n] = \text{Re}(x[n])$ (not homogeneous, hence not linear; Sec. 2.8.1).
3. $y[n] = \begin{cases} \dfrac{x^2[n]}{x[n-1]} & x[n-1] \neq 0 \\ 0 & \text{otherwise} \end{cases}$ (not additive, hence not linear; Sec. 2.8.1).
4. $y[n] = nx[n]$ (time varying; Sec. 2.8.3).
5. $y[n] = x[n] + (-1)^n x[n-1]$ (time varying; Sec. 2.8.3).
6. $y[n] = x[2n]$ (time varying; Sec. 2.8.3).

It can be shown that the first three systems above are time invariant (although nonlinear). The last three systems are linear (although time varying). We prove a couple of these here. For the rest, see Problem 2.16.

Sample proofs. For example, take the system $y[n] = x^2[n]$ and consider the shifted input $x_s[n] = x[n-K]$. Then the output is

$$y_s[n] = x_s^2[n] = x^2[n-K] = y[n-K], \tag{2.125}$$

proving time invariance. Next consider $y[n] = nx[n]$ and let $x_i[n] \longmapsto y_i[n]$. Consider the new input $x[n] = c_1x_1[n] + c_2x_2[n]$. This produces the output

$$y[n] = nx[n] = n\Big(c_1x_1[n] + c_2x_2[n]\Big) = c_1\big(nx_1[n]\big) + c_2\big(nx_2[n]\big), \tag{2.126}$$

which is precisely $c_1y_1[n] + c_2y_2[n]$, proving linearity. ▽ ▽ ▽

2.10 Causality

A system is said to be **causal** if the output at time m does not depend on the input at future times $m + K$, $K > 0$. It may depend only on the past and present input samples. See Fig. 2.26(a). A system that is not causal is said to be **noncausal**. For example, the system

$$y[n] = x[n] + x[n+1] \tag{2.127}$$

is noncausal because the output $y[n]$ depends on the future input sample $x[n+1]$, whereas the system

$$y[n] = x[n] + x[n-1] \tag{2.128}$$

is causal.

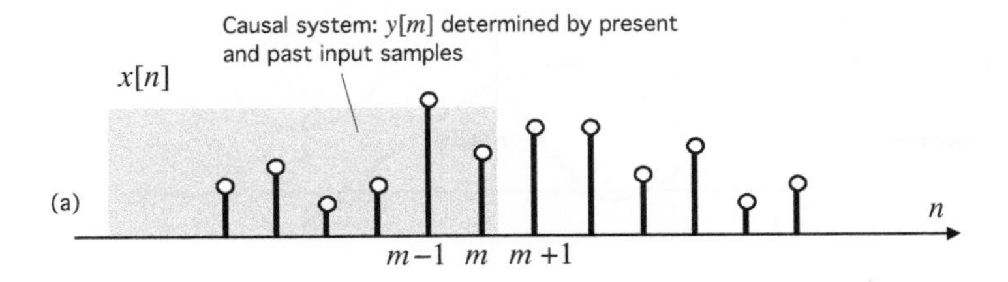

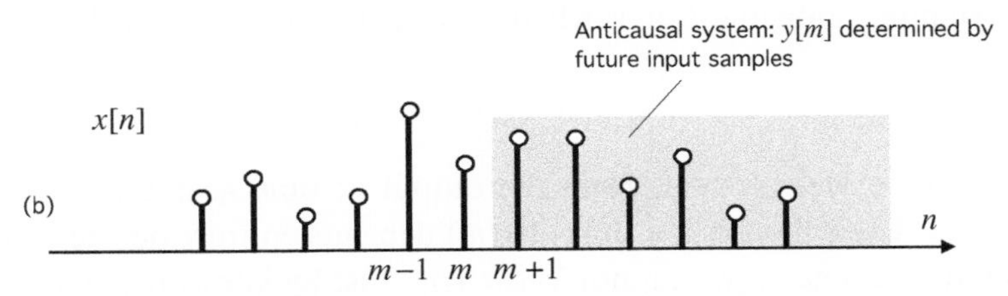

Figure 2.26 The type of dependency of the output sample $y[m]$ on input, for (a) causal systems and (b) anticausal systems.

Anticausal systems. For the sake of completeness, we also mention anticausal systems, although these are encountered only rarely. A system is said to be **anticausal** if the output at time m does not depend on the input at the *present or past* times $m + K$, $K \leq 0$. It may depend only on future input samples. See Fig. 2.26(b). For example, the system

$$y[n] = x[n+1] + x[n+2] \tag{2.129}$$

is anticausal. But the system (2.127) is neither causal nor anticausal because the output depends on future and on present input samples. According to the above definitions, if a system is both causal and anticausal, then the output does not depend on any input sample at all. The output is fully predetermined without help from the input! For example, the system

$$y[n] = (-1)^n \tag{2.130}$$

is both causal and anticausal. Such systems are degenerate in the sense that the output is independent from the input. Notice that the system (2.129) is both non-causal and anticausal.

▽ ▽ ▽

Next, a system is said to be **memoryless** if the output at any given time m can be fully determined by knowing the input at the same time. No knowledge of the past or future of the input is required. Examples of memoryless systems include

$$y[n] = x^2[n], \quad y[n] = nx[n], \quad y[n] = (-1)^n, \tag{2.131}$$

and so forth.[6] It can be shown that a system which is both causal and anticausal is also memoryless, but the converse is not true (Problem 2.13).

[6] A slightly different definition for anticausal systems is that $y[m]$ does not depend on the input at the *past* times $m + K$, $K < 0$. Under this definition, "memoryless" becomes synonymous with "causal and anticausal." We do not use this definition in this book.

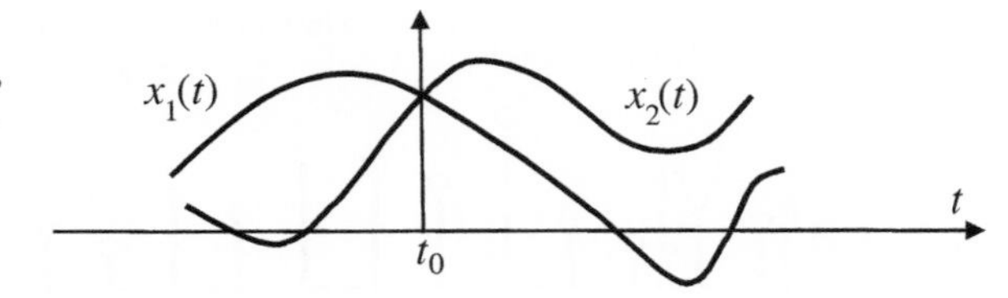

Figure 2.27 Two signals with the same value at t_0, but with different slopes.

An interesting question is whether the continuous-time differentiator

$$y(t) = \frac{dx(t)}{dt} \tag{2.132}$$

is a memoryless system. Here the output at time t_0 is the slope of the input at time t_0. Even though the output $y(t_0)$ depends entirely on the behavior of the input $x(t)$ at time t_0, we cannot know $y(t_0)$ just by knowing $x(t_0)$. Thus, two signals $x_1(t)$ and $x_2(t)$ with the same value at t_0 can have different slopes at t_0, producing two different outputs (Fig. 2.27). So it cannot be said that the differentiator is memoryless.

Electrical circuit elements such as capacitors and inductors are differentiators. For example, the capacitor differentiates the voltage to produce current (Sec. 2.11.2). Such devices store electrical quantities. Thus, the capacitor stores electric charge. This storage property gives a physical interpretation as to why differentiators are not memoryless.

2.11 Stable Systems

A system is said to be stable if the output remains bounded for all time, whenever the input is bounded. To be more precise, let us first define bounded signals. Suppose there exists a constant B_x with $0 \leq B_x < \infty$ such that

$$|x[n]| \leq B_x \quad \forall n. \tag{2.133}$$

Then $x[n]$ is said to be a **bounded** signal. For example, the unit step $\mathcal{U}[n]$ (Fig. 2.4) is a bounded signal whereas the exponential a^n (Fig. 2.14) is not, unless $|a| = 1$.

Definition 2.4 *Stable systems.* Suppose the system is such that any bounded input $x[n]$ produces a bounded output $y[n]$. That is, whenever (2.133) holds for some $0 \leq B_x < \infty$, then

$$|y[n]| \leq B_y \quad \forall n, \tag{2.134}$$

for some constant $B_y < \infty$. Then the system is said to be stable. ◇

This is also called *bounded-input bounded-output* or **BIBO** stability, to distinguish it from many other types of stability that are defined in the literature. A system that is not stable is said to be **unstable**.

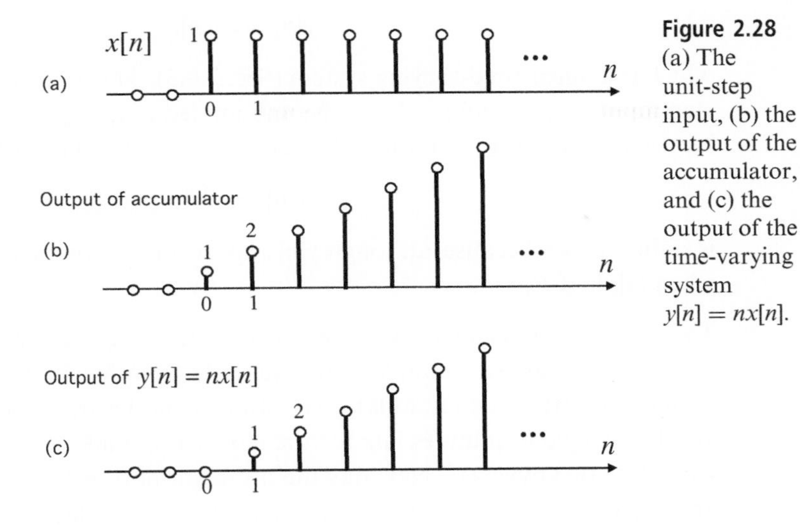

Figure 2.28 (a) The unit-step input, (b) the output of the accumulator, and (c) the output of the time-varying system $y[n] = nx[n]$.

2.11.1 Stable and Unstable Examples

The system

$$y[n] = x[n] + x[n-1] \tag{2.135}$$

is stable because $|x[n]| \leq B_x$ implies $|y[n]| \leq 2B_x$. Next consider the system

$$y[n] = \sum_{k=0}^{\infty} x[n-k] = x[n] + x[n-1] + x[n-2] + \cdots . \tag{2.136}$$

Thus, at time n the system output is simply the sum of all input samples up until time n. In other words, the system accumulates all the past and present sample values. It is therefore called an **accumulator**. It is clear that the system can also be described by the equation

$$y[n] = \sum_{k=-\infty}^{n} x[k], \tag{2.137}$$

and is analogous to the continuous-time **integrator**

$$y(t) = \int_{-\infty}^{t} x(\tau)d\tau. \tag{2.138}$$

The accumulator is an example of an unstable system. To prove that it is unstable, simply consider the unit-step input $x[n] = \mathcal{U}[n]$, which is bounded. Then the output is

$$y[n] = \begin{cases} 0 & n < 0, \\ n+1 & n \geq 0. \end{cases} \tag{2.139}$$

So the output grows unboundedly as n increases (Figs. 2.28(a) and (b)), proving that the system is unstable. In the same way, the continuous-time integrator is also unstable. All the above examples are LTI systems. Next consider the example

$$y[n] = nx[n], \tag{2.140}$$

which is a linear time-varying system (Sec. 2.8.3). This is unstable because the unit-step input $x[n] = \mathcal{U}[n]$ produces the unbounded output $y[n] = n\mathcal{U}[n]$ (Fig. 2.28(c)). Notice finally that even though the system $y[n] = e^{n}x[n]$ is unstable, the system

$$y[n] = e^{x[n]} \tag{2.141}$$

is stable. This is because, although $x[n]$ appears in the exponent, we have $|y[n]| \leq e^{B_x}$ when $|x[n]| \leq B_x$.

The professor's dilemma. A professor teaches classes at 1:00 pm and notices that most students arrive a couple of minutes late from lunch. The professor therefore decides to start two minutes late. The students notice this, and some of them come another couple of minutes late for the next class, thinking "Where is the hurry? It will start late anyways." Next day the professor notices that some students are still arriving only after the lecture starts. So from the next class the professor decides to start *another* couple of minutes late. Again some students notice this and decide to arrive even later, by another couple of minutes. The professor notices it, and again decides to starts even later. This goes on and on until ... well, we don't need to finish the story, but can you retell it using the language of unstable systems? ▽ ▽ ▽

2.11.2 The Capacitor and the Inductor

In electrical engineering, circuit elements such as the capacitor and the inductor are important building blocks. If $i(t)$ is the current flowing in a capacitor with voltage $v(t)$ across it (Fig. 2.29(a)), then these are related by

$$i(t) = C\frac{dv(t)}{dt}, \tag{2.142}$$

where the constant $C > 0$ is called the *capacitance*. If we view $v(t)$ as the input and $i(t)$ as the output, then the capacitor can be regarded as a differentiator. Since the above relation can also be written as

$$v(t) = \frac{1}{C}\int_{-\infty}^{t} i(\tau)d\tau, \tag{2.143}$$

the capacitor can also be regarded as an integrator if the current $i(t)$ is taken as the input and the voltage $v(t)$ as the output. Since the integrator is unstable, it follows that the capacitor is unstable. For example, a unit-step current source as in

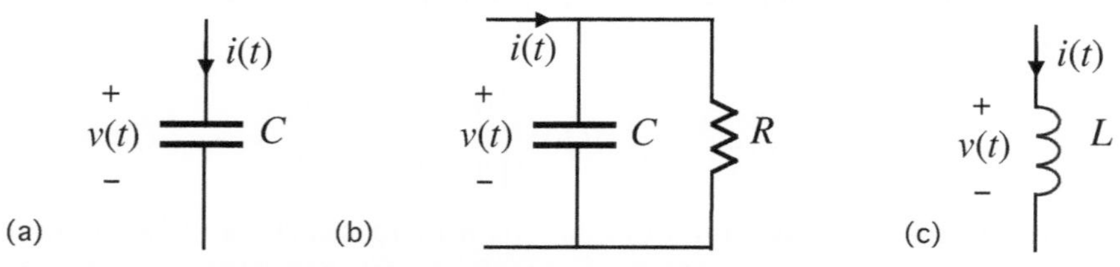

Figure 2.29 Some electrical circuit elements. (a) The ideal capacitor, (b) model for a practical capacitor with electrical loss, and (c) the ideal inductor.

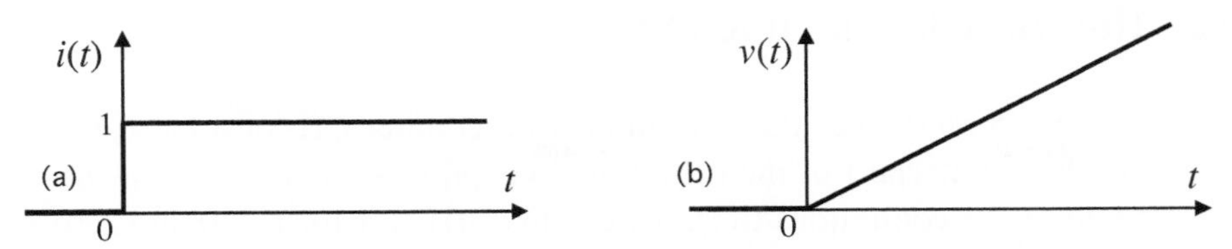

Figure 2.30 For an ideal capacitor, the unit-step input current as in (a) produces an unbounded output voltage as in (b).

Fig. 2.30(a) produces a linearly increasing voltage output, which is unbounded as in Fig. 2.30(b).

Another way to see the instability is to use Eq. (2.142): suppose

$$v(t) = \sin(\omega_0 t^2). \tag{2.144}$$

Clearly $|v(t)| \leq 1$, and this is a bounded signal. The corresponding output is

$$i(t) = 2t\omega_0 C \cos(\omega_0 t^2), \tag{2.145}$$

which is unbounded! This proves instability. In this example, the input $v(t) = \sin(\omega_0 t^2) = \sin((\omega_0 t)t)$ can be regarded as a signal with *instantaneous frequency* $\omega_0 t$, which increases linearly with time.

The real-world capacitor. A capacitor satisfying Eq. (2.142) is called an *ideal capacitor*. For a practical capacitor, the input–output equation is not exactly Eq. (2.142) because of losses in the capacitor arising due to dielectric loss, resistance in the connecting wires, and even radiation loss. A practical capacitor, with its losses, is actually stable. For such a system, Eq. (2.142) is replaced with

$$i(t) = C\frac{dv(t)}{dt} + \frac{v(t)}{R}, \tag{2.146}$$

where $R > 0$ is a resistor in parallel with C, representing electrical losses (Fig. 2.29(b)). $\triangledown\triangledown\triangledown$

Next, an ideal inductor (Fig. 2.29(c)) is a device which forces the following relation between the voltage and current:

$$v(t) = L\frac{di(t)}{dt}, \tag{2.147}$$

or equivalently

$$i(t) = \frac{1}{L}\int_{-\infty}^{t} v(\tau)d\tau, \tag{2.148}$$

where $L > 0$ is called the *inductance* of the inductor. So, like the ideal capacitor, an ideal inductor is unstable. A practical inductor turns out to be stable because of the inevitable electrical losses; its mathematical description is a corrected version of Eq. (2.147), similar to Eq. (2.146): $v(t) = Ldi(t)/dt + Ri(t), R > 0$.

2.12 The Dirac Delta Function

In Fig. 2.5 we showed the continuous-time counterparts of some signals in Fig. 2.4, but the counterpart of the delta function $\delta[n]$ was not shown. This has to be done carefully. In continuous time, if we define $\delta(t) = 1$ for $t = 0$ and zero everywhere else, then its integral is zero (although it is non-negative everywhere). This turns out to be not a useful way to define the delta function in continuous time, as will become clear in Sec. 3.9 and Chapter 9.

2.12.1 Defining the Continuous-Time Delta Function

A more useful approach is to define the continuous-time delta function as the limit of rectangular pulses $p_\Delta(t)$ of the form shown in Fig. 2.31(a) as $\Delta \to 0$. For any fixed $\Delta > 0$, the integral of $p_\Delta(t)$ is unity. But as $\Delta \to 0$, the height of the pulse increases in an unbounded manner. The continuous-time impulse, which is also called the **Dirac delta** function, is defined as follows:

$$\delta_c(t) = \lim_{\Delta \to 0} p_\Delta(t). \tag{2.149}$$

This should be interpreted carefully. What it means is that

$$\delta_c(t) = \begin{cases} 0 & t \neq 0, \\ \text{undefined} & t = 0, \end{cases} \tag{2.150}$$

and furthermore

$$\int_{-\infty}^{\infty} \delta_c(t)dt = 1, \tag{2.151}$$

because the integral of $p_\Delta(t)$ is unity for all $\Delta > 0$. Equations (2.150) and (2.151) together constitute the definition of the Dirac delta, although Eq. (2.149) is useful to provide insight. Note that we have used the subscript c in $\delta_c(t)$ as a second reminder that it is the continuous-time delta function (although the round brackets in $\delta_c(t)$ already distinguish it from $\delta[n]$).

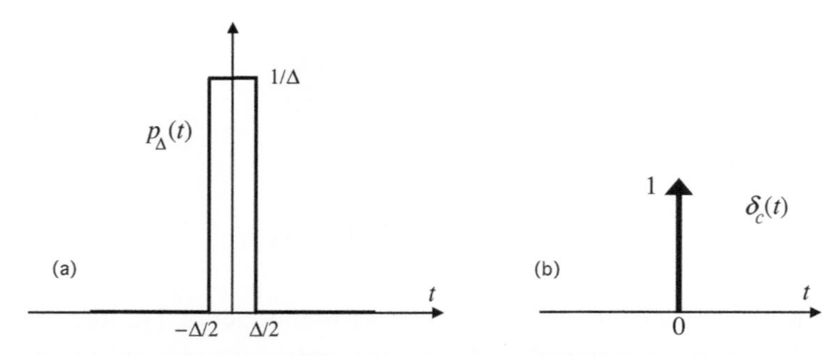

Figure 2.31 (a) A rectangular pulse with unit area. The Dirac delta function can be regarded as the limit of this as $\Delta \to 0$. (b) The notation used in figures for the Dirac delta function.

Figure 2.31(b) shows the notation used for the Dirac delta function. We simply draw a thick vertical arrow at $t = 0$. Even though the height of this arrow is labeled as "1," the meaning of this is that the *area* or *integral* of this function is unity. The height at $t = 0$ is actually infinite. Notice finally that, instead of regarding $\delta_c(t)$ as the limit of the rectangular pulses in Fig. 2.31(a), we could also have started with other shapes such as a *triangular* pulse or a *Gaussian* pulse (Fig. 2.5), to arrive at the same result.

2.12.2 Some Properties of the Dirac Delta

Since the function $\delta_c(-t)$ also satisfies properties similar to Eqs. (2.150) and (2.151), we say that $\delta_c(t)$ is an even function, that is,

$$\delta_c(-t) = \delta_c(t). \tag{2.152}$$

It is easy to show that for any real $\alpha \neq 0$,

$$\delta_c(\alpha t) = \frac{1}{|\alpha|}\delta_c(t), \tag{2.153}$$

which is a generalization of Eq. (2.152). To prove this, we only have to verify that

$$\int_{-\infty}^{\infty} \delta_c(\alpha t)dt = \frac{1}{|\alpha|}\int_{-\infty}^{\infty} \delta_c(t)dt, \tag{2.154}$$

which is simple (Problem 2.19). Next, notice that the definition given by Eqs. (2.150) and (2.151) also implies

$$\int_{-\alpha}^{\alpha} \delta_c(t)dt = 1, \tag{2.155}$$

for any $\alpha > 0$. Furthermore,

$$\int_{t_0-\alpha}^{t_0+\alpha} \delta_c(t - t_0)dt = 1, \tag{2.156}$$

for any real t_0 and any $\alpha > 0$. Here, $\delta_c(t - t_0)$ is the Dirac delta function shifted to the point t_0.

The picky function. Since $\delta_c(t - t_0) = 0$ everywhere except at $t = t_0$, it is clear that in the product $x(t)\delta_c(t - t_0)$, only the value of $x(t)$ at $t = t_0$ matters, assuming that it is well defined. So, we can write

$$x(t)\delta_c(t - t_0) = x(t_0)\delta_c(t - t_0), \tag{2.157}$$

whenever $x(t)$ is continuous at t_0. This is nothing but the Dirac delta shifted to time t_0 and scaled by the constant $x(t_0)$. In particular, note that

$$\int_{-\infty}^{\infty} x(t)\delta_c(t - t_0)dt = \int_{-\infty}^{\infty} x(t_0)\delta_c(t - t_0)dt = x(t_0), \tag{2.158}$$

where Eq. (2.151) has been used. Thus, the Dirac delta can be used to "pick" the value of $x(t)$ at any point t_0 where it is continuous. Even though the integral or area under $x(t_0)\delta_c(t - t_0)$ is $x(t_0)$, we refer to $x(t_0)$ as the height or amplitude or strength

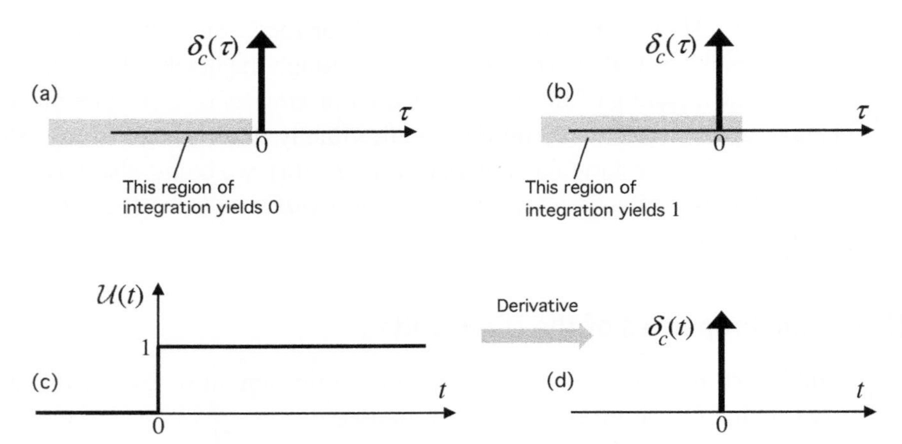

Figure 2.32 (a), (b) Integrating the Dirac delta function over different regions. (c), (d) The unit-step function $\mathcal{U}(t)$ and its derivative $\delta_c(t)$.

of the scaled Dirac delta in Eq. (2.157). This is a slight misuse of language, but is commonly used. ▽ ▽ ▽

2.12.3 Relation between the Unit Step and the Dirac Delta

From the properties in Eqs. (2.151) and (2.155) of the integrals involving the Dirac delta function, it should now be clear that

$$\int_{-\infty}^{t} \delta_c(\tau)d\tau = \begin{cases} 0 & t < 0, \\ 1 & t > 0. \end{cases} \tag{2.159}$$

The right-hand side above is nothing but the definition of the unit step, so

$$\int_{-\infty}^{t} \delta_c(\tau)d\tau = \mathcal{U}(t). \tag{2.160}$$

From this it follows that

$$\frac{d\mathcal{U}(t)}{dt} = \delta_c(t). \tag{2.161}$$

Figure 2.32 summarizes these relations between the unit step and the Dirac delta. These are analogous to Eqs. (2.11) and (2.12). Note that $\mathcal{U}(t)$ has a discontinuity at $t = 0$, and is not differentiable in the strict sense. However, if we allow the Dirac delta to enter our mathematics, we can say that the derivative of $\mathcal{U}(t)$ is $\delta_c(t)$.

2.13 Two-Dimensional Frequencies*

In two dimensions, the complex sinusoid or single-frequency signal has the form $x(\mathbf{t}) = e^{j\mathbf{\Omega}^T\mathbf{t}}$, where

$$\mathbf{\Omega} = \begin{bmatrix} \omega_1 \\ \omega_2 \end{bmatrix}, \quad \mathbf{t} = \begin{bmatrix} t_1 \\ t_2 \end{bmatrix} \tag{2.162}$$

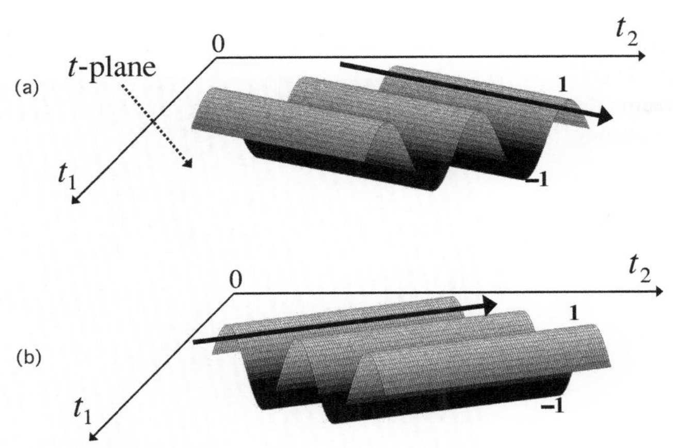

Figure 2.33 Plots of Eq. (2.164) in the (t_1, t_2) plane, for two different values of the frequency vector $\mathbf{\Omega}$. In the (t_1, t_2) plane, the frequency vector is orthogonal to the thick arrow.

are real-valued column vectors. Here $\mathbf{\Omega}$ is a constant, called the **frequency** vector, and $\mathbf{t}$ is the 2D version of "time." Similarly we have real sinusoids like $x_1(\mathbf{t}) = \cos(\mathbf{\Omega}^T\mathbf{t})$, $x_2(\mathbf{t}) = \sin(\mathbf{\Omega}^T\mathbf{t})$, and so on. We can regard t_1 and t_2 as spatial coordinates. In some examples, t_1 can be time and t_2 can represent a spatial coordinate as in Eq. (2.26). In any case, we often refer to $\mathbf{t}$ as "time" for convenience. Note that

$$\mathbf{\Omega}^T\mathbf{t} = \omega_1 t_1 + \omega_2 t_2, \tag{2.163}$$

so that

$$\cos(\mathbf{\Omega}^T\mathbf{t}) = \cos(\omega_1 t_1 + \omega_2 t_2). \tag{2.164}$$

How does this behave as a function of $\mathbf{t}$? To understand this, notice two things:

1. If we fix t_2 at a constant value, then this behaves like a 1D sinusoid in the variable t_1 with frequency ω_1.
2. Similarly for fixed t_1, it is a 1D sinusoid in the variable t_2 with frequency ω_2.

It is insightful to show this as a surface plot in the (t_1, t_2) plane. Figure 2.33 shows examples for two different frequency vectors $\mathbf{\Omega}$. Such a plot looks like a carpet unrolled like a cosine! In the direction shown by the thick arrow in the (t_1, t_2) plane, the signal has no variation at all. And in the direction orthogonal to the thick arrow in the (t_1, t_2) plane, the variation is maximum, that is, the cosine behavior is most prominent. This is the direction of the $\mathbf{\Omega}$ vector (as we shall see in Sec. 2.13.2).

2.13.1 Gray-Level Plots

To get a different perspective, one often plots $\cos(\mathbf{\Omega}^T\mathbf{t})$ as a gray level in the 2D plane (t_1, t_2). Cosines take both positive and negative values, so in order to plot this as a gray level, one might be tempted to plot $\cos^2(\mathbf{\Omega}^T\mathbf{t})$ (or just the absolute value).

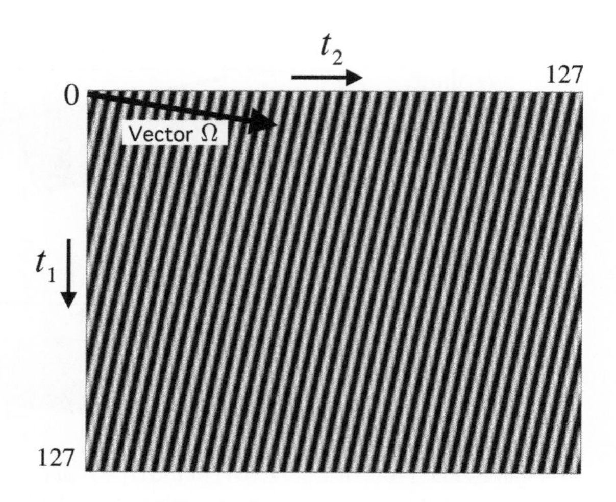

Figure 2.34 A plot of $[1 + \cos(\mathbf{\Omega}^T\mathbf{t})]/2$ for $\mathbf{\Omega} = 2\pi[0.05\ 0.3]^T$. The thick arrow represents the direction of the frequency vector $\mathbf{\Omega}$.

But this varies twice as fast, because each -1 becomes $+1$ and the distance between peaks is thereby halved. So, to get an accurate representation of the peak-to-peak distances, we plot

$$\frac{1 + \cos(\mathbf{\Omega}^T\mathbf{t})}{2}. \tag{2.165}$$

This is demonstrated in Fig. 2.34 for the frequency

$$\mathbf{\Omega} = 2\pi[0.05 \quad 0.3]^T. \tag{2.166}$$

The dark lines represent points where the intensity is zero and the bright lines represent unit intensity. We will see that these lines of constant intensity are orthogonal to the orientation of the frequency vector $\mathbf{\Omega}$, shown as the thick arrow in the figure. Although the eye perceives the plot mostly as a bunch of parallel lines, there is smooth variation of intensity between the lines. Note that in the t_1 direction the variation of intensity is slower than in the t_2 direction because $\omega_1 = 2\pi \times 0.05$ is smaller than $\omega_2 = 2\pi \times 0.3$.

2.13.2 Understanding the Frequency Vector Geometrically

We mentioned that the thick arrow in Fig. 2.34 represents the direction of the frequency vector $\mathbf{\Omega}$. To understand this, first note that in these sinusoids, $\mathbf{t}$ arises only through the expression $\mathbf{\Omega}^T\mathbf{t}$. In the (t_1, t_2) plane, the quantity $\mathbf{\Omega}^T\mathbf{t}$ has a constant value along any line with a certain special slope (like the thick arrows in Fig. 2.33). To see this, first note that any $\mathbf{t}$ can be uniquely written as

$$\mathbf{t} = \mathbf{t}_a + \mathbf{t}_\perp, \tag{2.167}$$

where $\mathbf{t}_a$ is aligned to (i.e., parallel to) $\mathbf{\Omega}$ and $\mathbf{t}_\perp$ is orthogonal to it, that is, $\mathbf{\Omega}^T\mathbf{t}_\perp = 0$. So,

$$\mathbf{\Omega}^T\mathbf{t} = \mathbf{\Omega}^T\mathbf{t}_a + \mathbf{\Omega}^T\mathbf{t}_\perp = \mathbf{\Omega}^T\mathbf{t}_a. \tag{2.168}$$

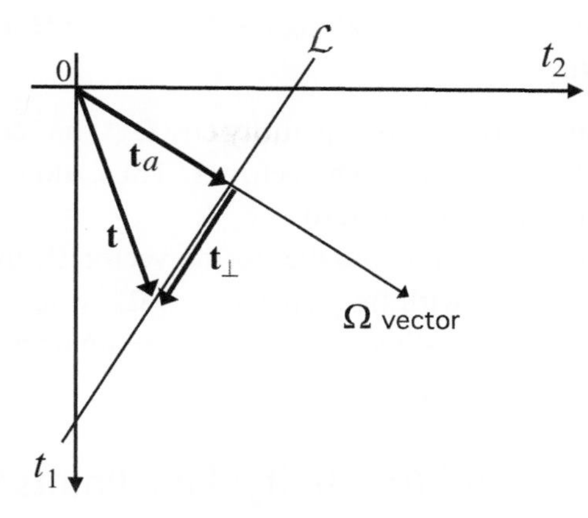

Figure 2.35 The 2D frequency vector $\boldsymbol{\Omega}$. A plot of $\cos(\boldsymbol{\Omega}^T\mathbf{t})$ has constant value along any line $\mathcal{L}$ orthogonal to $\boldsymbol{\Omega}$. See text.

Now, in the **t**-plane, draw a vector representing $\boldsymbol{\Omega}$ and consider any line $\mathcal{L}$ orthogonal (i.e., perpendicular) to it, as shown in Fig. 2.35. The vector **t** representing any point on this line $\mathcal{L}$ can be written as $\mathbf{t} = \mathbf{t}_a + \mathbf{t}_\perp$, where $\mathbf{t}_a$ is *fixed* and only $\mathbf{t}_\perp$ changes, as we move along this line. So, from Eq. (2.168) it follows that $\boldsymbol{\Omega}^T\mathbf{t}$ has the same value for all **t** on the line $\mathcal{L}$, and this value is determined by $\mathbf{t}_a$. Hence, functions of the form $e^{j\boldsymbol{\Omega}^T\mathbf{t}}$, $\cos(\boldsymbol{\Omega}^T\mathbf{t})$, and $\sin(\boldsymbol{\Omega}^T\mathbf{t})$ are **constant along lines** orthogonal to $\boldsymbol{\Omega}$. This is clearly seen in the plot of Fig. 2.34. Next, we can always write

$$\mathbf{t}_a = \tau\mathbf{u}_a, \tag{2.169}$$

where $\mathbf{u}_a$ is the unit vector along the direction of $\mathbf{t}_a$ (i.e., along the direction of $\boldsymbol{\Omega}$). That is,

$$\mathbf{u}_a = \frac{\boldsymbol{\Omega}}{\sqrt{\omega_1^2 + \omega_2^2}} = \frac{\boldsymbol{\Omega}}{\|\boldsymbol{\Omega}\|}, \tag{2.170}$$

where

$$\|\boldsymbol{\Omega}\| = \sqrt{\omega_1^2 + \omega_2^2} \tag{2.171}$$

is the norm or "length" of the vector $\boldsymbol{\Omega}$. Then it follows that

$$\boldsymbol{\Omega}^T\mathbf{t} = \boldsymbol{\Omega}^T\mathbf{t}_a = \boldsymbol{\Omega}^T\mathbf{u}_a\tau = \frac{\boldsymbol{\Omega}^T\boldsymbol{\Omega}}{\|\boldsymbol{\Omega}\|}\tau = \|\boldsymbol{\Omega}\|\tau, \tag{2.172}$$

which depends only on the scalar τ and the norm of the frequency $\boldsymbol{\Omega}$. Thus,

$$\cos(\boldsymbol{\Omega}^T\mathbf{t}) = \cos(\boldsymbol{\Omega}^T\mathbf{u}_a\tau) = \cos(\|\boldsymbol{\Omega}\|\tau) = \cos(\omega\tau), \tag{2.173}$$

where

$$\omega = \boldsymbol{\Omega}^T\mathbf{u}_a = \|\boldsymbol{\Omega}\|. \tag{2.174}$$

In short, if we move along the direction of the vector $\boldsymbol{\Omega}$, we simply see an ordinary 1D cosine $\cos(\omega\tau)$ in the variable τ, with 1D frequency $\omega = \boldsymbol{\Omega}^T\mathbf{u}_a = \|\boldsymbol{\Omega}\|$. Here, τ represents the amount of movement along $\mathbf{u}_a$ (i.e., along $\boldsymbol{\Omega}$). For example, if $\tau = 0.1$

secs then the movement is 0.1 along the direction of $\mathbf{\Omega}$. Here is the summary of what we have shown.

1. In the 2D plane (t_1, t_2), the quantity $\cos(\mathbf{\Omega}^T\mathbf{t})$ has constant value along any line orthogonal to the frequency vector $\mathbf{\Omega}$. Thus, along lines orthogonal to $\mathbf{\Omega}$, the perceived 1D frequency is zero.
2. If we move along a line parallel to the vector $\mathbf{\Omega}$, then $\cos(\mathbf{\Omega}^T\mathbf{t})$ behaves like a 1D cosine $\cos(\omega\tau)$ with frequency $\omega = \|\mathbf{\Omega}\|$, where τ represents the amount of movement along the direction of $\mathbf{\Omega}$. So, along lines parallel to $\mathbf{\Omega}$, the perceived 1D frequency is $\omega = \|\mathbf{\Omega}\|$.

2.14 Homogeneity and Additivity: Fine Points*

We know that there exist systems which are homogeneous but not additive, as shown by the following example (Sec. 2.8):

$$y[n] = \begin{cases} \dfrac{x^2[n]}{x[n-1]} & x[n-1] \neq 0, \\ 0 & x[n-1] = 0. \end{cases} \tag{2.175}$$

On the other hand, it was also shown that the system

$$y[n] = \text{Re}\Big(x[n]\Big) \tag{2.176}$$

is additive but not homogeneous. Thus, homogeneity does not imply additivity and vice versa. Note that even if we restrict the input $x[n]$ to be real valued, the system of Eq. (2.175) is homogeneous and non-additive. So, homogeneity does not imply additivity whether we consider the real or complex case. On the other hand, if we restrict the input to be real in Eq. (2.176), then the system reduces to the identity system $y[n] = x[n]$, which is trivially homogeneous (in addition to being additive). This therefore raises the following question:

For real systems, does additivity imply homogeneity automatically?

The term "real systems" refers to systems that take only real inputs, and produce real outputs in response to them. Carefully note that homogeneity here means that if $x[n] \longmapsto y[n]$ then $cx[n] \longmapsto cy[n]$ for any **real** c. Complex c need not be considered because the system is "real" in this discussion.

To examine the above question, we consider a real additive system $x[n] \longmapsto y[n]$ and make a number of observations.

1. First, $x[n] + x[n] \longmapsto y[n] + y[n]$ by additivity. By repeated application of this we see that $Nx[n] \longmapsto Ny[n]$ for any positive integer N. But since additivity implies subtractivity (Sec. 2.8.1), we see that $-Nx[n] \longmapsto -Ny[n]$ as well. By additivity again, this means that $Nx[n] - Nx[n] \longmapsto Ny[n] - Ny[n]$, that is zero input produces zero output! Combining these properties, it follows that for an additive system $x[n] \longmapsto y[n]$ we also have

$$Nx[n] \longmapsto Ny[n], \tag{2.177}$$

for *any* integer N (positive, negative, or zero).

2. We now claim that the additive system also satisfies

$$x[n]/M \longmapsto y[n]/M \quad \text{for any nonzero integer } M. \tag{2.178}$$

To prove this, let the output in response to $x[n]/M$ be denoted as $y_1[n]$. So we have

$$x[n] \longmapsto y[n] \quad \text{and} \quad x[n]/M \longmapsto y_1[n]. \tag{2.179}$$

So, by (2.177) we have $Mx[n]/M \longmapsto My_1[n]$, that is, $x[n] \longmapsto My_1[n]$, which proves $y[n] = My_1[n]$ or equivalently $y_1[n] = y[n]/M$.

3. By combining the mappings (2.177) and (2.178), it can therefore be concluded that for an additive system $x[n] \longmapsto y[n]$,

$$\frac{N}{M}x[n] \longmapsto \frac{N}{M}y[n] \quad \text{for any integer } N \text{ and nonzero integer } M. \tag{2.180}$$

Now recall that a **rational number** is a ratio of integers, like N/M. So Eq. (2.180) says that, for an additive system $x[n] \longmapsto y[n]$, we certainly have

$$rx[n] \longmapsto ry[n] \quad \text{for any rational number } r. \tag{2.181}$$

But it is well known that any real number c can be approximated arbitrarily accurately by a rational number $r = N/M$ [Haaser and Sullivan, 1991]. Can we therefore conclude from Eq. (2.181) that $cx[n] \longmapsto cy[n]$ for any real c? For all practical purposes the answer is yes, but we have to be careful about some hypothetical situations. So, let us refine this a bit more.

4. Since Eq. (2.181) is already proved, we only have to consider real irrational c now (like $c = \pi$, $c = \sqrt{2}$, etc.). Any real irrational number c can be approximated, as closely as we please, by a rational number. That is, there exists a sequence of rationals r_k such that $r_k \to c$ as $k \to \infty$. Given $x[n] \longmapsto y[n]$, we have

$$r_k x[n] \longmapsto r_k y[n]. \tag{2.182}$$

The question is, can we also say $cx[n] \longmapsto cy[n]$? For now, let us denote the output by $y_1[n]$, that is,

$$cx[n] \longmapsto y_1[n]. \tag{2.183}$$

Then, from the mappings (2.182) and (2.183),

$$(r_k - c)x[n] \longmapsto r_k y[n] - y_1[n], \tag{2.184}$$

in view of additivity (which also implies subtractivity). But we already know that additivity also implies that zero input ($x[n] = 0 \ \forall n$) produces zero output:

$$\mathbf{0} \longmapsto \mathbf{0}. \tag{2.185}$$

The use of bold $\mathbf{0}$ is a reminder that it is a sequence $\{\cdots 0, 0, 0, \cdots\}$ and not just a number. Now, the left-hand side of (2.184) goes to zero as $k \to \infty$. Suppose the right-hand side does not. This means that we can find arbitrarily small input $(r_k - c)x[n]$ such that the output $\epsilon[n]$ is strictly away from zero, that is, $|\epsilon[n]| = \delta > 0$ for some n. By comparing this conclusion with the result (2.185), it follows that the system $x[n] \longmapsto y[n]$ has **discontinuity at the origin**. That is, when $x[n] = 0\ \forall n$, we also have $y[n] = 0\ \forall n$, but there exists an input arbitrarily close to $x[n] = 0\ \forall n$, such that the output abruptly moves to a value $\epsilon[n]$ strictly away from zero.

5. We therefore conclude that, as long as the real system $x[n] \longmapsto y[n]$ does not have such discontinuous behavior at $x[n] = 0$, we can say that additivity implies homogeneity.

Here is a summary of what we have shown.

1. Homogeneity does not imply additivity, whether the system is real or complex (see example (2.175)).
2. For complex systems, additivity does not imply homogeneity (see example (2.176)).
3. For real systems (which have input restricted to be real, and output also real), additivity implies homogeneity in a restricted sense, that is, the property (2.181) holds true. For all practical purposes this means that the system is homogeneous. The only hypothetical case where this can fail is when the system $x[n] \longmapsto y[n]$ has discontinuity at the origin in the sense explained above.

2.15 Summary of Chapter 2

This chapter first introduced continuous-time signals $x(t)$ and discrete-time signals $x[n]$. Various examples, such as the exponential signal, the unit step, and the rectangular pulse, were given. We introduced the unit pulse or discrete-time impulse $\delta[n]$ and the Dirac delta or continuous-time impulse $\delta_c(t)$. A number of real-world signals were also mentioned. We discussed even and odd signals, periodic signals, and single-frequency signals (or complex sinusoids). Single-frequency signals can be represented by rotating vectors whose speed of rotation represents frequency. We also explained the meaning of two-dimensional frequencies and demonstrated the idea using two-dimensional sinusoids. There are some differences between continuous-time and discrete-time frequencies.

1. The frequency range in discrete time is $-\pi \leq \omega < \pi$ whereas that in continuous time is $-\infty < \Omega < \infty$.
2. The single-frequency signal $e^{j\omega n}$ in discrete time is periodic in n only when ω is a rational multiple of π. But the single-frequency signal $e^{j\Omega t}$ in continuous time is always periodic in t.
3. The single-frequency signal $e^{j\omega n}$ in discrete time is periodic in ω.

Then the chapter introduced the meaning of "systems." A system takes an input $x[n]$ (or $x(t)$) to produce a unique output $y[n]$ (or $y(t)$). It is also called a mapping or an operator. Several types of systems were presented, as summarized below.

1. Homogeneous: $x[n] \longmapsto y[n]$ implies $cx[n] \longmapsto cy[n]$.
2. Additive: $x_k[n] \longmapsto y_k[n]$ implies $x_1[n] + x_2[n] \longmapsto y_1[n] + y_2[n]$.
3. Linear: $x_k[n] \longmapsto y_k[n]$ implies $c_1x_1[n] + c_2x_2[n] \longmapsto c_1y_1[n] + c_2y_2[n]$.
4. Time invariant: $x[n] \longmapsto y[n]$ implies $x[n-N] \longmapsto y[n-N]$.
5. Causal: Output does not depend on the input at future times.
6. Anticausal: Output does not depend on the input in the present or past times.
7. Memoryless: Output at any time is determined if input at that time is known.
8. Stable: Bounded input produces bounded output.

Many examples were presented to demonstrate the behaviors of various types of systems. Several subtle points relating to homogeneity and addivity were also discussed. We showed that homogeneity and addivity do not in general imply each other. A real-time system is required to be causal, and most useful systems are required to be stable. It should be mentioned that some unstable systems can be useful starting points for certain applications, for example, in the design of an oscillator or waveform generator (Sec. 8.5).

Moving forward, we will see that many of the concepts introduced in this chapter form the backbone for the next few chapters, where we go deeper into linear time-invariant systems and digital filters.

PROBLEMS

Note: If you are using MATLAB for the computing assignments at the end, the "stem" command is convenient for plotting sequences.

2.1 Let $x[n]$ be a discrete-time signal such that

$$x[-2] = 1, \quad x[-1] = 2, \quad x[0] = 3, \quad x[1] = -1, \quad x[2] = 4, \quad x[3] = -2,$$

and let $x[n] = 0$ for all other n. Plot the following sequences:
(a) $x[n]$, $x[-n]$, $x[n-2]$, $x[-n+2]$, and $x[-n-2]$;
(b) $x[2n]$;
(c) $x[n]\mathcal{U}[n]$ and $x[n]\mathcal{U}[-n+2]$ where $\mathcal{U}[n]$ is the unit step;
(d) $x[n]\delta[n+1]$ and $x[n]\delta[n-2]$.

2.2 For the signal $x[n]$ in Problem 2.1, find the even part $x_e[n]$ and the odd part $x_o[n]$, and plot them.

2.3 For the signal $x(t)$ shown in Fig. P2.3, find the even part $x_e(t)$ and the odd part $x_o(t)$.

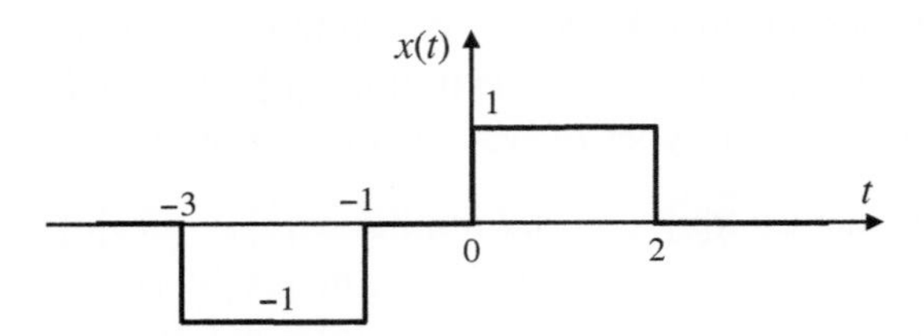

Figure P2.3 Signal $x(t)$ for Problem 2.3.

2.4 We know that cosines and sines are superpositions of two single-frequency signals. For each of the following signals, find these two frequencies in Hertz, assuming t is expressed in seconds. (a) $\cos(\pi t)$, (b) $\cos(t)$, (c) $\sin(2\pi t + 2)$, and (d) $\sin(-3t + 2)$. For $\cos(t)$, what are the two frequencies in rads/sec?

2.5 Which of the following discrete-time signals are periodic? (a) $e^{j0.1\pi n}$, (b) $e^{j0.1n}$, (c) e^{jn^2} and (d) $e^{j\pi n^2}$. For those that are periodic, find the period.

2.6 Find an example of an unstable, non-memoryless, linear time-varying system in discrete time.

2.7 Show that the system $y[n] = 1 + x[n]$ is not homogeneous. Is the system additive?

2.8 For the signal $x[n]$ in Fig. 2.22(a), plot the signal $x[2 - 3n]$.

2.9 For the signal $x(t)$ in Fig. 2.25(a) we obtained a plot of $y(t) = x(1 - 2t)$ by performing a dilation, then a flip, and then a shift. Instead, we can first flip $x(t)$, then shift by an appropriate amount, and then dilate to obtain $y(t) = x(1 - 2t)$. Show the plots of these intermediate signals, analogous to what we plotted in Fig. 2.25.

2.10 For the signal $x(t)$ shown in Fig. P2.10, plot the signal $x(1 + 3t)$.

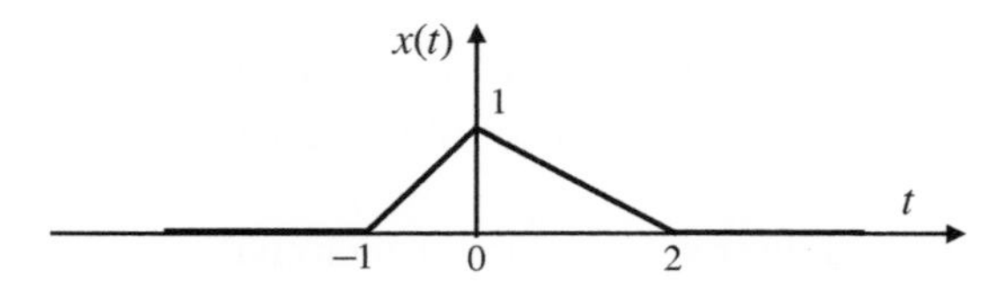

Figure P2.10 Signal $x(t)$ for Problem 2.10.

2.11 For each of the systems below, examine which of these properties are satisfied: linearity, time invariance, stability, causality, and memorylessness:
(a) $y[n] = x[n^2]$;
(b) $y[n] = 2^{-n}x[n]x[n - 1]$;
(c) $y[n] = \sin(\pi n/4)x[n]$;
(d) $y[n] = x[n + 1] + x[n - 20]$.

2.12 In this chapter we saw some examples of computational graphs for systems (Figs. 2.18, 2.20, and 2.21). Based on these, draw a computational graph for the system $y[n] = n\big(x[n] + ax[n - 1]\big) + 2$.

2.13 Show that a discrete-time system which is both causal and anticausal is also memoryless. Show also that the converse is not true.

2.14 Consider the continuous-time system

$$y(t) = \begin{cases} x(t)/t & t > 0, \\ 0 & t \le 0. \end{cases}$$

Which of the following properties is satisfied by this system? (a) Linearity, (b) time invariance, (c) stability. Also, find an unbounded input such that the output is bounded.

2.15 Recall Eq. (2.130), which shows a system that is both causal and anticausal.

(a) Show that this is a time-varying system.

(b) Next, find a discrete-time system that is both causal and anticausal, and also time invariant. Just to avoid trivial answers make sure the output is not the zero signal when $x[n] = \delta[n]$.

2.16 In Sec. 2.9.2 we listed six examples and claimed that the first three systems are time invariant, whereas the last three systems are linear. We proved a couple of these in the chapter. Prove the remainder.

2.17 In this problem we consider a number of systems and check some of their basic properties.

(a) Consider the system

$$y[n] = \begin{cases} x[n]/n & n \neq 0, \\ 1 & n = 0. \end{cases} \tag{P2.17a}$$

Which of the following properties are satisfied by the system? (i) Linearity, (ii) time invariance, (iii) causality, (iv) memoryless property, and (v) stability.

(b) Consider the system

$$y[n] = \begin{cases} x[n]/n & n \neq 0, \\ 0 & n = 0. \end{cases} \tag{P2.17b}$$

Is this linear? Is this stable?

(c) Consider the system

$$y[n] = \begin{cases} nx[n] & n \neq 0, \\ 0 & n = 0. \end{cases} \tag{P2.17c}$$

Is this linear? Is this stable? *Note:* The second line in Eq. (P2.17c) is redundant, but we added it for uniformity with Eqs. (P2.17a) and (P2.17b).

2.18 Suppose a discrete-time system is additive. Show that the system is also subtractive. That is, if $x_1[n] \mapsto y_1[n]$ and $x_2[n] \mapsto y_2[n]$, then $x_1[n] - x_2[n] \mapsto y_1[n] - y_2[n]$. (This was mentioned in Sec. 2.8 but the proof was omitted.) *Warning:* Do *not* assume the system is homogeneous.

2.19 Show that $\int_{-\infty}^{\infty} \delta_c(\alpha t)dt = \int_{-\infty}^{\infty} \delta_c(t)dt/|\alpha|$, for any real nonzero α.

2.20 The pulse in Fig. 2.5(b) is discontinuous at $t = 0$ and $t = 4$, and is not differentiable in the traditional sense. However, we can obtain a "derivative" if we use the Dirac delta function in the answer, like we did for the unit-step function $\mathcal{U}(t)$ in Figs. 2.32(c) and (d). Find the derivative of the pulse in this sense, and plot it.

2.21 *Period and repetition interval.* Let $x[n]$ have period P (i.e., P is the smallest positive integer such that $x[n] = x[n+P]$, $\forall n$). Suppose the positive integer R is also a **repetition interval** for $x[n]$, that is, $x[n] = x[n+R]$, $\forall n$. Show that R is necessarily a multiple of P, that is, $R = R_1 P$, for some integer R_1.

2.22 *Period and repetition interval, continuous time.* Let $x(t)$ have period T (i.e., T is the smallest positive number such that $x(t) = x(t+T)$, $\forall t$). Suppose $R > 0$ is also a repetition interval for $x(t)$, that is, $x(t) = x(t+R)$, $\forall t$. Show that R is necessarily a multiple of T, that is, $R = mT$, for some integer m.

2.23 *Decimating a periodic signal.* Let $x[n]$ be periodic with period P (i.e., P is the smallest positive integer such that $x[n] = x[n+P]$, $\forall n$). Define the decimated version $y[n] = x[Mn]$, where $M > 0$ is an integer.

(a) Show that $y[n]$ is periodic.

(b) Show that $y[n] = y[n+Q]$, where $Q = P/(M,P)$. Here, (M,P) denotes the greatest common divisor (gcd) of M and P. This shows that the period of $y[n]$ is Q or a divisor of Q.

2.24 We know that a continuous-time signal $x_c(t)$ is said to be periodic if $x_c(t) = x_c(t+\tau)$ for all t, for some $\tau > 0$, and that a discrete-time signal $x[n]$ is said to be periodic if $x[n] = x[n+P]$ for all n, for some **integer** $P > 0$.

(a) Recall that $x_c(t) = e^{j\Omega_0 t}$ is periodic with period $\tau = 2\pi/|\Omega_0|$. Let $x[n] = x_c(nT)$ be the sampled version. Show that $x[n]$ is periodic if and only if τ/T is rational, that is, $\tau/T = P/Q$ for some positive integers P and Q. (*Hint:* Use the fact that $e^{j\omega_0 n}$ is periodic if and only if ω_0 is a rational multiple of π, that is, $\omega_0 = (K/L)\pi$ where K and L are integers.)

(b) More generally, let $x_c(t)$ be an arbitrary continuous-time signal with period τ, and let $x[n] = x_c(nT)$ be the sampled version. Let τ/T be rational. Show that $x[n]$ is periodic.

In part (b), it can be shown that rationality of τ/T is in fact *necessary*, and not merely *sufficient*, for periodicity of $x[n]$. But we are not asking you to prove it here.

2.25 (Computing assignment) *Difference operators.* Let $x[n] = (1.1)^n \mathcal{U}[n]$. This is a one-sided exponential starting at $n = 0$. Let $x_1[n]$ and $x_2[n]$ be the first difference and second difference of $x[n]$, respectively (Sec. 2.5.2). Plot $x[n], x_1[n]$ and $x_2[n]$ for $0 \le n \le 40$. Do the plots of $x_1[n]$, and $x_2[n]$ look the same as $x[n]$ except for some scale factors? If not, is there an n_0 such that all three plots look the same for $n \ge n_0$ (except for a scale factor)?

2.26 (Computing assignment) *A time-varying operator.* Let $x[n] = (1.1)^n \mathcal{U}[n]$ and let

$$y[n] = x[n] + (-1)^n x[n-1],$$

which is a time-varying system. Plot $x[n]$ and $y[n]$ for $0 \le n \le 40$. You will find that $y[n]$ never looks like an exponential (even approximately), no matter how large n gets.

2.27 (Computing assignment) *A non-additive operator.* Consider again the system (2.86) which is not additive (hence not linear). For $x[n] = a^n\mathcal{U}[n]$, we easily see from the system definition that $y[n] = a \cdot a^n\mathcal{U}[n-1]$. So, one-sided exponential inputs produce one-sided exponential outputs for this system. But if we apply $x[n] = \cos(\omega_0 n)\mathcal{U}[n]$, then the output is not necessarily a cosine. For example, let $\omega_0 = \pi\sqrt{2}/10$. Then plot $x[n]$ and $y[n]$ for $0 \leq n \leq 40$. You will see that $x[n]$ looks like a cosine (because it is), whereas $y[n]$ does not look like a cosine even approximately.

3 Linear Time-Invariant Systems

3.1 Introduction

In the previous chapter we introduced many different types of systems. Of these, linear time-invariant (LTI) systems have considerable importance in several areas of engineering, including signal processing, because a number of real-world systems can be approximated by LTI systems. In this chapter we look into discrete-time LTI systems in greater detail. It will be seen that the input–output behavior of an LTI system is completely characterized by the so-called impulse response.

An operation called convolution arises in the study of LTI systems and is discussed in detail. Using this it will be shown that exponential signals such as a^n and $e^{j\omega_0 n}$ retain their functional form when passed through an LTI system. That is, the shape is unaltered, although they may get scaled by a constant. This property leads to the important ideas of transfer functions, frequency responses, and eigenfunctions of LTI systems. The frequency response is fundamental because it describes how the system handles various frequency components of the input. It also gives a systematic meaning to the term "filtering."

The discrete-time Fourier transform (DTFT) also arises naturally in the discussion. This allows us to represent a signal as a superposition of single-frequency signals. Such a representation is called the Fourier representation. We will discuss the properties of DTFT along with several examples. The z-transform, which is of great importance in the study of LTI systems, is also introduced. It will be seen that the DTFT is a special case of the z-transform. The continuous-time counterparts of the main results are then summarized, along with several examples. A summary of properties of the DTFT is given in Sec. 3.13, and a table of DTFT pairs is given in Sec. 3.14 for quick reference.

Note: In this chapter we use the terms time-invariant and shift-invariant interchangeably.

3.2 Input–Output Description

Let $x[n]$ and $y[n]$ denote the input and output of an LTI system. First consider the simple input signal

$$x[n] = \delta[n], \tag{3.1}$$

which is nothing but the impulse or unit pulse. We will use $h[n]$ to denote the corresponding output. This output is called the **impulse response** of the system. It turns out that if the impulse response $h[n]$ of an LTI system is known, we can find the output $y[n]$ for any arbitrary input $x[n]$. To understand this, observe that $x[n]$ is nothing but a sum of the form

$$x[n] = \cdots + x[-1]\delta[n+1] + x[0]\delta[n] + x[1]\delta[n-1] + x[2]\delta[n-2] + \cdots, \quad (3.2)$$

that is, a sum of shifted and scaled impulses. If the input were $\delta[n-k]$, that is, the impulse shifted by k samples, then the output would be

$$h[n-k], \quad (3.3)$$

because of **time invariance** (Sec. 2.8.3). So the output in response to the kth term $x[k]\delta[n-k]$ in Eq. (3.2) is

$$x[k]h[n-k], \quad (3.4)$$

by the **homogeneity** property of linear systems (Sec. 2.8.1). This is demonstrated in Fig. 3.1 for $k = 0, 1, 2$. So the output in response to the input (3.2) is

$$y[n] = \cdots x[-1]h[n+1] + x[0]h[n] + x[1]h[n-1] + x[2]h[n-2] + \cdots \quad (3.5)$$

by the **additivity** property of linear systems (Sec. 2.8.1). Thus, the output is the sum of all these scaled and shifted versions demonstrated in Fig. 3.1. This can be written more compactly as

$$y[n] = \sum_{k=-\infty}^{\infty} x[k]h[n-k]. \quad (3.6)$$

Notice that, in obtaining this expression, we have used linearity (to be specific, both homogeneity and additivity) and time invariance. The right-hand side of Eq. (3.6) is called the **convolution** sum of $x[n]$ with $h[n]$. Thus, the output $y[n]$ of *any* discrete-time LTI system can be expressed as a convolution of the input $x[n]$ with the impulse response $h[n]$ of the system. Conversely, suppose an arbitrary system is such that the output is given by Eq. (3.6) for any input $x[n]$. Does this mean that the system is LTI? This is indeed so, as we show next.

Proof of the converse. Assume a system is described by Eq. (3.6). Assume $x_1[n]$ and $x_2[n]$ produce outputs $y_1[n]$ and $y_2[n]$, so that

$$y_1[n] = \sum_{k=-\infty}^{\infty} x_1[k]h[n-k], \quad y_2[n] = \sum_{k=-\infty}^{\infty} x_2[k]h[n-k]. \quad (3.7)$$

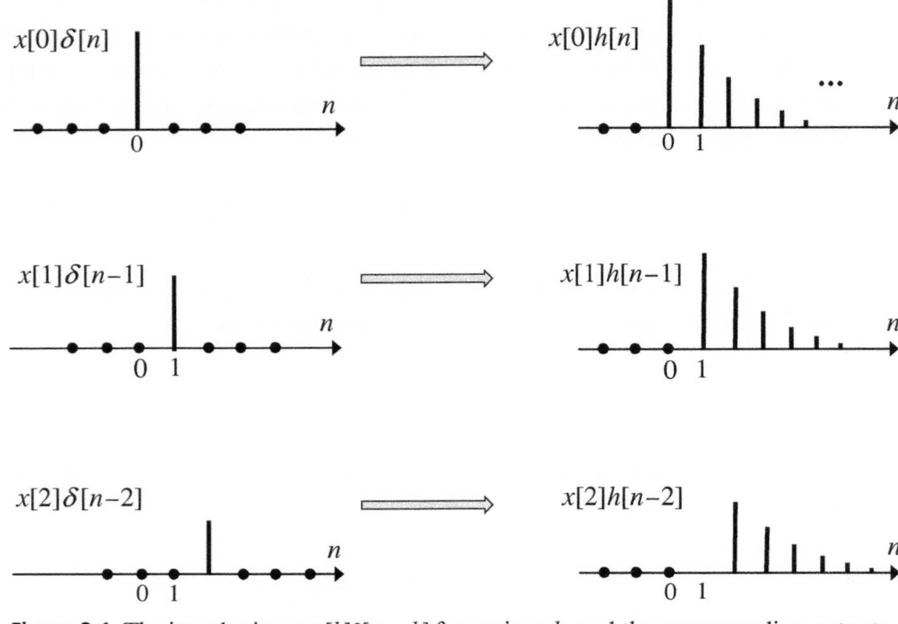

Figure 3.1 The impulse input $x[k]\delta[n-k]$ for various k, and the corresponding outputs of an LTI system. Here, $h[n]$ is the impulse response of the system.

Now apply the input $x[n] = c_1x_1[n] + c_2x_2[n]$. Then the output is

$$\begin{aligned}
y[n] &= \sum_{k=-\infty}^{\infty} x[k]h[n-k] \\
&= \sum_{k=-\infty}^{\infty} \Big(c_1x_1[k] + c_2x_2[k]\Big)h[n-k] \\
&= c_1 \sum_{k=-\infty}^{\infty} x_1[k]h[n-k] \;+\; c_2 \sum_{k=-\infty}^{\infty} x_2[k]h[n-k] \\
&= c_1y_1[n] + c_2y_2[n],
\end{aligned} \tag{3.8}$$

which proves linearity. Next, suppose we define the shifted version $\widehat{x}[n] = x[n-l]$ as the input. Then the output is

$$\begin{aligned}
\widehat{y}[n] &= \sum_{k=-\infty}^{\infty} \widehat{x}[k]h[n-k] \\
&= \sum_{k=-\infty}^{\infty} x[k-l]h[n-k] \\
&= \sum_{m=-\infty}^{\infty} x[m]h[n-l-m] = y[n-l],
\end{aligned} \tag{3.9}$$

which proves shift invariance. ▽ ▽ ▽

By making the change of variables $m = n - k$, we see that Eq. (3.6) becomes

$$y[n] = \sum_{m=-\infty}^{\infty} h[m]x[n-m], \tag{3.10}$$

which shows that the output of an LTI system can be written in two ways:

$$y[n] = \sum_{k=-\infty}^{\infty} x[k]h[n-k] = \sum_{k=-\infty}^{\infty} h[k]x[n-k]. \tag{3.11}$$

Thus, the operands $x[n]$ and $h[n]$ can be interchanged, that is, convolution is a **commutative** operation. The notation $(x * h)[n]$ is often used to denote convolution of $x[\cdot]$ and $h[\cdot]$. Thus

$$\sum_{k=-\infty}^{\infty} x[k]h[n-k] = (x * h)[n] = (h * x)[n]. \tag{3.12}$$

3.3 Impulse Response and LTI Properties

Given the impulse response coefficients $h[n]$ of an LTI system, we can find the output $y[n]$ for *any* input, which shows that an LTI system is completely characterized by its impulse response, as far as all input–output properties are concerned. In particular, all of the standard system properties can be judged from $h[n]$, as discussed next.

1. *Causality*. First, from Eq. (3.10), we see that $y[n]$ depends on the *past* input samples $x[n-1], x[n-2]$, and so on because of the coefficients $h[1], h[2]$, and so on. Similarly, $y[n]$ depends on the *future* input samples $x[n+1], x[n+2]$, and so on because of

$$h[-1], h[-2], h[-3] \cdots. \tag{3.13}$$

This shows that an LTI system is causal (Sec. 2.10) if and only if

$$h[n] = 0, \quad n < 0. \tag{3.14}$$

Similarly, from our definition of anticausal systems (Sec. 2.10), it follows that an LTI system is anticausal if and only if

$$h[n] = 0, \quad n \geq 0. \tag{3.15}$$

See Figs. 3.2(a)–(c) for examples of causal, noncausal, and anticausal impulse responses. Because of this property of LTI systems, one often says that a signal $x[n]$ (say, an input sequence) is "causal" or otherwise, according to the following:

$$x[n] = \begin{cases} 0 & n < 0 \quad \text{(causal)}, \\ 0 & n \geq 0 \quad \text{(anticausal)}, \end{cases} \tag{3.16}$$

and noncausal if it is not causal. While this is a misuse of terminology, it is often used.

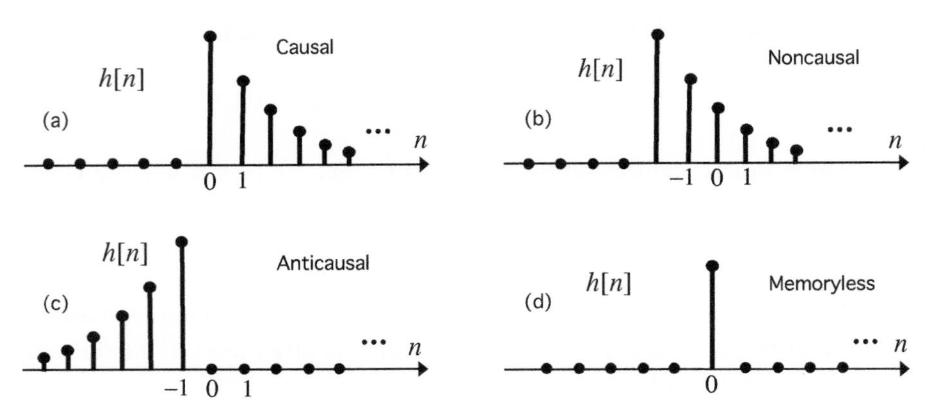

Figure 3.2 Impulse responses of discrete-time LTI systems that are (a) causal, (b) noncausal, (c) anticausal, and (d) memoryless.

2. *Memoryless property*. The system is memoryless if $y[n]$ does not depend on $x[n-k]$ for $k > 0$ or $k < 0$. From Eq. (3.10) it follows that an LTI system is memoryless if and only if

$$h[n] = 0, \quad n \neq 0, \tag{3.17}$$

   or equivalently $h[n] = c\delta[n]$, that is, if and only if the impulse response is itself an impulse! See Fig. 3.2(d).

3. *Stability*. Recall that a system is said to be stable if the output in response to any bounded input is bounded (Sec. 2.11). We now show that an LTI system is **stable if and only if**

$$\sum_{n=-\infty}^{\infty} |h[n]| < \infty, \tag{3.18}$$

   that is, the impulse response is **absolutely summable**. Thus, the condition (3.18) is both necessary and sufficient for stability. When a sequence $h[n]$ satisfies condition (3.18), we say that it is an ℓ_1 *sequence*. This is indicated as $h[n] \in \ell_1$, which is read as "the sequence $h[n]$ belongs to the space of ℓ_1 signals."

Proof that condition (3.18) is necessary and sufficient for stability. If the input is bounded, that is, if $|x[n]| \leq B < \infty$ for some B, then from Eq. (3.11)

$$|y[n]| \leq \sum_{k=-\infty}^{\infty} |h[k]x[n-k]| \leq B \sum_{k=-\infty}^{\infty} |h[k]| < \infty, \tag{3.19}$$

when condition (3.18) holds. This proves that $y[n]$ is bounded as well. So, the system is stable when condition (3.18) holds. Next, for the converse, we have to show that if condition (3.18) is not satisfied then the system is unstable. For this, let us write $h[n]$ in its polar form

$$h[n] = |h[n]|e^{j\theta[n]}. \tag{3.20}$$

Now consider the sample $y[0] = \sum_{k=-\infty}^{\infty} h[k]x[-k]$. We can easily make this infinite by choosing a bounded input as follows:

$$x[n] = e^{-j\theta[-n]}. \tag{3.21}$$

With this input, we have

$$y[0] = \sum_{k=-\infty}^{\infty} h[k]x[-k] = \sum_{k=-\infty}^{\infty} |h[k]|e^{j\theta[k]}e^{-j\theta[k]} = \sum_{k=-\infty}^{\infty} |h[k]|, \tag{3.22}$$

which is not finite when condition (3.18) is not satisfied. That is, $y[n]$ is unbounded although the input (3.21) is bounded, proving the system is unstable. ▽ ▽ ▽

Notice, finally, that the impulse response can be defined for any system, whether LTI or not. For this we simply apply the input $x[n] = \delta[n]$ and measure the output $h[n]$. But this is not useful unless the system is LTI. That is, just from a knowledge of the impulse response we cannot compute the response of the system to an arbitrary input, unless the system is LTI. In fact, there are non-LTI systems whose impulse response is zero for all n. An example is the system (2.86) discussed earlier, and there are many other examples (Problem 3.11). This does not mean that the output is zero for arbitray input, because the convolution property does not hold for non-LTI systems.

3.4 Exponentials are Eigenfunctions of LTI Systems

Now assume that an exponential input signal

$$x[n] = a^n, \quad -\infty < n < \infty \tag{3.23}$$

is applied to an LTI system. Then the output is, from Eq. (3.11),

$$y[n] = \sum_{k=-\infty}^{\infty} h[k]x[n-k] = \sum_{k=-\infty}^{\infty} h[k]a^{n-k} = a^n \sum_{k=-\infty}^{\infty} h[k]a^{-k}. \tag{3.24}$$

The summation on the very right is independent of time n, and depends only on the system $h[k]$ and the input parameter a. We denote it as

$$H(a) = \sum_{k=-\infty}^{\infty} h[k]a^{-k}. \tag{3.25}$$

Assuming this infinite summation converges, the output is therefore

$$y[n] = H(a)a^n, \quad -\infty < n < \infty. \tag{3.26}$$

Thus, for an LTI system,

$$x[n] = a^n, \ \forall n \quad \Longrightarrow \quad y[n] = H(a)a^n, \ \forall n, \tag{3.27}$$

where the notation ∀ means "for all." That is, an exponential input produces the *same* exponential as the output, except that it is scaled by a constant $H(a)$ that does not depend on n. By "same exponential" we mean the output is not proportional to b^n for some other $b \neq a$, as might happen for some non-LTI systems. So the exponential shape is completely preserved by the LTI system, except for a scale factor. We therefore say that *exponentials are* **eigenfunctions** *of LTI systems*, and the scale factor $H(a)$ is said to be the *eigenvalue* corresponding to the eigenfunction.[1] More commonly, $H(a)$ is called the transfer function of the system, as we shall see.

How about non-LTI systems? For a *non-LTI* system, if the input is $x[n] = a^n$, the output is not necessarily an exponential. For example, consider the nonlinear system

$$y[n] = x[n] + x^2[n]. \tag{3.28}$$

If $x[n] = a^n$ then this produces the output $y[n] = a^n + a^{2n} = a^n(1 + a^n)$, which is not an exponential – rather, it is a sum of *two* different exponentials. As another example, the time-varying system

$$y[n] = x[n] + x[3n] \tag{3.29}$$

produces $y[n] = a^n + a^{3n}$, which is also a sum of two different exponentials. How about the time-varying system $y[n] = nx[n]$? This produces the output $y[n] = na^n$, which is not even a sum of exponentials. It is a linearly modulated exponential!

In short, the property that exponential inputs a^n produce exponential outputs $H(a)a^n$ is indeed special for LTI systems. But we will see later that there exist rare non-LTI systems for which exponentials *are* eigenfunctions! ▽ ▽ ▽

Notice finally that, even for LTI systems, if the exponential input is one-sided, as in

$$x[n] = a^n \mathcal{U}[n], \tag{3.30}$$

then the output is **not** just a one-sided exponential like $H(a)a^n, n \geq 0$, except in some trivial cases. This will be discussed in Sec. 7.12.

3.4.1 Transfer Function and Frequency Response

The scale factor $H(a)$ in Eq. (3.27) is given by Eq. (3.25). It depends on the system through $h[n]$, and also depends on the possibly complex number a. It is more convenient to use a generic variable z, and rewrite Eq. (3.25) as

$$H(z) = \sum_{n=-\infty}^{\infty} h[n]z^{-n}. \tag{3.31}$$

[1] More generally, a function $v[n]$ is an eigenfunction of a transformation $\mathcal{T}$ if $v[n]$ is not the zero signal, and $\mathcal{T}(v[n]) = \lambda v[n]$ for some λ. The constant λ is called the eigenvalue corresponding to this eigenfunction.

The above summation is called the **z-transform of $h[n]$**. Since it is an infinite summation, it converges only for certain values of the complex quantity z. We shall discuss this convergence in Chapter 6. $H(z)$ is also called the **transfer function** of the LTI system. The physical meaning of the transfer function $H(z)$ is that, if an input a^n is applied, then the output is $H(a)a^n$. So $H(z)$ is like the "gain" of the system for exponential inputs. Notice, however, that $H(z)$ is in general complex. For the special case where $a = e^{j\omega_0}$, the input becomes

$$x[n] = e^{j\omega_0 n}, \tag{3.32}$$

and the corresponding output is $y[n] = H(e^{j\omega_0})e^{j\omega_0 n}$ so that, for an LTI system,

$$x[n] = e^{j\omega_0 n}, \quad \forall n \quad \Longrightarrow \quad y[n] = H(e^{j\omega_0})e^{j\omega_0 n}, \quad \forall n. \tag{3.33}$$

Here

$$H(e^{j\omega}) = \sum_{n=-\infty}^{\infty} h[n]e^{-j\omega n}. \tag{3.34}$$

This summation is called the **Fourier transform** (FT) or, more specifically, the discrete-time Fourier transform or **DTFT** of the sequence $h[n]$. It is also called the **frequency response** of the LTI system.

Equation (3.33) states that if we apply a single-frequency signal $x[n] = e^{j\omega_0 n}$ to an LTI system then the output is the *same single-frequency signal*, except for a scale factor $H(e^{j\omega_0})$. This is a beautiful property special to LTI systems. The scale factor $H(e^{j\omega_0})$ depends on the system $h[n]$, and on the frequency ω_0. Equation (3.33) assumes that the infinite sum (3.34) converges for $\omega = \omega_0$. We shall discuss convergence in Chapter 6. For now we just mention that if the stability condition (3.18) is satisfied, then the frequency response summation converges for all ω, although the converse is not true (Sec. 6.9.4).

For non-LTI systems a single-frequency input $x[n] = e^{j\omega_0 n}$ can produce multiple frequencies at the output. For example, the nonlinear system defined by $y[n] = x[n] + x^2[n]$ produces the ouput

$$y[n] = e^{j\omega_0 n} + e^{j2\omega_0 n}, \tag{3.35}$$

which contains two frequencies ω_0 and $2\omega_0$. Notice also that, even for LTI systems, if the input is one-sided, as in

$$x[n] = e^{j\omega_0 n}\mathcal{U}[n], \tag{3.36}$$

then the output is in general not of the form $y[n] = H(e^{j\omega_0})e^{j\omega_0 n}, n \geq 0$ (Sec. 7.12).

3.4.2 The z-Transform, z-Plane, and Unit Circle

In the definition of the transfer function $H(z)$, the variable z is in general complex. The complex z-plane is therefore very useful in the discussion of LTI systems. This is shown in Fig. 3.3. Points of the form $z = e^{j\omega}$ have unit magnitude, and therefore lie on a circle with unit radius, centered at $z = 0$. This is called the **unit circle** and

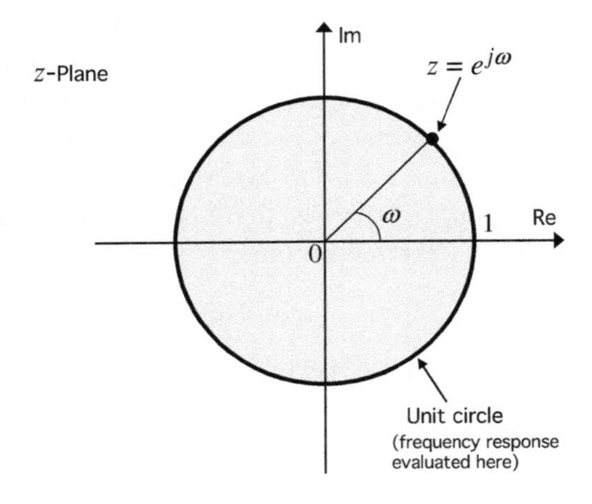

Figure 3.3 The complex z-plane and the unit circle in it. The frequency response $H(e^{j\omega})$ is the transfer function evaluated on the unit circle, that is, for $z = e^{j\omega}$. More generally, the Fourier transform $X(e^{j\omega})$ is the z-transform $X(z)$ evaluated on the unit circle.

is indicated in the figure. The transfer function $H(z)$ can be evaluated for any z in the complex z-plane, as long as the defining summation (3.31) converges. The frequency response $H(e^{j\omega})$ is nothing but the transfer function evaluated on the unit circle, that is, for $z = e^{j\omega}$.

In the same way that we defined the z-transform of the impulse response in Eq. (3.31), one can define a z-**transform** for any discrete-time signal $x[n]$ as follows:

$$X(z) = \sum_{n=-\infty}^{\infty} x[n]z^{-n}. \tag{3.37}$$

This summation converges for certain values of z, depending on what $x[n]$ is. And, just like Eq. (3.34), the z-transform $X(z)$ evaluated on the unit circle

$$X(e^{j\omega}) = \sum_{n=-\infty}^{\infty} x[n]e^{-j\omega n} \qquad \text{(DTFT)}, \tag{3.38}$$

is called the discrete-time **Fourier transform** or **DTFT** of $x[n]$. Notice that $X(e^{j\omega})$ can be complex even if $x[n]$ is real. Readers familiar with continuous-time signals and systems may already notice some differences. The continuous-time counterpart of the above will be summarized in Sec. 3.9 for convenience.

The DTFT is periodic! Notice that if we replace ω with $\omega + 2\pi$ on the right-hand side of Eq. (3.38), there is no change because

$$e^{-j(\omega+2\pi)n} = e^{-j\omega n}e^{-j2\pi n} = e^{-j\omega n}. \tag{3.39}$$

This shows that $X(e^{j\omega})$ **repeats periodically** as a function of ω with a repetition interval 2π:

$$X(e^{j(\omega+2\pi)}) = X(e^{j\omega}). \tag{3.40}$$

This is demonstrated in Fig. 3.4 for the case where $X(e^{j\omega})$ is real, and is consistent with the fact that in discrete time, the frequencies ω and $\omega + 2\pi$ cannot be

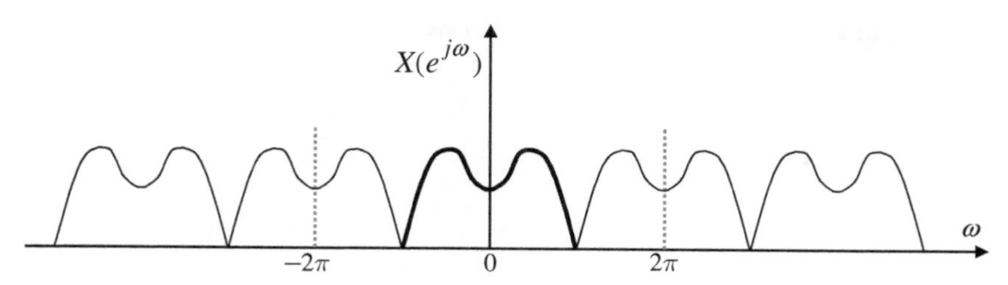

Figure 3.4 Demonstrating that the discrete-time Fourier transform $X(e^{j\omega})$ is periodic in ω, with repetition interval 2π.

distinguished (Sec. 2.6.1). Thus, $X(e^{j\omega})$ needs to be specified only over a finite interval such as

$$0 \leq \omega < 2\pi \qquad \text{or} \qquad -\pi \leq \omega < \pi. \tag{3.41}$$

The notation $X(e^{j\omega})$ (instead of $X(j\omega)$) is therefore very apt because it naturally reflects the fact that $X(e^{j\omega})$ is periodic. Similarly, the frequency response $H(e^{j\omega})$ of a discrete-time LTI system (Eq. (3.34)) also repeats periodically with repetition interval 2π, and needs to be specified only in some finite region like (3.41). This is a major departure from the continuous-time case, where the frequency range is infinite and Fourier transforms are not necessarily periodic (Sec. 3.9). ▽ ▽ ▽

3.4.3 The Convolution Theorem

The definition of the transfer function $H(z)$ for LTI systems was motivated by the fact that the input a^n produces the output $H(a)a^n$, and the definition of the frequency response $H(e^{j\omega})$ was similarly motivated. But what is the motivation for defining the z-transform (3.37) for an *arbitrary* signal $x[n]$? We know that if $x[n]$ is input to an LTI system with impulse response $h[n]$, then the output is the convolution

$$y[n] = \sum_{k=-\infty}^{\infty} x[k]h[n-k]. \tag{3.42}$$

With $X(z)$ and $Y(z)$ denoting the z-transforms of $x[n]$ and $y[n]$, we will show that Eq. (3.42) implies

$$Y(z) = H(z)X(z). \tag{3.43}$$

That is, convolution in the **time domain** is equivalent to pointwise multiplication of the two z-transforms. This is a very important result, and is called the *convolution theorem*. Convolution in the time domain is equivalent to multiplication in the z-**domain**. In particular on the unit circle, that is, for $z = e^{j\omega}$, this means that

$$Y(e^{j\omega}) = H(e^{j\omega})X(e^{j\omega}), \tag{3.44}$$

which is pointwise multiplication in the **frequency domain**.

Proof of Eq. (3.43). From Eq. (3.42) we have

$$\begin{aligned} Y(z) = \sum_{n=-\infty}^{\infty} y[n]z^{-n} &= \sum_{n=-\infty}^{\infty} z^{-n} \sum_{k=-\infty}^{\infty} x[k]h[n-k] \\ &= \sum_{n=-\infty}^{\infty} \sum_{k=-\infty}^{\infty} x[k]z^{-k}h[n-k]z^{-(n-k)} \\ &= \sum_{k=-\infty}^{\infty} x[k]z^{-k} \sum_{n=-\infty}^{\infty} h[n-k]z^{-(n-k)}. \end{aligned} \tag{3.45}$$

The third equality is obtained by inserting $1 = z^{-k}z^{k}$, and the fourth equality is obtained by interchanging the summations. Since the inner sum in the last line is an infinite summation, the integer k can be dropped so that

$$Y(z) = \sum_{k=-\infty}^{\infty} x[k]z^{-k} \sum_{n=-\infty}^{\infty} h[n]z^{-n} = X(z)H(z), \tag{3.46}$$

which proves the desired result. ▽ ▽ ▽

It was observed earlier that convolution is commutative, that is, $(x*h)[n] = (h*x)[n]$. This property is now obvious in view of the convolution theorem, because it simply says $X(z)H(z) = H(z)X(z)$.

3.5 The Inverse Fourier Transform

The definition (3.38) of the discrete-time Fourier transform arose naturally when we introduced LTI systems and derived the property that exponential inputs produced exponential outputs. Based only on this, it is difficult to appreciate what Eq. (3.38) physically means, or why it is so useful. The beauty and significance of Eq. (3.38) can be appreciated from the inverse formula, which tells us that $x[n]$ can be recovered for all n in the infinite range $-\infty < n < \infty$, if $X(e^{j\omega})$ is known in the finite range $0 \leq \omega < 2\pi$. More precisely, the inverse discrete-time Fourier transform formula is

$$x[n] = \frac{1}{2\pi} \int_0^{2\pi} X(e^{j\omega})e^{j\omega n}d\omega \qquad \text{(IDTFT)}. \tag{3.47}$$

This expression is also called the **Fourier integral**. We will first prove this, and then discuss its significance. The proof is based on the following simple result: for any integer k

$$\int_0^{2\pi} e^{j\omega k}d\omega = 2\pi\delta[k] = \begin{cases} 2\pi & \text{if } k = 0, \\ 0 & \text{if } k \neq 0. \end{cases} \tag{3.48}$$

For $k = 0$ this is obvious. When $k \neq 0$ the left-hand side is

$$\int_0^{2\pi} e^{j\omega k} d\omega = \frac{e^{j\omega k}}{jk}\bigg|_0^{2\pi} = \frac{e^{j2\pi k} - 1}{jk} = 0, \tag{3.49}$$

proving Eq. (3.48). It is often useful to rewrite Eq. (3.48) as

$$\int_0^{2\pi} e^{j\omega l} e^{-j\omega m} d\omega = 2\pi\delta[l - m]. \tag{3.50}$$

Note that this has the form

$$\int_0^{2\pi} f_l(\omega) f_m^*(\omega) d\omega = 2\pi\delta[l - m] \quad \text{(orthogonality)}. \tag{3.51}$$

Whenever such a property holds, we say that $f_l(\omega)$ and $f_m(\omega)$ are orthogonal in the interval $0 \leq \omega < 2\pi$. Thus, Eq. (3.50) is the **orthogonality property** of the family of functions $f_k(\omega) = e^{jk\omega}$ in the interval $[0, 2\pi)$.

Proof of Eq. (3.47). Substituting from Eq. (3.38), the right-hand side of Eq. (3.47) becomes

$$\begin{aligned} \frac{1}{2\pi}\int_0^{2\pi}\Big(\sum_{m=-\infty}^{\infty} x[m] e^{-j\omega m}\Big) e^{j\omega n} d\omega &= \frac{1}{2\pi}\sum_{m=-\infty}^{\infty} x[m] \int_0^{2\pi} e^{j\omega(n-m)} d\omega \\ &= \sum_{m=-\infty}^{\infty} x[m]\delta[n - m] = x[n]. \end{aligned} \tag{3.52}$$

On the far left, we use $x[m]$ instead of $x[n]$ in the summation, because n is a reserved symbol in Eq. (3.47). The first equality follows from interchanging the summation with the integral, and the second equality follows from Eq. (3.48). ▽ ▽ ▽

Since $X(e^{j\omega})$ is periodic as in Eq. (3.40), the IDFT can also be written as

$$x[n] = \frac{1}{2\pi}\int_{-\pi}^{\pi} X(e^{j\omega}) e^{j\omega n} d\omega \qquad \text{(IDTFT)}, \tag{3.53}$$

or as an integral over any region of the form $\alpha \leq \omega < \alpha + 2\pi$.

3.5.1 Approximating $x[n]$ as a Sum of Complex Sinusoids

To understand the importance of Eq. (3.47), let us appoximate the Fourier integral with a summation as follows:

$$x[n] = \frac{1}{2\pi}\int_0^{2\pi} X(e^{j\omega}) e^{j\omega n} d\omega \approx \frac{1}{2\pi}\sum_{k=0}^{K-1} X(e^{j\omega_k}) e^{j\omega_k n} \Delta\omega, \tag{3.54}$$

where K is a large integer and

$$\omega_k = k\Delta\omega \tag{3.55}$$

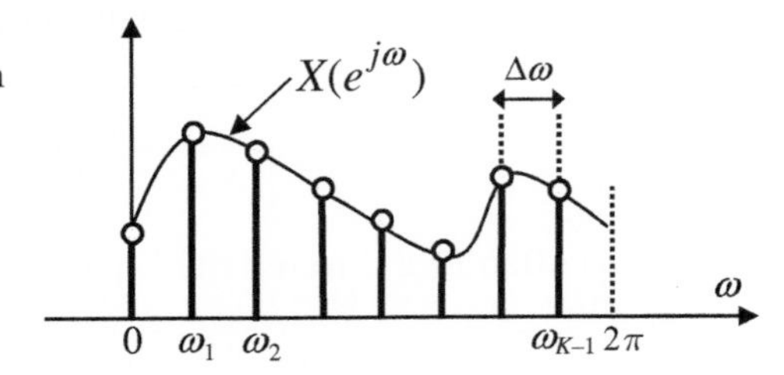

Figure 3.5 Pertaining to the approximation of $x[n]$ as a sum of single-frequency components. See Eq. (3.56).

is a set of closely spaced frequencies separated by the small interval $\Delta\omega = 2\pi/K$. This is demonstrated in Fig. 3.5, with the spacing $\Delta\omega$ exaggerated. Thus, the IDTFT formula says that we can approximate $x[n]$ with a sum

$$x[n] \approx \frac{1}{K}\sum_{k=0}^{K-1} X(e^{j\omega_k})e^{j\omega_k n}. \tag{3.56}$$

Each term in the summation is a single-frequency signal, with frequency ω_k and amplitude $X(e^{j\omega_k})$, which is possibly complex. Thus, the expression (3.56) approximates $x[n]$ by a linear combination of single-frequency signals $e^{j\omega_k n}$ or **complex sinusoids**. The closer the frequencies ω_k are, the better the approximation. The amplitude or "amount" of the frequency component ω_k present in $x[n]$ is proportional to $X(e^{j\omega_k})$. This gives a "physical" meaning to the Fourier transform $X(e^{j\omega})$, which was defined in Sec. 3.4.1, motivated purely by the behavior of LTI systems. The integral (3.53) is of course an *exact* expression for $x[n]$. It says that $x[n]$ can be represented by a "linear combination" of a *continuum* of single-frequency signals $e^{j\omega n}$.

3.5.2 The Filtering Interpretation

Equation (3.44) is what makes LTI systems so important in linear filtering applications. Since

$$Y(e^{j\omega}) = H(e^{j\omega})X(e^{j\omega}), \tag{3.57}$$

it follows that, where $|H(e^{j\omega})|$ is large, the system "passes" the corresponding frequency component $X(e^{j\omega})$ of the input, and where $|H(e^{j\omega})|$ is small, the system attenuates that frequency component from the input. For example, if $H(e^{j\omega_0}) = 0$, then the input $x[n] = e^{j\omega_0 n}$ produces zero output (from Eq. (3.33)). In this case we say that the system has a **transmission zero** at $\omega = \omega_0$. Note that $H(e^{j\omega_0}) = 0$ is equivalent to

$$H(z) = 0, \quad z = e^{j\omega_0}. \tag{3.58}$$

That is, if there is a transmission zero at $\omega = \omega_0$, then $H(z)$ is zero at $z = e^{j\omega_0}$ in the z-plane. Since this is a point on the unit circle, it follows that there is a transmission zero at ω_0 if and only if $H(z)$ has a zero on the **unit circle** at $z = e^{j\omega_0}$.

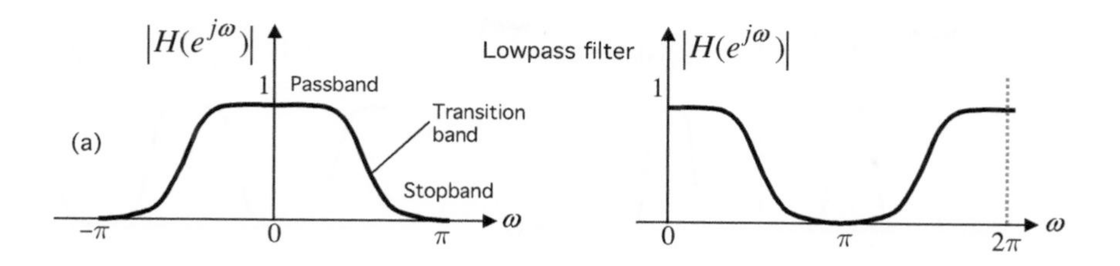

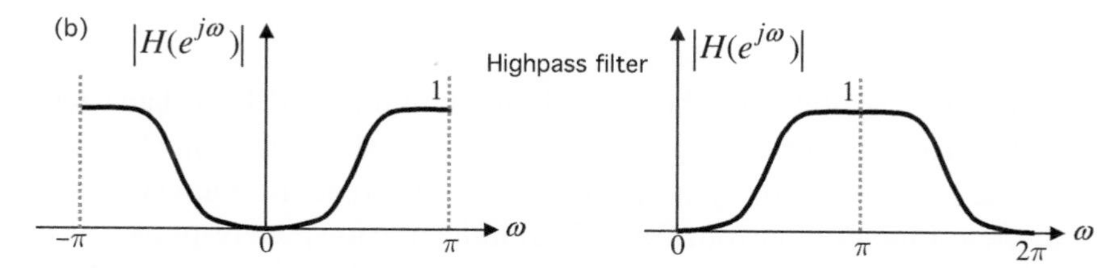

Figure 3.6 The magnitude response of (a) a lowpass digital filter and (b) a highpass digital filter. Both are shown for $-\pi \leq \omega < \pi$ and for $0 \leq \omega < 2\pi$ for clarity.

Note that the frequency response $H(e^{j\omega})$ can be complex even when $h[n]$ is real, but it can always be written in polar form as

$$H(e^{j\omega}) = |H(e^{j\omega})|e^{j\phi(\omega)}, \tag{3.59}$$

where $|H(e^{j\omega})|$ is the magnitude (absolute value) and $\phi(\omega)$ is the phase. We say that $|H(e^{j\omega})|$ is the **magnitude response** and $\phi(\omega)$ is the **phase response**. Figure 3.6(a) shows an example of a magnitude response. Both the regions $-\pi \leq \omega < \pi$ and $0 \leq \omega < 2\pi$ are shown for clarity.

3.5.2.1 Lowpass Digital Filters

Recall from Sec. 2.6.1 that for the discrete-time case the region around $\omega = 0$ is regarded as low frequency and the region around $\omega = \pm\pi$ is regarded as high frequency. So the system shown in Figure 3.6(a) transmits low frequencies and attenuates high frequencies. Such a system is called a **lowpass** filter (LPF) because it "passes" low frequencies in preference to high frequencies. The region that is "passed" is called the **passband** and the region that is attenuated is the **stopband**. In between these there is a gray area where the response transitions from 1 to 0, so it is called the **transition band**. As these are discrete-time systems, these filters are also referred to as **digital filters**.

3.5.2.2 Other Types of Digital Filters

Next, Fig. 3.6(b) shows a filter which passes high frequencies in preference to low. So this is called a **highpass** filter (HPF). Figure 3.7(a) shows the magnitude response of a **bandpass** filter (BPF). Such a filter passes frequencies that are close to a nonzero frequency ω_0 (and $-\omega_0$ usually), and attenuates all other frequencies. The opposite

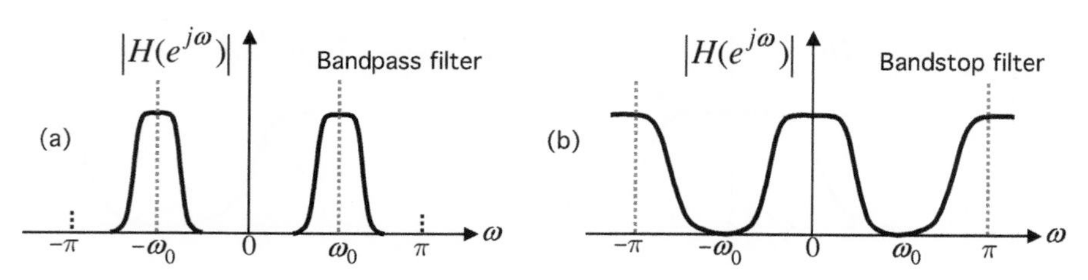

Figure 3.7 The magnitude response of (a) a bandpass digital filter and (b) a bandstop (or band elimination) digital filter.

behavior is exhibited by the band elimination filter (BEF) or **bandstop** filter, which eliminates frequencies close to $\pm\omega_0$ and passes all others (Fig. 3.7(b)).

The responses $H(e^{j\omega})$ are periodic in ω with repetition interval 2π, just like any discrete-time Fourier transform. Examining the lowpass response for the range $0 \leq \omega < 2\pi$ (Fig. 3.6), we get the impression that there are two passbands, one near $\omega = 0$ and one near $\omega = 2\pi$. This is because $\omega = 2\pi$ is equivalent to $\omega = 0$ in discrete time (Sec. 2.6.1). The region near 2π shown in the figure just corresponds to negative frequencies close to $\omega = 0$. Similarly, the highpass response plotted for $-\pi \leq \omega < \pi$ shows two passbands because $\omega = \pi$ and $\omega = -\pi$ refer to the same frequency.

Real-world filters. A healthy human ear can hear frequencies in the range 20 Hz to 20 kHz and this range gradually decreases as we age. In any case the ear can be considered to be approximately a bandpass filter, whose bandwidth decreases with age. In some applications such as audio signal processing, it is convenient to model the ear as a collection of bandpass filters with non-overlapping passbands and increasing center frequencies. This will be elaborated in Sec. 11.8.2 (see Fig. 11.14). Next, the human visual system (HVS) also cannot perceive frequencies that are arbitrarily high. For example, we cannot visually perceive the roughness of a metallic surface, or even the roughness of a whiteboard surface. Individual strands of hair on the head can be perceived from close quarters, but not from far away. The HVS is commonly regarded as a two-dimensional lowpass (or bandpass) filter. Thus, filtering is inherently involved in Nature. These are of course continuous-time (or continuous-space) filters, so the frequency variable can be arbitrarily large, unlike in discrete-time systems where we restrict it to $-\pi \leq \omega < \pi$ (Sec. 2.6.1) without loss of generality. $\triangledown\triangledown\triangledown$

3.6 Real Signals and Real Filters

If $x[n]$ is real valued for all n, we say that it is a real signal. Similarly, an LTI system or a filter is said to be real if the impulse response coefficients $h[n]$ are real, so that real inputs $x[n]$ always produce real outputs. Real filters are also called *real-coefficient filters*. For the case of real signals and systems, some simplification of notation is possible based on a property of the Fourier transform. To understand this, observe that given an arbitrary $x[n]$ (possibly complex) with DTFT $X(e^{j\omega})$, the DTFT of its complex conjugate $x^*[n]$ is given by

$$\sum_{n=-\infty}^{\infty} x^*[n]e^{-j\omega n} = \Big(\sum_{n=-\infty}^{\infty} x[n]e^{j\omega n}\Big)^* = \big(X(e^{-j\omega})\big)^* = X^*(e^{-j\omega}), \qquad (3.60)$$

where $S^*(e^{j\omega})$ is the simplified notation for the complex conjugate $(S(e^{j\omega}))^*$. Thus the DTFT of $x^*[n]$ is given by $X^*(e^{-j\omega})$, which is the *frequency-reversed* and *conjugated* version of $X(e^{j\omega})$. If $x[n]$ is real, then $x[n] = x^*[n]$, which proves that

$$X(e^{j\omega}) = X^*(e^{-j\omega}), \quad \text{or equivalently} \quad X(e^{-j\omega}) = X^*(e^{j\omega}). \qquad (3.61)$$

That is, the FT at $-\omega$ is the conjugate of the FT at ω. So, even though $X(e^{j\omega})$ can be complex for real $x[n]$, it has to satisfy Eq. (3.61). Similarly, if an LTI system or filter is real (i.e., $h[n]$ is real), then the frequency response satisfies

$$H(e^{j\omega}) = H^*(e^{-j\omega}), \quad \text{that is} \quad H(e^{-j\omega}) = H^*(e^{j\omega}) \quad \text{(for real filters)}. \qquad (3.62)$$

Equation (3.62) is called **Hermitian symmetry**. If $H(e^{j\omega})$ also happens to be real (in addition to $h[n]$ being real), then Eq. (3.62) is equivalent to ordinary symmetry or evenness in ω.

We know that the frequency response $H(e^{j\omega})$ can always be written as $H(e^{j\omega}) = |H(e^{j\omega})|e^{j\phi(\omega)}$ (see Eq. (3.59)). So the property (3.62), which holds for real $h[n]$, is equivalent to

$$|H(e^{j\omega})| = |H(e^{-j\omega})|, \quad \phi(\omega) = -\phi(-\omega). \qquad (3.63)$$

That is, when $h[n]$ *is real*, the magnitude response $|H(e^{j\omega})|$ is a *symmetric* (or even) function and the phase response $\phi(\omega)$ is an *antisymmetric* (or odd) function. This is demonstrated in Figs. 3.8(a) and (b) for a lowpass filter. Because of this symmetry property, we usually plot these responses only in the range $0 \leq \omega \leq \pi$, as demonstrated in Figs. 3.8(c) and (d).

The reader will notice that in Figs. 3.6 and 3.7, the examples of $|H(e^{j\omega})|$ are shown as even functions. This is because of the implicit assumption that $h[n]$ is real, although we did not mention it. For complex $h[n]$ a plot of $|H(e^{j\omega})|$ may not be even, as shown by the examples of lowpass and bandpass filters in Fig. 3.9. It should be mentioned that in the digital world there are no difficulties in implementing complex filters, as complex computations can be readily performed. But most of the digital filters we encounter in practice have real $h[n]$, so we mostly focus on them.

Further remarks on frequency responses of real systems can be found in Sec. 5.13. In particular, the ouput of a real system in response to a real sinusoid like $\cos(\omega_0 n)$ is discussed in Sec. 5.13.1.

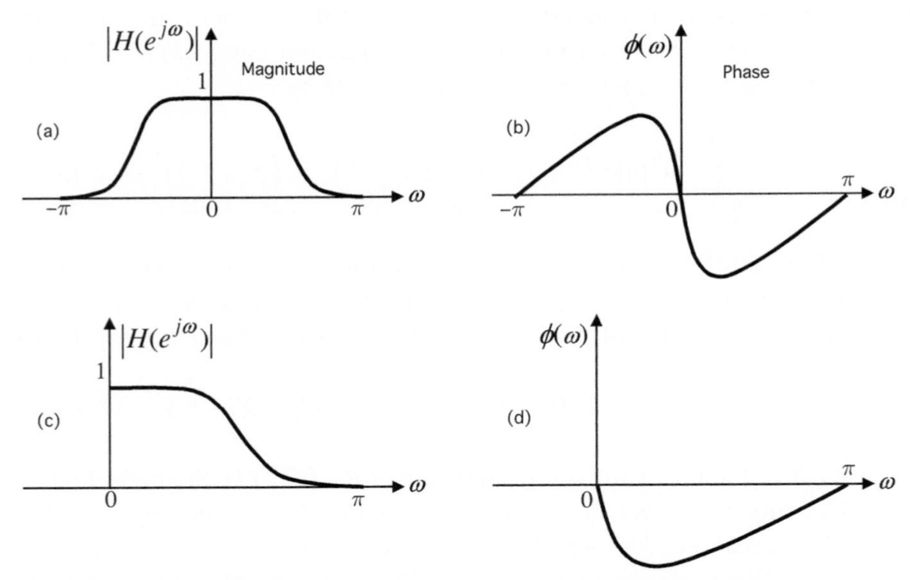

Figure 3.8 (a), (b) Typical magnitude and phase responses for a real-coefficient digital filter. The magnitude is an even function and the phase is an odd function. In (c) and (d) we demonstrate the usual convention of showing these plots only for $0 \leq \omega \leq \pi$.

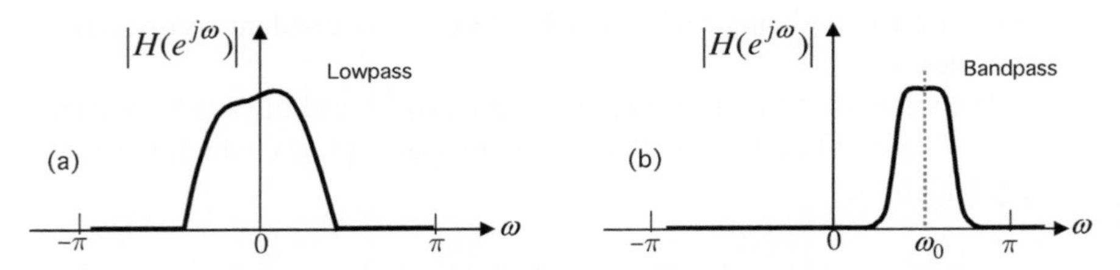

Figure 3.9 Examples of magnitude responses for filters with complex $h[n]$. (a) Lowpass and (b) bandpass. These are not symmetric functions because $h[n]$ is complex.

3.7 Filtering an Image

Figure 3.10(a) shows a grayscale (i.e., black-and-white) image $x[n_1, n_2]$ where the integers $[n_1, n_2]$ are the spatial coordinates and $x \geq 0$ is the intensity. If we think of the image as a matrix, then n_1 is the row index and n_2 the column index. As mentioned at the end of Sec. 2.2, images are two-dimensional (2D) signals. Imagine we wish to filter this image using LTI filters. To explain how this can be done, note that each row in the image is a sequence of numbers and similarly each column is a sequence of numbers. Suppose $H(z)$ is a lowpass filter with impulse response $h[n]$, that is, a one-dimensional (1D) filter like we have been discussing in this chapter. We can then filter each row with this filter as follows:

$$u[n_1, n_2] = \sum_{m_2} x[n_1, n_2 - m_2]h[m_2] \text{ (horizontal filtering for each row } n_1\text{)}. \quad (3.64)$$

We can then filter each column of this result $u[n_1, n_2]$ with this same filter:

$$y_{LL}[n_1, n_2] = \sum_{m_1} u[n_1 - m_1, n_2]h[m_1] \text{ (vertical filtering for each col. } n_2). \quad (3.65)$$

This is a lowpass/lowpass (LL) filtered version (i.e., it is filtered in both directions using a lowpass filter), and the subscript on y is a reminder of this. Figure 3.10(b) shows such a filtered image. Note that lowpass filtering creates a smoothing effect. The sharp edges get blurred because edges represent high-frequency information (fast variations) and are not "passed" favorably by the lowpass filter.

It is possible to filter this image in other ways. For example, one can filter every row using a lowpass filter $H(z)$ as before, take the result $u[n_1, n_2]$, and filter every column using a *highpass* filter $G(z)$:

$$y_{LH}[n_1, n_2] = \sum_{m_1} u[n_1 - m_1, n_2]g[m_1] \text{ (vertical filtering for each col. } n_2). \quad (3.66)$$

This is a lowpass/highpass (LH) filtered version (i.e., lowpass horizontally and high-pass vertically). Figure 3.10(c) shows such a filtered version. Since highpass filters pass high frequencies (rapid changes like edges), we see that edge information is clearly visible as we move vertically along the image, but not as we move horizontally. Thus, highpass filters are useful to extract edge information and lowpass filters are useful to extract smooth portions of images. Splitting a signal into multiple frequency bands like this is a very important operation and is used widely in a data compression scheme called subband coding [Sayood, 2000].

Figure 3.11 shows the magnitude responses of the 1D lowpass and highpass filters used to generate the above filtered images. This is just a preview, as the design of such filters comes only later (Sec. 4.3). $H(z)$ is the lowpass filter given in Eq. (4.32) with $L = 11$, and the highpass filter $G(z)$ is the transformed version in Eq. (4.50). The behaviors of these filters will be explained in detail later; here we just want to give a flavor for image filtering. Since the filters are real, only the region $0 \leq \omega < \pi$ is shown in Fig. 3.11.

There are some subtleties involved in image filtering. Since the impulse responses can have some negative coefficients, the filtered image which is obtained by convolution with impulse responses may sometimes have negative pixel values. Since negative intensities are meaningless, one often adds a small positive constant to the entire filtered image to make it non-negative, so that the image can be displayed. We have done so in all image renditions when necessary.

3.8 The Discrete-Time Fourier Transform

In the previous sections, some properties of the DTFT were proved. We now discuss more properties and examples. A complete list of properties is given in the table in Sec. 3.13. The definition of the DTFT, and the expression for its inverse (proved in Sec. 3.5), are reproduced below:

Figure 3.10 (a) Original image, (b) image lowpass filtered in both directions, (c) image lowpass filtered horizontally and highpass filtered vertically.

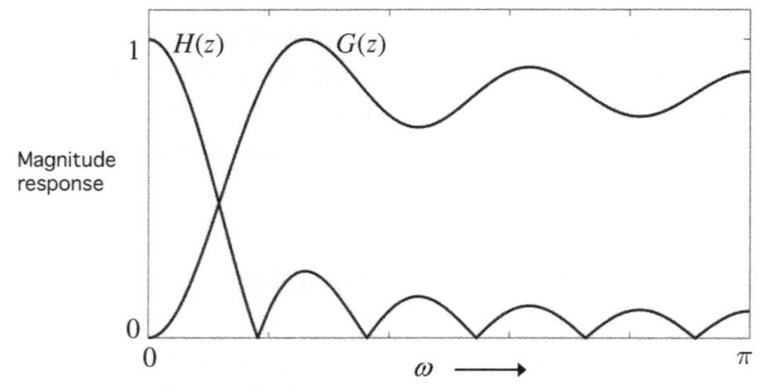

Figure 3.11 Magnitude responses of the 1D lowpass and highpass filters used to generate the 2D filtered images in Figs. 3.10(b) and (c).

$$X(e^{j\omega}) = \sum_{n=-\infty}^{\infty} x[n]e^{-j\omega n} \qquad \text{(DTFT)}, \tag{3.67}$$

$$x[n] = \frac{1}{2\pi}\int_0^{2\pi} X(e^{j\omega})e^{j\omega n}d\omega \qquad \text{(IDTFT)}. \tag{3.68}$$

The quantities $X(e^{j\omega})$ and $x[n]$ are together referred to as a *Fourier transform pair* or **DTFT pair** and denoted as

$$x[n] \longleftrightarrow X(e^{j\omega}). \tag{3.69}$$

Let us recapitulate the properties we have seen so far. As shown in Sec. 3.4.2, $X(e^{j\omega})$ is periodic in ω with repetition interval 2π. In general, $X(e^{j\omega})$ is complex even when $x[n]$ is real, and can be written in polar form as

$$X(e^{j\omega}) = |X(e^{j\omega})|e^{j\phi(\omega)}. \tag{3.70}$$

In Sec. 3.6 it was shown that if $x[n]$ is real valued, then $X(e^{j\omega})$ has the Hermitian symmetry property (3.61), which means that $|X(e^{j\omega})|$ is an even function and the phase of $\phi(\omega)$ is odd. The convolution theorem was proved in Sec. 3.4.3. Most of the remaining properties listed in Sec. 3.13 are proved in this section.

3.8.1 Proofs of Selected Properties

The linearity property mentioned in Sec. 3.13 follows readily from the definition (3.67). It is also clear from Eqs. (3.67) and (3.68) that

$$X(1) = X(e^{j0}) = \sum_{n=-\infty}^{\infty} x[n], \tag{3.71}$$

$$x[0] = \frac{1}{2\pi}\int_0^{2\pi} X(e^{j\omega})d\omega. \tag{3.72}$$

So, $x[0]$ is the average of $X(e^{j\omega})$, whereas the zero-frequency value $X(1)$ is the sum of the samples $x[n]$.

3.8.1.1 Shifting and Modulation

Next, the shifted version $x[n-n_0]$ has FT

$$\sum_{n=-\infty}^{\infty} x[n-n_0]e^{-j\omega n} = \sum_{n=-\infty}^{\infty} x[n]e^{-j\omega(n+n_0)} = e^{-j\omega n_0}\sum_{n=-\infty}^{\infty} x[n]e^{-j\omega n},$$

which proves that

$$x[n-n_0] \longleftrightarrow e^{-j\omega n_0}X(e^{j\omega}). \tag{3.73}$$

Similarly, from Eq. (3.67), it follows that

$$e^{j\omega_0 n}x[n] \longleftrightarrow X\left(e^{j(\omega-\omega_0)}\right). \tag{3.74}$$

The quantity $e^{j\omega_0 n}x[n]$ is called the **amplitude modulated** (AM) version of $x[n]$. Thus, modulation in time results in a shift in the frequency, whereas a shift in time results in "modulation in frequency." Amplitude modulation is the basis for AM radio, and is used to shift the frequency band of a low-frequency signal (say, speech) into a very high-frequency signal, which can then be transmitted using radio waves over long distances. In Problem 11.25 we will see another type of modulation called angle modulation. The familiar frequency modulation (FM) is a special case of that.

3.8.1.2 Convolution in the Frequency Domain

In Sec. 3.4.3 we proved that convolution in the time domain is equivalent to multiplication in the frequency domain. Now consider multiplication $x_1[n]x_2[n]$ of two signals in time. The DTFT is

$$\begin{aligned}\sum_{n=-\infty}^{\infty} x_1[n]x_2[n]e^{-j\omega n} &= \sum_{n=-\infty}^{\infty}\left(\frac{1}{2\pi}\int_0^{2\pi} X_1(e^{ju})e^{jun}du\right)x_2[n]e^{-j\omega n} \\ &= \frac{1}{2\pi}\int_0^{2\pi} X_1(e^{ju})\left(\sum_{n=-\infty}^{\infty} x_2[n]e^{-j(\omega-u)n}\right)du \\ &= \frac{1}{2\pi}\int_0^{2\pi} X_1(e^{ju})X_2(e^{j(\omega-u)})du. \end{aligned} \tag{3.75}$$

The first equality follows by writing $x_1[n]$ in terms of its FT, and the second equality is obtained by interchanging the integral with the summation. The last integral is called the **convolution** of the two Fourier transforms. This shows that multiplication in time corresponds to convolution in frequency (and scaling by $1/2\pi$):

$$x_1[n]x_2[n] \longleftrightarrow \frac{1}{2\pi}\int_0^{2\pi} X_1(e^{ju})X_2(e^{j(\omega-u)})du = \frac{(X_1 * X_2)(e^{j\omega})}{2\pi}. \tag{3.76}$$

The term "convolution" is used for the above integral because it is similar in principle to the convolution sum. The integral above has finite limits, covering one period of the integrand (which is periodic). We will have a more detailed discussion of convolution integrals for functions of continuous arguments in Sec. 3.9 later.

3.8.1.3 Symmetry and Realness

It was already proved in Sec. 3.6 that $x^*[n] \longleftrightarrow X^*(e^{-j\omega})$, which proves that if $x[n]$ is real then $X(e^{j\omega})$ has the Hermitian symmetry property (3.61). Furthermore, for any $x[n]$ it follows readily from Eq. (3.67) that $x[-n] \longleftrightarrow X(e^{-j\omega})$. Combining these it follows that $x^*[-n] \longleftrightarrow X^*(e^{j\omega})$. Summarizing, if $x[n] \longleftrightarrow X(e^{j\omega})$, then we have

$$x[-n] \longleftrightarrow X(e^{-j\omega}), \quad x^*[n] \longleftrightarrow X^*(e^{-j\omega}), \quad x^*[-n] \longleftrightarrow X^*(e^{j\omega}). \tag{3.77}$$

In short, *reversal in the time domain* is equivalent to *reversal in the frequency domain*. Furthermore, *complex conjugation* in one domain is equivalent to *complex conjugation and reversal* in the other domain. Next, a signal $x[n]$ is said to be an even (or symmetric) function if

$$x[n] = x[-n]. \tag{3.78}$$

It is clear from the first property in Eq. (3.77) that $x[n]$ is even if and only if $X(e^{j\omega})$ is even as well. Furthermore, by combining the properties in Eq. (3.77) it follows that

$$x[n] \text{ is real and even if and only if } X(e^{j\omega}) \text{ is real and even.} \tag{3.79}$$

Similar properties can be proved for imaginary and odd sequences, and details can be found in Problem 3.24.

3.8.1.4 Correlations, Energy, and Parseval's Relation

The quantity

$$r_{xy}[k] \stackrel{\Delta}{=} \sum_{n=-\infty}^{\infty} x[n]y^*[n-k] \tag{3.80}$$

is called the **cross-correlation** between the signals $x[n]$ and $y[n]$. It is a sequence by definition, as it depends on the integer k, which is the shift index in $y^*[n-k]$. This argument k in $r_{xy}[k]$ is also called the **lag** variable (since $y[n-k]$ lags behind $y[n]$ by k samples). We can think of the cross-correlation (3.80) as the *convolution* between $x[n]$ and the time-reversed and conjugated version of $y[n]$, that is, $y^*[-n]$. Since the DTFT of $y^*[-n]$ is $Y^*(e^{j\omega})$ (from Eq. (3.77)), it follows from the convolution theorem that the DTFT of $r_{xy}[k]$ is the product $X(e^{j\omega})Y^*(e^{j\omega})$, that is,

$$r_{xy}[k] \longleftrightarrow X(e^{j\omega})Y^*(e^{j\omega}). \tag{3.81}$$

For the special case where $x[n] = y[n]$, the cross-correlation (3.80) reduces to

$$r_{xx}[k] \stackrel{\Delta}{=} \sum_{n=-\infty}^{\infty} x[n]x^*[n-k], \tag{3.82}$$

which is called the **autocorrelation** of $x[n]$ (i.e., correlation with itself). It follows from the mapping (3.81) that

$$r_{xx}[k] \longleftrightarrow |X(e^{j\omega})|^2. \tag{3.83}$$

Using the property (3.72), it follows from (3.81) that

$$r_{xy}[0] = \frac{1}{2\pi}\int_0^{2\pi} X(e^{j\omega})Y^*(e^{j\omega})d\omega. \tag{3.84}$$

But since $r_{xy}[0] = \sum_{n=-\infty}^{\infty} x[n]y^*[n]$, the above is equivalent to

$$\sum_{n=-\infty}^{\infty} x[n]y^*[n] = \frac{1}{2\pi}\int_0^{2\pi} X(e^{j\omega})Y^*(e^{j\omega})d\omega. \tag{3.85}$$

This is called **Parseval's relation**. The sum on the left is the **inner product** of the two sequences $x[n]$ and $y[n]$, whereas the integral on the right is the inner product of $X(e^{j\omega})$ with $Y(e^{j\omega})$ (Sec. 18.3.2). So, Parseval's relation says that the inner product is preserved by the Fourier transform operation (except for the scale factor $1/2\pi$). For the special case where $x[n] = y[n]$, Eq. (3.85) reduces to

$$\sum_{n=-\infty}^{\infty} |x[n]|^2 = \frac{1}{2\pi}\int_0^{2\pi} |X(e^{j\omega})|^2 d\omega. \tag{3.86}$$

The sum on the left is called the **energy** of the signal $x[n]$, and the quantity on the right is the energy in $X(e^{j\omega})$. The above version (3.86) of Parseval's relation states that the energy of a signal evaluated in the time domain is the same as the energy evaluated in the frequency domain. This is also called the **energy conservation** property of the Fourier transform. It can be shown (Problem 3.26) that this is a direct consequence of the orthogonality property (3.50).

In Problem 3.28 we will consider some symmetry properties of correlations. Autocorrelations and cross-correlations have important applications in digital communications and in radar signal processing. In fact these are at the heart of **matched filtering** techniques, which are crucial for the detection of signals buried in noise (Sec. 9.10.2). For example, one of the key signal processing elements in the detection of **gravity waves** in the LIGO project was the matched filter.

3.8.2 Examples of DTFT Pairs

If $x[n] = \delta[n]$ then, from Eq. (3.67), we have $X(e^{j\omega}) = 1$ for all ω. So we write

$$x[n] = \delta[n] \longleftrightarrow X(e^{j\omega}) = 1, \quad \forall\omega. \tag{3.87}$$

This is shown in Fig. 3.12. Thus, the impulse signal has *all frequency components present in it*, in equal measure. We next present some examples which require more effort.

3.8.2.1 The Rectangular Pulse

Consider the signal

$$x[n] = \begin{cases} 1 & -M \leq n \leq M, \\ 0 & \text{otherwise.} \end{cases} \tag{3.88}$$

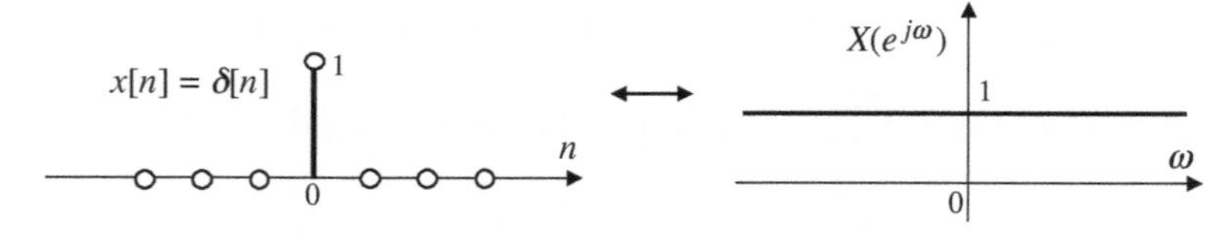

Figure 3.12 The discrete-time impulse signal $\delta[n]$ and its Fourier transform.

This is called a rectangular pulse signal, and its duration or length is $2M+1$ samples. To compute its Fourier transform, it is useful to know a mathematical identity.

A simple identity. In signal processing, one often encounters summations of the form $\sum_{n=0}^{L-1} \alpha^n$. There is a closed-form expression for this, namely

$$\sum_{n=0}^{L-1} \alpha^n = \frac{1-\alpha^L}{1-\alpha}. \tag{3.89}$$

To prove this, simply observe that

$$(1-\alpha)\sum_{n=0}^{L-1} \alpha^n = 1+\alpha+\alpha^2+\cdots+\alpha^{L-1} - (\alpha+\alpha^2+\cdots+\alpha^L) = 1-\alpha^L, \tag{3.90}$$

which proves Eq. (3.89). Note that if $\alpha = 1$, the numerator and denominator of the right-hand side of Eq. (3.89) are both zero, but the limiting value as $\alpha \to 1$ is exactly L. Since this is the correct answer (as verified by setting $\alpha = 1$ on the left-hand side of Eq. (3.89)), we take the expression (3.89) to be valid for *all* α. An important extension of Eq. (3.89) is

$$\sum_{n=0}^{\infty} \alpha^n = \frac{1}{1-\alpha}, \quad |\alpha| < 1. \tag{3.91}$$

The infinite sum above does not coverge to a finite value if $|\alpha| \geq 1$. $\nabla\nabla\nabla$

Returning to the pulse (3.88), its Fourier transform is

$$X(e^{j\omega}) = \sum_{n=-M}^{M} e^{-j\omega n} = e^{j\omega M}\sum_{n=0}^{2M} e^{-j\omega n} = e^{j\omega M}\left(\frac{1-e^{-j\omega(2M+1)}}{1-e^{-j\omega}}\right), \tag{3.92}$$

where we have used Eq. (3.89) to get the last equality. Now, we rewrite the quanity in brackets by factoring out $e^{-j\omega/2}$ from its denominator and $e^{-j\omega(2M+1)/2}$ from its numerator. This gives a factor $e^{-j\omega(2M+1)/2}/e^{-j\omega/2} = e^{-j\omega M}$, which cancels $e^{j\omega M}$ in Eq. (3.92), so that

$$X(e^{j\omega}) = \frac{e^{j\omega(2M+1)/2} - e^{-j\omega(2M+1)/2}}{e^{j\omega/2} - e^{-j\omega/2}} = \frac{2j\sin(\omega(2M+1)/2)}{2j\sin(\omega/2)}, \tag{3.93}$$

which proves that the pulse (3.88) has the Fourier transform

$$X(e^{j\omega}) = \frac{\sin(\omega(2M+1)/2)}{\sin(\omega/2)}. \tag{3.94}$$

The pulse (3.88) and its DTFT (3.94) are plotted in Fig. 3.13. Notice that the pulse is even ($x[n] = x[-n]$) and real, which explains why $X(e^{j\omega})$ is real and even as well (see property (3.79)). Notice also that for $M = 0$ we have $x[n] = \delta[n]$, and Eq. (3.94) indeed reduces to $X(e^{j\omega}) = 1$, as shown earlier in Eq. (3.87).

In mathematics, the function (3.94) is called the **Dirichlet** kernel [Haaser and Sullivan, 1991]. It is somewhat similar to the so-called sinc function (Eq. (4.44)), which we shall encounter later. The main difference is that function (3.94) is periodic in

(a)

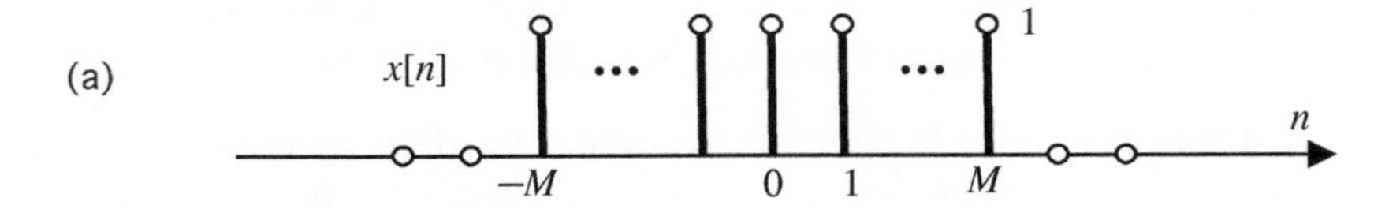

(b)

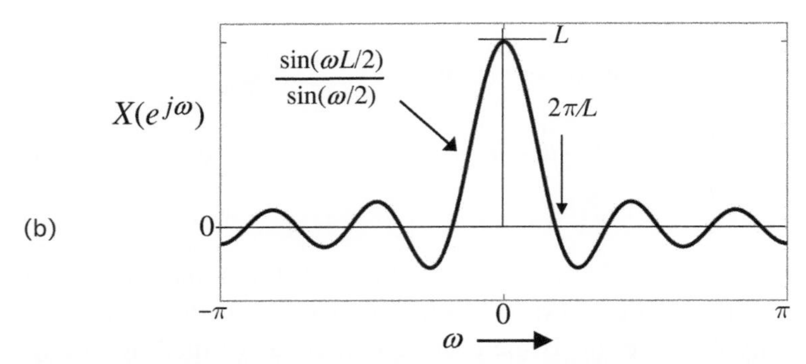

Figure 3.13 (a) The rectangular pulse signal $x[n]$ and (b) its Fourier transform $X(e^{j\omega}) = \sin(\omega L/2)/\sin(\omega/2)$, which is called the Dirichlet kernel. Here, $L = 2M + 1$.

ω with period 2π, because it is a discrete-time FT, whereas the sinc function is not periodic.

Further properties of the Dirichlet kernel. Since $x[0] = 1$ in our example, it follows from $x[0] = \int_{-\pi}^{\pi} X(e^{j\omega})d\omega/2\pi$ that

$$\int_{-\pi}^{\pi} \frac{\sin(\omega(2M+1)/2)}{\sin(\omega/2)} d\omega = 2\pi, \tag{3.95}$$

no matter what the integer $M \geq 0$ is. We will find this a useful result. To discuss the properties of Eq. (3.94) further, it is convenient to introduce the notation

$$L = 2M + 1, \tag{3.96}$$

which is the duration or length of the sequence $x[n]$. Then

$$X(e^{j\omega}) = \left(\frac{\sin(\omega L/2)}{\sin(\omega/2)} \right). \tag{3.97}$$

Notice that if $\omega = 0$, the above takes the form $0/0$, but we can take the limit as $\omega \to 0$. Thus, since $\sin\theta \to \theta$ for $\theta \to 0$, it follows that for $\omega \to 0$,

$$\frac{\sin(\omega L/2)}{\sin(\omega/2)} \longrightarrow \frac{\omega L/2}{\omega/2} = L. \tag{3.98}$$

This is consistent with the fact that

$$X(e^{j0}) = \sum_{n=-M}^{M} x[n] = 2M + 1 = L. \tag{3.99}$$

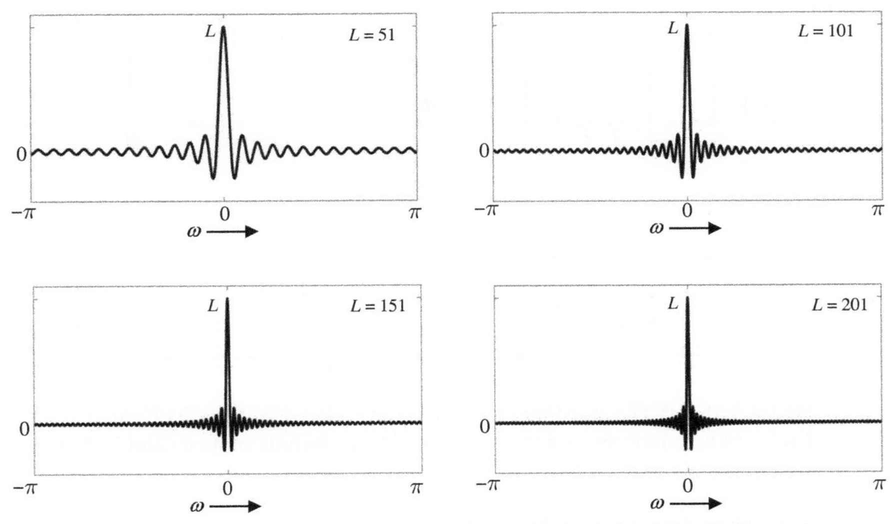

Figure 3.14 Plots of the Dirichlet kernel $X(e^{j\omega}) = \sin(\omega L/2)/\sin(\omega/2)$ for various odd values of L. As $L \to \infty$, the plot approaches $2\pi\delta_c(\omega)$. See text.

The plot in Fig. 3.13(b) has maximum value of L at $\omega = 0$. Also, $\sin(\omega L/2) = 0$ at $\omega L/2 = k\pi$ for integer k. So $X(e^{j\omega})$ is zero at

$$\omega = \frac{2\pi k}{L}, \quad k \neq 0, \tag{3.100}$$

as indicated in the plot in Fig. 3.13. The zero crossings are therefore uniformly spaced, with spacing $2\pi/L$. $\bigtriangledown\bigtriangledown\bigtriangledown$

Figure 3.14 shows plots of the Dirichlet kernel (3.97) for increasing values of L. As L increases, the pulse spreads out more in the time domain, but its Fourier transform gets more and more localized in frequency. In fact, the plots in Fig. 3.14 look more and more like the Dirac delta (Sec. 2.12) as L increases. Since the integral of this function is 2π for all L (see Eq. (3.95)), it follows that the limit is actually $2\pi\delta_c(\omega)$, that is,

$$\lim_{M\to\infty} \frac{\sin(\omega(2M+1)/2)}{\sin(\omega/2)} \longrightarrow 2\pi\delta_c(\omega). \tag{3.101}$$

This is a correct statement in the range $-\pi \leq \omega < \pi$, but since the left-hand side is periodic with period 2π, a more complete way to write the above would be

$$\lim_{M\to\infty} \frac{\sin(\omega(2M+1)/2)}{\sin(\omega/2)} \longrightarrow 2\pi \sum_{k=-\infty}^{\infty} \delta_c(\omega + 2\pi k), \quad \forall\omega. \tag{3.102}$$

The curious reader might wonder about the properties of Eq. (3.97) when L is even. In this case, it turns out that Eq. (3.97) is not a valid Fourier transform of any discrete-time signal! This is discussed further in Problem 3.27.

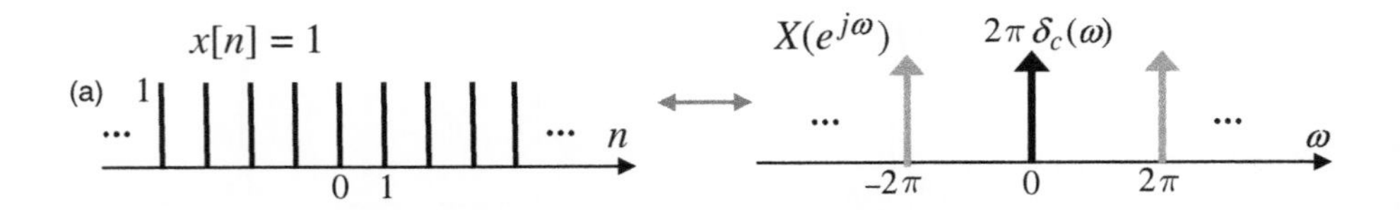

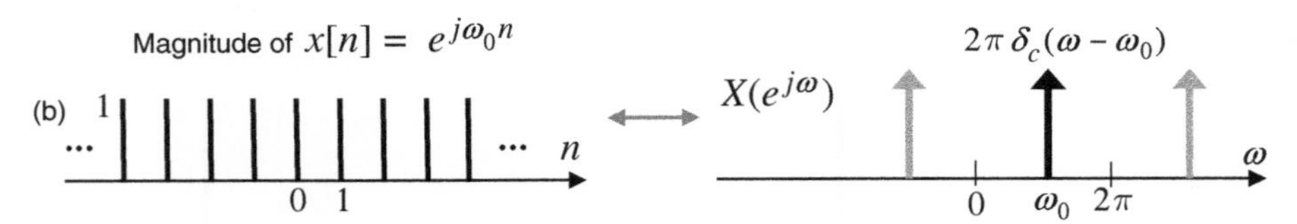

Figure 3.15 (a) The constant (zero-frequency) signal $x[n]$ and its Fourier transform. (b) The single-frequency signal $x[n] = e^{j\omega_0 n}$ (magnitude plotted) and its Fourier transform.

3.8.2.2 Single-Frequency Signals

Next consider the signal $x[n] = 1$ (zero-frequency signal or constant signal). From the definition of the DTFT, we have

$$X(e^{j\omega}) = \sum_{n=-\infty}^{\infty} e^{-j\omega n}. \tag{3.103}$$

This summation does not converge in the traditional sense. However, we can regard $x[n]$ as the limit of the pulse (3.88) as $M \to \infty$. The pulse has the DTFT given by Eq. (3.94), and it has the limit (3.101) as $M \to \infty$. Thus, if we allow Dirac delta functions into the answer, it is possible to get a valid answer for $X(e^{j\omega})$, namely, $X(e^{j\omega}) = 2\pi\,\delta_c(\omega)$ in the range $-\pi \leq \omega < \pi$, repeating periodically with period 2π. That is,

$$x[n] = 1, \ \forall n \quad \longleftrightarrow \quad X(e^{j\omega}) = 2\pi\,\delta_c(\omega). \tag{3.104}$$

Thus, the zero-frequency signal has its Fourier transform concentrated at zero frequency, as expected. This is demonstrated in Fig. 3.15(a), and is called a **line spectrum**. More generally, a signal whose Fourier transform is a sum of shifted and scaled Dirac delta functions (e.g., as we shall see in Fig. 3.16) is called a line spectral signal.

Another way to prove the result (3.104) is to substitute $X(e^{j\omega}) = 2\pi\,\delta_c(\omega)$ into the right-hand side of Eq. (3.68). Then

$$\text{RHS of Eq. (3.68)} = \frac{1}{2\pi}\int_0^{2\pi} X(e^{j\omega})e^{j\omega n}d\omega = \int_0^{2\pi} \delta_c(\omega)e^{j\omega n}d\omega = 1 \tag{3.105}$$

for all n, proving that indeed $x[n] = 1, \forall n$. The two examples (3.87) and (3.104) represent *opposite extremes*. That is, in Fig. 3.12 the signal is *localized* at one point in time and the FT is *spread out* in frequency completely. In Fig. 3.15(a) the opposite is true. The pulse (3.88) is a generalization of these two examples (which correspond to $M = 0$ and $M = \infty$, respectively).

In Fig. 3.15(a) the light-gray arrows in $X(e^{j\omega})$ serve as a reminder that $X(e^{j\omega})$ is periodic in ω with repetition interval 2π (Sec. 3.4.2). So, it is more accurate to write the result (3.104) as

$$x[n] = 1, \ \forall n \quad \longleftrightarrow \quad X(e^{j\omega}) = 2\pi \sum_{k=-\infty}^{\infty} \delta_c(\omega + 2\pi k), \quad \forall \omega, \tag{3.106}$$

although we usually use (3.104) and implicitly understand that $X(e^{j\omega})$ repeats periodically. Since $X(e^{j\omega}) = \sum x[n]e^{-j\omega n} = \sum e^{-j\omega n}$ in this case, it follows from (3.106) that

$$\sum_{n=-\infty}^{\infty} e^{-j\omega n} = 2\pi \sum_{k=-\infty}^{\infty} \delta_c(\omega + 2\pi k), \quad \forall \omega, \tag{3.107}$$

which can be regarded as a "mathematical" identity. We will find this useful later.

Next, by using the modulation property (3.74) in the example (3.106), it follows that

$$x[n] = e^{j\omega_0 n}, \ \forall n \quad \longleftrightarrow \quad X(e^{j\omega}) = 2\pi \sum_{k=-\infty}^{\infty} \delta_c(\omega + 2\pi k - \omega_0). \tag{3.108}$$

That is, the single-frequency signal $e^{j\omega_0 n}$ has its Fourier transform concentrated at ω_0 as expected, and repeating periodically with period 2π. This is shown in Fig. 3.15(b).

Remark on convergence. So we saw that the DTFT of the constant signal $x[n] = 1$ (zero-frequency signal) can be regarded as the limit

$$\lim_{M \to \infty} \sum_{n=-M}^{M} e^{-j\omega n}, \tag{3.109}$$

where the sum is the DTFT of the rectangular pulse of duration $2M + 1$. This sum approaches the Dirac delta function as $M \to \infty$, as demonstrated in Fig. 3.14. The constant signal therefore has the DTFT given in Eq. (3.104).

One thing should be noticed carefully. Even though the result (3.104) suggests that the DTFT sum $\sum_{n=-\infty}^{\infty} e^{-j\omega n}$ converges to zero for $\omega \neq 0$, we cannot prove this directly. For example, take $\omega = \pi$. Then

$$\sum_{n=-\infty}^{\infty} e^{-j\omega n} = \cdots - 1 + 1 - 1 + 1 - 1 + 1 - 1 \cdots, \tag{3.110}$$

which does not converge although it is bounded between 1 and -1 because of cancellation of adjacent terms. As we increase the number of terms in the sum

$$\sum_{n=-M}^{M} e^{-j\omega n}, \tag{3.111}$$

even though we cannot prove that this "becomes zero," we can show that it becomes negligibly small *compared to* the zero-frequency value $\sum_{n=-M}^{M} e^{-j0n}$. It is in this sense that the sum approaches the Dirac delta, and yields the answer (3.104). In fact, whenever the expression for the Fourier transform involves Dirac delta functions, we have to interpret the result carefully like this. $\bigtriangledown \bigtriangledown \bigtriangledown$

We will see in Sec. 6.3.1 that the constant signal $x[n] = 1$ (and more generally $x[n] = e^{j\omega_0 n}$) does not have a z-transform. In particular, the Fourier transform "does not exist," in the sense that there is no bounded answer. But if we accept unbounded things like the Dirac delta, then an expression for the FT can be found as in (3.108).

3.8.2.3 Cosines and Sines

There exist many examples for which the Fourier transform summation does not converge to a finite answer, but one can get an expression for $X(e^{j\omega})$ by allowing the Dirac delta function into our mathematics. Since

$$\cos(\omega_0 n) = \frac{e^{j\omega_0 n} + e^{-j\omega_0 n}}{2}, \quad \sin(\omega_0 n) = \frac{e^{j\omega_0 n} - e^{-j\omega_0 n}}{2j}, \tag{3.112}$$

it follows from (3.108) that

$$\cos(\omega_0 n) \longleftrightarrow \pi \sum_{k=-\infty}^{\infty} \Big(\delta_c(\omega + \omega_0 + 2\pi k) + \delta_c(\omega - \omega_0 + 2\pi k)\Big), \tag{3.113}$$

$$\sin(\omega_0 n) \longleftrightarrow j\pi \sum_{k=-\infty}^{\infty} \Big(\delta_c(\omega + \omega_0 + 2\pi k) - \delta_c(\omega - \omega_0 + 2\pi k)\Big). \tag{3.114}$$

Thus, the real cosine is a superposition of two frequencies ω_0 and $-\omega_0$ in $-\pi \leq \omega < \pi$. So its DTFT has two Dirac delta functions concentrated at these two frequencies. The real sine is similar. See Fig. 3.16. Note that $x[n] = \sin(\omega_0 n)$ is real as well as odd (i.e., $x[n] = -x[-n]$), whereas its FT is imaginary and odd. This is a general property of Fourier transforms (Problem 3.24).

Now consider the truncated cosine signal

$$x[n] = \begin{cases} \cos(\omega_0 n) & -M \leq n \leq M, \\ 0 & \text{otherwise.} \end{cases} \tag{3.115}$$

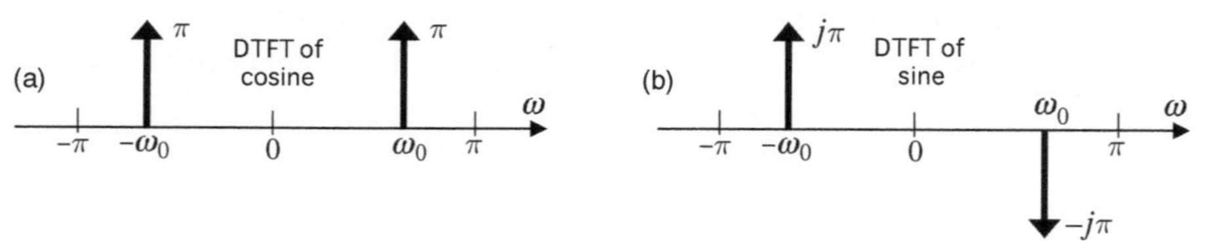

Figure 3.16 (a) The Fourier transform of $\cos(\omega_0 n)$ and (b) the Fourier transform of $\sin(\omega_0 n)$. Each of these has two Dirac delta functions. The heights shown indicate the integrals or "areas" under the Dirac deltas.

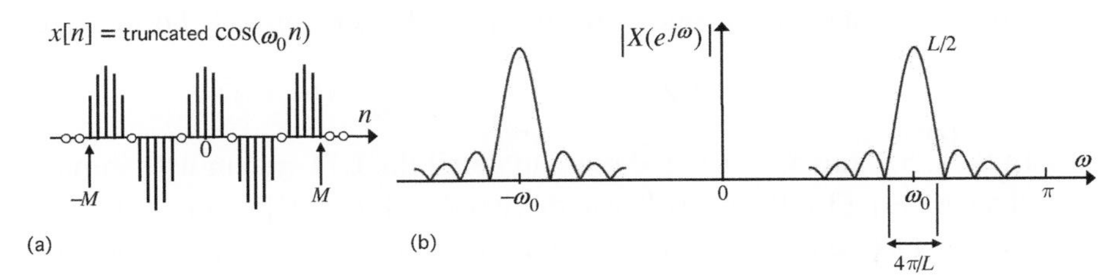

Figure 3.17 (a) A truncated cosine signal $x[n]$, with $L = 2M + 1$ samples retained. (b) Magnitude of its Fourier transform $X(e^{j\omega})$.

Only $L = 2M + 1$ samples are retained, and the rest discarded. See Fig. 3.17(a). We can think of this as the product $\cos(\omega_0 n)p[n]$, where $p[n]$ is the pulse

$$p[n] = \begin{cases} 1 & -M \leq n \leq M, \\ 0 & \text{otherwise,} \end{cases} \tag{3.116}$$

and its DTFT $P(e^{j\omega})$ is the Dirichlet kernel shown in Fig. 3.13(b). Thus

$$x[n] = \cos(\omega_0 n)p[n] = 0.5e^{j\omega_0 n}p[n] + 0.5e^{-j\omega_0 n}p[n]. \tag{3.117}$$

Since $e^{j\omega_0 n}p[n] \longleftrightarrow P(e^{j(\omega-\omega_0)})$ it follows that

$$X(e^{j\omega}) = 0.5P\left(e^{j(\omega-\omega_0)}\right) + 0.5P\left(e^{j(\omega+\omega_0)}\right). \tag{3.118}$$

This is a superposition of two copies of the Dirichlet kernel, shifted to the left and right by ω_0. Thus, the Dirac delta functions in Fig. 3.16(a) are "smeared" or spread out in the frequency domain, because of the truncation in the time domain, as shown in Fig. 3.17(b). The width of spread is $4\pi/L$, as shown in the figure, where L is the number of samples of $\cos(\omega_0 n)$ retained.

Another way to derive $X(e^{j\omega})$ is as follows: since $x[n]$ is the product of $p[n]$ with $\cos(\omega_0 n)$, its DTFT is proportional to the convolution of $P(e^{j\omega})$ with the twin Diracs in Fig. 3.16(a) (see Eq. (3.76)). Since the convolution of $P(e^{j\omega})$ with $\delta_c(\omega - \omega_0)$ is simply the shifted version $P\left(e^{j(\omega-\omega_0)}\right)$ (as we shall see in Sec. 3.10.3), the result follows readily.

3.8.2.4 One-Sided Exponential

Now consider the one-sided exponential

$$h[n] = p^n \mathcal{U}[n]. \tag{3.119}$$

Assume $|p| < 1$ so that $h[n] \to 0$ as $n \to \infty$. The DTFT is

$$\begin{aligned} H(e^{j\omega}) &= \sum_{n=-\infty}^{\infty} h[n]e^{-j\omega n} = \sum_{n=0}^{\infty} p^n e^{-j\omega n} \\ &= \sum_{n=0}^{\infty} (pe^{-j\omega})^n = \frac{1}{1 - pe^{-j\omega}}, \end{aligned} \tag{3.120}$$

where the last equality follows from Eq. (3.91) using $|p| < 1$. So we have proved

$$p^n \mathcal{U}[n] \quad \longleftrightarrow \quad \frac{1}{1 - pe^{-j\omega}} \quad (|p| < 1). \tag{3.121}$$

In Secs. 5.2 and 5.3 we will discuss in detail the LTI system with impulse response given by Eq. (3.119). We will also discuss the plots of $|h[n]|$ and $|H(e^{j\omega})|$ for specific values of p (see Figs. 5.2 and 5.3 for a preview). For now, we will mention one more closely related example, namely

$$(n+1)p^n \mathcal{U}[n] \quad \longleftrightarrow \quad \frac{1}{(1 - pe^{-j\omega})^2} \quad (|p| < 1). \tag{3.122}$$

This will become clear later; its proof follows from Eqs. (6.29) and (6.31) by setting $z = e^{j\omega}$ in Eq. (6.29).

3.8.3 The Unit Step

We now derive the Fourier transform of the unit-step signal $x[n] = \mathcal{U}[n]$. This is a tricky one, because

$$\widehat{\mathcal{U}}(e^{j\omega}) = \sum_{n=-\infty}^{\infty} \mathcal{U}[n] e^{-j\omega n} = \sum_{n=0}^{\infty} e^{-j\omega n}, \tag{3.123}$$

which does not converge. This is similar in spirit to the FT (3.103) of the constant signal $x[n] = 1$, which does not converge, although we were able to obtain an expression for the DTFT by admitting the Dirac delta into our mathematics (see Eq. (3.104)). Similarly, there are some tricks which can be used to derive an expression for $\widehat{\mathcal{U}}(e^{j\omega})$, which will have some unbounded terms. First observe that

$$\mathcal{U}[n] - \mathcal{U}[n-1] = \delta[n]. \tag{3.124}$$

Taking the Fourier transform on both sides, it follows that

$$\widehat{\mathcal{U}}(e^{j\omega}) - e^{-j\omega}\widehat{\mathcal{U}}(e^{j\omega}) = 1, \tag{3.125}$$

where the shift property (3.73) has been used. This proves that

$$\widehat{\mathcal{U}}(e^{j\omega}) = \frac{1}{1 - e^{-j\omega}}, \quad \omega \neq 0. \tag{3.126}$$

The reason for excluding $\omega = 0$ is that the denominator of the right-hand side is zero when $\omega = 0$, making the expression unbounded. Equation (3.126) shows that we have found the answer "almost everywhere." For $\omega = 0$, note that

$$\widehat{\mathcal{U}}(e^{j0}) = \sum_{n=0}^{\infty} e^{-j0n} = \sum_{n=0}^{\infty} 1 = \infty. \tag{3.127}$$

The situation was the same for the constant signal, although the behavior at $\omega = 0$ was captured well by the Dirac delta. To make progress along similar lines, note that the final expression for $\widehat{\mathcal{U}}(e^{j\omega})$ should be such that

$$\int_{-\pi}^{\pi} \widehat{\mathcal{U}}(e^{j\omega}) d\omega = 2\pi, \tag{3.128}$$

because $\mathcal{U}[0] = 1$ (use the property (3.72)). Now, it turns out (as proved below) that the integral of the right-hand side of Eq. (3.126) is

$$\int_{-\pi}^{\pi} \frac{1}{1 - e^{-j\omega}} d\omega = \pi, \tag{3.129}$$

where it is understood that an infinitesimally small neighborhood $-\epsilon < \omega < \epsilon$ is excluded from the range of integration, to avoid the singularity at $\omega = 0$. The correct expression for $\widehat{\mathcal{U}}(e^{j\omega})$, which accounts for the behavior at $\omega = 0$ while at the same time satisfying Eqs. (3.126) and (3.128), is therefore

$$\widehat{\mathcal{U}}(e^{j\omega}) = \pi \delta_c(\omega) + \frac{1}{1 - e^{-j\omega}} \,. \tag{3.130}$$

Indeed, the first term is such that (a) the value of Eq. (3.126) at $\omega \neq 0$ is not disturbed, (b) at $\omega = 0$ the FT is infinite as required by Eq. (3.127), and (c) the integral property (3.128) holds. This assures us that Eq. (3.130) is the correct answer for the Fourier transform of the unit step $\mathcal{U}[n]$. So, we have proved that

$$\mathcal{U}[n] \quad \longleftrightarrow \quad \pi \delta_c(\omega) + \frac{1}{1 - e^{-j\omega}} \,. \tag{3.131}$$

Note that we cannot obtain the result (3.131) by substituting $p = 1$ in Eq. (3.121). This is not surprising, because (3.121) is valid only when $|p| < 1$.

Proof of Eq. (3.129). It remains to prove Eq. (3.129). For this, observe that

$$\frac{1}{1 - e^{-j\omega}} = \frac{1}{1 - \cos\omega + j\sin\omega} = \frac{1 - \cos\omega - j\sin\omega}{(1 - \cos\omega)^2 + \sin^2\omega}, \tag{3.132}$$

which can be rewritten (Problem 3.25) as follows:

$$\frac{1}{1 - e^{-j\omega}} = 0.5 - j0.5\cot(\omega/2). \tag{3.133}$$

The imaginary term above is unbounded at $\omega = 0$. Notice, however, that since $\cot(\omega/2)$ is an odd function,

$$\int_{-\pi}^{-\epsilon} \cot(\omega/2) + \int_{\epsilon}^{\pi} \cot(\omega/2) = 0, \tag{3.134}$$

for any ϵ in $0 < \epsilon < \pi$. Thus, for infinitesimally small $\epsilon > 0$, Eq. (3.133) yields

$$\int_{-\pi}^{-\epsilon} \frac{1}{1 - e^{-j\omega}} d\omega + \int_{\epsilon}^{\pi} \frac{1}{1 - e^{-j\omega}} d\omega = 0.5 \int_{-\pi}^{\pi} d\omega = \pi, \tag{3.135}$$

which proves Eq. (3.129). ▽ ▽ ▽

3.8.3.1 A Corollary of the Unit-Step Example

Now consider the signal

$$s[n] = \begin{cases} 1 & n > 0, \\ -1 & n < 0, \\ 0 & n = 0. \end{cases} \tag{3.136}$$

To find $S(e^{j\omega})$, observe that $s[n]$ can be written as

$$s[n] = 2\Big(\mathcal{U}[n] - 0.5 - 0.5\delta[n]\Big). \tag{3.137}$$

Taking the DTFT and using Eqs. (3.130), (3.104), and (3.108), we get

$$S(e^{j\omega}) = 2\Big(\pi\delta_c(\omega) + \frac{1}{1-e^{-j\omega}} - \pi\delta_c(\omega) - 0.5\Big). \tag{3.138}$$

This simplifies to

$$S(e^{j\omega}) = \frac{1+e^{-j\omega}}{1-e^{-j\omega}} = \frac{\cot(\omega/2)}{j}, \quad \omega \neq 0. \tag{3.139}$$

We can take $S(e^{j0}) = 0$ because $\sum_n s[n] = 0$.

3.8.3.2 Interpreting Eq. (3.130) Carefully

The expression (3.130) was obtained by imposing the constraints (3.126)–(3.128), which should be satisfied by $\widehat{\mathcal{U}}(e^{j\omega})$. Careful thought shows that some questions still remain unanswered. For example, as mentioned right at the beginning, the expression (3.123) does not converge for any ω. So, what does it mean to say that $\widehat{\mathcal{U}}(e^{j\omega})$ is given by Eq. (3.126) when $\omega \neq 0$? For example, let us try $\omega = \pi$. Then, according to the definition (3.123), we have

$$\widehat{\mathcal{U}}(e^{j\pi}) = \sum_{n=0}^{\infty} e^{-j\pi n} = 1 - 1 + 1 - 1 + 1 - \cdots, \tag{3.140}$$

which does not converge. As the number of terms in the partial sum $\sum_{n=0}^{m-1} e^{-j\pi n}$ increases, the result simply oscillates between 1 and 0 without converging to anything. On the other hand, substituting $\omega = \pi$ in Eq. (3.126), we get

$$\widehat{\mathcal{U}}(e^{j\pi}) = \frac{1}{1-e^{-j\pi}} = \frac{1}{1+1} = 0.5, \tag{3.141}$$

which is the average of successive partial sums (1 and 0). Thus, the defining sum (3.140) and the advertised answer (3.126) do not seem to be saying the same thing at all! Consider one more example. Let $\omega = \pi/2$. Then

$$\widehat{\mathcal{U}}(e^{j\pi/2}) = \sum_{n=0}^{\infty} e^{-j\pi n/2} = 1 - j - 1 + j + 1 - j \cdots \tag{3.142}$$

whereas, if we substitute $\omega = \pi/2$ into Eq. (3.126), we get

$$\widehat{\mathcal{U}}(e^{j\pi/2}) = \frac{1}{1-e^{-j\pi/2}} = \frac{1}{1-\cos(\pi/2)+j\sin(\pi/2)} = \frac{1}{1+j} = \frac{1-j}{2}. \tag{3.143}$$

The summation (3.142) does not converge, so in particular it is not equal to what is claimed by Eq. (3.143). However, notice something miraculous here: if we consider the partial sums $\sum_{n=0}^{m-1} e^{-j\pi n/2}$, then as we increase the number of terms m, the first four results are

$$1,\ 1-j,\ -j,\ 0, \tag{3.144}$$

and then the sequence repeats! Thus, the average of the partial sums converges to

$$\frac{1+1-j-j}{4} = \frac{1-j}{2}, \tag{3.145}$$

which is precisely the result of Eq. (3.143)! Thus, at least in the above two examples, the average of the partial sums in the nonconvergent series $\sum_{n=0}^{\infty} e^{-j\omega n}$ seems to agree with the formula (3.126)! Is this a coincidence? Certainly not, as explained next.

3.8.4 Average of Partial Sums*

Consider the partial sum

$$S_m(e^{j\omega}) = \sum_{n=0}^{m-1} e^{-j\omega n}, \tag{3.146}$$

where $m \geq 1$. This is a truncation to m terms of the infinite sum which represents the Fourier transform of $\mathcal{U}[n]$. Define the average of the first L partial sums as follows:

$$A_L(e^{j\omega}) = \frac{1}{L}\sum_{m=1}^{L} S_m(e^{j\omega}) = \frac{1}{L}\sum_{m=1}^{L}\sum_{n=0}^{m-1} e^{-j\omega n}. \tag{3.147}$$

For example,

$$\begin{aligned} A_5(e^{j\omega}) = \frac{1}{5}\Big(& 1 \\ & + 1 + e^{-j\omega} \\ & + 1 + e^{-j\omega} + e^{-2j\omega} \\ & + 1 + e^{-j\omega} + e^{-2j\omega} + e^{-3j\omega} \\ & + 1 + e^{-j\omega} + e^{-2j\omega} + e^{-3j\omega} + e^{-4j\omega}\Big). \end{aligned} \tag{3.148}$$

We claim that for $\omega \neq 0$,

$$\lim_{L\to\infty} A_L(e^{j\omega}) = \frac{1}{1-e^{-j\omega}}. \tag{3.149}$$

Proof. We can rewrite Eq. (3.147) as

$$\begin{aligned} A_L(e^{j\omega}) = \frac{1}{L}\sum_{m=1}^{L}\frac{1-e^{-j\omega m}}{1-e^{-j\omega}} &= \frac{1}{L(1-e^{-j\omega})}\left(L - \sum_{m=1}^{L} e^{-j\omega m}\right) \\ &= \frac{1}{L(1-e^{-j\omega})}\left(L - e^{-j\omega}\sum_{m=0}^{L-1} e^{-j\omega m}\right) \\ &= \frac{1}{L(1-e^{-j\omega})}\left(L - e^{-j\omega}\frac{1-e^{-j\omega L}}{1-e^{-j\omega}}\right), \end{aligned}$$

which can be rewritten as

$$A_L(e^{j\omega}) = \frac{1}{1-e^{-j\omega}}\left(1 - e^{-j\omega}e^{-j\omega(L-1)/2}\frac{\sin(\omega L/2)}{L\sin(\omega/2)}\right). \tag{3.150}$$

Now, we know that $\sin(\omega L/2)/\sin(\omega/2)$ is the Dirchlet function plotted earlier in Fig. 3.13, and has value L at $\omega = 0$. So, the magnitude of the second term in brackets in Eq. (3.150) is unity at $\omega = 0$, and has its zero crossings at $\omega = 2\pi k/L, k \neq 0$. Thus, for any fixed $\omega \neq 0$ in $[-\pi, \pi]$, as $L \to \infty$, the second term in Eq. (3.150) goes to zero. This proves the result (3.149). $\nabla\nabla\nabla$

Summarizing, for $\omega \neq 0$, the expression (3.126) (hence Eq. (3.130)) should be interpreted as the limit of the average (3.147) of the partial sums (3.146). This is the only way we can reconcile an expression like (3.130) with the nonconvergent summation (3.123) from which it came!

In the theory of Fourier series, such averages of partial sums are called Cesàro sums [Apostol, 1974]. Summations like $\sum_{n=0}^{\infty} e^{-j\omega n}$ for which the average of partial sums converges are said to be **Cesàro summable**. Thus, the Fourier transform expression (3.130) for the unit step should be interpreted as a Cesàro sum.

As a final remark, we note that the unit step $x[n] = \mathcal{U}[n]$ has z-transform

$$\widehat{\mathcal{U}}(z) = \frac{1}{1 - z^{-1}}, \tag{3.151}$$

as long as $|z| > 1$. This is readily proved using Eq. (3.91). Note that the z-transform does not exist on the unit circle $|z| = 1$ because the summation (3.123) does not converge. This is consistent with the fact that the Fourier transform expression (3.130) is unbounded.

3.9 Overview of Continuous-Time Counterpart

All the developments in this chapter so far have been based on discrete-time signals and systems. We now provide a brief summary of the continuous-time counterpart, to serve as a convenient reference. Because of the resemblance to the discrete-time case, details will be omitted. As mentioned in Chapter 2, the definitions of linearity and time invariance are similar for continuous and discrete-time systems. If we apply an impulse input $\delta_c(t)$ (i.e., Dirac delta, Sec. 2.12) to a continuous-time LTI system, the output $h(t)$ is called an **impulse response**. Knowing this response $h(t)$, it is possible to express the output $y(t)$ in response to any arbitray input $x(t)$, using the convolution integral

$$y(t) = \int_{-\infty}^{\infty} x(\tau)h(t-\tau)d\tau = \int_{-\infty}^{\infty} h(\tau)x(t-\tau)d\tau = (h * x)(t), \tag{3.152}$$

similar to the convolution sum for discrete-time systems. You can verify that the LTI system is causal if and only if $h(t) = 0, t < 0$, and stable if and only if $h(t)$ is absolutely integrable, that is,

$$\int_{-\infty}^{\infty} |h(t)|dt < \infty \quad \text{(stability)}. \tag{3.153}$$

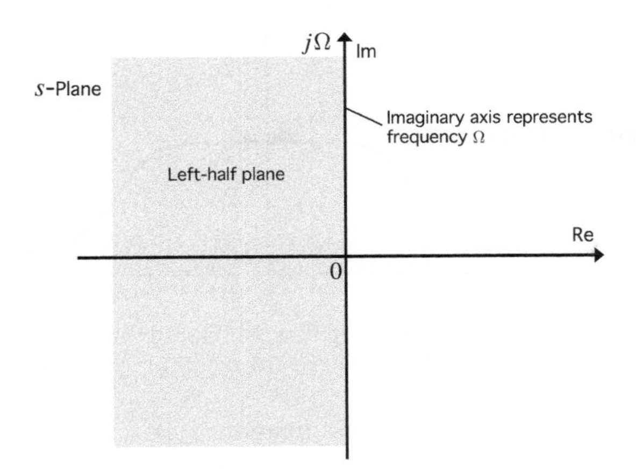

Figure 3.18 The complex s-plane. The imaginary axis $s = j\Omega$ represents the frequency variable Ω.

Next, if an exponential input $x(t) = e^{at}$ is applied, the output is the same exponential scaled, that is, $y(t) = H(a)e^{at}$, where

$$H(s) = \int_{-\infty}^{\infty} h(t)e^{-st}dt. \tag{3.154}$$

This is called the *transfer function* of the LTI system. The above integral is also called the **Laplace transform** of $h(t)$. Here, s is a complex variable and the integral in general converges only in a certain region of the complex s-plane, the region depending on the function $h(t)$. Figure 3.18 shows the complex s-plane which plays a role similar to the complex z-plane in the discrete-time case (Fig. 3.3).

In particular, note that the single-frequency signal $x(t) = e^{j\Omega_0 t}$ produces output $y(t) = H(j\Omega_0)e^{j\Omega_0 t}$, where

$$H(j\Omega) = \int_{-\infty}^{\infty} h(t)e^{-j\Omega t}dt \tag{3.155}$$

is said to be the **frequency response** of the LTI system. The above integral is also called the continuous-time Fourier transform (CTFT) of $h(t)$. Note that the Fourier transform $H(j\Omega)$ is nothing but the Laplace transform $H(s)$ evaluated on the **imaginary axis** of the s-plane. In discrete time, the unit circle of the z-plane represented frequencies with $0 \leq \omega < 2\pi$ (Fig. 3.3). In the continuous-time case, the frequency variable has the range

$$-\infty < \Omega < \infty, \tag{3.156}$$

and is represented by the imaginary axis. With the frequency range as above, a typical lowpass filter response would be as in Fig. 3.19. We will see later (Sec. 7.7) that the left half of the s-plane, shown shaded in Fig. 3.18, has a connection to the region inside the unit circle of the z-plane. The continuous-time Fourier transform is discussed in considerable detail in Chapter 9.

Next, for the input $x(t)$ and the output $y(t)$, we can similarly define the Laplace transforms

$$X(s) = \int_{-\infty}^{\infty} x(t)e^{-st}dt, \quad Y(s) = \int_{-\infty}^{\infty} y(t)e^{-st}dt. \tag{3.157}$$

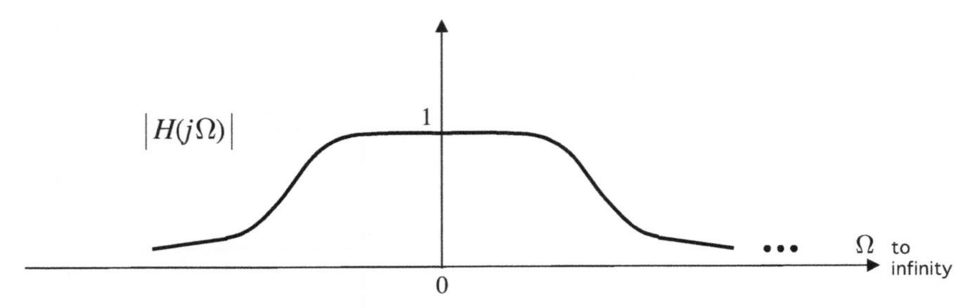

Figure 3.19 The magnitude response of a typical lowpass filter in the continuous-time case. Note that the frequency range is $-\infty < \Omega < \infty$.

These quantities evaluated on the imaginary axis, that is, $X(j\Omega)$ and $Y(j\Omega)$, are the Fourier transforms of $x(t)$ and $y(t)$, respectively. The convolution theorem, whose proof is similar to the discrete-time case, says that if $y(t)$ is given by Eq. (3.152) then

$$Y(s) = H(s)X(s) \quad \text{and} \quad Y(j\Omega) = H(j\Omega)X(j\Omega). \tag{3.158}$$

Thus, convolution in time becomes multiplication in the transform domain. The Laplace transform is discussed in greater detail in Chapter 13.

3.10 Convolution Examples

The convolution $y(t) = \int_{-\infty}^{\infty} x(\tau)h(t-\tau)d\tau$ can be interpreted graphically as follows:

1. First time-reverse the signal $h(\tau)$ to obtain $h(-\tau)$.
2. Shift the time-reversed signal $h(-\tau)$ to the right by t to get $h(-(\tau - t))$ or $h(t-\tau)$. (For negative t this is actually a left shift.)
3. Multiply this right-shifted signal $h(t-\tau)$ by $x(\tau)$ and integrate with respect to τ. The answer is $y(\cdot)$ at time t. If this is repeated for all t (positive and negative), we get the entire plot of $y(t)$.

This **graphical approach** to convolution is demonstrated in Fig. 3.20 for an example, and is self-explanatory. For a discrete-time signal, the procedure is similar except that the integrals are replaced with summations:

$$y[n] = \sum_{k=-\infty}^{\infty} x[k]h[n-k]. \tag{3.159}$$

An example is shown in Fig. 3.21.

3.10.1 Convolving the Pulse with Itself

Another good example in continuous time is the convolution of the pulse $b(t)$ shown in Fig. 3.22(a) with itself. This can be derived using the graphical approach to convolution described above. The result, denoted

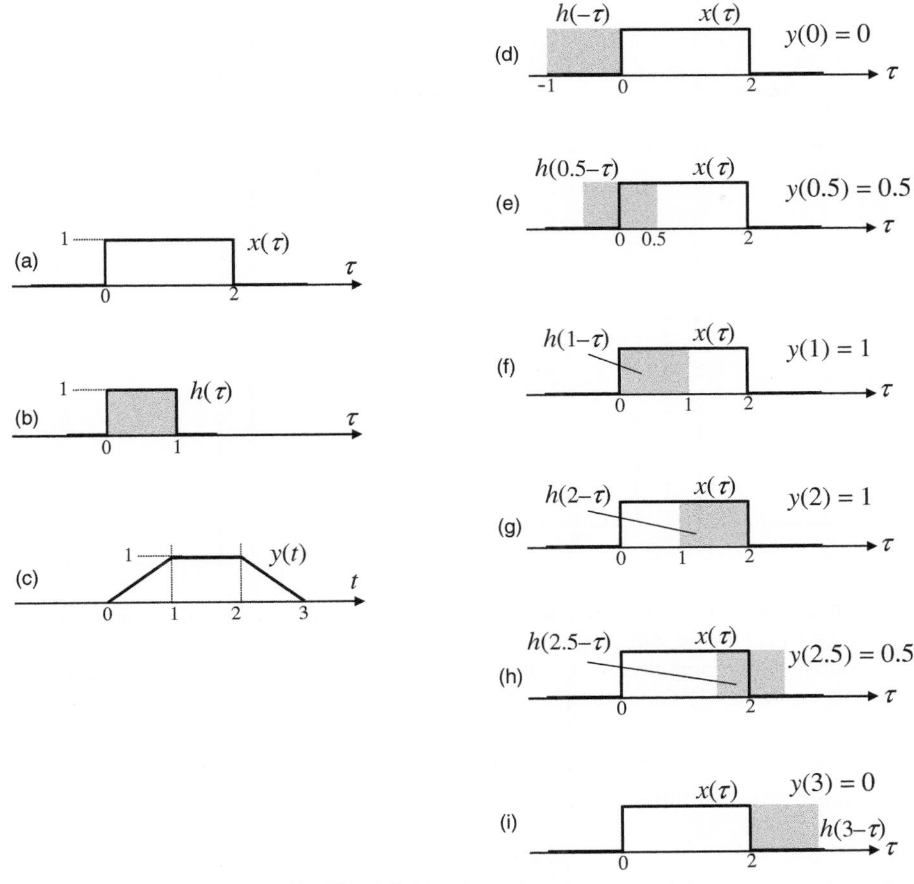

Figure 3.20 Demonstration of the graphical approach to convolution. (a), (b) The two signals to be convolved, (c) the result of convolution $y(t)$, and (d)–(i) relative positions of $x(\tau)$ and $h(t-\tau)$ for various values of t. The area of the region of overlap is the integral of the product, and represents the result of convolution $y(t)$ for that specific t (which is numerically indicated on the right).

$$b_1(t) = (b * b)(t), \tag{3.160}$$

is a triangle, twice as long. See Fig. 3.22(b). If we convolve the result with the pusle again, the result is

$$b_2(t) = (b * b * b)(t). \tag{3.161}$$

This has duration three and is shown in Fig. 3.22(c). This has the mathematical expression (Problem 3.20)

$$b_2(t) = \begin{cases} 0.5t^2 & 0 \le t < 1, \\ 0.75 - (t-1.5)^2 & 1 \le t < 2, \\ 0.5(t-3)^2 & 2 \le t < 3, \\ 0 & \text{otherwise.} \end{cases} \tag{3.162}$$

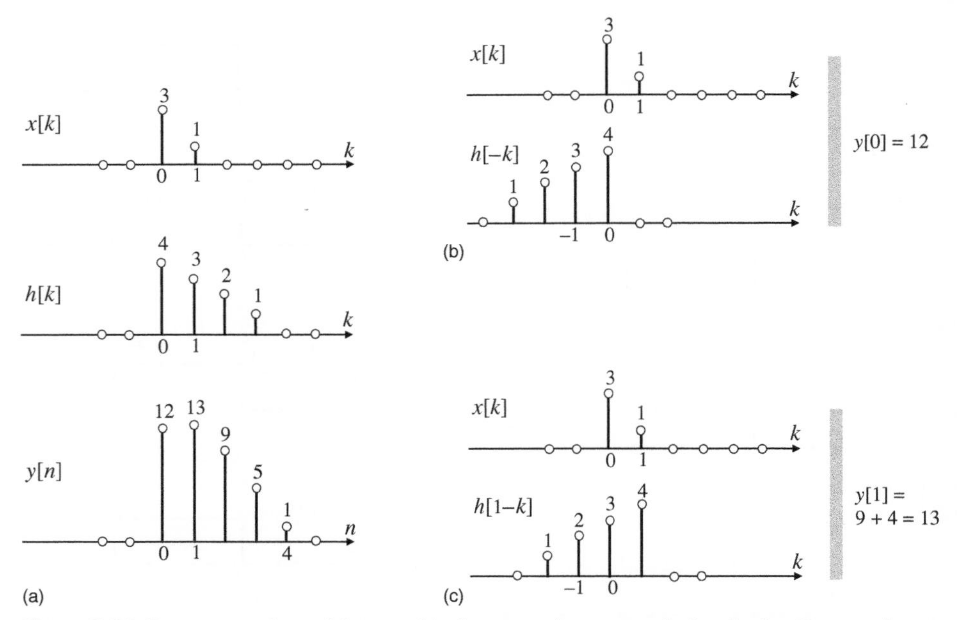

Figure 3.21 Demonstration of the graphical approach to convolution in the discrete-time case. (a) Plots of $x[k]$, $h[k]$, and their convolution $y[n]$, (b) $x[k]$ and the flipped version $h[-k]$ involved in computing $y[0]$, and (c) $x[k]$ and the shifted version $h[1-k]$ involved in computing $y[1]$. Other samples of $y[n]$ are computed similarly.

Notice that $b_2(t)$ is made of three polynomial pieces and each polynomial is a quadratic. Similarly, we can define $b_N(t)$ for any integer $N > 0$ as

$$b_N(t) = \Big(\underbrace{b * b * \cdots * b}_{b \text{ occurs } N+1 \text{ times}} \Big)(t), \tag{3.163}$$

where the b occurs $N+1$ times on the right-hand side. The subscript N is the number of convolutions ("$*$" signs) involved, for example, $b_2(t)$ in Eq. (3.161) has two convolutions. In the mathematics literature, $b_N(t)$ is known as the Nth-order ***B*-spline**. In Sec. 9.11.1 we will discuss splines in detail. It will be explained that splines have certain smoothness properties which make them useful in signal interpolation applications. For now, we just mention that $b_N(t)$ has duration $N+1$ and is made of $N+1$ polynomial pieces, each of degree N.

3.10.2 Duration of the Result of a Convolution

Notice that if $x(t)$ and $h(t)$ have finite durations D_x and D_h, then the duration of their convolution is $D_x + D_h$. In discrete time, if $x[n]$ and $h[n]$ have finite durations N_x and N_h samples, then the duration of convolution is N_x+N_h-1 samples. Notice also that if one of the durations is infinite, then the duration of the convolution is not necessarily infinite. For example, suppose $x[n]$ is the finite-duration signal

$$x[n] = \cdots 0, 0, 1, -a, 0, 0 \cdots \tag{3.164}$$

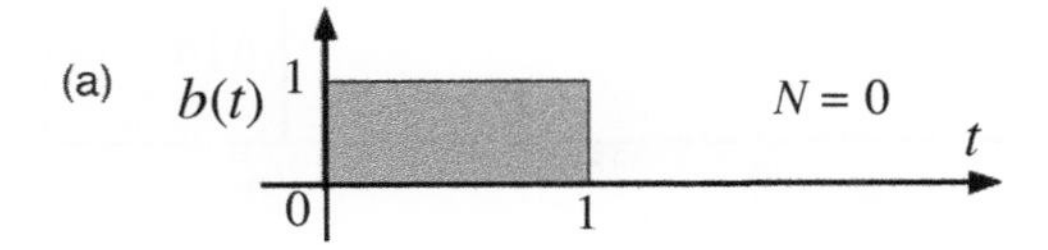

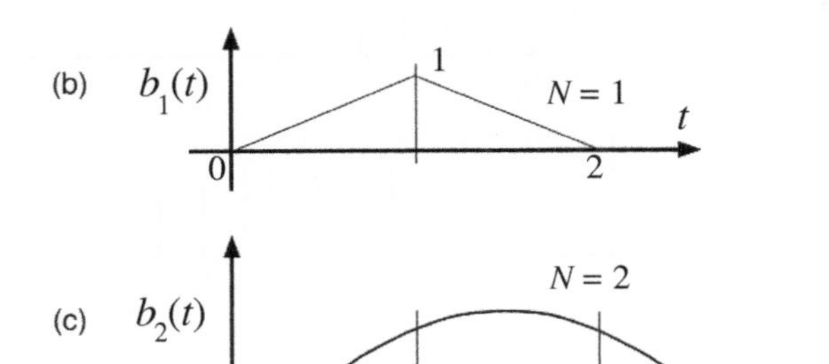

Figure 3.22 (a) A rectangular pulse, (b) result of convolution of the pulse with itself, and (c) result of further convolution of the triangle in (b) with the pulse in (a). If the pulse is convolved with itself repeatedly, we get the Nth-order B-spline $b_N(t) = (b * b * \cdots * b)(t)$. See text.

and $h[n]$ is the infinite-duration, one-sided, exponential signal

$$h[n] = \cdots 0, 0, 1, a, a^2, a^3, \cdots \tag{3.165}$$

where the "1" occurs at time $n = 0$ in both cases. Then their convolution is $y[n] = \delta[n]$ (Problem 3.3), which has duration one!

3.10.3 Convolving with an Impulse is Like Shifting

From the definition of the impulse $\delta[n]$, it readily follows that

$$\sum_{k=-\infty}^{\infty} \delta[k]x[n-k] = x[n]. \tag{3.166}$$

The left-hand side is a convolution, which shows that if we convolve a signal with the impulse, we get back the same signal. That is, the impulse signal is the **identity element** for the operation of convolution. Similarly, if we convolve $x[n]$ with the shifted impulse $\delta[n - n_0]$, we get

$$\sum_{k=-\infty}^{\infty} \delta[k - n_0]x[n-k] = x[n - n_0], \tag{3.167}$$

which is the right-shifted version of $x[n]$ by n_0 samples. This is demonstrated in Fig. 3.23. Similarly, in continuous time we have

$$\int_{-\infty}^{\infty} \delta_c(\tau)x(t-\tau)d\tau = x(t) \tag{3.168}$$

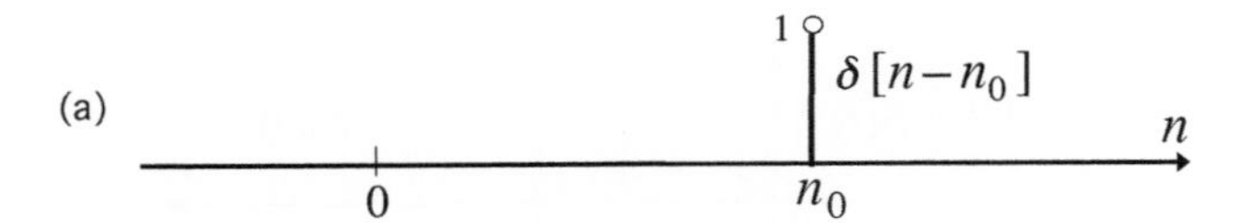

Figure 3.23 (a) The shifted impulse $\delta[n - n_0]$, (b) a signal $x[n]$, and (c) the convolution of these two signals, which is the shifted version $x[n - n_0]$.

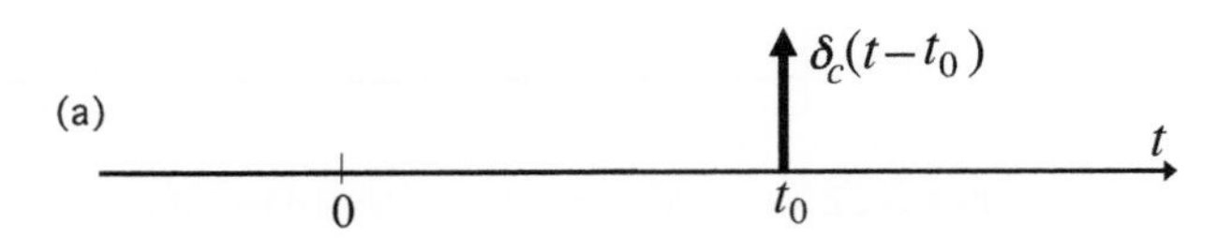

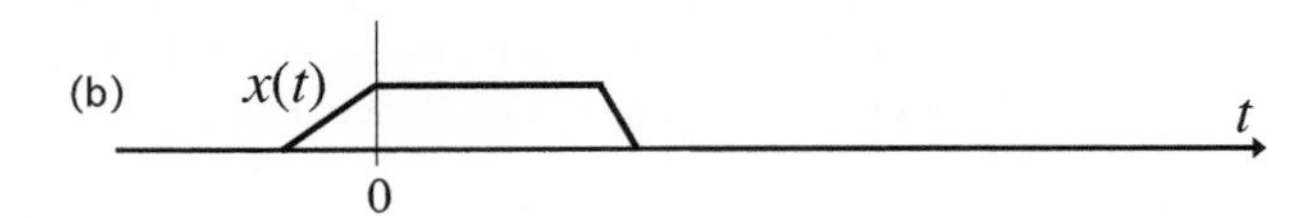

Figure 3.24 (a) The shifted Dirac delta $\delta_c(t - t_0)$, (b) a signal $x(t)$, and (c) the convolution of these two signals, which is the shifted version $x(t - t_0)$.

and

$$\int_{-\infty}^{\infty} \delta_c(\tau - t_0)x(t - \tau)d\tau = x(t - t_0). \tag{3.169}$$

So, convolution of $x(t)$ with $\delta_c(t - t_0)$ shifts $x(t)$ to the right by t_0. This is demonstrated in Fig. 3.24.

3.11 Some Fine Points about Eigenfunctions*

Some interesting points about the eigenfunctions of LTI systems will now be discussed. In Sec. 3.4 we showed that exponentials a^n are eigenfunctions of LTI systems, that is,

$$x[n] = a^n, \ \forall n \quad \Longrightarrow \quad y[n] = H(a)a^n, \ \forall n, \tag{3.170}$$

where

$$H(z) = \sum_{n=-\infty}^{\infty} h[n]z^{-n}. \tag{3.171}$$

We now mention a number of points in connection with this property.

1. First, the summation (3.171) does not converge for all z, and there is usually a region in the z-plane where it converges (Chapter 6). If the constant a is not in this region of convergence, then $H(a)$ is undefined, and the statement (3.170) is not meaningful for that a.
2. One natural question that arises when we look at (3.170) is this: if $v[n]$ is an eigenfunction of a particular LTI system, does it *have* to be an exponential? The answer is "not necessarily." For example, consider the identity system $H(z) = 1$. Since $y[n] = x[n]$, any signal is an eigenfunction for this system! But are there nontrivial examples, other than the identity system? Surely it is easy to create many examples. All we have to do is construct $H(z)$ such that $H(a) = H(b)$ for two distinct constants a and b. Then $a^n \longmapsto H(a)a^n$ and $b^n \longmapsto H(b)b^n$, so that

$$a^n + b^n \longmapsto H(a)a^n + H(b)b^n = H(a)(a^n + b^n), \tag{3.172}$$

since $H(a) = H(b)$. Thus $a^n + b^n$, which is not an exponential for $a \neq b$, is an eigenfunction. That is, if the eigenvalues (transfer functions) at $z = a$ and $z = b$ are identical, then the sum $a^n + b^n$ is also an eigenfunction! Here is an example of how to create an LTI system such that $H(a) = H(b)$. Let

$$H(z) = h[0] + h[1]z^{-1} + h[2]z^{-2}. \tag{3.173}$$

One can easily choose the impulse response coefficients $h[n]$ such that $H(e^{j\omega_1}) = H(e^{j\omega_2})$ for $\omega_2 = -\omega_1$, for some ω_1. This can be achieved by choosing (Problem 3.18)

$$h[0] = 0, \quad h[1] = 1, \quad h[2] = -\frac{\sin \omega_1}{\sin 2\omega_1}. \tag{3.174}$$

Extending this idea, suppose we have a good filter with $H(e^{j\omega}) \approx 1$ in the passband. If the input signal already belongs in the passband (i.e., $X(e^{j\omega}) \approx 0$ in the filter's stopband), then $Y(e^{j\omega}) \approx X(e^{j\omega})$, that is $y[n] \approx x[n]$, which means that such an $x[n]$ is (approximately) an eigenfunction of such a filter.
3. In fact, one can create more examples like this: suppose the input $x[n]$ to the LTI system is a sequence for which the Fourier transform $X(e^{j\omega})$ exists. Then $Y(e^{j\omega}) = H(e^{j\omega})X(e^{j\omega})$. If $x[n]$ happens to be an eigenfunction for this system, we have $Y(e^{j\omega}) = cX(e^{j\omega})$ for some constant c. So we have

$$Y(e^{j\omega}) = H(e^{j\omega})X(e^{j\omega}), \quad \text{and} \quad Y(e^{j\omega}) = cX(e^{j\omega}) \text{ as well}, \tag{3.175}$$

which shows that $H(e^{j\omega})X(e^{j\omega}) = cX(e^{j\omega})$, that is,

$$\Big(H(e^{j\omega}) - c\Big)X(e^{j\omega}) = 0. \tag{3.176}$$

The above equation simply says that for each ω, we *have* to have

$$\text{either } X(e^{j\omega}) = 0, \quad \text{or } H(e^{j\omega}) = c. \tag{3.177}$$

That is, where $H(e^{j\omega}) = c$, the quantity $X(e^{j\omega})$ can be arbitrary, and where $H(e^{j\omega}) \neq c$, we have to have $X(e^{j\omega}) = 0$. We can create many examples of

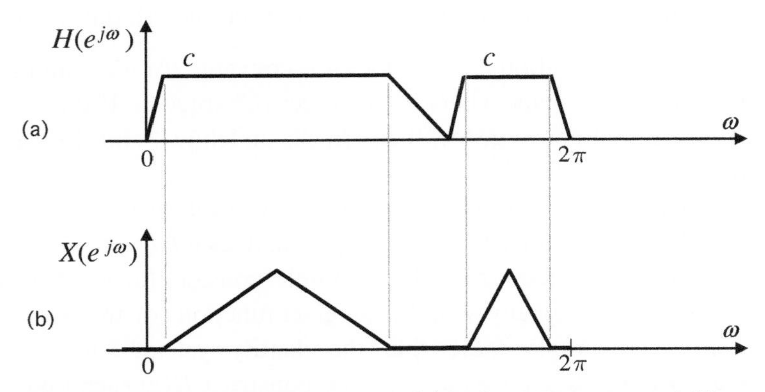

Figure 3.25 (a) Frequency response $H(e^{j\omega})$ of an LTI system and (b) Fourier transform of an eigenfunction $x[n]$, which is not an exponential. Where $H(e^{j\omega}) = c$, $X(e^{j\omega})$ can be arbitrary, and where $H(e^{j\omega}) \neq c$, $X(e^{j\omega}) = 0$.

eigenfunctions like this. Figure 3.25 shows one. Inspection shows that $Y(e^{j\omega}) = cX(e^{j\omega})$, which explains the eigenfunction property. By taking the inverse transform of this $X(e^{j\omega})$ we have therefore constructed an eigenfunction $x[n]$ for this $H(e^{j\omega})$, and this eigenfunction is not an exponential!

4. When encountering the eigenfunction property (3.170) for the first time, one might think "If $X(z)$ is the z-transform of a^n, then (3.170) implies that $Y(z) = H(a)X(z)$. But we also have $Y(z) = H(z)X(z)$ by the convolution theorem. So we have both

$$Y(z) = H(z)X(z) \quad and \quad Y(z) = H(a)X(z), \tag{3.178}$$

which implies $H(z) = H(a)$ for all z, that is, $H(z)$ is a constant. So unless a system has a constant transfer function, there are no eigenfunctions!" The flaw in this argument arises from the fact that for an exponential signal $x[n] = a^n$, the z-transform does not exist, that is, $\sum_{n=-\infty}^{\infty} x[n]z^{-n}$ does not converge for any z, as explained in Sec. 6.3.1. So arguments such as the above, based in the z-domain, are not valid!

Next, consider a very different question. Suppose we happen to know that a certain system is such that exponentials are eigenfunctions:

$$x[n] = a^n, \ \forall n \quad \Longrightarrow \quad y[n] = c_a a^n, \ \forall n. \tag{3.179}$$

Assuming we do not have any further knowledge about the system, is the property (3.179) itself enough to say that the system is LTI? The answer is no. To prove this, we only have to revisit an example given in Sec. 2.8.1, namely Eq. (2.86) reproduced below:

$$y[n] = \begin{cases} \dfrac{x^2[n]}{x[n-1]} & x[n-1] \neq 0, \\ 0 & \text{otherwise.} \end{cases} \tag{3.180}$$

It was seen earlier that this system is homogeneous and time invariant, or **HTI**, but it is not linear (because it is not additive). However, it is clear that if $x[n] = a^n$ then $y[n] = a \cdot a^n$, so Eq. (3.179) is satisfied. For example, if $x[n] = e^{j\omega_0 n}$ then

$$y[n] = e^{j\omega_0} e^{j\omega_0 n}. \tag{3.181}$$

Does this mean that we can even define a frequency response

$$H(e^{j\omega}) = e^{j\omega} \tag{3.182}$$

for this system, although it is not LTI? Let us accept that this is so, and examine it a bit further. As noted in Sec. 2.8.1, if we apply the input $x[n] = \delta[n]$ then the output is $y[n] = 0$, so that this system has impulse response

$$h[n] = 0, \quad \forall n. \tag{3.183}$$

That is, $H(e^{j\omega})$ is *not* the Fourier transform of $h[n]$ in this case! The system (3.180) has impulse response (3.183) but the frequency response (3.182) is nonzero for all ω. In fact, it is "allpass" in the sense that $|H(e^{j\omega})| = 1$ for all ω. Equation (3.183) does not imply that the output is zero for arbitrary inputs, because it is not an LTI system anyway (i.e., the output is not a convolution of $x[n]$ and $h[n]$). The main point of the above example is that there may exist a system for which the frequency response can be defined, but it is not useful unless the system is LTI. For example, if the input is a linear combination of several signals of the form $e^{j\omega_i n}$, the output of such a system is not necessarily a similar linear combination of $H(e^{j\omega_i})e^{j\omega_i n}$ (because the system is not additive). So the definition of a frequency response is not meaningful for non-LTI systems.

HTI systems and exponentials. In fact, we can generalize the above example as follows [Vaidyanathan, 1999]. Given *any* homogeneous and time-invariant (HTI) system, suppose the input $x[n] = a^n$ produces an output $y[n]$. Then by homogeneity we have

$$aa^n \longmapsto ay[n], \tag{3.184}$$

and by time invariance we have

$$a^{n+1} \longmapsto y[n+1]. \tag{3.185}$$

Since $aa^n = a^{n+1}$ we conclude from the above that $y[n+1] = ay[n]$. This shows that for any n the output has the form

$$y[n] = y[0]a^n. \tag{3.186}$$

That is, the ouput is also the same exponential a^n, scaled by a constant. In short, for any HTI system, exponentials are eigenfunctions. However, as seen from the example (3.180), this does not mean that the concept of a frequency response is useful for such systems. $\bigtriangledown \bigtriangledown \bigtriangledown$

As already mentioned in Sec. 3.4, for non-LTI systems, we cannot make any general statements about the output in response to an exponential input. For example, with

$x[n] = e^{j\omega_0 n}$, which is a single-frequency signal, the outputs for a number of non-LTI examples are listed below.

1. The system (3.180) (nonlinear): Now $y[n] = e^{j\omega_0}e^{j\omega_0 n}$, which is indeed a single-frequency signal with the same frequency ω_0 as the input. This was already discussed in detail above.
2. $y[n] = x^2[n]$ (nonlinear) and $y[n] = x[2n]$ (time varying): For both these systems we have $y[n] = e^{j2\omega_0 n}$. So the output is also a single-frequency signal but the frequency has changed to $2\omega_0$.
3. $y[n] = x[n] + x^2[n]$ (nonlinear): We have $y[n] = e^{j\omega_0 n} + e^{j2\omega_0 n}$. So the output has a new frequency component $2\omega_0$ in addition to ω_0.
4. $y[n] = e^{x[n]}$ (nonlinear): Now the output is $y[n] = e^{e^{j\omega_0 n}}$, which is not even a sum of a finite number of single-frequency signals!
5. $y[n] = nx[n]$ (time varying): Now the output is $y[n] = ne^{j\omega_0 n}$, which is not a sum of a finite number of single-frequency signals.

3.12 Summary of Chapter 3

This chapter presented a detailed study of linear time-invariant (LTI) systems. One of the most important properties of LTI systems is that the impulse response $h[n]$ completely determines the output $y[n]$ in response to an input $x[n]$, via convolution:

$$y[n] = (h * x)[n] = \sum_{k=-\infty}^{\infty} h[k]x[n-k] \quad \text{(convolution)}. \tag{3.187}$$

The LTI system can therefore be characterized completely by $h[n]$. Properties such as causality and stability can be expressed in terms of $h[n]$. For example, an LTI system is stable if and only if

$$\sum_{n=-\infty}^{\infty} |h[n]| < \infty, \tag{3.188}$$

that is, the impulse response is absolutely summable. We introduced the discrete-time Fourier transform $X(e^{j\omega})$ and the z-transform $X(z)$ for any sequence $x[n]$:

$$X(z) = \sum_{n=-\infty}^{\infty} x[n]z^{-n}, \quad X(e^{j\omega}) = \sum_{n=-\infty}^{\infty} x[n]e^{-j\omega n}. \tag{3.189}$$

The inverse Fourier transform $x[n] = \int_0^{2\pi} X(e^{j\omega})e^{j\omega n}d\omega/2\pi$ expresses $x[n]$ as a superposition of single-frequency signals $e^{j\omega n}$. This is the Fourier integral representation for $x[n]$. The z-transform of $h[n]$, namely $H(z)$, is called the transfer function of the LTI system, and the Fourier transform of $h[n]$ is called the frequency response. The z-transform evaluated on the unit circle of the z-plane is the Fourier transform. We showed that any exponential signal a^n is an eigenfunction of any LTI system, that is,

$$x[n] = a^n \quad \Longrightarrow \quad y[n] = H(a)a^n \qquad \text{(LTI systems)}. \tag{3.190}$$

This gives a physical meaning for the transfer function $H(z)$. Similarly, if $x[n] = e^{j\omega_0 n}$ then $y[n] = H(e^{j\omega_0})e^{j\omega_0 n}$, that is, the frequency response $H(e^{j\omega_0})$ is the complex gain of the system for the single-frequency input $e^{j\omega_0 n}$. The Fourier transforms of many commonly occurring signals, such as the pulse, the unit step, and sinusoids, were derived (see summary in Sec. 3.14). Several properties of discrete-time Fourier transforms were derived in considerable detail (see summary in Sec. 3.13). The convolution theorem says that convolution in time is equivalent to multiplication in the frequency domain (or equivalently in the z-domain). Thus, if an LTI system with impulse response $h[n]$ has input $x[n]$ and output $y[n]$, then

$$Y(z) = H(z)X(z), \quad Y(e^{j\omega}) = H(e^{j\omega})X(e^{j\omega}). \tag{3.191}$$

This property gives rise to the concept of filtering. Several types of filters, such as lowpass, highpass, and bandpass filters, were also introduced. The filtering process for a two-dimensional image was explained in some detail. While the earlier part of the chapter focused on the discrete-time case, the continuous-time counterpart of the above results was also presented along with some examples. One of the interesting examples was the repeated convolution of the pulse with itself, which gives rise to the family of spline functions.

For the reader wishing to study more, there are many books on fundamentals of signals and systems [e.g., Lathi, 1992; Lee and Varaiya, 2003; Mitra, 2015; Oppenheim and Willsky, 1997; Papoulis, 1977a, 1980; Vetterli et al., 2014].

3.13 Table of DTFT Properties

We summarize here some of the important properties of the discrete-time Fourier transform (DTFT, or just FT). Most of the proofs can be found in Secs. 3.4.3, 3.5, 3.6, and 3.8. Recall that the DTFT is defined as

$$X(e^{j\omega}) = \sum_{n=-\infty}^{\infty} x[n]e^{-j\omega n}, \tag{3.192}$$

and that the inverse Fourier transform (IFT, or IDTFT) is given by

$$x[n] = \frac{1}{2\pi}\int_0^{2\pi} X(e^{j\omega})e^{j\omega n}d\omega.$$

The notation

$$x[n] \longleftrightarrow X(e^{j\omega})$$

means that $x[n]$ and $X(e^{j\omega})$ constitute a Fourier transform pair, that is, they satisfy the above two equations. $X(e^{j\omega})$ is periodic in ω, and repeats with repetition interval 2π. So, $\int_0^{2\pi} d\omega$ can also be replaced with $\int_{-\pi}^{\pi} d\omega$ in Eq. (3.192) and in the properties below.

1. Fourier transformation is a *linear* operator, that is, if $x_1[n] \longleftrightarrow X_1(e^{j\omega})$ and $x_2[n] \longleftrightarrow X_2(e^{j\omega})$, then $c_1x_1[n] + c_2x_2[n] \longleftrightarrow c_1X_1(e^{j\omega}) + c_2X_2(e^{j\omega})$.
2. $X(1) = X(e^{j0}) = \sum_{n=-\infty}^{\infty} x[n]$ and $x[0] = \int_0^{2\pi} X(e^{j\omega})d\omega/2\pi$.
3. $x[n-n_0] \longleftrightarrow e^{-j\omega n_0}X(e^{j\omega})$; $x[n-n_0]$ is called the *shifted* version of $x[n]$.
4. $e^{j\omega_0 n}x[n] \longleftrightarrow X(e^{j(\omega-\omega_0)})$; $e^{j\omega_0 n}x[n]$ is called the *amplitude modulated* version of $x[n]$.
5. $nx[n] \longleftrightarrow jdX(e^{j\omega})/d\omega$ (Problem 3.22).
6. $y[n] \triangleq \sum_{k=-\infty}^{\infty} x[k]h[n-k] \longleftrightarrow X(e^{j\omega})H(e^{j\omega})$ (convolution in time).
7. $x_1[n]x_2[n] \longleftrightarrow \int_0^{2\pi} X_1(e^{ju})X_2(e^{j(\omega-u)})du/2\pi$ (convolution in frequency).
8. $x[-n] \longleftrightarrow X(e^{-j\omega})$; time reversal is equivalent to frequency reversal.
9. $x^*[-n] \longleftrightarrow X^*(e^{j\omega})$ and $x^*[n] \longleftrightarrow X^*(e^{-j\omega})$. Conjugation in one domain is equivalent to conjugation and reversal in the other domain.
10. $x[n]$ real $\Longleftrightarrow X(e^{j\omega}) = X^*(e^{-j\omega})$. So $|X(e^{j\omega})| = |X(e^{-j\omega})|$ and furthermore $\arg(X(e^{j\omega})) = -\arg(X(e^{-j\omega}))$.
11. $x[n]$ real and even $\Longleftrightarrow X(e^{j\omega})$ real and even ($x[n]$ even means $x[n] = x[-n]$).
12. $r_{xy}[k] \triangleq \sum_{n=-\infty}^{\infty} x[n]y^*[n-k] \longleftrightarrow X(e^{j\omega})Y^*(e^{j\omega})$; $r_{xy}[k]$: cross-correlation between $x[n]$ and $y[n]$.
13. $r_{xx}[k] \triangleq \sum_{n=-\infty}^{\infty} x[n]x^*[n-k] \longleftrightarrow |X(e^{j\omega})|^2$; $r_{xx}[k]$: autocorrelation of $x[n]$.

14. $\sum_{n=-\infty}^{\infty} x[n]y^*[n] = \int_0^{2\pi} X(e^{j\omega})Y^*(e^{j\omega})d\omega/2\pi$; Parseval's relation: inner product preserved.

15. $\sum_{n=-\infty}^{\infty} |x[n]|^2 = \int_0^{2\pi} |X(e^{j\omega})|^2 d\omega/2\pi$; Parseval's relation: energy preserved.

3.14 Table of DTFT Pairs

Signal	Fourier transform (DTFT)
$x[n] = \delta[n]$	$X(e^{j\omega}) = 1$
$x[n] = 1$	$X(e^{j\omega}) = 2\pi \sum_{k=-\infty}^{\infty} \delta_c(\omega + 2\pi k)$
$x[n] = e^{j\omega_0 n}$	$X(e^{j\omega}) = 2\pi \sum_{k=-\infty}^{\infty} \delta_c(\omega + 2\pi k - \omega_0)$
$p[n] = \begin{cases} 1 & -M \leq n \leq M \\ 0 & \text{otherwise} \end{cases}$	$P(e^{j\omega}) = \frac{\sin(\omega(2M+1)/2)}{\sin(\omega/2)}$ (Dirichlet kernel)
$\cos(\omega_0 n)p[n]$	$0.5P(e^{j(\omega-\omega_0)}) + 0.5P(e^{j(\omega+\omega_0)})$
$\cos(\omega_0 n)$	$\pi \sum_{k=-\infty}^{\infty} \Big(\delta_c(\omega + \omega_0 + 2\pi k) + \delta_c(\omega - \omega_0 + 2\pi k)\Big)$
$\sin(\omega_0 n)$	$j\pi \sum_{k=-\infty}^{\infty} \Big(\delta_c(\omega + \omega_0 + 2\pi k) - \delta_c(\omega - \omega_0 + 2\pi k)\Big)$
$\mathcal{U}[n]$ (unit step)	$\widehat{\mathcal{U}}(e^{j\omega}) = \pi\,\delta_c(\omega) + \frac{1}{1 - e^{-j\omega}}$
$p^n\mathcal{U}[n], \quad \lvert p\rvert < 1$	$\frac{1}{1 - pe^{-j\omega}}$
$(n+1)p^n\mathcal{U}[n], \quad \lvert p\rvert < 1$	$\frac{1}{(1 - pe^{-j\omega})^2}$
$h[n] = \frac{\sin(\omega_c n)}{\pi n}$ (sinc function)	$H(e^{j\omega}) = \begin{cases} 1 & -\omega_c < \omega < \omega_c \\ 0 & \omega_c < \lvert\omega\rvert \leq \pi \end{cases}$ (Sec. 4.3.4) (ideal lowpass frequency response)

PROBLEMS

Note: If you are using MATLAB for the computing assignments at the end, the "stem" command is convenient for plotting sequences. The "plot" command is useful for plotting real-valued functions such as $|X(e^{j\omega})|$.

3.1 For each of the following, sketch $x[n]$ and $h[n]$. Also, compute the convolution $y[n] = (x * h)[n]$.
(a) $x[n] = a^n \mathcal{U}[n]$ and $h[n] = p^n \mathcal{U}[n-1]$, where $\mathcal{U}[n]$ is the unit step.
(b) $x[n] = \mathcal{U}[n] - \mathcal{U}[n-10]$ and $h[n] = \delta[n+2]$. Also, sketch a plot of $y[n]$.

3.2 Let $x[n] = \mathcal{U}[n] - \mathcal{U}[n-10]$ and $h[n] = x[n]$. Sketch $x[n]$ and plot the convolution $y[n] = (x * h)[n]$. Since this is the convolution of a rectangular pulse $x[n]$ with itself, you will find that the result is a triangle.

3.3 Show that the convolution of the two sequences (3.164) and (3.165) yields the result $y[n] = \delta[n]$.

3.4 Convolve the signals $x(t)$ and $h(t)$ in Fig. P3.4 and plot the result.

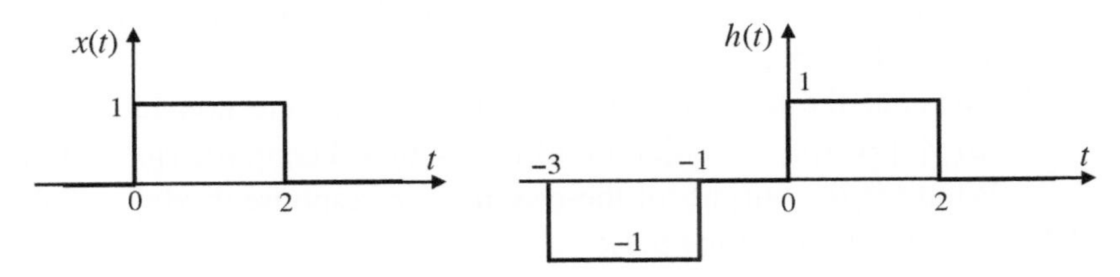

Figure P3.4 Signals for Problem 3.4.

3.5 Let $x_1(t)$ and $x_2(t)$ be rectangular pulses of unequal duration:

$$x_1(t) = \begin{cases} 1 & -1 < t < 1, \\ 0 & \text{otherwise,} \end{cases} \qquad x_2(t) = \begin{cases} 1 & -2 < t < 2, \\ 0 & \text{otherwise.} \end{cases}$$

What is the convolution of $x_1(t)$ with $x_2(t)$? Handsketch the result.

3.6 Given below are impulse responses of some LTI systems. For each system, check whether it is (i) causal and (ii) stable:
(a) $h_1[n] = (1/2)^n \mathcal{U}[n-1]$;
(b) $h_2[n] = (1/2)^n \mathcal{U}[-n]$;
(c) $h_3[n] = n\mathcal{U}[n+1]$;
(d) $h_4[n] = n(1/2)^n \mathcal{U}[n+1]$.
Also, find closed-form expressions for the transfer functions $H_1(z)$ and $H_2(z)$. (*Hint:* Where necessary, use the "ratio test" for convergence of an infinite sum.)

3.7 For the LTI system with $h[n] = \mathcal{U}[n] - \mathcal{U}[n-10]$, find a closed-form expression for the transfer function $H(z)$ and frequency response $H(e^{j\omega})$, and plot the magnitude response $|H(e^{j\omega})|$.

3.8 Find an example of a nonlinear system such that if the input is the single-frequency signal with frequency ω_0 (i.e., $x[n] = e^{j\omega_0 n}$), then the output is a superposition of signals with frequencies $2\omega_0, 5\omega_0$, and $-\omega_0$.

3.9 Figure P3.9(a) shows the unit delay z^{-1} (an LTI system) in cascade with a decimator (which is not LTI). Figure P3.9(b) shows the two systems interchanged in cascade. Show that these two cascades are not equivalent, that is, find an input $x[n]$ such that $y_1[n] \neq y_2[n]$.

Figure P3.9 Cascading a delay and a decimator in two ways.

3.10 Let $H(z) = 1-(1/3)z^{-1}$. Find an *infinite*-duration input $x[n]$ (with $x[n] = 0$ for $n < 0$) such that $y[n]$ has *finite* duration. What is the duration of the output in your construction? Thus, even simple LTI systems like the above $H(z)$ have the ability to annihilate certain infinite-duration inputs everywhere except in a very short region.

3.11 Consider the following systems, all of which are non-LTI: (a) $y[n] = nx[n]$, (b) $y[n] = x[n] - x^2[n]$, and (c) $y[n] = (n+1)x[n]$, (d) $y[n] = 1 + x[n] - x^2[n]$. What are the outputs of these systems in response to $x[n] = \delta[n]$?

3.12 For the discrete-time pulse

$$p[n] = \begin{cases} 1 & -1 \leq n \leq 1, \\ 0 & \text{otherwise,} \end{cases}$$

plot the convolutions $p_1[n] = (p * p)[n]$ and $p_2[n] = (p * p * p)[n]$. Find closed-form expressions for the Fourier transforms $P(e^{j\omega})$, $P_1(e^{j\omega})$, and $P_2(e^{j\omega})$, and handsketch these.

3.13 Suppose $p_1[n]$ in Problem 3.12 is the impulse response of an LTI system. Then what is the transfer function $P_1(z)$? Is this a causal system?

3.14 For convenience, let $p[n]$ denote the symmetric rectangular pulse shown in Fig. 3.13(a) and let $p_1[n] = (p * p)[n]$ (not to be confused with the notation $p_1[n]$ in other problems).

(a) Plot $p_1[n]$ and clearly indicate the locations and heights of the leftmost and rightmost nonzero samples. You must have found $p_1[n]$ to be a triangle.

(b) Find a closed-form expression for $P_1(e^{j\omega})$ and plot it. Clearly indicate the height at $\omega = 0$, and indicate the locations of a few zero crossings.

3.15 Convolve the signals $x[n]$ and $h[n]$ in Fig. P3.15 and plot the result.

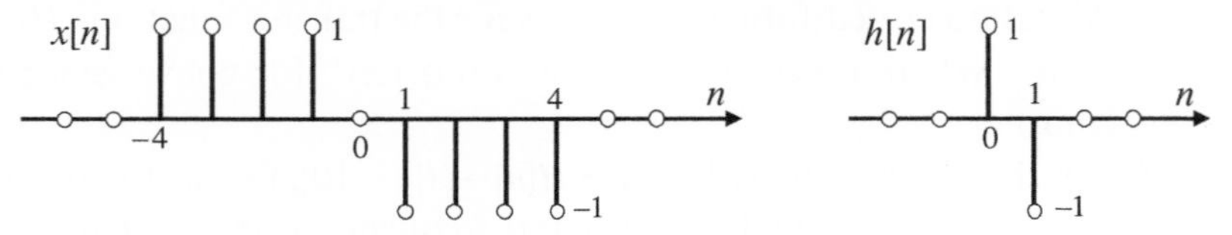

Figure P3.15 Signals for Problem 3.15.

3.16 For $x[n]$ in Problem 3.15, find $X(e^{j\omega})$ in closed form. Plot the magnitude.

3.17 Let $x[n] = \cos(\omega_0 n)p_1[n]$, where $p_1[n]$ is as in Problem 3.14. So $x[n]$ is a cosine, but truncated to a finite number of samples using a triangle instead of a rectangle as in Eq. (3.115).

(a) Handsketch $x[n]$. The triangle gradually tapers the cosine to zero, instead of abruptly truncating it.

(b) With $P(e^{j\omega})$ denoting the Dirichlet kernel (DTFT of $p[n]$ in Problem 3.14), find an expression for $X(e^{j\omega})$.

(c) Show that $X(e^{j\omega}) \geq 0$ and handsketch it.

Be sure to compare the plot of $X(e^{j\omega})$ with Fig. 3.17(b) and observe the differences.

3.18 For the system $H(z)$ in Eq. (3.173) show that if the impulse response is chosen as in Eq. (3.174), then $H(e^{j\omega_1}) = H(e^{j\omega_2})$ for $\omega_2 = -\omega_1$. Thus, the eigenvalues of this LTI system corresponding to the eigenfunctions $e^{j\omega_1 n}$ and $e^{j\omega_2 n}$ are equal.

3.19 Consider the LTI system $H(z) = 1 - z^{-2}$.

(a) Show that $x_1[n] = 1 + 0.5(-1)^n$ is an eigenfunction. Note that this is not an exponential. What is the corresponding eigenvalue?

(b) Find one more non-exponential example $x_2[n]$ of an eigenfunction for this system.

For part (b) make sure your answer is not just a shifted and scaled version $cx_1[n - n_0]$. Also, remember that the zero signal ($x[n] = 0, \forall n$) is not a valid eigenfunction.

3.20 With $b(t)$ denoting the rectangular pulse in Fig. 3.22, show that the repeated convolution $b_2(t) = (b * b * b)(t)$ has the form given in Eq. (3.162).

3.21 Convolve the signals $x(t)$ and $h(t)$ in Fig. P3.21 and plot the result. Note that the "height" indicated for each Dirac delta in $x(t)$ is the area under that Dirac delta, as usual.

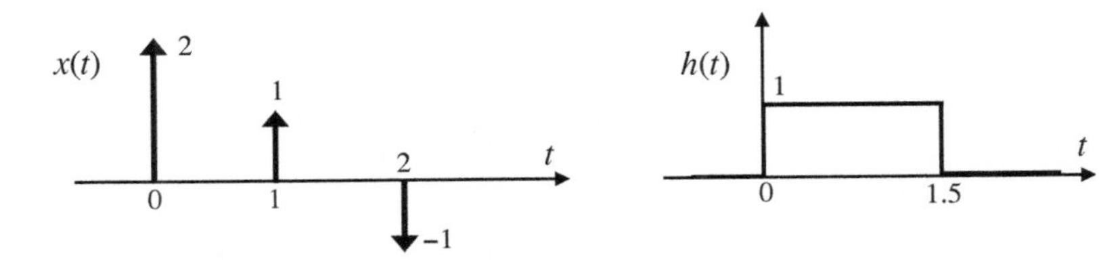

Figure P3.21 Signals for Problem 3.21.

3.22 Starting from the definition of the Fourier transform, prove that the Fourier transform of $nx[n]$ is $jdX(e^{j\omega})/d\omega$. This was tabulated in Sec. 3.13 without proof.

3.23 Find a discrete-time signal $x[n]$ with finite duration such that $X(e^{j\omega})$ is real and positive for all ω. To avoid trivial answers, make sure $x[n]$ has at least three nonzero samples.

3.24 *Odd symmetry*. In Sec. 3.8.1.3 we proved some symmetry properties for real signals. For example, if $x[n]$ is real and even (as demonstrated in Fig. P3.24), then $X(e^{j\omega})$ is also real and even. In this problem we explore symmetry properties further.

(a) If $x[n]$ is real and odd ($x[n] = -x[-n]$ as demonstrated in Fig. P3.24), then show that $X(e^{j\omega})$ is imaginary and odd.

(b) Hence show that if $x[n]$ is imaginary and odd, then $X(e^{j\omega})$ is real and odd.

(c) Show that if $X(e^{j\omega})$ is imaginary and odd, then $x[n]$ is real and odd. This is the converse of part (a).

It follows then that $x[n]$ is real and odd if and only if $X(e^{j\omega})$ is imaginary and odd.

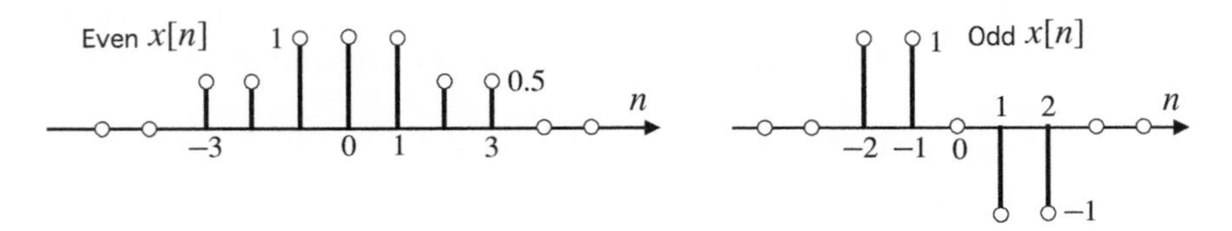

Figure P3.24 Even and odd signals.

3.25 Prove Eq. (3.133). This is an important step in the understanding of Eq. (3.130).

3.26 By substituting the definition of the Fourier transforms into the right-hand side of Eq. (3.85) and using the orthogonality property (3.50), show that this reduces to the left-hand side of Eq. (3.85). This is a direct proof of Parseval's relation based on orthogonality.

3.27 *A strange Fourier transform?* In Sec. 3.8.2 we found that the symmetric pulse (3.88) has Fourier transform $\sin(\omega L/2)/\sin(\omega/2)$, where $L = 2M + 1$ is odd. The curious student might wonder what the signal $x[n]$ would be, whose Fourier transform is

$$f(\omega) = \frac{\sin(\omega L/2)}{\sin(\omega/2)}, \quad L \text{ even.} \tag{P3.27}$$

It turns out that this is *not* a valid Fourier transform of any discrete-time signal $x[n]$! We now explore this. In what follows, assume L is even.

(a) Show that $f(\omega + 2\pi) = -f(\omega)$. This shows that $f(\omega)$ does not have period 2π. So it cannot be the Fourier transform of any discrete-time signal.

(b) Show that $f(\omega + 4\pi) = f(\omega)$. So the period is 4π.

(c) What is the value of $f(0)$, and where are the zero crossings of $f(\omega)$?

(d) Handsketch $f(\omega)$ in the region $-2\pi \le \omega \le 2\pi$.

In Problem 9.16 we will see a *continuous-time* signal $x(t)$ whose Fourier transform is precisely that given in Eq. (P3.27). So there is nothing strange about this!

3.28 We now look into some properties of correlations.

(a) Show that the cross-correlation $r_{xy}[k] = \sum_{n=-\infty}^{\infty} x[n]y^*[n-k]$ satisfies the property

$$r_{xy}^*[-k] = r_{yx}[k]. \quad \text{(P3.28)}$$

In particular, therefore, $r_{xx}^*[-k] = r_{xx}[k]$, so the autocorrelation is Hermitian symmetric. When it is real valued, $r_{xx}[k]$ is an even function.

(b) From the definition of $r_{xx}[k]$ it follows that $r_{xx}[0] = \sum_{n=-\infty}^{\infty} |x[n]|^2$, which is the energy of $x[n]$. Show that $|r_{xx}[k]| \leq r_{xx}[0]$ for all k. (*Hint*: Try the Cauchy–Schwarz inequality.)

3.29 Consider the discrete-time system described by

$$y[n] = \begin{cases} \dfrac{x^K[n]}{x^{K-1}[n-1]} & \text{if} x[n-1] \neq 0, \\ 0 & \text{otherwise,} \end{cases}$$

where $K > 1$ is an integer.

(a) Show that this system is homogeneous but not additive. So it is not linear.

(b) If you apply the impulse input $x[n] = \delta[n]$, what is the output $y[n]$? This is, by definition, the "impulse response" $h[n]$.

(c) If you apply the input $x[n] = e^{j\omega_0 n}$, show that the output has the form $y[n] = c_{\omega_0} e^{j\omega_0 n}$, where c_{ω_0} depends on ω_0 but not on n. So it is tempting to say that c_{ω_0} is the frequency response. Find c_{ω_0} and show that

$$c_{\omega_0} \neq \sum_n h[n] e^{-j\omega_0 n},$$

that is, c_{ω_0} is not the Fourier transform of $h[n]$. So the concepts of impulse response and frequency response are not useful for *non-LTI* systems.

(d) Just for completeness, plot $h[n]$ and $|c_\omega|$.

3.30 Suppose we apply the truncated exponential $a^n \mathcal{U}[-n]$ to a causal stable LTI system $H(z)$, where $\mathcal{U}[n]$ is the unit step. Assume $|a| > 1$ so the input remains bounded as $n \to -\infty$.

(a) What is the output $y[n]$ for $n \leq 0$?

(b) Suppose $y[n]$ is zero for $n > 0$, that is, the output becomes zero as soon as the input becomes zero. This leads us to suspect that the system might be *memoryless* [i.e., $H(z) = \text{const.}$]. This is indeed true. Prove it.

3.31 (Computing assignment) *FT of truncated cosine*. Consider the truncated cosine

$$x[n] = \begin{cases} \cos(\omega_0 n) & \text{for } -N \leq n \leq N, \\ 0 & \text{otherwise.} \end{cases}$$

Let $\omega_0 = 0.15\pi$. Plot $|X(e^{j\omega})|$ in $0 \leq \omega \leq 2\pi$ (or $-\pi \leq \omega \leq \pi$ if you prefer) for $N = 10$, $N = 20$, and $N = 50$. You will see that there are two peaks, and as N increases each peak gets sharper, approaching the Dirac delta behavior.

3.32 (Computing assignment) *Symmetry properties of* $X(e^{j\omega})$. Let $x[n] = \cos(\omega_0 n) + \sin(3\omega_0 n)$ for $0 \le n \le 10$, and zero otherwise. Assume $\omega_0 = 0.2\pi$. Since $x[n]$ is real, we know that

$$X(e^{j\omega}) = X^*(e^{-j\omega}),$$

so that $|X(e^{j\omega})|$ is an even function (Sec. 3.6). With $X(e^{j\omega}) = X_r(e^{j\omega}) + jX_i(e^{j\omega})$ where $X_r(e^{j\omega})$ and $X_i(e^{j\omega})$ are the real and imaginary parts of $X(e^{j\omega})$, this also implies that $X_r(e^{j\omega})$ is an even function and $X_i(e^{j\omega})$ an odd function. Plot $|X(e^{j\omega})|$, $X_r(e^{j\omega})$, and $X_i(e^{j\omega})$ in $0 \le \omega \le 2\pi$ (or $-\pi \le \omega \le \pi$ if you prefer) and convince yourself of these symmetry properties. (For all plots, you can plot 100 to 200 points in $0 \le \omega \le 2\pi$.)

3.33 (Computing assignment) *Spline-like sequences.* Let $x[n]$ be the following pulse signal:

$$x[n] = \begin{cases} 1 & \text{for } 0 \le n \le N-1, \\ 0 & \text{otherwise,} \end{cases}$$

and let

$$y_1[n] = (x * x)[n], \quad y_2[n] = (x * x * x)[n], \quad y_3[n] = (x * x * x * x)[n],$$

where $*$ denotes convolution as usual. For $N = 20$, plot $y_1[n], y_2[n]$, and $y_3[n]$. You will find that $y_k[n]$ gets longer and looks "smoother" as k increases. These $y_k[n]$ are discrete-time analogs of kth-order splines in continuous time (Sec. 3.10.1), which have some specific smoothness properties as k increases (Sec. 9.11.1).

3.34 (Computing assignment) *Autocorrelations.* Let $x[n]$ be the pulse signal

$$x[n] = \begin{cases} N-n & \text{for } 0 \le n \le N-1, \\ 0 & \text{otherwise,} \end{cases}$$

with $N = 20$. Let $r_{xx}[k]$ be the autocorrelation of $x[n]$. Plot $x[n]$ and $r_{xx}[k]$, clearly showing the regions where they are nonzero. Notice that $r_{xx}[0] \ge r_{xx}[k]$ for all k, and $r_{xx}[k] = r_{xx}[-k]$. Compute the energy of $x[n]$ from its definition and verify that it is equal to $r_{xx}[0]$.

4 Digital Filters and Filter Structures

4.1 Introduction

Digital filters are linear time-invariant (LTI) systems with selective frequency responses, as seen in the previous chapter. In this chapter we elaborate further on these. Digital filters can be represented in terms of time-domain difference equations, or by transfer functions and frequency responses in the transform domain. They can also be represented by computational graphs or structures. This chapter begins by introducing structures and structural interconnections for LTI systems. Then several examples of digital filters are presented. We also show how to convert lowpass filters into other types, such as highpass, bandpass, and so on, by use of simple transformations.

The meaning of phase distortion is explained and a family of filters called linear-phase digital filters are introduced, which do not create such distortions. Application of digital filters in noise removal (denoising) is also discussed, with examples for one-dimensional (1D) signals and two-dimensional (2D) images. An optional section at the end (marked with an asterisk) shows how convolution can be represented using Toeplitz matrices, and discusses some insights which result from this.

Things to review. In this chapter we will see many examples of frequency responses and transfer functions for discrete-time LTI systems. The frequency response $H(e^{j\omega})$ of an LTI system is the Fourier transform of the impulse response $h[n]$, whereas the transfer function $H(z)$ is the z-transform of the impulse response $h[n]$. These were defined in Secs. 3.4.1 and 3.4.2, and the student may want to review these at this time. While the z-transform is studied in greater detail in Chapter 6, those details are not required for this chapter.

4.2 Structures and Building Blocks

In earlier chapters we introduced a number of building blocks which are useful in the flowgraph representation of systems, as shown in Figs. 2.18 and 2.20. For the implementation of LTI systems, the three building blocks shown in Fig. 4.1 are sufficient. These are the constant multiplier a, the two-input adder, and the unit-delay element. On the left are shown the basic forms used in Chapter 2, and on the

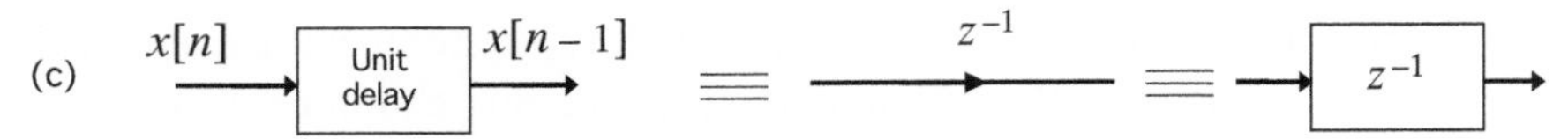

Figure 4.1 Notations for building blocks of discrete-time LTI systems. (a) Constant multiplier a, (b) adder, and (c) unit-delay element.

right we show equivalent notations which are frequently used. To understand why the notation z^{-1} is used for the unit delay, observe that the delayed version of $x[n]$ is

$$y[n] = x[n-1]. \tag{4.1}$$

Compared with the convolution sum $y[n] = \sum_k h[k]x[n-k]$, we see that Eq.(4.1) represents an LTI system with impulse response $h[n] = \delta[n-1]$. So its transfer function $H(z)$, which is the z-transform of $h[n]$, is given by

$$H(z) = \sum_{n=-\infty}^{\infty} h[n]z^{-n} = \sum_{n=-\infty}^{\infty} \delta[n-1]z^{-n} = z^{-1}. \tag{4.2}$$

So the unit-delay element indeed has transfer function $H(z) = z^{-1}$. Similarly, the system $y[n] = x[n-K]$ has transfer function $H(z) = z^{-K}$.

Figure 4.2 makes an important clarification. The notation in part (a) shows that $x[n]$ branches off into two copies. The same meaning is conveyed in part (b); the little circle at the branching point is just for aesthetics. In part (c) the two crossing lines do not touch. So $x[n]$ and $s[n]$ are conveyed to the ends of the respective lines without change. If crossing lines ever touch, there will be some special notation to indicate this, as in parts (d) and (e), where there is an adder at the crossing.

Next, Fig. 4.3 shows the many ways in which an LTI system is depicted in flowgraphs. In part (a) we write the impulse response in the box. This means that the output is $y[n] = (h * x)[n]$ (convolution). In part (b) we write $H(z)$ instead of $h[n]$ in the box, which is sometimes convenient. The meaning is still the same, namely, $y[n] = (h * x)[n]$. Finally, in part (c) the input and output are written in z-transform notation instead of in the time domain. Once again the output is $y[n] = (h * x)[n]$.

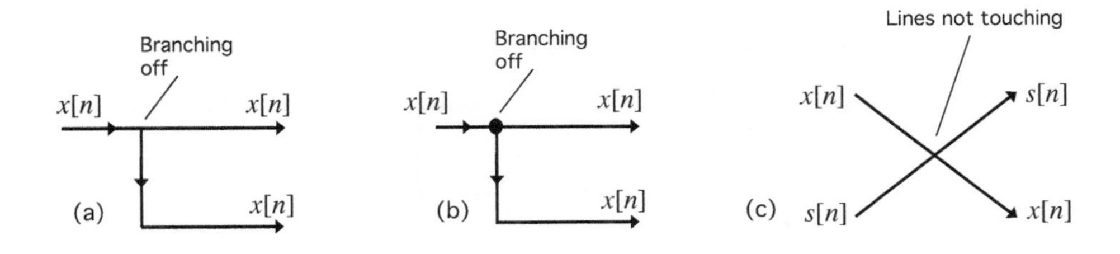

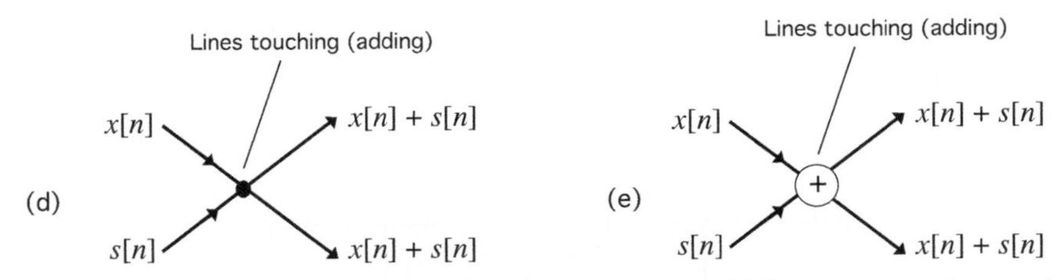

Figure 4.2 (a), (b) A signal $x[n]$ branching into two copies. (c) Two signal paths crossing but not touching. (d), (e) Notation when paths touch (i.e., signals add).

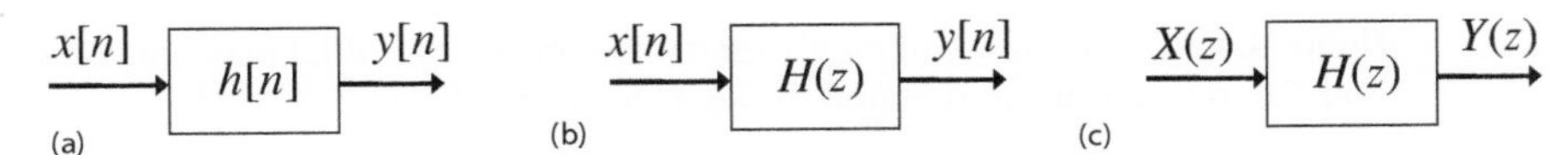

Figure 4.3 (a)–(c) Equivalent ways to represent an LTI system in signal flowgraphs. In all these figures, $y[n] = (h * x)[n]$, that is, $Y(z) = H(z)X(z)$.

4.2.1 Interconnections of LTI Systems

Figure 4.4 shows three commonly used interconnections of LTI systems. In each case the system from $x[n]$ to $y[n]$ is LTI and can be characterized by an overall transfer function $F(z) = Y(z)/X(z)$ and the corresponding impulse response $f[n]$. For the **parallel** connection it is clear that $f[n] = g[n] + h[n]$, so that the overall transfer function is the sum

$$F(z) = G(z) + H(z). \tag{4.3}$$

For the **cascade** connection, if $x[n] = \delta[n]$ then the output $s[n]$ of $H(z)$ is $h[n]$. So the output of $G(z)$ is

$$y[n] = (g * s)[n] = (g * h)[n], \tag{4.4}$$

so that $f[n] = (g * h)[n]$, and therefore the overall transfer function is the product

$$F(z) = G(z)H(z). \tag{4.5}$$

Finally, consider the connection in Fig. 4.4(c). Here the output of $H(z)$ is filtered again by $G(z)$, and the result is fed back, to be added to the input $x[n]$. So we have a feedback loop, and this is called a **feedback** connection. In this case, it is not easy to figure out the impulse response from $x[n]$ to $y[n]$ directly in the time domain. This is

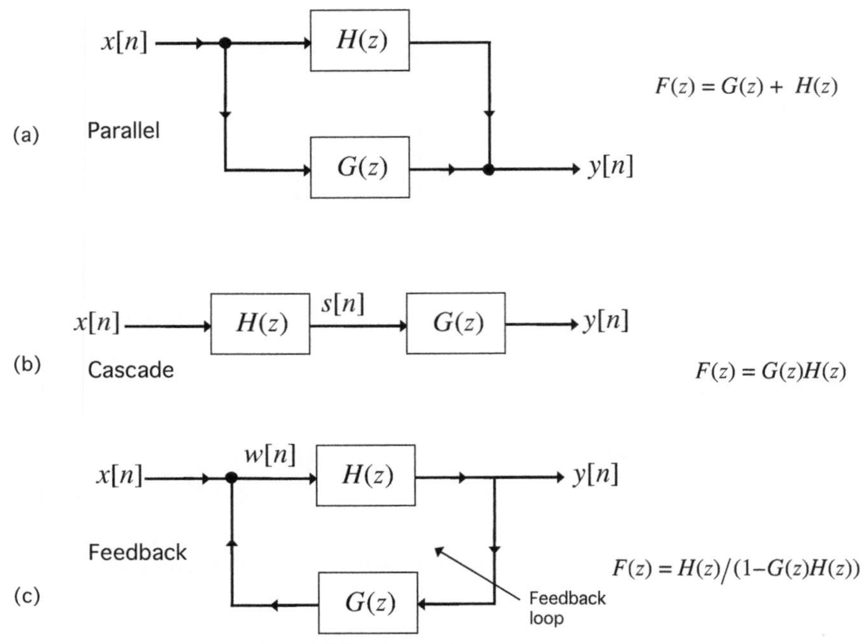

Figure 4.4 (a)–(c) Three commonly used interconnections of LTI systems. In each case the overall transfer function $F(z)$ is also indicated.

because the output of $H(z)$ gets convolved with $g[n]$ and this is again filtered by $H(z)$ and propagated back through $G(z)$, an infinite number of times! This is where the technique of z-transforms comes to our help. Thus, consider the signal $w[n]$ (output of the adder) and denote its z-transform as $W(z)$. Then we have the following two equations in the z-domain to describe the system:

$$W(z) = X(z) + G(z)Y(z), \quad Y(z) = H(z)W(z). \tag{4.6}$$

Substituting from the first into the second equation, we eliminate $W(z)$ and get

$$Y(z) = H(z)\Big(X(z) + G(z)Y(z)\Big), \tag{4.7}$$

which implies $\Big(1 - H(z)G(z)\Big)Y(z) = H(z)X(z)$. So the transfer function $F(z) = Y(z)/X(z)$ is given by

$$F(z) = \frac{H(z)}{1 - G(z)H(z)}. \tag{4.8}$$

So we have a closed-form expression for the transfer function, from which the impulse response $f[n]$ can be found by inverse z-transformation (Chapter 6). Figure 4.5 shows an example of a system with a feedback loop. Here $H(z) = 1$ and $G(z) = pz^{-1}$, so that

$$F(z) = \frac{1}{1 - pz^{-1}}. \tag{4.9}$$

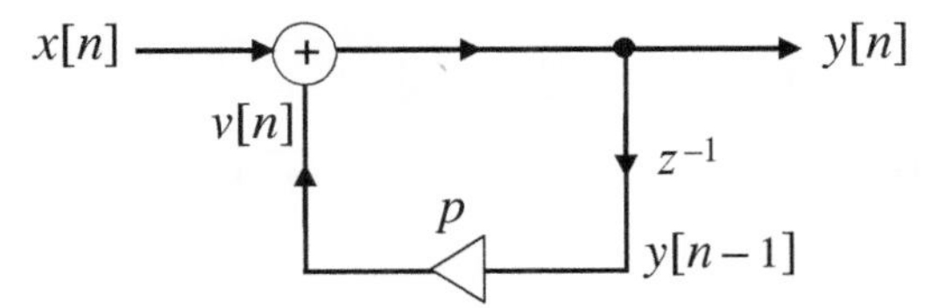

Figure 4.5 An example of a system with a feedback interconnection. The transfer function is $F(z) = 1/(1 - pz^{-1})$. Such a drawing is called a structure, computational graph, or signal flowgraph for the system.

It will be seen in Sec. 5.2 that an LTI system with impulse response

$$f[n] = p^n \mathcal{U}[n] \tag{4.10}$$

has the above transfer function, where $\mathcal{U}[n]$ is the unit-step function. As in Chapter 2, such a figure is called a **structure** for the transfer function (4.9). We also refer to it as a **computational graph**, or a signal flowgraph.

The three types of interconnections discussed above are also used for continuous-time LTI systems, and formulas for the overall transfer function $F(s)$ are similar to what we presented above. However, the internal details of the building blocks in the continuous-time case depend on the type of implementation. For example, the implementations can be based on passive electrical circuit elements (resistors, capacitors, and inductors) or active elements (e.g., operational amplifiers), or even mechanical components such as masses, springs, and frictional surfaces. We shall not go into these details here. The interested reader can refer to Oppenheim and Willsky [1997] and references therein.

LTI systems commute in cascade. Since the cascade connection in Fig. 4.4(b) has the transfer function $G(z)H(z)$, it follows that if we interchange the two systems in cascade, the transfer function does not change. Thus two LTI systems in cascade can be interchanged.[1] Notice, however, that unless both systems are LTI we **cannot** in general **interchange** them in cascade. For example, consider the two systems in Fig. 4.6(a). This is a cascade of the unit-delay element with the time-varying multiplier n. Thus, the two systems in cascade are $x_1[n] = x[n-1]$ (which is LTI) and $y_1[n] = nx_1[n]$ (which is time varying although linear). The final output is clearly

$$y_1[n] = nx[n-1]. \tag{4.11}$$

Now, if we interchange the two systems in cascade, we get Fig. 4.6(b). The first system in this cascade has output $x_2[n] = nx[n]$, so that

$$y_2[n] = x_2[n-1] = (n-1)x[n-1]. \tag{4.12}$$

[1] It turns out that this is not true for MIMO systems (Sec. 14.4). For such systems transfer functions are matrices, and the matrix product $\mathbf{G}(z)\mathbf{H}(z)$ is in general not equal to $\mathbf{H}(z)\mathbf{G}(z)$.

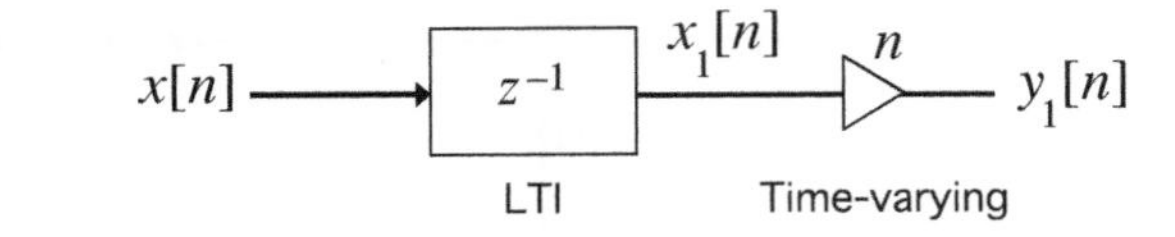

Figure 4.6 (a) Two systems in cascade and (b) the two systems interchanged in cascade.

Thus $y_2[n] \neq y_1[n]$, and the systems in Figs. 4.6(a) and (b) are not equivalent. For example, if $x[n] = \delta[n]$, then $y_1[n] = n\delta[n-1] = \delta[n-1]$, whereas $y_2[n] = (n-1)\delta[n-1] = 0$ for all n, so that $y_2[n] \neq y_1[n]$. For another example of this nature, see Problem 3.9. $\triangledown \triangledown \triangledown$

4.2.2 Delay-Free Loops

Notice in Fig. 4.5 that the feedback loop has a delay z^{-1}. In discrete-time systems, any feedback loop should contain at least one unit of delay like this, for otherwise the system is not realizable. For example, consider Fig. 4.7. Here we have a delay-free loop as indicated. The equations describing this system are

$$\begin{aligned} y[n] &= x[n] + v[n], \\ v[n] &= cy[n] + dy[n-1]. \end{aligned}$$

Thus, to compute $y[\cdot]$ at time n one needs to know $v[\cdot]$ at time n. But to compute $v[\cdot]$ at time n one needs to know $y[\cdot]$ at time n (in addition to $y[\cdot]$ at time $n-1$). So we are not able to proceed with this computation! Compare this with Fig. 4.5, where

$$\begin{aligned} y[n] &= x[n] + v[n], \\ v[n] &= py[n-1]. \end{aligned}$$

Now, to compute $y[\cdot]$ at time n one needs to know $v[\cdot]$ at time n. And to compute $v[\cdot]$ at time n we only need to know $y[\cdot]$ at time $n-1$, which we have available. So this is a perfectly realizable system. Systems with delay-free loops like Fig. 4.7 are unrealizable. Returning to the general feedback system in Fig. 4.4(c), we see that there is no delay-free loop if $H(z)$ is causal:

$$H(z) = h[0] + h[1]z^{-1} + h[2]z^{-2} + h[3]z^{-3} + \cdots \tag{4.13}$$

and $G(z)$ has the form

$$G(z) = g[1]z^{-1} + g[2]z^{-2} + g[3]z^{-3} + \cdots . \tag{4.14}$$

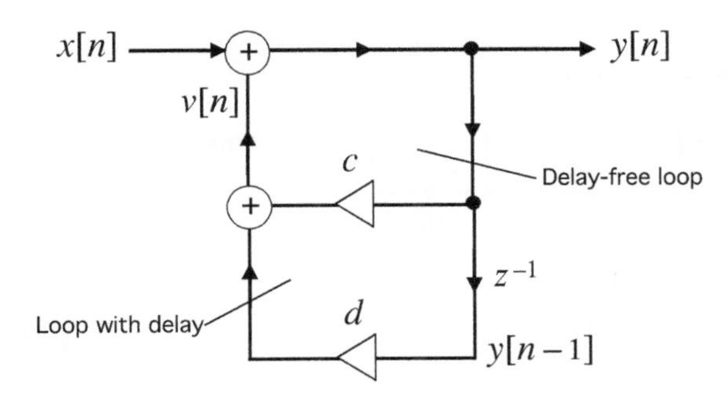

Figure 4.7 Example of a discrete-time structure with a delay-free loop. Such structures are unrealizable.

That is, the zeroth coefficient $g[0] = 0$ and $G(z)$ has the form $G(z) = z^{-1}G_1(z)$ where $G_1(z)$ is still a causal system. In this case there is no delay-free loop, and the structure is realizable.

For continuous-time systems the story is slightly different. Even if a delay element is not explicitly included in the loop, there are delays inherently present in the implementation of physical components (circuit elements, mechanical components, etc.). So, delay-free loops are not an issue.

4.3 Examples of Digital Filters

We now consider some specific examples of useful digital filters and examine their frequency responses.

4.3.1 The Moving Average Filter

First consider the system described by the input–output relation

$$y[n] = \frac{x[n] + x[n-1]}{2}. \tag{4.15}$$

For every time index n, this computes the average of the present and the past input samples. So this is called a **sliding** average or a **moving average** system. Figure 4.8(a) shows the structure for this system, using the notations introduced in Sec. 4.2. By comparing this with the convolution sum $y[n] = \sum_k h[k]x[n-k]$, we can draw a couple of conclusions. First, Eq. (4.15) is a convolution, and therefore represents an LTI system. Second, the impulse response satisfies

$$h[0] = h[1] = 0.5, \tag{4.16}$$

with $h[n] = 0$ for all other n, as shown in Fig. 4.8(b). That is,

$$h[n] = \frac{1}{2}\delta[n] + \frac{1}{2}\delta[n-1]. \tag{4.17}$$

The transfer function is simply the z-transform

$$H(z) = \sum_{n=-\infty}^{\infty} h[n]z^{-n} = \frac{1+z^{-1}}{2}, \tag{4.18}$$

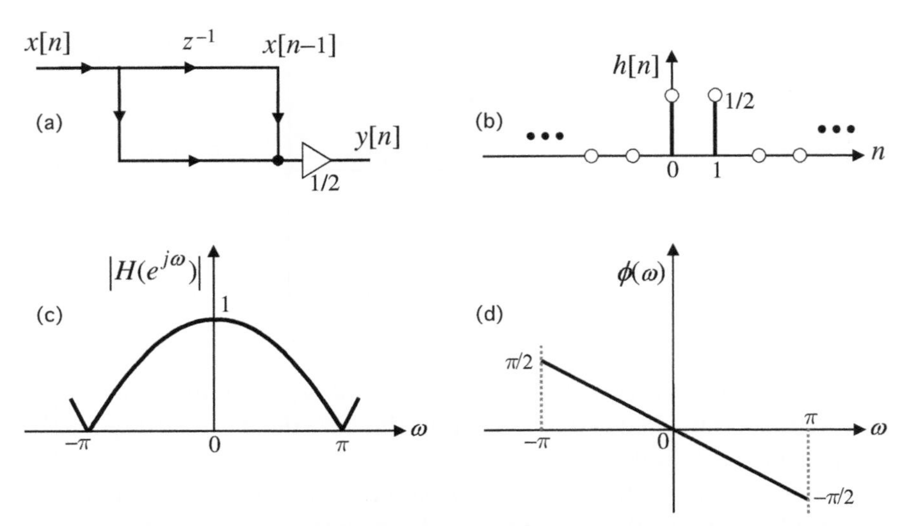

Figure 4.8 The moving average filter. (a) Structure or computational graph, (b) impulse response, (c) magnitude response, and (d) phase response. This is a simple lowpass filter.

so that the frequency response is

$$H(e^{j\omega}) = \frac{1 + e^{-j\omega}}{2}. \tag{4.19}$$

Another way to arrive at this frequency response is to apply an input $x[n] = e^{j\omega n}$ directly into Eq. (4.15) and find the output:

$$y[n] = \frac{e^{j\omega n} + e^{j\omega(n-1)}}{2} = \left(\frac{1 + e^{-j\omega}}{2}\right) e^{j\omega n}, \tag{4.20}$$

which proves, by definition, that the frequency response is (4.19). To get a physical feeling for this frequency response, let us rewrite it as

$$H(e^{j\omega}) = \frac{1 + e^{-j\omega}}{2} = e^{-j\omega/2}\left(\frac{e^{j\omega/2} + e^{-j\omega/2}}{2}\right) = e^{-j\omega/2}\cos(\omega/2). \tag{4.21}$$

This shows that the magnitude and phase response are

$$|H(e^{j\omega})| = |\cos(\omega/2)| \quad \text{and} \quad \phi(\omega) = -\omega/2 \tag{4.22}$$

in $-\pi \leq \omega < \pi$. These are plotted in Figs. 4.8(c) and (d). From this we see that this is a **lowpass** filter, with a transmission zero at $\omega = \pm\pi$. That is, $H(z) = 0$ at $z = e^{j\pi} = -1$. So the transmission zero at $\omega = \pi$ corresponds to a zero in the z-plane at $z = -1$, which is also clear from an inspection of Eq.(4.18).

Notice that the phase response is a straight line passing through the origin, so this is called a **linear-phase** filter. Summarizing, the moving average filter (4.15) has impulse response, transfer function, and frequency response given by

$$h[n] = \frac{\delta[n] + \delta[n-1]}{2}, \quad H(z) = \frac{1+z^{-1}}{2}, \quad H(e^{j\omega}) = e^{-j\omega/2}\cos(\omega/2), \quad (4.23)$$

and represents a very simple lowpass filter. More sophisticated examples of lowpass filters will be presented later (Sec. 4.3.4 and Chapter 7). In our example, since the response changes very slowly from 1 to 0, it is difficult to specify the passband and stopband regions exactly. This demarcation depends on the error tolerance we are willing to accept in our definitions of passband and stopband. One final remark is that a filter which performs an operation like

$$y[n] = \frac{x[n] + x[n+1]}{2} \quad (4.24)$$

(instead of (4.15)) is also a lowpass moving average filter (Problem 4.4) but it is noncausal because $y[n]$ depends on the future input $x[n+1]$.

4.3.2 The First Difference Operator or Filter

Now consider a simple variation of Eq. (4.15), namely,

$$y[n] = \frac{x[n] - x[n-1]}{2}. \quad (4.25)$$

The only modification is that we use the difference between adjacent samples in the numerator. So, this is called a first-difference operator. The name seems to hint that it is possible to define higher-order differences as well. This is indeed true, as we will see in Problem 4.5. In (2.49) we briefly demonstrated this system (but without the 2 in the denominator which makes it more convenient).

Figure 4.9(a) shows the structure for this system, using the notations introduced in Sec. 4.2. If we analyze Eq. (4.25) like we did Eq. (4.15), we find that

$$h[n] = \frac{\delta[n] - \delta[n-1]}{2}, \quad H(z) = \frac{1-z^{-1}}{2}, \quad H(e^{j\omega}) = je^{-j\omega/2}\sin(\omega/2), \quad (4.26)$$

where the last expression follows because

$$H(e^{j\omega}) = \frac{1-e^{-j\omega}}{2} = e^{-j\omega/2}\left(\frac{e^{j\omega/2} - e^{-j\omega/2}}{2}\right) = je^{-j\omega/2}\sin(\omega/2). \quad (4.27)$$

Note that, since $e^{j\pi/2} = j$, we have

$$je^{-j\omega/2}\sin(\omega/2) = e^{j(\pi/2-\omega/2)}\sin(\omega/2). \quad (4.28)$$

Since $\sin(\omega/2) < 0$ for $-\pi < \omega < 0$, this contributes to an additional phase of π, which can also be regarded as $-\pi$ (because $e^{\pm j\pi} = -1$). So the magnitude and phase responses should be written as

$$|H(e^{j\omega})| = |\sin(\omega/2)| \quad \text{and} \quad \phi(\omega) = \begin{cases} \pi/2 - \omega/2 & 0 < \omega < \pi, \\ -\pi/2 - \omega/2 & -\pi < \omega < 0. \end{cases} \quad (4.29)$$

Figure 4.9 shows these plots. It is clear that the first-difference operator is a **highpass** filter. Note that $H(e^{j\omega}) = 0$ at $\omega = 0$, so the filter has a transmission zero at zero frequency. That is, $H(z)$ has a zero at $z = 1$ in the z-plane.

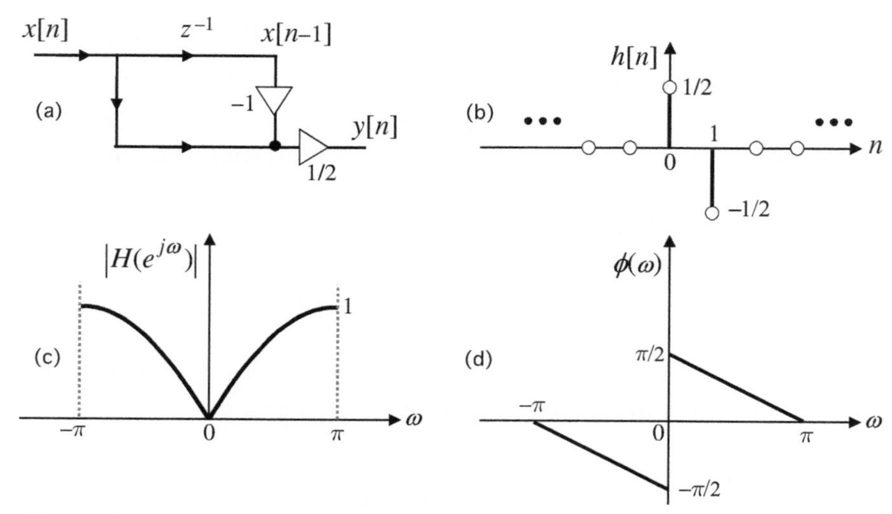

Figure 4.9 The first-difference filter. (a) Structure, (b) impulse response, (c) magnitude response, and (d) phase response. This is a simple highpass filter. The jump in the phase response at $\omega = 0$ is due to the change of sign of $\sin(\omega/2)$ in Eq. (4.27).

Comment on periodicity. The fact that $H(e^{j\omega})$ is periodic with repetition interval 2π implies in particular that the magnitude $|H(e^{j\omega})|$ and the phase $\phi(\omega)$ individually have this periodicity property. But in some of the examples, the plot of $\phi(\omega)$ does not give the impression of being periodic. However, if we plot $\phi(\omega)$ modulo 2π, that is, maintain it in the range $-\pi \leq \phi(\omega) < \pi$ or $0 \leq \phi(\omega) < 2\pi$ by adding an appropriate multiple of 2π, then the plot will look periodic in ω with repetition interval 2π. We should also remember to add $\pm\pi$ to $\phi(\omega)$ whenever there is a zero crossing of $H(e^{j\omega})$, as in Fig. 4.9(c) at $\omega = 0$.
▽ ▽ ▽

Problem 4.6 considers a generalization of the above filters of the form $y[n] = (x[n] + e^{j\omega_0}x[n-1])/2$, which can act as a simple tunable bandpass filter. Equations of the form (4.15) and (4.25), which express the output as a finite linear combination of input samples at different times, are also called **difference equations**, abbreviated as "**d.e.**" Thus, the more general moving average filter (4.30) to be discussed next is also a d.e. In Chapter 5 we will see a generalization of this called the **recursive** d.e.

4.3.3 A More General Moving Average Filter

Consider the system with input–output relation

$$y[n] = \frac{x[n] + x[n-1] + \cdots + x[n-L+1]}{L} = \frac{1}{L}\sum_{k=0}^{L-1} x[n-k]. \tag{4.30}$$

The output is the average of L consecutive input samples, namely the present sample and the past $L-1$ samples. This is clearly a generalization of Eq. (4.15). Figure 4.10(a) shows the structure for this system, using the notations introduced in

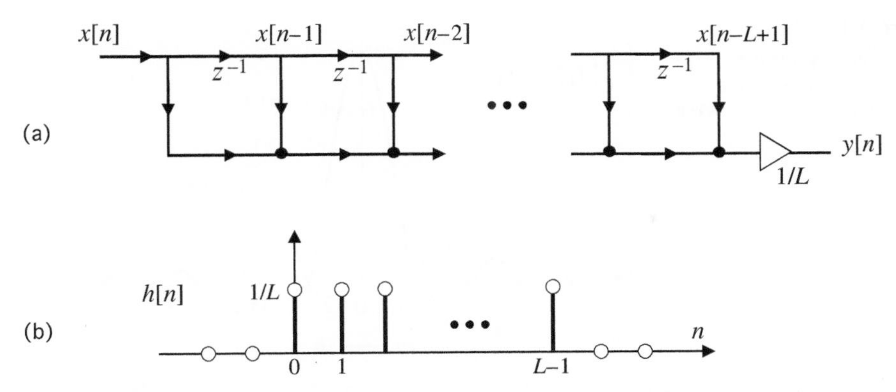

Figure 4.10 The moving average filter (4.30). (a) Computational graph and (b) impulse response.

Sec. 4.2. By comparing with the convolution sum $y[n] = \sum_k h[k]x[n-k]$, it follows that Eq. (4.30) is an LTI system with impulse response

$$h[n] = \begin{cases} \dfrac{1}{L} & 0 \le n \le L-1, \\ 0 & \text{otherwise.} \end{cases} \tag{4.31}$$

This is shown in Fig. 4.10(b). The transfer function is therefore

$$H(z) = \frac{1}{L}\sum_{n=0}^{L-1} z^{-n}. \tag{4.32}$$

We say that this is an $(L-1)$th-order moving average filter because the highest power of z^{-1} in Eq. (4.32) is $L-1$. By using the identity (3.89), we see that Eq. (4.32) can be rewritten as

$$H(z) = \frac{1}{L}\left(\frac{1-z^{-L}}{1-z^{-1}}\right), \tag{4.33}$$

so that the frequency response is

$$H(e^{j\omega}) = \frac{1}{L}\left(\frac{1-e^{-j\omega L}}{1-e^{-j\omega}}\right). \tag{4.34}$$

To understand the properties of this frequency response, we write it in the form

$$H(e^{j\omega}) = \frac{1}{L}\left(\frac{1-e^{-j\omega L}}{1-e^{-j\omega}}\right) = \frac{e^{-j\omega L/2}}{Le^{-j\omega/2}}\left(\frac{e^{j\omega L/2}-e^{-j\omega L/2}}{e^{j\omega/2}-e^{-j\omega/2}}\right), \tag{4.35}$$

which simplifies to

$$H(e^{j\omega}) = \frac{e^{-j\omega(L-1)/2}}{L}\left(\frac{\sin(\omega L/2)}{\sin(\omega/2)}\right). \tag{4.36}$$

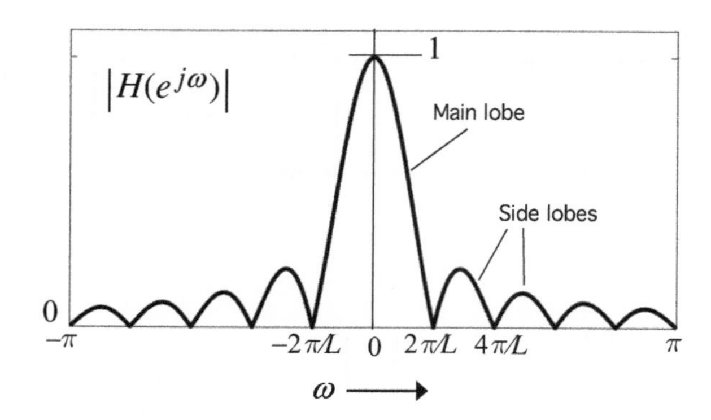

Figure 4.11 The magnitude response of the moving average filter (4.30).

The quantity in brackets is the Dirichlet kernel (3.97), whose properties were discussed earlier in Sec. 3.8.2. A plot of this was also shown in Fig. 3.13(b). Notice from Eq. (4.36) that $H(e^{j0}) = 1$, which is consistent with the fact that

$$H(e^{j0}) = \sum_{n=0}^{L-1} h[n] = \sum_{n=0}^{L-1} (1/L) = 1. \tag{4.37}$$

Since the Dirichlet kernel has zeros (3.100), the filter (4.36) has transmission zeros at

$$\omega = \frac{2\pi k}{L}, \quad k \neq 0 \quad \text{(transmission zeros of (4.36))}, \tag{4.38}$$

as indicated in the magnitude response plot in Fig. 4.11. The response has maximum value of unity at $\omega = 0$, and represents a lowpass filter. Thus, the *L-sample moving average system is a lowpass filter*. The magnitude response has a **main lobe** and several **side lobes**, as indicated. As the filter length L increases (i.e., the number of samples averaged in the filter (4.30) increases), the main lobe gets narrower. Thus, we can design this filter to pass an arbitrarily **narrow band** of frequencies around $\omega = 0$, simply by increasing L.

Since the transfer function (4.32) can be rewritten as in Eq. (4.33), there is a more efficient way to implement the filter. This is shown in Fig. 5.11 later, and will be explained in Chapter 5.

4.3.3.1 FIR and IIR Filters

The impulse responses in Eqs. (4.23), (4.26), and (4.31) all have finite duration. So, these systems are said to be **FIR** (finite impulse response) filters. By contrast, digital filters with infinitely long impulse response are called **IIR** (infinite impulse response) filters. An example of an IIR filter is Eq. (4.9), which has the impulse response (4.10). In general, LTI systems with feedback (as in Fig. 4.5) are IIR. If we examine the structures for the FIR filters in Eqs. (4.23), (4.26), and (4.31), we see that they do not have feedback loops (e.g., see Fig. 4.8(a)). IIR filters will be discussed in detail in Chapter 7. For now, we just mention that the frequency response of Eq. (4.9) is

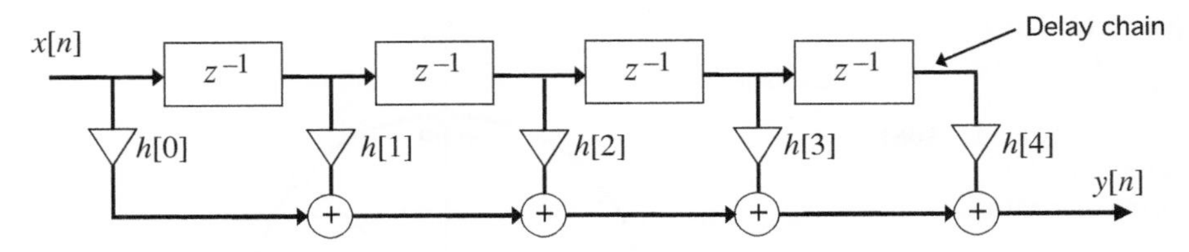

Figure 4.12 The direct-form structure for an FIR filter of order N. Here, $N = 4$.

as shown later in Fig. 5.3. Depending on the pole, this can be lowpass or highpass. For causal FIR filters,

$$H(z) = \sum_{n=0}^{N} h[n]z^{-n} \tag{4.39}$$

is a polynomial of order N in z^{-1}. Figure 4.12 shows a structure for implementing the FIR filter (4.39) when $N = 4$. This is often called the direct-form structure for $H(z)$, and is just a generalization of Fig. 4.10(a). The cascade of N delays in the structure is referred to as a delay chain. We say that N is the **order** of the filter, and the **length** of the filter is $L = N + 1$ (length of the impulse response).

For IIR filters, the length $L = \infty$ but the order can be finite. For example, the IIR filter in Eq. (4.9) is said to have order $N = 1$ because the highest power of z^{-1} that is required in the expression $H(z)$ is unity. Structures for IIR filters will be presented in Chapter 5. The order of a filter (FIR or IIR) is sometimes also called its degree.[2]

4.3.3.2 Transmission Zeros and Unit-Circle Zeros of $H(z)$

The transmission zeros (4.38) imply that $H(z)$ has zeros at the unit-circle points $z_k = e^{j2\pi k/L}$ in the z-plane for integers $k \neq 0$. These $L - 1$ zeros are equispaced on the unit circle as shown in Fig. 4.13. This can also be derived from the expression (4.33) for $H(z)$. The numerator has zeros at $z^{-L} = 1$, or equivalently $z^L = 1$. There are L solutions to this, namely the Lth roots of unity given by

$$z_k = e^{j2\pi k/L}, \quad 0 \leq k \leq L - 1. \tag{4.40}$$

Of these, $z_0 = 1$ is also a zero of the denominator of Eq. (4.33), and cancels off. The remaining $L - 1$ zeros are the zeros shown in Fig. 4.13.

4.3.4 The Ideal Lowpass Digital Filter

In the previous examples we defined the digital filter by its time-domain input–output description, for example as in Eq. (4.30), and then derived the impulse

[2] For systems with multiple inputs and outputs, there is a subtle difference between order and degree (Sec. 14.5.1), but we will not be concerned with this here.

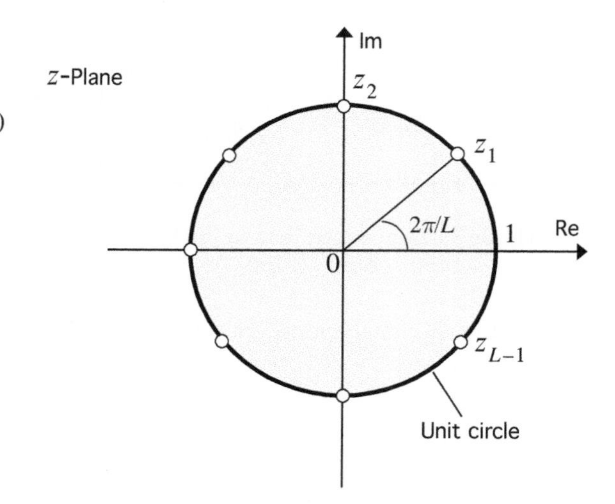

Figure 4.13 The $L-1$ zeros of $H(z)$ in Eq. (4.33). These are on the unit circle at angles (4.38) representing the transmission zeros of the filter.

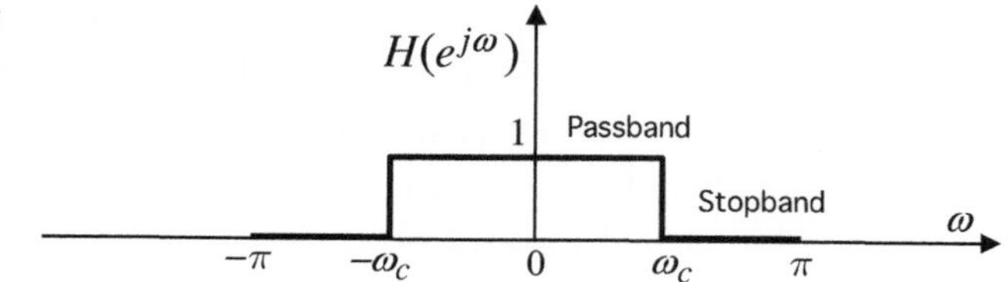

Figure 4.14 Frequency response of the ideal lowpass digital filter.

response, transfer function, and frequency response. An example is now presented with a different approach. Namely, we specify the digital filter frequency response $H(e^{j\omega})$ to be as shown in Fig. 4.14, and wish to derive the impulse response. This filter is called the ideal lowpass filter, and it satisfies

$$H(e^{j\omega}) = \begin{cases} 1 & -\omega_c < \omega < \omega_c, \\ 0 & \omega_c < |\omega| \leq \pi, \end{cases} \tag{4.41}$$

and repeating periodically with period 2π, like any digital filter. There are discontinuities at the band edges $\omega = \pm\omega_c$. Referring to the terminology in Fig. 3.6(a), we can describe this response as follows: it is exactly unity everywhere in the passband, exactly zero everywhere in the stopband, and the transition bandwidth is zero! This is called an ideal filter because it cannot be implemented in practice, as we shall see. However, it serves as a benchmark for practical designs, and is therefore useful to understand. Compared with the general form $H(e^{j\omega}) = |H(e^{j\omega})|\, e^{j\phi(\omega)}$, we see that the ideal lowpass filter has phase response $\phi(\omega) = 0$ for all ω. So it is also a **zero-phase** filter. The filter in Fig. 4.14 is also called a **brick-wall filter**, to distinguish it from more relaxed ideal filters as in Problem 4.25.

The filter impulse response $h[n]$ can be obtained by computing the inverse Fourier transform of $H(e^{j\omega})$. Thus, using Eq. (3.47), we have

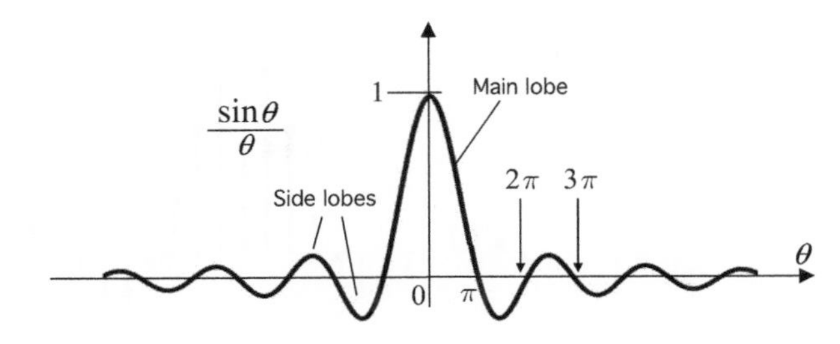

Figure 4.15 The sinc function $\sin\theta/\theta$, which arises frequently in our discussions.

$$h[n] = \frac{1}{2\pi}\int_{-\pi}^{\pi} H(e^{j\omega})e^{j\omega n}d\omega$$

$$= \frac{1}{2\pi}\int_{-\omega_c}^{\omega_c} e^{j\omega n}d\omega = \frac{e^{j\omega_c n} - e^{-j\omega_c n}}{j2\pi n} = \frac{2j\sin(\omega_c n)}{j2\pi n}, \quad (4.42)$$

which simplifies to

$$h[n] = \frac{\sin(\omega_c n)}{\pi n}. \quad (4.43)$$

To understand the behavior of this impulse response, we now introduce an important function, which arises frequently in the theory of signals and systems.

The sinc function. The function defined as

$$f(\theta) = \frac{\sin\theta}{\theta} \quad (4.44)$$

is very important in many of our discussions. It is called the sinc function and is plotted in Fig. 4.15. To understand this plot, recall first that $\sin\theta$ is a periodic function of θ with period 2π. So the decay of the sinc function for large θ comes entirely from the decay of $1/\theta$. To understand the behavior for small θ, note that $\sin\theta \approx \theta$ when $\theta \approx 0$, so that $f(\theta) \to 1$ as $\theta \to 0$. Thus, even though Eq. (4.44) looks like 0/0 for $\theta = 0$, the limit is well defined, and is equal to 1. Since $\sin\theta = 0$ for $\theta = k\pi$ for any integer k, the sinc function has periodic zero crossings at $\pi, 2\pi$, and so forth, as shown in the figure. Thus, the sinc function has a main lobe and infinitely many side lobes, as indicated. Finally, since $\sin\theta$ and θ are both odd functions, their ratio $f(\theta)$ is an even function of θ. Notice that the sinc function $\sin\theta/\theta$ has some resemblance to the Dirichlet function plotted in Fig. 3.13, which has the form $\sin(N\theta)/\sin(\theta)$. This is periodic in θ, whereas the sinc is not periodic. So although the Dirichlet function has a "sinc-ish" behavior, it is different. ▽ ▽ ▽

It is clear that the impulse response (4.43) can be regarded as a sampled version of the sinc function, appropriately scaled. Thus, rewriting it as

$$h[n] = \frac{\omega_c}{\pi}\frac{\sin(\omega_c n)}{\omega_c n}, \quad (4.45)$$

it follows that $h[0] = \omega_c/\pi$ and that there are zero crossings at $n = k\pi/\omega_c$ whenever $k\pi/\omega_c$ is a nonzero integer. Figure 4.16 shows the plots of $h[n]$ for $\omega_c = 0.26\pi$ and 0.4π. Note that as ω_c increases, the main lobe gets narrower. Thus, if the passband of the filter is wider, the main lobe of the impulse response is narrower.

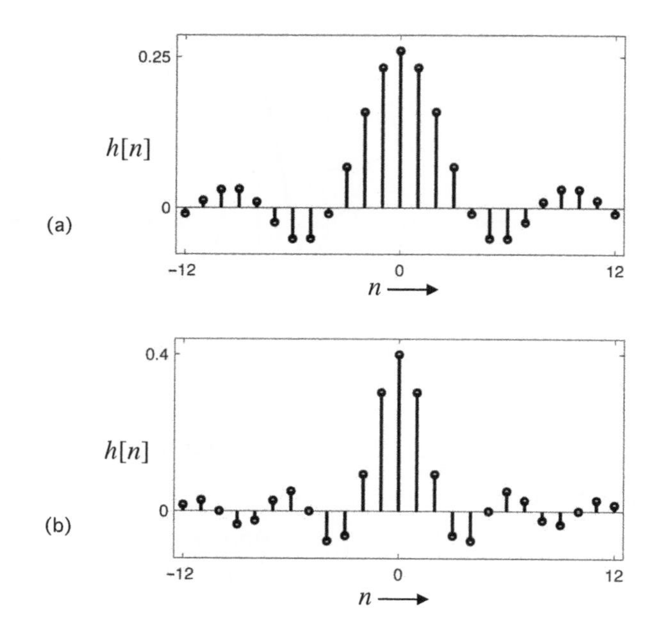

Figure 4.16 The impulse response (4.43) of the ideal lowpass digital filter for (a) $\omega_c = 0.26\pi$ and (b) $\omega_c = 0.4\pi$.

4.3.4.1 Remarks on the Ideal Lowpass Filter

From the impulse response (4.45), we can draw some conclusions about the ideal lowpass filter, based on the results of Sec. 3.3.

1. The filter (4.45) is **noncausal** since $h[n]$ can be nonzero for $n < 0$.
2. The impulse response (4.45) is infinitely long, so the filter is IIR.
3. The ideal lowpass filter is **unstable** because (4.45) is not absolutely summable. That is,

$$\sum_{n=-\infty}^{\infty} |h[n]| = \infty. \tag{4.46}$$

This is because $\sin \omega_c n$ keeps oscillating without decay as $n \to \infty$, and therefore $h[n]$ decays only as fast as $1/n$. Since

$$\sum_{n=1}^{\infty} \frac{1}{n} \tag{4.47}$$

does not converge, the sum in Eq. (4.46) does not converge either. A rigorous proof can be given as follows: if $\sum_n |h[n]|$ is finite, then $H(e^{j\omega})$ is a continuous function, as proved in Sec. 6.9.4. However, since $H(e^{j\omega})$ is discontinuous at $\omega = \omega_c$ (as seen from Fig. 4.14), it follows that $\sum_n |h[n]|$ cannot be finite.

Thus the ideal LPF is unstable and noncausal. So it is not realizable. There is yet another reason why this filter is unrealizable. In Chapter 5 we will study rational LTI systems. It will be seen (Sec. 5.8) that the frequency responses of such systems are continuous functions of ω. Thus, since the ideal LPF is discontinuous at ω_c, it is *not a rational* LTI system. It will be shown later (Sec. 14.4.4) that in the discrete-time case, only rational LTI systems are realizable (i.e., can be implemented with a *finite*

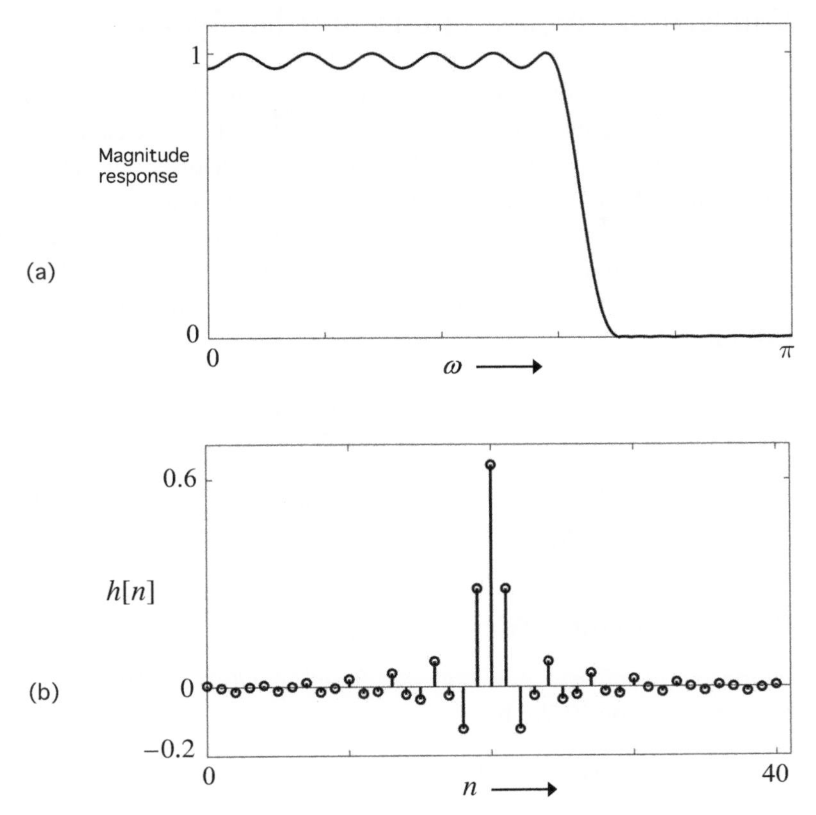

Figure 4.17 (a) Magnitude response of an FIR filter approximating the ideal lowpass filter and (b) its impulse response. The filter order is $N = 40$.

number of resources such as storage elements, multipliers, and adders). Summarizing, ideal lowpass filters are unrealizable because they are not rational LTI systems (although obviously LTI).

4.3.4.2 Approximating the Ideal Lowpass Filter

Even though the ideal lowpass filter is unrealizable, we will see in Chapter 7 that it can be approximated well by realizable filters. For example, a simple way to modify the ideal filter response to obtain FIR filters, called the window design method, is described in Sec. 7.9. Figure 4.17(a) shows the magnitude response of an FIR filter which approximates the ideal lowpass filter. Its impulse response is shown in Fig. 4.17(b). This is a filter with order $N = 40$ (see Eq. (4.39)), designed using a method called the McClellan and Parks method [McClellan and Parks, 1973; Oppenheim and Schafer, 2010]. This is called an **equiripple** filter because the local extrema of the errors in the passband are all equal (unlike in window-based designs), and the same holds in the stopband. While the description of such filters is beyond the scope of our discussion here, we mention that the equiripple property implies a kind of optimality called the **minimax** property, as explained in the above references.

4.4 Converting Lowpass Filters to Other Types

Given a lowpass filter $H(z)$, there are some simple ways to convert it to other types such as highpass, bandpass, and so forth. These are called frequency or spectral transformations of filters. They are useful because they save us the trouble of redesigning the filter. Also, some of these transformations can be directly implemented on the structure or flowgraph, which is useful in hardware implementations. There exist more general transformations [Constantinides, 1970], but we only give some simple examples here. In what follows we assume that we are given a causal FIR filter

$$H(z) = \sum_{n=0}^{N} h[n]z^{-n}, \tag{4.48}$$

with real $h[n]$ and a lowpass response as in Fig. 4.18(a). Here $|H(e^{j\omega})|$ is an even function because $h[n]$ is real.

4.4.1 The Transformation $G(z) = H(-z)$

Suppose we define a new transfer function $G(z) = H(-z)$. Then what is the frequency response? We have

$$G(e^{j\omega}) = H(-e^{j\omega}) = H(e^{-j\pi}e^{j\omega}) = H(e^{j(\omega-\pi)}). \tag{4.49}$$

So the frequency response has been shifted to the right by π as shown in Fig. 4.18(b), yielding a highpass filter. To understand this plot, remember that the response in Fig. 4.18(a) repeats with period 2π, so in particular there is a passband at $\omega = -2\pi$. This copy of the passband moves to $-\pi$ due to the right shift, as seen

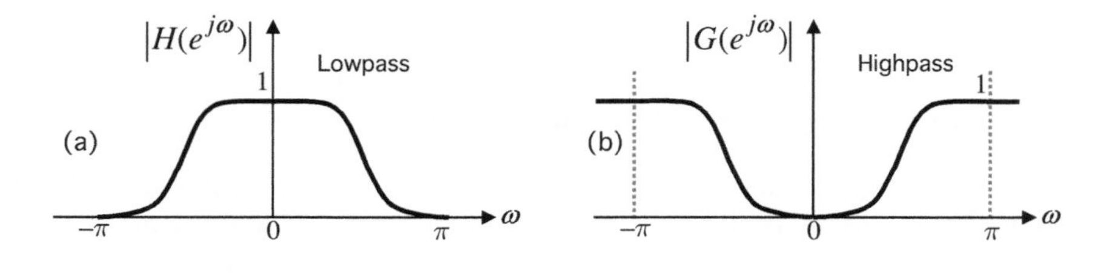

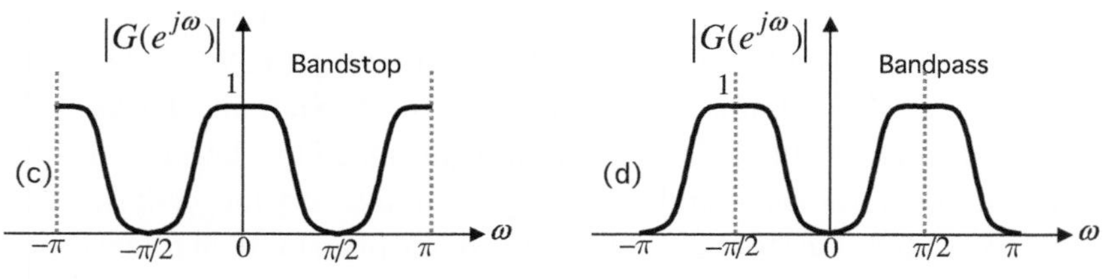

Figure 4.18 Converting a lowpass filter to other types of filter. (a) A lowpass filter and (b)–(d) other types of filters derived from it.

in Fig. 4.18(b). Thus, $G(z) = H(-z)$ is a **lowpass to highpass** transformation. Next, how is the impulse response $g[n]$ of $G(z)$ related to that of $H(z)$? Note that

$$G(z) = H(-z) = \sum_{n=0}^{N} h[n](-z)^{-n} = \sum_{n=0}^{N} (-1)^n h[n] z^{-n}, \tag{4.50}$$

which proves that

$$g[n] = (-1)^n h[n]. \tag{4.51}$$

That is, we simply change the sign of every odd-numbered coefficient in the impulse response. A more general frequency shift $H(e^{j(\omega-\omega_0)})$ can be achieved by modifying $h[n]$ in other appropriate ways (Problem 4.6).

4.4.2 The Transformation $G(z) = H(z^2)$

Now let us define a new transfer function $G(z) = H(z^2)$. Then

$$G(e^{j\omega}) = H(e^{j2\omega}). \tag{4.52}$$

Thus, the response $H(e^{j\omega})$ is simply "squeezed" by a factor of two, as demonstrated in Fig. 4.18(c). Remember that the response in Fig. 4.18(a) repeats with period 2π. When squeezed by two, the copy at 2π comes to π and the copy at -2π comes to $-\pi$. The resulting filter $G(e^{j\omega})$ is therefore a bandstop filter, which passes all frequencies except a band around $\pm\pi/2$. In short, $G(z) = H(z^2)$ is a **lowpass to bandstop** transformation. To find what the new impulse response $g[n]$ is, simply note that

$$G(z) = H(z^2) = \sum_{n=0}^{N} h[n] z^{-2n} = h[0] + h[1]z^{-2} + h[2]z^{-4} + \cdots. \tag{4.53}$$

That is, the impulse response $g[n]$ is

$$h[0],\ 0,\ h[1],\ 0,\ h[2],\ 0,\ h[3],\ \ldots, \tag{4.54}$$

or

$$g[n] = \begin{cases} h[n/2] & n \text{ even,} \\ 0 & \text{otherwise.} \end{cases} \tag{4.55}$$

One simply inserts a zero between adjacent samples of the original impulse response to get the bandstop filter.

4.4.3 The Transformation $G(z) = H(-z^2)$

If we define a new transfer function $G(z) = H(-z^2)$, then

$$G(e^{j\omega}) = H(-e^{j2\omega}) = H(e^{-j\pi} e^{j2\omega}) = H(e^{j(2\omega-\pi)}) = H(e^{j(2(\omega-\pi/2))}). \tag{4.56}$$

Thus, the response is the right-shifted version of Fig. 4.18(c), by the amount $\pi/2$. This produces a bandpass filter as shown in Fig. 4.18(d). In short, $G(z) = H(-z^2)$ is

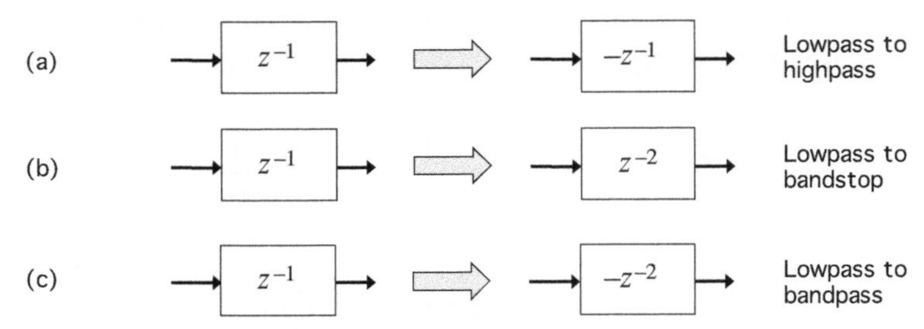

Figure 4.19 (a)–(c) Frequency transformations in a filter structure by replacement of each delay element z^{-1} by another building block.

a **lowpass to bandpass** transformation. To find what the new impulse response $g[n]$ is, note that

$$G(z) = H(-z^2) = \sum_{n=0}^{N} h[n](-z^2)^{-n} = h[0] - h[1]z^{-2} + h[2]z^{-4} - h[3]z^{-6} \cdots .$$

That is, the impulse response $g[n]$ is

$$h[0],\ 0,\ -h[1],\ 0,\ h[2],\ 0,\ -h[3],\ \ldots, \tag{4.57}$$

or

$$g[n] = \begin{cases} (-1)^{n/2} h[n/2] & n \text{ even}, \\ 0 & \text{otherwise}. \end{cases} \tag{4.58}$$

In all three transformations given above, the magnitude $|G(e^{j\omega})|$ continues to be an even function of ω, consistent with the fact that the filter coefficients $g[n]$ remain real (because $h[n]$ was assumed real). Figure 4.19 summarizes the transformations. In a structure that implements a lowpass filter, if we replace each delay unit as shown, then we achieve the transformation indicated.

Examples of frequency transformations are covered through computing assignments such as Problems 4.10 and 4.23 in this chapter, and Problems 7.17 and 7.20 in Chapter 7.

4.5 Phase Distortion

In the general form of the frequency response (3.59), it is easy to understand what the magnitude response $|H(e^{j\omega})|$ does. It simply gives relative weights to different frequency components of the input, producing a "filtering" effect. But what is the role of the phase response $\phi(\omega)$? In what way does it "physically" change the input signal $x[n]$? In this section we discuss this. The reader can use this section as a reference, and come back to it later, as needed.

4.5.1 Effect of Phase Response

The frequency response (3.59), expressed in polar form, is reproduced below:

$$H(e^{j\omega}) = |H(e^{j\omega})|e^{j\phi(\omega)}. \tag{4.59}$$

Similarly, $X(e^{j\omega})$ can be expressed as

$$X(e^{j\omega}) = |X(e^{j\omega})|e^{j\phi_x(\omega)}. \tag{4.60}$$

Then from Eq. (3.57) we have

$$Y(e^{j\omega}) = H(e^{j\omega})X(e^{j\omega}) = |H(e^{j\omega})X(e^{j\omega})|e^{j(\phi(\omega)+\phi_x(\omega))}. \tag{4.61}$$

So the output $y[n]$ has Fourier transform magnitude $|Y(e^{j\omega})| = |H(e^{j\omega})X(e^{j\omega})|$, which gives the filtering interpretation we already know. Moreover, from Eq. (4.61) the phase of $Y(e^{j\omega})$ is given by

$$\phi_y(\omega) = \phi(\omega) + \phi_x(\omega). \tag{4.62}$$

Thus, the phase response $\phi(\omega)$ of the filter is simply added to the phase of the input $\phi_x(\omega)$. To understand what this really means, it is easier to first consider the example of a single-frequency input signal $x[n] = e^{j\omega_0 n}$. Then

$$\begin{aligned} y[n] = H(e^{j\omega_0})\, e^{j\omega_0 n} &= |H(e^{j\omega_0})|\, e^{j\phi(\omega_0)} e^{j\omega_0 n} \\ &= |H(e^{j\omega_0})|\, e^{j(\phi(\omega_0)+\omega_0 n)} \\ &= |H(e^{j\omega_0})|\, e^{j\omega_0(n+\phi(\omega_0)/\omega_0)} \\ &= |H(e^{j\omega_0})|\, e^{j\omega_0(n-D_0)}, \end{aligned} \tag{4.63}$$

where $D_0 = -\phi(\omega_0)/\omega_0$. The factor $|H(e^{j\omega_0})|$ simply scales the amplitude of the input signals (filtering effect). The last line above shows that the effect of the filter phase $\phi(\omega_0)$ is to *delay* the signal by the amount D_0. Thus, if D_0 is an integer, we can write

$$y[n] = |H(e^{j\omega_0})|\, x[n - D_0]. \tag{4.64}$$

Non-integer D_0 will be addressed in Sec. 4.5.3.

4.5.2 Two Input Frequencies Applied Together

To generalize the above, assume next that $x[n]$ has two single-frequency terms as in

$$x[n] = e^{j\omega_1 n} + e^{j\omega_2 n}. \tag{4.65}$$

If these frequencies are in the passband, that is, $|H(e^{j\omega_1})| \approx |H(e^{j\omega_2})| \approx 1$, it follows that

$$y[n] \approx e^{j\omega_1(n-D_1)} + e^{j\omega_2(n-D_2)}, \tag{4.66}$$

where $D_i = -\phi(\omega_i)/\omega_i$. If the delays at the two frequencies are equal ($D_1 = D_2$), then this implies

$$y[n] \approx x[n - D_1], \tag{4.67}$$

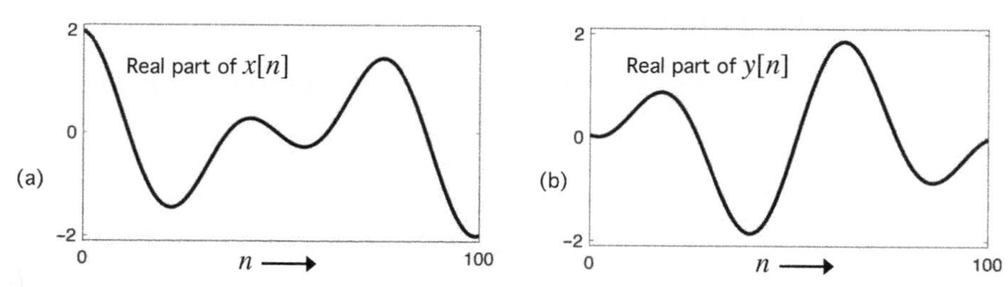

Figure 4.20 Demonstration of phase distortion. (a) Real part of Eq. (4.65) and (b) real part of Eq. (4.66), when $\omega_1 = 0.03\pi$ and $\omega_2 = 0.05\pi$, $D_1 = 3$ and $D_2 = 23$.

assuming D_1 is an integer. That is, the output is a simple delayed version of the input. On the other hand, if $D_1 \neq D_2$ then $y[n]$ is **not** a delayed version of the input. There is distortion in the shape of the output (compared to the shape of the input), in spite of the fact that the "filtering action" did not affect the amplitudes of the two sinusoids (i.e., $|H(e^{j\omega_1})| \approx |H(e^{j\omega_2})| \approx 1$). This distortion is called **phase distortion.** It is created by the fact that the delay

$$D(\omega) = -\phi(\omega)/\omega \tag{4.68}$$

due to the phase response is not identical at different frequencies. To demonstrate, we plot the real parts of Eqs. (4.65) and (4.66) in Fig. 4.20, for $\omega_1 = 0.03\pi$ and $\omega_2 = 0.05\pi$, when the delays are $D_1 = 3$ and $D_2 = 23$. It is clear that the two plots have completely different shapes. Thus, phase distortion is a visible distortion in the signal shape. In image processing applications it can readily be perceived (Sec. 7.10), and we prefer to avoid it.

If the filter has a phase response of the form $\phi(\omega) = -K\omega$, then the delay $-\phi(\omega)/\omega = K$ is the same at all frequencies and the filter is said to be a **linear-phase** filter. Such a filter does not create phase distortion. We will discuss such filters in some detail in Secs. 4.6 and 7.10.

4.5.3 Case when *D* is Not an Integer*

If $D(\omega) = -\phi(\omega)/\omega$ is *not* an integer, then we cannot interpret it as a time shift because in discrete time, the shift has to be an integer. However, a meaningful interpretation is still possible. As explained above, the input $x[n] = e^{j\omega_0 n}$ produces the output

$$y[n] = H(e^{j\omega_0})e^{j\omega_0 n} = |H(e^{j\omega_0})|e^{j\omega_0(n-D_0)},$$

where $D_0 = -\phi(\omega_0)/\omega_0$. Assume $|H(e^{j\omega_0})| = 1$ for convenience. Now, we can always regard the input signal $x[n] = e^{j\omega_0 n}$ as a sampled version of the continuous-time signal

$$x_c(t) = e^{j\omega_0 t},$$

with sample spacing $\Delta = 1$. That is, $x[n] = x_c(n)$. Similarly, $y[n] = e^{j\omega_0(n-D_0)}$ can be regarded as the sampled version of the shifted signal

$$y_c(t) = e^{j\omega_0(t-D_0)} = x_c(t - D_0). \tag{4.69}$$

That is, $y[n] = y_c(n)$. Thus, even though $y[n]$ cannot be regarded as an integer-shifted version of $x[n]$, we can regard $y[n]$ as the sampled version of a shifted version $y_c(t)$ of $x_c(t)$. Here, $x_c(t)$ is the "underlying" or hypothetical continuous-time signal from which $x[n]$ can be considered to have been obtained by sampling. Thus, when D_0 is not an integer, it can be interpreted as a "fractional delay" in the above sense.

Fractional delay for more general inputs. We now generalize the above interpretation to arbitrary $x[n]$. This part requires knowledge of continuous-time Fourier transforms (CTFTs) (Chapter 9) and the concept of sampling (Chapter 10), so the reader can skip this and return to it later. Any discrete-time signal $x[n]$ can be regarded as a sampled version of a continuous-time signal $x_c(t)$ bandlimited to $-\pi < \omega < \pi$ (Chapter 10), whose continuous-time Fourier transform is

$$X_c(j\omega) = \begin{cases} X(e^{j\omega}) & -\pi < \omega < \pi, \\ 0 & \text{otherwise.} \end{cases} \tag{4.70}$$

(For convenience, we assume $X(e^{j\omega}) = 0$ at $\omega = \pm\pi$.) The signal $x_c(t)$ is related to its Fourier transform by

$$x_c(t) = \frac{1}{2\pi}\int_{-\infty}^{\infty} X_c(j\omega)e^{j\omega t}d\omega \tag{4.71}$$

$$= \frac{1}{2\pi}\int_{-\pi}^{\pi} X_c(j\omega)e^{j\omega t}d\omega \tag{4.72}$$

$$= \frac{1}{2\pi}\int_{-\pi}^{\pi} X(e^{j\omega})e^{j\omega t}d\omega. \tag{4.73}$$

Since $x[n] = \int_{-\pi}^{\pi} X(e^{j\omega})e^{j\omega n}d\omega/2\pi$, it follows that

$$x[n] = x_c(n). \tag{4.74}$$

That is, $x[n]$ is a uniformly sampled version of $x_c(t)$ with sample spacing $\Delta = 1$. Now consider a discrete-time LTI system with frequency response

$$H(e^{j\omega}) = e^{-jK\omega} \tag{4.75}$$

in $|\omega| < \pi$, and repeating periodically with period 2π (so that it is a valid frequency response of a discrete-time system even if K may not be an integer). Then an input $x[n]$ produces an output $y[n]$ with

$$Y(e^{j\omega}) = e^{-jK\omega}X(e^{j\omega}) \tag{4.76}$$

in $|\omega| < \pi$, and repeating periodically with period 2π. If K is an integer, then we have $y[n] = x[n - K]$, so the output is a delayed version of the input. If K is not an integer, then it is more interesting. It is tempting to interpret Eq. (4.75) as a "fractional delay" but we have to do so carefully. Observe that

$$\begin{aligned} y[n] &= \frac{1}{2\pi}\int_{-\pi}^{\pi} Y(e^{j\omega})e^{j\omega n}d\omega \\ &= \frac{1}{2\pi}\int_{-\pi}^{\pi} e^{-jK\omega}X(e^{j\omega})e^{j\omega n}d\omega = \frac{1}{2\pi}\int_{-\pi}^{\pi} X_c(j\omega)e^{j\omega(n-K)}d\omega, \end{aligned}$$

where we have used Eq. (4.70). Define the continuous-time signal

$$y_c(t) = \frac{1}{2\pi}\int_{-\pi}^{\pi} X_c(j\omega)e^{j\omega(t-K)}d\omega. \tag{4.77}$$

By comparing with Eq. (4.72), we see that

$$y_c(t) = x_c(t-K). \tag{4.78}$$

Clearly

$$y[n] = y_c(n) = x_c(n-K), \tag{4.79}$$

so that $y[n]$ is the sampled version of $y_c(t)$ with sample spacing $\Delta = 1$. Thus, while $y[n]$ cannot be directly related to $x[n]$ by integer shift, we can regard $y[n]$ as the sampled version of a shifted version $y_c(t)$ of $x_c(t)$ (the shift being the non-integer amount K). As before, $x_c(t)$ is the "underlying" continuous-time signal from which $x[n]$ was obtained by sampling. ▽ ▽ ▽

Now consider an even more general situation where a discrete-time filter has frequency response

$$H(e^{j\omega}) = e^{j\phi(\omega)}, \tag{4.80}$$

in $|\omega| < \pi$, and repeating periodically with period 2π. Since $|H(e^{j\omega})| = 1$, this is said to be an allpass filter and it will be discussed in some detail in Sec. 7.4. The output has Fourier transform $Y(e^{j\omega}) = e^{j\phi(\omega)}X(e^{j\omega})$. Then

$$\begin{aligned} y[n] &= \frac{1}{2\pi}\int_{-\pi}^{\pi} Y(e^{j\omega})e^{j\omega n}d\omega \\ &= \frac{1}{2\pi}\int_{-\pi}^{\pi} X(e^{j\omega})e^{j(\omega n+\phi(\omega))}d\omega = \frac{1}{2\pi}\int_{-\pi}^{\pi} X_c(j\omega)e^{j\omega(n-D(\omega))}d\omega, \end{aligned}$$

where we have used Eq. (4.70). Here $D(\omega) = -\phi(\omega)/\omega$, as before. Define

$$y_c(t) = \frac{1}{2\pi}\int_{-\pi}^{\pi} X_c(j\omega)e^{j\omega(t-D(\omega))}d\omega. \tag{4.81}$$

The interpretation of this is that the frequency component of $x_c(t)$ with frequency ω, that is $X_c(j\omega)e^{j\omega t}$, is delayed by $D(\omega)$, and these are superposed to obtain $y_c(t)$.

Clearly $y[n] = y_c(n)$, which is the sampled version of $y_c(t)$. Thus, we can regard $y[n]$ as the sampled version of a modified version $y_c(t)$ of $x_c(t)$, where $x_c(t)$ is the underlying continuous-time signal from which $x[n]$ was obtained by sampling. The modified version $y_c(t)$ is such that each frequency component of $x_c(t)$ is delayed by (a possibly different, non-integer) amount $D(\omega)$, and then superimposed to obtain $y_c(t)$.

In this section we discussed phase responses $\phi(\omega)$ and their interpretations in some detail. A closely related quantity called the group delay $\tau(\omega) = -d\phi(\omega)/d\omega$ is described later in Sec. 7.10.

4.6 Linear-Phase Digital Filters

Recall that the frequency response of a digital filter can always be written in the polar form

$$H(e^{j\omega}) = |H(e^{j\omega})|e^{j\phi(\omega)}. \tag{4.82}$$

If the phase response $\phi(\omega)$ is linear, or more specifically

$$\phi(\omega) = -K\omega \tag{4.83}$$

for some real K, we say that the filter has linear phase. As explained in Sec. 4.5, such filters do not create phase distortion. Sometimes one extends the definition of linear phase to include responses like that of Eq. (4.29), where the phase response is piecewise linear and does not pass through the origin. We will not get into a detailed discussion of this, but some examples are covered in Problems 4.14 to 4.16. Suffice it to say that linear-phase filters are important in applications such as image processing (Sec. 7.10).

4.6.1 How to Construct Linear-Phase Filters

The ideal lowpass filter (Sec. 4.3.4) evidently has linear phase because $\phi(\omega) = 0$, but it is unrealizable. In this section we will show that it is easy to design realizable digital filters which enjoy an exactly linear phase. We already know this from the example of the moving average filter in Eq. (4.23). More generally, consider an Nth-order FIR filter with transfer function and frequency response

$$H(z) = \sum_{n=0}^{N} h[n]z^{-n}, \quad H(e^{j\omega}) = \sum_{n=0}^{N} h[n]e^{-j\omega n}. \tag{4.84}$$

Now assume that the filter is real (i.e., $h[n]$ is real valued) and furthermore that $h[n]$ satisfies the symmetry property

$$h[n] = h[N-n]. \tag{4.85}$$

This symmetry implies $h[0] = h[N], h[1] = h[N-1]$, and so forth. This is demonstrated in Fig. 4.21. For a system satisfying Eq. (4.85), we can rewrite the frequency response in a special form, as demonstrated below for $N = 4$:

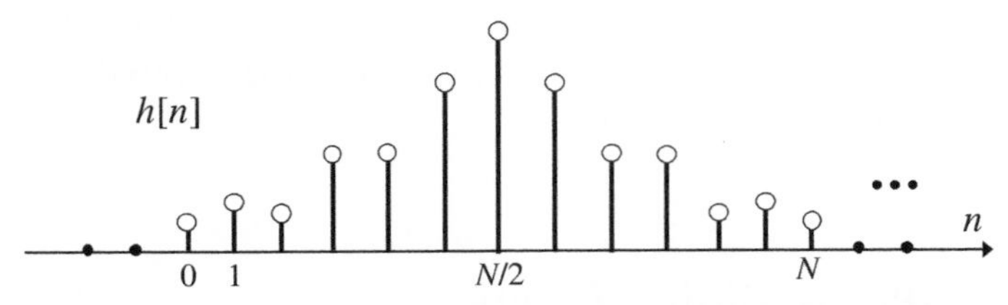

Figure 4.21 A real-valued impulse response with the symmetry property $h[n] = h[N - n]$. This represents a linear-phase filter. See text.

$$\begin{aligned} H(e^{j\omega}) &= h[0] + h[1]e^{-j\omega} + h[2]e^{-j2\omega} + h[3]e^{-j3\omega} + h[4]e^{-j4\omega} \\ &= h[0] + h[1]e^{-j\omega} + h[2]e^{-j2\omega} + h[1]e^{-j3\omega} + h[0]e^{-j4\omega} \\ &= e^{-j2\omega}\Big(h[0]e^{j2\omega} + h[1]e^{j\omega} + h[2] + h[1]e^{-j\omega} + h[0]e^{-j2\omega}\Big) \\ &= e^{-j2\omega}\underbrace{\Big(2h[0]\cos(2\omega) + 2h[1]\cos(\omega) + h[2]\Big)}_{\text{call this } H_R(\omega)}. \end{aligned} \tag{4.86}$$

Since $h[n]$ is real, the quantity $H_R(\omega)$ indicated above is real valued for all ω, and the subscript R is used to signify this. Thus, because of the symmetry in Eq. (4.85), we could combine pairs of terms in the summation within the large brackets, resulting in a real-valued sum $H_R(\omega)$. Now, any real function $H_R(\omega)$ has a phase equal to either 0 or π. If $H_R(\omega)$ can be guaranteed to be non-negative for all ω, then its phase is zero for all ω, in which case the phase of the filter (4.86) is

$$\phi(\omega) = -2\omega, \tag{4.87}$$

which is exactly linear. More generally, for arbitrary even N, if $h[n]$ is real and satisfies the symmetry (4.85), then we can readily verify that $H(e^{j\omega})$ has the form

$$H(e^{j\omega}) = e^{-j\omega N/2}H_R(\omega), \tag{4.88}$$

where $H_R(\omega)$ is real valued and can be gauranteed to be non-negative simply by making $h[N/2]$ a large enough positive number! This is readily seen from Eq. (4.86): by making $h[2]$ large enough we can make the quantity in brackets non-negative for all ω. With the definition of linear phase slightly generalized, it is possible to achieve linear phase even when $h[n]$ has more generalized forms of symmetry, such as antisymmetry or Hermitian symmetry (Problems 4.14 to 4.16). The evenness of N can also be relaxed.

We have shown how FIR linear-phase filters can be constructed. The curious student might wonder if it is also possible to design IIR filters with linear phase. This is addressed in Sec. 6.8.

Transfer Function Implied by Eq. (4.85)

The symmetry property (4.85) implies that

$$H(z) = z^{-N}H(z^{-1}). \tag{4.89}$$

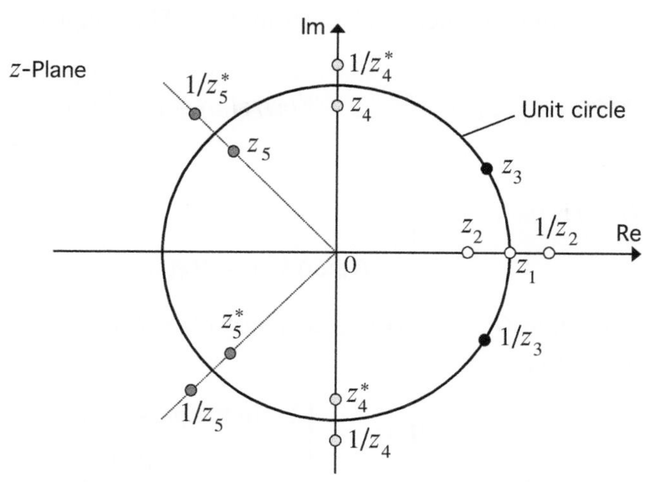

Figure 4.22 Pattern of the zeros of a linear-phase filter $H(z)$. Here, $h[n]$ is real and $h[n] = h[N - n]$.

To prove this, simply note that

$$H(z) = \sum_{n=0}^{N} h[n]z^{-n} = \sum_{n=0}^{N} h[N - n]z^{-n}$$
$$= \sum_{m=0}^{N} h[m]z^{-(N-m)} = z^{-N} \sum_{m=0}^{N} h[m]z^{m},$$

which proves Eq. (4.89). Equation (4.89) shows that if z_k is a zero of $H(z)$, then so is $1/z_k$. Thus, the zeros come in **reciprocal pairs**. Since $h[n]$ has real coefficients, it follows that z_k^* and $1/z_k^*$ are also zeros. Depending on the value of z_k, the four quantities $(z_k, 1/z_k, z_k^*, 1/z_k^*)$ may or may not be all distinct. Figure 4.22 shows some examples. Since $z_1 = 1$, it is its own reciprocal and conjugate. z_2 and $1/z_2$ form a pair. Since z_3^* and $1/z_3$ are identical, we just have the pair $(z_3, 1/z_3)$. Next, $(z_4, 1/z_4, z_4^*, 1/z_4^*)$ are distinct, and form a quadruple. Similarly, $(z_5, 1/z_5, z_5^*, 1/z_5^*)$ form a quadruple.

4.6.2 The Complement of a Linear-Phase Filter

Let $H(z)$ in Eq. (4.84) be a lowpass filter with real $h[n]$ satisfying the symmetry Eq. (4.85), and assume the order N is even so that the mid-term $h[N/2]$ exists. Then $H(e^{j\omega})$ satisfies Eq. (4.88). Now define

$$G(z) = z^{-N/2} - H(z), \tag{4.90}$$

that is,

$$g[n] = \delta\left[n - \frac{N}{2}\right] - h[n], \tag{4.91}$$

which is meaningful since $N/2$ is an integer. The frequency response of $G(z)$ is therefore

$$G(e^{j\omega}) = e^{-j\omega N/2} - e^{-j\omega N/2} H_R(\omega) = e^{-j\omega N/2} \underbrace{\Big(1 - H_R(\omega)\Big)}_{G_R(\omega)}. \tag{4.92}$$

Since $H_R(\omega)$ is real, so is

$$G_R(\omega) \triangleq 1 - H_R(\omega), \tag{4.93}$$

which shows that $G(z)$ is also a linear-phase filter. More interestingly, suppose $H(z)$ is lowpass so that

$$H_R(\omega) \approx \begin{cases} 1 & \text{in the passband,} \\ 0 & \text{in the stopband.} \end{cases} \tag{4.94}$$

Then

$$G_R(\omega) \approx \begin{cases} 0 & \text{in the passband of } H(z), \\ 1 & \text{in the stopband of } H(z). \end{cases} \tag{4.95}$$

That is, $G(z)$ is a highpass filter! Thus, merely by performing the subtraction operation (4.90), we are able to obtain a highpass filter $G(z)$. This has been possible because (a) $H(z)$ has linear-phase property (4.88) and (b) N is even. Note that if $H(z)$ were an arbitrary lowpass filter not satisfying these two properties, then $G(z)$ in Eq. (4.90) would not be a good highpass filter.

In the above construction, since $G_R(\omega) = 1 - H_R(\omega)$, we say that $G(e^{j\omega})$ is a complement of $H(e^{j\omega})$. More precisely, since they satisfy Eq. (4.90), $H(z)$ and $G(z)$ are said to be **delay complementary**. That is, $H(z)$ and $G(z)$ add up to a delay $z^{-N/2}$.

As a specific example, let $h[n]$ be as in Eq. (4.31), where the order $N = L-1 = 10$. Then $H(z)$ has linear phase, and Fig. 4.23 shows its magnitude response, which is

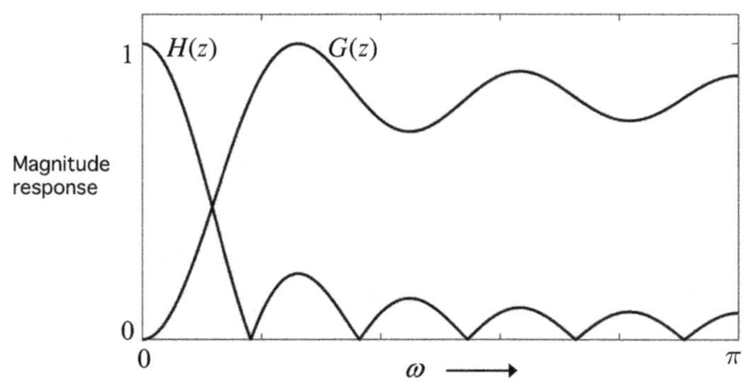

Figure 4.23 Magnitude responses of the lowpass and highpass linear-phase filters, which satisfy the delay-complementary property (4.90). Here $h[n]$ is as in Eq. (4.31), with $N = L-1 = 10$. In the plot, the peak of $|G(e^{j\omega})|$ is normalized to 1 for convenience.

lowpass. Since N is even, $G(z)$ satisfies the complementary property (4.95). The magnitude response of $G(z)$ is also shown, and $G(z)$ is indeed highpass as expected. In the plot, the peak of $|G(e^{j\omega})|$ is normalized to unity for convenience.

4.6.3 Image-Filtering Examples

Section 3.7 presented some examples of 2D images filtered by lowpass and highpass filters. The reader may want to review Sec. 3.7 at this time, because we will use the same language here to present a more detailed example. We use the linear-phase delay-complementary filters $H(z)$ (lowpass) and $G(z)$ (highpass) shown in Fig. 4.23, to filter the image $x[n_1, n_2]$ shown in Fig. 4.24(a). There are four filtered versions shown in this figure, and they are generated as follows.

1. Part (b) shows the lowpass/lowpass (LL) filtered version $y_{LL}[n_1, n_2]$. That is, the filter $H(z)$ in Fig. 4.23 is used for horizontal as well as vertical filtering. The filtered image is a highly smoothed version of the original – all sharp edges have been smoothed out. This is a common effect of lowpass filtering.
2. Part (c) shows the lowpass/highpass (LH) filtered version $y_{LH}[n_1, n_2]$. That is, $H(z)$ is used for horizontal filtering and $G(z)$ (highpass) for vertical filtering. Thus, if we move in the vertical direction, we are able to see sharp outlines of the objects, which carry "high-frequency information."
3. Part (d) shows the highpass/lowpass (HL) filtered version $y_{HL}[n_1, n_2]$. That is, $G(z)$ (highpass) is used for horizontal filtering and $H(z)$ for vertical filtering. In this case we are able to see sharp outlines of objects if we move in the horizontal direction.
4. Part (e) shows the highpass/highpass (HH) filtered version $y_{HH}[n_1, n_2]$. That is, $G(z)$ is used for horizontal filtering as well as vertical filtering. So we are able to see sharp outlines of objects in both directions.

This example demonstrates the effects of lowpass and highpass filtering of an image. As mentioned in Sec. 3.7, whenever impulse responses have some negative values, the pixels of filtered images can have occasional negative values as well. A small positive constant has been added to the entire filtered image to make it non-negative, so that the image can be displayed.

Now comes the interesting part. Suppose we add the four filtered images. Then it can be shown that the resulting sum image is exactly the original image (except for an inconsequential delay $N/2$ in both directions). That is,

$$y_{LL}[n_1, n_2] + y_{LH}[n_1, n_2] + y_{HL}[n_1, n_2] + y_{HH}[n_1, n_2] = x[n_1 - M, n_2 - M], \quad (4.96)$$

where $M = N/2$. This sum image is shown in part (f). The reason for this wonderful behavior is the complementary property (4.93), or equivalently (4.90). If the lowpass/highpass pair $\{H(z), G(z)\}$ did not exactly satisfy this, then Eq. (4.96) would not be true.

More real-world applications: audio and image signals. Dividing an image into frequency subbands and adding back to get the original signal is certainly fun when we

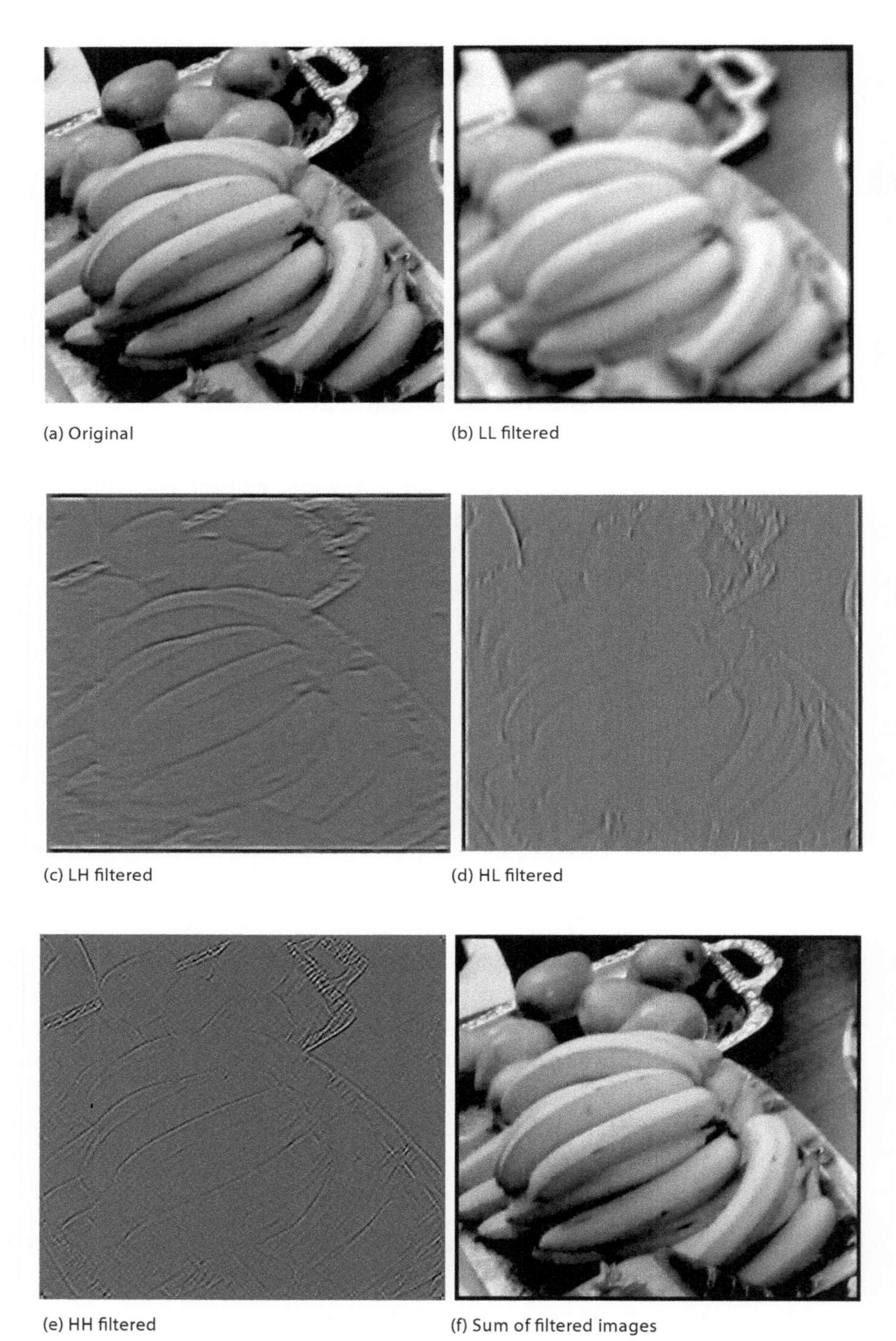

(a) Original (b) LL filtered

(c) LH filtered (d) HL filtered

(e) HH filtered (f) Sum of filtered images

Figure 4.24 (a) Original image, (b) LL filtered, (c) LH filtered, (d) HL filtered, (e) HH filtered, and (f) sum of four filtered images. The filters $H(z)$ and $G(z)$ in Fig. 4.23 are used for lowpass and highpass filtering, respectively.

are learning the basics. But is there any use for this exercise? Indeed, there are practical applications where it is actually useful to divide a signal into various subbands like this.

A famous technique called **subband coding**, which is used for data compression, comes to mind. In this technique, the four subband images y_{LL}, y_{LH}, y_{HL}, and y_{HH} are **quantized** to a small number of bits. Notice that most of the filtered images in Fig. 4.24 have very little energy. Such subband signals are assigned very few bits, and this is where the compression aspect comes in. After such bit allocation, the collection of subband signals requires much fewer bits for storage than the original signal $x[n_1, n_2]$. Later on, when we are ready to view the image again, the stored compressed subband signals are combined back to obtain the full image, which is then displayed. Techniques similar to this are at the heart of compression systems such as JPEG, MPEG, and MP3 systems. Be aware that our description here is oversimplified. In practice, the subband signals are downsampled before quantization (otherwise no compression can be achieved). Moreover, the quantized subband signals are interpolated with filters before being added back. Subband compression has been used for both 2D and 1D signals.

Another application is in audio engineering, where the aesthetic quality of a signal is edited by using a **graphic equalizer**. For this, the sound signal (music or speech) is divided into subbands, and the subbands are weighted by different constants before being added back. This redistributes the relative energies in the bass, treble, and other components, according to the listener's personal aesthetic preference. ▽ ▽ ▽

4.7 Denoising or Noise Removal by Filtering

When signals are obtained by various means (sensing, acquisition, and so on), they often suffer from additive noise. Thus, instead of the clean signal $x[n]$, we only have access to its noisy version

$$x_{noisy}[n] = x[n] + e[n], \tag{4.97}$$

where $e[n]$ is an additive noise component. For example, a clean signal $x[n]$ and its noisy version $x_{noisy}[n]$ are shown in Figs. 4.25(a) and (b). Here the clean signal has the form

$$x[n] = a_1 \sin(\omega_1 n) + a_2 \cos(\omega_2 n) + a_3 \cos(\omega_3 n), \tag{4.98}$$

where the frequencies of the sinusoids are

$$\omega_1 = 0.06\pi, \quad \omega_2 = 0.15\pi, \quad \omega_3 = 0.3\pi, \tag{4.99}$$

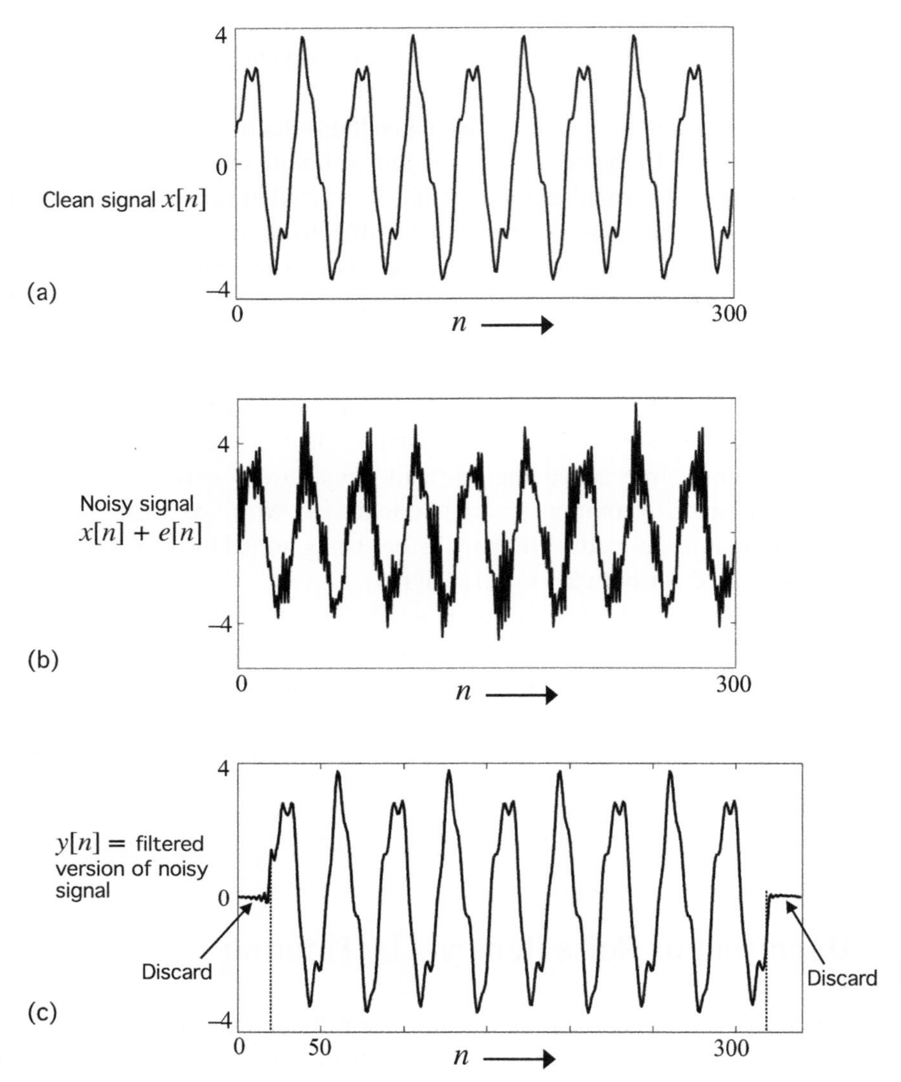

Figure 4.25 (a) A clean signal $x[n]$, (b) the noisy version $x[n] + e[n]$, where $e[n]$ is high-frequency noise, and (c) the lowpass filtered version of the noisy version.

and their amplitudes are $a_1 = 3$, $a_2 = 0.5$, $a_3 = 0.4$. The noise term is $e[n] = 0.9\cos(\omega_4 n) + 1.1\cos(\omega_5 n)$, where $\omega_4 = 0.9\pi$ and $\omega_5 = 0.95\pi$. Thus, the clean signal $x[n]$ has low-frequency components, and the additive noise term $e[n]$ has high-frequency components. This is also clear from the plots in Figs. 4.25(a) and (b).

4.7.1 Denoising Example

We now use the lowpass filter shown in Fig. 4.17 to filter out the noise in the above image. The filtered version $y[n]$ is shown in Fig. 4.25(c). It is clear that $y[n]$ closely resembles the input signal $x[n]$, except for a short initial segment of about 20 samples and a short segment in the end of similar duration, marked as "discard" in the figure. The removal of noise from a signal is often called **denoising** and is a major topic in signal processing.

The short segments denoted as "discard" in the figure are unavoidable in LTI filtering. To understand why they arise, consider a finite-duration input

$$x[n], \quad 0 \leq n \leq M, \tag{4.100}$$

applied to an FIR filter $H(z) = \sum_{n=0}^{N} h[n]z^{-n}$. As explained in Sec. 4.8.2 later, the output $y[n]$ has a *steady-state* part of duration $M - N + 1$ and two *transient* parts of duration N, one at the beginning and one at the end. Returning to Fig. 4.25 we have $M = 300$. $H(z)$ is a linear-phase filter of order $N = 40$ (see Fig. 4.17(b)). So the steady state starts at $n = 40$ and ends at $n = 300$ (see Eq. (4.109)). But since the impulse response of the linear-phase filter $h[n]$ has most of its energy concentrated near $n = 20$ (Fig. 4.17), the output $y[n]$ is close to steady state as soon as n exceeds 20 slightly. Similarly, it remains in steady state for almost 20 samples after $y[300]$. Thus, the portion of $y[n]$ between the two dotted lines is as good as the steady state, and it agrees well with the clean input signal $x[n]$.

In our example, denoising has been especially easy because the signal components have low frequencies and the noise has high frequency. So a simple lowpass filter could do the magic. But in practice the noise can be more complicated and have its energy distributed in a wide range of frequencies. Furthermore, the noise $e[n]$ often has random characteristics, and can only be described as a *random process*. Removing random noise from a signal comes under statistical signal processing, which is a major topic in advanced signal processing [Sayed, 2023]. There are many sophisticated, often nonlinear, methods for denoising [Krim et al., 1999].

4.7.2 Image-Denoising Example

We now present an example of image denoising. Figure 4.26(a) shows a clean image $x[n_1, n_2]$ and Fig. 4.26(b) shows a noisy version $x[n_1, n_2] + e[n_1, n_2]$. The additive noise $e[n_1, n_2]$ is basically a high-frequency pattern and gives the impression that the image has been painted on coarse cardboard. So we call this "cardboard noise." Figure 4.27 shows a filtered version of the noisy image; it looks almost like the original image, although there is some loss of sharpness and contrast due to lowpass filtering. The filter used is a linear-phase FIR lowpass filter of order $N = 40$, having magnitude response $|H(e^{j\omega})|$ as in Fig. 4.28(a). The 2D filtering is performed by performing filtering with $H(z)$ along horizontal and vertical directions, as described in Sec. 3.7.

Figure 4.28(b) shows the so-called decibel or dB response of the filter. This is nothing but a plot of $20 \log_{10} |H(e^{j\omega})|$. Such a plot is useful to view the stopband response in greater detail. We will have more to say about dB responses in Sec. 7.6.2. It should be noted that in image filtering also, there are transients at the borders of the signal similar to the portions in Fig. 4.25 marked as "discard." In

(a)

(b)

Figure 4.26 (a) A clean original image and (b) the same image contaminated with high-frequency noise (cardboard noise).

Figure 4.27 The denoised or filtered version of the noisy image in Fig. 4.26. A linear-phase FIR filter with magnitude response as in Fig. 4.28 is used.

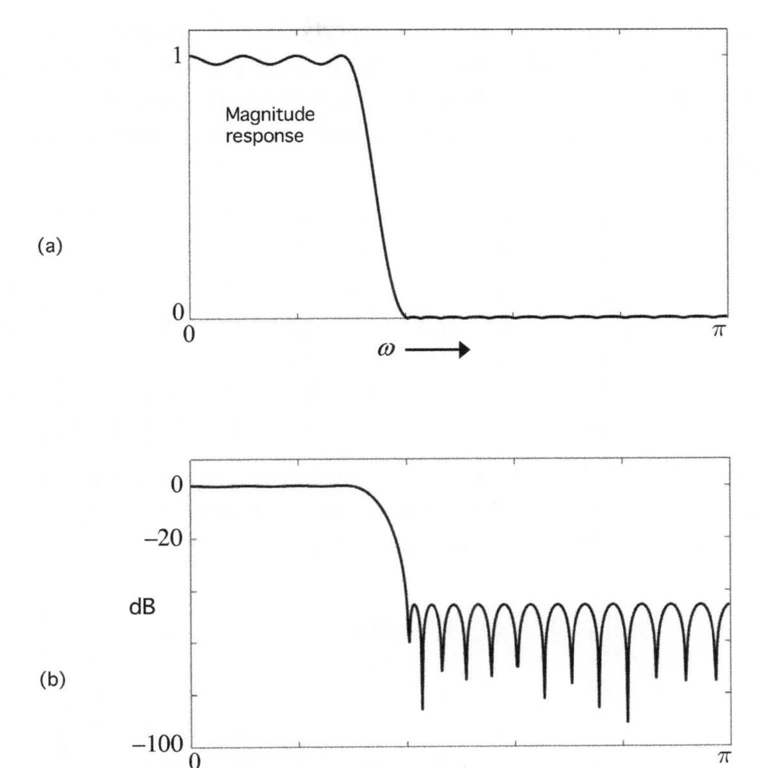

Figure 4.28 Responses of the filter used in image denoising. (a) Magnitude response $|H(e^{j\omega})|$ and (b) plot of $20\log_{10}|H(e^{j\omega})|$ (decibel or dB plot, see Sec. 7.6.2).

Fig. 4.27 we do not see them because they have been discarded before plotting. The same is true in most of the filtered images we show.

4.8 Toeplitz Matrices and Convolution*

It is sometimes convenient to express the convolution sum $y[n] = \sum_k h[k]x[n-k]$ as a matrix vector multiplication. For example, when $h[n]$ and $x[n]$ are both causal, we have $y[n] = 0$ for $n < 0$ and

$$y[n] = \sum_{k=0}^{n} h[k]x[n-k], \quad n \geq 0. \tag{4.101}$$

If we want to show the first five samples of the output, the matrix equation is

$$\begin{bmatrix} y[0] \\ y[1] \\ y[2] \\ y[3] \\ y[4] \end{bmatrix} = \underbrace{\begin{bmatrix} h[0] & 0 & 0 & 0 & 0 \\ h[1] & h[0] & 0 & 0 & 0 \\ h[2] & h[1] & h[0] & 0 & 0 \\ h[3] & h[2] & h[1] & h[0] & 0 \\ h[4] & h[3] & h[2] & h[1] & h[0] \end{bmatrix}}_{\mathbf{H}} \begin{bmatrix} x[0] \\ x[1] \\ x[2] \\ x[3] \\ x[4] \end{bmatrix}. \tag{4.102}$$

Note that the matrix $\mathbf{H}$ has the property that all elements along the diagonal are identical. Similarly, all elements along any line parallel to the diagonal are identical. Such a matrix is called a **Toeplitz** matrix. It has the property that $[\mathbf{H}]_{km}$ is identical for all (k, m) such that $k - m$ is a constant. Here is another example where $h[n]$ is FIR with length three:

$$\begin{bmatrix} y[0] \\ y[1] \\ y[2] \\ y[3] \\ y[4] \\ y[5] \\ y[6] \end{bmatrix} = \begin{bmatrix} h[0] & 0 & 0 & 0 & 0 & 0 & 0 \\ h[1] & h[0] & 0 & 0 & 0 & 0 & 0 \\ h[2] & h[1] & h[0] & 0 & 0 & 0 & 0 \\ 0 & h[2] & h[1] & h[0] & 0 & 0 & 0 \\ 0 & 0 & h[2] & h[1] & h[0] & 0 & 0 \\ 0 & 0 & 0 & h[2] & h[1] & h[0] & 0 \\ 0 & 0 & 0 & 0 & h[2] & h[1] & h[0] \end{bmatrix} \begin{bmatrix} x[0] \\ x[1] \\ x[2] \\ x[3] \\ x[4] \\ x[5] \\ x[6] \end{bmatrix}. \tag{4.103}$$

Notice that all the samples of $h[k]$ participate in the expression for $y[n]$, starting only from $n = 2$. So $y[n], n \geq 2$ is called the steady-state output (Sec. 7.12.3). If we write out only this steady-state part, then

$$\begin{bmatrix} y[2] \\ y[3] \\ y[4] \\ y[5] \\ y[6] \end{bmatrix} = \begin{bmatrix} h[2] & h[1] & h[0] & 0 & 0 & 0 & 0 \\ 0 & h[2] & h[1] & h[0] & 0 & 0 & 0 \\ 0 & 0 & h[2] & h[1] & h[0] & 0 & 0 \\ 0 & 0 & 0 & h[2] & h[1] & h[0] & 0 \\ 0 & 0 & 0 & 0 & h[2] & h[1] & h[0] \end{bmatrix} \begin{bmatrix} x[0] \\ x[1] \\ x[2] \\ x[3] \\ x[4] \\ x[5] \\ x[6] \end{bmatrix}. \tag{4.104}$$

Now the Toeplitz matrix has an equal number of nonzero elements in each row. This is called a **banded Toeplitz** matrix. Finally, consider an example where $h[n]$ is causal and FIR with length three, but $x[n]$ is not restricted to be causal. Then the output need not be zero for $n \leq 0$. The output for the finite interval $-1 \leq n \leq 2$ is shown below:

$$\begin{bmatrix} y[-1] \\ y[0] \\ y[1] \\ y[2] \end{bmatrix} = \begin{bmatrix} x[-1] & x[-2] & x[-3] \\ x[0] & x[-1] & x[-2] \\ x[1] & x[0] & x[-1] \\ x[2] & x[1] & x[0] \end{bmatrix} \begin{bmatrix} h[0] \\ h[1] \\ h[2] \end{bmatrix}. \tag{4.105}$$

Notice that for convenience we have put the input samples in the matrix and the filter samples in the vector. Since convolution is commutative, we can always interchange the roles of $x[n]$ and $h[n]$ in this manner.

4.8.1 Deconvolution and Applications

Since the matrix (4.102) is triangular, it is nonsingular as long as $h[0] \neq 0$ (Sec. 18.7.2). We can therefore find the input $x[n]$ from the output $y[n]$ simply by inverting this matrix. Finding the input of an LTI system from measurements of the output is called **deconvolution**. For example, in digital communications [Haykin, 1988; Lathi, 1989], the input $x[n]$ is a symbol stream to be sent from a transmitter to a receiver. In the process of transmission, the signal $x[n]$ gets distorted by a **channel** (the medium of communication). This distortion can often be modeled as a convolution. So the received symbol stream $y[n]$ is as in Eq. (4.101). Identifying the transmitted symbols $x[n]$ is therefore a deconvolution problem. It should be added here that in practice there is additive noise, which complicates this symbol detection. Thus, instead of Eq. (4.101) we have

$$y[n] = \sum_k h[k]x[n-k] + e[n], \tag{4.106}$$

where $e[n]$ is additive noise. A more sophisticated version of deconvolution is therefore necessary. A final remark is that in many situations the channel $h[n]$ is unknown, and a **channel identification** algorithm is often needed. For this, one sends a *known* input symbol stream $x[n]$ (a pilot stream) for a short duration. By measuring the output $y[n]$ in response to this, we can identify $h[n]$ using deconvolution. Again, in the presence of additive noise, the algorithm is more involved.

4.8.2 Steady State and Transient State

Now consider a finite-duration input

$$x[n], \quad 0 \le n \le M, \tag{4.107}$$

applied to a causal FIR filter $H(z) = \sum_{n=0}^{N} h[n]z^{-n}$. For example, if $M = 4$ and $N = 2$, then $y[n]$ has at most $M + N + 1 = 7$ nonzero samples, and these are

$$\begin{bmatrix} y[0] \\ y[1] \\ y[2] \\ y[3] \\ y[4] \\ y[5] \\ y[6] \end{bmatrix} = \begin{bmatrix} h[0] & 0 & 0 & 0 & 0 \\ h[1] & h[0] & 0 & 0 & 0 \\ h[2] & h[1] & h[0] & 0 & 0 \\ 0 & h[2] & h[1] & h[0] & 0 \\ 0 & 0 & h[2] & h[1] & h[0] \\ 0 & 0 & 0 & h[2] & h[1] \\ 0 & 0 & 0 & 0 & h[2] \end{bmatrix} \begin{bmatrix} x[0] \\ x[1] \\ x[2] \\ x[3] \\ x[4] \end{bmatrix}. \tag{4.108}$$

Notice that only $y[2], y[3], y[4]$ are affected by *all* the filter coefficients $h[n]$. So we call these the **steady-state** output samples or the "steady-state response." The samples $y[0], y[1]$, which are affected only by some of the $h[n]$, can be regarded as **transients** or the "transient response." Similarly, $y[5], y[6]$ can also be regarded as transients. To distinguish these, we can call $y[0], y[1]$ "leading transients" and $y[5], y[6]$ "trailing transients." Thus, there are N leading transients and N trailing transients:

$$\underbrace{y[0], \ldots, y[N-1]}_{\text{leading transients}}, \; \underbrace{y[N], \ldots, y[M]}_{\text{steady state}}, \; \underbrace{y[M+1], \ldots, y[M+N]}_{\text{trailing transients}}. \tag{4.109}$$

Therefore,

$$\text{Number of steady-state output samples} = M - N + 1, \tag{4.110}$$

which is 3 in our example. Clearly, if $N > M$ there are no steady-state samples at all! When $N << M$, which is the typical situation, there are many more samples in the steady state than in the transient state.

The true significance of the steady state will be explained in Sec. 7.12. Here we merely borrow from Sec. 7.12 and say that if $x[n] = e^{j\omega_0 n}, 0 \leq n \leq M$ (and zero outside), then the eigenfunction behavior

$$y[n] = H(e^{j\omega_0})e^{j\omega_0 n} \tag{4.111}$$

is valid only in the steady state. You can readily verify this by substituting $x[n] = e^{j\omega_0 n}$ into Eq. (4.108) and using

$$H(e^{j\omega}) = h[0] + h[1]e^{-j\omega} + h[2]e^{-j2\omega}. \tag{4.112}$$

Thus, frequency responses and the filtering interpretation (lowpass, highpass, and so forth) make sense only for the steady state. Here are a couple of other points to note:

1. The output $y[n]$ transitions into and out of the steady state rather "gradually" (especially for large N). For example, referring to Eq. (4.108), $y[N-1]$ is almost a steady-state sample because almost all samples of $h[n]$ are involved in its computation, whereas $y[N-2]$ is less so. Similarly, $y[M+1]$ is almost a steady-state sample, and $y[M+2]$ less so.
2. For IIR filters, the filter length is $L = \infty$, although the filter order N is some finite number. For example, $H(z) = 1/(1-pz^{-1})$ has order 1 but filter length ∞. Thus, strictly speaking, the steady state never arrives (as seen from Eq. (4.109)). However, in practice, $h[n]$ decays to zero exponentially for realizable filters. So, as long as the input length $M+1$ is large enough, the behavior of the output $y[n]$ will eventually approximate the steady-state behavior admirably well (Sec. 7.12).

4.9 Summary of Chapter 4

This chapter first introduced computational graphs, which are useful for representing systems in the form of structures or signal flowgraphs. Then three types of interconnections for LTI systems were presented, namely the cascade, parallel, and feedback interconnections, along with their transfer functions. We then presented the first-order moving average filter and first-order difference filter:

$$y[n] = \begin{cases} \dfrac{x[n] + x[n-1]}{2} & \text{(moving average)}, \\ \dfrac{x[n] - x[n-1]}{2} & \text{(first difference)}, \end{cases} \tag{4.113}$$

and showed that they are, respectively, lowpass and highpass. Higher-order moving average filters were also presented, and shown to be narrow band lowpass filters. For each filter the impulse response, transfer function, frequency response, and computational graph were derived. We also distinguished between FIR and IIR filters. The ideal lowpass filter was then presented, and it was shown that its impulse response is a sinc function. This filter is noncausal, unstable, and unrealizable. Following these, simple frequency transformations were presented, for converting a lowpass filter $H(z)$ to other types of filters $G(z)$:

$$G(z) = \begin{cases} H(-z) & \text{(lowpass to highpass)}, \\ H(z^2) & \text{(lowpass to bandstop)}, \\ H(-z^2) & \text{(lowpass to bandpass)}. \end{cases} \tag{4.114}$$

Next, the meaning of phase distortion was explained in detail. Thus, if different frequency components of an input signal are delayed by different amounts by a filter, this results in phase distortion. It was shown that filters with the linear-phase property do not create phase distortion, and that linear phase can be achieved by constraining the impulse response to be symmetric.

The effect of filtering a 2D image with lowpass and highpass filters was demonstrated. Lowpass filters smooth out the edges, whereas highpass filters enhance the edges. The application of filters for signal denoising was demonstrated for both a 1D signal and a 2D image. Finally, it was shown that the convolution operation can be represented as a matrix vector multiplication where the matrix has the Toeplitz structure. The matrix representation also shows us how to undo a filtering operation through a process called deconvolution. The matrix representation was also useful to explain the meaning of steady-state and transient components of filter outputs. A more direct approach to steady states and transients will be presented in Sec. 7.12.

PROBLEMS

Note: If you are using MATLAB for the computing assignments at the end, the "stem" command is convenient for plotting sequences. The "plot" command is useful for plotting real-valued functions such as $|X(e^{j\omega})|$.

4.1 Consider the LTI system described by

$$y[n] = \frac{x[n+1] + 4x[n] + x[n-1]}{6}.$$

This is a moving average system which averages three samples.

(a) What is the impulse response $h[n]$?

(b) What is the frequency response $H(e^{j\omega})$? Handsketch the magnitude response $|H(e^{j\omega})|$ for $-\pi \leq \omega \leq \pi$.

(c) What is the phase response of the system?

(d) Is the system causal?

(e) Is the system stable?

4.2 For causal LTI systems described by the following difference equations, compute the frequency response $H(e^{j\omega})$, transfer function $H(z)$, and impulse response $h[n]$. In each case, specify what kind of filters are described by these equations (lowpass, highpass, bandpass, bandstop, multiband, allpass, etc.). Also specify whether these are FIR or IIR.

(a) $y[n] = \Big[x[n] - x[n-2]\Big]/2.$

(b) $y[n] = \Big[x[n] + x[n-10]\Big]/2.$

(c) $y[n] + 0.5y[n-1] = 0.5x[n] + x[n-1].$

4.3 Figure P4.3 shows the frequency response of a bandpass filter $H(e^{j\omega})$. There is a passband in the negative-frequency region also, because the filter has real impulse response $h[n]$.

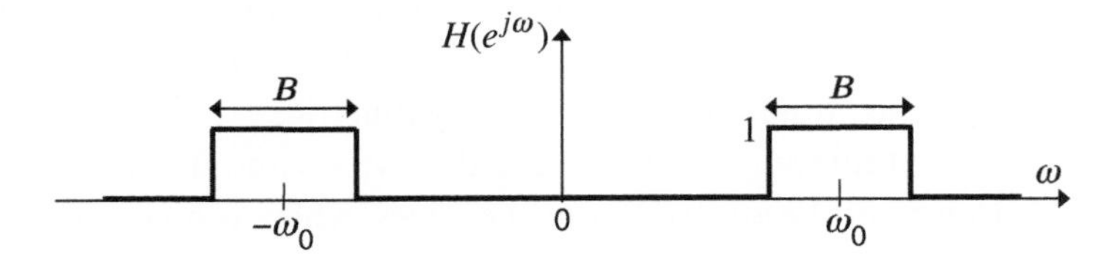

Figure P4.3 A bandpass filter.

(a) Find a closed-form expression for $h[n]$. (*Hint:* Save work by modifying the impulse response of the lowpass filter in Fig. 4.14 using simple properties of DTFT.)

(b) Is this a linear-phase filter?

4.4 For the noncausal LTI system $y[n] = (x[n] + x[n+1])/2$, what are the impulse response $h[n]$ and transfer function $H(z)$? Plot the magnitude response $|H(e^{j\omega})|$ and the phase response $\phi(\omega)$. You will find that this is also a lowpass filter similar to the causal sliding average system (4.15).

4.5 *The second-difference operator or filter.* We know that $y_1[n] = (x[n] - x[n-1])/2$ is the first-difference operator. Now define

$$y_2[n] = \frac{y_1[n] - y_1[n-1]}{2}. \qquad \text{(P4.5)}$$

This is the first difference of the first difference, and is therefore called the second difference of $x[n]$. The system with input $x[n]$ and output $y_2[n]$ is called the second-difference operator.

(a) Write $y_2[n]$ in terms of $x[n]$ by eliminating $y_1[n]$ from the above equation.

(b) What is the impulse response $h_2[n]$ of the second-difference operator, and what is its transfer function $H_2(z)$?

(c) Plot the magnitude response $|H_2(e^{j\omega})|$. You will find it a highpass filter (but with a sharper peak at $\omega = \pi$ than the first-difference operator).

The idea can readily be generalized to obtain third and higher-order differences.

4.6 Consider an LTI system defined by

$$y[n] = \frac{x[n] + e^{j\omega_0}x[n-1]}{2}.$$

What are the impulse response $h[n]$, transfer function $H(z)$, and frequency response? Plot the magnitude response $|H(e^{j\omega})|$ in the range $0 \leq \omega < 2\pi$ for $\omega_0 = 0.4\pi$ and 1.3π. (Note that the first-order moving average filter ($\omega_0 = 0$) and the first-order difference operator ($\omega_0 = \pi$) discussed in Sec. 4.3 are special cases of the above system.)

4.7 For the LTI system in Fig. P4.7, do the following:

(a) Find the (i) impulse response $h[n]$, (ii) transfer function $H(z)$, and (iii) frequency response $H(e^{j\omega})$.

(b) Give a plot of the magnitude response $|H(e^{j\omega})|$. What kind of filter is this (lowpass, highpass, bandpass, bandstop, etc.)?

(c) What is the phase response of the system? Plot it.

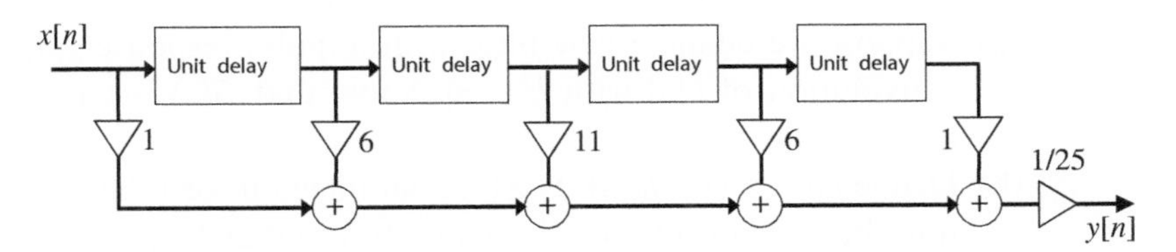

Figure P4.7 An LTI filter.

4.8 For the LTI system in Fig. P4.8, write the transfer function $G(z)$ in terms of $H(z)$ in Fig. P4.7. Give a plot of the magnitude response $|G(e^{j\omega})|$. What kind of filter is this (lowpass, highpass, bandpass, bandstop, etc.)?

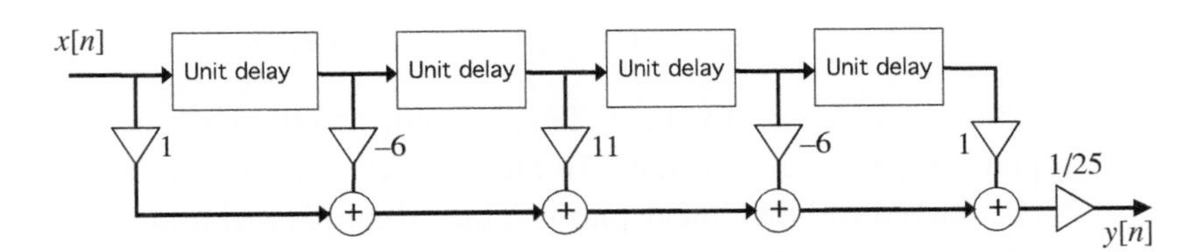

Figure P4.8 A modified LTI filter.

4.9 Find the energy of the signal $x[n] = \sin(\sigma n)/n$, where $0 < \sigma < \pi$.

4.10 Let $H(z) = 1 + 4z^{-1} + z^{-2}$ and consider the following frequency transformed filters: $H(-z)$, $H(z^2)$, and $H(-z^2)$. (a) Plot the magnitude response $|H(e^{j\omega})|$ and the magnitude responses of all the transformed filters. (b) Plot the impulse response of $H(z)$ and that of all the transformed filters.

4.11 Let $G(z)$ be an ideal transfer function such that $G(e^{j\omega}) = 1$ for $0 \leq |\omega| < \pi/4$ and $G(e^{j\omega}) = 0$ for $\pi/4 \leq |\omega| \leq \pi$.

(a) Consider the new system $H(z) = G(z^2)$. Plot $|H(e^{j\omega})|$ for $-\pi \leq \omega \leq \pi$. What kind of filter is this (i.e., lowpass, highpass, etc.)?

(b) Now consider the LTI system constructed according to the flowgraph in Fig. P4.11. What is the transfer function $F(z) = Y(z)/X(z)$?

(c) Plot the magnitude response of $F(z)$ for $-\pi \leq \omega \leq \pi$. What kind of filter is this?

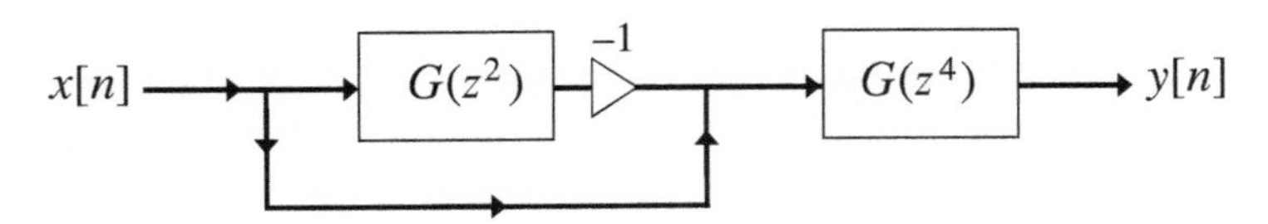

Figure P4.11 An interconnection of LTI systems.

(*Note:* We know that the frequency response of $G(z)$ has period 2π. So the frequency response of $G(z^L)$ has period $2\pi/L$. So draw things carefully.)

4.12 Plot the phase responses of the following filters:

(a) $H_1(z) = 2z^2 + z + 7 + z^{-1} + 2z^{-2}$;

(b) $H_2(z) = 2 + z^{-1} + 7z^{-2} + z^{-3} + 2z^{-4}$.

4.13 Let $h[n]$ be the impulse response of some filter $H(z)$.

(a) Suppose we define a new filter with impulse response $g[n]$, which is the convolution of $h[n]$ with $h^*[-n]$. Show that $G(z)$ is guaranteed to be a linear-phase filter.

(b) Define $f[n] = (h * h)[n]$. Is $F(z)$ guaranteed to be a linear-phase filter? If not, then what is the condition on $H(z)$ so that $F(z)$ has linear phase?

4.14 Consider an FIR filter $H(z) = \sum_{n=0}^{N} h[n]z^{-n}$ with real coefficients $h[n]$ satisfying the antisymmetry property

$$h[n] = -h[N-n]. \tag{P4.14a}$$

Show that the frequency response has the form

$$H(e^{j\omega}) = je^{-j\omega N/2} H_R(\omega), \tag{P4.14b}$$

where $H_R(\omega)$ is real valued. This is similar to the response (4.88) of a linear phase filter, except for the j. In fact, a filter with the response (P4.14b) is also considered to have linear phase. Thus, if $h[n]$ is real and antisymmetric as in (P4.14a) (instead of symmetric as in Sec. 4.6), it also leads to linear phase.

Remarks. The response of the first-order difference operator (4.26) is a special case of Eq. (P4.14b) with $H_R(\omega) = \sin(\omega/2)$. The phase response for this was plotted in Fig. 4.9(d). The discontinuity of the phase at $\omega = 0$ is because $H_R(\omega)$ has a zero crossing there, and the sign change alters the phase by π. The phase response of (P4.14b) is $\pi/2 - N\omega/2$ where $H_R(\omega) > 0$ and $-\pi/2 - N\omega/2$ where $H_R(\omega) < 0$. Linear-phase filters like (P4.14b) with the j-factor are required in digital approximations of functions called differentiators and Hilbert transformers (Sec. 9.9.4).

4.15 From Sec. 4.6 and Problem 4.14 we know that if $h[n]$ is real, and has symmetry or antisymmetry, it leads to linear phase. Now consider the FIR filter $H(z) = \sum_{n=0}^{N} h[n]z^{-n}$ and assume $h[n]$ can be complex. If $h[n]$ is Hermitian symmetric, that is,

$$h[n] = h^*[N-n], \tag{P4.15a}$$

then show that

$$H(e^{j\omega}) = e^{-j\omega N/2} H_R(\omega), \tag{P4.15b}$$

where $H_R(\omega)$ is real. Thus, linear phase can be achieved by imposing (P4.15a).

4.16 The most general definition of linear phase is that the frequency response has the form

$$H(e^{j\omega}) = e^{-j\theta} e^{-jK\omega} H_R(\omega) \tag{P4.16a}$$

in $-\pi \leq \omega < \pi$ (and repeating periodically with period 2π), where $H_R(\omega)$ is real and K and θ are real constants. The filters in Sec. 4.6 and in Problems 4.14 and 4.15 are clearly special cases of this.

(a) Show that Eq. (P4.16a) is equivalent to

$$H(e^{j\omega}) = e^{-2j\theta} e^{-2jK\omega} H^*(e^{j\omega}). \tag{P4.16b}$$

(b) Now assume that $2K$ is an integer (because it can be shown that for rational filters Eq. (P4.16b) implies this). Prove now that Eq. (P4.16b) is equivalent to

$$h[n] = ch^*[2K - n], \tag{P4.16c}$$

where $c = e^{-2j\theta}$. The above equation can be regarded as a *generalized Hermitian* property. Real-coefficient filters with $h[n] = \pm h[N - n]$, and the filter satisfying Eq. (P4.15), are clearly special cases of this.

Summarizing, the time-domain symmetry (P4.16c) is equivalent to the linear-phase property (P4.16a) for any *rational* digital filter.

Remark. In this problem we have *not* assumed the filter to be FIR. The result is valid for rational IIR filters as well. It can be shown based on Eq. (P4.16c) that if the filter is IIR then linear phase in particular necessitates that the filter be noncausal (Sec. 6.8). So *there are no causal, IIR, linear-phase filters.*

4.17 *Reciprocal conjugate zeros.* Consider an FIR filter satisfying the generalized Hermitian property $h[n] = ch^*[N - n]$, where $|c| = 1$. From Problem 4.16 we know that this has linear phase in a generalized sense. Show that the transfer function satisfies

$$H(z) = cz^{-N}[H(1/z^*)]^*. \tag{P4.17}$$

This is a generalization of the property (4.89), which was valid only when $h[n]$ was real and $h[n] = h[N - n]$. Equation (P4.17) shows that if z_k is a zero of $H(z)$, then so is $1/z_k^*$. That is, the zeros come in reciprocal conjugate pairs.

4.18 We know that the impulse response $h[n]$ can be defined for any system, whether it is LTI or not (although it is useful only for LTI systems). We also know that the frequency response can be defined only if $x[n] = e^{j\omega n}$ produces $y[n] = c(\omega)e^{j\omega n}$ for some $c(\omega)$ which possibly depends on ω, but not on n. Thirdly, we know that if the system is not LTI, then even if the frequency response can be defined, $c(\omega)$ may not be the Fourier transform of $h[n]$ (see the discussion following Eq. (3.180)). Now assume that, for a particular system under analysis, $c(\omega)$ does indeed happen to be the Fourier transform of $h[n]$. Does this mean that the system is LTI? The answer is "not necessarily," as shown by the following example. Consider a system (call it System 1) defined as follows:

$$y[n] = \begin{cases} 0 & \text{if } x[n-1] = 0, \\ 0 & \text{if } x[n-2] = 0, \\ 0 & \text{if } x[n]x[n-2] = [x[n-1]]^2, \\ 1 & \text{otherwise.} \end{cases}$$

(a) If $x[n] = \delta[n-i]$ for some integer i, what is the output $y[n]$? In particular, what is the impulse response $h[n]$?
(b) If $x[n] = e^{j\omega_0 n}$, show that $y[n] = c(\omega_0)e^{j\omega_0 n}$ for some function $c(\omega)$. Is this $c(\omega)$ the Fourier transform of $h[n]$ calculated in part (a)?
(c) If $x[n] = \delta[n] + \delta[n-1] + \delta[n-2]$, then what is the output $y[n]$?
(d) Based on your answers to the preceding, argue that System 1 is nonlinear.
(e) Now consider the parallel connection of System 1 with an LTI system with impulse response $g[n]$ and transfer function $G(z)$ (see Fig. P4.18). (i) What are the impulse response $f[n]$ and frequency response (if it can be defined) of this new system? Is the latter the Fourier transform of $f[n]$? (ii) Argue that the system in Fig. P4.18 is nonlinear.

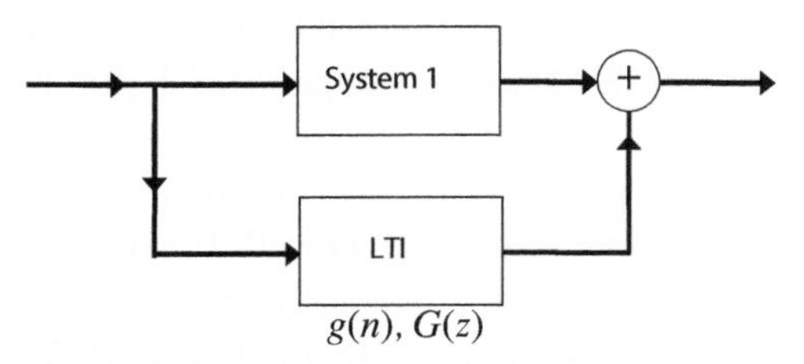

Figure P4.18 A parallel connection of systems for Problem 4.18

4.19 (Computing assignment) *Unusual linear-phase filters.* Consider a digital filter with frequency response

$$H(e^{j\omega}) = e^{-j\alpha\omega} \quad \text{(P4.19)}$$

in $-\pi \leq \omega < \pi$ (and repeating periodically with period 2π), where α is some real number. This is clearly a linear-phase digital filter.

(a) Compute a closed-form expression for the impulse response $h[n]$, in terms of α.
(b) Plot $h[n]$ for $\alpha = 0.0, 0.5$, and 0.75.
(c) Which of these three plots satisfies the symmetry property (P4.16c)?

You will see that not all of these plots satisfy the symmetry (P4.16c). This shows that the linear-phase property (P4.16a) is not in general equivalent to the time-domain symmetry. Such equivalence can be claimed only when $H(z)$

is rational, as proved in Problem 4.16. It turns out that for some choices of α, Eq. (P4.19) represents an irrational transfer function.

4.20 (Computing assignment) *Effect of filtering*. Consider the lowpass filter (4.32). We now perform some actual filtering with this filter, for some inputs. You can use computational tools (e.g., MATLAB) for this. Let $L = 10$.

(a) Assume that the input is $x[n] = \cos(\omega_0 n)\mathcal{U}[n]$, where $\mathcal{U}[n]$ is the unit step. Compute $y[n]$ for the examples $\omega_0 = \pi/20, \pi/10$, and $\pi/5$.

(b) Plot $y[n]$ for $0 \leq n \leq 200$, so you can clearly see the cosine features of the output.

You will find that $y[n]$ looks like the input cosine, but has smaller and smaller amplitudes as ω_0 is increased as above. This is consistent with the lowpass feature of the filter. In fact, for $\omega_0 = \pi/5$ the output is nearly zero because $H(e^{j\omega})$ has a zero at $\omega = \pm\pi/5$. Note also that $y[n]$ does not look like a cosine for the first few samples; this is the "transient response" region (Sec. 7.12).

4.21 (Computing assignment) *Smoothing effect of lowpass filters*. Let $x[n]$ be defined by

$$x[n] = \begin{cases} 1 & \text{for } 0 \leq n \leq 39, \\ 2 & \text{for } 40 \leq n \leq 79, \\ 0.5 & \text{for } 80 \leq n \leq 119, \\ 0 & \text{otherwise,} \end{cases}$$

and let $h[n]$ be the FIR lowpass filter (4.31) with length $L = 10$. Define $f[n] = (h * h)[n]$, which is still a lowpass filter because $F(e^{j\omega}) = H^2(e^{j\omega})$. Let $y[n]$ be the output of $F(z)$ in response to $x[n]$. Plot $x[n]$ and $y[n]$. You will see that the sharp edges in $x[n]$ are smoothed out by this lowpass filter. This is one effect of lowpass filtering: edges get smoothed out. (If using MATLAB, use the "plot" command for this rather than the "stem" command.)

4.22 (Computing assignment) *Edge detectors and difference operators*. Let

$$x[n] = \begin{cases} 1 & \text{for } 0 \leq n \leq 39, \\ 2 & \text{for } 40 \leq n \leq 79, \\ 0.5 & \text{for } 80 \leq n \leq 119, \\ 0 & \text{otherwise,} \end{cases}$$

and let $H_k(z) = (1-z^{-1})^k$, which is the kth-order difference operator. Let $y_k[n]$ be the output of $H_k(z)$ in response to $x[n]$. Plot $x[n]$ and $y_k[n]$ for $k = 1, 2$, and 3. You will see that the filters $H_k(z)$ extract the sharp edges in $x[n]$. That is, $y_k[n]$ are zero in most places, but have large amplitudes near the edges of $x[n]$. Edge detectors are useful in image processing. There exist more sophisticated edge detectors; this problem shows only basic examples.

4.23 (Computing assignment) *Frequency transformations*. Consider the lowpass filter $h[n]$ in Eq. (4.31) and assume $L = 10$. Plot the frequency responses of $H(z)$ and the frequency-transformed filters $H(-z)$, $H(z^2)$, and $H(-z^2)$. Clearly label the filter types (lowpass, highpass, etc.).

4.24 (Computing assignment) *Phase distortion.* Verify the plots given in Fig. 4.20 by starting from Eqs. (4.65), (4.66) and using $\omega_1 = 0.03\pi, \omega_2 = 0.05\pi$, $D_1 = 3$, $D_2 = 23$. Also, generate these plots when D_2 is changed to $D_2 = 11$.

4.25 (Computing assignment) *Relaxed lowpass filter.* Consider the relaxed ideal lowpass digital filter shown in Fig. P4.25. Here, $0 < \alpha < \beta < \pi$ and the response is symmetric with respect to zero frequency. The transition from passband to stopband is gradual (and not discontinuous as in a brick-wall lowpass filter (Fig. 4.14), which has $\alpha = \beta$).

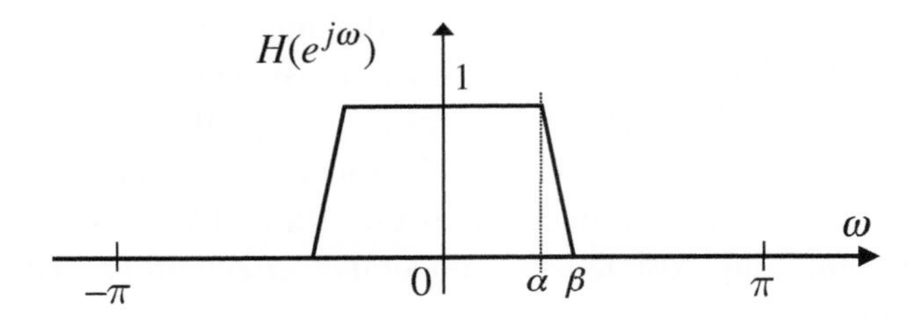

Figure P4.25 A relaxed lowpass filter.

(a) Derive a closed-form expression for the impulse response $h[n]$ of this filter. (*Hint:* You can regard $H(e^{j\omega})$ as a convolution of two rectangular pulses of different widths.)

(b) Show that this is a stable filter, that is, $\sum_n |h[n]| < \infty$. (Remember that $0 < \alpha < \beta < \pi$. If $\alpha = \beta$, $h[n]$ is a sinc, which is unstable.)

(c) Let $\alpha = 0.3$. Plot $h[n]$ for $\beta = 0.31$, $\beta = 0.35$, and $\beta = 0.4$. Make each plot for the range $-150 \leq n \leq 150$. (You can use any platform, say MATLAB, to do this.)

5 Recursive Difference Equations

5.1 Introduction

In previous chapters we studied the properties of linear time-invariant (LTI) systems, and showed how simple instances of lowpass and highpass filtering can be implemented with the help of difference equations (d.e.). In most cases these filters had finite-duration impulse responses, so the systems were referred to as FIR systems. The difference equation describing such an **FIR system** has the form

$$y[n] = \sum_{k=0}^{N} h[k]x[n-k], \tag{5.1}$$

assuming the system is causal for simplicity. That is, the output $y[n]$ at any time n is a linear combination of the present and the past N samples of the input $x[n]$, for some finite integer N. This chapter discusses a generalization of this system, namely

$$y[n] = -\sum_{k=1}^{N} b_k y[n-k] + \sum_{k=0}^{N} a_k x[n-k]. \tag{5.2}$$

In this system, the output at time n depends not only on the present and past N input samples $x[n-k]$, but also on the past N output samples

$$y[n-1], y[n-2], \ldots, y[n-N].$$

Because of the dependence of $y[n]$ on past output samples, this system is said to be a **recursive** difference equation or **recursive d.e.**, with **coefficients** a_k and b_k. Equation (5.1), on the other hand, is **nonrecursive**, that is, $y[n]$ does not depend on past outputs. Recursive systems have infinitely long impulse responses as we shall see, so they are **IIR** filters.

The integer N is said to be the **order** of the difference equation, assuming that Eq. (5.2) is **irreducible** (i.e., it cannot be rewritten using fewer coefficients, say due to some cancellations (Sec. 5.8.1)). Recursive difference equations have some beautiful properties, which make them very useful in digital signal processing, as we shall see throughout the book.

5.2 First-Order Recursive Difference Equations

We begin with an example of the first-order recursive difference equation

$$y[n] = py[n-1] + x[n]. \tag{5.3}$$

Here, $N = 1, b_1 = -p$, and $a_0 = 1$. All other coefficients a_k and b_k are zero. Equation (5.3) states that if $x[n]$ and $y[n-1]$ are known for some n, we can calculate the current output $y[n]$. Thus, starting from any initial time, say $n = 0$, it follows that

$$\begin{aligned} y[0] &= py[-1] + x[0], \\ y[1] &= py[0] \quad + x[1], \\ y[2] &= py[1] \quad + x[2], \end{aligned} \tag{5.4}$$

and so on. Thus, given any value for $y[-1]$, and any set of values $x[n], n \geq 0$, we can compute $y[n]$ for $n \geq 0$. We usually assume that the input starts at $n = 0$, that is,

$$x[n] = \begin{cases} 0 & \text{for } n < 0, \\ \text{possibly nonzero} & \text{for } n \geq 0, \end{cases} \tag{5.5}$$

and start computing the output $y[n]$ from $n = 0$.

5.2.1 The Initial Condition

The initial output $y[-1]$ is referred to as the **initial condition (IC)** or **initial state**. From Eq. (5.3) it is clear that the output $y[n]$ gets nonzero contributions from two sources. One is the input (5.5) and the other is the initial condition $y[-1]$. Most importantly, notice that even if the input signal is zero,

$$x[n] = 0, \quad -\infty < n < \infty, \tag{5.6}$$

the output can be nonzero if the initial condition $y[-1] \neq 0$. This is because

$$y[0] = py[-1], \quad y[1] = py[0] = p^2 y[-1], \quad y[2] = py[1] = p^3 y[-1], \quad \ldots, \tag{5.7}$$

so that $y[n] = p^{n+1} y[-1] \neq 0$ for all $n \geq 0$. So a nonzero output can be sustained without the help of any input!

Now recall from Sec. 2.9 that if a system $x[n] \longmapsto y[n]$ is *linear*, then *zero input produces zero output*. So, if we want the recursive system (5.3) to behave as a linear system, it is necessary to set

$$y[-1] = 0. \tag{5.8}$$

This is called the **zero initial condition** assumption. It will be shown later (Sec. 5.7.1) that the Nth-order recursive difference equation (5.2) behaves as a linear system $x[n] \longmapsto y[n]$ for an input satisfying Eq. (5.5) if and only if

$$y[-1] = y[-2] = \cdots = y[-N] = 0 \quad \text{(zero IC assumption).} \tag{5.9}$$

Note that there are N initial conditions $y[-k]$ in the Nth-order case. We will see (Sec. 5.7.2) that the system is also *time invariant* under the assumption (5.9), so that under the zero IC assumption, the system (5.2) is LTI. For now, we will accept this conclusion, and proceed with our calculations for the first-order example (5.3).

5.2.2 Impulse Response and Transfer Function of Eq. (5.3)

Assuming then that the input starts from zero time as in Eq. (5.5) and that $y[-1] = 0$, we now compute the impulse response of the LTI system (5.3). Setting $x[n] = \delta[n]$ and $y[-1] = 0$ into Eq. (5.4), we have

$$\begin{aligned} y[0] &= 1, \\ y[1] &= py[0] + 0 = p, \\ y[2] &= py[1] + 0 = p^2, \\ y[3] &= py[2] + 0 = p^3, \end{aligned} \tag{5.10}$$

and so on, which shows that the impulse response is

$$h[n] = p^n \mathcal{U}[n]. \tag{5.11}$$

Since $h[n] = 0$ for $n < 0$, the system is **causal** (Sec. 3.3). This is not surprising because of the very nature of computation dictated by Eq. (5.3). Namely, $y[n]$ never depends on $x[n + k]$ for $k > 0$. Since the impulse response (5.11) has infinite duration, the LTI system (5.3) is an **IIR** filter.

Is causality inherent? An important point here is that the recursive system of Eq. (5.3) does not "inherently" represent a causal system. Causality of Eq. (5.3) arose because we assumed that the input was zero before a certain time and computed the output recursively for *future* times, starting from that time. On the other hand, Eq. (5.3) can be rearranged in such a way that the computations are carried out backwards in time, leading to an anticausal system (Eq. (6.76)). This will be explained in Chapter 6 (Sec. 6.7). By contrast, the nonrecursive system of Eq. (5.1) is inherently causal, as $y[n]$ depends only on present and past inputs $x[n - k]$.

The beauty of a recursion. For a causal IIR filter with input $x[n]$ starting from $n = 0$, the convolution sum for the output gives

$$y[n] = \sum_{k=0}^{n} h[k]x[n-k], \quad n \geq 0. \tag{5.12}$$

It follows from the right-hand side that as n increases, the computation of each output sample $y[n]$ requires more and more multiplications and additions. However, for IIR filters described by recursive difference equations such as Eq. (5.2), we can use recursive computation instead of Eq. (5.12). This requires only about $2N$ multiplications and additions per output sample! The recursion (5.2) does the clever job of *storing the past output samples* $y[n-1], y[n-2], \ldots, y[n-N]$, to save computations. The stored samples neatly summarize information about the infinite past required to compute the output $y[n]$. ▽ ▽ ▽

5.2.2.1 Transfer Function of Eq. (5.3)

With the impulse response given by Eq. (5.11), we can now find the transfer function $H(z)$. Since $H(z)$ is the z-transform of $h[n]$, it follows that

$$H(z) = \sum_{n=-\infty}^{\infty} h[n]z^{-n} = \sum_{n=0}^{\infty} p^n z^{-n} = \sum_{n=0}^{\infty} (p/z)^n = \frac{1}{1 - pz^{-1}}, \tag{5.13}$$

as long as $|p/z| < 1$, that is,

$$|z| > |p|. \tag{5.14}$$

In Eq. (5.13) we have used the fact that $\sum_{n=0}^{\infty} \alpha^n$ converges if and only if $|\alpha| < 1$ (Eq. (3.91)). We now notice a striking thing. Unlike most examples in Chapter 4, the transfer function (5.13) is not a polynomial in z or z^{-1}. Rather, it is a *ratio of two polynomials* in z^{-1}, namely,

$$H(z) = \frac{A(z)}{B(z)}, \tag{5.15}$$

where $A(z) = 1$ (a trivial polynomial) and $B(z) = 1 - pz^{-1}$. So we say that $H(z)$ is a **rational** function of z^{-1}. Equivalently, it is a rational function of z because we can always rewrite

$$H(z) = \frac{z}{z - p}. \tag{5.16}$$

For convenience, we also refer to Eq. (5.13) as a **rational LTI** system.

5.2.2.2 Poles and Zeros of Eq. (5.13)

The denominator of $H(z)$ above becomes zero at the point $z = p$, and we say that p is a **pole** of the LTI system. The rational transfer functiton $H(z)$ "blows up" at the pole $z = p$. Next, since the numerator in Eq. (5.16) becomes zero at $z = 0$, we say that the system has a **zero** at $z = 0$. We shall explain the physical meanings of poles and zeros later.

For discrete-time systems, we regard a pole p to be "**nontrivial**" only when $p \neq 0$ and $p \neq \infty$. The same remark holds for zeros. For example, the FIR system $H(z) = z^{-1}$ has a pole at $z = 0$, and the FIR system

$$H(z) = z^{-1} + z \tag{5.17}$$

has a pole at $z = 0$ and a pole at $z = \infty$. These are counted as poles, but they are regarded as trivial poles. FIR filters have no "nontrivial" poles. So we say that FIR LTI systems are **all-zero** filters.

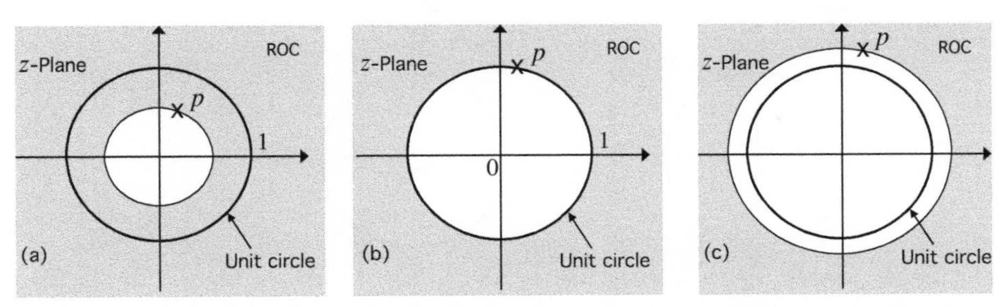

Figure 5.1 (a) – (c) The region of convergence for a causal system with transfer function $H(z) = 1/(1 - pz^{-1})$, for three pole positions (indicated by crosses).

5.2.2.3 Region of Convergence

The region (5.14) in which the infinite sum for $H(z)$ converges to a rational expression is called the **region of convergence** or **ROC** of the transfer function. The ROC is shown in Fig. 5.1 for various values of the pole p (indicated with a cross) and is the region outside the indicated circle with radius $|p|$. The unit circle is also shown for reference in each case. Recall that for FIR filters, $H(z)$ is meaningful for all z except possibly $z = 0$ and $z = \infty$. But the IIR transfer function expression (5.13) is valid only in the ROC (5.14).

5.2.2.4 Stability

Recall that an LTI system is stable if and only if the impulse response $h[n]$ is absolutely summable (Eq. (3.18)). For the example above, since $h[n]$ is given by Eq. (5.11), we have

$$\sum_{n=-\infty}^{\infty} |h[n]| = \sum_{n=0}^{\infty} |p|^n,$$

which converges to a finite value if and only if

$$|p| < 1. \tag{5.18}$$

Under this condition,

$$\sum_{n=-\infty}^{\infty} |h[n]| = \sum_{n=0}^{\infty} |p|^n = \frac{1}{1 - |p|}, \tag{5.19}$$

where we have used Eq. (3.91). Thus, the LTI system described by Eq. (5.3) is causal, and is stable if and only if Eq. (5.18) holds, that is, if and only if the **pole is inside the unit circle** of the z-plane. Figure 5.2 shows three pole positions in the z-plane, with the unit circle indicated for reference. The impulse response decays to zero as $n \to \infty$ only when $|p| < 1$ (pole is inside the unit circle). When the pole is outside, $h[n]$ is unbounded and is clearly unstable. When the pole is *on* the unit circle, although $h[n]$ is bounded, the sum $\sum_n |h[n]|$ does not converge.

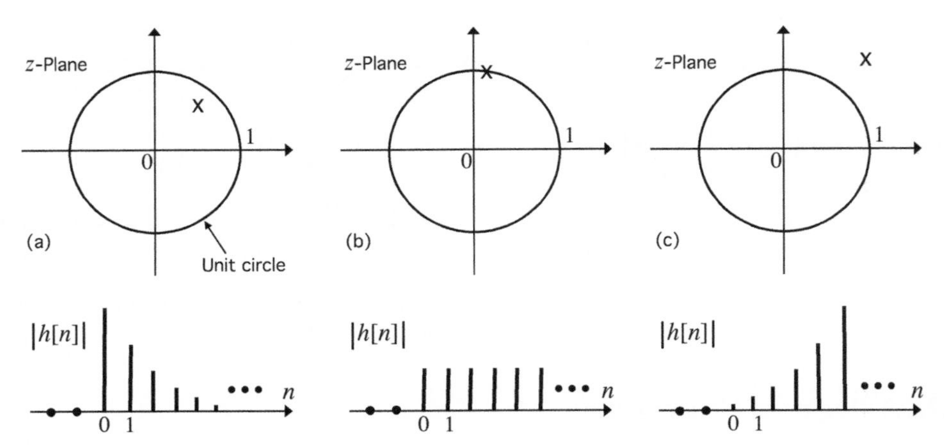

Figure 5.2 Three examples of the pole p for the recursive difference equation (5.3) which has $h[n] = p^n\mathcal{U}[n]$. The pole (indicated by a cross) is (a) inside the unit circle, (b) on the unit circle, and (c) outside the unit circle. Only the first one corresponds to a stable system.

5.3 Frequency Response for Eq. (5.13)

For the transfer function (5.13), the ROC includes the unit circle (Fig. 5.1(a)) if $|p| < 1$, otherwise it does not. Since $H(z)$ evaluated on the unit circle is the frequency response $H(e^{j\omega})$ (Sec. 3.4.2), it follows from the above that the *frequency response exists if and only if the pole is inside the unit circle*, that is, if and only if the system is stable. This is a very important conclusion and will be generalized later (Secs. 6.4.3 and 6.4.4). Since $H(z) = 1/(1 - pz^{-1})$, it follows that the frequency response, when it exists, is

$$H(e^{j\omega}) = \frac{1}{1 - pe^{-j\omega}}, \tag{5.20}$$

which is in general complex even if p is real. For the special case where p is real, we can rewrite this as

$$H(e^{j\omega}) = \frac{1}{1 - p\cos\omega + jp\sin\omega} = \frac{e^{j\phi(\omega)}}{\sqrt{1 + p^2 - 2p\cos\omega}}, \tag{5.21}$$

where

$$\phi(\omega) = -\tan^{-1}\left(\frac{p\sin\omega}{1 - p\cos\omega}\right). \tag{5.22}$$

Thus the magnitude response is

$$|H(e^{j\omega})| = \frac{1}{\sqrt{1 + p^2 - 2p\cos\omega}}, \tag{5.23}$$

and the phase response is as in Eq. (5.22). Since $|p| < 1$ for stability, the pole in the real case is restricted to be $-1 < p < 1$. Figure 5.3 shows plots of the magnitude response $|H(e^{j\omega})|$ and phase response $\phi(\omega)$ for the two cases $0 < p < 1$ and

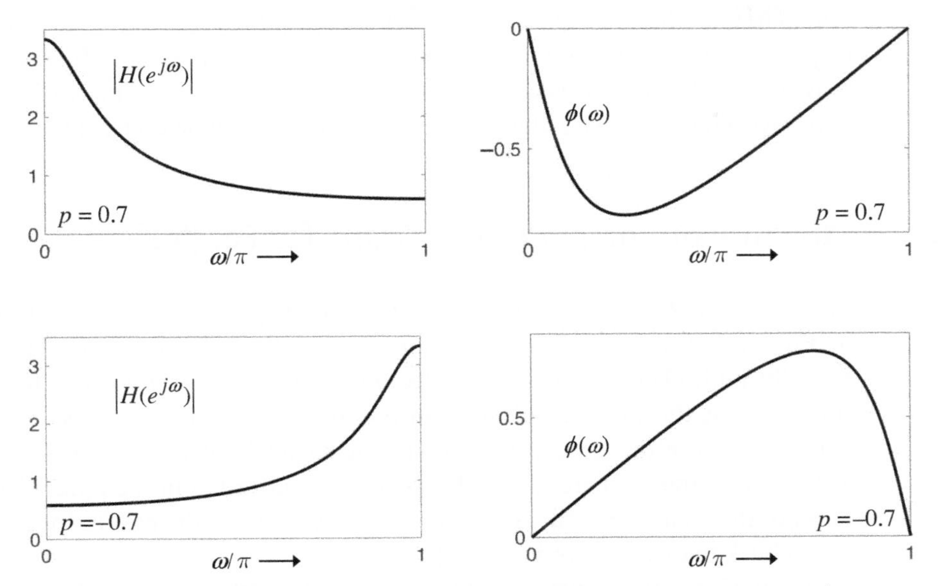

Figure 5.3 The magnitude response and phase responses of $H(z) = 1/(1 - pz^{-1})$ for two different values of the pole p.

$-1 < p < 0$. It is readily verified (Problem 5.7) that Eq. (5.23) has extrema only at $\omega = 0, \pi$. It follows that

$$H(z) = \begin{cases} \text{lowpass} & \text{for } 0 < p < 1, \\ \text{highpass} & \text{for } -1 < p < 0. \end{cases} \tag{5.24}$$

Since $h[n]$ is real for real p, the magnitude response $|H(e^{j\omega})|$ is even and $\phi(\omega)$ is odd (Sec. 3.6), so we have plotted them only for $0 \leq \omega \leq \pi$. Notice also that we can always scale the filter to *normalize* the peak passband response to unity, if necessary, by defining $H(z) = c/(1 - pz^{-1})$ for appropriate c.

Non-existence of frequency response. What does it mean for the frequency response to "not exist"? This could be confusing to some readers, and deserves an explanation. The infinite sum in Eq. (5.13) converges to the expression $1/(1 - pz^{-1})$ only in the ROC. If the unit circle ($z = e^{j\omega}$) is not in the ROC, then $\sum_n h[n]e^{-j\omega n}$ does not converge, so the frequency response does not exist. True, even in this case, if we substitute $z = e^{j\omega}$ into $1/(1 - pz^{-1})$ we will get the expression

$$\frac{1}{1 - pe^{-j\omega}}, \tag{5.25}$$

which can be evaluated for all ω. But this expression does *not* represent the infinite sum $\sum_n h[n]e^{-j\omega n}$ in this case, for the latter does not converge anyway. So, Eq. (5.25) is not the frequency response of the causal LTI system (5.3) when $|p| > 1$. We will provide more insights into this in Sec. 6.4.4. Also see "Connection to stability of $H(z)$" in Sec. 7.12.1.1. ▽ ▽ ▽

Summarizing, the causal IIR filter described by $y[n] = py[n-1] + x[n]$ has impulse response $h[n] = p^n\mathcal{U}[n]$ and transfer function $H(z) = 1/(1 - pz^{-1})$. It is stable if and only if $|p| < 1$, and in this case the frequency response is as in Eq. (5.20). For real p, the system is a lowpass filter for $0 < p < 1$, and a highpass filter for $-1 < p < 0$.

5.4 Structure or Computational Graph for Eq. (5.3)

Figure 5.4 shows the structure or computational graph for the system described by Eq. (5.3). The past value of the output, $y[n-1]$, is stored by the delay element z^{-1}. The multiplier p therefore produces the output $py[n-1]$. Adding together the current input $x[n]$ and the multiplier output therefore produces $y[n]$ as shown. Compared to most of the structures for LTI systems in Chapter 4, there is an important difference. Namely, there is a **feedback loop**, as indicated in the figure. This is because of the recursive nature of Eq. (5.3), which requires the past output $y[n-1]$ to be fed back in order to compute $y[n]$.

Summarizing, the recursive system of Eq. (5.3) is different from the nonrecursive systems presented in Chapter 4 in many ways:

1. It has an infinite impulse response $h[n] = p^n\mathcal{U}[n]$.
2. The transfer function $H(z) = 1/(1 - pz^{-1})$ is a rational function rather than a polynomial.
3. The structure for its implementation has a feedback loop.
4. $H(z)$ has a nontrivial pole p in the z-plane.

From Fig. 5.4 we see that the delay element stores the previous output $y[n-1]$. So, delay elements are like *storage* or **memory** elements. At zero time, the output of the delay element is $y[-1]$. That is, the initial condition is initially stored in the delay element. Zero IC simply means that the content of the storage element is zero when we start the system. We will see (Sec. 5.9.1) that when the structure gets more complicated than Fig. 5.4, the contents of the delay elements are not necessarily equal to $y[n-k]$. However, zero IC (i.e., Eq. (5.9)) can still be achieved by ensuring that *all the delay elements have zero content* when we start the system.

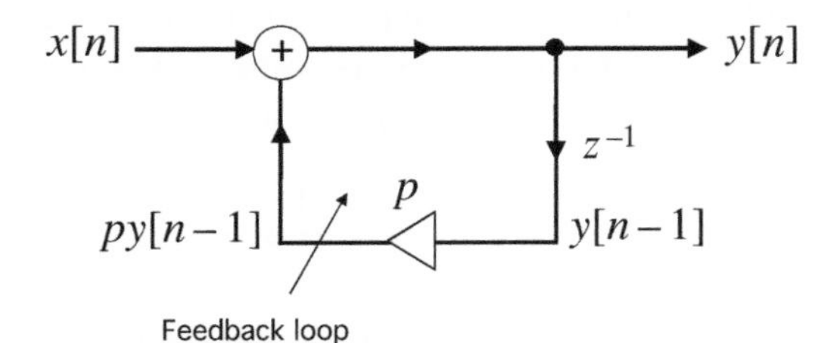

Figure 5.4 The computational graph or "structure" for the first-order recursive difference equation (5.3). With zero initial conditions this behaves as an LTI system with transfer function $H(z) = 1/(1 - pz^{-1})$.

5.4.1 Revisiting the Accumulator

For the special case of $p = 1$, the structure of Fig. 5.4 corresponds to the difference equation

$$y[n] = y[n-1] + x[n]. \tag{5.26}$$

This is the first-order recursive system (5.3), with pole $p = 1$. So the impulse response and transfer function are given by

$$h[n] = \mathcal{U}[n], \quad H(z) = \frac{1}{1 - z^{-1}}. \tag{5.27}$$

Using the convolution sum $y[n] = \sum_k h[k]x[n-k]$, it is clear that the output can also be expressed as

$$y[n] = \sum_{k=0}^{\infty} x[n-k]. \tag{5.28}$$

That is, $y[n]$ is the sum of the present and all the past samples of the input. This is precisely the *accumulator* discussed earlier (see Eq. (2.136)). The recursive implementation (5.26) (equivalently, Fig. 5.4 with $p = 1$) is efficient because it takes only one addition per output sample, unlike Eq. (5.28), which takes an infinite number of additions.

Leaky accumulator. Since Eq. (5.27) is an LTI system with a pole on the unit circle, the accumulator is unstable. We already know this from Sec. 2.11.1, where we saw examples of bounded inputs which produce unbounded accumulator outputs. Another undesirable property of this accumulator is that it remembers all past samples with equal weight, as seen from Eq. (5.28). A more practical accumulator would incorporate a **forgetting factor** so that progressively older samples are forgotten more and more. This is precisely what is done by Eq. (5.3), reproduced below:

$$y[n] = py[n-1] + x[n]. \tag{5.29}$$

Since $h[n] = p^n \mathcal{U}[n]$, Eq. (5.28) is replaced with

$$y[n] = \sum_{k=0}^{\infty} p^k x[n-k] = x[n] + px[n-1] + p^2 x[n-2] + p^3 x[n-3] + \cdots. \tag{5.30}$$

Assuming $|p| < 1$, the system is stable. In this case the weights p^k decay as k increases. So the older inputs $x[n-k]$ get less and less weight as k increases. This is the forgetting factor achieved by the pole p. This accumulator with a forgetting factor can be regarded as a leaky accumulator (i.e., the memory of the past leaks away slowly). ▽ ▽ ▽

From Sec. 2.11.2 you will recall that an ideal capacitor is an unstable integrator but a lossy (practical) capacitor is a stable integrator. The leaky accumulator is analogous to the lossy capacitor.

5.5 Does the Starting Time have to be $n = 0$?

In our analysis of the recursive difference equation (5.3), we assumed that the input starts at $n = 0$ (see Eq. (5.5)), and that the output $y[-1]$ before that was zero (zero IC assumption). However, the choice $n = 0$ is arbitrary, and there is nothing special about it. We can choose the starting time to be an arbitrary integer, say n_0, so that

$$x[n] = \begin{cases} 0 & \text{for } n < n_0, \\ \text{possibly nonzero} & \text{for } n \geq n_0. \end{cases} \tag{5.31}$$

The "zero initial condition assumption" now becomes

$$y[n_0 - 1] = 0 \tag{5.32}$$

or, more generally for the Nth-order difference equation (5.2),

$$y[n_0 - 1] = y[n_0 - 2] = \cdots = y[n_0 - N] = 0. \tag{5.33}$$

Once the impulse response $h[n]$ has been found by applying $\delta[n]$ with the zero IC $y[-k] = 0$ for $1 \leq k \leq N$, the convolution expression $y[n] = \sum h[k]x[n-k]$ holds for input $x[n]$ starting at any n_0 as long as Eq. (5.33) holds. In particular, with $n_0 \to -\infty$ we can say that the input $x[n] = a^n$ produces the output $H(a)a^n$ provided, of course, that $a \in$ ROC of $H(z)$. Thus, once $h[n]$ and $H(z)$ have been found we can apply the theory to inputs like

$$x[n] = e^{j\omega_0 n}, \tag{5.34}$$

which start at $n = -\infty$, and claim that the output is

$$y[n] = H(e^{j\omega_0})e^{j\omega_0 n} \tag{5.35}$$

for *all* n, as we did in Chapter 3. Since the starting time $n_0 = 0$ simplifies the notation, we will implicitly assume that this is the case whenever we refer to "starting time." As there is no loss of generality in this, no further reference to n_0 will be made.

5.6 Nth-Order Recursive Difference Equations

We now generalize the ideas of the preceding sections to the case of the Nth-order recursive difference equation reproduced below:

$$y[n] = -\sum_{k=1}^{N} b_k y[n-k] + \sum_{k=0}^{N} a_k x[n-k]. \tag{5.36}$$

If we know the initial conditions

$$y[-1], y[-2], \ldots, y[-N], \tag{5.37}$$

then we can compute the output $y[n]$ uniquely for $n \geq 0$ recursively, assuming the input $x[n]$ starts at $n = 0$. For example, with $N = 2$, we compute

$$\begin{aligned}
y[0] &= -b_1 y[-1] - b_2 y[-2] \;+\; a_0 x[0], \\
y[1] &= -b_1 y[0] - b_2 y[-1] \;+\; a_0 x[1] + a_1 x[0], \\
y[2] &= -b_1 y[1] - b_2 y[0] \;+\; a_0 x[2] + a_1 x[1] + a_2 x[0], \\
&\vdots
\end{aligned} \tag{5.38}$$

and so on. Note that starting from the Nth output $y[N]$, two things happen:

1. The initial conditions $y[-k], k > 0$ do not make an explicit appearance in the computations any more, although they implicitly affect the output because they affected $y[N-1], y[N-2]$, and so on.
2. All the $N + 1$ input terms are present.

For convenience, we sometimes define a vector $\mathbf{v}$ containing the initial conditions:

$$\mathbf{v} = [y[-1] \quad y[-2] \quad \cdots \quad y[-N]\,]^T. \tag{5.39}$$

This is also called the initial **state vector** (a column vector by convention). It is clear that the **output** $y[n], n \geq 0$ **is unique** for a given $x[n], n \geq 0$ and given initial conditions $\mathbf{v}$. So the system (5.36) is schematically denoted as a mapping:

$$\{x[n], \mathbf{v}\} \longmapsto y[n], \tag{5.40}$$

where it is implicitly assumed that $n \geq 0$. Recall that if the initial conditions are nonzero, then zero input can produce nonzero output, violating linearity (see the remarks following Eq. (5.7)). We often assume, as in Sec. 5.2, that the input starts at zero time:

$$x[n] = \begin{cases} 0 & \text{for } n < 0, \\ \text{possibly nonzero} & \text{for } n \geq 0, \end{cases} \tag{5.41}$$

and that the initial conditions are zero:

$$y[-1] = y[-2] = \cdots = y[-N] = 0, \quad \text{that is,} \quad \mathbf{v} = \mathbf{0} \quad \text{(zero IC assumption).} \tag{5.42}$$

We will also see interesting applications with nonzero IC (Chapter 8). A number of remarks are in order.

1. Equation (5.36) is sometimes called a **linear**, **constant-coefficient**, recursive difference equation; here the term "linear" just refers to the fact that the equation does not contain nonlinear expressions such as $x^2[n]$, $x[n]x[n-1]$, $y[n-1]x[n]$, and so forth. This also means that the mapping (5.40) is linear, as we shall see in Sec. 5.7.1.
2. But the above does not mean that the mapping

   $$x[n] \longmapsto y[n] \tag{5.43}$$

   is linear in the sense defined in Sec. 2.8.2, because the initial conditions $\mathbf{v}$ may be nonzero.

3. In short, we will see in Sec. 5.7.1 that if we think of Eq. (5.36) as a mapping from $\{x[n], \mathbf{v}\}$ to $y[n]$, it is linear regardless of what the IC is. If we think of it as a mapping from $x[n]$ to $y[n]$, it is linear if and only if the zero IC assumption holds.
4. The term "constant coefficient" means that a_k and b_k do not depend on the time index n; we will see in Sec. 5.7.2 that this property implies time invariance of the system.

5.7 Linearity and Time Invariance*

Under the zero IC assumption, the system (5.36) behaves like an LTI system, and its transfer function will be derived in Sec. 5.8. In this section we explain how linearity and time-invariance properties come about. If the reader prefers to go directly to Sec. 5.8 at this point, with the intention of returning to this section later, that will be fine.

5.7.1 Linearity of the Recursive d.e.

Because linearity can have two subtle shades of meaning, we would like to clarify these here. Consider the recursive difference equation (5.36) again. With the notation (5.40), assume that $\{x_i[n], \mathbf{v}_i\}$ produces the output $y_i[n]$ for $i = 1, 2$:

$$\{x_1[n], \mathbf{v}_1\} \longmapsto y_1[n], \quad \{x_2[n], \mathbf{v}_2\} \longmapsto y_2[n]. \tag{5.44}$$

More explicitly, the following equations are satisfied:

$$y_1[n] = -\sum_{k=1}^{N} b_k y_1[n-k] + \sum_{k=0}^{N} a_k x_1[n-k], \quad n \geq 0, \tag{5.45}$$

$$y_2[n] = -\sum_{k=1}^{N} b_k y_2[n-k] + \sum_{k=0}^{N} a_k x_2[n-k], \quad n \geq 0, \tag{5.46}$$

where $\mathbf{v}_i = [y_i[-1] \;\; y_i[-2] \;\; \cdots \;\; y_i[-N]\,]^T$ are the initial state vectors. If we compute the linear combination $c_1 y_1[n] + c_2 y_2[n]$, we therefore find

$$\begin{aligned} c_1 y_1[n] + c_2 y_2[n] = &-\sum_{k=1}^{N} b_k\Big(c_1 y_1[n-k] + c_2 y_2[n-k]\Big) \\ &+\sum_{k=0}^{N} a_k\Big(c_1 x_1[n-k] + c_2 x_2[n-k]\Big), \; n \geq 0. \end{aligned} \tag{5.47}$$

This shows that Eq. (5.36) is satisfied by the input–output pair

$$x[n] = c_1 x_1[n] + c_2 x_2[n], \quad y[n] = c_1 y_1[n] + c_2 y_2[n], \quad n \geq 0, \tag{5.48}$$

as long as the new initial condition is also similarly adjusted to be the same linear combination:

$$y[-k] = c_1 y_1[-k] + c_2 y_2[-k], \; 1 \leq k \leq N, \quad \text{that is,} \quad \mathbf{v} = c_1 \mathbf{v}_1 + c_2 \mathbf{v}_2. \tag{5.49}$$

In short, Eq. (5.44) implies

$$c_1\{x_1[n], \mathbf{v}_1\} + c_2\{x_2[n], \mathbf{v}_2\} \longmapsto c_1 y_1[n] + c_2 y_2[n], \tag{5.50}$$

for any c_1, c_2, any initial states $\mathbf{v}_1, \mathbf{v}_2$, and any inputs $x_1[n], x_2[n], n \geq 0$ (where $c_i\{x_i[n], \mathbf{v}_i\}$ stands for $\{c_i x_i[n], c_i\mathbf{v}_i\}$). Equation (5.50) shows that the mapping

$$\{x[n], \mathbf{v}\} \longmapsto y[n] \tag{5.51}$$

is linear for any initial condition $\mathbf{v}$. This is why Eq. (5.36) is called a "linear" difference equation. In particular, this implies that *if the IC is zero* then the mapping

$$x[n] \longmapsto y[n] \tag{5.52}$$

is *also* linear.

Is zero IC necessary for linearity of (5.52)? We now show that zero IC is necessary for the mapping (5.52) to be linear. For this, it is enough to show that, with $x[n] = 0$ for all n, we will have $y[n] = 0$ for all $n \geq 0$ only if the zero IC assumption holds. Without loss of generality, assume $N \geq 1$ and $b_N \neq 0$. From Eq. (5.36) we have, in particular,

$$y[N-1] = -b_1 y[N-2] - b_2 y[N-3] - \cdots - b_{N-1} y[0] - b_N y[-1], \tag{5.53}$$

when $x[n] = 0$ for all n. Assume $y[n] = 0$ for all $n \geq 0$. Then the above implies $y[-1] = 0$ as well. Next

$$y[N-2] = -b_1 y[N-3] - b_2 y[N-4] - \cdots - b_{N-1} y[-1] - b_N y[-2]. \tag{5.54}$$

Since $y[N-2] = y[N-3] = \cdots = y[-1] = 0$, it follows that $y[-2] = 0$ as well. Repeating this, we finally obtain $y[-1] = y[-2] = \cdots = y[-N] = 0$. Summarizing, we have shown that when $x[n] = 0$ for all n, we have $y[n] = 0$ for all $n \geq 0$ only if zero IC prevails. So, for linearity of (5.52), zero IC is necessary. ▽ ▽ ▽

Summarizing, if we think of (5.36) as a mapping from $\{x[n], \mathbf{v}\}$ to $y[n]$, it is linear regardless of what the IC is. If we think of it as a mapping from $x[n]$ to $y[n]$, it is linear if and only if the zero IC assumption holds.

5.7.2 Time Invariance of the Recursive d.e.

It will now be shown that the system described by Eq. (5.36) is time invariant in a certain sense. Thus, let the system with input signal

$$x_1[n] = \begin{cases} 0 & \text{for } n < 0, \\ \text{possibly nonzero} & \text{for } n \geq 0, \end{cases} \tag{5.55}$$

produce the output $y_1[n], n \geq 0$ when the initial conditions are specified by a vector $\mathbf{v}$. That is,

$$y_1[n] = -\sum_{k=1}^{N} b_k y_1[n-k] + \sum_{k=0}^{N} a_k x_1[n-k], \quad n \geq 0, \tag{5.56}$$

when

$$[y_1[-1] \quad y_1[-2] \quad \cdots \quad y_1[-N]\,]^T = \mathbf{v}. \tag{5.57}$$

Now define the shifted signal $x_2[n] = x_1[n-K]$, which starts at time $n = K$ (where K is not necessarily positive), and define

$$y_2[n] = y_1[n-K], \; n \geq K. \tag{5.58}$$

Then the pair $(x_2[n], y_2[n])$ satisfies

$$y_2[n] = -\sum_{k=1}^{N} b_k y_2[n-k] + \sum_{k=0}^{N} a_k x_2[n-k], \quad n \geq K, \tag{5.59}$$

provided the new "initial conditions" are given by the same vector $\mathbf{v}$, that is,

$$[y_2[K-1] \quad y_2[K-2] \quad \cdots \quad y_2[K-N]\,]^T = \mathbf{v}. \tag{5.60}$$

To prove this, simply substitute the definitions of $y_2[n]$ and $x_2[n]$ into Eq. (5.59) to show that it just reduces to Eq. (5.56), which we know holds. So Eq. (5.59) must be true as well.

In summary, if we shift the input by K and start the system at time K with the same initial conditions, then the output is also shifted by K. This is the time-invariance property of the difference equation (5.36), and springs from the fact that the coefficients a_k and b_k do not depend on time n. In particular, this proves that under zero IC,

$$x[n] \longmapsto y[n] \qquad \Longrightarrow \qquad x[n-K] \longmapsto y[n-K]. \tag{5.61}$$

That is, the recursive d.e. (5.36) is time invariant in the traditional sense (Sec. 2.8.3) under the zero IC assumption.

5.8 Transfer Function

Assuming zero initial conditions, the recursive d.e. is an LTI system as proved in Sec. 5.7. So we can find its transfer function $H(z)$. For the first-order case ($N = 1$), we found the impulse response first, and took its z-transform to find the transfer function (Sec. 5.2.2). For $N > 1$, however, it is not so easy to find an expression for $h[n]$ by running the recursive d.e. For example, consider Eq. (5.38) where $N = 2$. Setting $x[n] = \delta[n]$ (and $y[-1] = y[-2] = 0$), we can compute the impulse response numerically for as many samples as we want but it is not easy to get a closed-form expression for $h[n]$ from Eq. (5.38).

However, it will be seen next that it is quite easy to find the transfer function $H(z)$ directly from the difference equation, which is reproduced below:

$$y[n] = -\sum_{k=1}^{N} b_k y[n-k] + \sum_{k=0}^{N} a_k x[n-k]. \tag{5.62}$$

Once $H(z)$ has been found, we can find $h[n]$ as explained in Sec. 5.10. To find $H(z)$, assume we apply $x[n] = a^n$ for all n. Then we know that the LTI system has the exponential output $y[n] = H(a)a^n$. Substituting these expressions for $x[n]$ and $y[n]$ into Eq. (5.62), it follows that

$$\begin{aligned} H(a)a^n &= -\sum_{k=1}^{N} b_k H(a)a^{n-k} + \sum_{k=0}^{N} a_k a^{n-k} \\ &= a^n \left(-H(a)\sum_{k=1}^{N} b_k a^{-k} + \sum_{k=0}^{N} a_k a^{-k} \right). \end{aligned} \tag{5.63}$$

Cancelling a^n from both sides and rearranging, we get

$$H(a)\left(1 + \sum_{k=1}^{N} b_k a^{-k}\right) = \sum_{k=0}^{N} a_k a^{-k}, \tag{5.64}$$

so that $H(a) = \sum_{k=0}^{N} a_k a^{-k} / \left(1 + \sum_{k=1}^{N} b_k a^{-k}\right)$. Summarizing, $H(z)$ takes the form

$$H(z) = \frac{\sum_{k=0}^{N} a_k z^{-k}}{1 + \sum_{k=1}^{N} b_k z^{-k}} = \frac{A(z)}{B(z)}, \tag{5.65}$$

where

$$A(z) = \sum_{k=0}^{N} a_k z^{-k} \quad \text{and} \quad B(z) = 1 + \sum_{k=1}^{N} b_k z^{-k}. \tag{5.66}$$

Thus the transfer function of the Nth-order recursive difference equation (5.62) is a ratio of two polynomials in z^{-1}. So we say that it is a **rational function** in z^{-1} (equivalently in z). These systems are therefore called **rational LTI** systems. So an LTI system is rational if the transfer function is as in Eq. (5.65) or equivalently, it can be implemented as a recursive difference equation (5.62). Assuming that Eq. (5.65) is **irreducible** (i.e., cannot be rewritten using smaller N, see Sec. 5.8.1), the integer N is said to be the **order** of the rational LTI system. We now make a few important observations regarding Eq. (5.65).

1. The frequency response of a rational LTI system can readily be found by setting $z = e^{j\omega}$ in Eq. (5.65). Thus

$$H(e^{j\omega}) = \frac{\sum_{k=0}^{N} a_k e^{-j\omega k}}{1 + \sum_{k=1}^{N} b_k e^{-j\omega k}}. \tag{5.67}$$

This exists only when the ROC of $H(z)$ includes the unit circle, similar to what we explained in Sec. 5.3.

2. *Role of coefficients* a_k, b_k. The numerator coefficients of $H(z)$ come from a_k (weights of past inputs $x[n-k]$ in Eq.(5.62)), and the denominator coefficients of $H(z)$ come from b_k (weights of past outputs $y[n-k]$). Thus, the denominator of $H(z)$ is entirely due to the recursive part of Eq. (5.62).
3. *FIR case.* Any causal and FIR (nonrecursive) system

$$H(z) = \sum_{n=0}^{N} h[n]z^{-n} \tag{5.68}$$

is a special case of Eq. (5.65) with $b_k = 0$ and $a_k = h[k]$. Thus, for FIR systems $B(z) = 1$. When $B(z) \neq 1$ the system is IIR (as in Sec. 5.2.2) but it takes some more work to find an expression for the impulse response (Sec. 5.10).
4. *Computation and storage.* Any rational LTI system (5.65) can be implemented as in Eq. (5.62). It therefore requires multiplications (like $b_k y[n-k]$ and $a_k x[n-k]$), additions, and storage (to store past samples like $x[n-k]$ and $y[n-k]$). We will present the computational graph or structure for this later (Sec. 5.9), but it is clear that to implement a rational LTI system it takes only a finite amount of computation and storage (proportional to N) *per output sample.*
5. *Realizable LTI systems.* What is less obvious is this: if a discrete-time LTI system is *realizable*, that is, can be implemented with finite computation and storage (per output sample), then it is necessarily a rational system! That is, it has a transfer function of the form (5.65), or equivalently can be implemented with a difference equation like (5.62). This will be proved in Sec. 14.4.4.
6. *Smoothness of* $H(e^{j\omega})$. For a rational LTI system, since $H(z)$ is a ratio of two polynomials, the frequency response $H(e^{j\omega})$ given by Eq. (5.67) is a **continuous** function of ω, and is **differentiable** any number of times for all ω. For example, if a filter has a response $H(e^{j\omega})$ with some discontinuity (like the ideal lowpass filter), then it is nonrational.

5.8.1 Poles and Zeros

The transfer function $H(z) = A(z)/B(z)$ of the rational LTI system (5.65) is a ratio of the polynomials $A(z)$ and $B(z)$. Now, an Nth-order polynomial has N roots (or zeros). Denoting the zeros of $A(z)$ by z_m, we can write it as a product of first-order factors:

$$A(z) = \sum_{k=0}^{N} a_k z^{-k} = a_0 \prod_{m=1}^{N} (1 - z_m z^{-1}). \tag{5.69}$$

Similarly, denoting the zeros of $B(z)$ by p_m, we can write

$$B(z) = 1 + \sum_{k=1}^{N} b_k z^{-k} = \prod_{m=1}^{N} (1 - p_m z^{-1}). \tag{5.70}$$

So the transfer function of an Nth-order rational LTI system can be written

$$H(z) = \frac{A(z)}{B(z)} = \frac{\sum_{k=0}^{N} a_k z^{-k}}{1 + \sum_{k=1}^{N} b_k z^{-k}} = \frac{a_0 \prod_{m=1}^{N} (1 - z_m z^{-1})}{\prod_{m=1}^{N} (1 - p_m z^{-1})}. \tag{5.71}$$

The zeros z_m of the numerator are called the **zeros** of $H(z)$, and the zeros p_m of the denominator are called the **poles** of $H(z)$. So, at a zero we have $H(z_m) = 0$. At a pole we have $H(p_m) = \infty$, that is, $H(z)$ "blows up." We will see later (Sec. 8.7) that the poles have a number of interesting time-domain interpretations, which offer more physical insights. Notice right away that if b_k are real, then the poles are either real or come in **complex-conjugate pairs**. If a_k are real, the same is true of the zeros. When a_k and b_k are both real, then the impulse response of this LTI system is also real, and we say that it is a real system or a **real filter**.

Notice that we have not assumed z_m to be distinct, and similarly p_m need not be distinct; for example, it is possible that $p_3 = p_4$. One slight loss of generality is that if $a_0 = 0$, then Eq. (5.69) is not valid. Instead, we have to write

$$A(z) = a_M z^{-M} \prod_{m=1}^{N-M} (1 - z_m z^{-1}), \tag{5.72}$$

where a_M is the first nonzero coefficient in $A(z)$, and modify Eq. (5.71) accordingly.

Irreducible form and system order. If we have a pole and a zero at the same place ($z_k = p_m$), then the factors $(1 - z_k z^{-1})$ and $(1 - p_m z^{-1})$ cancel, and N can be reduced. So we normally assume that there are no common uncancelled factors in the rational form $A(z)/B(z)$. In this case we say that $A(z)$ and $B(z)$ are **relatively prime** or **coprime** polynomials, and that $A(z)/B(z)$ is **irreducible**. When $H(z) = A(z)/B(z)$ is irreducible, and at least one of a_N, b_N is nonzero, N is the **order** of $H(z)$. ▽ ▽ ▽

5.8.2 Examples of Poles and Zeros

Consider the rational function $H(z) = (1 + z^{-1})/(1 + 0.5z^{-1})$, which can be rewritten as

$$H(z) = \frac{z + 1}{z + 0.5}. \tag{5.73}$$

This has the pole $p_1 = -0.5$ and the zero $z_1 = -1$, as shown in Fig. 5.5(a). Note that poles and zeros are indicated with small crosses and circles, respectively. If the numerator has higher order in z^{-1} as in

$$H(z) = \frac{1 + z^{-2}}{1 + 0.5z^{-1}} = \frac{z^2 + 1}{z^2 + 0.5z} = \frac{z^2 + 1}{z(z + 0.5)}, \tag{5.74}$$

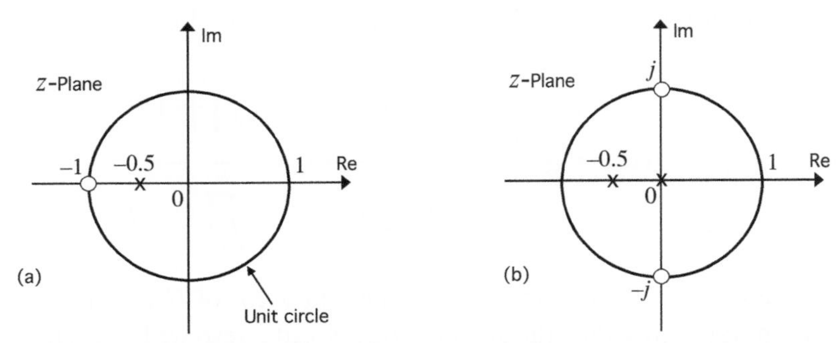

Figure 5.5 Poles and zeros for (a) the system (5.73) and (b) the system (5.74). A pole is indicated by a cross and a zero by a small circle.

then there is a pole at $z = 0$. In this example there is another pole at $z = -0.5$, and zeros at $z = \pm j$, as shown in Fig. 5.5(b). If the denominator has higher order in z^{-1} as in

$$H(z) = \frac{z^{-1} + 2}{1 + 0.5z^{-2}} = \frac{2z^2 + z}{z^2 + 0.5} = \frac{z(2z + 1)}{z^2 + 0.5}, \tag{5.75}$$

then we have a zero at $z = 0$. There could be multiple poles at the same point, and similarly multiple zeros. For example, consider

$$H(z) = \frac{z^{-1}(1 + 2z^{-1})^3(1 + 5z^{-1})}{(1 + 4z^{-1})^2(1 + 0.5z^{-1})}. \tag{5.76}$$

This has the following poles and zeros:

1. Pole of order (or multiplicity) two at $z = -4$.
2. Pole of order one at $z = -0.5$. Also called a "single pole."
3. Pole of order two at $z = 0$ (since $H(z) \to 5z^{-2}$ as $z \to 0$).
4. Zero of order three at $z = -2$.
5. Zero of order one at $z = -5$.
6. Zero of order one at $z = \infty$ (since $H(z) \to z^{-1}$ as $z \to \infty$).

In this example, the number of poles is five, and so is the number of zeros, provided we count multiple occurences of poles and zeros. More generally, *the number of poles and number of zeros of a rational transfer function are always equal* provided we account for multiplicities, and remember to count those at $z = 0$ and $z = \infty$ as well.

FIR case. Notice that causal FIR systems have transfer functions of the form

$$H(z) = \sum_{n=0}^{N} h[n]z^{-n}. \tag{5.77}$$

Thus all the N poles are at $z = 0$. Noncausal FIR systems can have poles at $z = \infty$ as well. Thus

$$H(z) = z^3 + 1 + z^{-2} \tag{5.78}$$

has two poles at $z = 0$ and three at $z = \infty$. As mentioned earlier, we regard a pole p to be "nontrivial" only when $p \neq 0$ and $p \neq \infty$. FIR systems are called all-zero filters because they do not have poles except at $z = 0$ and $z = \infty$. That is, they do not have nontrivial poles. $\nabla\nabla\nabla$

5.8.2.1 A Subtle Point about Zeros

We now mention a subtle point that arises with the definition of zeros of IIR systems. Consider $h[n] = p^n \mathcal{U}[n]$, which has

$$H(z) = h[0] + h[1]z^{-1} + h[2]z^{-2} + \cdots . \tag{5.79}$$

In the ROC $|z| > |p|$, this reduces to the rational form $H(z) = 1/(1 - pz^{-1})$ (see Eq. (5.13)), or equivalently

$$H(z) = \frac{z}{z - p}. \tag{5.80}$$

From Eq. (5.80) it is clear that there is a zero at $z = 0$ and a pole at $z = p$. But if we set $z = 0$ in Eq. (5.79), the right-hand side "blows up." So there seems to be a contradiction! Does it mean that there is also a pole at $z = 0$? The fact is that poles and zeros are defined with respect to the rational form (5.80), and not with respect to the original infinite series (5.79). (In the FIR case this does not make a difference because the series $\sum h[n]z^{-n}$ itself is finite, and becomes the rational form.)

So we should examine Eq. (5.80) (and not Eq. (5.79)) to find out about poles and zeros. Notice that these two expressions are identical only in the ROC. In the non-ROC we simply cannot evaluate Eq. (5.79), as it does not converge. The poles obtained by examining rational forms like (5.80) are clearly not in the ROC. The zeros obtained by examining the rational form (5.80) may or may not be in the ROC, depending on the example. If such a zero ($z_0 = 0$ in our example) is not in the ROC, then we have this strange situation: at $z = z_0$, the rational expression (5.80) becomes zero. But at $z = z_0$ the infinite series "blows up" because z_0 is not in the ROC.

When we are asked to find the poles and zeros, we should therefore use the rational expression (e.g., Eq. (5.80)) rather than the infinite series (5.79).

5.9 Structures or Computational Graphs

For the general rational transfer function (5.65), we now derive the computational graphs or structures similar to the first-order case shown in Fig. 5.4.

5.9.1 Direct-Form Structure

First let us rewrite Eq. (5.65) as

$$H(z) = \frac{\sum_{k=0}^{N} a_k z^{-k}}{1 + \sum_{k=1}^{N} b_k z^{-k}} = \frac{1}{B(z)} \times A(z). \tag{5.81}$$

This can be regarded as a **cascade** of an LTI system with transfer function $1/B(z)$ followed by an LTI system with transfer function $A(z)$. We will therefore derive the structure for $1/B(z)$ and cascade it with that of $A(z)$. Since $A(z)$ is FIR, its structure is easily obtained, as demonstrated in Fig. 5.6(b) for $N = 3$. For $1/B(z)$ note that its input–output description is the recursive difference equation

$$w[n] = -\sum_{k=1}^{N} b_k w[n-k] + x[n], \tag{5.82}$$

where $w[n]$ denotes the output of $1/B(z)$. This is an IIR system because of the polynomial $B(z)$ in the denominator. In the same way that Fig. 5.4 represented the structure of Eq. (5.3), we can draw the computational graph for Eq. (5.82) as shown in Fig. 5.6(a) for $N = 3$. Note the feedback loops, which arise due to the denominator coefficients b_k (i.e., due to the recursive part of the difference equation).

The output $w[n]$ of $1/B(z)$ can be taken as the input to $A(z)$, to produce the final output $y[n]$. That is, we simply have to connect the two structures in parts (a) and (b) to obtain the final structure as shown in Fig. 5.7. Note that the two **delay**

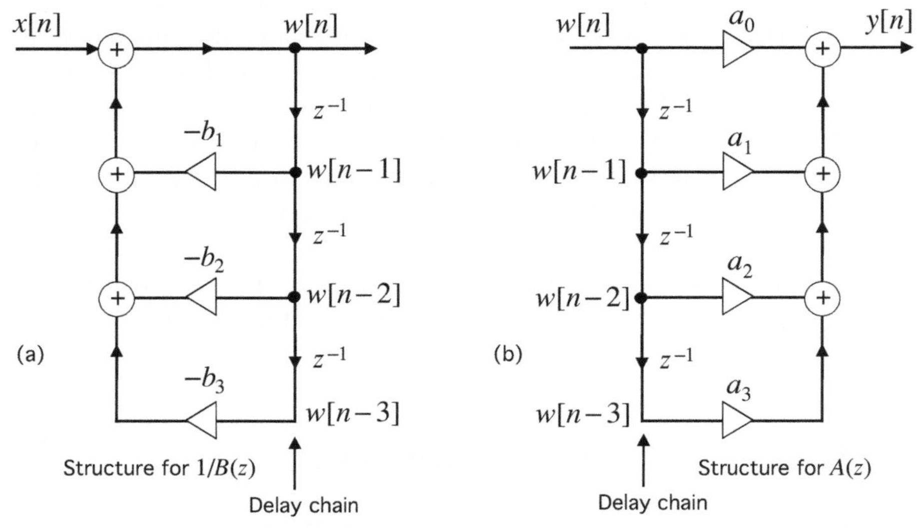

Figure 5.6 The computational graph or structure (a) for $1/B(z)$ and (b) for $A(z)$. See text for details.

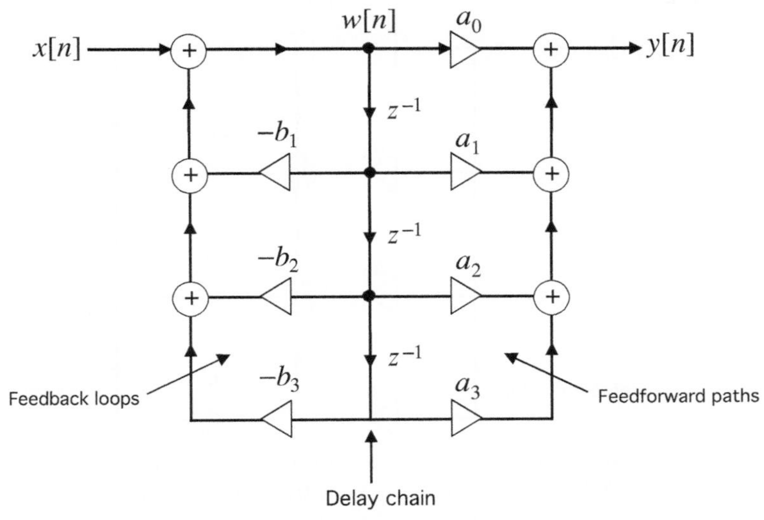

Figure 5.7 The computational graph or structure for the Nth-order rational LTI system (5.81), shown for $N = 3$. This is called the direct-form structure.

chains (cascades of z^{-1}) can be shared because the same signal $w[n]$ and its delayed versions appear in Figs. 5.6(a) and (b). The final structure in Fig. 5.7 is called the **direct-form** structure, because the coefficients a_k and b_k which appear in the transfer function appear directly in the structure as multipliers. Note that the direct-form structure uses

$$2N + 1 \text{ multipliers, } 2N \text{ adders, and } N \text{ delays (storage elements).}$$

This is the minimum **computational** and **storage** complexity required to implement an arbitrary Nth-order rational LTI system with nonzero coefficients a_k, b_k. It should be mentioned that when a multiplier has values such as ± 1 or $\pm j$, it is not counted as a multiplication. Similarly, powers of 2 such as $\pm 2^m$ or $\pm j2^m$, where m is an integer of either sign, are not counted as multiplers, as these can be realized with binary shifts in digital implementations. Such multipliers are referred to as **trivial multipliers**; only nontrivial multiplications are counted towards computational complexity.

Another remark is worthwhile here. We do not see an explicit appearance of the delayed signals $x[n-k]$ and $y[n-k]$ in the structure of Fig. 5.7, even though it implements the recursive difference equation (5.2), which has $x[n-k]$ and $y[n-k]$. Only the delayed versions of the internal signal $w[n]$ make explicit appearence in the structure. This is not unusual; most of the implementations of difference equations do not have $x[n-k]$ and $y[n-k]$ appearing directly in the structure, because of all sorts of rearrangements and simplifications in the flowgraphs. As mentioned earlier, zero IC (i.e., Eq. (5.9)) can still be achieved by ensuring that all the delay elements have zero content when we start the system.

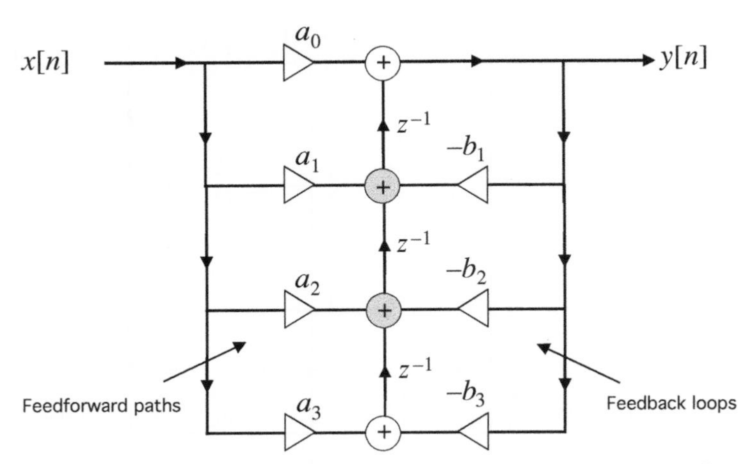

Figure 5.8 The modified direct-form structure. This is a variation of Fig. 5.7, and implements the same transfer function (5.81).

5.9.2 Modified Direct-Form Structure

A modified version of the direct-form structure is shown in Fig. 5.8. It is not hard to verify (Problem 5.19) that this implements the same transfer function (5.81), or equivalently the difference equation (5.2). Notice that the delay chain (cascade of delay elements) is oriented upwards in this modified structure. Unlike Fig. 5.7, the feedforward loops are on the left and the feedback loops on the right. It appears that we only have four adders in this structure, whereas Fig. 5.7 had six adders. But we have to be careful. All the adders in Fig. 5.7 are two-input adders, whereas there are some adders in Fig. 5.8 with three inputs (shown in gray). Each three-input adder is equivalent to two two-input adders, so both structures have the same number of two-input adders. Thus, both structures have the same computational and storage complexity. In Sec. 5.14 we will present an application which shows the advantage of this structure over the structure of Fig. 5.7.

5.9.3 Cascade-Form Structures

Returning to Eq. (5.71), it is clear that $H(z)$ can be written as the product

$$H(z) = a_0 H_1(z) H_2(z) \cdots H_N(z), \tag{5.83}$$

where

$$H_m(z) = \frac{1 - z_m z^{-1}}{1 - p_m z^{-1}}. \tag{5.84}$$

Thus, $H(z)$ can also be implemented as a cascade of first-order sections $H_m(z)$, where each $H_m(z)$ is implemented using a direct-form structure. This is called a cascade-form structure. See Fig. 5.9 for an example with $N = 2$.

Real filters. For *real-coefficient* filters, the impulse response $h[n]$ is real. A real-coefficient rational filter can be realized using real-valued a_n and b_n. In this case, if

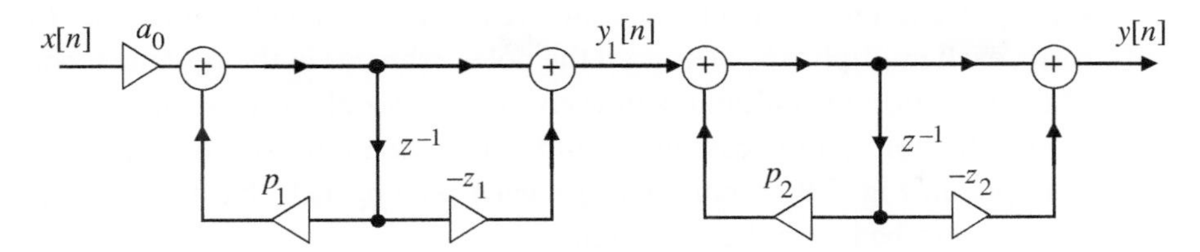

Figure 5.9 Cascade-form structure for Eq. (5.83) with two first-order sections (5.84).

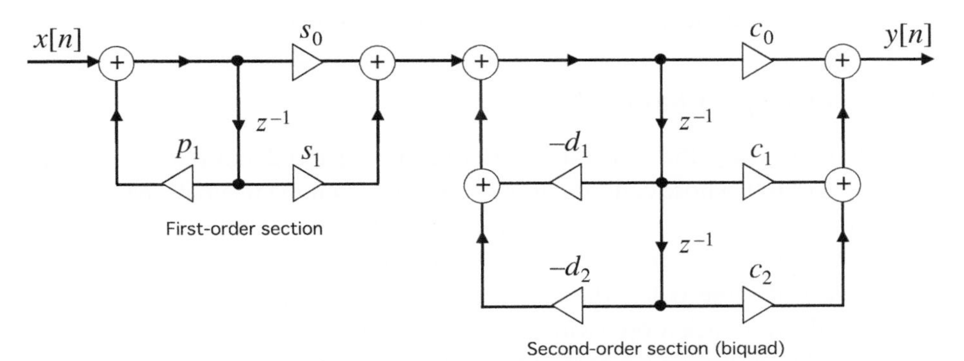

Figure 5.10 Cascade-form structure for a third-order real-coefficient filter. All the multipliers are real. The second-order real-multiplier section is called a biquadratic section (biquad).

a pole p_m is complex, its conjugate p_m^* is also a pole. So we can combine factors as follows:

$$(1 - pz^{-1})(1 - p^* z^{-1}) = 1 + d_1 z^{-1} + d_2 z^{-2}, \tag{5.85}$$

where d_1 and d_2 are real. The same is true for zeros. Thus, we can implement a real-coefficient filter as a cascade of first and second-order sections, all having real coefficients (multipliers). Second-order building blocks

$$\frac{c_0 + c_1 z^{-1} + c_2 z^{-2}}{1 + d_1 z^{-1} + d_2 z^{-2}}, \tag{5.86}$$

with real c_i, d_i, are also called **biquads** (biquadratic sections). For example, a typical third-order real-coefficient filter can be factored into the form

$$\left(\frac{s_0 + s_1 z^{-1}}{1 - p_1 z^{-1}} \right) \left(\frac{c_0 + c_1 z^{-1} + c_2 z^{-2}}{1 + d_1 z^{-1} + d_2 z^{-2}} \right), \tag{5.87}$$

where p_1, s_i, c_i, and d_i are real. This leads to the cascade-form implementation shown in Fig. 5.10. More generally, it can be verified that for any filter order, a real-coefficient rational transfer function can be implemented in cascade form, with only real multipliers. ▽ ▽ ▽

The student may wonder why we discuss different types of structures to implement the same transfer function. The fact is, some structures can have advantages over others for practical reasons. For example, in a digital implementation, the multipliers in the structure have to be quantized to a finite number of bits. This may cause some poles to move outside the unit circle, making the filter unstable. In the direct form, each multiplier b_k affects all the poles, whereas in the cascade form the effects of quantization of multipliers in a given section only affects the poles represented by that section. So the cascade form tends to be more robust to quantization. For example, in Fig. 5.9 we can quantize each pole p_i such that the quantized version continues to be inside the unit circle.

There are many other undesirable effects due to quantization (e.g., see Sec. 8.5.1), and some structures behave better than others under these circumstances. The details are quite involved and will not be considered further.

5.9.4 Summary for Rational LTI Systems

Summarizing, the LTI system described by Eq. (5.62) has the rational transfer function (5.65), whose denominator $B(z)$ comes from the recursive part of the difference equation (5.62) (which makes $y[n]$ depend on past outputs $y[n-k]$). The system therefore has nontrivial poles p_i (zeros of $B(z)$) because of the recursive part. Because of these poles, the system is IIR. The recursive part $1/B(z)$ is responsible for the feedback loops (with multipliers b_k) in the implementation, as shown in Fig. 5.7. The nonrecursive part $A(z)$ is FIR, and is responsible for the feedforward paths (with multipliers a_k) in the figure. In short, the following statements are equivalent:

1. The difference equation has a recursive part.
2. The filter is IIR.
3. $H(z)$ has a denominator $B(z)$.
4. $H(z)$ has nontrivial poles ($p_i \neq 0$).
5. The computational graph has feedback loops.

It should be mentioned, however, that sometimes it is economical to implement *FIR filters* using *recursive structures*. This will now be demonstrated. Thus, recall the FIR transfer function (4.32). Since it can be rewritten as in Eq. (4.33), it can be implemented as in Fig. 5.11 (Problem 5.21). This structure has only two adders, unlike Fig. 4.10(a), and is therefore more economical. It has a feedback loop because of the denominator in Eq. (4.33), and is called a **recursive** implementation of the FIR filter. The filter is therefore sometimes called a *recursive running sum*. The feedback loop, however, does not mean that the filter is IIR because the denominator $(1-z^{-1})$ cancels with a factor $(1-z^{-1})$ in the numerator, that is, there is a pole–zero cancellation in Eq. (4.33), which yields the original FIR filter (4.32) (Problem 5.20). In practice, this cancellation may not be perfect because of numerical errors in the

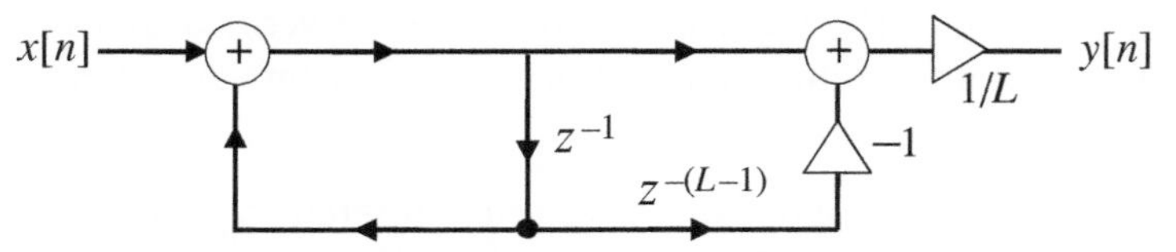

Figure 5.11 A recursive implementation (i.e., a structure with a feedback loop) of the FIR filter (4.32).

implementation, such as quantization. This can create a problem if the pole to be cancelled is on or outside the unit circle, as in this example. Unless the cancellation is exact, the system is, strictly speaking, an unstable IIR filter. Therefore, one ought to be careful with IIR-like (i.e., recursive) implementations of FIR filters.

Benefits of the z-transform approach. By now it should be clear to the student why we use z-transforms for the study of LTI systems. The advantages of the transform-domain method are so implicit and ingrained in our discussion that we hardly pause to ask why we use them at all. So, let us note some advantages explicitly. (a) We can study stability by examining the zeros of a polynomial. (b) We can write an Nth-order difference equation as a cascade of first-order difference equations. (c) We can plot a frequency response which tells us how the system handles every single input frequency. (d) We can analyze feedback systems which are hard to analyze directly in the time domain, as explained in Sec. 4.2.1. And so on. $\triangledown\triangledown\triangledown$

5.10 Impulse Response and Stabiliity

It remains to find the impulse response $h[n]$ of the LTI system described by Eq. (5.62). Since $h[n]$ is the inverse z-transform of $H(z)$, we have to find the inverse z-transform of the transfer function reproduced below:

$$H(z) = \frac{A(z)}{B(z)} = \frac{\sum_{k=0}^{N} a_k z^{-k}}{1 + \sum_{k=1}^{N} b_k z^{-k}} = \frac{a_0 \prod_{m=1}^{N} (1 - z_m z^{-1})}{\prod_{m=1}^{N} (1 - p_m z^{-1})}. \tag{5.88}$$

Once an expression for $h[n]$ is found, it is easy to study stability, as we shall see. In this section we will assume that $b_N \neq 0$, so that *denominator order* $\geq$ *numerator order*. We will also assume that the **poles are distinct**, that is, $p_i \neq p_j$ for $i \neq j$. Under these two assumptions, it can be shown that $H(z)$ can be written as follows:

$$H(z) = A_0 + \frac{A_1}{1 - p_1 z^{-1}} + \frac{A_2}{1 - p_2 z^{-1}} + \cdots + \frac{A_N}{1 - p_N z^{-1}}. \tag{5.89}$$

This is called a **partial fraction expansion** or **PFE** and will be explained in detail in Sec. 6.6.

Finding A_i. The coefficients A_i can readily be evaluated as follows. First, by comparing Eqs. (5.88) and (5.89) as $z \to 0$, it follows that

$$A_0 = \frac{a_N}{b_N}. \tag{5.90}$$

Next, we see readily from Eq. (5.89) that

$$A_i = \left(1 - z^{-1} p_i\right) H(z)\Big|_{z=p_i}, \qquad 1 \leq i \leq N. \tag{5.91}$$

Using Eq. (5.88), the right-hand side of Eq. (5.91) can be simplified to obtain

$$A_i = \frac{A(p_i)}{\prod_{\substack{m=1 \\ m \neq i}}^{N} (1 - p_m p_i^{-1})}, \qquad 1 \leq i \leq N. \tag{5.92}$$

Thus, all the A_i can be calculated. ▽ ▽ ▽

Since the z-transform is a *linear* operator (Sec. 6.12), it follows that the inverse transform of $H(z)$ is the *sum of the inverse transforms* of individual terms. In Sec. 5.2.2 we showed that the transfer function $1/(1 - pz^{-1})$ has the causal impulse response $p^n \mathcal{U}[n]$. Thus, the term $A_k/(1 - p_k z^{-1})$ in Eq. (5.89) has the causal impulse response

$$A_k p_k^n \mathcal{U}[n]. \tag{5.93}$$

Since the z-transform of $A_0\delta[n]$ is the constant A_0 for all z, it follows that $A_0\delta[n]$ is the inverse transform of the constant term A_0. Thus, the causal impulse response corresponding to Eq. (5.89) is given by

$$h[n] = A_0\delta[n] + \left(A_1 p_1^n + A_2 p_2^n + \cdots + A_N p_N^n\right)\mathcal{U}[n]. \tag{5.94}$$

This is a sum of right-sided (causal) exponentials, plus a delta function. This is a simple and insightful expression indeed. It shows that the dynamics (i.e., time-domain shape) of the impulse response is determined by the pole locations. If

$$|p_i| < 1 \tag{5.95}$$

for all i, then the exponentials in Eq. (5.94) decay to zero as $n \to \infty$, so $h[n]$ itself decays to zero as $n \to \infty$. In fact more is true, namely, condition (5.95) is necessary and sufficient for stability:

Theorem 5.1 *Stability of causal rational LTI systems.* Consider the causal rational LTI system with transfer function (5.89), where the poles p_i are distinct. This system is stable, that is,

$$\sum_{n=-\infty}^{\infty} |h[n]| < \infty, \tag{5.96}$$

if and only if condition (5.95) is true for all i, that is, if and only if *all poles are inside* the unit circle. ◇

(The result is true even when the poles are not distinct, see Theorem 6.1.) It is easy to prove that condition (5.95) is *sufficient* for stability. Indeed, assuming (5.95) holds, we have

$$|h[n]| \leq |A_0|\delta[n] + \sum_{i=1}^{N} |A_i||p_i|^n \mathcal{U}[n], \tag{5.97}$$

so that

$$\sum_{n=-\infty}^{\infty} |h[n]| \leq |A_0| + \sum_{i=1}^{N} |A_i| \sum_{n=0}^{\infty} |p_i|^n$$
$$= |A_0| + \sum_{i=1}^{N} \frac{|A_i|}{1 - |p_i|} < \infty, \tag{5.98}$$

where the sum $\sum_{n=0}^{\infty} |p_i|^n$ converges because of condition (5.95). This shows that $H(z)$ is stable when condition (5.95) holds for all i.

Conversely, the proof that condition (5.95) is also *necessary* for stability is a bit more involved. The basic idea is that if $|p_i| \geq 1$ for some i, then p_i^n does not decay to zero, and this prevents $\sum_n |h[n]|$ from being finite. But there may exist more than one pole violating condition (5.95), and some of them might even have identical magnitudes. So, how do we know that the sum $\sum_i A_i p_i^n$ with a few $|p_i| \geq 1$ cannot add up to something that is still absolutely summable? The intuitive reason why this cannot happen is that the functions p_i^n for distinct p_i are linearly independent, so the nondecaying behavior of p_i^n cannot be cancelled by the behavior of other terms p_j^n. In what follows, this argument is made more precise.

Proof that condition (5.95) is necessary for stability*. There will be no loss of continuity if the reader skips this proof during first reading. A review of notations from Chapter 18 will be helpful before reading on. Assume condition (5.95) is not true, say, $|p_N| \geq 1$. Without loss of generality, assume $A_N \neq 0$, for otherwise, the summation (5.89) would not yield the transfer function (5.88) with p_N present in the denominator (i.e., p_N would not be a pole). From Eq. (5.94) we see that

$$\underbrace{\begin{bmatrix} h[n] \\ h[n+1] \\ h[n+2] \\ \vdots \\ h[n+N-1] \end{bmatrix}}_{\text{call this } \mathbf{h}} = \underbrace{\begin{bmatrix} p_1^n & p_2^n & p_3^n & \cdots & p_N^n \\ p_1^{n+1} & p_2^{n+1} & p_3^{n+1} & \cdots & p_N^{n+1} \\ p_1^{n+2} & p_2^{n+2} & p_3^{n+2} & \cdots & p_N^{n+2} \\ \vdots & \vdots & \vdots & \cdots & \vdots \\ p_1^{n+N-1} & p_2^{n+N-1} & p_3^{n+N-1} & \cdots & p_N^{n+N-1} \end{bmatrix}}_{\text{call this } \mathbf{P}} \underbrace{\begin{bmatrix} A_1 \\ A_2 \\ A_3 \\ \vdots \\ A_N \end{bmatrix}}_{\text{call this } \mathbf{a}}, \tag{5.99}$$

for $n \geq 1$. This can be rewritten as

$$\mathbf{h} = \underbrace{\begin{bmatrix} 1 & 1 & 1 & \cdots & 1 \\ p_1 & p_2 & p_3 & \cdots & p_N \\ p_1^2 & p_2^2 & p_3^2 & \cdots & p_N^2 \\ \vdots & \vdots & \vdots & \cdots & \vdots \\ p_1^{N-1} & p_2^{N-1} & p_3^{N-1} & \cdots & p_N^{N-1} \end{bmatrix}}_{\text{call this } \mathbf{V}} \underbrace{\begin{bmatrix} p_1^n & 0 & 0 & \cdots & 0 \\ 0 & p_2^n & 0 & \cdots & 0 \\ 0 & 0 & p_3^n & \cdots & 0 \\ \vdots & \vdots & \vdots & \cdots & \vdots \\ 0 & 0 & 0 & \cdots & p_N^n \end{bmatrix} \begin{bmatrix} A_1 \\ A_2 \\ A_3 \\ \vdots \\ A_N \end{bmatrix}}_{\text{call this vector } \mathbf{d}}, \tag{5.100}$$

for $n \geq 1$. The matrix $\mathbf{V}$ is a Vandermonde matrix, and it is *nonsingular because the poles are assumed distinct* (see Sec. 5.16). So $\mathbf{V}^H\mathbf{V}$ is positive definite. Let $\epsilon^2 > 0$ be its smallest eigenvalue.

Then

$$\mathbf{h}^H\mathbf{h} = \mathbf{d}^H\mathbf{V}^H\mathbf{V}\mathbf{d} \geq \epsilon^2\mathbf{d}^H\mathbf{d}, \tag{5.101}$$

where the inequality follows from the Rayleigh–Ritz principle [Horn and Johnson, 1985]. So

$$\mathbf{h}^H\mathbf{h} \geq \epsilon^2\mathbf{d}^H\mathbf{d} = \epsilon^2\sum_{i=1}^{N}|p_i^n A_i|^2 \geq \epsilon^2|p_N^n A_N|^2 \geq \epsilon^2|A_N|^2 > 0, \tag{5.102}$$

since $A_N \neq 0$ and $|p_N| \geq 1$. Since the energy $\mathbf{h}^H\mathbf{h}$ has a lower bound that does not decrease with increasing n (where n is the time index in Eq. (5.99)), it follows in particular that $|h[m]| \geq \delta > 0$ for some δ for infinitely many m. So $\sum_{n=0}^{\infty}|h[n]|$ cannot be finite, which proves that the system is unstable. $\nabla\nabla\nabla$

In this section we assumed that the poles p_i are distinct in (5.88). We will see later that condition (5.95) is necessary and sufficient for the stability of a causal rational LTI system even when the **poles are not distinct** (Theorem 6.1). But the expressions (5.89) and (5.94) have to be modified in this case.

5.10.1 Stability, ROC, and Frequency Response

The ROC of the z-transform of the causal system (5.94) is the intersection of the ROCs of the individual terms in (5.89), which have the form $|z| > |p_i|$ (see Eqs. (5.13) and (5.14)). So, the ROC is the region

$$|z| > |p_N|, \tag{5.103}$$

assuming that p_N is the pole with the largest magnitude. Since the system is stable if and only if condition (5.95) holds, that is, $|p_N| < 1$, it follows that the causal rational LTI system is *stable if and only if the unit circle is in the ROC.*[1] The frequency response is well defined only in this case. This is similar to the discussion we had in Sec. 5.3.

5.10.2 Parallel-Form Structures

Recall from Sec. 4.2.1 that when two LTI systems are connected in parallel, the overall transfer function is the sum of the individual transfer functions. The partial fraction expansion (5.89) therefore leads to a structure for the implementation of $H(z)$, in which each term of (5.89) is a separate section. This is demonstrated in Fig. 5.12 for $N = 2$, and is called the parallel-form structure. Like the cascade form, the poles of the parallel form remain inside the unit circle under quantization, guaranteeing stability.

[1] We will see in Chapter 6 that the conclusion generalizes to any rational system (causal or not) but not to nonrational systems.

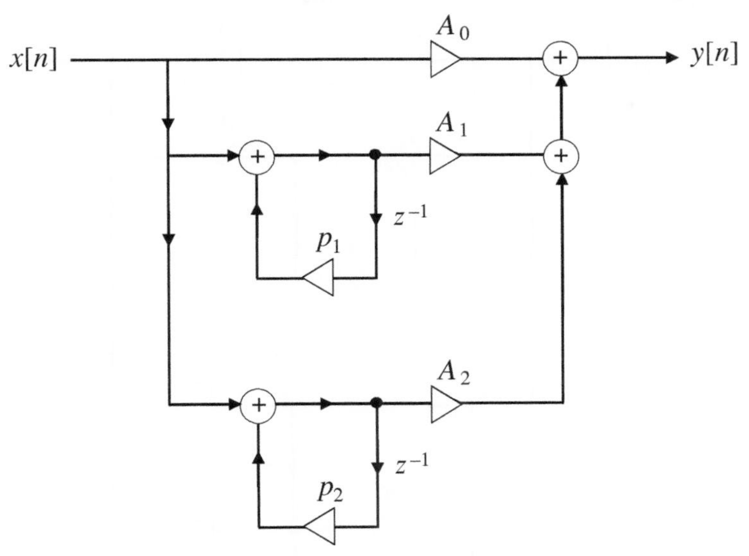

Figure 5.12 The parallel-form structure for a second-order digital filter.

For real-coefficient filters, the poles p_i are either real or occur in complex-conjugate pairs. Consider a third-order system where p_1 is real, p_2 is complex, and $p_3 = p_2^*$. Thus

$$H(z) = A_0 + \frac{A_1}{1 - p_1 z^{-1}} + \frac{A_2}{1 - p_2 z^{-1}} + \frac{A_3}{1 - p_2^* z^{-1}}. \tag{5.104}$$

Since $H(z)$ has real coefficients, $A(z)$ in Eq. (5.88) has real coefficients as well. Thus, using Eqs. (5.90) and (5.91) it can be shown that A_0 and A_1 are real, and (with some more effort) that $A_3 = A_2^*$. Thus, in this case we can rewrite Eq. (5.104) as

$$\begin{aligned} H(z) &= A_0 + \frac{A_1}{1 - p_1 z^{-1}} + \frac{A_2}{1 - p_2 z^{-1}} + \frac{A_2^*}{1 - p_2^* z^{-1}} \\ &= A_0 + \frac{A_1}{1 - p_1 z^{-1}} + \frac{g_0 + g_1 z^{-1}}{1 + d_1 z^{-1} + d_2 z^{-2}}, \end{aligned} \tag{5.105}$$

where A_i, g_i, d_i, and p_1 are real. This leads to the parallel-form structure shown in Fig. 5.13, where all the multiplier coefficients are real. More generally, it can be verified that for any filter order, a real-coefficient rational transfer function can be implemented in parallel form, with only real multipliers.

For real-coefficient filters, the impulse response is real, and the expressions can be simplified to show this. The causal system (5.105) has impulse response

$$h[n] = A_0 \delta[n] + \left(A_1 p_1^n + A_2 p_2^n + A_2^* (p_2^*)^n \right) \mathcal{U}[n]. \tag{5.106}$$

The first two terms are real. Now examine the sum of the last two terms, namely $g[n] = \left(A_2 p_2^n + A_2^* (p_2^*)^n \right) \mathcal{U}[n]$. We can always write the complex numbers A_2 and p_2 in the form $A_2 = \rho e^{j\phi}$ and $p_2 = R e^{j\theta}$, so that

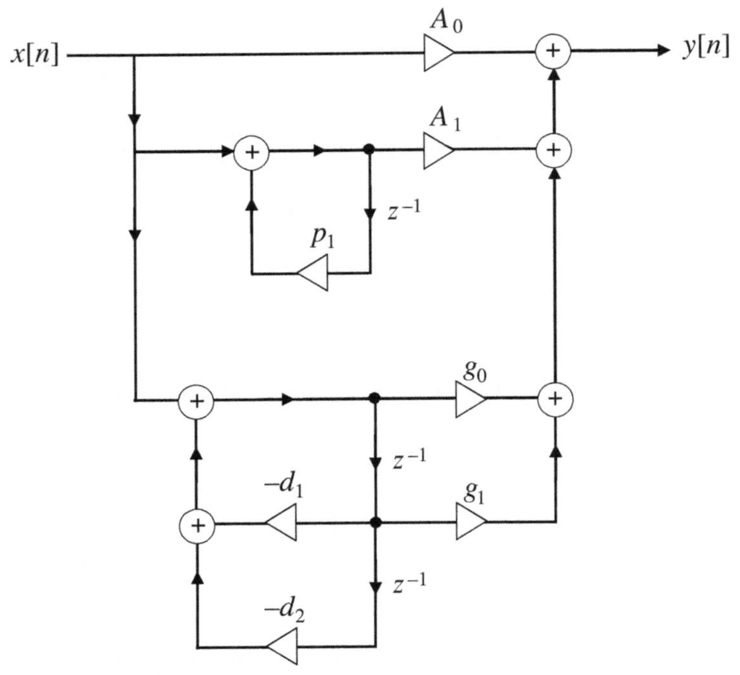

Figure 5.13 The parallel-form structure for a third-order digital filter with real coefficients. All multipliers (p_i, d_i, A_i, g_i) are real.

$$\begin{aligned} g[n] &= \left(\rho e^{j\phi} R^n e^{j\theta n} + \rho e^{-j\phi} R^n e^{-j\theta n}\right)\mathcal{U}[n] \\ &= 2\rho R^n \cos(\theta n + \phi)\mathcal{U}[n]. \end{aligned} \tag{5.107}$$

Thus

$$h[n] = A_0\delta[n] + \left(A_1 p_1^n + 2\rho R^n \cos(\theta n + \phi)\right)\mathcal{U}[n]. \tag{5.108}$$

All quantities in this expression are real.

5.11 Some Real-World Applications

Discrete-time FIR and IIR filters arise in many applications. One application which impacts our lives every day is **digital communications**. Here we send a sequence of symbols $s[n]$ from a transmitter (like your mobile phone) to a possibly remote receiver (like a base station). This is shown on the left in Fig. 5.14 where the symbol spacing is normalized to unity for simplicity. These symbols are typically sampled and coded versions of signals such as speech or image, or even coded versions of text messages. The transmission medium between the transmitter and the receiver is called a **channel**. The channel includes the electronics at the transmitter and receiver, the propagation characteristics in space including obstacles and multipaths, and additive noise. It also includes discrete to continuous-time converters at the

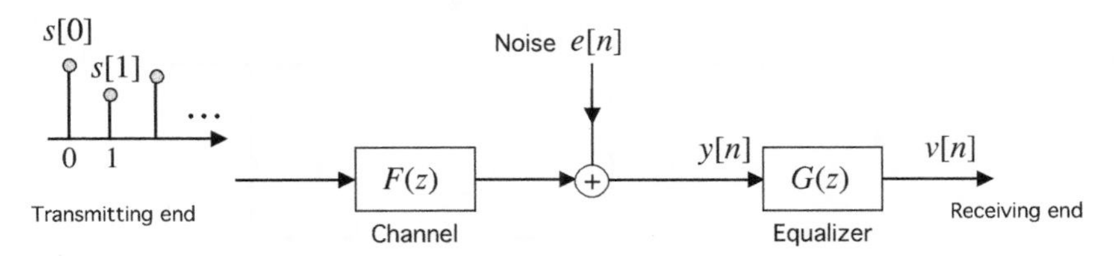

Figure 5.14 A simple model for a digital communication system.

transmitter, and continuous to discrete-time converters at the receiver [Lathi, 1989; Proakis and Salehi, 2008]. In many cases the channel can be approximately modeled as a discrete-time LTI system followed by an additive noise source $e[n]$ as shown in Fig. 5.14. The LTI system $F(z)$ in the model distorts the transmitted symbol stream $s[n]$ because of the convolution of $s[n]$ with $f[n]$. So the received signal $y[n]$ is this convoluted signal, which is in addition contaminated by noise $e[n]$:

$$y[n] = (s * f)[n] + e[n]. \tag{5.109}$$

The goal at the receiver is to satisfactorily recover the transmitted symbol stream $s[n]$ from $y[n]$. One way to do this is to pass the received signal $y[n]$ through the inverse filter $G(z) = 1/F(z)$, called the channel **equalizer**. This clearly results in

$$v[n] = (g * y)[n] = s[n] + (g * e)[n]. \tag{5.110}$$

Assuming that the noise $e[n]$ is not too strong, it is therefore possible to obtain an approximation of the transmitted symbol stream $s[n]$. In practice, $s[n]$ belongs to a quantized set of values (which could simply be 0 and 1). For each n, $v[n]$ is also first quantized to the nearest quantization level before use. The channel model $F(z)$ is often FIR, so the equalizer $G(z)$ used at the receiver is IIR. For example, if $F(z) = 1 - 2z^{-1}$ then

$$G(z) = \frac{1}{1 - 2z^{-1}}, \tag{5.111}$$

which is IIR. More importantly, the equalizer $G(z)$ is unstable in this example because the pole $p = 2$ is outside the unit circle. In general, since the channel $F(z)$ is not in our control, it is possible to end up with unstable equalizers like this, which is not good news because unstable filters can amplify the noise severely. One solution in practice is to use an FIR filter $G(z)$ which only approximately equalizes the channel, that is, $F(z)G(z) \approx 1$. Another solution is to use a more sophisticated system called the *fractionally spaced equalizer* (FSE); such equalizers can be FIR even for FIR channels. The details are beyond the scope of our discussion, and the interested student can refer to Treichler et al. [1996] and Sec. 4.8 of Vaidyanathan et al. [2010] for more details.

Another application we would like to mention is in feedback control systems. Consider again Fig. 4.4(c), reproduced in Fig. 5.15. There are applications where $H(z)$ is an unstable system which cannot be changed, although we have access to its input and output nodes. In such cases it is possible to feed the output $y[n]$ back to

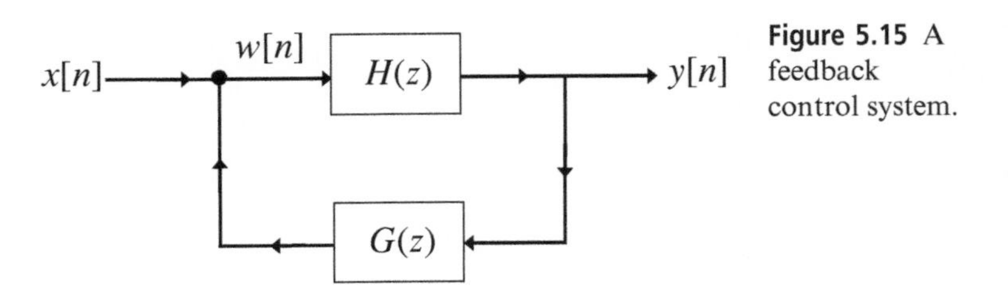

Figure 5.15 A feedback control system.

the input using another system $G(z)$ such that the closed-loop system is stable. This comes under the topic of feedback control systems, and has been studied extensively for many decades.

For example, $H(z)$ could be the discrete-time model of a power plant or some subsystem of a driverless vehicle, and the goal could be to keep the system stable by designing $G(z)$ appropriately. In such applications, $H(z)$ is called the "plant" and $G(z)$ the "controller." We shall be content with showing a basic toy example of stabilization. Thus, consider the case where

$$H(z) = \frac{1}{1 - z^{-1}}, \tag{5.112}$$

and $G(z) = az^{-1}$ where a is a constant. Clearly, $H(z)$ is unstable because there is a pole on the unit circle at $z = 1$. Now the transfer function (4.8) of the feedback system becomes

$$F(z) = \frac{Y(z)}{X(z)} = \frac{H(z)}{1 - G(z)H(z)} = \frac{\dfrac{1}{1 - z^{-1}}}{1 - \dfrac{az^{-1}}{1 - z^{-1}}} = \frac{1}{1 - (1 + a)z^{-1}}. \tag{5.113}$$

It is clear that if we choose a to be a small negative number such as $a = -0.01$, then the pole of the closed-loop system moves to $1 + a = 1 - 0.01 = 0.99$, resulting in a stable system. Of course, we could choose $a = -1$ and completely remove the pole. But this is not useful if the goal is to choose the controller such that the new system $F(z)$ is still a useful model for the original plant $H(z)$. Thus, in this example it is more appropriate to making a a negative number close to zero.

5.12 Finding a_k and b_k from Impulse Response*

Let $H(z)$ be an Nth-order causal LTI system with transfer function

$$H(z) = \frac{\sum_{k=0}^{N} a_k z^{-k}}{1 + \sum_{k=1}^{N} b_k z^{-k}}, \tag{5.114}$$

and impulse response $h[n]$, $n \geq 0$. Even though this is IIR, the first $2N+1$ coefficients of $h[n]$, namely,

$$h[0],\ h[1],\ h[2], \ldots, h[2N], \tag{5.115}$$

can be used to uniquely determine $h[n]$ for all n. We prove this by showing how a_k and b_k can be identified from the sequence (5.115). The result is useful to identify IIR LTI systems from a measurement of a finite number of samples $h[n]$. To prove the claim, consider the difference equation corresponding to $H(z)$:

$$y[n] = -\sum_{k=1}^{N} b_k y[n-k] + \sum_{k=0}^{N} a_k x[n-k]. \tag{5.116}$$

With $x[n] = \delta[n]$, we get

$$\begin{aligned}
h[0] &= a_0, \\
h[1] &= -b_1 h[0] + a_1, \\
h[2] &= -b_1 h[1] - b_2 h[0] + a_2, \\
h[3] &= -b_1 h[2] - b_2 h[1] - b_3 h[0] + a_3, \\
&\vdots \\
h[N] &= -b_1 h[N-1] - b_2 h[N-2] - \cdots - b_N h[0] + a_N.
\end{aligned} \tag{5.117}$$

If we continue to the next equation, we find that a_k does not enter any more:

$$h[N+1] = -b_1 h[N] - b_2 h[N-1] - \cdots - b_N h[1]. \tag{5.118}$$

In fact, the next N equations can be neatly expressed using matrix/vector notation as follows:

$$\underbrace{\begin{bmatrix} h[N+1] \\ h[N+2] \\ \vdots \\ h[2N] \end{bmatrix}}_{\text{call this } \mathbf{h}} = -\underbrace{\begin{bmatrix} h[N] & h[N-1] & \cdots & h[1] \\ h[N+1] & h[N] & \cdots & h[2] \\ & \vdots & & \\ h[2N-1] & h[2N-2] & \cdots & h[N] \end{bmatrix}}_{\text{call this } -\mathbf{H}} \underbrace{\begin{bmatrix} b_1 \\ b_2 \\ \vdots \\ b_N \end{bmatrix}}_{\text{call this } \mathbf{b}}. \tag{5.119}$$

For example, when $N = 4$ we have

$$\begin{bmatrix} h[5] \\ h[6] \\ h[7] \\ h[8] \end{bmatrix} = -\begin{bmatrix} h[4] & h[3] & h[2] & h[1] \\ h[5] & h[4] & h[3] & h[2] \\ h[6] & h[5] & h[4] & h[3] \\ h[7] & h[6] & h[5] & h[4] \end{bmatrix} \begin{bmatrix} b_1 \\ b_2 \\ b_3 \\ b_4 \end{bmatrix}. \tag{5.120}$$

It can be shown (see Sec. 14.5.5.2) that the $N \times N$ matrix in Eq. (5.119) is nonsingular as long as $H(z)$ is truly of order N, that is, at least one of a_N, b_N is nonzero, and there are no common factors between the numerator and the denominator of Eq. (5.114). Thus, from the coefficients $h[1], h[2], \ldots, h[2N]$, we can uniquely find the N coefficients b_k by solving Eq. (5.119) through matrix inversion:

$$\mathbf{b} = -\mathbf{H}^{-1}\mathbf{h}. \tag{5.121}$$

Once the b_k have been found, we can find the a_k from the kth equation in (5.117) for each k individually. No matrix inversion is required for this. An example is worked out in Problem 5.18. Notice that the $N \times N$ matrix in (5.119) is Toeplitz, that is, all elements on any line parallel to the diagonal are identical.

5.13 Remarks on Estimating Frequency Response*

Given a black box representing an LTI system $H(z)$, we can measure the frequency response $H(e^{j\omega})$, at least conceptually, as follows: apply the input signal $x[n] = e^{j\omega_i n}$ and measure the output, which has the form $y[n] = H(e^{j\omega_i})e^{j\omega_i n}$. Thus, by observing the samples of the output, we can estimate $H(e^{j\omega_i})$ and repeat this on a dense grid of frequncies ω_i. While there are better ways to measure the frequency response, even conceptually we have to be careful. Namely, in practice we can apply $e^{j\omega_i n}$ only starting from some initial time, say, $n = 0$. In this case the output has an initial "transient portion" during which it does not look like $H(e^{j\omega_i})e^{j\omega_i n}$ (Sec. 7.12). The transients eventually decay to insignificant values for stable systems; only then does the output begin to look like $H(e^{j\omega_i})e^{j\omega_i n}$. In this section we will discuss some other aspects of interest, assuming for simplicity that the input starts at $n = -\infty$ so we don't have to worry about transients. The main focus in this section is on some interesting points which arise in the case of real systems.

The discussions in this section are not restricted to rational systems. If the system is rational we can estimate a_k and b_k from the first few coefficients of $h[n]$ as in Sec. 5.12, and from these we can find $H(e^{j\omega})$ immediately.

5.13.1 Response of Real LTI Systems to Real Sinusoids

Recall that an LTI system or filter is said to be real if $h[n]$ is real. In this case the frequency response can still be complex, and has the general form

$$H(e^{j\omega}) = |H(e^{j\omega})|e^{j\phi(\omega)} = H_r(e^{j\omega}) + jH_i(e^{j\omega}), \tag{5.122}$$

where

$$H_r(e^{j\omega}) = |H(e^{j\omega})| \cos \phi(\omega), \quad H_i(e^{j\omega}) = |H(e^{j\omega})| \sin \phi(\omega) \tag{5.123}$$

are the real part and the imaginary part of $H(e^{j\omega})$. Since $h[n]$ is real, it follows that $H(e^{-j\omega}) = H^*(e^{j\omega})$ (Sec. 3.6), so that $H_r(e^{j\omega})$ is even and $H_i(e^{j\omega})$ is odd. For real systems we will see that we can measure the complex quantity $H(e^{j\omega})$, just by applying real inputs like $\cos(\omega_0 n)$. First observe that since $\cos(\omega_0 n) = 0.5(e^{j\omega_0 n} + e^{-j\omega_0 n})$, the output in response to $x[n] = \cos(\omega_0 n)$ is

$$\begin{aligned} y[n] &= 0.5\Big(H(e^{j\omega_0})e^{j\omega_0 n} + H(e^{-j\omega_0})e^{-j\omega_0 n}\Big) \\ &= 0.5\Big(H(e^{j\omega_0})e^{j\omega_0 n} + H^*(e^{j\omega_0})e^{-j\omega_0 n}\Big) \\ &= 0.5|H(e^{j\omega_0})|\Big(e^{j\phi(\omega_0)}e^{j\omega_0 n} + e^{-j\phi(\omega_0)}e^{-j\omega_0 n}\Big). \end{aligned} \tag{5.124}$$

Thus the output in response to $x[n] = \cos(\omega_0 n)$ is

$$y_c[n] = |H(e^{j\omega_0})| \cos(\omega_0 n + \phi(\omega_0)) \tag{5.125}$$

$$= H_r(e^{j\omega_0}) \cos(\omega_0 n) - H_i(e^{j\omega_0}) \sin(\omega_0 n), \tag{5.126}$$

where the subscript c is added as a reminder that the input is a cosine, $x[n] = \cos(\omega_0 n)$. Note that the output is not a cosine, but a linear combination of $\cos(\omega_0 n)$ and $\sin(\omega_0 n)$. This shows that $\cos(\omega_0 n)$ is in general *not an eigenfunction* of an LTI system, even if $h[n]$ is real. Similarly, if we apply $x[n] = \sin(\omega_0 n)$ then we get

$$y_s[n] = H_r(e^{j\omega_0}) \sin(\omega_0 n) + H_i(e^{j\omega_0}) \cos(\omega_0 n). \tag{5.127}$$

5.13.2 Estimating $H(e^{j\omega})$ of a Real LTI System

First let us focus on the output (5.126) obtained by using the input $x[n] = \cos(\omega_0 n)$, where ω_0 is known. By measuring $y_c[n]$ for some n where $\sin(\omega_0 n) = 0$ (like $n = 0$), we can readily find $H_r(e^{j\omega_0}) = y_c[n]$ (since $\cos(\omega_0 n) = 1$). Similarly, by measuring $y_c[n]$ for some n where $\cos(\omega_0 n) \approx 0$, we can estimate $H_i(e^{j\omega_0})$. But such methods based on a single sample are not robust to noise and other errors, so in practice it is better to estimate these based on several samples (or snapshots) of the output. Say we have K successive samples of $y_c[n]$. Then[2]

$$\underbrace{\begin{bmatrix} y_c[0] \\ y_c[1] \\ \vdots \\ y_c[K-1] \end{bmatrix}}_{\text{call this } \mathbf{y}_c} = \underbrace{\begin{bmatrix} 1 & 0 \\ \cos\omega_0 & -\sin\omega_0 \\ \vdots & \vdots \\ \cos(K-1)\omega_0 & -\sin(K-1)\omega_0 \end{bmatrix}}_{K\times 2 \text{ matrix } \mathbf{A}} \begin{bmatrix} H_r(e^{j\omega_0}) \\ H_i(e^{j\omega_0}) \end{bmatrix}. \tag{5.128}$$

As long as the $K \times 2$ matrix $\mathbf{A}$ above has rank 2 (which will happen for almost all ω_0), we can estimate $H_r(e^{j\omega_0})$ and $H_i(e^{j\omega_0})$ (hence $H(e^{j\omega_0})$) as follows:

$$\begin{bmatrix} H_r(e^{j\omega_0}) \\ H_i(e^{j\omega_0}) \end{bmatrix} = (\mathbf{A}^H\mathbf{A})^{-1}\mathbf{A}^H\mathbf{y}_c. \tag{5.129}$$

A second method would be to observe that Eq. (5.126) implies

$$\sum_{n=0}^{K-1} y_c[n]\cos(\omega_0 n) = H_r(e^{j\omega_0}) \sum_{n=0}^{K-1} \cos^2(\omega_0 n) - H_i(e^{j\omega_0}) \sum_{n=0}^{K-1} \cos(\omega_0 n)\sin(\omega_0 n). \tag{5.130}$$

In discrete time, sines and cosines may or may not be periodic, depending on whether ω_0 is a rational multiple of π or not. Nevertheless, if K is large enough,

[2] Chapter 18 reviews matrix notations and concepts.

$\cos(\omega_0 n)$ and $\sin(\omega_0 n)$ tend to be almost orthogonal in the interval $0 \leq k \leq K-1$, in the sense that

$$\frac{\sum_{n=0}^{K-1} \cos(\omega_0 n)\sin(\omega_0 n)}{\sum_{n=0}^{K-1} \cos^2(\omega_0 n)} << 1. \tag{5.131}$$

Then the second summation on the right-hand side of Eq. (5.130) is negligible compared to the first. So we obtain the estimate

$$H_r(e^{j\omega_0}) \approx \frac{\sum_{n=0}^{K-1} y_c[n]\cos(\omega_0 n)}{\sum_{n=0}^{K-1} \cos^2(\omega_0 n)}, \tag{5.132}$$

and similarly we can estimate

$$H_i(e^{j\omega_0}) \approx -\frac{\sum_{n=0}^{K-1} y_c[n]\sin(\omega_0 n)}{\sum_{n=0}^{K-1} \sin^2(\omega_0 n)}. \tag{5.133}$$

There are similar ways to estimate $H(e^{j\omega})$ from $y_s[n]$. Finally, it is also possible to estimate $H(e^{j\omega})$ by taking advantage of both the measurements $y_c[n]$ and $y_s[n]$. For this, notice that the measurements can be written as

$$\begin{bmatrix} y_c[n] \\ y_s[n] \end{bmatrix} = \begin{bmatrix} \cos(\omega_0 n) & -\sin(\omega_0 n) \\ \sin(\omega_0 n) & \cos(\omega_0 n) \end{bmatrix} \begin{bmatrix} H_r(e^{j\omega_0}) \\ H_i(e^{j\omega_0}) \end{bmatrix}. \tag{5.134}$$

The 2×2 matrix above is unitary and can be inverted to obtain

$$\begin{bmatrix} H_r(e^{j\omega_0}) \\ H_i(e^{j\omega_0}) \end{bmatrix} = \begin{bmatrix} \cos(\omega_0 n) & \sin(\omega_0 n) \\ -\sin(\omega_0 n) & \cos(\omega_0 n) \end{bmatrix} \begin{bmatrix} y_c[n] \\ \\ y_s[n] \end{bmatrix}. \tag{5.135}$$

Again, with noisy or imperfect measurements in practice, one does not rely on a single sample measurement. There are ways to obtain more robust estimates that depend on K snapshots. One approach is to rewrite Eq. (5.134) as

$$\underbrace{\begin{bmatrix} y_c[n] \\ y_s[n] \end{bmatrix}}_{\mathbf{y}[n]} = \underbrace{\begin{bmatrix} H_r(e^{j\omega_0}) & -H_i(e^{j\omega_0}) \\ H_i(e^{j\omega_0}) & H_r(e^{j\omega_0}) \end{bmatrix}}_{\mathbf{H}} \underbrace{\begin{bmatrix} \cos(\omega_0 n) \\ \sin(\omega_0 n) \end{bmatrix}}_{\mathbf{x}[n]}. \tag{5.136}$$

We have available the measurements $\mathbf{y}[n], 0 \leq n \leq K-1$. And the input $\mathbf{x}[n], 0 \leq n \leq K-1$ is known as well. So we can obtain a matrix equation

$$\mathbf{Y} = \mathbf{HX}, \tag{5.137}$$

where $\mathbf{Y}$ and $\mathbf{X}$ are the $2 \times K$ matrices

$$\mathbf{Y} = [\mathbf{y}[0] \quad \mathbf{y}[1] \quad \cdots \quad \mathbf{y}[K-1]\,], \quad \mathbf{X} = [\mathbf{x}[0] \quad \mathbf{x}[1] \quad \cdots \quad \mathbf{x}[K-1]\,].$$

Assuming that $\mathbf{X}$ has rank 2, the 2×2 matrix $\mathbf{X}\mathbf{X}^H$ is invertible, and we obtain

$$\mathbf{H} = \mathbf{Y}\mathbf{X}^H(\mathbf{X}\mathbf{X}^H)^{-1}. \tag{5.138}$$

Since this is based on imperfect or noisy data, the diagonal elements of $\mathbf{H}$ may not be equal, in which case we can take their average to estimate $H_r(e^{j\omega_0})$. Similarly, $H_i(e^{j\omega_0})$ can be estimated from the nondiagonal elements of $\mathbf{H}$.

5.14 Polynomial Division Using IIR Structure*

Polynomial division is an operation which arises in many disciplines including computer science, coding theory, and digital communications. In this section we show that there is a clever way to achieve this just by running an IIR filter (i.e., a recursive difference equation) for a finite amount of time. Classically, a polynomial of degree K has non-negative powers of z, as in $P(z) = p_0 + p_1 z + \cdots + p_K z^K$. So let us say we have two polynomials

$$\begin{aligned} X_1(z) &= x[L] + x[L-1]z + \cdots + x[1]z^{L-1} + x[0]z^L, \\ B_1(z) &= b_N + b_{N-1}z + \cdots + b_1 z^{N-1} + z^N, \end{aligned} \tag{5.139}$$

with $L \geq N$. To avoid trivialities assume $x[0] \neq 0$, so the polynomial degrees are L and N, respectively. The notational convention above may look strange at first, but it will be convenient when we relate these to IIR filtering.

When we divide $X_1(z)$ by $B_1(z)$ there is a quotient polynomial $Q_1(z)$ and a remainder polynomial $R_1(z)$ (which are unique). To be more specific,

$$\frac{X_1(z)}{B_1(z)} = Q_1(z) + \frac{R_1(z)}{B_1(z)}, \tag{5.140}$$

where

$$\begin{aligned} Q_1(z) &= q_{L-N} + q_{L-N-1}z + \cdots + q_0 z^{L-N}, \\ R_1(z) &= r_{N-1} + r_{N-2}z + \cdots + r_0 z^{N-1}. \end{aligned} \tag{5.141}$$

Given $X_1(z)$ and $B_1(z)$, the goal of polynomial division is to find $Q_1(z)$ and $R_1(z)$. Note that the quotient has degree $L - N$ and the remainder has degree $< N$. The remainder is zero if and only if $X_1(z)$ is a polynomial multiple of $B_1(z)$, that is, $X_1(z) = Q_1(z)B_1(z)$. To relate Eq. (5.140) to filtering, define a causal IIR filter

$$H(z) = 1/B(z), \tag{5.142}$$

where

$$B(z) = 1 + b_1 z^{-1} + \cdots + b_N z^{-N}, \tag{5.143}$$

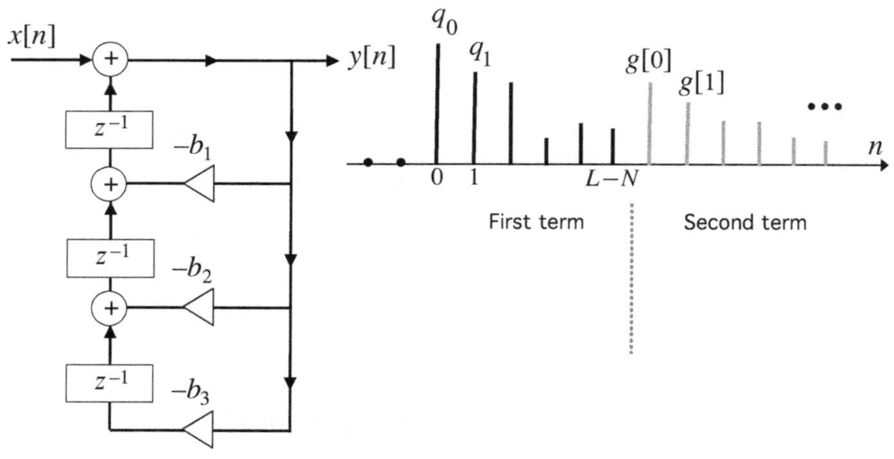

Figure 5.16 Filter $H(z) = 1/B(z)$ with input $x[n]$. The output signal $y[n]$, given by Eq. (5.146), is also sketched. The first few samples of $y[n]$ yield the quotient coefficients q_k and the second part $g[n]$ can be used to find the remainder coefficients r_k.

and consider a finite-duration causal input to this filter, with z-transform

$$X(z) = x[0] + x[1]z^{-1} + \cdots + x[L]z^{-L}. \tag{5.144}$$

Figure 5.16 shows this filter implemented using the modified direct-form structure of Fig. 5.8. $X(z)$ and $B(z)$ are related to $X_1(z)$ and $B_1(z)$ in Eqs. (5.139) by

$$X(z) = z^{-L}X_1(z), \quad B(z) = z^{-N}B_1(z). \tag{5.145}$$

Given $X(z)$ and $B(z)$, it will be shown that we can find $Q_1(z)$ and $R_1(z)$ from the ouput of $H(z)$ in response to the input $X(z)$. The output is

$$Y(z) = \frac{X(z)}{B(z)} = z^{-(L-N)}\frac{X_1(z)}{B_1(z)} = z^{-(L-N)}Q_1(z) + z^{-(L-N)}\frac{R_1(z)}{B_1(z)},$$

from Eq. (5.140). From Eqs. (5.139) and (5.141), this can be rewritten as

$$Y(z) = q_0 + q_1z^{-1} + \cdots + q_{L-N}z^{-(L-N)}$$
$$+z^{-(L-N+1)}\underbrace{\left(\frac{r_0 + r_1z^{-1} + \cdots + r_{N-1}z^{-(N-1)}}{B(z)}\right)}_{\text{call this } G(z)}. \tag{5.146}$$

The first term produces an output that lasts only in $0 \le n \le L - N$. And because of the factor $z^{-(L-N+1)}$, the second term produces an output starting only at time $L-N+1$. Thus, in the time domain, the first and second terms are non-overlapping (Fig. 5.16). So the quotient can be identified from the first $L - N + 1$ samples of the output:

$$q_n = y[n], \quad 0 \le n \le L - N. \tag{5.147}$$

To find the remainder coefficients r_k, consider $g[n]$ (the causal inverse transform of $G(z)$ defined in Eq. (5.146)). This can be found from $y[n]$ as follows:

$$g[n] = y[n + L - N + 1], n \geq 0. \tag{5.148}$$

Since $g[n]$ can be regarded as the impulse response of a hypothetical causal filter $G(z)$, it follows that $g[n], b_k$, and r_k are related by an equation similar to (5.117). Since $g[n]$ and b_k are known, we can therefore find the N remainder coefficients r_k from the first N equations of (5.117) with appropriate change of notations:

$$\begin{aligned}
g[0] &= r_0, \\
g[1] &= -b_1 g[0] + r_1, \\
g[2] &= -b_1 g[1] - b_2 g[0] + r_2, \\
g[3] &= -b_1 g[2] - b_2 g[1] - b_3 g[0] + r_3, \\
&\vdots \\
g[N-1] &= -b_1 g[N-2] - b_2 g[N-3] - \cdots - b_{N-1} g[0] + r_{N-1}.
\end{aligned} \tag{5.149}$$

Each r_k can be obtained trivially from its equation, as the equations are decoupled. This completes the task of finding the quotient $Q_1(z)$ and the remainder $R_1(z)$.

An efficient way to compute the remainder. It turns out that r_k can be found even more easily, without using Eq. (5.149), in fact without any additional computation at all! For this, imagine we implement $G(z)$ using the modified direct-form structure (Fig. 5.8). This is shown in Fig. 5.17(a) for $N = 3$. (Remember, $r_N = 0$.) The sequence $g[n]$ is the impulse response of this structure, as indicated in the figure. The impulse at the input creates the signal $r_k\delta[n]$ at the output of the multiplier r_k. Since these are directly connected to the delay elements as shown, we can generate

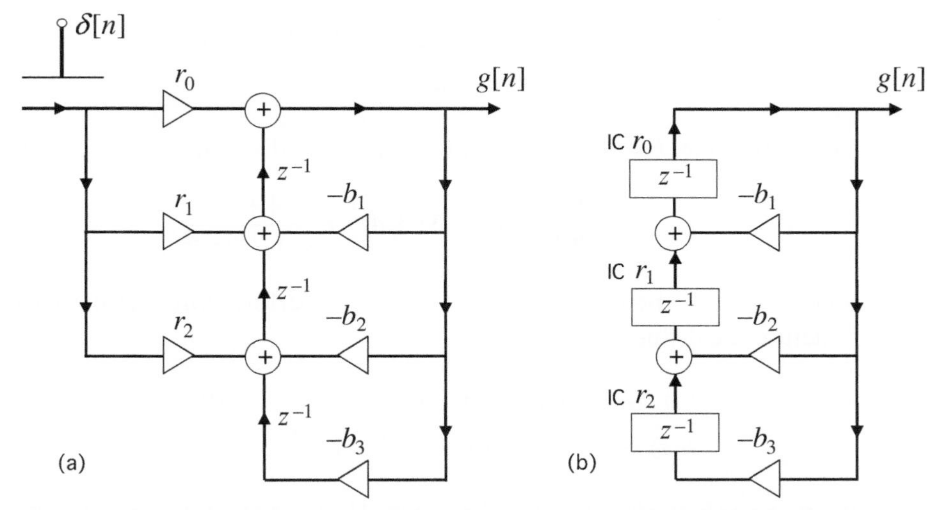

Figure 5.17 (a) The modified direct-form structure (Fig. 5.8) for the LTI system $G(z)$ in Eq. (5.146) with input $\delta[n]$; (b) equivalent way to generate $g[n]$ by storing r_k as initial values in the delay elements. The figure is for $N = 3$.

the output $g[n]$ in a different way, as shown in Fig. 5.17(b). Namely, we simply store the multiplier values r_k in the delay elements as *initial conditions*, and start the system with *zero input*. The output due to this will *also be* equal to the output $g[n]$ generated in Fig. 5.17(a).

Notice, however, that Fig. 5.17(b) is nothing but the structure we used to implement $1/B(z)$ with input $X(z)$ (Fig. 5.16), to generate $y[n]$. What our above discussion shows is that, in Fig. 5.16, after we have noted down the quotient samples q_k from the output $y[n]$, if we examine the contents of the delay elements at time $n = L-N+1$, we will find that they are precisely equal to the remainder coefficients r_k, because the output $y[n]$ starting at time $n=L-N+1$ is $g[0], g[1], \ldots$.

A subtle point here is that $g[n]$ uniquely determines r_k; there cannot be two sets of r_k yielding the same $g[n]$. This follows from the theory of Sec. 5.12. In short, in Fig. 5.16(a) we can find the quotients q_k by reading out the first $L-N+1$ output samples, and then find the remainder coefficients r_k by reading out the contents of the delay elements at time $L-N+1$.

An elegant way to perform polynomial division indeed! $\bigtriangledown \bigtriangledown \bigtriangledown$

It should be noted that if we had used the direct-form structure of Fig. 5.7 for the implementation of $1/B(z)$ instead of Fig. 5.16, then the contents of the delay elements at $n = L-N+1$ would not match r_k; it would take some computation to identify r_k from those. The fact that the input in Fig. 5.17(a) is directly connected to the outputs of the delays through r_k is what makes the modified direct form especially suitable in this application.

Polynomial Division with Filter Structure: An Example

Let the polynomials $X_1(z)$ and $B_1(z)$ be

$$\begin{aligned} X_1(z) &= 8 + 17z + 16z^2 + 11z^3 + 4z^4, \\ B_1(z) &= 2 + 2z + z^2. \end{aligned} \tag{5.150}$$

In this example, $L = 4$ and $N = 2$. By direct long division we can verify that

$$\frac{X_1(z)}{B_1(z)} = 2 + 3z + 4z^2 + \frac{4+7z}{2+2z+z^2}.$$

So $Q_1(z) = 2+3z+4z^2$ and $R_1(z) = 4+7z$. To perform this operation using a filter structure, we define

$$\begin{aligned} X(z) &= z^{-4}X_1(z) = 4 + 11z^{-1} + 16z^{-2} + 17z^{-3} + 8z^{-4}, \\ B(z) &= z^{-2}B_1(z) = 1 + 2z^{-1} + 2z^{-2}. \end{aligned} \tag{5.151}$$

We then apply the input $x[n]$ to the system $H(z) = 1/B(z)$ shown in Fig. 5.16. (Since $B(z)$ has degree 2, there are only two delays in the structure now, and there is no b_3.) The coefficients of the quotient $Q_1(z)$ can be obtained from the output samples as

$$q_0 = y[0] = 4, \quad q_1 = y[1] = 3, \quad q_2 = y[2] = 2. \tag{5.152}$$

At the time instant $L - N + 1 = 3$, the contents of the delay elements reveal the remainder coefficients $r_0 = 7$ and $r_1 = 4$, as indicated clearly in Fig. 5.17(b). There is no r_2 in this example, since there is no b_3. Note that $1/B(z)$ is unstable in this example (because the highest coefficient of $B(z)$ is 2, which shows that at least one pole is outside the unit circle). But since the system is implemented only for a short duration of time $L - N + 1 = 3$, this is not of any concern.

5.15 Summary of Chapter 5

This chapter started off with a study of the first-order recursive difference equation $y[n] = py[n-1] + x[n]$. Given an input $x[n], n \geq 0$ and an initial condition $y[-1]$, it was shown at the beginning of this chapter that the output can be computed recursively for all $n, n \geq 0$. When the initial condition $y[-1] = 0$, this represents an LTI system, and its impulse response and transfer function were derived, and frequency response plotted for various p. This system is causal, and stable when $|p| < 1$. The frequency response expression shows that for real p, this system is a lowpass filter when $0 < p < 1$ and a highpass filter when $-1 < p < 0$. Recursive difference equations offer a computationally efficient way to implement systems whose outputs may depend on an infinite number of past inputs, like the accumulator. The recursive property allows the infinite past to be remembered by remembering only a finite number of past outputs. After the above first-order example, the Nth-order recursive difference equation

$$y[n] = -\sum_{k=1}^{N} b_k y[n-k] + \sum_{k=0}^{N} a_k x[n-k] \tag{5.153}$$

was considered. Again, given $x[n], n \geq 0$ and initial condition $y[-1], y[-2], \ldots, y[-N]$, the output can be computed for all $n \geq 0$. It was shown that this is a causal LTI system when the initial conditions are zero, and the transfer function is

$$H(z) = \frac{\sum_{k=0}^{N} a_k z^{-k}}{1 + \sum_{k=1}^{N} b_k z^{-k}} = \frac{A(z)}{B(z)}. \tag{5.154}$$

This is called a rational transfer function, and the system is a rational LTI system. Assuming that the above $H(z)$ cannot be written in terms of lower-degree polynomials, the integer N is called the order of the system. This is an IIR system unless $B(z) = 1$. The zeros z_m of the numerator polynomial $A(z)$ are called the zeros of the system, and the zeros p_m of the denominator polynomial $B(z)$ are called the poles of the system. A computational graph was presented for this LTI system and is called the direct-form structure. It has $2N + 1$ multipliers, $2N$ two-input adders, and N delay elements. We derived the causal impulse response of the rational LTI system $H(z)$ using the partial fraction expansion (PFE) method. This causal system is stable if and only if

$$|p_m| < 1 \quad \text{(stability, causal rational } H(z)\text{)}.$$

Next, it was shown that given the $2N + 1$ measurements $h[0], h[1], \ldots, h[2N]$ of the causal infinite impulse response $h[n]$, the $2N + 1$ coefficients a_k and b_k can be computed. Also presented were methods to estimate the frequency response of an LTI system by making appropriate input–output measurements. Finally, it was shown that the operation of polynomial division can be implemented efficiently by running a recursive difference equation with appropriately chosen coefficients a_k and b_k.

We will see later that recursive difference equations can be used to design sophisticated digital filters such as sharp-cutoff lowpass filters, notch filters, anti-notch filters, and so on (Chapter 7). Recursive difference equations with *nonzero* initial conditions also have interesting applications (Chapter 8). The definitions of poles and zeros given in this chapter do not directly tell us what they "do" in the time domain. There are interesting time-domain meanings for poles and zeros, and these will be discussed in Sec. 8.7 and in the Problems section of Chapter 8.

5.16 Appendix: Vandermonde Matrices

Consider a matrix of the form

$$\mathbf{V} = \begin{bmatrix} 1 & 1 & \cdots & 1 \\ \alpha_1 & \alpha_2 & \cdots & \alpha_N \\ \alpha_1^2 & \alpha_2^2 & \cdots & \alpha_N^2 \\ \vdots & \vdots & \vdots & \vdots \\ \alpha_1^{N-1} & \alpha_2^{N-1} & \cdots & \alpha_N^{N-1} \end{bmatrix}. \tag{5.155}$$

Here each column is of the form

$$\begin{bmatrix} 1 & \alpha & \alpha^2 & \cdots & \alpha^{N-1} \end{bmatrix}^T, \tag{5.156}$$

which looks like an exponential. These columns are called *Vandermonde vectors*, and the matrix (5.155) is called a Vandermonde matrix. The transpose of such a matrix is also called a Vandermonde matrix. An important result is that this matrix is nonsingular if and only if the α_i are distinct. We now prove this.

First assume the α_i are not distinct, say $\alpha_1 = \alpha_2$. Then the first two columns are identical, so the rank is less than N and the matrix is singular. This proves that distinctness of the α_i is necessary. For the converse, assume the α_i *are* distinct. We will assume that $\mathbf{V}$ is singular and bring about a contradiction: singularity of $\mathbf{V}$ means that the rows are linearly *dependent*, so there exist scalars $w_0, w_1, \ldots, w_{N-1}$, not all zero, such that

$$[w_0 \quad w_1 \quad \cdots \quad w_{N-1}] \begin{bmatrix} 1 & 1 & \cdots & 1 \\ \alpha_1 & \alpha_2 & \cdots & \alpha_N \\ \alpha_1^2 & \alpha_2^2 & \cdots & \alpha_N^2 \\ \vdots & \vdots & \vdots & \vdots \\ \alpha_1^{N-1} & \alpha_2^{N-1} & \cdots & \alpha_N^{N-1} \end{bmatrix} = [0 \quad 0 \quad \cdots \quad 0]. \quad (5.157)$$

Now define a polynomial

$$W(x) = \sum_{n=0}^{N-1} w_n x^n. \quad (5.158)$$

Equation (5.157) simply says that $W(\alpha_i) = 0$ for $1 \leq i \leq N$. Since the α_i are distinct, this means that $W(x)$ has N distinct roots. This is a contradiction because $W(x)$ is a polynomial with degree only $N - 1$, and cannot have more than $N - 1$ roots! This proves that $\mathbf{V}$ is indeed nonsingular when the α_i are distinct.

PROBLEMS

Note: If you are using MATLAB for the computing assignments at the end, the "stem" command is convenient for plotting sequences. The "plot" command is useful for plotting real-valued functions such as $|X(e^{j\omega})|$.

5.1 Consider the recursive difference equation $y[n] = py[n-1] + x[n]$.
  (a) Assume $x[n] = \delta[n]$, and that the initial condition is nonzero, namely, $y[-1] = 1/p$. Find the output $y[n]$ for $n \geq 0$.
  (b) In part (a), if we change the initial condition to $y[-1] = -1/p$, then what is the output for $n \geq 0$?

5.2 Consider the second-order recursive system shown in Fig. P5.2.
  (a) Write down the difference equation describing the system.
  (b) Assume $x[n] = \delta[n]$, and that the initial conditions are zero, namely, $y[-1] = y[-2] = 0$. Find the output $y[n]$ for $n = 0, 1, 2$.
  (c) In part (a), if we change the initial condition to $y[-1] = y[-2] = 1$, and set $x[n] = 0$ for all n, then find the output for $n = 0, 1, 2$.

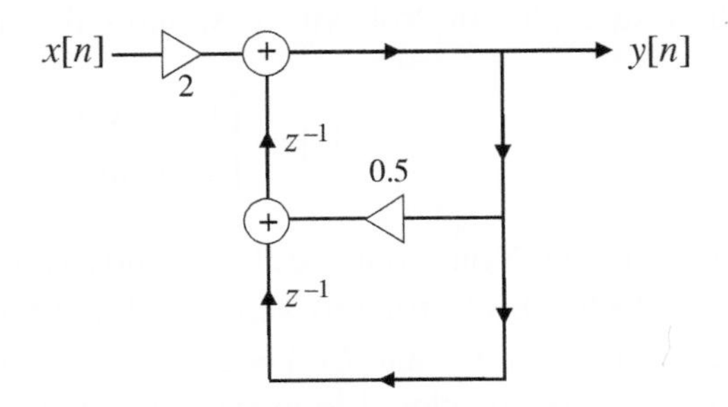

Figure P5.2 A second-order recursive system.

5.3 Consider the causal LTI system described by

$$y[n] = y[n-1] + 5y[n-2] + 29y[n-3] + y[n-4] + x[n].$$

(a) Find an expression for the transfer function.
(b) Draw the direct-form structure for the system.
(c) Without explicitly calculating the locations of the poles, determine whether the system is stable or not.

5.4 Let a causal rational discrete-time LTI system have the transfer function

$$H(z) = \frac{1 - z^{-1} + z^{-2}}{1 - (3/4)z^{-1} + (1/8)z^{-2}}.$$

(a) What are the poles p_i and zeros z_i of the system? Indicate them approximately in the z-plane. Be sure to draw the unit circle for reference.
(b) Write the *recursive causal difference equation* that represents this system in the time domain. Assuming the initial conditions are $y[-1] = y[-2] = 0$, and letting $x[n] = \delta[n]$, compute the impulse response $h[n]$ for $n = 0, 1, 2, 3$.
(c) Draw the direct-form structure for the system. Use no more than two memory elements (unit delays).

5.5 For the system $H(z)$ in Problem 5.4, draw a cascade-form structure with first-order sections. Are all the multipliers real?

5.6 Find the poles and zeros of the transfer functions given below.
(a) $H_1(z) = (1 + 2z^{-1})(1 - z^{-1})^3/(1 + z^{-2})^2$.
(b) $H_2(z) = (1 - 2z^{-1})^2(1 + z^{-1})^2/(1 + z^{-4})^3$.
(c) $H_3(z) = z^2 + z^{-3}$.
Be sure to specify multiplicities (orders) of poles and zeros, and do not forget to examine $z = 0$ and $z = \infty$.

5.7 For the magnitude response (5.23) (which assumes $-1 < p < 1$), show that the derivative of $|H(e^{j\omega})|^2$ with respect to ω has zeros at $\omega = 0$ and π, but no zeros in $0 < \omega < \pi$. This shows that $|H(e^{j\omega})|^2$ has no extrema in $0 < \omega < \pi$.

5.8 Let $H(z) = (1 + z^{-1})/(1 - pz^{-1})$ represent a causal LTI system. (a) Write the difference equation, (b) find an expression for the impulse response $h[n]$, and (c) plot the magnitude responses of $H(z)$ and $H(-z)$ for $p = 0.9$.

5.9 For the system $H(z)$ in Problem 5.8, suppose the input is

$$x[n] = \begin{cases} 1 & n \text{ even}, \\ 0 & n \text{ odd}. \end{cases}$$

Then what is $y[n]$? You should be able to find a simple closed-form expression.

5.10 Consider the feedback structure shown in Fig. P5.10.
(a) Let $F(z) = 1 + 2z^{-1}$ and $G(z) = z^{-1}$. (i) Find the transfer function $H(z) = Y(z)/X(z)$ of the closed-loop system. (ii) Write the recursive difference equation corresponding to this transfer function. (iii) Is $H(z)$ stable?

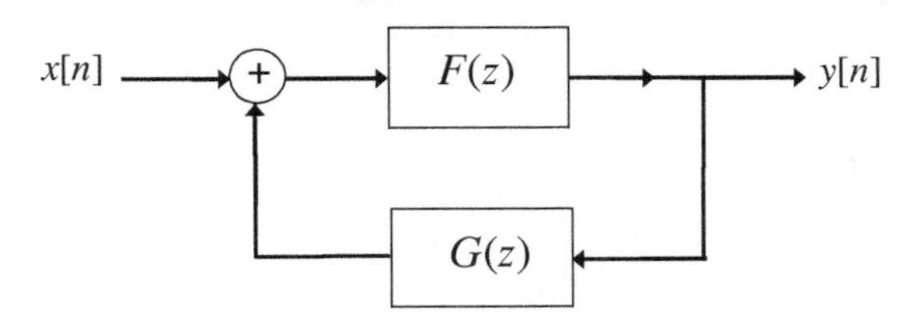

Figure P5.10 A feedback system.

(b) Now assume that $G(z)$ is changed to $z^{-1}/4$. Then what is $H(z)$? Is it stable?

5.11 If p_i are the poles of a rational LTI system $H(z)$, where are the poles of the transformed filters $H(-z), H(z^2)$, and $H(-z^2)$? Assuming $H(z)$ is causal and stable, show that the transformed filters are also causal and stable. As a specific example, assume that $p = 1/4$ is the only pole of $H(z)$; indicate the pole locations of $H(-z), H(z^2)$, and $H(-z^2)$ using a z-plane diagram.

5.12 Consider a causal LTI system $x[n] \longmapsto y[n]$ described by

$$w[n] = w[n-1] + x[n] + x[n-1],$$
$$y[n] = 0.5y[n-1] + w[n] - w[n-2].$$

What is the transfer function $H(z) = Y(z)/X(z)$? Is the system stable?

5.13 We now consider Fourier transforms of some modified exponentials.

(a) For the truncated exponential $x[n] = p^n(\mathcal{U}[n] - \mathcal{U}[n-6])$ shown in Fig. P5.13(a), find $X(e^{j\omega})$ in closed form.

(b) For the symmetric filter $h[n] = p^{|n|}$ shown in Fig. P5.13(b), where $0 < p < 1$, find $H(e^{j\omega})$ in closed form and plot it for $-\pi \le \omega \le \pi$. Clearly mark the values of $H(e^{j\omega})$ at $\omega = 0, \pi, -\pi$. What kind of filter is this (lowpass, highpass, etc.)? Show that $H(e^{j\omega})$ is positive in this example.

(c) How does the plot for $H(e^{j\omega})$ change if $-1 < p < 0$?

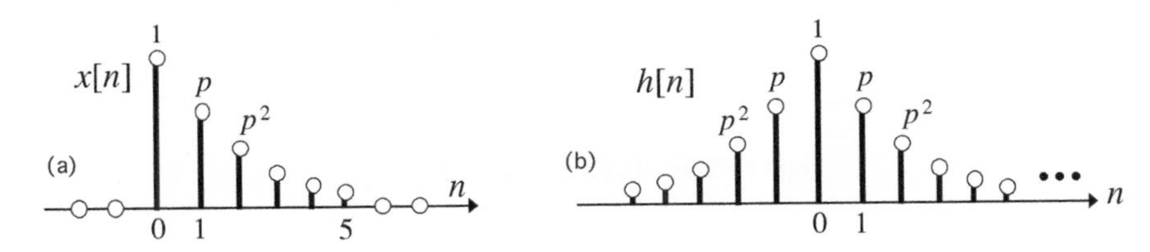

Figure P5.13 Modified exponentials.

5.14 Suppose a system is described by the recursive difference equation

$$y[n] = ny[n-1] + x[n].$$

Assume zero initial conditions $y[-1] = 0$. What is the output $y[n], n \ge 0$ for (a) $x[n] = \delta[n]$ and (b) $x[n] = \delta[n-2]$? Based on the results, argue that the system is time varying.

5.15 Consider a system described by the recursive difference equation

$$y[n] = y[n-1]x[n] + x[n].$$

(a) Assume zero initial condition $y[-1] = 0$. What is the output $y[n], n \geq 0$ for (i) $x[n] = \delta[n]$, (ii) $x[n] = \delta[n-1]$, and (iii) $x[n] = \delta[n] + \delta[n-1]$? Based on this, argue that the system is nonlinear.

(b) Now consider the modified system $y[n] = y[n-1]x[n]$. Assume zero initial condition $y[-1] = 0$. What is the output $y[n], n \geq 0$ for arbitrary input (with $x[n]$ finite for all n)?

5.16 This problem develops some interesting manifestations of stability.

(a) Consider a causal IIR filter with transfer function $H(z) = 1/D(z)$, where

$$D(z) = 1 + d_1 z^{-1} + d_2 z^{-2} + \cdots + d_N z^{-N},$$

with d_n real for all n. Show that this is BIBO stable *only if* $D(1) > 0$ and $D(-1) > 0$.

(b) As a special case, consider a second-order causal IIR filter with transfer function $H(z) = 1/(1 + az^{-1} + bz^{-2})$, with a and b real. Show that if $H(z)$ is BIBO stable then the values of a and b are restricted to be *strictly* inside the triangular area in Fig. P5.16 (called the *stability triangle*).

(c) Finally, show that, in the above second-order case, the converse is also true, that is, if a and b are strictly inside the triangular region, then $H(z)$ is BIBO stable.

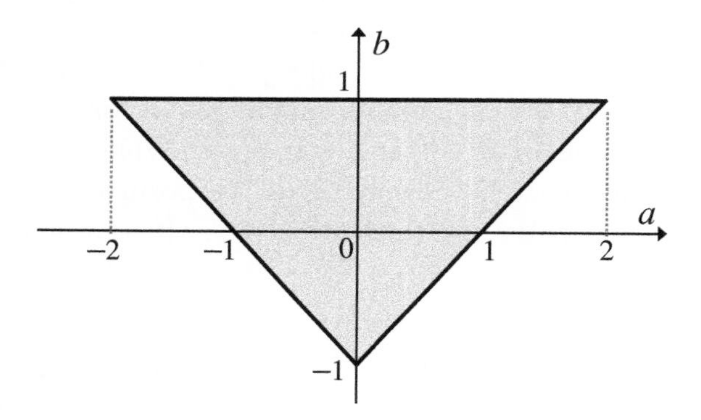

Figure P5.16 The stability triangle.

5.17 A causal LTI system has impulse response $h[n]$ equal to zero for all n except $0 \leq n \leq 3$. When the following input with duration four

$$x[0] = 2, x[1] = 1, x[2] = 1, x[3] = -1$$

is applied to the system, the first four samples of the output are

$$y[0] = 2, y[1] = -1, y[2] = 2, y[3] = 1.$$

(a) Find $h[n]$.

(b) Find $y[n]$ for $n = 4$.

5.18 Consider the causal IIR filter $H(z) = (a_0 + a_1 z^{-1} + a_2 z^{-2})/(1 + b_1 z^{-1} + b_2 z^{-2})$.
(a) Find the coefficients a_n and b_n, if the first few impulse response coefficients are $h[0] = 4, h[1] = 3, h[2] = 2, h[3] = 1$, and $h[4] = 1$.
(b) What are $h[5]$ and $h[6]$?
(c) Is the filter stable?

5.19 Show that the structure in Fig. 5.8 implements $H(z)$ in Eq. (5.65) for $N = 3$.

5.20 Consider the LTI system $G(z) = (1 - z^{-L})/(1 - z^{-1})$. Show that the zeros of $1 - z^{-L}$ have the form $z_k = e^{j2\pi k/L}, 0 \le k \le L - 1$, as shown in Fig. P5.20. Thus $z_0 = 1$ is a zero, and it cancels with the zero of the denominator $1 - z^{-1}$. So $G(z)$ is FIR, and in fact

$$\frac{1 - z^{-L}}{1 - z^{-1}} = 1 + z^{-1} + z^{-2} + \cdots + z^{-(L-1)}.$$

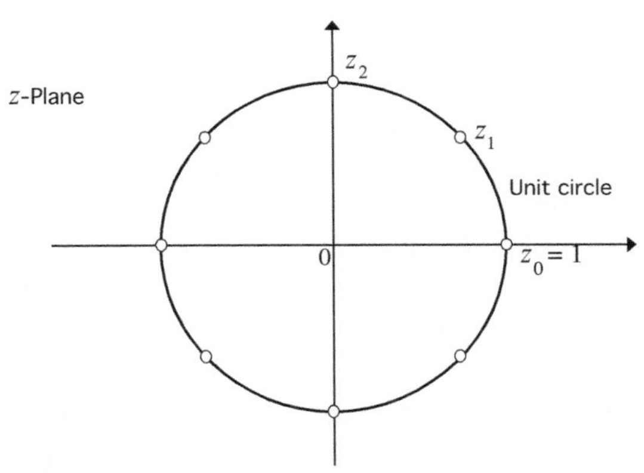

Figure P5.20 Zeros distributed uniformly on the unit circle.

5.21 Show that the transfer function of the system in Fig. 5.11 is given by Eq. (4.33), or equivalently (4.32).

5.22 Consider the LTI system

$$y[n] = \frac{1}{L}\Big(x[n] - x[n-1] + x[n-2] - x[n-3] + \cdots + (-1)^{L-1}x[n-L+1]\Big).$$

(a) Find the impulse response $h[n]$.
(b) What is the transfer function $H(z)$?
(c) Where are the poles and zeros?
(d) Derive a closed-form expression for the frequency response $H(e^{j\omega})$. Handsketch the magnitude response $|H(e^{j\omega})|$, indicating where the maximum is and where the transmission zeros are. Is it lowpass or highpass?

5.23 Consider the system in Problem 5.22. We know this FIR system can always be implemented with $L - 1$ adders per output sample. Draw a “clever” recursive structure which requires only *two* adders. How many unit delays are used by this recursive structure?

5.24 Consider two exponential signals a^n and b^n, where a and b are nonzero with $a \neq b$.

(a) Show that their sum cannot be an exponential, that is, we cannot satisfy

$$a^n + b^n = \beta c^n, \quad \forall n,$$

for any c and β.

(b) In fact, more is true. Show that we cannot satisfy

$$\alpha_0 a^n + \alpha_1 b^n = \beta c^n, \quad \forall n,$$

for any choice of $\alpha_0, \alpha_1, \beta$, and c whatsoever, unless $\alpha_0 = \alpha_1 = \beta = 0$, or c equals a or b.

5.25 (Computing assignment) *IIR lowpass filter*. Consider a causal LTI system with transfer function $H(z) = 1/(1 - pz^{-1})$ where the pole $p = 0.9$. Let the input to this system be

$$x[n] = \cos(\omega_1 n) + 0.3\cos(\omega_2 n), \quad n \geq 0,$$

and zero otherwise. Assume $\omega_1 = 0.05\pi$ (low frequency) and $\omega_2 = 0.3\pi$ (mid frequency).

(a) Plot the magnitude response $|H(e^{j\omega})|$.

(b) Plot $x[n]$ and the system output $y[n]$ for $0 \leq n \leq 200$.

You will find that $y[n]$ is mostly dominated by the low-frequency sinusoid $\cos(\omega_1 n)$. This is because $|H(e^{j\omega_1})|$ is significantly larger than $|H(e^{j\omega_2})|$.

5.26 (Computing assignment) *IIR highpass filter*. In Problem 5.25 change p and ω_2 to the following values: $p = -0.9$ and $\omega_2 = 0.95\pi$. Plot $x[n]$ and $y[n]$ for $0 \leq n \leq 200$. Also plot $|H(e^{j\omega})|$. You will find that $y[n]$ is mostly dominated by the high-frequency sinusoid $\cos(\omega_2 n)$. Explain this behavior.

5.27 (Computing assignment) *Oscillatory impulse response*. Consider a causal LTI system with transfer function given by

$$H(z) = \frac{1}{1 - 2R\cos\theta z^{-1} + R^2 z^{-2}} = \frac{1}{(1 - Re^{j\theta}z^{-1})(1 - Re^{-j\theta}z^{-1})},$$

where $R = 0.8$ and $\theta = 0.2\pi$. This is a stable system with complex poles at $Re^{j\theta}$ and $Re^{-j\theta}$. Implement the second-order difference equation to compute the impulse response $h[n]$ numerically, and plot it for $0 \leq n \leq 40$. You will find that $h[n]$ has an oscillatory behavior.

6 The z-Transform and its Inverse

6.1 Introduction

The z-transform was defined in Sec. 3.4.1 when we introduced basic properties of discrete-time LTI systems. A number of examples were also presented in Chapters 3–5. In particular, there was a brief discussion of the region of convergence (ROC) of the z-transform in Sec. 5.2.2.

This chapter goes deeper into the properties of the z-transform. First, it is shown that given a z-transform expression $X(z)$, the inverse transform $x[n]$ is not unique, and depends on the ROC of $X(z)$. Given an $X(z)$, some of the inverse transforms can be causal, some noncausal, some bounded, some unbounded, and so forth. The chapter gives a detailed explanation of how the ROC, causality, and stability of an LTI system are interrelated. One outcome of this is a detailed discussion of the stability conditions for different types of LTI systems. The partial fraction expansion (PFE) is discussed in its generality, as it is a powerful tool for finding the inverse transforms for rational $X(z)$.

Some special topics are also discussed, such as the implementation of IIR linear-phase filters, anticausal rational filters, and so forth. We also make connections between the z-transform and the theory of complex variables and analytic functions in mathematics (Sec. 6.9). It will be seen that this leads to many useful insights. A summary of the properties of the z-transform can be found in Sec. 6.12. A table of commonly used z-transform pairs is given in Sec. 6.13. In Chapter 13 we shall briefly discuss the Laplace transform, which is the continuous-time counterpart of the z-transform.

6.2 Basic Properties of the z-Transform

From the definition of the z-transform

$$X(z) = \sum_{n=-\infty}^{\infty} x[n]z^{-n}, \tag{6.1}$$

a number of basic properties can be derived, similar to what we did in Sec. 3.8 for the discrete-time Fourier transform (DTFT). Section 6.12 provides a detailed summary. Most derivations are similar to those for the DTFT. So we present derivations only for a few of them here. As done in Chapter 3 for Fourier transforms, the z-transform pair is denoted as

$$x[n] \longleftrightarrow X(z). \tag{6.2}$$

First consider the time-reversed version $x[-n]$. From the definition of $X(z)$ it readily follows that $\sum_{n=-\infty}^{\infty} x[-n]z^{-n} = \sum_{n=-\infty}^{\infty} x[n]z^{n} = X(1/z)$, so that

$$x[-n] \longleftrightarrow X(1/z). \tag{6.3}$$

Next consider a time-shifted version $x[n-K]$. Then

$$\sum_{n=-\infty}^{\infty} x[n-K]z^{-n} = \sum_{n=-\infty}^{\infty} x[n]z^{-(n+K)} = z^{-K}X(z),$$

so that

$$x[n-K] \longleftrightarrow z^{-K}X(z). \tag{6.4}$$

Next, how about the modulated version $a^n x[n]$? We have

$$\sum_{n=-\infty}^{\infty} a^n x[n]z^{-n} = \sum_{n=-\infty}^{\infty} x[n](z/a)^{-n} = X(z/a), \tag{6.5}$$

so that

$$a^n x[n] \longleftrightarrow X(z/a). \tag{6.6}$$

For the conjugated version $x^*[n]$ it can be shown that

$$x^*[n] \longleftrightarrow X^*(z^*) \tag{6.7}$$

because

$$\sum_{n=-\infty}^{\infty} x^*[n]z^{-n} = \Big(\sum_{n=-\infty}^{\infty} x[n](z^*)^{-n}\Big)^* = \big(X(z^*)\big)^* = X^*(z^*). \tag{6.8}$$

Derivations of the remaining properties are given in the various sections and problems indicated in Sec. 6.12. Since $X(z)$ is an infinite summation, it converges only in certain regions of the z-plane, and this depends on $x[n]$. In the next section we will derive the ROC of the z-transform for several examples. Given the ROC for $X(z)$, the ROCs of the z-transforms of modified signals such as $a^n x[n]$, $x[-n]$, and so forth can be readily derived, and are also summarized in Sec. 6.12.

6.3 Finding the Inverse z-Transform

A common procedure for finding the impulse response of $H(z)$ in Eq. (5.88) is to write it as a sum of simple first-order terms like $A_k/(1-p_k z^{-1})$ and add their inverse transforms to get the answer. This was done in Sec. 5.10. However, some subtleties are involved in this, which we shall explain in this section.

6.3.1 First-Order z-Transform

From Sec. 5.2.2 we know that the right-sided exponential signal

$$x[n] = a^n \mathcal{U}[n] \tag{6.9}$$

has the z-transform

$$X(z) = \sum_{n=0}^{\infty} a^n z^{-n} = \frac{1}{1 - az^{-1}}, \quad |z| > |a|. \tag{6.10}$$

(See the derivation of Eq. (5.13) from Eq. (5.11).) Since the summation converges only when $|az^{-1}| < 1$, the ROC is $|z| > |a|$ (Fig. 5.1). Now consider

$$x[n] = -a^n \mathcal{U}[-n-1], \tag{6.11}$$

which is a *left-sided* sequence. This has z-transform

$$X(z) = -\sum_{n=-\infty}^{-1} a^n z^{-n} = -\sum_{n=1}^{\infty} a^{-n} z^n = -a^{-1} z \sum_{n=0}^{\infty} (a^{-1} z)^n. \tag{6.12}$$

The summation on the right converges only when $|a^{-1}z| < 1$. Since $\sum_{n=0}^{\infty} \alpha^n = 1/(1-\alpha)$ if $|\alpha| < 1$, we see that Eq. (6.12) simplifies to

$$X(z) = \frac{1}{1 - az^{-1}}, \quad |z| < |a|. \tag{6.13}$$

Notice that the ROC is $|z| < |a|$ now. Thus, the two signals (6.9) and (6.11) have exactly the same z-transform expression $1/(1 - az^{-1})$, but the ROCs are different! Putting it another way, the *inverse* z-transform of $X(z) = 1/(1 - az^{-1})$ has two answers, depending on the ROC:

$$X(z) = \frac{1}{1 - az^{-1}} \implies x[n] = \begin{cases} a^n \mathcal{U}[n] & \text{if ROC is } |z| > |a|, \\ -a^n \mathcal{U}[-n-1] & \text{if ROC is } |z| < |a|. \end{cases} \tag{6.14}$$

So, given the z-transform expression $X(z) = 1/(1 - az^{-1})$, the inverse transform can be determined uniquely only if we are told what the ROC is. Figure 6.1 shows the ROCs for both cases, along with plots of $|x[n]|$. The **right-sided** sequence $a^n \mathcal{U}[n]$ has ROC everywhere **outside** a circle of radius $|a|$ and the **left-sided** sequence $-a^n \mathcal{U}[-n-1]$ has ROC everywhere **inside** a circle of radius $|a|$. A number of important remarks are now in order.

1. *Case when* $|a| < 1$. In Fig. 6.1 we have assumed $|a| < 1$ just for demonstration. So the unit circle is in the ROC for $a^n \mathcal{U}[n]$ but *not* in the ROC for $-a^n \mathcal{U}[-n-1]$. This means that the Fourier transform exists for $a^n \mathcal{U}[n]$ but not for $-a^n \mathcal{U}[-n-1]$ when $|a| < 1$ (because the Fourier transform exists for rational $X(z)$ only if the ROC includes the unit circle). This is consistent with the fact that $-a^n \mathcal{U}[-n-1]$ is an unbounded sequence (examine $n \to -\infty$).
2. *Case when* $|a| > 1$. Figure 6.2 shows the situation when $|a| > 1$. Now the unit circle is *not* in the ROC for $a^n \mathcal{U}[n]$, although *it is* in the ROC for $-a^n \mathcal{U}[-n-1]$. This means that the Fourier transform does not exist for $a^n \mathcal{U}[n]$ but it exists for

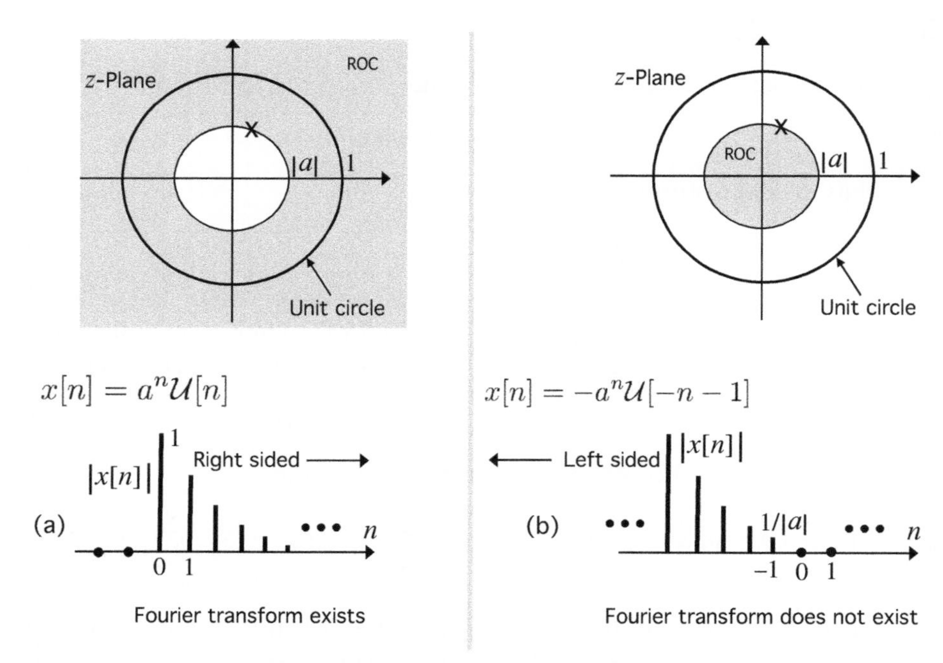

Figure 6.1 The ROC of the z-transform of (a) $x[n] = a^n\mathcal{U}[n]$ and (b) $x[n] = -a^n\mathcal{U}[-n-1]$. Plots of $|x[n]|$ are also shown. The pole $z = a$ is indicated by the cross. We assume $|a| < 1$.

$-a^n\mathcal{U}[-n-1]$ when $|a| > 1$. This is consistent with the fact that $a^n\mathcal{U}[n]$ is an unbounded sequence (examine $n \to \infty$).

3. *z-Transform may exist even if FT does not.* The preceding examples show that the z-transform often exists even when the Fourier transform does not. For example, $a^n\mathcal{U}[n]$ does not have an FT when $|a| > 1$ but its z-transform still exists for all z in the ROC $|z| > |a|$. So the z-transform is more generally applicable than the FT.
4. *Exponentials do not have a z-transform!* There exist signals for which even the z-transform does not exist. Thus, consider the exponential signal

$$x[n] = a^n, \tag{6.15}$$

that is, the usual *two-sided* exponential. We can write this as

$$x[n] = a^n\mathcal{U}[n] + a^n\mathcal{U}[-n-1]. \tag{6.16}$$

The expression for the z-transform is

$$X(z) = \sum_{n=0}^{\infty} a^n z^{-n} + \sum_{n=-\infty}^{-1} a^n z^{-n}. \tag{6.17}$$

As shown earlier, the first term converges for $|z| > |a|$ and the second term converges for $|z| < |a|$. So the ROC of $X(z)$ is the intersection (overlap) of these two regions, which is empty! The z-transform of $x[n]$ does not converge anywhere, that is, the exponential $x[n] = a^n$ does not have a z-transform.

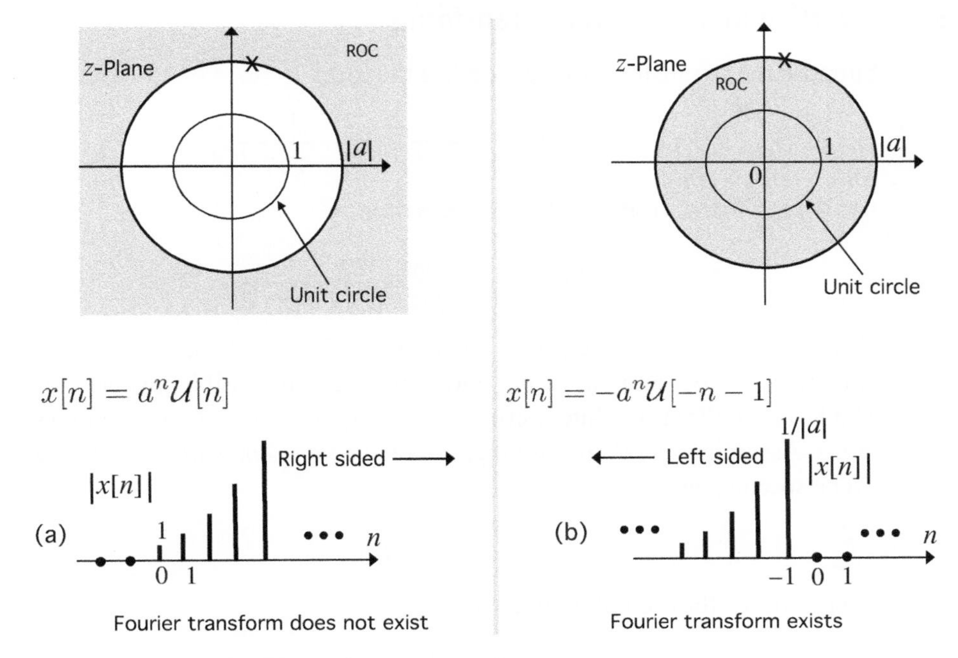

Figure 6.2 Similar to Fig. 6.1, but now $|a| > 1$.

5. *The paradox of* $e^{j\omega_0 n}$. Now consider the special case where $a = e^{j\omega_0}$. Then $x[n] = a^n = e^{j\omega_0 n}$, which has the Fourier transform (see Eq. (3.108))

$$X(e^{j\omega}) = 2\pi\delta_c(\omega - \omega_0) \tag{6.18}$$

in $-\pi \leq \omega < \pi$ (repeating periodically outside this). This might be puzzling upon first sight, because the z-transform of $x[n] = e^{j\omega_0 n}$ does not exist since it is an exponential, but the FT (6.18) evidently "exists." The fact is that in z-transform theory, $X(z)$ is required to be finite everywhere in the ROC. Since Eq. (6.18) has a Dirac delta, it is not regarded as a valid FT in z-transform theory. But engineers accept it as a valid FT because they regard Dirac delta as the limiting case of the pulse (Sec. 2.12). See Sec. 6.9.4 for other examples of this kind.

Next, by starting from the first z-transform pair in Eq. (6.14), we can readily verify that

$$R^n \cos(\theta n)\mathcal{U}[n] \quad \longleftrightarrow \quad \frac{1 - R\cos\theta\, z^{-1}}{1 - 2R\cos\theta\, z^{-1} + R^2 z^{-2}}, \quad |z| > |R| \tag{6.19}$$

and similarly

$$R^n \sin(\theta n)\mathcal{U}[n] \quad \longleftrightarrow \quad \frac{R\sin\theta\, z^{-1}}{1 - 2R\cos\theta\, z^{-1} + R^2 z^{-2}}, \quad |z| > |R|. \tag{6.20}$$

6.3.2 Inverting More General z-Transforms

Suppose the z-transform expression has the form

$$X(z) = \frac{1}{1 - p_1 z^{-1}} + \frac{1}{1 - p_2 z^{-1}}, \tag{6.21}$$

where $p_i \neq 0$ and $p_1 \neq p_2$. This can be written as

$$X(z) = \frac{a_0 + a_1 z^{-1}}{(1 - p_1 z^{-1})(1 - p_2 z^{-1})}. \tag{6.22}$$

Clearly, p_1 and p_2 are the poles. Each term in Eq. (6.21) has two possible ROCs, namely, $|z| > |p_i|$ and $|z| < |p_i|$, and therefore two possible inverse z-transforms. The ROC of $X(z)$ is the **intersection** (or overlap) of the ROCs of the two individual terms. Assuming $|p_1| \neq |p_2|$, we therefore have *four combinations* of the two ROCs. To be specific, let

$$|p_1| < |p_2|. \tag{6.23}$$

Then three of these combinations are

$$|p_1| < |p_2| < |z|, \quad |p_1| < |z| < |p_2|, \quad |z| < |p_1| < |p_2|. \tag{6.24}$$

These three choices for the ROC of $X(z)$ are shown in Fig. 6.3. The fourth combination, $|z| > |p_2|$ and $|z| < |p_1|$, is evidently not possible because $|p_1| < |p_2|$. Using Eq. (6.14) it follows that the three answers for the inverse z-transform of Eq. (6.21) are

$$x[n] = \begin{cases} p_1^n \mathcal{U}[n] + p_2^n \mathcal{U}[n] & \text{if ROC is } |z| > |p_2|, \\ p_1^n \mathcal{U}[n] - p_2^n \mathcal{U}[-n-1] & \text{if ROC is } |p_1| < |z| < |p_2|, \\ -p_1^n \mathcal{U}[-n-1] - p_2^n \mathcal{U}[-n-1] & \text{if ROC is } |z| < |p_1|. \end{cases} \tag{6.25}$$

More generally, if we have a rational $X(z)$ with N **distinct pole magnitudes** $|p_i|$, then there are $N + 1$ possible ROCs for $X(z)$ and each of these gives a different inverse z-transform $x[n]$. For example, if $N = 3$ with $|p_1| < |p_2| < |p_3|$, the ROCs are

$$|z| < |p_1|, \quad |p_1| < |z| < |p_2|, \quad |p_2| < |z| < |p_3|, \quad |z| > |p_3|. \tag{6.26}$$

The $N + 1$ ROCs (which are open regions not including the circles which form the boundaries) are usually written in the form[1]

$$0 < |z| < |p_1|, \quad |p_i| < |z| < |p_{i+1}|, \quad |p_N| < |z| < \infty, \tag{6.27}$$

where $i = 1, 2, \ldots, N - 1$. Thus, the ROCs are open annuli, bounded by two concentric circles centered at the origin. Some thought shows that the following conclusions are true:

[1] Note that whether the z-transform converges at $z = 0$ and/or $z = \infty$ depends on some trivial details, so we have excluded these points. For example, examine $X(z) = z$, $X(z) = z^{-1}$, and $X(z) = z + z^{-1}$.

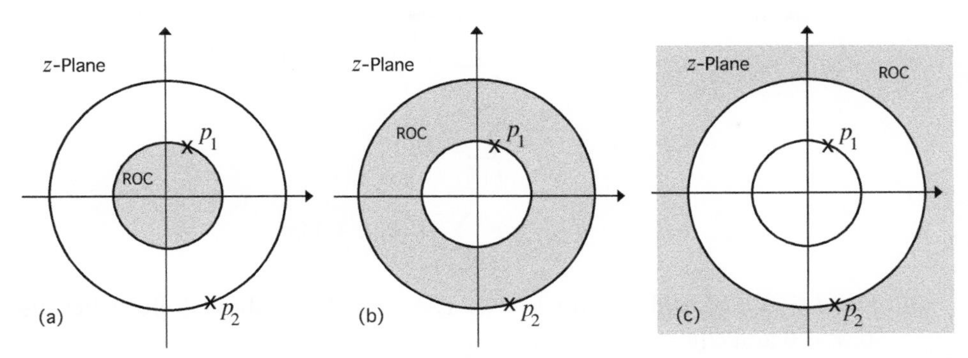

Figure 6.3 The three possible ROCs of the z-transform, when there are two finite poles with distinct magnitudes. The poles are indicated by crosses.

1. *Left and right-sided answers.* The ROC $|p_N| < |z| < \infty$ gives rise to a right-sided answer for the inverse z-transform. The ROC $0 < |z| < |p_1|$ gives rise to a left-sided answer. The remaining $N - 1$ answers are two sided. For example, when $N = 2$ we have $N - 1 = 1$, and precisely one answer is two sided as shown in Eq. (6.25).
2. *Unit circle poles.* It is clear that the unit circle will not be in any ROC if there is a pole on the unit circle.
3. *Unit circle in ROC.* When there are no poles on the unit circle, there is **exactly one ROC** *which contains the unit circle.* This is the only answer for which the Fourier transform exists. In this case the left-sided terms $p_i^n\mathcal{U}[-n-1]$ have $|p_i| > 1$ and the right-sided terms $p_i^n\mathcal{U}[n]$ have $|p_i| < 1$, so the answer is **bounded**. This is also the only answer for which

$$\sum_n |x[n]| < \infty. \tag{6.28}$$

   Note that this bounded answer is itself right sided if and only if $|p_i| < 1$ for all i, and left sided if and only if $|p_i| > 1$ for all i.

6.3.3 Handling Multiple Poles

We now consider the example

$$X(z) = \frac{1}{(1 - pz^{-1})^2}, \tag{6.29}$$

which has a **double pole** at p (pole of multiplicity two). There are only two possible ROCs, namely, $|z| < |p|$ and $|z| > |p|$. If the ROC is $|z| > |p|$ then the inverse z-transform is the convolution of the right-sided inverse z-transform $p^n\mathcal{U}[n]$ of $1/(1 - pz^{-1})$ with itself. So the answer is

$$x[n] = \sum_{k=-\infty}^{\infty} p^k \mathcal{U}[k] p^{n-k} \mathcal{U}[n-k]$$
$$= \sum_{k=0}^{n} p^k p^{n-k}$$
$$= p^n \sum_{k=0}^{n} 1 = (n+1)p^n \tag{6.30}$$

for $n \geq 0$, and $x[n] = 0$ for $n < 0$. The second line above follows because $\mathcal{U}[k] = 0, k < 0$ and $\mathcal{U}[n-k] = 0, k > n$. So

$$x[n] = (n+1)p^n \mathcal{U}[n]. \tag{6.31}$$

In particular, if we set $z = e^{j\omega}$, this leads to the DTFT pair (3.122). For the ROC $|z| < |p|$, we will leave it to the student to figure out the details (Problem 6.11). If there is a pole of multiplicity K (Kth-order pole) at a point, that is, if

$$X(z) = \frac{1}{(1 - pz^{-1})^K}, \tag{6.32}$$

then assuming the ROC is $|z| > |p|$, the inverse z-transform $x[n]$ takes the form

$$x[n] = \left(\alpha_0 + \alpha_1 n + \cdots + \alpha_{K-1} n^{K-1}\right) p^n \mathcal{U}[n], \tag{6.33}$$

for appropriate constants α_i. See Problem 6.12. When $K = 1$, we say that p is a “single” pole or a “simple” pole.

6.4 Stability of Rational LTI Systems

For a rational LTI system with transfer function $H(z)$ as in Eq. (6.21), the impulse response $h[n]$ can have three possible answers (6.25), depending on the ROC. Thus, the LTI system is completely specified only when $H(z)$ *and* its ROC are specified. We now examine the stability of this system for different cases.

6.4.1 Causal Rational LTI Systems

In practice, we are most interested in **causal** systems which have

$$h[n] = 0, \quad n < 0. \tag{6.34}$$

Thus, if we know that our system is causal, then only one of the answers in Eq. (6.25) is valid, namely

$$h[n] = p_1^n \mathcal{U}[n] + p_2^n \mathcal{U}[n], \tag{6.35}$$

which corresponds to the ROC

$$|z| > |p_i| \tag{6.36}$$

for all i. That is, the ROC is everywhere outside a circle whose radius is the largest pole magnitude (Fig. 6.3(c)). More generally, given a rational LTI system $H(z)$ with any number of poles and zeros, the only ROC which corresponds to a causal $h[n]$ is (6.36), and the causal impulse response $h[n]$ has terms of the form

$$p_i^n \mathcal{U}[n]. \tag{6.37}$$

It may also have terms of the form

$$p_i^n \mathcal{U}[n], \quad np_i^n \mathcal{U}[n], \quad n^2 p_i^n \mathcal{U}[n], \quad \ldots \quad , n^{K_i - 1} p_i^n \mathcal{U}[n], \tag{6.38}$$

where K_i is the multiplicity of pole p_i. This follows from the theory of partial fraction expansions (Sec. 6.6). In addition to the above, there may also be terms of finite duration in the impulse response (which do not affect stability). In short, for any **causal rational LTI** system the impulse response has the form

$$h[n] = \sum_{i=1}^{P} \sum_{k=0}^{K_i - 1} c_{ik} n^k p_i^n, \tag{6.39}$$

for $n \geq n_0$, where n_0 is some finite non-negative integer. (Although it is not critical, Sec. 6.6 explains why the first few samples of $h[n]$ are not of the form (6.39)). Here P is the number of **distinct** poles, that is, $p_i \neq p_j$ in Eq. (6.39).[2] K_i is the multiplicity of pole p_i.

The first conclusion from Eq. (6.39) is that the impulse response of a causal rational LTI system **decays to zero** as $n \to \infty$, if $|p_i| < 1$ for all i. This is because

$$\lim_{n \to \infty} n^k p_i^n = 0, \tag{6.40}$$

for any finite $k \geq 0$. In fact we will prove the following, which is a generalization of Theorem 5.1 (that assumed the poles to be distinct).

Theorem 6.1 *Stability of causal rational LTI systems.* A causal rational LTI system is stable, that is,

$$\sum_{n=-\infty}^{\infty} |h[n]| < \infty, \tag{6.41}$$

if and only if

$$|p_i| < 1 \tag{6.42}$$

for all i, that is, if and only if **all poles are inside** the unit circle. ◇

[2] Note that in Sec. 6.3.2 the number of distinct pole *magnitudes* determined the number of ROCs. But in the current discussion P is the number of distinct poles, not pole magnitudes. For example, if the poles are $\{1, 1, 2, -1\}$ then there are three distinct poles $\{1, 2, -1\}$ and two distinct magnitudes $\{1, 2\}$. The pole $p_1 = 1$ has multiplicity 2.

Sketch of the proof. Since $h[n]$ is causal, and can be expressed as in Eq. (6.39) for $n \geq n_0$, we have

$$\sum_{n=-\infty}^{\infty} |h[n]| = \sum_{n=0}^{\infty} |h[n]| \leq \sum_{n=0}^{n_0-1} |h[n]| + \left(\sum_{i=1}^{P} \sum_{k=0}^{K_i-1} |c_{ik}| \sum_{n=n_0}^{\infty} |n^k p_i^n| \right). \quad (6.43)$$

Using the fact that

$$\sum_{n=n_0}^{\infty} |n^k p_i^n| < \infty \quad (6.44)$$

if $|p_i| < 1$, it then follows that $\sum_{n=-\infty}^{\infty} |h[n]| < \infty$. This proves that the condition (6.42) is sufficient for stability. The proof that it is necessary is similar to the proof of necessity in Theorem 5.1, although it is more involved. We can obtain an equation similar to Eq. (5.100) but the matrix $\mathbf{V}$ is not exactly Vandermonde anymore, because the terms p_i^n are replaced with more complicated terms like $n^k p_i^n$ due to multiplicity of poles. We shall skip the details here. $\nabla \nabla \nabla$

6.4.2 Anticausal Rational LTI Systems

We now make some remarks on noncausal LTI systems. Although causal systems are the most desirable, it is sometimes possible to simulate noncausal discrete-time systems on a computer. So the following discussion has more than mere academic value. Returning again to the example of Eq. (6.21), consider now the inverse z-transform corresponding to the ROC $|z| < |p_1|$. This ROC is everywhere *inside* the circle with radius equal to the smallest pole magnitude (Fig. 6.3(a)). In this case we have

$$h[n] = -p_1^n \mathcal{U}[-n-1] - p_2^n \mathcal{U}[-n-1], \quad (6.45)$$

which is noncausal. In fact, it is **anticausal** because $h[n] = 0$ for $n \geq 0$. So, the behavior of p_i^n as $n \to -\infty$ (and not $n \to \infty$) determines the stability of this system. It therefore follows that this system is stable if and only if

$$|p_i| > 1 \quad (6.46)$$

for all i. By using arguments similar to those leading up to Theorem 6.1, we can prove that an anticausal rational LTI system is stable if and only if all its poles satisfy the condition (6.46), that is, are **outside the unit circle**. Just like Theorem 6.1, this is valid no matter how many poles there are, and whether they are distinct or not.

As mentioned at the beginning of this subsection, anticausal systems have some practical value. They can actually be implemented on a computer, provided the entire input signal is already available in stored form (as in image processing). In some image compression applications, a combination of causal and anticausal filters is used to achieve linear phase, which prevents phase distortion (see Sec. 6.8 and Problem 6.15). Another example is in situations where one finds that a causal system has to be inverted (like a channel transfer function in digital communications). In such a situation, the inverse may not have a stable implementation unless it is noncausal. This happens if the system to be inverted has some zeros outside the unit circle, because these zeros become poles of the inverse. The details of the above applications are beyond the scope of our discussion. The interested reader can refer to Chapter 4 of Vaidyanathan et al. [2010]. For more results on anticausal inverses, see Vaidyanathan and Chen [1995a, 1995b, 1998], and references therein.

6.4.3 General Statements on Stability

Let us pause for a minute to recollect the main points about stability. For a rational LTI system with poles p_i, the impulse response contains causal (or right-sided) terms of the form

$$p_i^n \mathcal{U}[n], \tag{6.47}$$

and minor variations like $n^k p_i^n \mathcal{U}[n]$ (for multiple poles). It also contains anticausal (or left-sided) terms of the form

$$p_j^n \mathcal{U}[-n-1], \tag{6.48}$$

and minor variations. The causal terms decay to zero as $n \to \infty$ if $|p_i| < 1$, and the anticausal terms decay to zero as $n \to -\infty$ if $|p_j| > 1$. Since these are one-sided exponentials (sometimes weighted by polynomials n^k), it follows that whenever they *decay to zero*, they are also *absolutely summable*.

Now let us think of the ROC. The causal terms $p_i^n \mathcal{U}[n]$ arise when the ROC satisfies $|z| > |p_i|$ and the anticausal terms $p_j^n \mathcal{U}[-n-1]$ arise when the ROC satisfies $|z| < |p_j|$. So, the ROC

$$|p_{i_0}| < |z| < |p_{j_0}| \tag{6.49}$$

for which all the terms are decaying is such that $|p_{i_0}| < 1$ and $|p_{j_0}| > 1$. In other words, such an ROC includes the unit circle. Conversely, whenever the ROC contains the unit circle, it follows that $|p_i| < 1$ for all right-sided exponentials and $|p_j| > 1$ for all left-sided exponentials, so that all these one-sided exponentials are decaying.

The impulse response of a rational LTI system is therefore absolutely summable, that is, the system is **stable, if and only if the ROC contains the unit circle**, that is, the frequency response $H(e^{j\omega})$ exists (because $H(e^{j\omega})$ is the z-transform evaluated on the unit circle). In short, therefore, for a given rational LTI transfer function $H(z)$, the unique choice of ROC which gives a stable inverse transform $h[n]$ is the one which includes the unit circle. Of course, such an ROC does not exist if there is a pole on the unit circle ($|p_k| = 1$ for some k), in which case there is *no* stable inverse transform.

6.4.4 Summary on Stability

1. *Absolute summability and stability.* An LTI system $H(z)$ is stable if and only if $\sum_n |h[n]| < \infty$.
2. *Decay and stability.* For a rational LTI system $H(z)$, stability is equivalent to saying that $h[n] \to 0$ as $n \to \pm\infty$, because $h[n]$ is a linear combination of terms of the form $p_i^n \mathcal{U}[n], p_j^n \mathcal{U}[-n-1]$, and some minor variations.
3. *Stability, unit circle, and frequency response.* For a rational LTI system $H(z)$, stability is equivalent to saying that the ROC of $H(z)$, which has the form $R_1 < |z| < R_2$, includes the unit circle. This is equivalent to saying that the frequency response $H(e^{j\omega})$ exists.
4. *Causality and stability.* A causal rational LTI system $H(z)$ is stable if and only if all poles are inside the unit circle ($|p_i| < 1$).
5. *Anticausality and stability.* An anticausal rational LTI system $H(z)$ is stable if and only if all poles are outside the unit circle ($|p_j| > 1$).
6. *Two-sided $h[n]$ and stability.* If a rational LTI system $H(z)$ has nontrivial poles both inside and outside the unit circle, then it can be stable only if the impulse response is two-sided. Of course, not all possible two-sided impulse repsonses are stable. There is at most one ROC that can contain the unit circle, and the specific two-sided impulse response corresponding to this ROC is stable.

We will see some numerical examples in the next few sections and in the problems.

6.4.5 Unit-Circle Poles

If the rational transfer function $H(z)$ has a pole on the unit circle, then no choice of the ROC will result in a stable system. To give such an example, consider

$$H(z) = \frac{1}{1 - e^{j\omega_0} z^{-1}}, \tag{6.50}$$

which has the unit-circle pole $p = e^{j\omega_0}$. If we take the ROC to be $|z| > 1$, then the impulse response is

$$h[n] = e^{j\omega_0 n} \mathcal{U}[n]. \tag{6.51}$$

This does not decay to zero, though it is bounded. Evidently it is not absolutely summable, consistent with the instability property. In fact, a bounded input of exactly the same form $x[n] = e^{j\omega_0 n} \mathcal{U}[n]$ produces the output

$$y[n] = \sum_{k=0}^{n} h[k]x[n-k] = (n+1)e^{j\omega_0 n} \mathcal{U}[n], \tag{6.52}$$

which is unbounded, verifying instability. Next consider the rational transfer function

$$G(z) = \frac{1}{(1 - e^{j\omega_0} z^{-1})^2}. \tag{6.53}$$

This has a **double pole** on the unit circle at $z = e^{j\omega_0}$. The impulse response is now exactly the expression (6.52), and is not even bounded. Thus, *single poles* on the unit circle yield bounded (but unstable) impulse responses. *Multiple poles* on the unit circle yield unbounded (hence unstable) impulse responses. Unstable filters with single poles on the unit circle are useful to design digital oscillators and more general waveform generators (Sec. 8.5).

6.5 A Numerical Example of the Inverse z-Transform

Consider the z-transform

$$H(z) = \frac{4 - \frac{5}{4}z^{-1}}{(1 - \frac{1}{2}z^{-1})(1 - \frac{1}{4}z^{-1})} = \frac{4 - \frac{5}{4}z^{-1}}{1 - \frac{3}{4}z^{-1} + \frac{1}{8}z^{-2}}, \tag{6.54}$$

which has poles of multiplicity one at $z = 1/2$ and $z = 1/4$. It is readily verified that this can be rewritten as

$$H(z) = \frac{3}{1 - \frac{1}{2}z^{-1}} + \frac{1}{1 - \frac{1}{4}z^{-1}}. \tag{6.55}$$

A formal approach to obtain such a decomposition is the PFE technique described in Sec. 6.6. To find the inverse z-transform $h[n]$, we have to compute the inverse z-transforms of the first term and the second term and add the results. For this we require further information about the ROC. The first term can have two possible ROCs, namely $|z| < 1/2$ and $|z| > 1/2$. The second term has the two possible ROCs $|z| < 1/4$ and $|z| > 1/4$. Since $H(z)$ converges in the intersection of the ROCs of the two terms, we can identify three possibilities for the ROC of $H(z)$, namely

$$|z| < 1/4, \quad 1/4 < |z| < 1/2, \quad \text{and} \quad |z| > 1/2. \tag{6.56}$$

Depending on the choice of ROC, there are three possible answers for the inverse z-transform, as summarized below:

$$\begin{aligned}
h_a[n] &= -3(1/2)^n\mathcal{U}[-n-1] - (1/4)^n\mathcal{U}[-n-1] && (\text{ROC } |z| < 1/4),\\
h_m[n] &= -3(1/2)^n\mathcal{U}[-n-1] + (1/4)^n\mathcal{U}[n] && (\text{ROC } 1/4 < |z| < 1/2),\\
h_c[n] &= 3(1/2)^n\mathcal{U}[n] + (1/4)^n\mathcal{U}[n] && (\text{ROC } |z| > 1/2).
\end{aligned}$$

Needless to say, all of these are IIR systems. The subscripts a, m, and c stand for *anticausal*, *mixed*, and *causal*, respectively. Figure 6.4 summarizes the three possible ROCs, and also the one-sided exponential components in $h[n]$ in each case. There is one left-sided (anticausal) solution, one right-sided (causal) solution, and one two-sided solution. Only one of the solutions decays to zero as $|n| \to \infty$, namely the right-sided sequence. This is also the only stable solution satisfying $\sum_n |h[n]| < \infty$. It is the only solution for which the unit circle is in the ROC, so that the frequency response $H(e^{j\omega})$ exists.

Is there a mixed difference equation? In Sec. 6.7 we will explain that LTI systems with anticausal impulse responses can be implemented by running the difference equation backwards in time, that is, by rewriting Eq. (6.80) as in Eq. (6.81). Thus, $h_c[n]$ can be implemented using Eq. (6.80) whereas $h_a[n]$ can be implemented using Eq. (6.81). But how about $h_m[n]$? Is there a "mixed difference equation" to realize

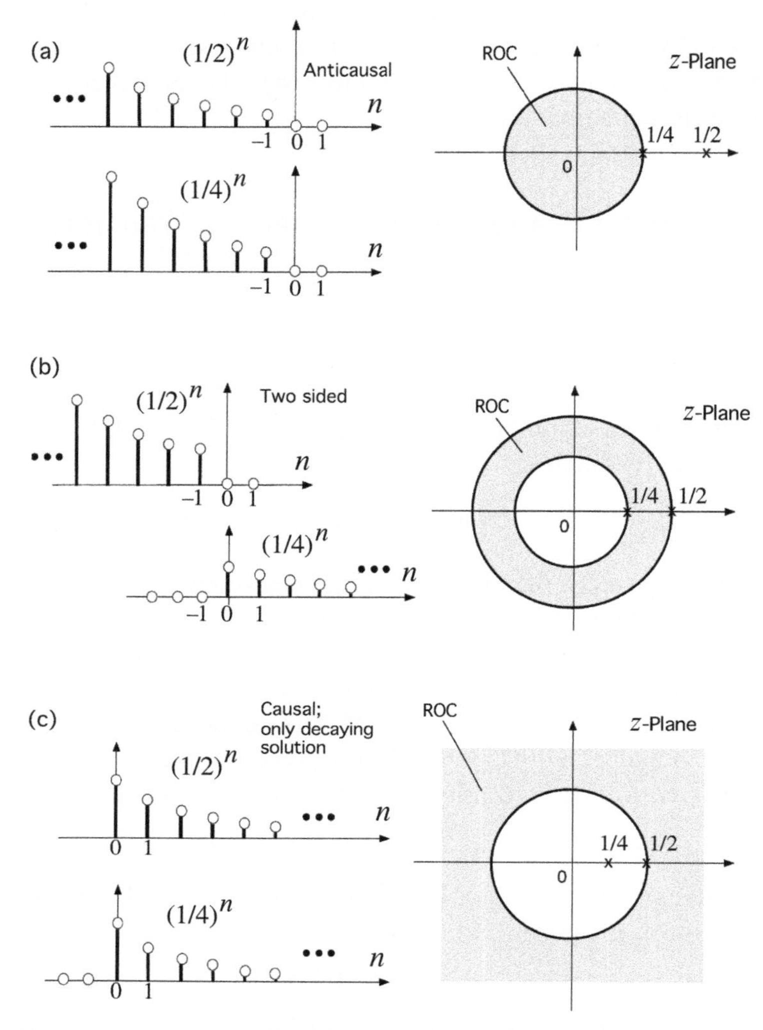

Figure 6.4 (a)–(c) Three possible regions of convergence for $H(z)$ in Eq. (6.54). In each case the one-sided exponentials which make up $h[n]$ are also shown.

it? More generally, if we have a system with N distinct pole magnitudes, then there are $N - 1$ such mixed responses. How do we figure out the right difference equation for each?

The answer is, we *do not need* to figure this out! Instead of working directly with the difference equation, we just go to the partial fraction expansion. For example, referring to Eq. (6.55), we can implement each term independently. Each term corresponds to a first-order difference equation. We implement the first term backwards in time (i.e., anticausally), the second term forward in time (i.e., causally), and add the results. The result is that we get precisely the mixed-impulse response $h_m[n]$ shown above!

▽ ▽ ▽

Notice that instead of specifying the ROC in order to get the right answer for the inverse z-transform, one can specify some side information indirectly. For example, "find the right-sided solution" or "find the two-sided solution" will produce a unique answer in the above example. Indeed, the discussion in Secs. 6.3.2 and 6.4 shows that for any rational LTI system with any number of poles there is precisely one inverse transform that is right sided and one that is left sided. But in general there can be $N-1$ distinct two-sided answers if there are N distinct pole magnitudes. So, "find the two-sided solution" does not produce a unique answer unless $N = 2$. Notice finally that the statement "find the stable solution" produces a unique answer (if one exists) because there is at most one ROC which contains the unit circle, and that yields the stable $h[n]$. Of course if there is a pole on the unit circle, then there is no such ROC and no stable inverse transform!

Pictorial interpretation. To obtain a clear understanding of ROC and time-domain behavior, consider

$$x[n] = a^n \mathcal{U}[n] + b^n \mathcal{U}[-n-1]. \tag{6.57}$$

In Fig. 6.5 we have shown four different combinations of values for a and b (assumed real for simplicity). For the first three cases there is a nonempty ROC $|a| < |z| < |b|$, whereas there is no ROC for the fourth case (because $|a| > |b|$). Of the first three cases, only the third case includes the unit circle in the ROC, and this is the only case where $x[n]$ decays to zero for both $n \to \infty$ and $n \to -\infty$. For each of the four cases, the thin lines indicate the envelopes of the plots of a^n and b^n. The signal $x[n]$ is obtained by choosing one of these envelopes for $n \geq 0$ and the other envelope for $n < 0$. We see that there is a nonempty ROC if and only if the chosen envelope for each side is the **smaller** of the two choices (assuming the envelopes are normalized to unity at the origin). If the chosen envelopes also decay to zero as $n \to \pm\infty$, then the ROC includes the unit circle. ▽ ▽ ▽

6.6 Partial Fraction Expansion

The PFE is a useful method to decompose a rational transfer function of arbitrary order into a sum of simple terms whose impulse responses can be found readily by inspection. We have already seen applications of this in Secs. 5.10 and 6.5. We now present the general theory of this decomposition. Let the rational transfer function be of the form

$$H(z) = \frac{\sum_{k=0}^{L} e_k z^{-k}}{1 + \sum_{k=1}^{N} b_k z^{-k}}. \tag{6.58}$$

It will be assumed that $e_L \neq 0$ and $b_N \neq 0$, so that the numerator order is L and the denominator order is N. This notation is more convenient than Eq. (5.65) when

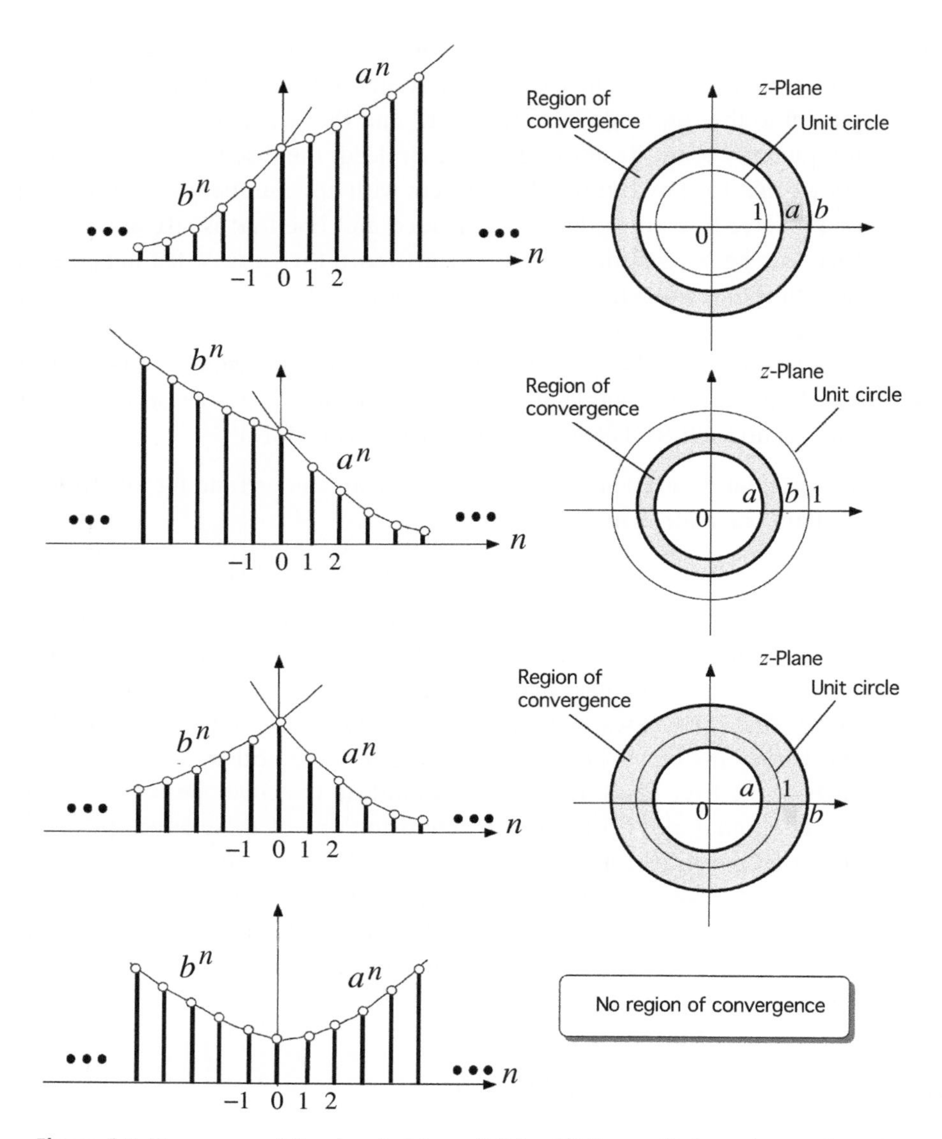

Figure 6.5 Four cases of the signal $x[n] = a^n\mathcal{U}[n] + b^n\mathcal{U}[-n-1]$. In each case the ROC $|a| < |z| < |b|$ of the z-transform is shown. In the last case $|a| > |b|$, so the ROC is empty. We assume a and b are real for simplicity.

one wants to distinguish between $L \geq N$ and $L < N$. When $L \geq N$, we can use long division and rewrite

$$H(z) = \sum_{n=0}^{L-N} c[n]z^{-n} + \frac{\sum_{k=0}^{N-1} a_k z^{-k}}{1 + \sum_{k=1}^{N} b_k z^{-k}} = C(z) + \frac{A(z)}{B(z)}, \tag{6.59}$$

where the **FIR part** $C(z)$ has the unique inverse transform $c[n]$. Here is an example:

$$\frac{5 - \frac{11}{8}z^{-2} + \frac{1}{4}z^{-3}}{1 - \frac{3}{4}z^{-1} + \frac{1}{8}z^{-2}} = 1 + 2z^{-1} + \frac{4 - \frac{5}{4}z^{-1}}{1 - \frac{3}{4}z^{-1} + \frac{1}{8}z^{-2}}. \tag{6.60}$$

Note that for $L = N$ the FIR part is just a constant term $c[0]$. If, on the other hand, $L < N$, there is no FIR term $C(z)$ at all. In any case, in order to find the general form of $h[n]$, it only remains to find the inverse transform of $A(z)/B(z)$ where $A(z)$ has order *less* than $B(z)$. We have

$$\frac{A(z)}{B(z)} = \frac{\sum_{k=0}^{N-1} a_k z^{-k}}{1 + \sum_{k=1}^{N} b_k z^{-k}} = \frac{\sum_{k=0}^{N-1} a_k z^{-k}}{\prod_{i=1}^{N}(1 - p_i z^{-1})}, \tag{6.61}$$

where p_i are the poles of $H(z)$, and hence the poles of $A(z)/B(z)$.

6.6.1 Case where Poles are Distinct

For the case where the poles are distinct, that is, when

$$p_i \neq p_j, \quad \text{for} \quad i \neq j, \tag{6.62}$$

Eq. (6.61) can be written in the form

$$\frac{\sum_{k=0}^{N-1} a_k z^{-k}}{\prod_{i=1}^{N}(1 - p_i z^{-1})} = \frac{A_1}{1 - p_1 z^{-1}} + \frac{A_2}{1 - p_2 z^{-1}} + \cdots + \frac{A_N}{1 - p_N z^{-1}}. \tag{6.63}$$

The right-hand side is called the partial fraction expansion of the left-hand side, and the constants A_i are called the **residues** of $A(z)/B(z)$ at the poles p_i. To prove Eq. (6.63), note that the right-hand side can be rewritten as

$$\frac{\sum_{k=1}^{N} A_k \prod_{i \neq k}(1 - p_i z^{-1})}{\prod_{i=1}^{N}(1 - p_i z^{-1})}. \tag{6.64}$$

Thus, given the left-hand side of Eq. (6.63), we can obtain the PFE by finding the N constants A_k such that

$$\sum_{k=0}^{N-1} a_k z^{-k} = \sum_{k=1}^{N} A_k \prod_{i \neq k}(1 - p_i z^{-1}). \tag{6.65}$$

This can be rearranged as a set of N linear equations in the N unknowns A_k, which can be solved for to obtain the partial fraction expansion (6.63). There is a simpler way to solve for A_k, which we explain next. Substituting Eq. (6.63) into Eq. (6.59) it follows that

$$H(z) = \sum_{n=0}^{L-N} c[n]z^{-n} + \frac{A_1}{1 - p_1 z^{-1}} + \frac{A_2}{1 - p_2 z^{-1}} + \cdots + \frac{A_N}{1 - p_N z^{-1}}. \tag{6.66}$$

It can easily be verified from Eq. (6.66) that

$$A_i = \left(1 - z^{-1}p_i\right)H(z)\Big|_{z=p_i}. \tag{6.67}$$

The residues A_i can readily be evaluated from this formula.

6.6.2 Case where Poles are Not Distinct

For the case where the poles are not distinct, the PFE is more complicated. For example, suppose $p_1 = p_2$. Then we say there is a double pole at p_1. In this case

$$\frac{A_1}{1 - p_1 z^{-1}} + \frac{A_2}{1 - p_2 z^{-1}} = \frac{A_1 + A_2}{1 - p_1 z^{-1}}. \tag{6.68}$$

So, this cannot contribute to a second-order transfer function. This shows that Eq. (6.66) is incapable of giving the correct order for $H(z)$, and is therefore not the correct form of the PFE when there is a multiple pole somewhere. It can be shown that when $p_1 = p_2$, the two terms which involve p_1 and p_2 in Eq. (6.66) should actually be replaced with

$$\frac{A_1}{1 - p_1 z^{-1}} + \frac{A_2}{(1 - p_1 z^{-1})^2}. \tag{6.69}$$

Similarly, if $p_1 = p_2 = p_3$ (i.e., there is a triple pole at p_1) then the three terms which involve p_1, p_2, and p_3 in Eq. (6.66) should be replaced with

$$\frac{A_1}{1 - p_1 z^{-1}} + \frac{A_2}{(1 - p_1 z^{-1})^2} + \frac{A_3}{(1 - p_1 z^{-1})^3}. \tag{6.70}$$

There are ways to identify the constants A_i in these modified expressions, using slightly more complicated formulas than (6.67).

The advantage of the PFE is that we can easily find the inverse z-transform of any rational $H(z)$ from the PFE. This is because the inverse transform of each of the terms in Eq. (6.66) can readily be found using the methods described in Sec. 6.3.1. Similarly, the inverse transforms of the terms in Eq. (6.70) can be found using the methods in Sec. 6.3.3. For example, if we are looking for the causal (right-sided) impulse response of Eq. (6.66), the answer is

$$h[n] = c[n] + \sum_{k=1}^{N} A_k p_k^n \mathcal{U}[n]. \tag{6.71}$$

The first term $c[n]$ is zero outside $0 \leq n \leq L - N$ and represents the FIR part. The second term is the IIR term due to the nontrivial poles p_k. If the PFE has a term of the form (6.69) due to a double pole, then $h[n]$ will have a term of the form

$$(\alpha_0 + \alpha_1 n)p_1^n \mathcal{U}[n], \tag{6.72}$$

for appropriate constants α_0 and α_1 (Sec. 6.3.3). More generally, when $H(z)$ has a pole of multiplicity K (i.e., the denominator of $H(z)$ has the factor $(1 - p_1 z^{-1})^K$), the impulse response of $H(z)$ contains a linear combination of terms

$$p_1^n,\ np_1^n,\ \ldots,\ n^{K-1}p_1^n, \tag{6.73}$$

in addition to contributions from the other poles (Sec. 6.3.3).

6.7 Implementation of Anticausal LTI Systems*

Consider again the first-order recursive difference equation (5.3) reproduced below:

$$y[n] = py[n-1] + x[n]. \tag{6.74}$$

In Sec. 5.2 we started from $n = 0$ and implemented this **forward** in time by computing $y[0], y[1], y[2], \ldots$, in that order. Because of this scheme of calculations we naturally got a causal system. It was LTI when the initial condition $y[-1] = 0$, with impulse response and transfer function given by

$$h[n] = p^n \mathcal{U}[n] \qquad \text{and} \qquad H(z) = \frac{1}{1 - pz^{-1}}, \quad |z| > |p|. \tag{6.75}$$

Now observe that the very same recursive difference equation can be rewritten as

$$y[n-1] = \frac{y[n]}{p} - \frac{x[n]}{p}. \tag{6.76}$$

Even though it is the same equation rearranged, it gives a different ordering of the computation. It is now **noncausal** because $y[k]$ depends on $x[n], n > k$. We now consider $y[0]$ to be the initial condition, assume the input is $x[0], x[-1], x[-2], \ldots$ (i.e., zero for positive time), and compute **backwards** in time. Thus

$$\begin{aligned} y[-1] &= \frac{y[0]}{p} - \frac{x[0]}{p}, \\ y[-2] &= \frac{y[-1]}{p} - \frac{x[-1]}{p}, \\ y[-3] &= \frac{y[-2]}{p} - \frac{x[-2]}{p}, \end{aligned}$$

and so on. In particular, with zero IC ($y[0] = 0$) we get an LTI system; with $x[n] = \delta[n]$ we can then find its impulse response as follows:

$$y[-1] = -\frac{1}{p}, \qquad y[-2] = -\frac{1}{p^2}, \qquad y[-3] = -\frac{1}{p^3}, \tag{6.77}$$

and so forth. That is,

$$h[n] = -p^n \mathcal{U}[-n-1], \tag{6.78}$$

which is precisely the anticausal inverse transform of $H(z) = 1/(1 - pz^{-1})$ obtained earlier by taking the ROC to be $|z| < |p|$ (Sec. 6.3.1). Summarizing, the difference equation (6.74) implemented forward in time is causal, with impulse response and transfer function (6.75). The same difference equation implemented backwards in time, that is, as in Eq. (6.76), has impulse response and transfer function

$$h[n] = -p^n \mathcal{U}[-n-1] \qquad \text{and} \qquad H(z) = \frac{1}{1 - pz^{-1}}, \quad |z| < |p|. \tag{6.79}$$

This is clearly anticausal.

Generalization. The above ideas can be generalized for arbitrary order N. We briefly summarize this for $N = 2$ here. The second-order recursive difference equation

$$y[n] = -b_1 y[n-1] - b_2 y[n-2] + a_0 x[n] + a_1 x[n-1] + a_2 x[n-2] \tag{6.80}$$

can be rearranged as

$$y[n-2] = \frac{1}{b_2}\Big(-y[n] - b_1 y[n-1] + a_0 x[n] + a_1 x[n-1] + a_2 x[n-2]\Big), \tag{6.81}$$

or equivalently

$$y[n] = \frac{1}{b_2}\Big(-b_1 y[n+1] - y[n+2] + a_2 x[n] + a_1 x[n+1] + a_0 x[n+2]\Big). \tag{6.82}$$

Once again, given an input

$$x[0], x[-1], x[-2], \ldots,$$

and any set of initial conditions $y[1], y[2]$, we can compute

$$y[0], y[-1], y[-2], \ldots$$

backwards in time. This gives a noncausal implementation of the difference equation. While Eq. (6.80) has a causal impulse response with terms like $p_i^n \mathcal{U}[n]$, Eq. (6.81) has an anticausal impulse response with terms like $-p_i^n \mathcal{U}[-n-1]$. Similarly, while the transfer function remains the same for both Eqs. (6.80) and (6.81), namely,

$$H(z) = \frac{a_0 + a_1 z^{-1} + a_2 z^{-2}}{1 + b_1 z^{-1} + b_2 z^{-2}}, \tag{6.83}$$

the ROCs are different in the two cases. With poles satisfying $0 < |p_1| < |p_2|$, the ROC for Eq. (6.80) (causal system) is $|z| > |p_2|$ (i.e., everywhere outside the circle whose radius is the largest pole magnitude). For Eq. (6.81) (anticausal system), the ROC is $|z| < |p_1|$ (i.e., everywhere inside the circle whose radius is the smallest pole magnitude).

▽ ▽ ▽

6.8 IIR Linear-Phase Filters*

The above implementation of anticausal systems, which uses a difference equation backwards in time, has an interesting application. Before explaining this, let us recall something about linear-phase filters. In Sec. 4.6 it was shown that FIR filters can achieve exact linear phase if the impulse response has certain symmetry, for example,

$$h[n] = h^*[K - n] \tag{6.84}$$

for some integer K (see Eq. (4.85)). Linear phase is important in some applications, as explained in Sec. 4.5.2. However, FIR linear-phase filters typically have much higher order than IIR filters with comparable magnitude responses (Sec. 7.10). Because of this, it is of interest to try and design IIR linear-phase filters.

But there is a catch. If an IIR filter has linear phase, then it has to satisfy the symmetry condition (6.84) or some variation of this. Since the filter is IIR, there exists nonzero $h[n]$ for arbitrarily large n. Because of the symmetry (6.84), this implies that there exists nonzero $h[n]$ for negative n with arbitrarily large $|n|$. This means the filter has infinite impulse response in both directions (i.e., as $n \to \infty$ and $n \to -\infty$). We cannot therefore obtain a shifted version $g[n] = h[n - n_0]$ to make it causal. It will necessarily be a noncausal filter.

In short, **IIR linear-phase filters are necessarily noncausal**. Such filters can, however, be made stable. For example, consider the IIR filter with impulse response

$$h[n] = p^n \mathcal{U}[n] + \left(\frac{1}{p}\right)^n \mathcal{U}[-n - 1], \tag{6.85}$$

where $-1 < p < 1$. The impulse response is plotted in Fig. 6.6(a) and satisfies $h[n] = h[-n]$, so that it has linear phase. The impulse response decays to zero as $n \to \pm\infty$. The filter is stable because Eq. (6.85) is absolutely summable when $|p| < 1$. The transfer function is

$$H(z) = \frac{1}{1 - pz^{-1}} - \frac{1}{1 - p^{-1}z^{-1}} \tag{6.86}$$

$$= \frac{az^{-1}}{1 - bz^{-1} + z^{-2}}, \tag{6.87}$$

where $a = p - p^{-1}$ and $b = p + p^{-1}$. Thus, there is one pole p inside the unit circle and one pole p^{-1} outside. The ROC is $p < |z| < p^{-1}$, which includes the unit circle. See Fig. 6.6(b).

It is possible to implement the above linear-phase IIR filter using recursive difference equations. The first term in Eq. (6.86) can just be implemented as a recursive difference equation running forward in time. The second term can be implemented as a recursive difference equation running backwards in time, as explained in Sec. 6.7. If the outputs generated by the two terms are added, the result is the desired output of $H(z)$.

Even though difference equations running backwards in time may seem impractical, this can be achieved if the *data to be filtered has already been acquired* and

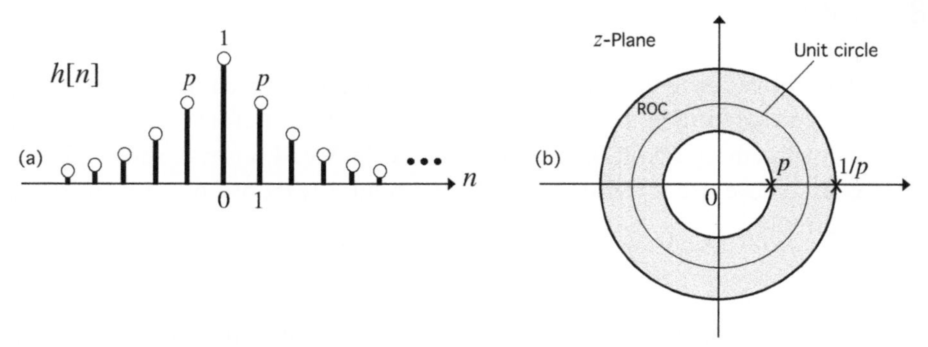

Figure 6.6 (a) The impulse response $h[n]$ of a linear-phase IIR filter; (b) ROC $p < |z| < 1/p$ of its transfer function $H(z)$.

stored in a computer, as in many digital signal processing applications, including image processing. Only in real-time processing do such noncausal filters become unrealizable.

6.9 Insights from Analytic Function Theory*

The z-transform, which is a popular tool in signal processing, was already a well-established topic in complex variable theory. In fact, it is nothing but a Laurent series which comes up in the discussion of analytic functions [Churchill and Brown, 1984]. We can gain more insights about the z-transform by looking at it from this point of view. The **Laurent series** is a doubly infinite series of the form[3]

$$X(z) = \sum_{n=-\infty}^{\infty} x[n]z^{-n}. \tag{6.88}$$

It is therefore nothing but the z-transform of $x[n]$. It is well known that this has ROC of the form

$$R_1 < |z| < R_2. \tag{6.89}$$

Thus, the ROC is the annulus bounded by two circles of radii R_1 and R_2 (Fig. 6.7), but it does not include the circles. So it is an open region. It is also known that the series necessarily diverges for all z in $|z| < R_1$ and $|z| > R_2$ [Apostol, 1974, p. 234]. For any z in the ROC, $X(z)$ is an **analytic** function, which implies in particular that it is **continuous** and **differentiable** any number of times. So it is infinitely smooth in the ROC. From the theory of analytic functions, it is also known that in the ROC, the convergence of Eq. (6.88) is **absolute** (i.e., $\sum |x[n]z^{-n}|$ also converges) as well as **uniform** [Apostol, 1974]. Whether the sum converges (absolutely or otherwise) at points on the bounding circles depends on the example under consideration. Even if it converges at all points on the circles, it is not guaranteed to be differentiable

[3] For causal or anticausal signals this becomes the McLaurin series in z or z^{-1} as appropriate.

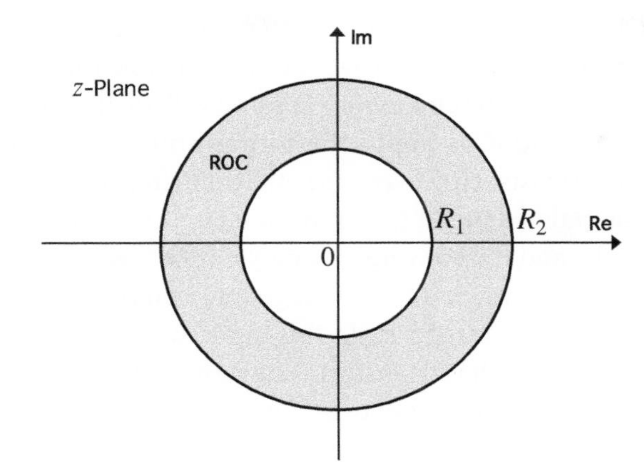

Figure 6.7 The general form of the ROC of the z-transform for a two-sided sequence.

there, as we shall see. The ROC is therefore formally defined as the **open region** which excludes the circles, even if the summation converges on the circles.

6.9.1 Expressions for Radii of Convergence

We can in fact express the radii of convergence R_1 and R_2 in terms of the limiting behavior of the signal $x[n]$. For this, let us rewrite

$$X(z) = \sum_{n=-\infty}^{-1} x_a[n]z^{-n} + \sum_{n=0}^{\infty} x_c[n]z^{-n}, \tag{6.90}$$

where $x_a[n]$ is anticausal and $x_c[n]$ causal. We can then show [Apostol, 1974, p. 234] that the radii of convergence R_1 and R_2 can be calculated from[4]

$$R_1 = \lim_{n\to\infty} \left|x_c[n]\right|^{1/n}, \qquad \frac{1}{R_2} = \lim_{n\to\infty} \left|x_a[-n]\right|^{1/n}. \tag{6.91}$$

For example, suppose

$$X(z) = \sum_{n=-\infty}^{\infty} (0.5)^{|n|} z^{-n}. \tag{6.92}$$

Then, since $x_c[n] = (0.5)^n$, we have $R_1 = \lim_{n\to\infty} [(0.5)^n]^{1/n} = 0.5$ and

$$\frac{1}{R_2} = \lim_{n\to\infty} \left|x_a[-n]\right|^{1/n} = 2^{-1}, \tag{6.93}$$

so that $R_2 = 2$. Thus, the ROC is $0.5 < |z| < 2$.

6.9.1.1 The Case of Empty ROC

If it turns out that $R_1 > R_2$ then the ROC is empty, that is, the z-transform does not converge anywhere. We will see that there are interesting examples where $R_1 =$

[4] When these limits do not exist, all the discussions of this section can be made valid by using the so-called "lim-sup" instead of the limit. For a detailed discussion, see, for example, Apostol [1974].

$R_2 = 1$. Clearly the ROC is empty in these cases too, as the ROC is defined to be the open region $R_1 < |z| < R_2$. However, it is possible for the z-transform summation to converge just on the circle $|z| = 1$, as we will see in examples later. In these cases, even though the ROC is empty, the z-transform exists on the isolated circle $|z| = 1$, that is, $X(e^{j\omega})$ exists. However, since the unit circle is not in the interior of an ROC, it cannot be claimed that $X(z)$ on the unit circle is analytic. Such $X(e^{j\omega})$ may not be differentiable, and the convergence may not be absolute.

For this reason, when $X(z)$ converges only on the unit circle, some books do not even consider it a z-transform. In this book, we will accept a more forgiving terminology and say that the z-transform converges only on the unit circle. Whenever we say the "z-transform does not exist," we mean not only that the ROC is empty but the z-transform does not converge on isolated circles either. An example is the exponential $x[n] = a^n$, for which the z-transform does not exist anywhere. Even when $a = 1$ it does not exist, even on the unit circle. True, the Fourier transform exists as a Dirac delta in that case, but it is unbounded, and is not a valid z-transform.

6.9.1.2 Magnitude Symmetry

Suppose a sequence satisfies magnitude symmetry, that is,

$$|x[-n]| = |x[n]|, \tag{6.94}$$

as in the case of sequences which are even or odd (or Hermitian symmetric). Then it is readily verified (Problem 6.16) that

$$R_2 = \frac{1}{R_1}, \tag{6.95}$$

so that the ROC has the form

$$R_1 < |z| < 1/R_1. \tag{6.96}$$

This is possible only if $R_1 < 1$. So we see that a sequence with magnitude symmetry has a nonempty ROC if and only if $R_1 < 1$. When this happens, the unit circle is in the interior of the ROC, so that the Fourier transform exists and is continuous and infinitely differentiable.

A nice corollary. Here is a consequence of the above result: suppose a sequence satisfies the magnitude symmetry (6.94), and has a Fourier transform which is not continuous or not differentiable. Then the ROC of the z-transform is empty, that is, the z-transform does not exist (except possibly on the unit circle). The reason is simple: if the ROC is not empty, it has to have the form (6.96), which means that the unit circle is in the ROC and the Fourier transform is continuous and infinitely differentiable, which contradicts the starting assumption. We will discuss examples in Sec. 6.9.4.
▽ ▽ ▽

6.9.2 More Examples of the ROC

Consider causal sequences of the form

$$x[n] = (1/n^k)\mathcal{U}[n-1], \tag{6.97}$$

where k is a fixed real number. Here are some special cases of this:

$$x[n] = \begin{cases} n\mathcal{U}[n-1] & (k=-1), \\ \mathcal{U}[n-1]/n & (k=1), \\ \mathcal{U}[n-1]/n^2 & (k=2), \\ \mathcal{U}[n-1]/\sqrt{n} & (k=0.5), \end{cases} \tag{6.98}$$

and so forth. Since these are causal, the ROC has the form $|z| > R_1$, that is, $R_2 = \infty$ in these cases. An easy way to find R_1 is to use the **ratio test** [Apostol, 1974]. To state this result, consider a series of the form $\sum_{n=1}^{\infty} a_n$ with nonzero terms a_n. Assume the limit

$$R = \lim_{n\to\infty} \left| \frac{a_{n+1}}{a_n} \right| \tag{6.99}$$

exists. Then the series $\sum_{n=1}^{\infty} a_n$ converges absolutely if $R < 1$ and diverges if $R > 1$. (If $R = 1$ there is no conclusion; further testing is necessary.) Returning to Eq. (6.97), we have

$$X(z) = \sum_{n=1}^{\infty} \frac{z^{-n}}{n^k}. \tag{6.100}$$

Letting $a_n = z^{-n}/n^k = 1/(n^k z^n)$, $n > 0$, we see that

$$R = \lim_{n\to\infty} \left| \frac{a_{n+1}}{a_n} \right| = \lim_{n\to\infty} \left| \left(\frac{n}{n+1}\right)^k \frac{z^n}{z^{n+1}} \right| = \frac{1}{|z|}. \tag{6.101}$$

Since convergence corresponds to $R < 1$, the z-transform (6.100) has the ROC

$$|z| > 1 \qquad \text{(ROC for Eq. (6.98))}, \tag{6.102}$$

regardless of the value of the real number k. Thus, all the examples in Eq. (6.98) have z-transforms converging outside the unit circle.

What is the intuition behind this? For example, why does $\sum_{n=1}^{\infty} n^{10} z^{-n}$ converge for $|z| > 1$? Because z^{-n} decays exponentially, but n^{10} grows only like a polynomial. The exponential decay more than makes up for the polynomial growth, and helps with convergence. Next, on the unit circle we have

$$X(e^{j\omega}) = \sum_{n=1}^{\infty} \frac{e^{-j\omega n}}{n^k}. \tag{6.103}$$

To examine whether this converges for a given ω, we use **Dirichlet's test** [Apostol, 1974], which says that a series of the form

$$\sum_{n=1}^{\infty} a_n b_n \tag{6.104}$$

converges if the sequences $\{a_n\}$ and $\{b_n\}$ have the following properties: (i) $\{b_n\}$ is a decreasing sequence with $b_n \to 0$ and (ii) the partial sum $p_L = \sum_{n=1}^{L} a_n$ is a bounded sequence, that is, $|p_L| < B$ for all $L > 0$, for some $B < \infty$. If we set

$$b_n = 1/n^k, \quad a_n = e^{-j\omega n}, \tag{6.105}$$

we can use this test to examine convergence of Eq. (6.103). Under the condition that $k > 0$, the sequence $\{b_n\}$ satisfies the requirements of the theorem. The partial sums are

$$p_L = \sum_{n=1}^{L} a_n = \sum_{n=1}^{L} e^{-j\omega n} = e^{-j\omega}\left(\frac{1 - e^{-j\omega L}}{1 - e^{-j\omega}}\right), \tag{6.106}$$

which shows that

$$|p_L| = \left|\frac{\sin(\omega L/2)}{\sin(\omega/2)}\right| \leq \frac{1}{|\sin(\omega/2)|}. \tag{6.107}$$

Thus, for any fixed $\omega \neq 2\pi i$, this has an upper bound independent of L. Dirichlet's condition is therefore satisfied. Thus, as long as $k > 0$, the z-transform of Eq. (6.97) converges at each point on the unit circle, except possibly at $\omega = 0$. That is, for $k > 0$, the Fourier transform (6.103) converges everywhere except possibly at zero frequency. For $\omega = 0$ (i.e., $z = 1$), the z-transform expression is

$$X(1) = \sum_{n=1}^{\infty} \frac{1}{n^k}, \tag{6.108}$$

which converges if and only if $k > 1$. Summarizing, the z-transform expression $X(z) = \sum_{n=1}^{\infty} z^{-n}/n^k$ has the following convergence properties:

1. For any real k, it converges in the region $|z| > 1$. For example, $X(z) = \sum_{n=1}^{\infty} nz^{-n}$ converges for $|z| > 1$.
2. For $k > 0$, it also converges on the unit circle, except possibly at $z = 1$. Thus, $X(z) = \sum_{n=1}^{\infty} (1/n)z^{-n}$ and $Y(z) = \sum_{n=1}^{\infty} (1/\sqrt{n})z^{-n}$ converge for $|z| \geq 1$ except at $z = 1$.
3. For $k > 1$, it also converges for $z = 1$. Thus, $X(z) = \sum_{n=1}^{\infty} (1/n^2)z^{-n}$ converges for all z such that $|z| \geq 1$.

As mentioned before, the ROC is formally taken to be the open region $|z| > 1$, even if the series converges everywhere on the circle $|z| = 1$. In the ROC $X(z)$ is analytic, but on the unit circle all we can say is that $X(e^{j\omega})$ converges; it may not be differentiable. We will see examples in Sec. 6.9.4 which demonstrate this subtle point.

6.9.2.1 Other Miscellaneous Examples

We can find several standard examples of power series in texts on mathematics [e.g., Churchill and Brown, 1984], serving as interesting z-transform pairs. For example, it is well known that e^w can be expressed as the power series $e^w = 1 + \sum_{n=1}^{\infty} w^n/n!$ with ROC $|w| < \infty$. Now consider the z-transform

$$X(z) = \exp(1/z) = 1 + z^{-1} + \frac{z^{-2}}{2!} + \frac{z^{-3}}{3!} + \cdots. \tag{6.109}$$

This has the ROC $|z| > 0$, and its inverse z-transform is the causal signal

$$x[n] = \frac{\mathcal{U}[n]}{n!}, \tag{6.110}$$

with 0! interpreted as unity. Another example of this kind is the logarithmic series

$$\text{Ln}(1 + z^{-1}) = z^{-1} - \frac{z^{-2}}{2} + \frac{z^{-3}}{3} - \frac{z^{-4}}{4} + \cdots, \tag{6.111}$$

which has the ROC $|z| > 1$ and yields the inverse transform

$$x[n] = \frac{-(-1)^n \mathcal{U}[n-1]}{n}. \tag{6.112}$$

Using basic properties of z-transforms, we therefore get the z-transform pair

$$\frac{\mathcal{U}[n-1]}{n} \longleftrightarrow -\text{Ln}(1 - z^{-1}), \quad |z| > 1. \tag{6.113}$$

Many such z-transform pairs can be generated from standard examples of analytic functions.

6.9.3 Inverse Z-Transform as a Contour Integral

While on the topic of analytic functions, we would like to point out that the inverse z-transform $x[n]$ can be expressed as a contour integral. Recall again the z-transform expression (6.88) and consider its values on a circle of radius R in the ROC (Fig. 6.8). On this circle we have $z = Re^{j\omega}$, so

$$X(z) = \sum_{n=-\infty}^{\infty} (x[n]R^{-n})e^{-j\omega n}, \tag{6.114}$$

which is nothing but the Fourier transform of the modified signal $x[n]R^{-n}$. Thus, using the standard inverse-FT expression, we have

$$x[n]R^{-n} = \frac{1}{2\pi} \int_0^{2\pi} X(z)e^{j\omega n} d\omega. \tag{6.115}$$

Thus, the inverse z-transform $x[n]$ can be computed just by computing the inverse Fourier transform $x[n]R^{-n}$ and multiplying the answer with R^n. Within the ROC, any radius R will yield the same answer for $x[n]$. But if we choose the circle to be in a different choice of ROC, then the answer $x[n]$ will be different, as expected. Now, Eq. (6.115) can also be rearranged as

$$\begin{aligned} x[n] &= \frac{1}{2\pi} \int_0^{2\pi} X(z)R^n e^{j\omega n} d\omega \\ &= \frac{1}{2\pi} \int_0^{2\pi} X(z)z^n d\omega = \frac{1}{2\pi j} \oint_{|z|=R} X(z)z^{n-1} dz, \end{aligned} \tag{6.116}$$

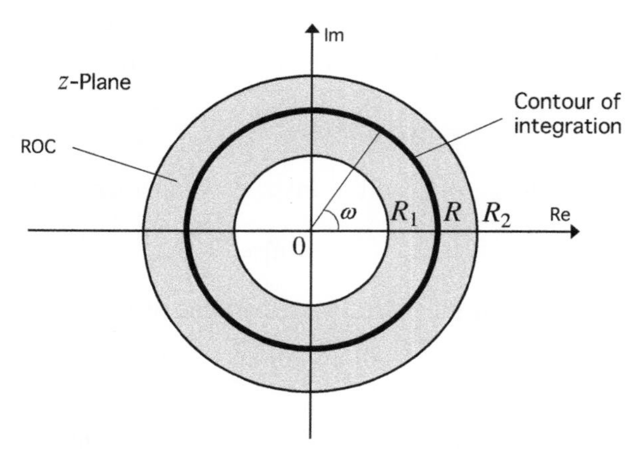

Figure 6.8 The inverse z-transformation can be interpreted as a contour integration in the ROC. See text.

where we have used the change of variables $z = Re^{j\omega}$. The notation in the last integral means that the integration is a **contour integral**, done counterclockwise, on the circle $|z| = R$ located in the ROC. Thus, the inverse z-transform $x[n]$ can be expressed as a contour integral involving $X(z)$.

The contour integral can be evaluated using standard techniques from complex variable theory [Churchill and Brown, 1984]. There is a theorem due to Cauchy which says that if $f(z)$ is single valued and analytic on and inside a closed contour C except for a finite number of poles $p_1, p_2, \ldots, p_N$ inside C, then

$$\frac{1}{2\pi j}\oint_C f(z)dz = \sum_{k=1}^{N} Q_k, \tag{6.117}$$

where Q_k is the so-called **residue** of $f(z)$ at the pole p_k. When p_k is a simple pole (i.e., has multiplicity one), the residue can be calculated as follows:

$$Q_k = \lim_{z\to p_k}(z - p_k)f(z). \tag{6.118}$$

If p_k is a pole of order M, then the residue is calculated using

$$Q_k = \lim_{z\to p_k}\frac{1}{(M-1)!}\frac{d^{M-1}}{dz^{M-1}}\left[(z-p_k)^M f(z)\right]. \tag{6.119}$$

So, given a rational function $X(z)$ and a specified ROC $R_1 < |z| < R_2$, we can compute the inverse z-transform $x[n]$ as follows:

1. Make a list of the poles of $z^{n-1}X(z)$ in the region $0 \le |z| \le R_1$.
2. Compute the residues of these poles and add them up. The result is precisely $x[n]$.

Students interested in pursuing the above approach for inverting the z-transform may benefit from the "residue" command in MATLAB. The contour integration method is insightful, but it can be somewhat cumbersome because of the poles created at $z = 0$ by the factor z^{n-1} in Eq. (6.116) (when $n < 1$). These are multiple poles and the multiplicity $n-1$ depends on the time index n. The methods based on partial fractions (Secs. 5.10 and 6.3.2) are often more convenient for inversion of rational $X(z)$.

6.9.4 Revisiting Unrealizable (Nonrational) Digital Filters

In Sec. 6.4.3 we found that for rational LTI systems the frequency response $H(e^{j\omega})$ exists if and only if the system is stable; furthermore, when $H(e^{j\omega})$ exists, it is continuous and infinitely differentiable (Sec. 5.8). But for nonrational LTI systems these statements are not necessarily true. We now elaborate on this for the sake of completeness. First recall that whether the LTI system is rational or not, the stability property is equivalent to

$$\sum_n |h[n]| < \infty. \tag{6.120}$$

A sequence satisfying Eq. (6.120) is called an ℓ_1 sequence, and we write $h[n] \in \ell_1$ (where ℓ_1 is the space of all such signals). The ℓ_1 norm of $h[n]$ is defined to be $\|h[n]\|_1 = \sum_{n=-\infty}^{\infty} |h[n]|$. Equation (6.120) implies the following properties:

1. The frequency response summation $\sum_n h[n]e^{-j\omega n}$ converges for all ω.
2. Furthermore, $|H(e^{j\omega})|$ is bounded by the ℓ_1 norm because

$$|H(e^{j\omega})| = \Big|\sum_{n=-\infty}^{\infty} h[n]e^{-j\omega n}\Big| \le \sum_{n=-\infty}^{\infty} |h[n]e^{-j\omega n}| = \sum_{n=-\infty}^{\infty} |h[n]|.$$

3. $H(e^{j\omega})$ is a continuous function.

Proof of continuity. We have to show that, given any $\epsilon > 0$, there exists a $\delta > 0$ such that $|H(e^{j\omega}) - H(e^{j(\omega-\delta)})| < \epsilon$. We have

$$\begin{aligned}\Big|H(e^{j\omega}) - H(e^{j(\omega-\delta)})\Big| &= \Big|\sum_{n=-\infty}^{\infty} h[n]e^{-j\omega n}(1-e^{j\delta n})\Big| \\ &\le 2\sum_{n=-\infty}^{\infty} \Big|h[n]\sin(\delta n/2)\Big|.\end{aligned}$$

Since $\sum_n |h[n]|$ converges, there exists N such that $2\sum_{|n|>N} |h[n]| < \epsilon/2$. Then

$$\Big|H(e^{j\omega}) - H(e^{j(\omega-\delta)})\Big| < 2\sum_{|n|\le N} \Big|h[n]\sin(\delta n/2)\Big| + \epsilon/2. \tag{6.121}$$

By making δ small enough, it can be ensured that $|\sin(\delta n/2)| \le |\sin(\delta N/2)|$ for $|n| \le N$. Thus

$$\Big|H(e^{j\omega}) - H(e^{j(\omega-\delta)})\Big| < |2\sin(\delta N/2)| \sum_{|n|\le N} |h[n]| + \epsilon/2. \tag{6.122}$$

Since $\sum_{|n|\le N} |h[n]|$ is finite, we can choose δ small enough to ensure that the first term is less that $\epsilon/2$, ensuring $|H(e^{j\omega}) - H(e^{j(\omega-\delta)})| < \epsilon$. ▽ ▽ ▽

6.9.4.1 Can the Sum for $H(e^{j\omega})$ Converge for Unstable Systems?

A subtle point is that although stability (Eq. (6.120)) implies that $H(e^{j\omega})$ exists, the converse is not in general true when the system is not rational. That is, even if $\sum_n h[n]e^{-j\omega n}$ converges pointwise for all $\omega \in [-\pi, \pi]$, it does not necessarily imply Eq. (6.120), unless the system is rational. Here is an example.[5] Let

$$h[n] = \frac{e^{j\sqrt{n}}}{n^{4/5}}\mathcal{U}[n-1], \tag{6.123}$$

which is a special case of an example on p. 200 (vol. 1) of Zygmund [1968], with $\alpha = 1/2$ and $\beta = 4/5$. Since $|h[n]| = 1/n^{4/5}$ for $n \geq 1$, it follows that Eq. (6.120) is not satisfied. The claim in Zygmund [1968], however, is that $\sum_n h[n]e^{-j\omega n}$ converges pointwise for each $\omega \in [-\pi, \pi]$ to a continuous function – in fact, the convergence is uniform. The ROC of the z-transform, which by definition is an open interval, is $|z| > 1$ (using the ratio test; see Sec. 6.9.2). Although this is a nonrational LTI system and is therefore unrealizable (Sec. 14.4.4), the example is certainly academically interesting. Another somewhat similar example is the ideal lowpass filter. See Eq. (6.127) below.

6.9.4.2 Nondifferentiable $H(e^{j\omega})$ and ROC

We showed above that stability (Eq. (6.120)) implies that $H(e^{j\omega})$ exists, is bounded by $\sum_n |h[n]|$, and is continuous. However, unless $H(z)$ is rational, $H(e^{j\omega})$ may still not be differentiable. To put this in context, consider the two lowpass filters shown in Fig. 6.9, namely the ideal lowpass filter and a filter with triangle-shaped passband. The impulse responses are $g[n] = \sin \omega_c n/\pi n$ and

$$h[n] = \left(\frac{\sin \omega_c n}{\pi n}\right)^2, \tag{6.124}$$

for appropriate choice of the height α in the figure.

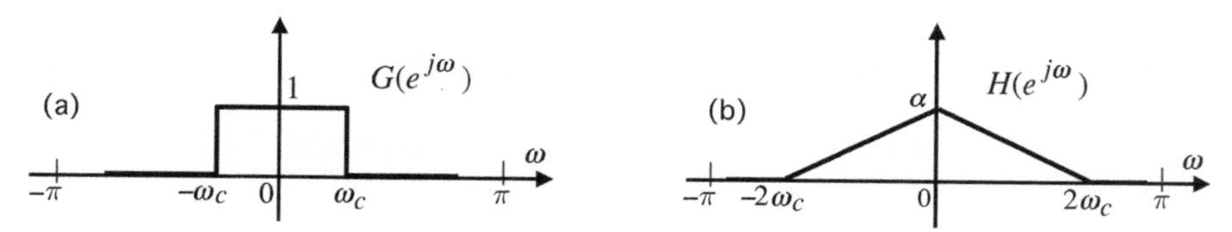

Figure 6.9 Frequency responses of (a) an ideal lowpass filter and (b) a triangular-passband filter.

[5] This example was brought to the author's attention by Professor Bhaskar Ramamurthi, Indian Institute of Technology, Chennai.

1. *Triangular filter*. Let us first think about $H(e^{j\omega})$. Since

$$\sum_{n=1}^{\infty} 1/n^2 \tag{6.125}$$

converges to a finite value ($= \pi^2/6$), it follows that $h[n]$ satisfies Eq. (6.120). As a result, $H(e^{j\omega})$ is continuous everywhere including at $\omega = 0$ and $\pm 2\omega_c$. But it is not differentiable at $\omega = 0$ and $\pm 2\omega_c$. Thus, here is an example of a stable filter whose frequency response exists and is continuous, but not differentiable! The filter $H(e^{j\omega})$ *does not have a rational transfer function*, for if it did, the frequency response would be differentiable (Sec. 5.8). Since the filter is non-rational, it is not realizable (Sec. 14.4.4).
2. *Rectangular filter*. Now return to the ideal lowpass filter (Fig. 6.9(a)), which has

$$g[n] = \frac{\sin \omega_c n}{\pi n}. \tag{6.126}$$

We know that this is unstable because $\sum_{n=1}^{\infty} 1/n$ does not converge, and therefore $\sum_n |g[n]|$ is not finite. So $g[n]$ is not an ℓ_1 sequence. And yet, the frequency response expression converges pointwise for all ω to a discontinuous function:

$$\sum_{n=-\infty}^{\infty} g[n]e^{-j\omega n} = \begin{cases} 1 & |\omega| < \omega_c, \\ 0.5 & \omega = \omega_c, \\ 0 & \omega_c < |\omega| < \pi. \end{cases} \tag{6.127}$$

This follows by application of Theorem 17.1 in Chapter 17.[6] Thus, it is quite possible for the sum $\sum_n g[n]e^{-j\omega n}$ to converge pointwise for all ω, even though the filter is unstable! This is unlike rational filters, for which the existence of a frequency response implies that the filter is stable.

Since the Fourier transforms of $h[n]$ and $g[n]$ are not differentiable, it is clear that the unit circle is not in the interior of the ROC of the z-transforms (because in the interior of the ROC, the transfer function is analytic, hence differentiable). In fact, we can go one step further and argue that the ROCs are empty! That is, the z-transforms do not exist *anywhere* except *on* the unit circle. To see this, simply note that $h[n]$ and $g[n]$ have the magnitude symmetry (6.94). So, if the z-transform had a nonempty ROC, it would have the form $R_1 < |z| < 1/R_1$ and would therefore include the unit circle, as explained in Sec. 6.9.1. This contradicts the above observation that the unit circle is not in the interior of the ROC.

6.10 History: Who Invented the *z*-Transform?

In data sciences we encounter many types of transforms, such as the Fourier transform, the Laplace transform, the Hadamard transform, and so forth. The idea for

[6] Essentially, $g[n]$ can be regarded as the Fourier series coefficients (Chapter 11) of the periodic function $G(e^{j\omega})$, so theorems on Fourier series convergence (Chapter 17) are applicable!

the Fourier transform originated from Fourier's early work on series representations of periodic signals. Similarly, we attribute the origins of the other transforms to those after whom these are named. But how about the z-transform? Surely, the name does not suggest an inventor!

The fact is that the z-transform has been there in the theory of complex variables even before the name z-transform got attached to it. As we explained in Sec. 6.9, the Laurent series is nothing but a z-transform, and its properties have been studied extensively in the literature on analytic functions of complex variables, by great minds such as Cauchy, Laurent, McLaurin, and so forth a few centuries ago [Churchill and Brown, 1984]. However, the credit for introducing the z-transform into data sciences goes back to a number of researchers of the twentieth century. For example, Ragazzini and Zadeh [1952] introduced it for sampled data systems, and Jury [1960] generalized it and proved many properties. It has also been traced back to earlier work by Hurewicz and even to Laplace [Ragazzini and Zadeh, 1952]. Today it is an integral part of signal processing and data sciences, and includes the discrete-time Fourier transform as a special case, as we saw.

Signal processing authors gave rebirth to the beauty of all this, by highlighting the interplay of causality, stability, and the region of convergence. In his student days, the current author used to be fascinated by this interplay, explained so beautifully in Oppenheim and Schafer [1975], through many examples and homework problems. Needless to say, in addition to its theoretical underpinnings, the z-transform has greatly impacted the practical side of signal processing as well.

6.11 Summary of Chapter 6

We started the chapter by deriving some basic properties of the z-transform, which are summarized in Sec. 6.12. Then it was shown that the first-order z-transform expression $X(z) = 1/(1 - az^{-1})$ has two possible inverse z-transforms, one right sided and one left sided. The correct answer depends on the region of convergence (ROC):

$$X(z) = \frac{1}{1 - az^{-1}} \quad \Longrightarrow \quad x[n] = \begin{cases} a^n \mathcal{U}[n] & \text{if ROC is } |z| > |a|, \\ -a^n \mathcal{U}[-n-1] & \text{if ROC is } |z| < |a|. \end{cases} \tag{6.128}$$

We demonstrated that the inverse z-transforms of more complicated rational functions can be found using partial fraction expansions (PFEs), and also presented the detailed theory of PFEs, including the case where there are multiple poles at one point. For the simple example

$$X(z) = \frac{1}{1 - p_1 z^{-1}} + \frac{1}{1 - p_2 z^{-1}},$$

which has two poles with $|p_1| < |p_2|$, we have

$$x[n] = \begin{cases} p_1^n \mathcal{U}[n] + p_2^n \mathcal{U}[n] & \text{if ROC is } |z| > |p_2|, \\ p_1^n \mathcal{U}[n] - p_2^n \mathcal{U}[-n-1] & \text{if ROC is } |p_1| < |z| < |p_2|, \\ -p_1^n \mathcal{U}[-n-1] - p_2^n \mathcal{U}[-n-1] & \text{if ROC is } |z| < |p_1|. \end{cases} \tag{6.129}$$

More generally, if there are poles with N distinct magnitudes then there are $N+1$ distinct ROCs, and $N+1$ answers for the inverse z-transforms, and we can say this:

1. One of the answers is right sided, and one is left sided. The rest are two sided.
2. There is at most one answer that is absolutely summable, and this has to have the unit circle in the interior of its ROC. Such an answer exists if and only if the rational $X(z)$ has no poles on the unit circle.
3. A rational LTI system with right-sided $h[n]$ (as in causal systems) is stable if and only if the poles satisfy $|p_m| < 1$. Similarly, a rational LTI system with left-sided $h[n]$ (as in anticausal systems) is stable if and only if the poles satisfy $|p_m| > 1$.

More generally, for a rational LTI system $H(z)$, stability is equivalent to saying that the ROC of $H(z)$, which has the form $R_1 < |z| < R_2$, includes the unit circle. This is equivalent to saying that the frequency response $H(e^{j\omega})$ exists. But for nonrational systems, there are examples where the frequency response $H(e^{j\omega})$ exists for all ω but the ROC of $H(z)$ is empty (see Sec. 6.9.4).

It was also shown in this chapter that IIR filters with linear phase cannot be causal and that, in the rational case, they can be implemented by running recursive difference equations using a combination of forward recursions and backward recursions.

In this chapter we also provided additional insights on z-transforms by mentioning a number of interesting facts from the theory of complex variables and analytic functions. For example, if $x[n]$ satisfies the magnitude symmetry $|x[-n]| = |x[n]|$, and has a Fourier transform which is not continuous or not differentiable, then the ROC of the z-transform is empty. That is, the z-transform does not exist (except possibly on the unit circle). Many examples of z-transform pairs were presented, some of which are summarized in Sec. 6.13.

6.12 Table of z-Transform Properties

A number of properties of the z-transform are summarized here. Where appropriate, section numbers are indicated in brackets, where details can be found. The notation $X(z) \longleftrightarrow x[n]$ means that $x[n]$ and $X(z)$ form a z-transform pair. The region of convergence (ROC) of $X(z)$, denoted (R_1, R_2), stands for $R_1 < |z| < R_2$. Uppercase letters indicate z-transforms of corresponding lower-case letters. For example, $x_k[n] \longleftrightarrow X_k(z)$. Most proofs are similar to those of Fourier transform properties. Some proofs are requested in the problems.

1. *Linearity.* $c_1x_1[n] + c_2x_2[n] \longleftrightarrow c_1X_1(z) + c_2X_2(z)$. The ROC of the result is at least the intersection of the ROCs of $X_1(z)$ and $X_2(z)$.
2. *Time-shift.* $x[n-K] \longleftrightarrow z^{-K}X(z)$; ROC (R_1, R_2) except possibly for behavior at $z = 0, z = \infty$.
3. *Exponential modulation.* $a^n x[n] \longleftrightarrow X(z/a)$; ROC $(|a|R_1, \quad |a|R_2)$. Special case: $(-1)^n x[n] \longleftrightarrow X(-z)$; ROC (R_1, R_2).
4. *Time reversal.* $x[-n] \longleftrightarrow X(1/z)$; ROC $(1/R_2, 1/R_1)$.
5. *Conjugation.* $x^*[n] \longleftrightarrow \triangleq X^*(z^*)$; ROC unchanged (Problem 6.18).
6. *Paraconjugation.* $x^*[-n] \longleftrightarrow \widetilde{X}(z)$; ROC $(1/R_2, 1/R_1)$. Here $\widetilde{X}(z) \triangleq X^*(1/z^*)$ is called the paraconjugate of $X(z)$, and $\widetilde{X}(e^{j\omega}) = X^*(e^{j\omega})$ (Problem 6.18).
7. *Derivative property.* $nx[n] \longleftrightarrow -zdX(z)/dz$; ROC (R_1, R_2) except for the behavior at $z = 0$ and $z = \infty$ (Problem 6.18).
8. *Convolution property.* $(h * x)[n] \triangleq \sum_k h[k]x[n-k] \longleftrightarrow H(z)X(z)$. The ROC of the result is at least the intersection of the ROCs of $H(z)$ and $X(z)$ (Sec. 3.4.3).
9. *Causal signals.* If $x[n] = 0, n < 0$ then $X(z)$ has ROC of the form $|z| > R$ (Sec. 6.9.1).
10. *Anticausal signals.* If $x[n] = 0, n \geq 0$ then $X(z)$ has ROC of the form $|z| < R$ (Sec. 6.9.1).
11. *Two-sided signals.* If the ROC of $X(z)$ is $R_1 < |z| < R_2$ with $R_1 > 0$ and $R_2 < \infty$, then $x[n]$ is neither causal nor anticausal.
12. *Stability.* Rational LTI $H(z)$ is stable $\Longleftrightarrow$ unit circle is in ROC $\Longleftrightarrow$ $H(e^{j\omega})$ exists for all ω. For the causal case this is equivalent to $|p_k| < 1$ (p_k being the poles); for the anticausal case it is equivalent to $|p_k| > 1$.

6.13 Table of z-Transform Pairs

Signal	z-Transform
$x[n] = \delta[n]$	$X(z) = 1$
$x[n] = \delta[n-K]$	$X(z) = z^{-K}$ (K units of delay)
$x[n] = a^n \mathcal{U}[n]$	$X(z) = \frac{1}{1-az^{-1}}, \quad \lvert z\rvert > \lvert a\rvert$
$x[n] = -a^n \mathcal{U}[-n-1]$	$X(z) = \frac{1}{1-az^{-1}}, \quad \lvert z\rvert < \lvert a\rvert$
$x[n] = a^n, \ \forall n$ (e.g., $x[n] = 1, \ \forall n$)	$X(z)$ does not exist
$x[n] = (n+1)a^n \mathcal{U}[n]$	$X(z) = \frac{1}{(1-az^{-1})^2}, \quad \lvert z\rvert > \lvert a\rvert$
$x[n] = R^n \cos(\theta n)\mathcal{U}[n]$	$X(z) = \frac{1 - R\cos\theta\, z^{-1}}{1 - 2R\cos\theta\, z^{-1} + R^2 z^{-2}}, \quad \lvert z\rvert > \lvert R\rvert$
$x[n] = R^n \sin(\theta n)\mathcal{U}[n]$	$X(z) = \frac{R\sin\theta\, z^{-1}}{1 - 2R\cos\theta\, z^{-1} + R^2 z^{-2}}, \quad \lvert z\rvert > \lvert R\rvert$
$x[n] = \begin{cases} 1 & 0 \le n \le L-1 \\ 0 & \text{otherwise} \end{cases}$	$X(z) = \frac{1-z^{-L}}{1-z^{-1}}, \quad \lvert z\rvert > 0$ (Sec. 4.3.3)
$x[n] = \frac{\mathcal{U}[n]}{n!}$ (Note: $0! = 1$)	$X(z) = \exp(1/z), \ \lvert z\rvert > 0$
$\frac{\mathcal{U}[n-1]}{n}$	$-\mathrm{Ln}(1-z^{-1}), \quad \lvert z\rvert > 1.$

PROBLEMS

6.1 Find the z-transforms of the following signals, and clearly indicate the region of convergence (ROC) and the pole–zero locations in each case. If the ROC is empty (i.e., if the z-transform does not exist), mention so. For each signal, determine whether the Fourier transform exists or not.
(a) $x[n] = 2^n\mathcal{U}[n]$ (b) $x[n] = 2^n\mathcal{U}[-n]$ (c) $x[n] = (1/3)^n\mathcal{U}[n]$

6.2 Which of the following signals has a nonempty ROC for the z-transform? For those that do, indicate the ROC and find $X(z)$.
(a) $x[n] = 1$ for all n (b) $x[n] = \mathcal{U}[n]$ (c) $(1 + j)^n\mathcal{U}[-n]$

6.3 Find the z-transforms of the following signals, and clearly indicate the ROC and the pole–zero locations in each case. If the ROC is empty (i.e., if the z-transform does not exist), mention so. For each signal, determine whether the Fourier transform exists or not.
(a) $x[n] = (1/2)^n\mathcal{U}[n] + 3^n\mathcal{U}[-n-1]$.
(b) $x[n] = 3^n\mathcal{U}[n] + (1/2)^n\mathcal{U}[-n-1]$.
(c) $x[n] = (1/2)^n\cos(\omega_0 n)\mathcal{U}[n]$.

6.4 Verify the validity of the z-transform pairs (6.19) and (6.20).

6.5 Let $X(z)$ be the z-transform of $x[n]$, and define

$$y[n] = \begin{cases} x[n/2] & \text{for } n \text{ even,} \\ 0 & \text{otherwise.} \end{cases}$$

Find $Y(z)$ in terms of $X(z)$.

6.6 Consider the discrete-time transfer function

$$H(z) = \frac{1 - z^{-1} + z^{-2}}{1 - (3/4)z^{-1} + (1/8)z^{-2}}.$$

This can be rewritten in the partial fraction form

$$H(z) = A + \frac{B}{1 - p_1 z^{-1}} + \frac{C}{1 - p_2 z^{-1}}. \quad \text{(P6.6)}$$

(a) What are the numerical values of p_1, p_2, and the constants A, B, C?
(b) Using the preceding, find a closed-form expression for the impulse response $h[n]$ of the causal LTI system described by $H(z)$.
(c) Is the system in (b) stable?

6.7 Consider the discrete-time transfer function

$$H(z) = \frac{4 - (11/4)z^{-1}}{1 - (9/4)z^{-1} + (1/2)z^{-2}}. \quad \text{(P6.7)}$$

(a) What are the poles p_i and zeros z_i of the system? Indicate them approximately in the z-plane (some of these might be complex). Be sure to draw the unit circle for reference.
(b) We know this $H(z)$ can be rewritten in the *partial fraction* form of Eq. (P6.6). What are the values of A, B, C?
(c) There are three possible regions of convergence (ROC) for the z-transform $H(z)$. Clearly indicate them in the z-plane.

(d) Depending on the ROC chosen, there are three possible answers for the inverse z-transform $h[n]$. Write down all these answers. So there are three LTI systems with the same transfer function (P6.7).

(e) Which of the above three answers represents a causal system? Which of the above three answers represents a stable system? Does there exist a causal stable solution in this case?

6.8 Consider

$$H(z) = \frac{1}{(1 - \frac{1}{2}z^{-1})(1 - z^{-1})}.$$

We know that this can have three possible inverse z-transforms. (a) Find the left-sided inverse z-transform $h[n]$. (b) Is this a bounded sequence? (c) Is this $h[n]$ the impulse response of a stable system?

6.9 Let $H(z) = 1/[1 - (9/2)z^{-1} + 2z^{-2}]$. We know the inverse transform has three possible answers. Find an answer $h[n]$ which is absolutely summable. Is the answer left sided, right sided, or both sided? Is there another answer which is also absolutely summable?

6.10 Consider a causal system with transfer function

$$H(z) = \frac{2 + (1/2)z^{-1} + (3/2)z^{-2} - z^{-3}}{1 - (1/2)z^{-1}}.$$

From Sec. 6.6 we know that this can be written as

$$H(z) = c_0 + c_1 z^{-1} + c_2 z^{-2} + \frac{a_0}{1 - (1/2)z^{-1}}.$$

Find c_0, c_1, c_2, and a_0. What is the impulse response $h[n]$?

6.11 For the z-transform $X(z) = 1/(1 - pz^{-1})^2$, if the ROC is $|z| > |p|$, then we showed that the inverse z-transform is given by Eq. (6.31). If the ROC is $|z| < |p|$, then what is the inverse z-transform?

6.12 Let $X(z) = 1/(1 - pz^{-1})^3$. Assuming the ROC is $|z| > |p|$, show that the inverse z-transform $x[n]$ takes the form

$$x[n] = \left(\alpha_0 + \alpha_1 n + \alpha_2 n^2\right) p^n \mathcal{U}[n], \tag{P6.12}$$

for appropriate constants α_i.

6.13 Consider the relaxed ideal lowpass digital filter shown in Fig. P6.13, where $0 < \alpha < \beta < \pi$ and the response is symmetric with respect to zero frequency. In Problem 4.25 you have derived the impulse response and verified that the filter is stable.

(a) What is the region of convergence of the z-transform of $h[n]$? *Hint*: Read Sec. 6.9.4.

(b) Although this filter is stable (Problem 4.25), it is still an "ideal filter" in the sense that it is unrealizable, that is, the transfer function is not rational. Explain why.

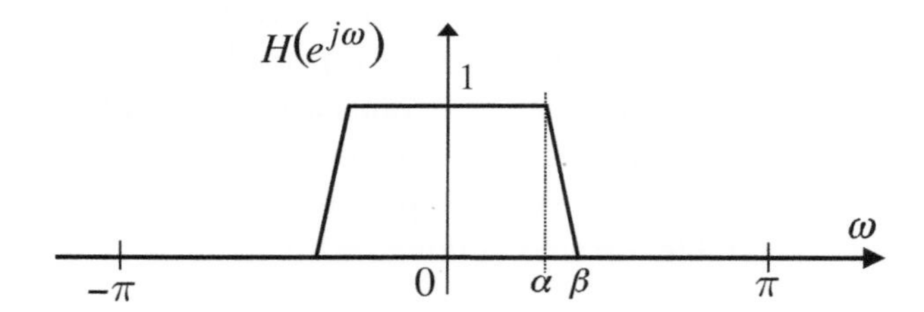

Figure P6.13 A relaxed lowpass filter.

6.14 Show that the filter $H(e^{j\omega})$ in Fig. 6.9(b) has the impulse response (6.124) for appropriate α. What is this value of α?

6.15 This problem demonstrates an IIR, stable, linear-phase filter. Consider

$$H(z) = \frac{-1.5z^{-1}}{1 - 2.5z^{-1} + z^{-2}}.$$

(a) Find the impulse response $h[n]$ by assuming a region of convergence that yields a stable solution. You will find that this solution is noncausal.

(b) What is the phase response of this solution?

(c) Show that $h[n] = h[N - n]$ for appropriate N. What is this N?

6.16 Suppose a sequence $x[n]$ satisfies magnitude symmetry (6.94). Using the closed-form expressions (6.91) for the radii of the circles bounding the ROC of $X(z)$, show that $R_2 = 1/R_1$.

6.17 Show that the ROC of the z-transform of Eq. (6.123) is $|z| > 1$.

6.18 Given a z-transform pair $x[n] \longleftrightarrow X(z)$ with ROC for $X(z)$ given by $R_1 < |z| < R_2$, show that

(a) $x^*[n] \longleftrightarrow X^*(z^*)$, with ROC unchanged.

(b) $x^*[-n] \longleftrightarrow \widetilde{X}(z)$, where $\widetilde{X}(z) \stackrel{\Delta}{=} X^*(1/z^*)$ is the paraconjugate of $X(z)$. Derive the ROC for $\widetilde{X}(z)$.

(c) $nx[n] \longleftrightarrow -zdX(z)/dz$.

6.19 Consider the decimator $y[n] = x[2n]$ which was discussed in Sec. 2.8.3. We now derive an expression for the z-transform $Y(z)$ in terms of $X(z)$. This is tricky because $Y(z) = \sum_n x[2n]z^{-n}$; we cannot blindly make a change of variables "$m = 2n$" to rewrite $Y(z) = \sum_m x[m]z^{-m/2} = X(z^{1/2})$ because m can only take even values. So we use a different approach. First define

$$s[n] = x[n]\left(\frac{1 + (-1)^n}{2}\right),$$

so that $s[2n] = x[2n]$ and $s[2n + 1] = 0$. Show that

$$S(z) = \frac{X(z) + X(-z)}{2}.$$

Now show that $Y(z) = S(z^{1/2})$ and therefore

$$Y(z) = \frac{X(z^{1/2}) + X(-z^{1/2})}{2}. \qquad \text{(P6.19)}$$

Notice in particular that $Y(e^{j\omega}) = [X(e^{j\omega/2}) + X(e^{j(\omega-2\pi)/2})]/2$. For insight on this, see the end of Problem 6.20.

6.20 Now consider the decimator $y[n] = x[Mn]$ for arbitrary integer $M > 1$. Define

$$s[n] = x[n]\left(\frac{1 + W^n + W^{2n} + \cdots + W^{(M-1)n}}{M}\right), \tag{P6.20a}$$

where $W = e^{-j2\pi/M}$ is an Mth root of unity (i.e., $W^M = 1$).

(a) Show that $s[Mn] = x[Mn]$, and $s[n] = 0$ otherwise.

(b) Hence show that $Y(z) = S(z^{1/M})$.

(c) From (P6.20a), also show that

$$S(z) = \sum_{k=0}^{M-1} X(zW^k)/M.$$

Hence show that

$$Y(z) = \frac{1}{M}\sum_{k=0}^{M-1} X(z^{1/M}W^k). \tag{P6.20b}$$

(d) Finally, show that $Y(e^{j\omega}) = \sum_{k=0}^{M-1} X(e^{j(\omega-2\pi k)/M})\,/M$. Thus, $Y(e^{j\omega})$ is a superposition of M copies

$$X(e^{j\omega/M}),\ X(e^{j(\omega-2\pi)/M}),\ X(e^{j(\omega-4\pi)/M}), \ldots.$$

Insights. The first copy is $X(e^{j\omega})$ stretched by M. The second copy is this stretched copy, shifted by 2π, and so forth. If there is any overlap between any two plots in the list of M copies above, we say that decimation creates aliasing. If $x[n]$ is bandlimited in the sense that $X(e^{j\omega})$ is zero everywhere in $[-\pi, \pi)$ except in the region $\omega_1 < \omega < \omega_1 + 2\pi/M$ (for some ω_1), then there is no aliasing due to decimation by M. This is an important concept in multi-rate signal processing theory. See Vaidyanathan [1993] for a detailed discussion.

7 More on Digital Filters

7.1 Introduction

Digital filtering is a major topic. Detailed books on digital signal processing usually contain multiple chapters on digital filters. In fact, there are many books dedicated to digital filters alone. Our purpose in this chapter is to give an introduction to a few simple methods and examples, so that the student can get a quick introduction. For detailed exposure to all well-known methods, there are excellent books such as Antoniou [1993], Diniz et al. [2010], Jackson [1996], Mitra [2011], Oppenheim and Schafer [2010], and Proakis and Manolakis [2007].

In Chapters 4 and 5, several simple examples of digital filters were presented, such as the moving average filter, the first-difference filter, the one-pole IIR filter, and so on. In this chapter we present more sophisticated types of digital filters such as notch and antinotch filters, and sharp-cutoff lowpass filters such as Butterworth filters. Also discussed are allpass filters and some of their amazing applications. It is explained how continuous-time filters can be transformed into discrete time by using appropriate mappings.

A simple method for the design of linear-phase FIR filters, called the window-based method, is also presented, and a comparative discussion of FIR and IIR filters is given. We demonstrate how nonlinear-phase filters can create visible phase distortion in images. Towards the end, a detailed discussion of steady-state and transient components of filter outputs is given.

This chapter gives exposure to a very important topic in digital signal processing. To give a deeper understanding, a number of optional sections are included for additional reading, marked with asterisks (Secs. 7.12, 7.14, and 7.15).

Note: Unless mentioned otherwise, all filters in this chapter are causal and rational. So stability is equivalent to the property that all poles p_i are inside the unit circle, that is, $|p_i| < 1$.

7.2 Adjustable Narrow-band Bandpass Filters

First consider the second-order transfer function

$$H(z) = \frac{1}{1 - pz^{-1}} + \frac{1}{1 - p^*z^{-1}}, \tag{7.1}$$

where p is a complex pole and the notation p^* refers to the complex conjugate of p. So we have a complex-conjugate pair of poles. We can always write

$$p = Re^{j\theta}, \quad 0 \le R < 1, \;\; 0 \le \theta < 2\pi, \tag{7.2}$$

where R is the pole radius and $\pm\theta$ are the angles of the two poles. It is assumed that $R < 1$, so that the causal impulse response

$$h[n] = (R^n e^{j\theta n} + R^n e^{-j\theta n})\mathcal{U}[n] = 2R^n \cos(\theta n)\mathcal{U}[n] \tag{7.3}$$

represents a stable system. Since

$$(1 - pz^{-1})(1 - p^* z^{-1}) = 1 - 2R\cos\theta\, z^{-1} + R^2 z^{-2}, \tag{7.4}$$

we can rewrite Eq. (7.1) as

$$H(z) = \frac{2 - (p + p^*)z^{-1}}{(1 - pz^{-1})(1 - p^* z^{-1})} = \frac{2(1 - R\cos\theta\, z^{-1})}{1 - 2R\cos\theta\, z^{-1} + R^2 z^{-2}}. \tag{7.5}$$

This represents a filter with real coefficients. The pole–zero plot is shown in Fig. 7.1; we have a complex-conjugate pair of poles, and a real zero at $z = R\cos\theta$ (besides a trivial zero at $z = 0$). The frequency response $H(e^{j\omega})$ is complex, and Fig. 7.2 shows plots of the magnitude response $|H(e^{j\omega})|$ for various (R, θ) combinations. This has a sharp peak at the frequency $\omega \approx \theta$, and the gain rapidly decays as ω moves away from θ. Filters of this form are *narrowband bandpass filters*. They are useful when we wish to pass a single frequency or a very narrowband of frequencies, and suppress the rest. The peak height of $|H(e^{j\omega})|$, which actually depends on (R, θ), has been normalized to unity for plotting convenience; this is equivalent to inserting a constant α in the numerator of the transfer function (7.5).

To explain why the response exhibits a peak at all, recall that the frequency response of the first term in Eq. (7.1) is

$$\frac{1}{1 - pe^{-j\omega}} = \frac{1}{1 - Re^{-j(\omega-\theta)}} = F(e^{j(\omega-\theta)}), \tag{7.6}$$

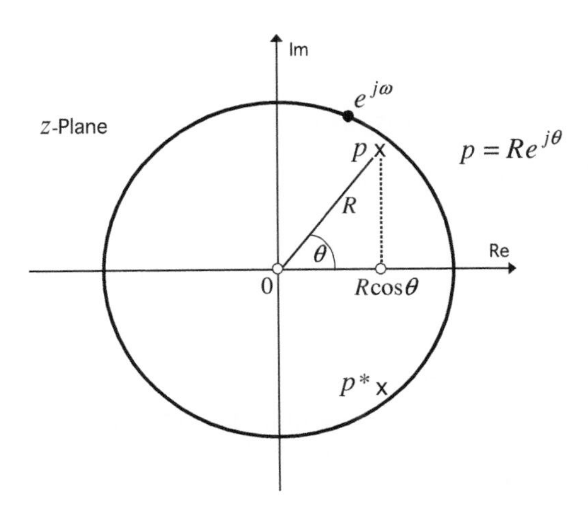

Figure 7.1 Poles and zeros of the transfer function (7.5), indicated by crosses and circles as usual.

where $F(z) = 1/(1 - Rz^{-1})$ is a first-order IIR filter as in Sec. 5.3, with pole R, where $0 < R < 1$. So $F(e^{j\omega})$ is lowpass with a peak at $\omega = 0$ (Fig. 5.3). Since Eq. (7.6) is the right-shifted version of this by an amount θ, the magnitude of Eq. (7.6) has a peak at $\omega = \theta$. Another way to see this is to write

$$\frac{1}{|1 - pe^{-j\omega}|} = \frac{1}{|e^{j\omega} - p|} = \frac{1}{|e^{j\omega} - Re^{j\theta}|}. \tag{7.7}$$

So, as $\omega \to \theta$, we have $e^{j\omega} \to e^{j\theta} \approx Re^{j\theta}$, when R is close to 1. The response (7.7) gets larger as $\omega \to \theta$. Pictorially, as the point $e^{j\omega}$ gets closer to the pole p in the complex plane, the quantity $1/|e^{j\omega} - p|$ grows (Fig. 7.1).

Similarly, the magnitude of the second term in Eq. (7.1) has a peak at $\omega = -\theta$. So the sum $H(e^{j\omega})$ has two peaks, approximately at θ and $-\theta$. Since we plot $|H(e^{j\omega})|$ only in $0 \leq \omega \leq \pi$, we only see the peak near $\omega = \theta$.

Peaks in the frequency response. The peak locations are not exactly at $\pm\theta$ because each term in Eq. (7.1) is a complex nonzero quantity at the peak of the other term, and these add up. Similarly, the peak heights are not exactly $1/(1 - R)$ as Eq. (7.6) would suggest. Nevertheless, the pole angle θ mostly determines the location of the peak, and the pole radius mostly determines the height of the peak $\approx 1/(1 - R)$. Thus, the peak gets sharper as R gets closer and closer to unity, as demonstrated in Fig. 7.2. ▽ ▽ ▽

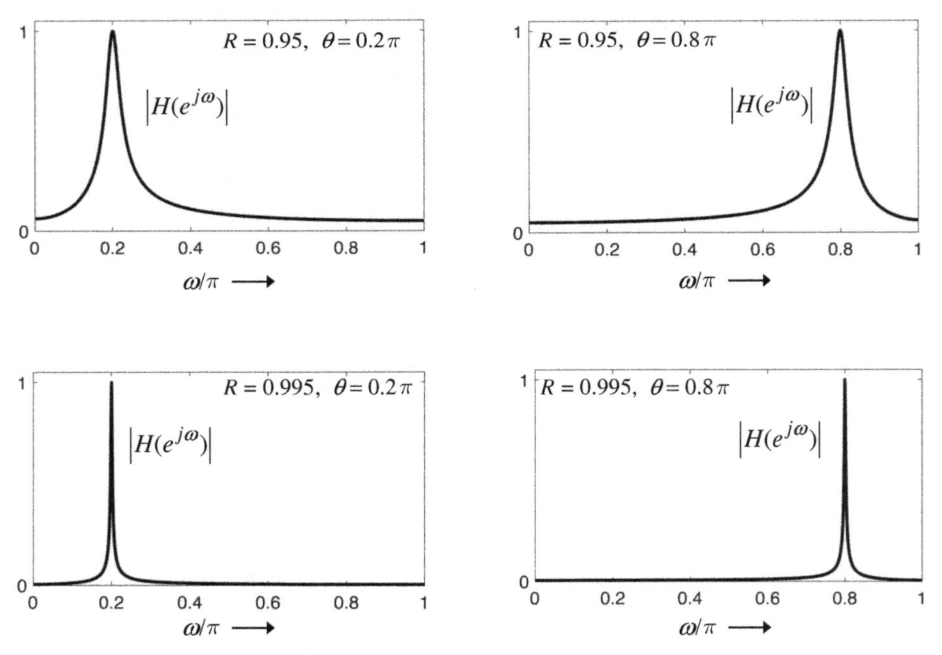

Figure 7.2 The magnitude response of the filter (7.5) for various values of the pole radius R and pole angle θ. The peak height of $|H(e^{j\omega})|$ has been normalized to unity for plotting convenience. This is a narrowband bandpass filter.

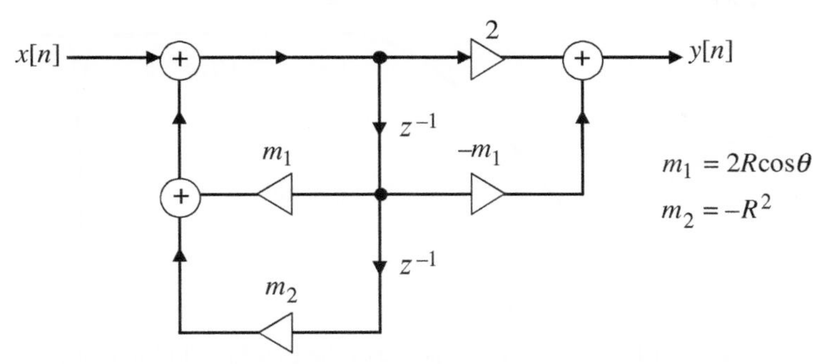

Figure 7.3 Computational graph for the narrowband bandpass filter (7.5). Multipliers m_1 and $-m_1$ can be shared, reducing the number of multipliers to two (Problem 7.1).

The filter can be implemented using the computational graph (direct-form structure) of Fig. 7.3. The parameters $m_1 = 2R\cos\theta$ and $m_2 = -R^2$ can be tuned to adjust the peak position and peak height of $|H(e^{j\omega})|$, respectively. By changing m_1 we adjust θ (i.e., the peak location), whereas by changing m_2 we can also make the peak sharper.

7.3 Adjustable Notch Filters

In some applications we require a digital filter to suppress a specific frequency, say $\omega = \phi$. The function of the filter is the opposite of the narrowband bandpass filter (7.5), which *passes* a specific frequency region around $\omega = \theta$ in preference to others. To suppress a specific frequency, all we need to do is to build a first-order filter

$$1 - e^{j\phi}z^{-1}. \tag{7.8}$$

This has a zero at $z = e^{j\phi}$ on the unit circle, that is, a transmission zero at $\omega = \phi$ in the plot of the frequency response. By multiplying filters like this, we can obtain zeros at a number of specified frequencies. If we are interested in real filters, then the zeros (and poles) are in complex-conjugate pairs. So we have to build a second-order filter section

$$A(z) = (1 - e^{j\phi}z^{-1})(1 - e^{-j\phi}z^{-1}) = 1 - 2\cos\phi\, z^{-1} + z^{-2}. \tag{7.9}$$

This has zeros at

$$z = e^{\pm j\phi}, \tag{7.10}$$

and suppresses frequencies $\omega = \pm\phi$. The magnitude response of this is shown in Fig. 7.4(a) for $0 \le \omega \le \pi$. This is called a **notch filter** and ϕ is called the **notch frequency**. Even though the filter achieves the goal of suppressing the notch frequency ϕ, it also attenuates a large neighbourhood of this frequency. A more desirable response of the notch filter would be as shown in Fig. 7.4(b), where the filter suppresses the frequency ϕ as before, and at the same time has gain close to unity

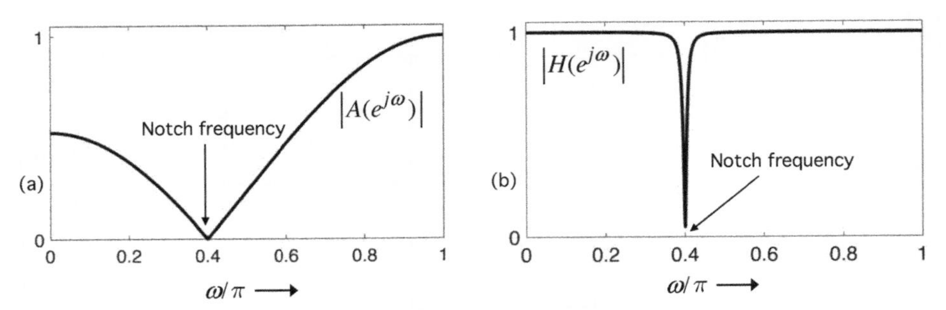

Figure 7.4 (a) The magnitude response of the FIR filter in Eq. (7.9); (b) the desired magnitude response of a good notch filter.

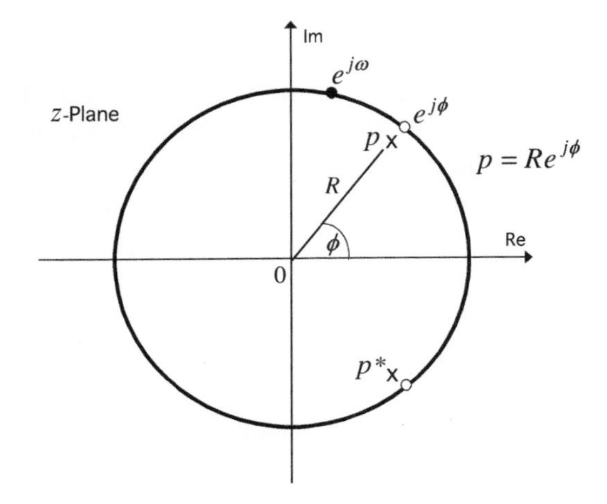

Figure 7.5 Designing a notch filter by positioning a pair of poles $Re^{\pm j\phi}$ close to the zeros $e^{\pm j\phi}$ representing the notch frequency ϕ.

at nearly all other frequencies. That is, the notch response is very sharp. We now show how to design good notch filters like this. It turns out that this can readily be achieved by using simple second-order IIR filters.

Figure 7.5 shows the two unit-circle zeros $z = e^{\pm j\phi}$ of the notch filter (7.9). To this filter, suppose we add poles $p = Re^{\pm j\phi}$ close to these zeros as shown. The poles have the same angles as the zeros ($\pm\phi$), but the pole radius $R < 1$ so that the filter is stable. Using Eq. (7.4), the transfer function becomes

$$H(z) = \frac{1 - 2\cos\phi\, z^{-1} + z^{-2}}{1 - 2R\cos\phi\, z^{-1} + R^2 z^{-2}}. \tag{7.11}$$

This is a second-order IIR filter with two poles and two zeros. Let us now see how this works:

1. At the notch frequency $\omega = \phi$, the numerator of Eq. (7.11) is zero, so $H(e^{j\omega}) = 0$ as expected. Understandably, at frequencies very close to the notch frequency $\omega = \phi$, the response $|H(e^{j\omega})|$ is small.
2. How about frequencies that are sufficiently away from the notch frequency $\omega = \phi$? Consider, for example, the black dot shown on the unit circle, which represents a frequency ω. Looking at the pole–zero pair from this frequency's

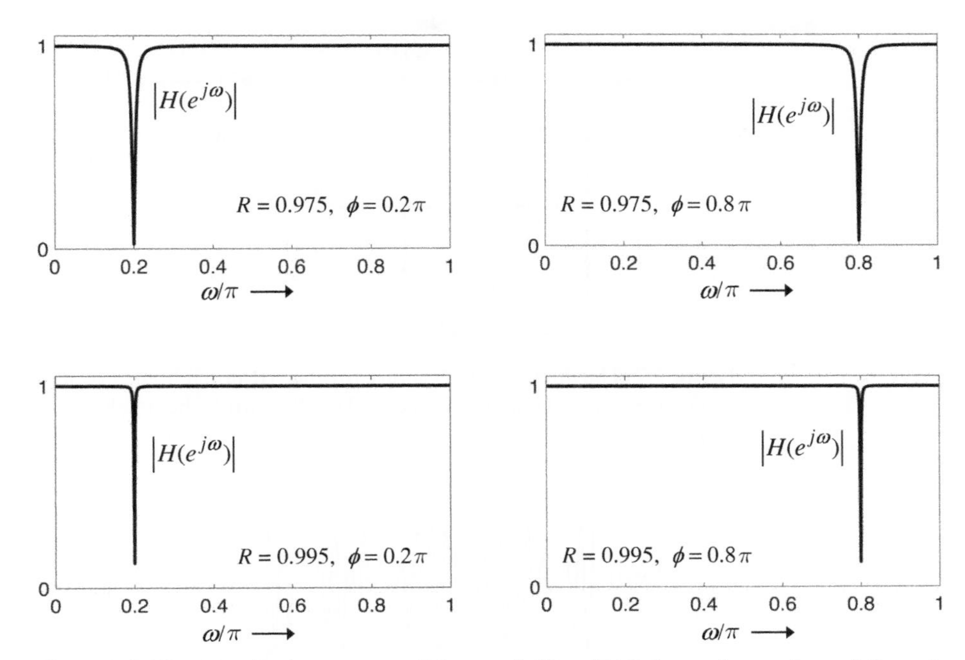

Figure 7.6 The magnitude response of the notch filter (7.11) for various values of the pole radius R and notch frequency ϕ.

viewpoint, the pole and zero almost cancel so that the response is close to unity! Thus, $|H(e^{j\omega})| \approx 1$ for frequencies that are not too close to ϕ. This means that the filter will indeed have a response somewhat like the one shown in Fig. 7.4(b).

Thus the transfer function (7.11) is expected to be a good notch filter. Furthermore, by making the pole radius R closer and closer to unity (while maintaining $R < 1$ for stability), the **quality of the notch** (i.e., sharpness) can be made better and better. To demonstrate, Fig. 7.6 shows the magnitude response of this notch filter for various values of the pole radius R and notch frequency ϕ. Thus, Eq. (7.11) provides a way to design notch filters with adjustable notch frequency and notch quality. Figure 7.7 shows a structure for the implementation of this IIR notch filter. The three multipliers can be tuned to achieve the desired notch frequency and notch quality.

7.3.1 A Notch Filtering Example

To see the notch filter in action, consider Fig. 7.8. Part (a) shows an input signal of the form

$$x[n] = \left(\frac{\cos(\omega_0 n) + \cos(\omega_1 n)}{2}\right)\mathcal{U}[n], \tag{7.12}$$

where $\omega_0 = 0.1\pi$ and $\omega_1 = 0.12\pi$. Consider the notch filter (7.11) with the notch frequency $\phi = \omega_0$, so that $\cos(\omega_0 n)$ can be suppressed at the output. Two situations will be considered:

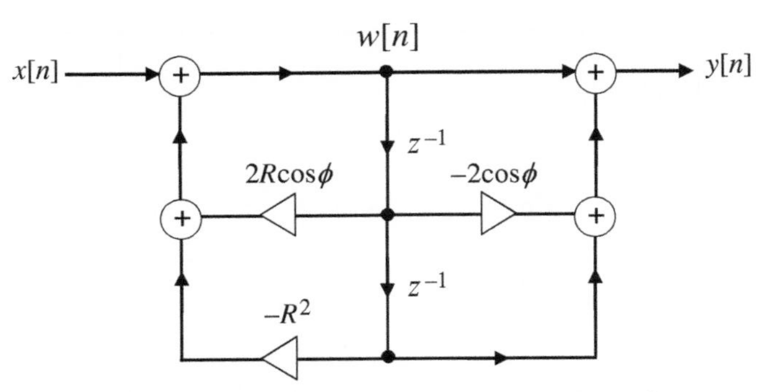

Figure 7.7 The direct-form structure for the IIR notch filter (7.11). Here, ϕ is the notch frequency and R controls the notch quality.

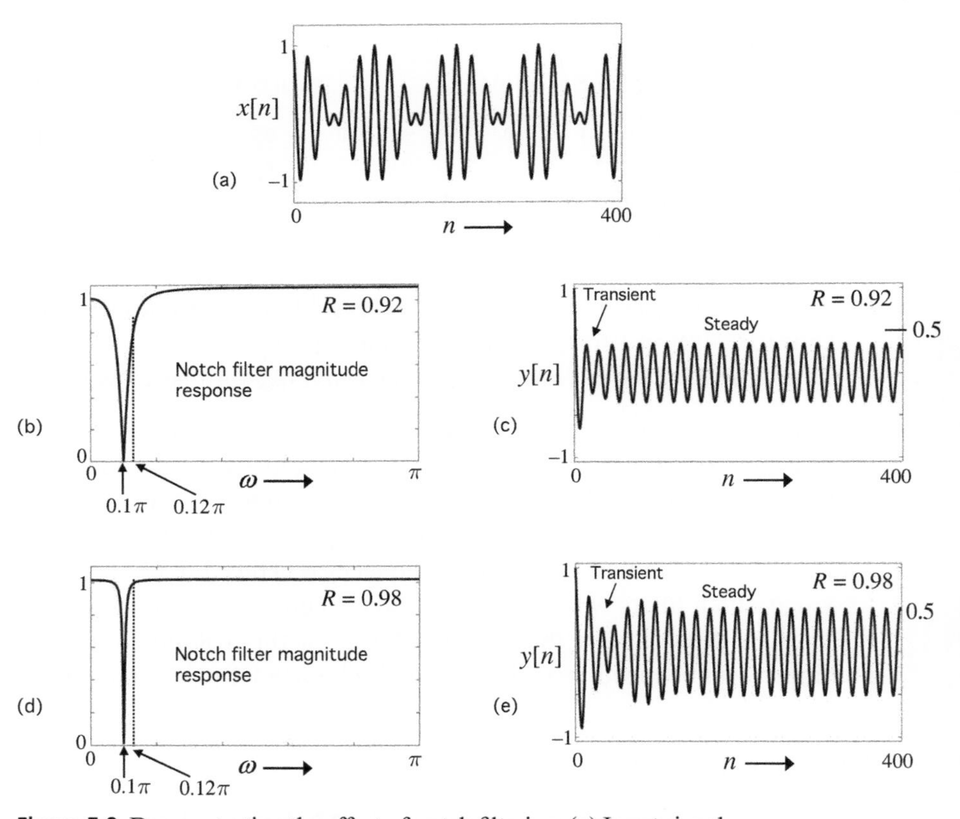

Figure 7.8 Demonstrating the effect of notch filtering. (a) Input signal $x[n] = 0.5\cos(\omega_0 n)\mathcal{U}[n] + 0.5\cos(\omega_1 n)\mathcal{U}[n]$; (b) and (c) notch filter magnitude response and filtered output for $R = 0.92$; (d) and (e) notch filter magnitude response and filtered output for $R = 0.98$. See text.

1. *Pole radius* $R = 0.92$. Figure 7.8(b) shows the magnitude response $|H(e^{j\omega})|$ of the notch filter, and Fig. 7.8(c) shows the filter output, when the pole radius $R = 0.92$. Notice that while $x[n]$ is a sum of sinusoids, $y[n]$ is almost a single sinusoid with frequency ω_1, since ω_0 has been suppressed. However, the first few

samples of the output do not look like a steady sinusoid. This is because of the so-called *transient response*, which will be explained in detail in Sec. 7.12. Notice also that even though the input cosines have amplitude 0.5, as seen from Eq. (7.12), the output cosine $\cos(\omega_1 n)$ has amplitude < 0.5. This is because $|H(e^{j\omega_1})|$ has not yet risen to unity, as ω_1 is very close to ω_0 (see dotted line in Fig. 7.8(b)).

2. *Pole radius* $R = 0.98$. In Fig. 7.8(d) and (e) we repeat the above experiment with $R = 0.98$ instead of 0.92. In this case the notch response $|H(e^{j\omega})|$ is sharper, so $|H(e^{j\omega_1})|$ is closer to unity (see dotted line in Fig. 7.8(d)). Thus, the steady part of the cosine in the output $y[n]$ has amplitude very close to 0.5, as expected. However, in this case the transient stage is longer. It will be seen in Sec. 7.12 that increasing the pole radius always has this disadvantage.

The above example demonstrates that even though the notch filter can be made arbitrarily sharp by increasing the pole radius R (subject to $R < 1$), there is a trade-off. If R is larger, the transient gets longer. This is especially undesirable when we are trying to filter short segments of data. This will become clearer in Sec. 7.12, especially when we explain Fig. 7.36. As you will learn in advanced classes, making R too close to unity is not desirable when quantization effects in the implementation are not negligible. For example, the quantization of filter coefficients to a finite number of bits may cause the pole to move outside the unit circle, making the filter unstable. Furthermore, any noise due to quantization of internal signals in a digital implementation (see Sec. 8.5.1) gets amplified as it reaches the output, and the amplification factor increases rapidly as $R \to 1$. For all these reasons, we cannot make R arbitrarily close to unity.

For more discussions on the design and applications of notch filters, see Hirano et al. [1974], Mitra and Hirano [1974], Pei et al. [2016], and Shiung [2022].

7.4 Allpass Filters

An allpass filter $H(z)$ is a filter such that

$$|H(e^{j\omega})| = 1, \quad \forall\omega. \tag{7.13}$$

The more general definition, $|H(e^{j\omega})| = c$ for some constant $c > 0$, is also sometimes used. The main point is that all frequencies are passed equally. The first impression would be that such a filter cannot be very useful because it is not filtering out any frequency. But it turns out that allpass filters have many applications in signal processing. In this section we study some basic properties of allpass filters, and in Secs. 7.5 and 7.14 we will see some applications.

7.4.1 Examples and Basic Properties

From the definition (7.13) we see that the frequency response of an allpass filter can be written as

$$H(e^{j\omega}) = e^{j\psi(\omega)}, \quad \forall\omega, \tag{7.14}$$

where $\psi(\omega)$ is the phase response. Since the magnitude response is unity for all ω, the phase response completely characterizes an allpass filter. It is readily verified that the delay element $H(z) = z^{-1}$, and more generally

$$H(z) = z^{-K}, \tag{7.15}$$

for integer K, are allpass. This is because $H(e^{j\omega}) = e^{-jK\omega}$ has unit magnitude. A more nontrivial example is the first-order filter

$$H(z) = \frac{b^* + z^{-1}}{1 + bz^{-1}}. \tag{7.16}$$

To prove that this is allpass, simply observe that

$$H(e^{j\omega}) = \frac{b^* + e^{-j\omega}}{1 + be^{-j\omega}} = e^{-j\omega}\left(\frac{1 + b^* e^{j\omega}}{1 + be^{-j\omega}}\right). \tag{7.17}$$

The quantity in brackets has unit magnitude because its numerator is the conjugate of the denominator. So $|H(e^{j\omega})| = 1$, proving that it is allpass. For convenience, we often rewrite Eq. (7.16) as

$$H(z) = \frac{-p^* + z^{-1}}{1 - pz^{-1}}, \tag{7.18}$$

where p is the pole. Notice that the zero is at $z = 1/p^*$. If two filters $H_1(z)$ and $H_2(z)$ are allpass, their cascade $H_1(z)H_2(z)$ is allpass because $|H_1(e^{j\omega})H_2(e^{j\omega})| = 1$. Thus, by multiplying first-order allpass filters of the form (7.18), we can get higher-order allpass filters:

$$H(z) = \alpha \prod_{i=1}^{N} \left(\frac{-p_i^* + z^{-1}}{1 - p_i z^{-1}}\right), \tag{7.19}$$

where $\alpha \neq 0$. A number of points are worth noting:

1. The constant α can be complex, and merely serves the purpose of generality. In nearly all applications we take $\alpha = 1$.
2. The allpass filter (7.19) has poles p_i and zeros $1/p_i^*$. Thus, poles and zeros are *reciprocal conjugate* pairs.
3. It is possible that some of the $p_i = 0$. For example, if $p_1 = 0$, then the first filter section reduces to z^{-1}. Thus, allpass filters of the form $H(z) = z^{-N}$ are special cases of Eq. (7.19).
4. More generally, it can be shown that any rational allpass filter in discrete time can always be written in the form (7.19) (see Sec. 3.4 of Vaidyanathan [1993]).
5. Multiplying out the first-order sections in Eq. (7.19), $H(z)$ takes the form

$$H(z) = \alpha\left(\frac{b_N^* + b_{N-1}^* z^{-1} + \cdots + b_1^* z^{-(N-1)} + z^{-N}}{1 + b_1 z^{-1} + b_2 z^{-2} + \cdots + b_N z^{-N}}\right). \tag{7.20}$$

   Thus, except for the scale factor α, the numerator has the same coefficients as the denominator but *conjugated, and in reverse order.* Denoting

$$B_N(z) = 1 + b_1 z^{-1} + b_2 z^{-2} + \cdots + b_N z^{-N}, \tag{7.21}$$

we have

$$z^{-N}\widetilde{B}_N(z) = z^{-N}[B_N(1/z^*)]^* = b_N^* + b_{N-1}^* z^{-1} + \cdots + b_1^* z^{-(N-1)} + z^{-N}, \tag{7.22}$$

where the tilde notation $\widetilde{B}_N(z) = [B_N(1/z^*)]^*$ follows the convention explained in Sec. 1.3. The quantity (7.22) is precisely the numerator polynomial in Eq. (7.20). Thus, we can write Eq. (7.20) as

$$H(z) = \alpha \frac{z^{-N}\widetilde{B}_N(z)}{B_N(z)}. \tag{7.23}$$

6. For real allpass filters (where b_k and α are real), Eq. (7.20) becomes

$$H(z) = \alpha \left(\frac{b_N + b_{N-1}z^{-1} + \cdots + b_1 z^{-(N-1)} + z^{-N}}{1 + b_1 z^{-1} + b_2 z^{-2} + \cdots + b_N z^{-N}} \right). \tag{7.24}$$

Everything mentioned so far is simple, and has been developed logically starting from the first-order allpass example (7.18). We now mention a couple of less obvious facts, for which proofs can be found in Vaidyanathan [1993]. First, *any rational allpass filter* has the form (7.20) (or equivalently (7.23)), and can be written in factored form (7.19). The poles and zeros are therefore always in **reciprocal conjugate pairs**

$$\{p_i, 1/p_i^*\}. \tag{7.25}$$

Second, it can also be shown that, as long as the poles are inside the unit circle, that is,

$$|p_i| < 1, \tag{7.26}$$

any rational allpass $H(z)$ satisfies the following curious property:

$$|H(z)| \begin{cases} = 1 & |z| = 1, \\ < 1 & |z| > 1, \\ > 1 & |z| < 1, \end{cases} \tag{7.27}$$

where we have assumed $|\alpha| = 1$ for convenience. (We have also assumed filter order $N \geq 1$, otherwise $|H(z)| = 1$ for all z.) Equation (7.27) is called the **modulus property**. For the special case of $H(z) = z^{-1}$, property (7.27) is very clear indeed! Since condition (7.26) is equivalent to the stability property for causal LTI systems, we can say that *any causal stable rational allpass filter* with $|H(e^{j\omega})| = 1$ satisfies the property (7.27).

The first property is obvious because it just says $|H(e^{j\omega})| = 1$. A proof of (7.27) for first-order allpass filters is easy to develop (Problem 7.5). For more general rational allpass filters we can use Eq. (7.19) to complete the proof. More properties of allpass filters can be found in Vaidyanathan [1993].

The modulus property. More generally, let $H(z)$ be any rational transfer function (not necessarily allpass) with all poles satisfying $|p_i| < 1$. If $|H(e^{j\omega})|$ has maximum value $M > 0$, then it can be shown that

$$|H(z)| \leq M \quad \text{for all } |z| > 1, \tag{7.28}$$

with equality for some z in $|z| > 1$ if and only if $H(z)$ is a constant, that is, $H(z) = Me^{j\theta}$ for some real constant θ, for all z. This follows from a result called the maximum modulus theorem in complex variable theory [Churchill and Brown, 1984]. Also see p. 75 of Vaidyanathan [1993]. ▽ ▽ ▽

7.4.2 Computational Graph or Structure

For the allpass filter

$$H(z) = \frac{b_2 + b_1 z^{-1} + z^{-2}}{1 + b_1 z^{-1} + b_2 z^{-2}} \tag{7.29}$$

with real coefficients, the direct-form structure is shown in Fig. 7.9. This requires four multipliers, but since the same coefficients b_k appear in numerator and denominator in different orders, there is a way to share these, as we shall see in this section. In fact, any Nth-order real coefficient allpass filter can be implemented with only N real multipliers.

To explain how multipliers can be shared, it is easier to first consider Fig. 7.10(a), which shows the direct-form structure for the first-order allpass filter (7.16). For the special case of real filters (i.e., when b is real), we have

$$H(z) = \frac{b + z^{-1}}{1 + bz^{-1}}. \tag{7.30}$$

Since the same coefficient b appears in the numerator and denominator, these can be shared as shown in Fig. 7.10(b). To understand this, simply observe that b and z^{-1} are interchangeable in Eq. (7.30). Although we now require only one multiplier,

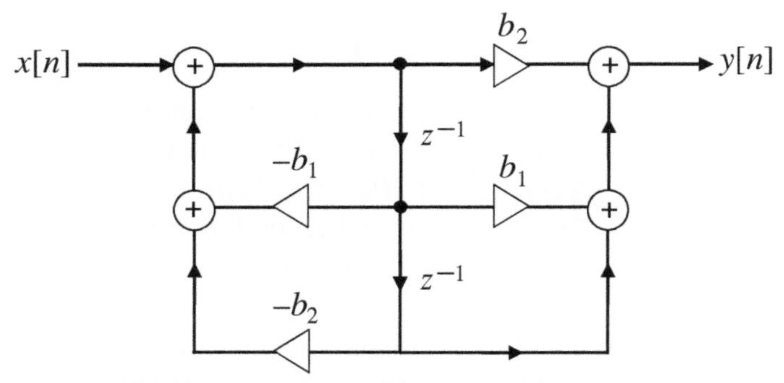

Figure 7.9 Direct-form structure for the second-order real-coefficient allpass filter given in Eq. (7.29). Each multiplier b_k appears twice, which makes this non-economic.

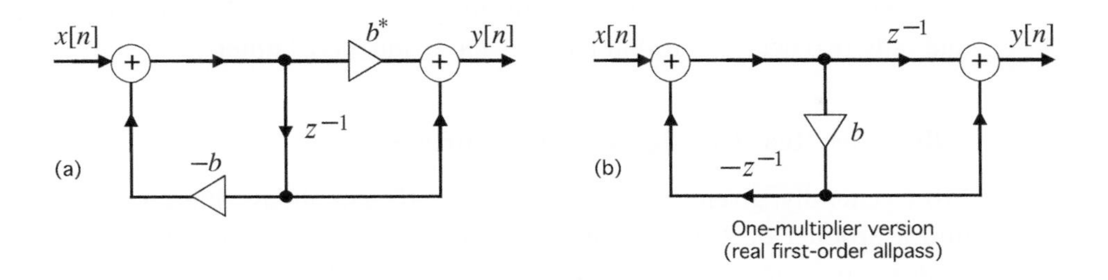

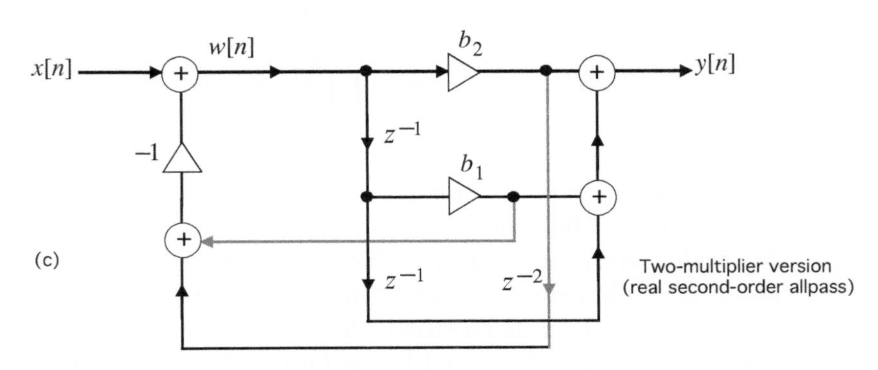

Figure 7.10 Computational graphs for allpass filters. (a) First-order allpass, (b) first-order allpass with real coefficients, and (c) second-order allpass with real coefficients.

unlike Fig. 7.10(a), two delay units are required. It is also possible to obtain a structure with one multiplier and one delay [Gray and Markel, 1973], but that requires three additions, instead of two as in the above structures. This is called the lattice structure, and it is reviewed in Chapter 14.

Next, for the real second-order allpass filter (7.29), we will show that Fig. 7.9 can be replaced with the more economic structure shown in Fig. 7.10(c); this requires only two multipliers and four adders, although it requires four delays.

Proof that Fig. 7.10(c) implements Eq. (7.29). The z-transform of the internal signal $w[n]$ is given by

$$W(z) = X(z) - \left(b_1 z^{-1} W(z) + b_2 z^{-2} W(z)\right). \tag{7.31}$$

The output signal, on the other hand, is given by

$$Y(z) = \left(b_2 + b_1 z^{-1} + z^{-2}\right) W(z). \tag{7.32}$$

From Eq. (7.31) we have $(1 + b_1 z^{-1} + b_2 z^{-2})W(z) = X(z)$. Substituting this into Eq. (7.32) and simplifying, we obtain $Y(z)/X(z) = (b_2 + b_1 z^{-1} + z^{-2})/(1 + b_1 z^{-1} + b_2 z^{-2})$, which indeed proves Eq. (7.29). ▽▽▽

Note that in Fig. 7.10(c), the multiplier with value -1 does not count as a multiplier. It is simply a reminder that we should reverse the sign of one of the signals that is going into the adder unit. It is possible to implement Eq. (7.29) with two multipliers and only two delays, but that requires more adders (Chapter 14).

Nth-Order Real-Coefficient Allpass Structure

More generally, given an Nth-order real-coefficient allpass filter, it can be factored into a product of N first-order allpass filters. If a pole is complex, its conjugate is also a pole and the two first-order sections with complex-conjugate poles can be combined into a second-order allpass section with real coefficients. Thus, any Nth-order real-coefficient allpass filter can be factored into the form

$$H(z) = c\left(\frac{b + z^{-1}}{1 + bz^{-1}}\right)^{l} \prod_{i=1}^{M} \left(\frac{b_{i,2} + b_{i,1}z^{-1} + z^{-2}}{1 + b_{i,1}z^{-1} + b_{i,2}z^{-2}}\right), \tag{7.33}$$

where $l = 0$ or 1 and the filter order is $N = 2M + l$. (If the filter order is even, the first-order section is not required, so we set $l = 0$.) Here, c, b, and $b_{i,k}$ are real coefficients. The first and second-order sections can be implemented with one and two multipliers, respectively, as in Figs. 7.10(b) and (c). The constant c is usually ± 1 and need not be regarded as a multiplier. This shows that a real-coefficient allpass filter can be implemented with *exactly N real multipliers*. But it requires more than N delays because of the extra delays in Figs. 7.10(b) and (c). While it is not obvious, it turns out that we can actually obtain a structure with N real multipliers and N delays only, achieving the minimum possible computational and storage complexity for real-coefficient allpass filters (Sec. 14.6.1).

7.5 Adjustable Notch Filters from Allpass Filters

This section shows that we can start from a second-order allpass filter and design notch filters like the ones in Sec. 7.3. It will be shown that this approach has several advantages. First, we require only two multipliers instead of three as in Fig. 7.7. Second, and more importantly, we get another filter at practically no additional cost, called the antinotch filter (which is a narrowband bandpass filter as in Sec. 7.2). Third, this second filter has better frequency response than those in Sec. 7.2 because it automatically has guaranteed transmission zeros at $\omega = 0$ and $\omega = \pi$. Let us begin by defining a second-order rational allpass filter[1]

$$G(z) = \frac{R^2 - 2R\cos\theta\, z^{-1} + z^{-2}}{1 - 2R\cos\theta\, z^{-1} + R^2 z^{-2}}, \tag{7.34}$$

where $0 < R < 1$ and θ is real. This is allpass because it has the form (7.20). Since $(1 - 2R\cos\theta\, z^{-1} + R^2 z^{-2}) = (1 - pz^{-1})(1 - p^* z^{-1})$, where

$$p = Re^{j\theta}, \quad p^* = Re^{-j\theta}, \tag{7.35}$$

[1] Here the notation $G(z)$ is used for the allpass filter instead of $H(z)$ as in Sec. 7.4. We shall use $H(z)$ for the notch filter.

the poles are given by the complex-conjugate pair (7.35). Since the pole radius is $R < 1$, $G(z)$ is stable. In short, *$G(z)$ is both stable and allpass*. Now define the filter

$$H(z) = \frac{1 + G(z)}{2}. \tag{7.36}$$

Substituting from Eq. (7.34) and simplifying, we get

$$H(z) = \kappa \left(\frac{1 - 2Qz^{-1} + z^{-2}}{1 - 2R\cos\theta z^{-1} + R^2 z^{-2}} \right), \tag{7.37}$$

where

$$Q = \left(\frac{2R}{1 + R^2} \right) \cos\theta, \quad \kappa = \frac{1 + R^2}{2}. \tag{7.38}$$

Since $(1 - R)^2 > 0$ it follows that $2R/(1 + R^2) < 1$. So $-1 < Q < 1$, and we can write $Q = \cos\phi$ for some real angle ϕ. Summarizing, Eq. (7.36) simplifies to

$$H(z) = \kappa \left(\frac{1 - 2\cos\phi z^{-1} + z^{-2}}{1 - 2R\cos\theta z^{-1} + R^2 z^{-2}} \right), \tag{7.39}$$

where

$$\cos\phi = \left(\frac{2R}{1 + R^2} \right) \cos\theta. \tag{7.40}$$

The pole–zero patterns for $H(z)$ and $G(z)$ are summarized in Fig. 7.11. $H(z)$ has the same poles as the allpass filter $G(z)$, namely $\{p, p^*\}$. The allpass filter has zeros at $\{1/p^*, 1/p\}$, which are outside the unit circle. But the filter $H(z)$ has zeros at $e^{\pm j\phi}$,

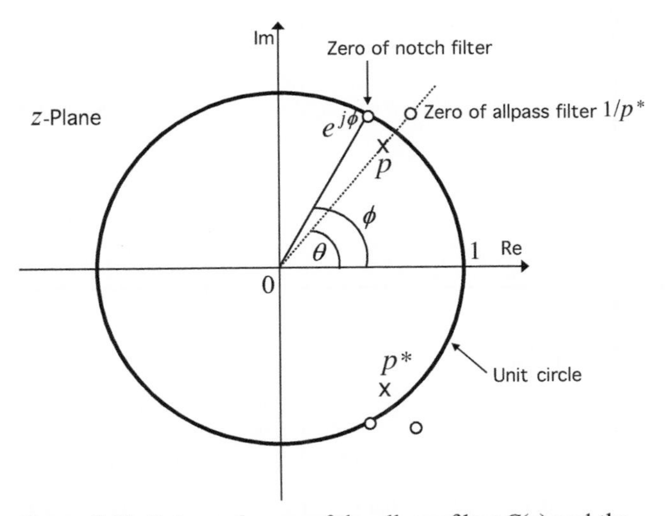

Figure 7.11 Poles and zeros of the allpass filter $G(z)$ and the notch filter $H(z) = (1 + G(z))/2$. The poles $\{p, p^*\}$ are common to both filters. The zeros of the allpass filter $G(z)$ are outside the unit circle, while those of the notch filter $H(z)$ are on the unit circle.

that is, *on the unit circle* at the angles $\pm\phi$. For $H(z)$, this is very similar to the pole–zero plot we devised in Fig. 7.5 to create a notch filter response. Indeed, $H(e^{j\omega}) = 0$ at $\omega = \phi$, so $H(z)$ behaves like a good notch filter when R is close to unity. For $R \approx 1$ we have $2R/(1 + R^2) \approx 1$, so that

$$\phi \approx \theta \tag{7.41}$$

from Eq. (7.40). Summarizing:

1. If we take the real second-order allpass filter (7.34) and define $H(z)$ as in Eq. (7.36), then $H(z)$ behaves like a notch filter when $0 < R < 1$ and R is close to 1.
2. The notch frequency ϕ is determined by Eq. (7.40) and depends on both the pole radius R and the pole angle θ. But for R close to the unit circle, the dependence on R is slight, and $\phi \approx \theta$.
3. As the pole moves closer to the unit circle ($R \to 1$) subject to the stability constraint $R < 1$, the notch quality improves because the poles of $H(z)$ almost cancel its zeros. Of course, the cancellation is never exact because the zeros are on the unit circle and the poles are inside.

Figure 7.12 shows the magnitude response of the notch filter (7.39) for various values of the pole radius R and pole angle θ. The notch frequency ϕ determined by Eq. (7.40) is also shown in each case, with four decimal digit accuracy. Note that ϕ is very close to θ because R is close to unity. So for all practical purposes we can

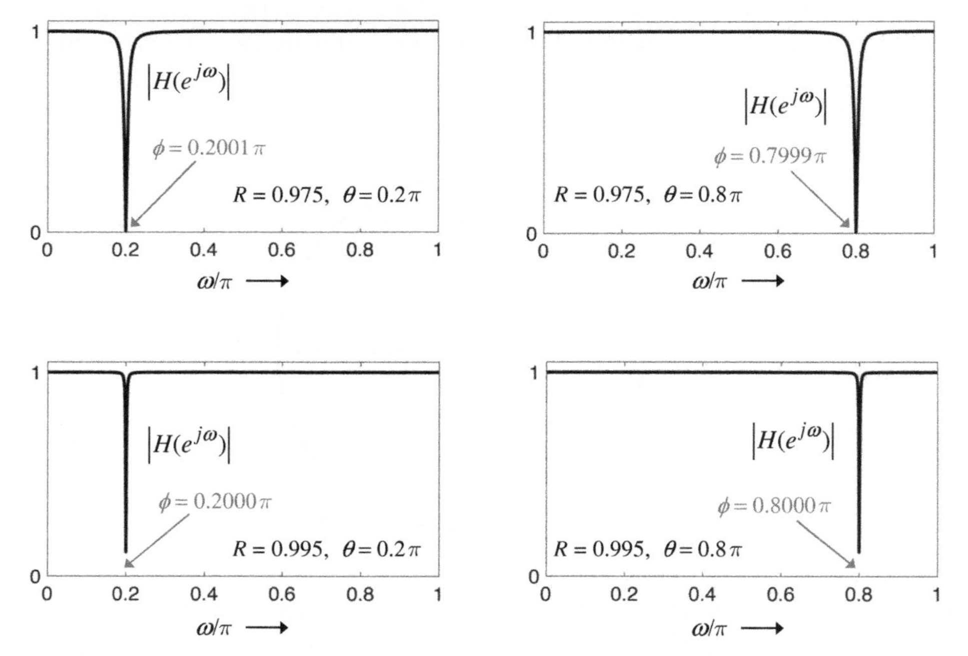

Figure 7.12 The magnitude response of the allpass-based notch filter (7.39) for various values of the pole radius R and pole angle θ. The notch frequency ϕ, determined by Eq. (7.40), is also indicated up to four-digit precision, and is very close to θ.

regard the pole angle θ to be the notch frequency ϕ. Notice from the plots that the notch gets sharper as R gets closer to unity. Thus, by controlling the pole radius R and pole angle θ of the allpass filter $G(z)$, we can control the notch quality and notch frequency.

Not suprisingly, these notch filter responses agree closely with those designed in Sec. 7.3 (Fig. 7.6). Since the second-order allpass filter (7.34) can be implemented with only two multipliers using a structure similar to Fig. 7.10(c), the notch filter (7.36) can be implemented using two multipliers, instead of three as in Sec. 7.3 (Fig. 7.7). If multipliers dominate the cost compared to adders and delays (as in hardware implementations), this is an advantage.

Starting from the allpass filter (7.34), we designed the notch filter using Eq. (7.36). Now consider a new filter defined as follows:

$$F(z) = \frac{1 - G(z)}{2}. \tag{7.42}$$

The magnitude response of this filter is shown in Fig. 7.13 for the same sets of values of R and θ (pole radius and angle) as in Fig. 7.12. The response of $F(z)$ appears to be complementary to that of $H(z)$: instead of suppressing the notch frequency ϕ, the filter now has the maximum magnitude at $\omega = \phi$. In Sec. 7.5.1 we will explain why this behavior arises.

Notice that the response is very sharp near ϕ when R is close to unity. Since the behavior of $F(z)$ is complementary to that of $H(z)$, we will call $F(z)$ an **antinotch** filter, and the notch frequency ϕ is sometimes referred to as the antinotch frequency.

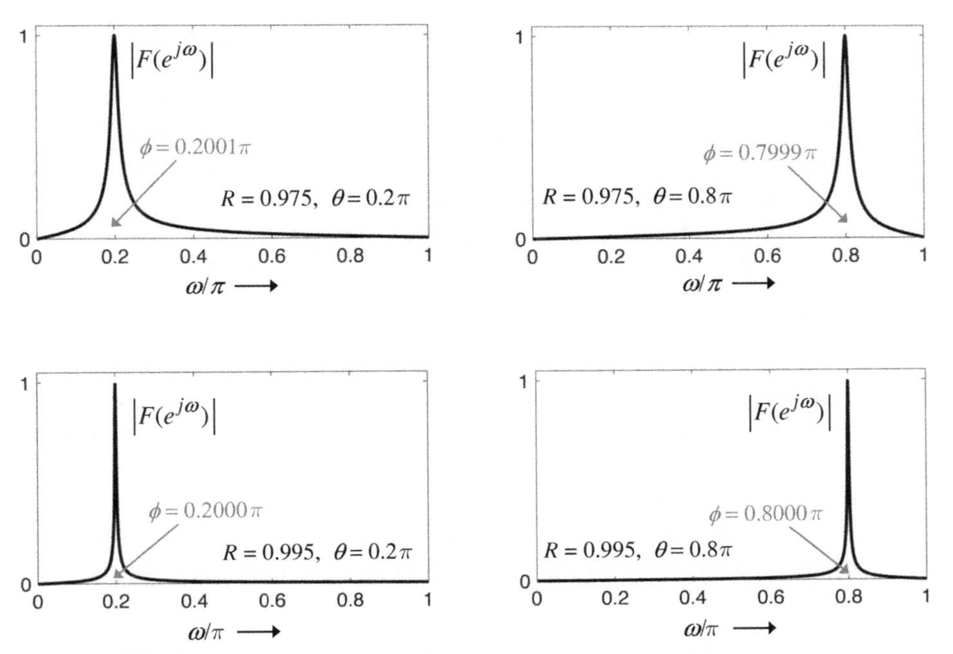

Figure 7.13 The magnitude response of the allpass-based antinotch filter (7.42) for various values of the pole radius R and pole angle θ. The antinotch frequency ϕ is determined by Eq. (7.40).

Notice that this is nothing but a narrowband bandpass filter. Indeed, the response resembles that in Fig. 7.2, although there are some differences which we shall explain below.

7.5.1 Power Complementary Property

We now prove a very important property of the notch–antinotch filter pair $\{H(z), F(z)\}$ given by Eqs. (7.36) and (7.42). The frequency responses are

$$H(e^{j\omega}) = \frac{1 + G(e^{j\omega})}{2}, \quad F(e^{j\omega}) = \frac{1 - G(e^{j\omega})}{2}, \tag{7.43}$$

where $G(z)$ is allpass. Thus

$$|H(e^{j\omega})|^2 = \frac{(1 + G(e^{j\omega}))(1 + G^*(e^{j\omega}))}{4} = \frac{2 + G(e^{j\omega}) + G^*(e^{j\omega})}{4},$$
$$|F(e^{j\omega})|^2 = \frac{(1 - G(e^{j\omega}))(1 - G^*(e^{j\omega}))}{4} = \frac{2 - G(e^{j\omega}) - G^*(e^{j\omega})}{4},$$

where we have used the allpass property $|G(e^{j\omega})| = 1$. Adding these, it follows that

$$|H(e^{j\omega})|^2 + |F(e^{j\omega})|^2 = 1, \qquad \forall\omega. \tag{7.44}$$

This is called the **power complementary** property. It says that the magnitude squares or "powers" of $H(e^{j\omega})$ and $F(e^{j\omega})$ exactly add up to unity for all ω, even though $|H(e^{j\omega})|$ and $|F(e^{j\omega})|$ are themselves not constants. In particular, therefore:

$$|H(e^{j\omega})| \approx 1 \quad \text{implies} \quad |F(e^{j\omega})| \approx 0, \quad \text{and vice versa.} \tag{7.45}$$

This explains why the notch behavior of $H(e^{j\omega})$ (Fig. 7.12) induces the antinotch behavior of $F(e^{j\omega})$ (Fig. 7.13). For further insight, observe from Eq. (7.34) that the allpass filter satisfies $G(1) = 1$ and $G(-1) = 1$. Since $z = 1$ and -1 correspond to $\omega = 0$ and π, respectively, this implies

$$H(e^{j0}) = H(e^{j\pi}) = \frac{1+1}{2} = 1, \quad F(e^{j0}) = F(e^{j\pi}) = \frac{1-1}{2} = 0, \tag{7.46}$$

which is also consistent with Eq. (7.44). Thus the antinotch filter $F(z)$ designed based on the allpass filter is *guaranteed to have zero response* at $\omega = 0$ and $\omega = \pi$, as seen clearly from the plots in Fig. 7.13. This is an advantage compared to the narrowband bandpass filters designed in Sec. 7.2; they were also antinotch filters, but they had nonzero responses at $\omega = 0$ and $\omega = \pi$, which were especially noticeable when R is not close to 1 (Fig. 7.2).

Finally, similar to the closed-form expression (7.39) for the notch filter, we can derive an expression for the antinotch filter $F(z)$. Thus

$$F(z) = \frac{1 - G(z)}{2} = 0.5\left(1 - \frac{R^2 - 2R\cos\theta\, z^{-1} + z^{-2}}{1 - 2R\cos\theta\, z^{-1} + R^2 z^{-2}}\right), \tag{7.47}$$

which simplifies to

$$F(z) = c\left(\frac{1 - z^{-2}}{1 - 2R\cos\theta\, z^{-1} + R^2 z^{-2}}\right), \tag{7.48}$$

where $c = 0.5(1 - R^2)$. Again, the numerator $(1 - z^{-2})$ shows that the antinotch filter $F(z)$ has its two zeros at $z = 1, -1$, consistent with the properties (7.46).

Consistency with the modulus property. In this section we started from a second-order allpass filter $G(z)$ and showed that $H(z) = (1 + G(z))/2$ is a notch filter and $F(z) = (1 - G(z))/2$ is antinotch. But what is the underlying principle behind such ideas? The fact is that both of these follow from the modulus property (7.27), which is satisfied by $G(z)$ since it is allpass with all poles inside the unit circle:

$$|G(z)| \begin{cases} = 1 & |z| = 1, \\ < 1 & |z| > 1, \\ > 1 & |z| < 1. \end{cases} \tag{7.49}$$

The modulus property implies that if

$$1 + G(z_0) = 0 \tag{7.50}$$

for some z_0, then z_0 is necessarily on the unit circle. This is because the above implies $|G(z_0)| = 1$, which is possible if and only if $|z_0| = 1$, according to Eq. (7.49). In short, given any Nth-order rational allpass $G(z)$ with all poles inside the unit circle, all the N zeros of $1 + G(z)$ are on the unit circle. The same is true for $1 - G(z)$, and more generally for $e^{j\alpha} + G(z)$ for any real constant α. This explains why, with our second-order allpass filter $G(z)$ with poles inside the unit circle, we found that $H(z) = (1 + G(z))/2$ is a notch filter with both its zeros $e^{\pm j\phi}$ on the unit circle. Similarly, $F(z) = (1 - G(z))/2$ has both zeros on the unit circle and they are at $z = 1, -1$.

More generally, therefore, if we wish to design a multinotch filter with notches at $\phi_1, \phi_2, \ldots$, this can be done by using a higher-order allpass filter $G(z)$, although this is not the only way to design it. We can also achieve this by multiplying two or more second-order notch filters. ▽ ▽ ▽

Figure 7.14(a) shows the implementation of the notch–antinotch pair

$$H(z) = \frac{1 + G(z)}{2}, \quad F(z) = \frac{1 - G(z)}{2}, \tag{7.51}$$

using a single allpass filter $G(z)$. The structure for $G(z)$, which is given by Eq. (7.34), is shown in Fig. 7.14(b); it is the allpass structure with only two multipliers and four adders, shown in Fig. 7.10(c). Thus, at the cost of only two multipliers and some adders, we obtain two filters $H(z)$ and $F(z)$ with complementary properties. Note that once $H(z)$ has been implemented, it takes just one more addition to implement $F(z)$. In that sense, the second filter $F(z)$ is practically free of cost.

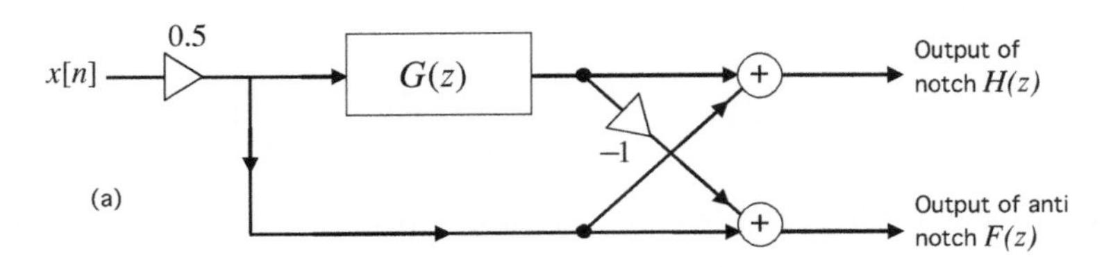

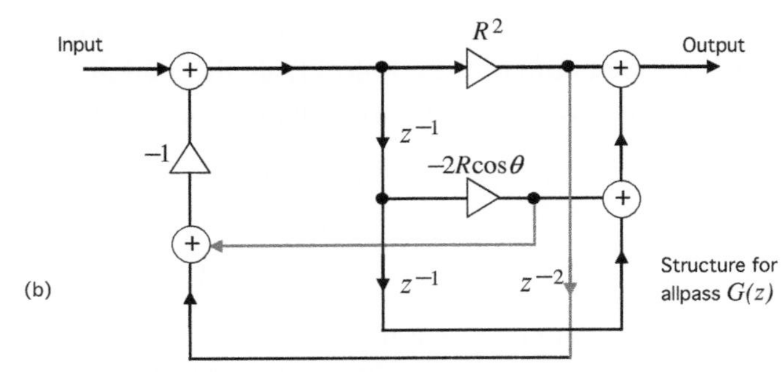

Figure 7.14 (a) Implementation of the notch–antinotch pair of filters using a single allpass filter $G(z)$; (b) details of the structure of $G(z)$.

7.6 Designing More General Digital Filters

In the previous section we showed how to design narrowband bandpass filters and notch filters based on second-order IIR transfer functions. We now discuss the design of lowpass filters without restricting the order to be two. In Sec. 4.3.4 it was seen that the ideal lowpass filter is unrealizable. A quick way to see this is that realizable digital filters $H(z)$ have to be *rational* transfer functions (Sec. 14.4.4), but we know that rational transfer functions have frequency responses of the form (5.67), which are *continuous in* ω. And since the ideal $H(e^{j\omega})$ is not continuous, it follows that it is not realizable. More generally, we claim the following. Consider any frequency response which is constant on some interval, that is,

$$H(e^{j\omega}) = c \quad \text{in} \quad \omega_a < \omega < \omega_b.$$

Such a filter is unrealizable (even if it is continuous, differentiable, etc.), unless $H(z) = c$ *everywhere*, in which case it is not an interesting filter.

Proof of the above claim. A rational $H(z)$ has the form $H(z) = A(z)/B(z)$ where $A(z)$ and $B(z)$ are polynomials in z^{-1}. If $H(e^{j\omega}) = c$ everywhere in an interval $\omega_a < \omega < \omega_b$, then $A(e^{j\omega}) = cB(e^{j\omega})$ there. With $A(z) = \sum_{k=0}^{N} a_k z^{-k}$ and $B(z) = \sum_{k=0}^{N} b_k z^{-k}$, this becomes

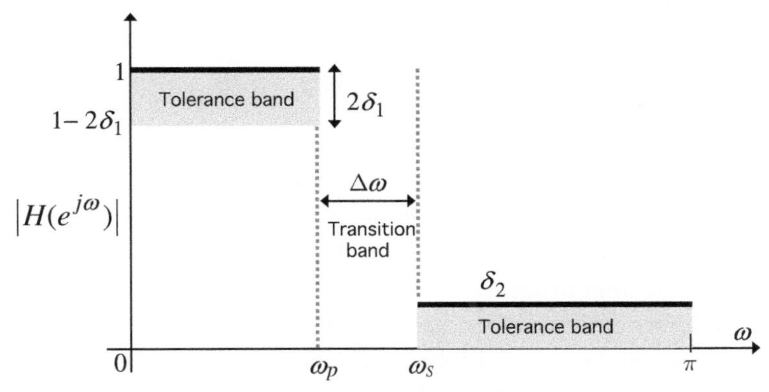

Figure 7.15 Design specifications for a digital lowpass filter magnitude. Here, $2\delta_1$ is the peak-to-peak passband error, δ_2 is the peak stopband error, ω_p and ω_s are the passband and stopband edges, and $\Delta\omega$ is the transition bandwidth.

$$\sum_{k=0}^{N}(a_k - cb_k)e^{-jk\omega} = 0, \tag{7.52}$$

for all ω in $\omega_a < \omega < \omega_b$. This implies that the Nth-order polynomial

$$\sum_{k=0}^{N}(a_k - cb_k)z^{-k} \tag{7.53}$$

has infinitely many zeros, namely the unit-circle points $z = e^{j\omega}$ with $\omega_a < \omega < \omega_b$. This is not possible unless $a_k - cb_k = 0, \ \forall k$, that is, unless $A(z) = cB(z)$ or equivalently $H(z) = c$ for all z. $\nabla\nabla\nabla$

So, with rational $H(z)$, we cannot satisfy $H(e^{j\omega}) = 1$ everywhere in the passband or $H(e^{j\omega}) = 0$ everywhere in the stopband. We can only approximate such responses. In practice, therefore, one specifies a tolerance region in these bands as shown in gray in Fig. 7.15. Furthermore, the transition of $H(e^{j\omega})$ from the passband to the stopband, cannot be abrupt because a rational $H(e^{j\omega})$ is a continuous (and differentiable) function. So, there has to be a nonzero transition bandwidth $\Delta\omega$ as shown. The extent to which the magnitude response $|H(e^{j\omega})|$ is allowed to deviate from the ideal is described by the tolerance regions in Fig. 7.15. The important quantities in this figure are:

- Peak-to-peak passband error $2\delta_1$ and peak stopband error δ_2. These determine the maximum **approximation errors** in the passband and stopband, respectively.
- Passband edge ω_p and stopband edge ω_s.
- Transition band width $\Delta\omega = \omega_s - \omega_p$, or equivalently $\Delta f = \Delta\omega/2\pi$.

We also call δ_1 the peak passband error or **ripple size** and δ_2 the peak stopband error or ripple size. As long as $|H(e^{j\omega})|$ is within the passband tolerance band and stopband tolerance band indicated in gray, the response is considered acceptable.

The four quantities

$$\omega_p, \ \omega_s, \ \delta_1, \ \delta_2 \tag{7.54}$$

are also called the filter **specifications** for the lowpass magnitude response. The filter order N required to meet the specifications depends on whether the filter is FIR or IIR, and also on the specific design method used. For a given design method, a smaller value of $\Delta\omega, \delta_1$, or δ_2 (with the other two fixed) usually results in a larger value of the filter order N.

The filter order usually determines the cost of its implementation (e.g., it determines the number of multipliers, adders, and delays, see Sec. 5.9). Typically FIR filters require much higher order than IIR filters (Sec. 7.10). This extra price paid in the case of FIR filters is offset by the fact that only FIR filters can achieve exact linear phase,[2] which is important in image processing applications (Sec. 7.10).

7.6.1 Classical Continuous-Time Filters

A well-known approach to designing an IIR digital filter $H(e^{j\omega})$ is to start from a continuous-time or analog lowpass filter[3] $H_c(j\Omega)$ and map it into a digital filter $H(e^{j\omega})$ using a mapping from Ω to ω. Since analog filters were well established decades before digital filters were even conceived, this approach became (and still is) popular, so we will describe it here. For this we first summarize some well-known analog filter responses. Figure 7.16 shows three classical lowpass filter approximations called the Butterworth, Chebyshev, and elliptic filters. These are cartoons, and the ripple sizes are exaggerated; they are much smaller in practice (see Fig. 7.31). All three filters have rational transfer functions

$$H_c(s) = \frac{\sum_{n=0}^{N} c_n s^n}{\sum_{n=0}^{N} d_n s^n} = \frac{A\prod_{k=1}^{N}(s-z_k)}{\prod_{k=1}^{N}(s-p_k)}. \tag{7.55}$$

Here, s is the Laplace transform variable, so that $s = j\Omega$ yields the frequency response $H_c(j\Omega)$ (see Sec. 3.9). Since these are real filters (i.e., c_n, d_n are real), only the responses in $0 \leq \Omega < \infty$ are shown in the figure. Notice that we have used the subscript c as an extra reminder that it is a "continuous-time" system.

The factored form on the right in Eq. (7.55) shows the poles p_k and zeros z_k of the transfer function. We assume $d_N \neq 0$ and that there are no uncancelled common factors in Eq. (7.55), so that N is actually the order of the filter. In a way similar to Theorem 6.1 for discrete-time systems, it can be shown that the continuous-time

[2] More precisely, IIR linear-phase filters are necessarily noncausal (Sec. 6.8).

[3] Note that "analog" is not exactly the same thing as "continuous time," as explained in Sec. 10.7. Here we use them interchangeably for convenience.

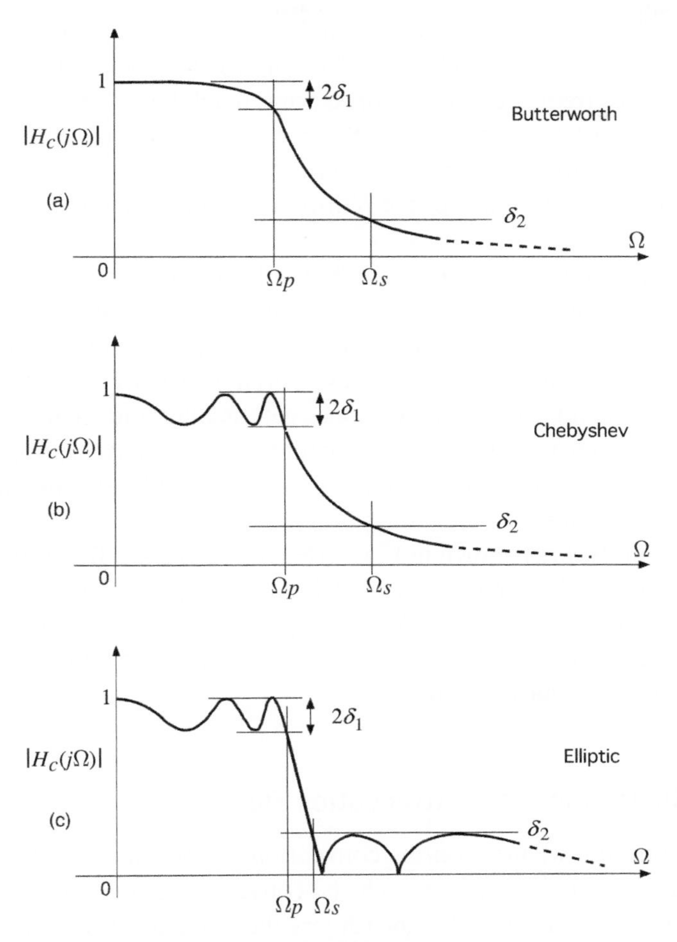

Figure 7.16 The three classical filter approximations for analog filters. (a) Butterworth filter, (b) Chebyshev filter, and (c) elliptic filter.

causal LTI system with transfer function (7.55) is stable (i.e., satisfies Eq. (3.153)) if and only if

$$\text{Re}\,(p_k) < 0 \quad \text{(stability condition)}, \tag{7.56}$$

that is, if and only if all the poles are in the **left-half plane**, that is, the left half of the s-plane.

Remark on notation. In this section we use Ω for continuous-time frequency and ω for discrete-time frequency, as this distinction is necessary to describe the mappings from Ω to ω.

Properties of the Classical Filters

The filters in Fig. 7.16 are supposed to approximate unity in the passband and zero in the stopband. Some of the features of these filters are summarized next.

1. The Butterworth filter shown in Fig. 7.16(a) has a **monotonically** decreasing magnitude response $|H_c(j\Omega)|$. The response is very accurate (i.e., close to unity) near zero frequency. Its performance is much better than the specified tolerance $2\delta_1$ in most of the passband. Similar comments apply in the stopband. So the Buttterworth filter is in some sense overdesigned for any given (δ_1, δ_2), and the approximation power of the Nth-order rational function is not most effectively utilized.
2. The Chebyshev filter shown in Fig. 7.16(b) has ripples in the passband. So the error is more nicely distributed throughout the passband. For this reason, this filter tends to have a smaller order N compared to the Butterworth filter, for fixed specifications (7.54). However, even the Chebyshev filter is overdesigned in the stopband because the response gets smaller and smaller than δ_2, as the frequency increases. The Chebyshev filter is said to have **equiripple** behavior in the passband (because the approximation error oscillates between a fixed maximum and minimum).
3. The elliptic filter shown in Fig. 7.16(c) has ripples in the stopband in addition to the passband, which helps to distribute the approximation error in the stopband also more uniformly. So, for fixed specifications (7.54), the elliptic filter has the smallest order N. Since it has equiripple behavior in both bands, the elliptic filter is called an **equiripple filter**.

7.6.2 The dB Response and Attenuation Function

For future use we mention the convention of dB responses here. We use discrete-time notation ($H(e^{j\omega})$), although the convention is the same for continuous-time filters ($H_c(j\Omega)$). In well-designed filters the stopband response of $|H(e^{j\omega})|$ is very close to zero. In order to see the behavior of the plot in detail, it is convenient to plot it on a log scale. The quantity

$$20 \log_{10} |H(e^{j\omega})|, \qquad \text{that is,} \qquad 10 \log_{10} |H(e^{j\omega})|^2 \tag{7.57}$$

is often plotted and is called the dB plot or dB response (where dB is read as "decibel"). For typical filters as in Fig. 7.16, this quantity is negative almost everywhere, since $|H(e^{j\omega})| \leq 1$. The quantity

$$\mathcal{A}(\omega) = -10 \log_{10} |H(e^{j\omega})|^2, \tag{7.58}$$

which is typically positive, is called the **attenuation** function. In particular,

$$A_s = -20 \log_{10} \delta_2 \tag{7.59}$$

is called the *minimum stopband attenuation* achieved by the filter. Here are some examples:

$\lvert H(e^{j\omega})\rvert^2$	$10\log_{10}\lvert H(e^{j\omega})\rvert^2$	Attenuation
1	0 dB	0 dB
0.5	−3.01 dB	3.01 dB
10^{-1}	−10 dB	10 dB
10^{-3}	−30 dB	30 dB
10^{-6}	−60 dB	60 dB
0	$-\infty$	∞

Note that $|H(e^{j\omega})|^2 = 0.5$ corresponds to about 3 dB, so the frequency where this happens is called the **three dB point**, often written as 3 dB point. In the passband, the dB response is close to 0, and in the stopband it is a large negative number. At the transmission zeros where $H(e^{j\omega}) = 0$, the dB response is $-\infty$, and shows up as very sharp dips in the dB plots (e.g., see Figs. 7.31(b) and (e)).

The dB plot for analog filters is usually shown with the frequency axis on a **logarithmic** scale, so that very high frequencies can be included. This is convenient because the frequency range is $-\infty < \Omega < \infty$. With such plots, both axes are therefore logarithmic (as in log–log plots). For digital filters the dB plot uses a normal (nonlogarithmic) frequency scale because of the finite range $0 \leq \omega \leq \pi$.

7.7 The Bilinear Transformation

One way to convert the analog filter (7.55) to a digital filter is to replace the Laplace variable s with

$$s = \frac{1 - z^{-1}}{1 + z^{-1}} \qquad \text{(bilinear transform).} \tag{7.60}$$

This is called the bilinear transform, and it has some interesting properties which make it especially attractive. First notice that this is an invertible mapping because we can rewrite it as

$$z^{-1} = \frac{1 - s}{1 + s}, \qquad \text{that is,} \qquad z = \frac{1 + s}{1 - s}. \tag{7.61}$$

Thus, every s is mapped to a unique z and vice versa. So this is called a "one-to-one and onto" mapping. To see how the mapping operates, consider the example

$$H_c(s) = \frac{s - z_1}{s - p_1}. \tag{7.62}$$

Then

$$H(z) = H_c(s)\Big|_{s = (1 - z^{-1})/(1 + z^{-1})} = \frac{1 - z_1 - z^{-1}(1 + z_1)}{1 - p_1 - z^{-1}(1 + p_1)}, \tag{7.63}$$

that is, the digital filter is

$$H(z) = \frac{1 - \widehat{z}_1 z^{-1}}{1 - \widehat{p}_1 z^{-1}}, \tag{7.64}$$

where

$$\widehat{z}_1 = \frac{1+z_1}{1-z_1} \quad \text{and} \quad \widehat{p}_1 = \frac{1+p_1}{1-p_1}. \tag{7.65}$$

Thus, the pole p_1 and zero z_1 of $H_c(s)$ are mapped into the pole $\widehat{p}_1$ and zero $\widehat{z}_1$ of $H(z)$ in accordance with Eq. (7.61).

7.7.1 Details of the *s*-Plane to *z*-Plane Mapping

Let us consider a point $s = \sigma + j\Omega$ in the s-plane. Here σ is the real part of s and Ω the imaginary part. Then, from Eq. (7.61) we have

$$z = \frac{1+s}{1-s} = \frac{1+\sigma+j\Omega}{1-\sigma-j\Omega}, \tag{7.66}$$

which proves that

$$|z|^2 = \frac{(1+\sigma)^2+\Omega^2}{(1-\sigma)^2+\Omega^2}. \tag{7.67}$$

Since $\sigma = \text{Re}[s]$, it follows that

$$|z| \begin{cases} < 1 & \text{if Re}[s] < 0, \\ > 1 & \text{if Re}[s] > 0, \\ = 1 & \text{if Re}[s] = 0. \end{cases} \tag{7.68}$$

We already mentioned that the bilinear transform maps every s to a unique z and vice versa. Equation (7.68) shows further details. It says that every point on the left half of the s-plane is mapped to a point **inside** the unit circle of the z-plane. Similarly, every point on the right half of the s-plane maps to a point **outside** the unit circle of the z-plane. And finally, every point $s = j\Omega$ on the imaginary axis of the s-plane maps to a point $z = e^{j\omega}$ *on* the unit circle of the z-plane. This is demonstrated in Fig. 7.17.

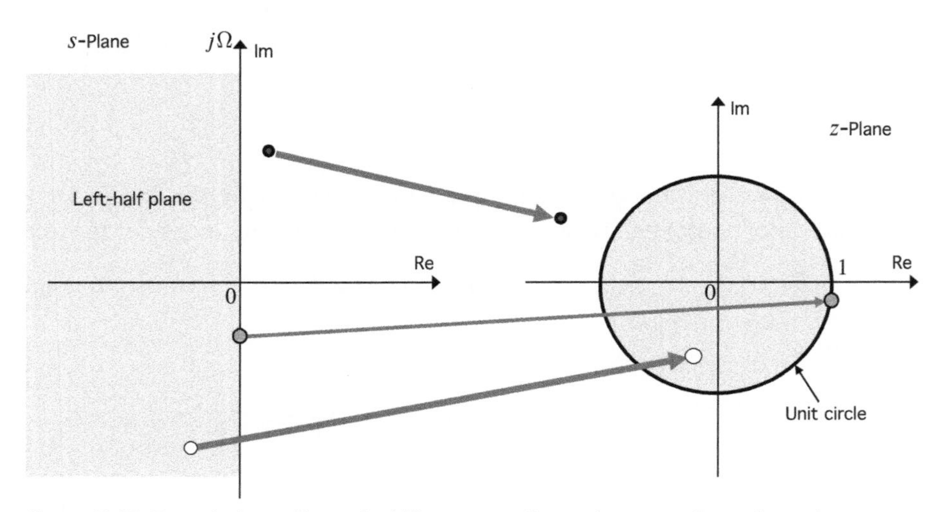

Figure 7.17 Description of how the bilinear transformation maps from the s-plane to the z-plane. The locations of the mapped points are not exact in this cartoon, the main point being that Eq. (7.68) is satisfied.

In particular, note that $H(z)$ has all its poles inside the unit circle if and only if $H_c(s)$ has all its poles in the left half of the s-plane. Assuming the filters are causal, it follows that

$$H(z) \text{ is stable if and only if } H_c(s) \text{ is stable.} \tag{7.69}$$

In short, the bilinear transform preserves stability properties.

7.7.2 The Ω to ω Mapping in Bilinear Transform

Since every point $s = j\Omega$ on the imaginary axis yields a point $z = e^{j\omega}$ on the unit circle, every frequency Ω maps to a unique frequency ω. To find the exact mapping, we substitute $s = j\Omega$ and $z = e^{j\omega}$ into Eq. (7.60). Then

$$j\Omega = \frac{1 - e^{-j\omega}}{1 + e^{-j\omega}} = \frac{e^{j\omega/2} - e^{-j\omega/2}}{e^{j\omega/2} + e^{-j\omega/2}} = j\frac{\sin(\omega/2)}{\cos(\omega/2)}, \tag{7.70}$$

so that

$$\Omega = \tan(\omega/2), \tag{7.71}$$

that is,

$$\omega = 2\tan^{-1}\Omega. \tag{7.72}$$

Figure 7.18(a) shows a plot of the function (7.71). Since it is an odd function, we have not shown the negative frequencies. Figure 7.18(b) shows the mapping (7.72) from Ω to ω. Thus, the infinite range of continuous-time frequencies $0 \leq \Omega \leq \infty$ is mapped to the finite range of discrete-time frequencies $0 \leq \omega \leq \pi$. The mapping is monotone increasing; in particular,

$$\Omega = 0 \text{ is mapped to } \omega = 0, \qquad \Omega = \infty \text{ is mapped to } \omega = \pi. \tag{7.73}$$

Figure 7.19 shows how the response $|H_c(j\Omega)|$ of an analog filter gets mapped into the digital filter response $|H(e^{j\omega})|$. Note that the ripple sizes δ_1 and δ_2 are preserved by the mapping. In particular, an *equiripple filter gets mapped into an equiripple filter*. The mapping from an infinite range $0 \leq \Omega \leq \infty$ to the finite range $0 \leq \omega \leq \pi$ has been possible because the mapping is nonlinear. The bilinear transformation is remarkable indeed. Other transformations have been tried in earlier days but they did not work very well (Problem 7.8).

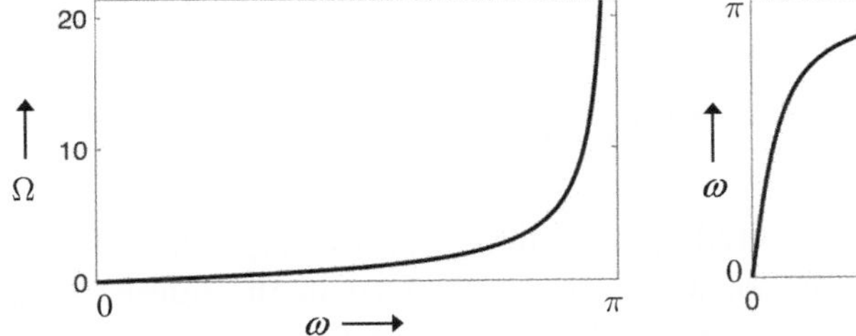

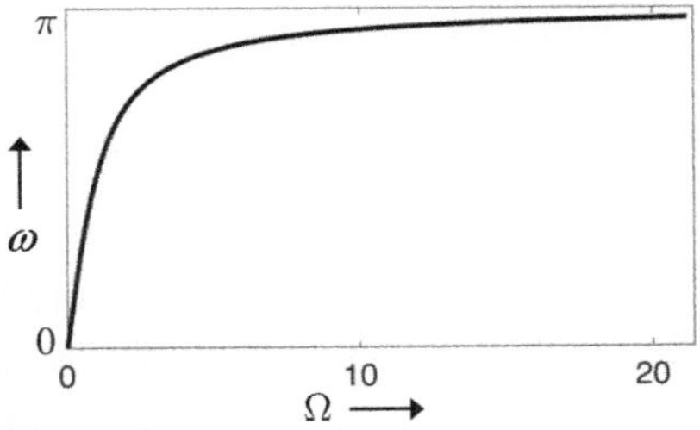

Figure 7.18 (a) The frequency mapping $\Omega = \tan(\omega/2)$ used in bilinear transformation; (b) the inverse mapping $\omega = 2\tan^{-1}\Omega$.

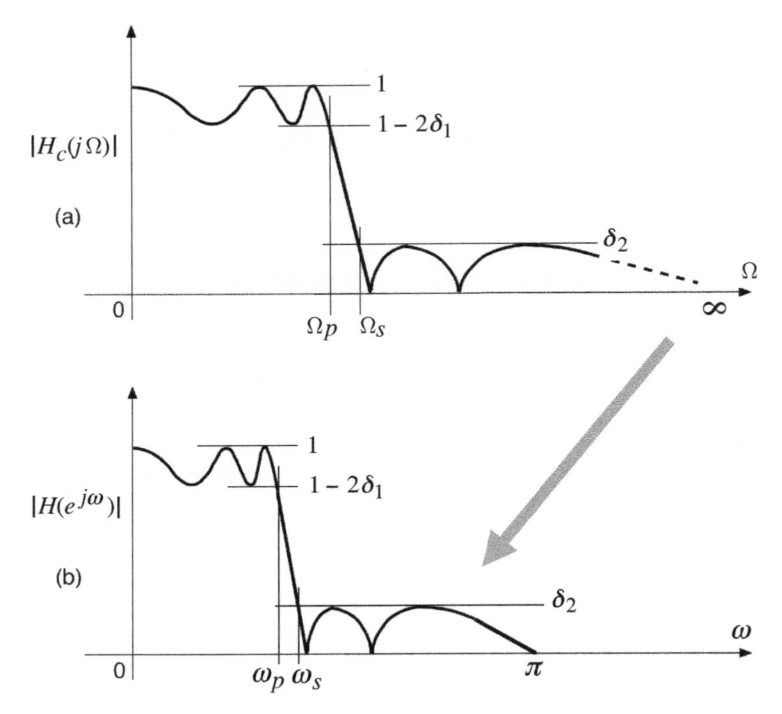

Figure 7.19 Illustration of how the frequency response is mapped under the bilinear transform $\Omega = \tan(\omega/2)$. (a) The response of the analog filter $H_c(s)$; (b) the response of the digital filter $H(z)$.

7.7.3 From Filter Specifications to Filter Coefficients

Given the specifications $\omega_p, \omega_s, \delta_1, \delta_2$ for the digital filter $H(z)$, the design proceeds as follows: we first map the specifications into the analog filter specifications using the inverse bilinear mapping formula $\Omega = \tan(\omega/2)$, that is,

$$\Omega_p = \tan(\omega_p/2), \quad \Omega_s = \tan(\omega_s/2). \tag{7.74}$$

We then design $H_c(s)$ with specifications $\Omega_p, \Omega_s, \delta_1, \delta_2$. There are standard methods for this. Once $H_c(s)$ is designed, we obtain the digital filter using the bilinear mapping

$$H(z) = H_c(s)\Big|_{s = (1-z^{-1})/(1+z^{-1})}. \tag{7.75}$$

This filter satisfies the desired specifications $\omega_p, \omega_s, \delta_1, \delta_2$. Of course, with standard digital filter design software, we simply feed in the numbers $\omega_p, \omega_s, \delta_1, \delta_2$ and obtain the filter coefficients $\{a_k, b_k\}$ of the transfer function $H(z)$. The bilinear transformation is internal to the code and we do not worry about it.

7.8 Butterworth Filters

The Nth-order Butterworth lowpass filter $H_c(s)$ has magnitude-squared response of the form

$$|H_c(j\Omega)|^2 = \frac{1}{1 + \left(\Omega/\Omega_c\right)^{2N}}. \tag{7.76}$$

The filter is completely determined by the two free parameters N and Ω_c. As shown in Fig. 7.16(a), the response is **monotone** decreasing, and decays from one to zero as Ω increases from 0 to ∞. Notice that

$$|H_c(j\Omega_c)|^2 = 1/2, \tag{7.77}$$

for all $N > 0$, so we call Ω_c the half-power point or the 3 dB point (since $10\log_{10}(1/2) \approx -3$ dB)). Figure 7.20(a) shows plots of Eq. (7.76) for $\Omega_c = 1$ and three values of N. As N increases the filter becomes sharper and sharper, approaching the ideal. The dB plot (Sec. 7.6.2) is shown in Fig. 7.20(b), using a logarithmic scale for the frequency axis, so that higher frequencies can be shown conveniently. Note that for $\Omega >> \Omega_c$ we have $|H_c(j\Omega)|^2 \approx (\Omega_c/\Omega)^{2N}$, so that

$$10\log_{10}|H_c(j\Omega)|^2 \approx 10\log_{10}\left(\Omega_c/\Omega\right)^{2N} \approx C - 20N\log_{10}\Omega, \tag{7.78}$$

where C is a constant independent of Ω. So for large frequencies the dB plot decreases linearly with $\log_{10}\Omega$ and the slope $-20N$ is proportional to the filter order N. This is clearly seen in Figure 7.20(b), where the frequency axis is logarithmic. For large Ω we see that the responses look like straight lines on this log–log plot, with negative slopes increasing with N.

7.8.1 Flatness at Zero Frequency

From the plots in Fig. 7.20(a) it is clear that the response of the Butterworth filter is very flat near zero frequency, in fact it looks flatter and flatter as N increases. This property can be quantified. The **degree of flatness** of a function at a point is measured by the number of consecutive derivatives (first derivative, second derivative, and so on) that become zero at that point. If the first N_f derivatives are zero at a point and the $(N_f + 1)$th derivative is nonzero, the degree of flatness there is N_f.

It will be shown next that the first $2N-1$ derivatives (but not the $2N$th derivative) of the function (7.76) are equal to zero at $\Omega = 0$. So we say that the degree of flatness is $2N-1$ at $\Omega = 0$. One way to show this is to simply differentiate Eq. (7.76) and examine the closed-form expressions for the higher derivatives. A perhaps more elegant way is to observe that for $\Omega < \Omega_c$ we can express the response in the form

$$|H_c(j\Omega)|^2 = 1 - (\Omega/\Omega_c)^{2N} + (\Omega/\Omega_c)^{4N} - (\Omega/\Omega_c)^{6N} + \cdots \tag{7.79}$$

by using $1/(1-x) = 1 + x + x^2 + \cdots$ for $|x| < 1$. Compare this with the Taylor series expansion around $x = 0$:

$$f(x) = f(0) + f^{(1)}(0)x + \frac{f^{(2)}(0)x^2}{2!} + \frac{f^{(3)}(0)x^3}{3!}\cdots, \tag{7.80}$$

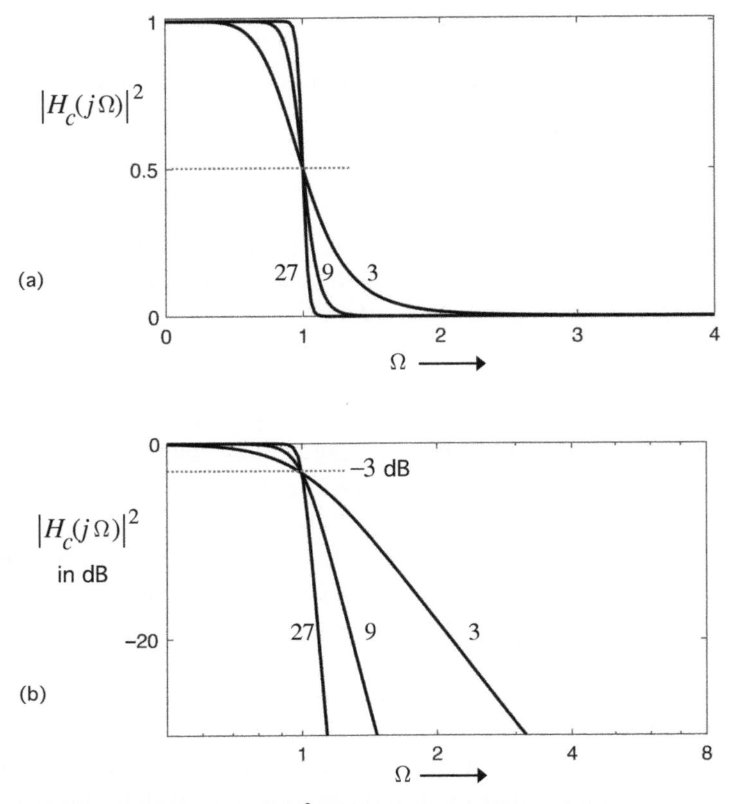

Figure 7.20 Plot of $|H_c(j\Omega)|^2$ for Butterworth filters with $\Omega_c = 1$. Filter orders $N = 3, 9$, and 27 are indicated next to the plots. (a) Normal plot; (b) dB plot.

where $f^{(n)}(0)$ is the nth derivative of $f(x)$ evaluated at $x = 0$. It is clear from Eq. (7.79) indeed that the first $2N-1$ derivatives of $|H_c(j\Omega)|^2$ are equal to zero at $\Omega = 0$.

Maximal flatness. In fact, it can be shown that the degree of flatness of $|H_c(j\Omega)|^2$ at zero frequency can be at most $2N - 1$ for an Nth-order filter $H_c(s)$. So, the Butterworth filter is said to be **maximally flat** at $\Omega = 0$. To see this, consider the rational function

$$\frac{P(x)}{Q(x)} = \frac{\sum_{n=0}^{K} p_n x^n}{\sum_{n=0}^{K} q_n x^n}, \tag{7.81}$$

and expand it in a Taylor series around $x = 0$ to write it as $\sum_{n=0}^{\infty} h_n x^n$. If this has the *first K derivatives* equal to zero at $x = 0$, then

$$\sum_{n=0}^{K} p_n x^n = \Big(\sum_{n=0}^{K} q_n x^n\Big)\Big(h_0 + h_{K+1}x^{K+1} + h_{K+2}x^{K+2} + \cdots\Big). \tag{7.82}$$

Equating like powers, we see indeed that $P(x) = h_0 Q(x)$, so that the rational function $P(x)/Q(x) = h_0$ is a constant! This shows that $P(x)/Q(x)$ cannot have more than $K - 1$ vanishing derivatives at $x = 0$, unless it is a constant. ▽ ▽ ▽

7.8.2 Poles of the Butterworth Filter

The Butterworth filter was defined based on its magnitude-squared response, Eq. (7.76). But we actually need to know the transfer function $H_c(s)$ itself, in order to convert it into the digital filter $H(z)$ with the help of the bilinear transform. Since Eq. (7.76) has a constant numerator, we can find an $H_c(s)$ of the form

$$H_c(s) = \frac{1}{D_c(s)}, \tag{7.83}$$

where

$$D_c(s) = 1 + d_1 s + d_2 s^2 + \cdots + d_N s^N, \tag{7.84}$$

such that Eq. (7.76) holds. In other words, we can take $H_c(s)$ to be an Nth-order all-pole filter. Since the quantity (7.76) is symmetric, that is, $|H_c(j\Omega)| = |H_c(-j\Omega)|$, we can further restrict the filter coefficients d_k to be real. To solve for the appropriate $D_c(s)$, consider the quantity $D_c(s)D_c(-s)$. When $s = j\Omega$ this becomes

$$D_c(j\Omega)D_c(-j\Omega) = |D_c(j\Omega)|^2, \tag{7.85}$$

because $D_c(-j\Omega) = (D_c(j\Omega))^*$ when the d_k are real. Thus we can achieve

$$|D_c(j\Omega)|^2 = 1 + (\Omega/\Omega_c)^{2N}, \tag{7.86}$$

by finding $D_c(s)$ such that

$$D_c(s)D_c(-s) = 1 + \left(\frac{s}{j\Omega_c}\right)^{2N}. \tag{7.87}$$

The right-hand side is a $2N$th-order polynomial and it has $2N$ zeros. Two more things should be noticed:

1. Since the right-hand side of Eq. (7.87) is an even function of s, it follows that if s_k is a zero then $-s_k$ is also a zero.
2. When $s = j\Omega$ the right-hand side of Eq. (7.87) is evidently strictly positive, so there are no zeros on the imaginary axis of the s-plane.

Summarizing, the $2N$ zeros of the right-hand side of Eq. (7.87) come as N pairs, $\{s_k, -s_k\}$, where s_k can be assumed, without loss of generality, to be in the left half of the s-plane, that is,

$$\text{Re}[s_k] < 0. \tag{7.88}$$

We can therefore find $D_c(s)$ satisfying Eq. (7.87) simply by defining

$$D_c(s) = c(s - s_0)(s - s_1)\cdots(s - s_{N-1}). \tag{7.89}$$

The constant c can readily be found by setting $s = 0$ on both sides of Eq. (7.87), or more easily by setting $D_c(0) = 1$ to ensure that $H_c(0) = 1$ (i.e., frequency repsonse $H_c(j\Omega) = 1$ at zero frequency).

This task of finding $D_c(s)$ such that Eq. (7.87) is satisfied is called the **spectral factorization** problem. Note that the spectral factor $D_c(s)$ is not unique. If we had replaced one of its roots s_k by $-s_k$, Eq. (7.87) would still be satisfied. However, Eq.

(7.89) is the unique solution such that all the zeros of $D_c(s)$ are in the left-half plane, ensuring that $H_c(s)$ is **stable**.

Expression for poles. We now find a simple expression for the $2N$ zeros s_k of the right-hand side of Eq. (7.87). The poles of $H_c(s)$, which are the subset of N values of s_k in the left-half plane, can then be found. Clearly, $\widehat{s}$ is a zero of Eq. (7.87) if and only if

$$1 + \Big(\frac{\widehat{s}}{j\Omega_c}\Big)^{2N} = 0, \quad \text{that is,} \quad \frac{\widehat{s}}{j\Omega_c} = (-1)^{1/2N} = e^{j(2k+1)\pi/2N}, \tag{7.90}$$

for some integer k. Here we have used the fact that $e^{j(2k+1)\pi} = -1$ for any integer k. It is clear that all the zeros have the form

$$s_k = j\Omega_c e^{j(2k+1)\pi/2N} = \Omega_c \exp\Big(j\Big(\frac{\pi}{2} + \frac{(2k+1)\pi}{2N}\Big)\Big). \tag{7.91}$$

This shows that $|s_k| = \Omega_c$, so all the zeros are on a circle in the s-plane with radius Ω_c, and centered at the origin. The angles of these zeros are

$$\theta_k = \frac{\pi}{2} + \frac{\pi}{2N} + \frac{k\pi}{N}, \quad k = 0, 1, 2, \ldots. \tag{7.92}$$

See Fig. 7.21, which shows the $2N$ zeros uniformly spaced on the circle of radius Ω_c. This is called the Butterworth circle. Since successive angles are separated uniformly by $\pi/N = 2\pi/2N$, it is clear that if we take $0 \le k \le 2N - 1$ we will get $2N$ distinct values of θ_k in $[0, 2\pi)$. As k is increased further, the values of θ_k start repeating modulo 2π, so we get $s_{2N} = s_0$, and so on. ▽ ▽ ▽

In short, there are exactly $2N$ distinct zeros s_k given by Eq. (7.91), as required. Exactly N of these are in the left-half plane, as shown by white circles in Fig. 7.21. These can be used to define $D_c(s)$ as in Eq. (7.89), and find the stable Butterworth filter $H_c(s) = 1/D_c(s)$. For example, if $N = 3$ and $\Omega_c = 1$, the first three pole angles (7.92) are

$$\theta_0 = 120°, \quad \theta_1 = 180°, \quad \theta_2 = -120°. \tag{7.93}$$

Since these are in the left-half plane, the poles of $H_c(s)$, given by $s_k = e^{j\theta_k}$, are

$$s_0 = e^{j120°}, \quad s_1 = e^{j180°} = -1, \quad s_2 = e^{-j120°}. \tag{7.94}$$

Thus, $D_c(s) = (s+1)(s - e^{j120°})(s - e^{-j120°}) = (s+1)(s^2 - 2s\cos 120° + 1) = s^3 + 2s^2 + 2s + 1$, so that

$$H_c(s) = \frac{1}{s^3 + 2s^2 + 2s + 1}. \tag{7.95}$$

This is the third-order Butterworth lowpass filter with $\Omega_c = 1$. For readers familiar with electrical circuits, Fig. 7.22 shows an LCR circuit implementing this filter. The actual transfer function is $H_c(s) = V_2(s)/V_1(s) = 0.5/(s^3 + 2s^2 + 2s + 1)$, which has an extra scale factor of 0.5. This is because, at zero frequency, the inductor L is a short circuit and the capacitor C an open circuit, so the system acts like a voltage divider $H_c(0) = R_2/(R_1 + R_2) = 1/2$.

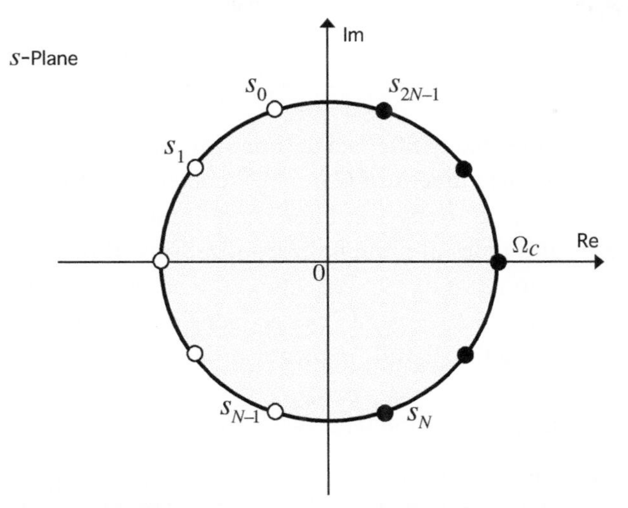

Figure 7.21 The $2N$ zeros of the right-hand side of Eq. (7.87) demonstrated for $N = 5$. These lie uniformly spaced on a circle of radius Ω_c. Those on the left-half plane are taken to be the poles of the Butterworth filter $H_c(s) = 1/D_c(s)$.

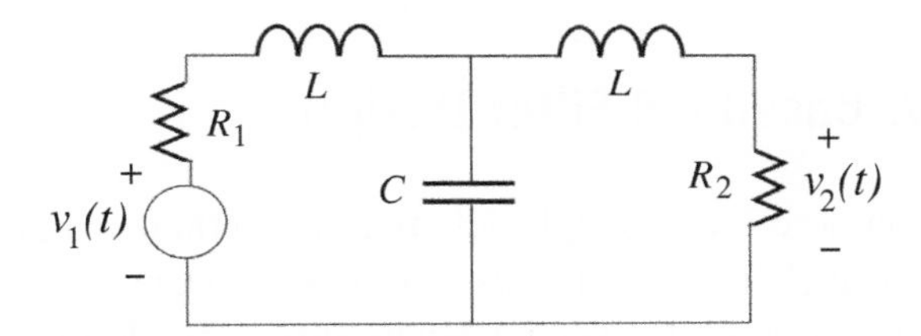

Figure 7.22 Third-order Butterworth filter $H_c(s) = V_2(s)/V_1(s)$, implemented using an LCR circuit. Here, $L = 1$ H, $C = 2$ F, and $R_1 = R_2 = 1\ \Omega$, so that $H_c(s) = 0.5/(s^3 + 2s^2 + 2s + 1)$.

7.8.3 Digital Butterworth Filters

We know the analog Butterworth transfer function has the form

$$H_c(s) = \frac{1}{1 + d_1 s + d_2 s^2 + \cdots + d_N s^N}, \tag{7.96}$$

so that the numerator is a constant. This is an all-pole filter, and the zeros are all at $s = \infty$. In fact, the frequency response $H_c(j\Omega)$ is nonzero everywhere except at $\Omega = \infty$. To convert $H_c(s)$ to a digital filter by using the bilinear transform, we substitute

$$s = \frac{1 - z^{-1}}{1 + z^{-1}} \tag{7.97}$$

into $H_c(s)$, and get the form

$$H(z) = \frac{c(1+z^{-1})^N}{1 + b_1 z^{-1} + b_2 z^{-2} + \cdots + b_N z^{-N}}. \tag{7.98}$$

The poles of the analog filter get mapped into the poles of the digital filter according to Eq. (7.97). The numerator $(1+z^{-1})^N$ comes from the denominator in Eq. (7.97). Thus, the digital Butterworth filter is not an all-pole filter. It is an IIR filter with both zeros and poles at finite nonzero locations. It has N zeros at $z = -1$, that is, N transmission zeros at the frequency $\omega = \pi$. Another way to see this is to recall that $H_c(j\Omega)$ has all N zeros at $\Omega = \infty$ (see Eq. (7.76)). Since the bilinear transform maps $\Omega = \infty$ to the digital frequency $\omega = \pi$ (see Fig. 7.19), the filter $H(z)$ has N zeros at π.

We conclude this section with two remarks:

1. It can be shown that the analog Chebyshev filter (Fig. 7.16(b)) is also an all-pole filter, so the digital Chebyshev filter is an IIR filter with the same numerator as the Butterworth filter (7.98).
2. The analog elliptic filter is not an all-pole filter because it has zeros at finite frequencies (see Fig. 7.16(c)). So the digital elliptic filter is an IIR filter with some zeros in $\omega_s < |\omega| < \pi$, besides a possible zero at $\omega = \pi$.

7.9 Window-Based FIR Filter Design

This section describes a simple yet effective method to approximate the ideal lowpass filter using FIR filters. In this section we will denote the impulse repsonse (4.43) of the ideal lowpass filter as $h_i[n]$, where subscript i denotes "ideal." Thus

$$h_i[n] = \frac{\sin(\omega_c n)}{\pi n} \quad \text{(ideal impulse response).} \tag{7.99}$$

Note that $h_i[n]$ decays to zero as $|n| \to \infty$. So we can obtain a finite-length (FIR) approximation simply by truncating $h_i[n]$ to a finite number of samples, say $-M \leq n \leq M$. The truncated filter can be expressed as

$$h[n] = w[n]h_i[n] \quad \text{(windowed impulse response),} \tag{7.100}$$

where

$$w[n] = \begin{cases} 1 & -M \leq n \leq M, \\ 0 & \text{otherwise,} \end{cases} \tag{7.101}$$

is the rectangular pulse, also called a **rectangular window**. Thus

$$h[n] = \begin{cases} \dfrac{\sin(\omega_c n)}{\pi n} & -M \leq n \leq M \\ 0 & \text{otherwise} \end{cases} \quad \text{(truncated impulse response).} \tag{7.102}$$

This is an FIR filter with length $L = 2M + 1$. It is clear that $H(e^{j\omega})$ approximates $H_i(e^{j\omega})$ better and better as M increases. Figure 7.23 shows plots of $|H(e^{j\omega})|$ for

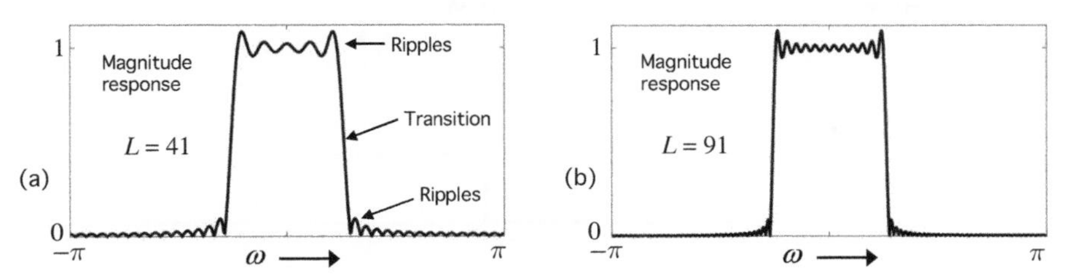

Figure 7.23 Magnitude responses $|H(e^{j\omega})|$ of truncated versions of the ideal lowpass filter for two filter lengths L.

two filter lengths $L = 41$ and 91, with filter cutoff frequency $\omega_c = 0.26\pi$. Note that there are two main effects due to truncation:

1. Unlike the ideal filter, the transition from passband to stopband is not abrupt anymore. There is a nonzero transition bandwidth $\Delta\omega$.
2. There are errors or ripples in the passband and stopband, that is, the response is not a constant in these bands.

As the filter length $L = 2M + 1$ increases, the transition bandwidth decreases and the ripples get crowded near the band edge, but the *peak ripple size* does not decrease, as seen from the figure. This is similar to the so-called **Gibbs phenomenon** in Fourier series theory (Sec. 11.6).[4]

7.9.1 Understanding the Effect of Windowing

To explain the frequency domain behavior of windowed filters, note that windowing (Eq. (7.100)) corresponds to convolution in frequency, or more precisely,

$$H(e^{j\omega}) = \int_{-\pi}^{\pi} H_i(e^{ju}) W(e^{j(\omega-u)}) \frac{du}{2\pi}. \tag{7.103}$$

If $L = \infty$ then $W(e^{j\omega})$ is the Dirac delta function (see Eq. (3.104)), that is,

$$W(e^{j\omega}) = 2\pi\delta_c(\omega) \tag{7.104}$$

in $|\omega| \leq \pi$, and the convolution (7.103) would just yield $H(e^{j\omega}) = H_i(e^{j\omega})$ as expected. But if the length $L = 2M + 1$ is finite (i.e., $w[n]$ has finite duration), then $W(e^{j\omega})$ is the Dirichlet function (3.97):

$$W(e^{j\omega}) = \frac{\sin(\omega(2M+1)/2)}{\sin(\omega/2)}, \tag{7.105}$$

which only approximates $2\pi\delta_c(\omega)$ for sufficiently large M (see examples in Fig. 3.14). Figure 7.24 shows $H_i(e^{j\omega})$ and $W(e^{j\omega})$, which are to be convolved to obtain $H(e^{j\omega})$. For finite L, $W(e^{j\omega})$ has a main lobe of nonzero width, and some side lobes or ripples as indicated. When the convolution (7.103) is performed, the sharp edges

[4] Since the filters are real, it is sufficient to show the plots in the region $0 \leq \omega \leq \pi$. In this section we use $-\pi \leq \omega \leq \pi$, just for added clarity.

(a)

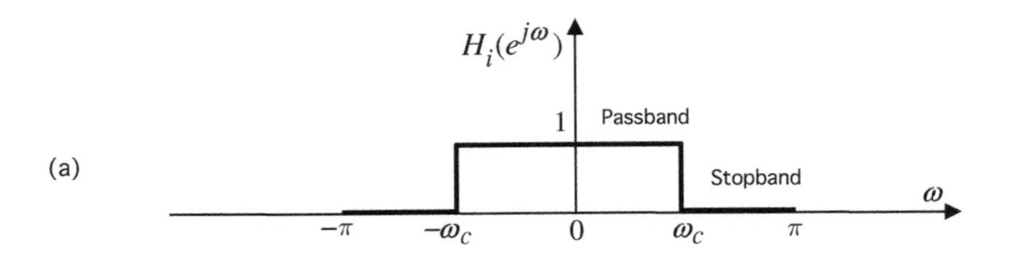

(b)

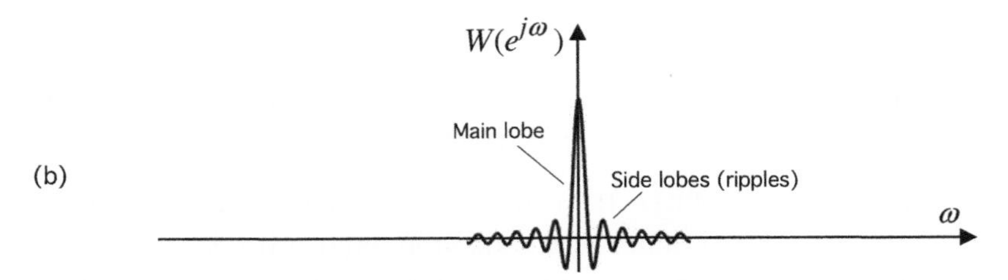

Figure 7.24 (a) The ideal lowpass filter and (b) the Fourier transform of a rectangular window $w[n]$.

of $H_i(e^{j\omega})$ at $\pm\omega_c$ get smeared because of the main lobe of $W(e^{j\omega})$ and furthermore, there are ripples in the passband and stopband of $H(e^{j\omega})$ because of the side lobes of $W(e^{j\omega})$. Summarizing, there are two effects:

1. Since the *main lobe* of $W(e^{j\omega})$ has nonzero width, the filter $H(e^{j\omega})$ has nonzero *transition band* $\Delta\omega$.
2. Since $W(e^{j\omega})$ has nonzero *side lobes*, there are *ripples* in the passbands and stopbands of $H(e^{j\omega})$, as we indicated in Fig. 7.23(a).

If the filter length (i.e., window length) $L = 2M+1$ is increased, it leads to a narrow main lobe for $W(e^{j\omega})$ but the *side lobe heights are not reduced.* So, as L increases, the plot $|H(e^{j\omega})|$ has narrower transition band $\Delta\omega$ but the ripples do not get smaller (Fig. 7.23).

7.9.2 More General Window Functions

If $w[n]$ is appropriately designed (instead of being taken to be rectangular), then the side lobes of $W(e^{j\omega})$ can be made smaller, which decreases the ripple sizes in $H(e^{j\omega})$. Thus, by choosing the shape of $w[n]$, the ripple sizes can be reduced, and by choosing the length L of $w[n]$, the transition bandwidth $\Delta\omega$ of $H(e^{j\omega})$ can be reduced. This is the *basic principle of window-based FIR filter* design. We will demonstrate this with several examples in this section. But first some general remarks:

1. *Symmetry*. Most windows used for filter design satisfy the symmetry property

$$w[n] = w[-n]. \tag{7.106}$$

The filter $h[n]$ in Eq. (7.100) is therefore also symmetric, since $h_i[n]$ is. Lowpass filters designed based on the window method are therefore linear-phase FIR filters (Sec. 4.6).

2. *Ripples*. Window-based methods produce filters with approximately equal ripple sizes in the passbands and stopbands:

$$\delta_1 \approx \delta_2. \tag{7.107}$$

Thus, the flexibility of having different ripple sizes in the two bands is not available. If δ_2 (or equivalently the minimum stopband attenuation $A_s = -20\log_{10}(\delta_2)$) is specified, then δ_1 is automatically determined. This is not surprising, because the passband and stopband ripples of $H(e^{j\omega})$ both have origin in the stopband ripples of $W(e^{j\omega})$.

3. *Causality*. In this section all windows $w[n]$ and filters $h[n]$ are nonzero in $-M \leq n \leq M$. But since the duration is finite, we can always obtain a causal filter $g[n]$ from the noncausal filter $h[n]$ simply by shifting it:

$$g[n] = h[n - M], \tag{7.108}$$

so that

$$G(e^{j\omega}) = e^{-j\omega M} H(e^{j\omega}). \tag{7.109}$$

Thus, $|G(e^{j\omega})| = |H(e^{j\omega})|$ and the filters have the same magnitude response. Since $H(e^{j\omega})$ has linear phase as explained above, $G(e^{j\omega})$ also has linear phase. Note that we can write

$$G(z) = \sum_{n=0}^{N} g[n]z^{-n}, \quad N = 2M. \tag{7.110}$$

This is a causal, linear-phase, FIR approximation of the ideal lowpass filter, and is therefore realizable. Clearly, $N = 2M$ is the filter order.

7.9.2.1 Example: The Triangular Window

To illustrate the idea of filter design using windows, consider the **triangular** window defined as

$$w[n] = \begin{cases} 1 - \dfrac{|n|}{M+1} & -M \leq n \leq M, \\ 0 & \text{otherwise.} \end{cases} \tag{7.111}$$

This is also known as the Bartlett window. While the rectangular window abruptly cuts off the ideal impulse response $h_i[n]$ for $|n| > M$, the triangular window gracefully decreases to zero as $|n|$ increases from 0 to M. Thus, for the triangular window there is less energy in higher frequencies, that is, $W(e^{j\omega})$ has smaller side lobes. For fixed window length $L = 2M + 1 = 41$, Fig. 7.25 compares plots of $|W(e^{j\omega})|$ (in normal scale and dB scale) for the rectangular and triangular windows. For ease of comparison we have normalized the plots to unity at $\omega = 0$.[5] Clearly the triangular window has much smaller side lobes, although its main lobe is wider. To understand this more quantitatively, observe that for even M we can regard the triangular window as the convolution of a *shorter* rectangular window with itself. Thus, define a rectangular window

$$w_r[n] = \begin{cases} \dfrac{1}{M+1} & -M/2 \leq n \leq M/2, \\ 0 & \text{otherwise.} \end{cases} \tag{7.112}$$

Then the triangular window is $w[n] = (w_r * w_r)[n]$, so that

$$W(e^{j\omega}) = W_r^2(e^{j\omega}), \tag{7.113}$$

where

$$W_r(e^{j\omega}) = \frac{\sin(\omega(M+1)/2)}{\sin(\omega/2)}. \tag{7.114}$$

On a dB scale it can be verified that for a rectangular window the peak side lobe is about 13 dB below the zero-frequency value, regardless of M (as long as M is not too small). This can be seen in Fig. 7.25(a). Equation (7.113) shows that the peak side lobe of the triangular window is the square of that of a rectangular window, so it is significantly smaller, and on a dB scale, 26 dB below the zero-frequency value (Fig. 7.25(b)). The main-lobe width of the triangular window $W(e^{j\omega})$ is comparable to that of $W_r(e^{j\omega})$, which is wider because $w_r[n]$ has only $M + 1$ (and not $2M + 1$) nonzero samples.

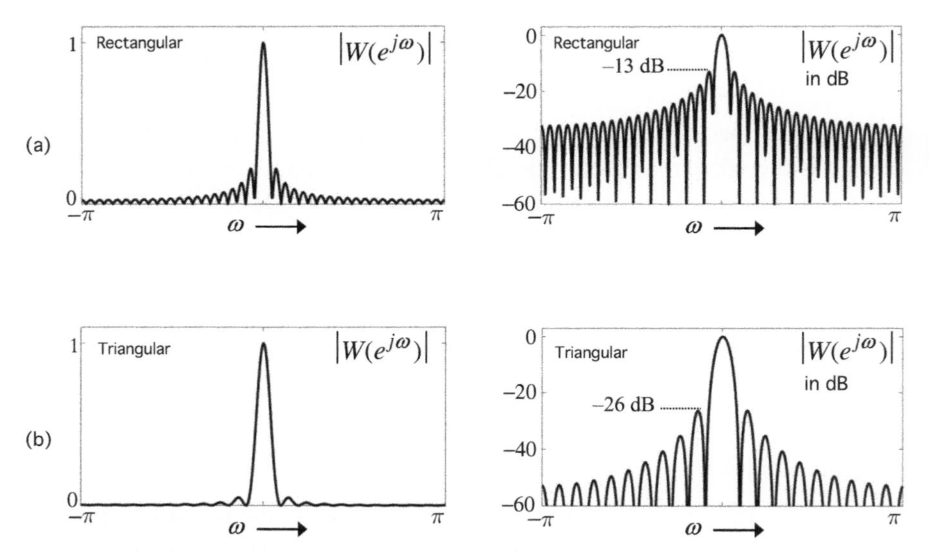

Figure 7.25 Plots of $|W(e^{j\omega})|$ in normal and dB scales for (a) rectangular window and (b) triangular (Bartlett) window. The length is $L = 41$ in both cases.

[5] This convention will often be followed, although we do not always mention it.

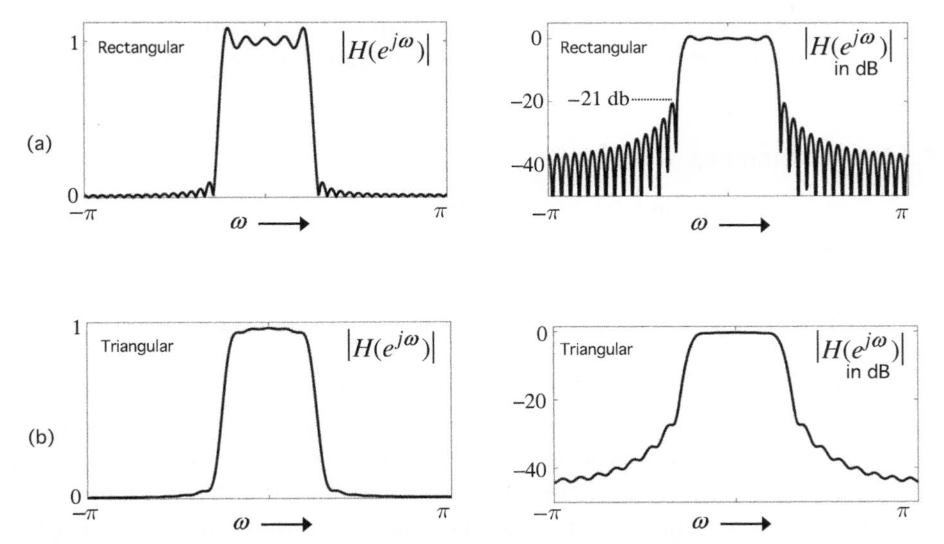

Figure 7.26 Plots of $|H(e^{j\omega})|$ in normal and dB scales for (a) rectangular window and (b) triangular (Bartlett) window. The length is $L = 41$ in both cases.

Figure 7.26 shows the responses $H(e^{j\omega})$ of the lowpass filters designed using the rectangular and triangular windows, for a fixed filter length $L = 41$, and $\omega_c = 0.26\pi$. The peak side lobe of $H(e^{j\omega})$ is 21 dB below the main lobe for the rectangular window-based filter, but it is significantly smaller for the triangle-based filter. As expected, the triangular window produces filters with smaller errors in the passbands and stopbands, but with a wider transition band. Since $W(e^{j\omega}) \geq 0$ for the triangular window (from Eq. (7.113)), the filter $H(e^{j\omega})$ is the convolution (7.103) of two non-negative functions. It therefore follows that $H(e^{j\omega}) > 0$ for all ω. This is also clear from the plots in Fig. 7.26(b), where $H(e^{j\omega})$ is never exactly zero anywhere in the stopband. Thus, the filter designed using the triangular window *cannot have any transmission zeros.*

It turns out that there are many other windows capable of producing better filters than this, so the triangular window is not widely used for filter design. We presented it here because it clearly demonstrates how a nonrectangular window can decrease the approximation errors in passbands and stopbands, at the expense of increased transition bandwidth.

7.9.3 Other Well-Known Windows

In the early days of digital signal processing many important windows were developed, such as the Hamming window, the Hann window, the Blackman window, the Kaiser window, and so forth [Harris, 1978; Oppenheim and Schafer, 2010; Proakis and Manolakis, 2007]. For fixed filter length, different windows provide different tradeoffs between stopband attenuation (ripple size δ_2) and transition bandwidth $\Delta\omega$. Details about these properties can be found in the above references. The paper by Harris [1978] discusses many more windows in great detail. About 17 different windows can be found in MATLAB by typing "help windows." In what follows we will mention only two windows, which are quite commonly used.

7.9.3.1 The Hamming Window

The Hamming window[6] is defined by

$$w[n] = \begin{cases} \alpha + (1-\alpha)\cos\left(\frac{2\pi}{N}n\right) & -M \leq n \leq M, \\ 0 & \text{otherwise,} \end{cases} \tag{7.115}$$

where $\alpha = 0.54$. Here, $N = 2M$ (filter order) and window length $L = 2M + 1$. (In some books you will see $v[n] = w[n - M]$ listed, which is nonzero in $0 \leq n \leq N$. Then the second term in Eq. (7.115) becomes $0.46\cos(2\pi(n - M)/N) = -0.46\cos(2\pi n/N)$, so it acquires a minus sign.) See Fig. 7.27, which summarizes the three windows mentioned so far. To understand the context in which Eq. (7.115) is defined, recall that the rectangular window abruptly cuts off the ideal impulse

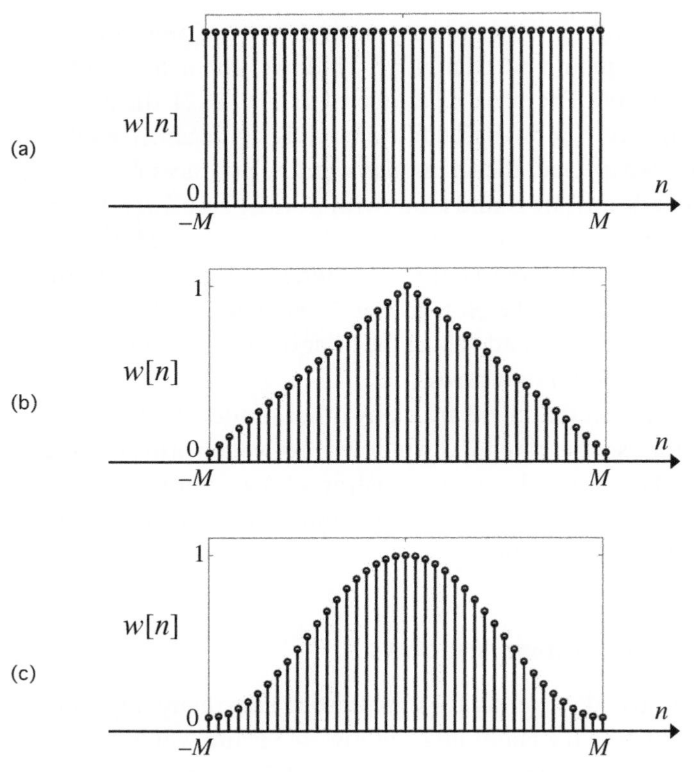

Figure 7.27 (a) The rectangular window, (b) the triangular window, and (c) the Hamming window.

[6] Named after the famous information and coding theorist Richard W. Hamming, who also toyed with digital filters during the early days of signal processing.

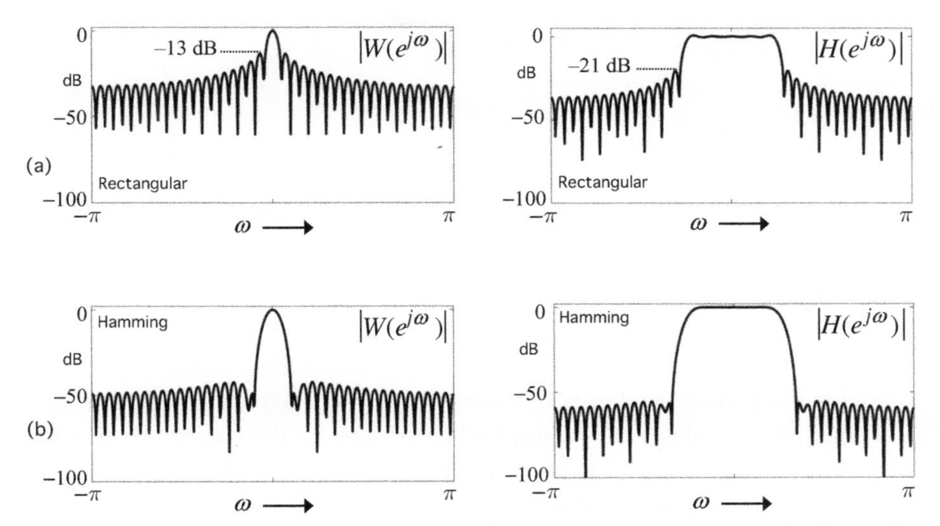

Figure 7.28 Plots of window magnitude $|W(e^{j\omega})|$ and filter magnitude $|H(e^{j\omega})|$ in dB, for (a) rectangular window and (b) Hamming window. The length is $L = 41$ in both cases.

response $h_i[n]$ for $|n| > M$, whereas the triangular window gracefully decreases towards zero as $|n|$ increases from 0 to M. So the triangular window has less energy at high frequencies compared to the rectangular window. The Hamming window also decreases gracefully. In fact it is even smoother than the triangular window near $n = 0$, as it does not have a sharp corner there. So there is even less energy at high frequencies, that is, the side lobes in $W(e^{j\omega})$ are even smaller for the Hamming window. There is another window called the **Hann window**, which is precisely as in Eq. (7.115), but with $\alpha = 0.5$. We do not elaborate on it, as Hamming's window has α optimized to produce better results. A generalization of the Hamming window, called the **Blackman window**, is discussed in Problem 7.26.

Figure 7.28 shows a comparison of the Hamming window design and the rectangular window design, in terms of the dB plots in the frequency domain. The Hamming window has much smaller side lobes. From the plots of $|W(e^{j\omega})|$ we see that for a rectangular window, the largest side lobe is 13 dB below the main lobe, whereas for a Hamming window, it is 50 dB below the main lobe. The filter plots $|H(e^{j\omega})|$ in the figure also show amazing improvement. For the rectangular window-based filter, the peak stopband ripple is 21 dB below the zero-frequency response, and for the Hamming-based filter it is more than 50 dB below!

The non-dB plots of the magnitude responses of the filters are shown in Fig. 7.29 for $\omega_c = 0.26\pi$. The filter based on the Hamming window has negligibly small ripples indeed, although it has a wider transition band, which is the price we pay for achieving smaller ripple sizes.

7.9.3.2 The Kaiser Window

The Kaiser window is unique in the sense that it comes with two degrees of freedom, namely (a) the window length $L = 2M + 1$ to control $\Delta\omega$ and (b) a parameter

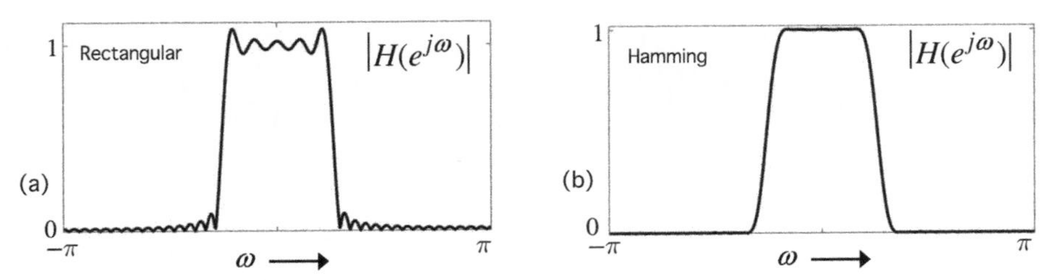

Figure 7.29 Plots of $|H(e^{j\omega})|$ for (a) rectangular window and (b) Hamming window. Here $\omega_c = 0.26\pi$ and length $L = 41$ in both cases.

β to control the stopband attenuation A_s of the filter $H(z)$. The other windows we described above had only one freedom, namely the window length. Since many of the other windows can be approximated well by the Kaiser window for an appropriate choice of β, the Kaiser window is the most commonly used window for filter design. The Kaiser window is given by

$$w(n) = \begin{cases} I_0\left[\beta\sqrt{1-(n/0.5N)^2}\,\right] \Big/ I_0(\beta) & -M \leq n \leq M, \\ 0 & \text{otherwise,} \end{cases} \tag{7.116}$$

where $N = 2M$ (filter order). Here, $I_0(x)$ is the so-called modified zeroth-order Bessel function. It can be computed from the infinite sum

$$I_0(x) = 1 + \sum_{k=1}^{\infty}\left(\frac{(0.5x)^k}{k!}\right)^2. \tag{7.117}$$

Clearly, $I_0(x) > 0$ for real x. In practice, only a few terms in the summation (7.117) (nearly 20) need to be retained to get a good approximation of $I_0(x)$. In Eq. (7.116), the parameter β can be used to control the stopband attenuation of the filter $H(e^{j\omega})$, and $N = 2M$ can be controlled to adjust the transition bandwidth of $H(e^{j\omega})$. Figure 7.30 shows examples of the window and the lowpass filter for two values of β, for $\omega_c = 0.26\pi$. Here $M = 20$, so the filter length $L = 41$. Plots of $w[n]$, $|W(e^{j\omega})|$, and $|H(e^{j\omega})|$ are shown. Note that for the larger β ($\beta = 7$) the window $w[n]$ is more localized in time, has a smaller side lobe in the frequency domain, and the filter $H(e^{j\omega})$ has a correspondingly larger stopband attenuation. The price paid is the larger main-lobe width for $W(e^{j\omega})$, and the correspondingly wider transition band for $H(e^{j\omega})$. As usual, the plots are normalized such that the response at zero frequency is unity (0 dB).

Through extensive experimentation, Kaiser came up with formulas for estimating the parameters β and N to achieve specified ripple size δ_2 and transition bandwidth $\Delta\omega$ [Kaiser, 1974]. Thus, β can be found using

$$\beta = \begin{cases} 0.1102(A_s - 8.7) & \text{if } A_s > 50, \\ 0.5842(A_s - 21)^{0.4} + 0.07886(A_s - 21) & \text{if } 21 < A_s < 50, \\ 0 & \text{if } A_s < 21, \end{cases} \tag{7.118}$$

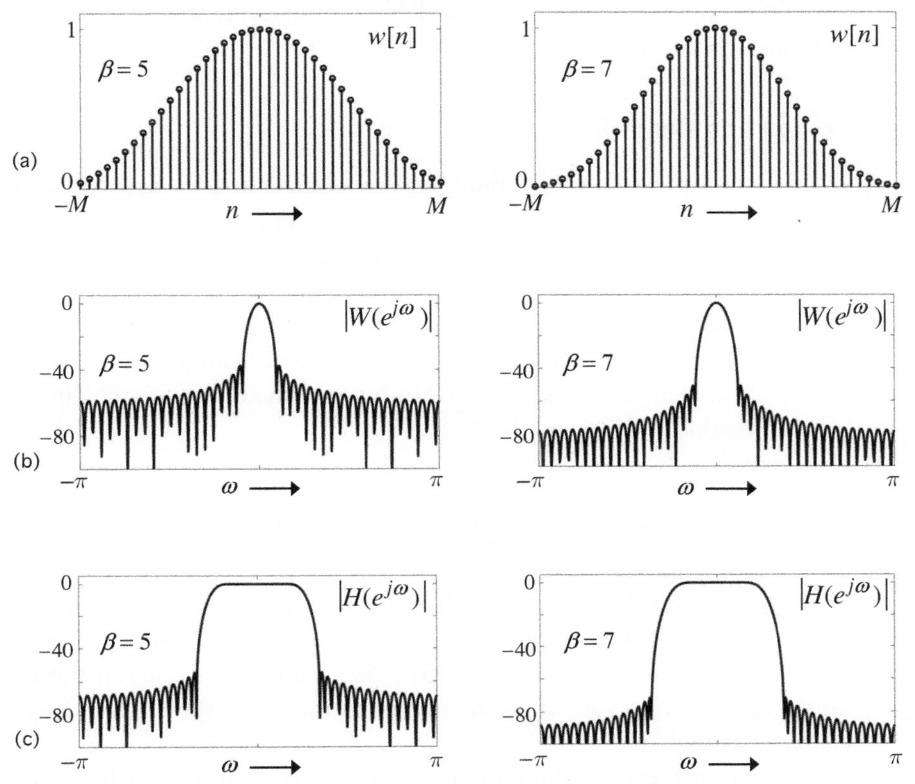

Figure 7.30 The Kaiser window and the corresponding lowpass filter for $\beta = 5$ (left) and $\beta = 7$ (right). The length is $L = 41$ (i.e., $M = 20$) and $\omega_c = 0.26\pi$ in both cases. (a) The windows $w[n]$, (b) the magnitudes $|W(e^{j\omega})|$ in dB, and (c) the filter magnitudes $|H(e^{j\omega})|$ in dB.

where $A_s = -20\log_{10}(\delta_2)$ is the minimum stopband attenuation that the filter $H(e^{j\omega})$ is required to achieve. The quantity $N = 2M$ (filter order) can be estimated from

$$N \approx \frac{A_s - 7.95}{14.36(\Delta\omega/2\pi)}. \tag{7.119}$$

There is no independent control over δ_1. As mentioned earlier, it turns out to be close to δ_2.

The curious student is no doubt wonderstruck as to how Kaiser came up with the above formulas, especially the expression (7.116). While there may be no systematic way to arrive at this expression through theory, there is another wonder waiting for the reader. Namely, the Kaiser window turns out to be a good approximation of a theoretically optimal window $v[n]$ called the prolate spheroidal sequence. This sequence has the property that the energy of $V(e^{j\omega})$ in the region $\sigma \leq |\omega| \leq \pi$ is minimized (for any fixed $\sigma > 0$) subject to the constraint that the total energy in $0 \leq |\omega| \leq \pi$ is fixed. Details can be found in Sec. 3.2.2 of Vaidyanathan [1993]. There are other types of optimal windows, such as the Dolph–Chebyshev window.

This finds application not only in FIR filter design but also in antenna arrays. We shall not discuss the details here.

7.9.4 Concluding Remarks

Even though the window method provides excellent FIR approximations to the ideal filter, the resulting filters are not optimal in any way. There exist a number of optimal FIR filter design methods, which minimize appropriate error objectives. For example, given the filter order and the band edges, they can minimize a linear combination of δ_1 and δ_2. The details are beyond the scope of our discussion here, but there are excellent books which the student can pursue, such as Antoniou [2006], Diniz et al. [2010], Mitra [2011], Oppenheim and Schafer [2010], and Proakis and Manolakis [2007].

7.10 Comparing Responses of FIR and IIR Filters

In order to give a flavor for how filters are designed, we discussed the analog Butterworth filter in some detail in Sec. 7.8. Chebyshev and elliptic filters will not be discussed in detail, and the interested reader can refer to one of the books mentioned above. However, we present some examples to show what the actual filter responses look like.

As mentioned earlier, digital filters designed from Butterworth, Chebyshev, and elliptic filters using bilinear transformation can be obtained directly from digital filter software packages. For this, one simply inputs the specifications $\omega_p, \omega_s, \delta_1, \delta_2$ and obtains the coefficients $\{a_k, b_k\}$ of the transfer function $H(z)$. These are **IIR filters** with nontrivial poles and zeros. Figure 7.31(a) shows the magnitude response $|H(e^{j\omega})|$ of an IIR digital elliptic filter of order $N = 5$, which satisfies the specifications

$$\omega_p = 0.1\pi,\ \omega_s = 0.22\pi,\ \delta_1 = 0.01, \delta_2 = 0.001. \tag{7.120}$$

The ripples in the passband are visible but the stopband ripples are very small. The dB plot of $|H(e^{j\omega})|$, shown in Fig. 7.31(b), clearly shows the ripples in the stopband. The peak stopband ripple in dB is

$$10 \log_2 \delta_2^2 = -60 \text{ dB}, \tag{7.121}$$

which is also evident from the plot. At the points where $H(e^{j\omega}) = 0$, the dB plot has value $-\infty$, which explains the three deep dips. The passband details are also shown in the inlay of Fig. 7.31(b), and we see clearly that $\delta_1 = 0.01$. The phase response $\phi(\omega)$ of the elliptic filter is shown in Fig. 7.31(c) and is seen to be very nonlinear. The negative derivative of the phase response, that is,

$$\tau(\omega) = \frac{-d\phi(\omega)}{d\omega}, \tag{7.122}$$

is called the **group delay** and is plotted in Fig. 7.31(d). The group delay can be interpreted as the amount of delay experienced by an input whose frequency

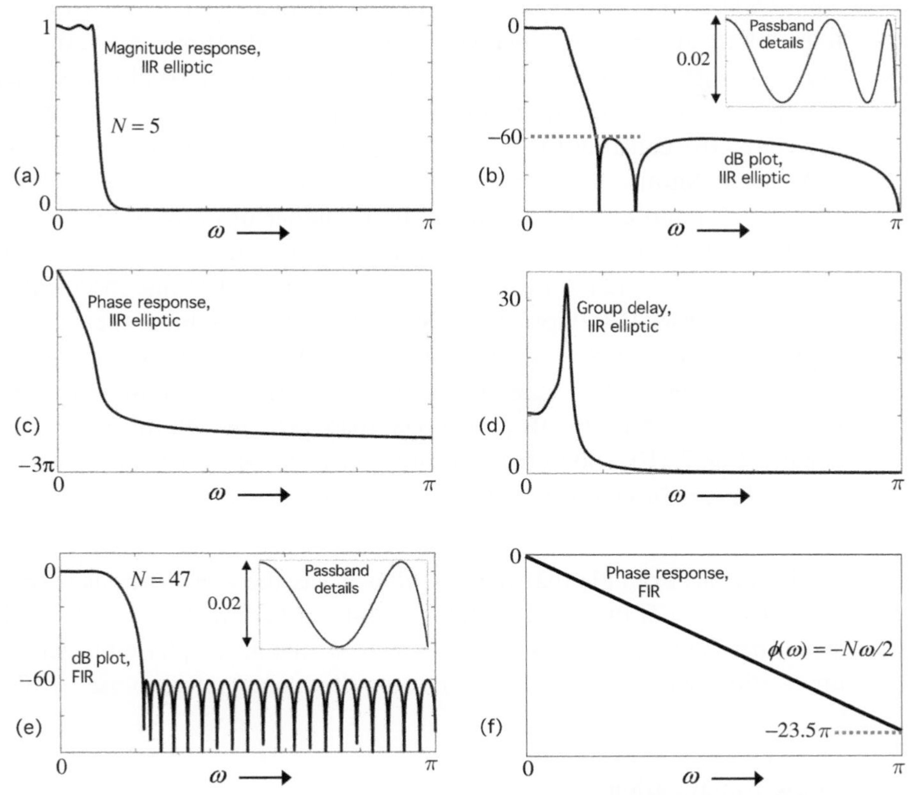

Figure 7.31 (a)–(d) The magnitude response $|H(e^{j\omega})|$, dB response $10\log_{10}|H(e^{j\omega})|^2$, phase response $\phi(\omega)$, and group delay $\tau(\omega)$ for an elliptic IIR lowpass filter (order $N = 5$). (e)–(f) The magnitude and phase responses for a linear-phase FIR filter (order $N = 47$). Both filters meet the same set of magnitude specifications (7.120).

components are concentrated around a frequency ω. For linear-phase filters, the group delay is the same at all frequencies, but for nonlinear-phase filters it is not, and this causes phase distortion (Sec. 4.5). The damaging effects of phase distortion can be quite visible in image processing applications (see below), but in other applications such as speech processing, the effect is not perceivable by the human ear.

Linear-phase FIR filters. It is generally true that IIR filters have very nonlinear phase (as in Fig. 7.31(c)), and create phase distortion. On the other hand, it is possible to design FIR filters of the form

$$\sum_{n=0}^{N} h[n]z^{-n}, \tag{7.123}$$

with perfectly linear phase. As shown in Sec. 4.6, this is achieved simply by imposing the *symmetry property*

$$h[n] = h[N-n], \tag{7.124}$$

where $h[n]$ is real. (For generalizations, see Problem 4.16.) Such FIR filters with linear phase are often used in image processing applications. Now, these FIR filters are not designed by using bilinear transformation on analog filters. Rather, they are designed by directly obtaining the filter coefficients $h[n]$ in the discrete-time domain subject to the constraint (7.124). One approach is the window method described in Sec. 7.9. Another is the equiripple design method developed by McClellan and Parks [1973] and described in detail in a number of books (see, e.g., Oppenheim and Schafer [2010]). Figure 7.31(e) shows the magnitude response $|H(e^{j\omega})|$ of such a linear-phase FIR filter of order $N = 47$, which meets the same specifications (7.120) as the fifth-order IIR elliptic filter. Since the order is so high, there are many more ripples, and many more transmission zeros in the stopband. But the ripple sizes are the same for both filters: the inlay shows the passband details, from which we see that $\delta_1 = 0.01$. And the dB plot shows $10\log_{10}\delta_2^2 = -60$ dB, that is, $\delta_2 = 0.001$. Finally, Fig. 7.31(f) shows the phase-response plot of the FIR filter, which is indeed exactly linear. In fact, it has the expression $\phi(\omega) = -N\omega/2$, where N is the filter order (Sec. 4.6).[7] $\nabla\nabla\nabla$

Notice in this example that the IIR filter meets the specifications with a small filter order $N = 5$, but the FIR filter requires a very high order ($N = 47$) for the same specifications. This is the price paid for the linear phase property. Thus, for applications where linear-phase is not required, one prefers to use IIR filters.

Phase Equalization

As explained in Sec. 6.8, it is not possible to design causal IIR filters with exact linear phase. However, given an IIR filter $H(z)$ with nonlinear-phase response, it is possible to design an approximately linear-phase filter $F(z)$ with the same magnitude response, by cascading it with an allpass filter $G(z)$:

$$F(z) = G(z)H(z). \tag{7.125}$$

Since $G(z)$ is allpass, we have $|F(e^{j\omega})| = |H(e^{j\omega})|$. However, the phase response of $F(z)$ is

$$\phi_f(\omega) = \phi_g(\omega) + \phi_h(\omega), \tag{7.126}$$

where $\phi_g(\omega)$ and $\phi_h(\omega)$ are the phase responses of $G(e^{j\omega})$ and $H(e^{j\omega})$, respectively. The nonlinearity in the phase $\phi_h(\omega)$ can be compensated for, by designing $\phi_g(\omega)$ appropriately, so that $\phi_f(\omega)$ approximates a straight line in the passband. This is called "phase equalization," and is sometimes used in applications where linear phase is required.

[7] Strictly speaking, this ignores the fact that whenever $H(e^{j\omega})$ crosses zero in the stopband, the phase jumps by π. Since the phase is not important in the stopband anyway, this jump does not matter.

7.11 Demonstration of Phase Distortion in Images

In Sec. 4.5 we explained in some detail the effect of phase distortion on signals. It was shown that phase distortion is visually noticeable (see the discussion around Fig. 4.20). This will now be demonstrated on a 2D image pattern. Consider Fig. 7.32(a), which shows a checkerboard image. Figure 7.32(b) shows a lowpass-filtered version of this, where the filter $H(z)$ is a linear-phase filter with magnitude response as in Fig. 7.33(a). The lowpass operation has smoothed out the sharp edges between the black and white squares. If we look carefully we also see some faint stripes in Fig. 7.32(b). These "filtering artifacts" are created by the fact that every point (pixel) in the original image spreads in both directions according to the impulse response $h[n]$, which is also known as the **point-spread function** in two dimensions. So, the strong lines in the center of the symmetric impulse response (Fig. 7.34(a)) can be "seen" in the filtered version.[8]

Next, Fig. 7.32(c) shows the same checkerboard image filtered by a filter $H_m(z)$ with the same magnitude response $|H_m(e^{j\omega})| = |H(e^{j\omega})|$ as before. But $H_m(z)$ is *not* a linear-phase filter. It is called a minimum-phase filter (explained below), and its phase response is as in Fig. 7.33(b). Since only the phase response in the passband

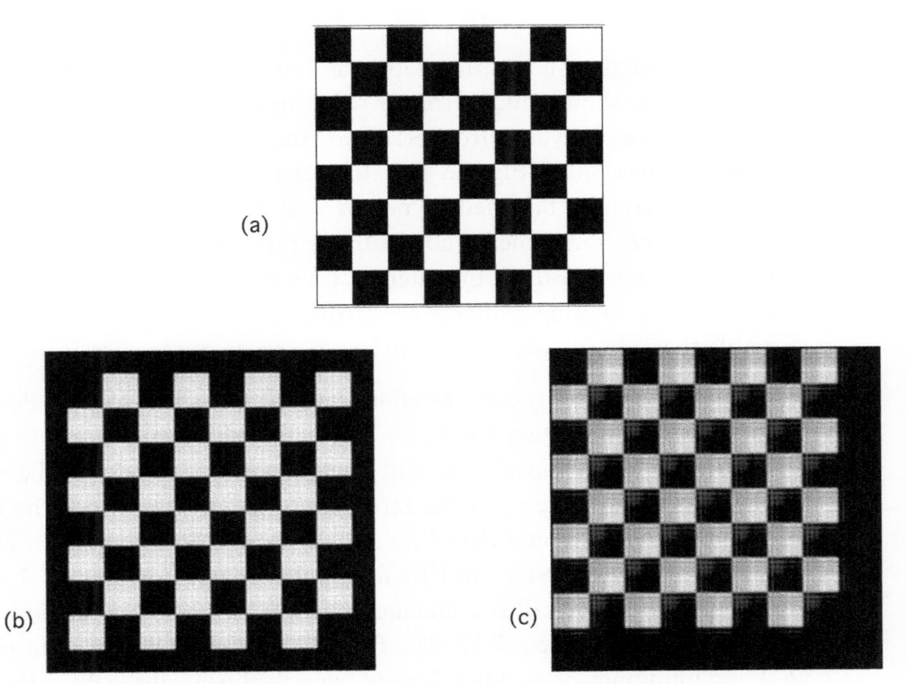

Figure 7.32 (a) Checkerboard image, (b) image filtered with a linear-phase lowpass filter $H(z)$, and (c) image filtered with a nonlinear-phase (minimum-phase) lowpass filter $H_m(z)$ with the same magnitude response $|H(e^{j\omega})|$.

[8] Since $h[n]$ has some negative samples, the filtered image usually has some negative pixels. One way to fix this problem is to add a small positive constant to all pixels to convert the most negative pixel to zero. This is what we have done here. There are other ways to handle the issue; all have pros and cons.

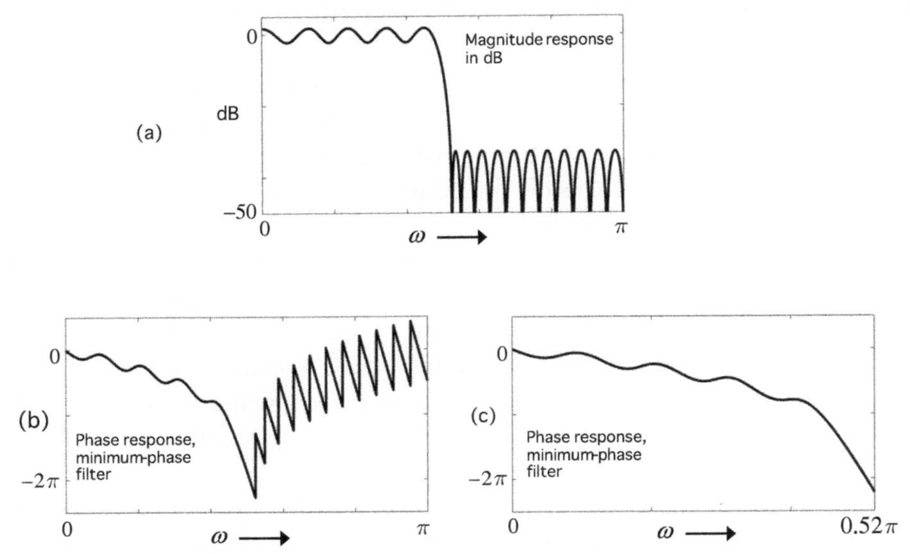

Figure 7.33 Responses of FIR filters $H(z)$ and $H_m(z)$ used for image filtering in Fig. 7.32. (a) Magnitude response $|H(e^{j\omega})| = |H_m(e^{j\omega})|$ in dB. (b) Phase response of nonlinear-phase filter $H_m(z)$. (c) Phase response of $H_m(z)$ in the passband.

matters, we have also shown that separately in Fig. 7.33(c).[9] Notice that the filtered version now shows much stronger "filtering artifacts." That is, the stripes are not faint anymore. They are stronger. If we examine the impulse response $h_m[n]$ of the nonlinear-phase filter (Fig. 7.34(b)), the reason for this becomes clear. Namely, there are large samples bunched up near $n = 0$. Even though the linear-phase filter $h[n]$ also shows some energy concentration near the middle (Fig. 7.34(a)), the concentration near $n = 0$ for the filter $H_m(z)$ seems stronger. To some extent this explains why the filtering artifacts are stronger. For reasons like this, it is often preferrable to use linear-phase filters in image-processing applications.

Minimum phase and energy concentration. Recall from Sec. 4.6 and Problem 4.17 that for linear-phase filters the zeros occur in reciprocal conjugate pairs, like $\{z_k, 1/z_k^*\}$. Thus, if z_k is inside the unit circle, then $1/z_k^*$ is outside the unit circle. It can be shown that if we move the zero $1/z_k^*$ to the point z_k inside the unit circle, then the magnitude response $|H(e^{j\omega})|$ does not change (Problem 7.12). The new filter has two zeros at z_k. Now, an FIR filter with all its zeros in $|z| \leq 1$ (i.e., on or inside the unit circle) is called a **minimum-phase** FIR filter. This topic is discussed in greater detail in Problem 7.12. The filter $H_m(z)$, which was used to obtain Fig. 7.32(c), is a minimum-phase filter derived from the linear-phase filter $H(z)$ by moving all the zeros outside the unit circle to reciprocal conjugate points inside the unit circle. That is how $|H_m(e^{j\omega})| = |H(e^{j\omega})|$ was achieved in this example. And since the minimum-phase filter $H_m(z)$ has all zeros in $|z| \leq 1$, it does not satisfy the

[9] The discontinuities in the phase response in the stopband are mere reflections of the fact that there are zero crossings of the filter in the stopband. At each zero crossing the phase jumps by π.

linear-phase property. Furthermore, it can be shown that its impulse response has a strong "energy concentration property" near $n = 0$, as explained more quantitatively in Problem 7.12. This property to some extent explains the strong filtering artifacts in Fig. 7.32(c). ▽ ▽ ▽

7.12 Steady-State and Transient Terms*

We have discussed many examples of digital filters in this chapter. Recall now how an LTI filter works: if the frequency response is $H(e^{j\omega})$ then the output in response to a single-frequency input $x[n] = e^{j\omega_0 n}$ is $y[n] = H(e^{j\omega_0})e^{j\omega_0 n}$. For example, if $H(e^{j\omega}) = 0$ at $\omega = \omega_0$ then the frequency ω_0 is completely attenuated, whereas if $H(e^{j\omega}) = 1$ at $\omega = \omega_0$ then ω_0 is passed with no distortion at all. But this situation assumes that the input $x[n] = e^{j\omega_0 n}$ starts at $n = -\infty$. In practice, we start the input of the causal filter at a certain time, say, $n = 0$. Then $x[n] = e^{j\omega_0 n}\mathcal{U}[n]$, and the output is **not** of the form $y[n] = H(e^{j\omega_0})e^{j\omega_0 n}$ for all $n \geq 0$. Instead, what we really have is

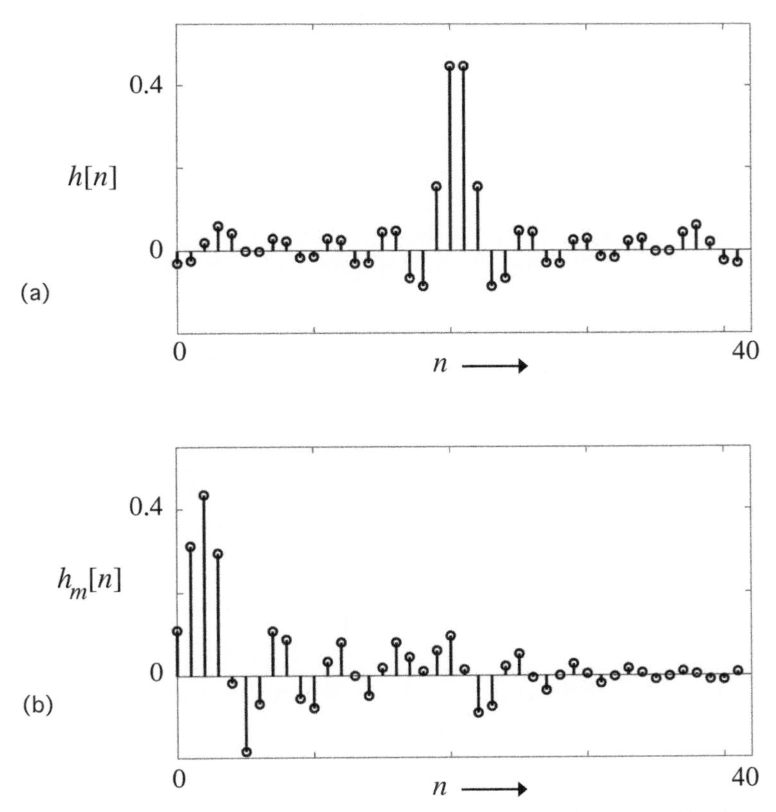

Figure 7.34 Impulse responses of the FIR filters used for image filtering in Fig. 7.32. (a) Linear-phase filter $H(z)$ and (b) nonlinear-phase filter $H_m(z)$.

$$y[n] = y_{tr}[n] + H(e^{j\omega_0})e^{j\omega_0 n}, \quad n \geq 0, \tag{7.127}$$

where the term $y_{tr}[n]$ decays to zero as $n \to \infty$ (assuming the filter is stable). Thus, $y[n] \approx H(e^{j\omega_0})e^{j\omega_0 n}$ for large n but not for small $n \geq 0$. The first term $y_{tr}[n]$ is called the transient response, and the second term is called the steady-state response. In all our earlier discussions we implicitly focused on the steady-state response. In this section we will study the properties of transients.

In Sec. 4.8.2 we briefly discussed steady-state and transient parts of the output of an LTI system. In this section we consider an important situation where the input is a one-sided exponential, namely, $x[n] = a^n\mathcal{U}[n]$. First, we focus on causal rational IIR filters and then discuss FIR filters in detail. Further generalizations are included towards the end of the section. This section can be read independently of Sec. 4.8.2 because we use a fresh approach.

7.12.1 One-Sided Exponential Input to an LTI System

Consider the one-sided exponential

$$x[n] = a^n\mathcal{U}[n], \tag{7.128}$$

applied to the first-order causal LTI system

$$H(z) = \frac{1}{1 - pz^{-1}}. \tag{7.129}$$

Clearly the system has a pole at p, and its impulse response is $h[n] = p^n\mathcal{U}[n]$. Thus, the output is the convolution

$$\begin{aligned} y[n] = \sum_{k=-\infty}^{\infty} h[k]x[n-k] &= \sum_{k=0}^{n} h[k]x[n-k] \\ &= \sum_{k=0}^{n} p^k a^{n-k} \\ &= a^n \sum_{k=0}^{n} (p/a)^k \end{aligned} \tag{7.130}$$

for $n \geq 0$, and by causality $y[n] = 0$ for $n < 0$. When $a \neq p$ this can be rewritten in the form

$$y[n] = a^n\left(\frac{1 - (p/a)^{n+1}}{1 - p/a}\right). \tag{7.131}$$

We will consider the case $a = p$ later. We can rewrite Eq. (7.131) as

$$y[n] = \frac{a^n - (p/a)p^n}{1 - pa^{-1}} = \frac{a^n}{1 - pa^{-1}} + cp^n = H(a)a^n + cp^n, \tag{7.132}$$

where

$$c = -\frac{pa^{-1}}{1 - pa^{-1}}. \tag{7.133}$$

Summarizing, when $a \neq p$, the output is

$$y[n] = \begin{cases} H(a)a^n + cp^n & n \geq 0, \\ 0 & n < 0, \end{cases} \tag{7.134}$$

where c is a constant, that is, independent of time n, although it depends on p (system parameter) and a (input parameter).

7.12.1.1 Remarks

A number of observations can be made about Eq. (7.134).

1. *One-sided exponentials are not eigenfunctions.* If the input were $x[n] = a^n$ (a true exponential starting at $n = -\infty$ instead of at $n = 0$), then we know that the output would be $y[n] = H(a)a^n$ (assuming a is in the region of convergence of $H(z)$; see Chapter 6). This is merely the statement that *exponentials are eigenfunctions* of LTI systems. Now, when $x[n] = a^n\mathcal{U}[n]$, which is an exponential starting at $n = 0$, the output is no longer of the form $H(a)a^n\mathcal{U}[n]$, that is, the eigenfunction property is not valid anymore. Referring to Eq. (7.134), we say that $H(a)a^n$ is the **steady-state** term, and cp^n is the **transient** term. So, when the exponential input a^n starts at a finite time, then the output has these two terms, instead of just the steady-state term $H(a)a^n$.
2. *Connection to ROC of $H(z)$.* If $|a| > |p|$, then Eq. (7.134) shows that

$$y[n] \approx H(a)a^n, \tag{7.135}$$

for large enough n. That is, the "eigenfunction behavior" of exponentials becomes more and more prominent as n increases. On the other hand, if $|a| \leq |p|$ then the steady state term $H(a)a^n$ can never dominate the transient term cp^n, that is, the steady-state "never arrives." From Sec. 6.3.1 we know that the region of convergence (ROC) of the impulse response $h[n] = p^n\mathcal{U}[n]$ is $|z| > |p|$. Thus, the above observation merely says that *when the input parameter a is in the ROC, then the steady state eventually dominates the transient*; otherwise, it does not.
3. *Connection to stability of $H(z)$.* Notice in particular that if $a = e^{j\omega_0}$, we have

$$x[n] = e^{j\omega_0 n}\mathcal{U}[n], \tag{7.136}$$

which is like the single-frequency signal, except that it starts at $n = 0$ instead of at $n = -\infty$. In this case the output does not have the form $y[n] = H(e^{j\omega_0})e^{j\omega_0 n}\mathcal{U}[n]$. Instead, the output is

$$y[n] = \left(H(e^{j\omega_0})e^{j\omega_0 n} + cp^n\right)\mathcal{U}[n]. \tag{7.137}$$

This has the steady-state term $H(e^{j\omega_0})e^{j\omega_0 n}$ and transient cp^n for all $n \geq 0$. If the causal system is stable then the pole satisfies $|p| < 1$, so that $p^n \to 0$ for $n \to \infty$.

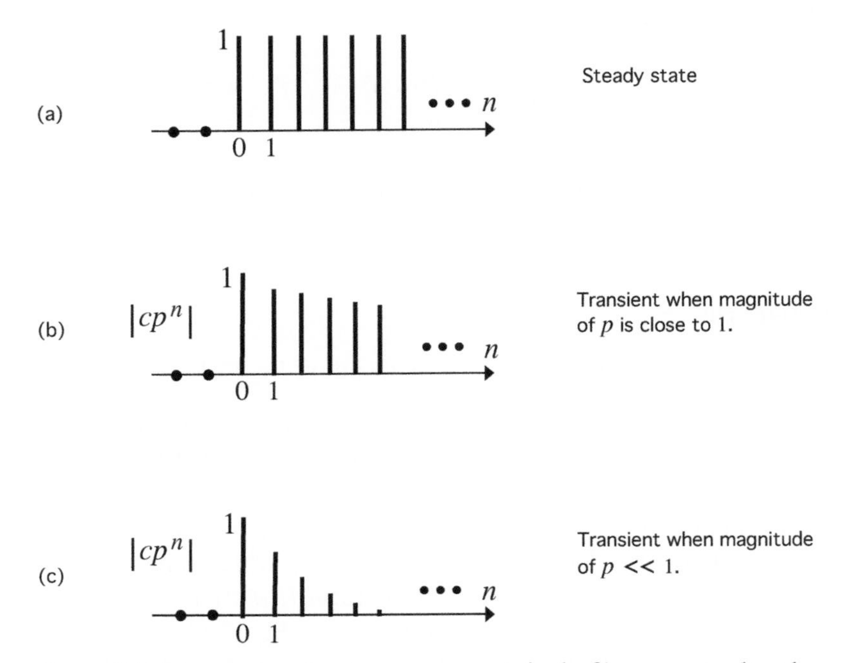

Figure 7.35 Magnitudes of various components in the filter output, when the input is $x[n] = e^{j\omega_0 n}\mathcal{U}[n]$. (a) Steady-state component, (b) transient component when $|p| < 1$ but close to unity, and (c) transient component when $|p| << 1$.

Thus, for a causal **stable** system, the steady-state term dominates more and more for large n and

$$y[n] \approx H(e^{j\omega_0})e^{j\omega_0 n}, \tag{7.138}$$

for large enough n. That is, the frequency response behavior can be "realized" for large n. For unstable systems we have $|p| \geq 1$, and the *frequency response cannot be "realized"* because $H(e^{j\omega_0})e^{j\omega_0 n}$ cannot dominate p^n, no matter how large n gets. Thus, in this case the frequency response has no meaning.

4. *Rate of decay of transients.* Assuming then that the system is stable, how long should we wait for the transient cp^n to die down, that is, become negligible compared to $H(e^{j\omega_0})e^{j\omega_0 n}$? The answer depends upon p. If it is close to the unit circle (i.e., $|p| \approx 1$), it takes a long time for the transient p^n to die down and the steady state to dominate. If $|p| << 1$ then the transient p^n dies downs quickly! See Fig. 7.35. For example, in Sec. 7.3 we found that the quality of a notch filter can be improved by moving the poles closer and closer to the unit circle. The **price paid** for this is that the transients get longer and longer! See the example in Sec. 7.3.1. A similar situation can arise with IIR filters of high quality, such as Butterworth, Chebyshev, and elliptic filters (Sec. 7.6), which tend to have several poles close to the unit circle.
5. *Short input segments.* Suppose the input is only a short segment of a sinusoid in $0 \leq n \leq n_0$ (and zero elsewhere). If n_0 is comparable to the duration where

the transient is strong, then by the time the steady state "arrives," the input has already disappeared! So, statements such as "the frequency ω_0 is attenuated completely by a transmission zero" are not very meaningful in such situations.

6. *Case when input is matched to the pole*. When $a = p$, we see from Eq. (7.130) that

$$y[n] = a^n(1+n) = a^n + na^n, \quad n \geq 0. \tag{7.139}$$

Compared to the first term (exponential), the second term (non-exponential) gets larger and larger as n grows. That is, the output never looks like an exponential a^n no matter how large n gets. The steady state "never arrives." The reason for this is that when $h[n] = p^n\mathcal{U}[n]$, the ROC of the transfer function $H(z)$ is $|z| > |p|$. It does not converge at $z = p$ (pole location). So when $a = p$, the quantity $H(a)$ is not defined, and it is not meaningful to expect a steady state $H(a)a^n$ to arrive.

7.12.2 More General LTI Systems

If the system has a higher order with more than one pole, then how do we calculate the steady state and transient? It is best to demonstrate this with an example. Thus, consider the second-order causal rational LTI system

$$H(z) = \underbrace{\frac{A_1}{1 - p_1 z^{-1}}}_{H_1(z)} + \underbrace{\frac{A_2}{1 - p_2 z^{-1}}}_{H_2(z)}, \tag{7.140}$$

so the poles are p_1 and p_2. Since the system is causal, the impulse response is

$$h[n] = \left(A_1 p_1^n + A_2 p_2^n\right)\mathcal{U}[n]. \tag{7.141}$$

If we apply an input $x[n] = a^n\mathcal{U}[n]$, then the output has the form

$$y[n] = \left(H_1(a)a^n + A_1 c_1 p_1^n \ + \ H_2(a)a^n + A_2 c_2 p_2^n\right)\mathcal{U}[n], \tag{7.142}$$

assuming that none of the poles p_i is equal to a. Here, c_i is as in Eq. (7.133), that is,

$$c_i = -\frac{p_i a^{-1}}{1 - p_i a^{-1}}. \tag{7.143}$$

Equation (7.142) can be rewritten as

$$y[n] = \Bigg(\underbrace{H(a)a^n}_{\text{steady state}} \ + \ \underbrace{A_1 c_1 p_1^n + A_2 c_2 p_2^n}_{\text{transient}}\Bigg)\mathcal{U}[n]. \tag{7.144}$$

Thus, the transient term is an appropriate linear combination of transients due to the individual poles. So, all discussions are similar to the first-order case. When $x[n] = e^{j\omega_0 n}\mathcal{U}[n]$, we have

$$y[n] = \Bigg(\underbrace{H(e^{j\omega_0})e^{j\omega_0 n}}_{\text{steady state}} \ + \ \underbrace{A_1 c_1 p_1^n + A_2 c_2 p_2^n}_{\text{transient}}\Bigg)\mathcal{U}[n]. \tag{7.145}$$

The steady state $H(e^{j\omega_0})e^{j\omega_0 n}$ dominates for large n if all poles are inside the unit circle ($|p_i| < 1$), that is, if the causal system is stable. In this case the frequency response term $H(e^{j\omega_0})e^{j\omega_0 n}$ begins to dominate more and more, that is, the frequency response gets "realized" for large n. Note that for large n, the transient due to the **dominant pole** (i.e., the pole with largest $|p_i|$) is more important than the other poles. See Fig. 7.35.

Case when $H(e^{j\omega_0}) = 0$. Suppose $H(z)$ has a zero on the unit circle at $z = e^{j\omega_0}$. Then $H(e^{j\omega_0}) = 0$ (transmission zero at ω_0), which means that the input $x[n] = e^{j\omega_0 n}$ starting from $n = -\infty$ produces $y[n] = 0, \forall n$. This perfect filtering behavior is not realizable when $x[n] = e^{j\omega_0 n}\mathcal{U}[n]$ (which starts at $n = 0$) because the output is as in Eq. (7.145). However, the steady-state component is certainly still zero, so the ouput is

$$y[n] = A_1 c_1 p_1^n + A_2 c_2 p_2^n, \quad n \geq 0. \tag{7.146}$$

The constants c_i depend on ω_0 because they are determined by Eq. (7.143), where $a = e^{j\omega_0}$. But the important point is that the time dependence or "shape" of the output signal is determined entirely by the exponentials p_i^n defined by the poles. The input component $e^{j\omega_0 n}$ is completely absent. In this sense, if the filter has a zero at $z = e^{j\omega_0}$, then the input $x[n] = e^{j\omega_0 n}$ is "suppressed," but the transient terms will still linger!

The curious reader might wonder what happens when the input sinusoid has finite duration, say, $0 \leq n \leq M - 1$. That is,

$$x[n] = e^{j\omega_0 n}\Big(\mathcal{U}[n] - \mathcal{U}[n - M]\Big). \tag{7.147}$$

This is a linear combination of two signals of the form $e^{j\omega_0 n}$, one starting at $n = 0$ and the other at $n = M$. Using the LTI property of the filter, the transient components can easily be found. In short, there are two transient components, one starting at $n = 0$ and the other at $n = M$. If the duration M of $x[n]$ is not long enough, then the input ceases even before the first transient has decayed satisfactorily! ▽ ▽ ▽

7.12.2.1 Real Filters with Real Inputs

It is easy to extend Eq. (7.145) for more complicated inputs. First consider the real-valued input $x[n] = \cos(\omega_0 n)$ (starting at $n = -\infty$), and a filter like Eq. (7.140) with real impulse response $h[n]$. This happens, for example, when $p_2 = p_1^*$ and $A_2 = A_1^*$. When $h[n]$ is real we have $H(e^{-j\omega}) = H^*(e^{j\omega})$ (Sec. 3.6). Since $\cos(\omega_0 n) = 0.5(e^{j\omega_0 n} + e^{-j\omega_0 n})$, the output is

$$\begin{aligned} y[n] &= 0.5\Big(H(e^{j\omega_0})e^{j\omega_0 n} + H(e^{-j\omega_0})e^{-j\omega_0 n}\Big) \\ &= 0.5\Big(H(e^{j\omega_0})e^{j\omega_0 n} + H^*(e^{j\omega_0})e^{-j\omega_0 n}\Big) \\ &= 0.5|H(e^{j\omega_0})|\Big(e^{j\psi(\omega_0)}e^{j\omega_0 n} + e^{-j\psi(\omega_0)}e^{-j\omega_0 n}\Big), \end{aligned} \tag{7.148}$$

where $H(e^{j\omega}) = |H(e^{j\omega})|e^{j\psi(\omega)}$. Thus, the output has the form

$$y[n] = |H(e^{j\omega_0})| \cos(\omega_0 n + \psi(\omega_0)), \quad \forall n. \tag{7.149}$$

While the output is also sinusoidal as expected, it is clear that $\cos(\omega_0 n)$ is not an eigenfunction of the LTI system (even though the input starts from $n = -\infty$). We noticed this also in Sec. 5.13.1. Next assume we apply $x[n] = \cos(\omega_0 n)\mathcal{U}[n]$ (i.e., starting from $n = 0$), then there will be transients before we observe a sinusoidal behavior like in Eq. (7.149). To analyze this, write

$$x[n] = \cos(\omega_0 n)\mathcal{U}[n] = \left(\frac{e^{j\omega_0 n} + e^{-j\omega_0 n}}{2}\right)\mathcal{U}[n]. \tag{7.150}$$

The first term in Eq. (7.150) creates a transient similar to the transient terms in Eq. (7.145). And the second term in Eq. (7.150) creates similar transient terms, but the constants c_i are replaced with possibly different constants c_i' because the constants depend on the frequencies $\pm\omega_0$ (see Eq. (7.143)). So the transient term created by Eq. (7.150) is

$$y_{trans}[n] = \Big(A_1(c_1 + c_1')p_1^n + A_2(c_2 + c_2')p_2^n\Big)\mathcal{U}[n]. \tag{7.151}$$

Summarizing, using Eq. (7.149) for the steady-state term, the output in response to $x[n] = \cos(\omega_0 n)\mathcal{U}[n]$ has the form

$$y[n] = y_{trans}[n] \quad + \quad |H(e^{j\omega_0})| \cos(\omega_0 n + \psi(\omega_0))\, \mathcal{U}[n], \tag{7.152}$$

where $y_{trans}[n]$ is as in Eq. (7.151).

7.12.2.2 An Example: How Transients Linger for Larger Pole Radius

We now demonstrate the effects of transients with an example. Consider a second-order causal IIR filter

$$H(z) = \frac{1 - 2\cos\phi\, z^{-1} + z^{-2}}{1 - 2R\cos\theta\, z^{-1} + R^2 z^{-2}}, \tag{7.153}$$

where $0 < R < 1$ is the pole radius, and ϕ and θ are real. Thus, the poles and zeros are at

$$z = Re^{\pm j\theta} \text{ (poles)} \qquad \text{and} \qquad z = e^{\pm j\phi} \text{ (zeros)}. \tag{7.154}$$

The zeros are on the unit circle, so that there are transmission zeros at $\omega = \pm\phi$ in the frequency response $H(e^{j\omega})$. We take the input to be the sinusoid

$$x[n] = \cos(\omega_0 n)\mathcal{U}[n], \tag{7.155}$$

where the input frequency is $\omega_0 = 0.03\pi$ (Fig. 7.36(a)). With the pole angle and zero angle fixed at $\theta = 0.003\pi$ and $\phi = 2\omega_0$, we plot the ouput $y[n]$ for two values of the pole radius: $R = 0.95$ and $R = 0.999$. These are shown in Figs. 7.36(b) and (c). For $R = 0.95$ we find that the output has a short transient followed by a nice sinusoid. For $R = 0.999$ the transient is much longer because $p^n = R^n e^{j\theta n}$ decays much slower; for clarity, we show this plot again in part (d) for many more samples.

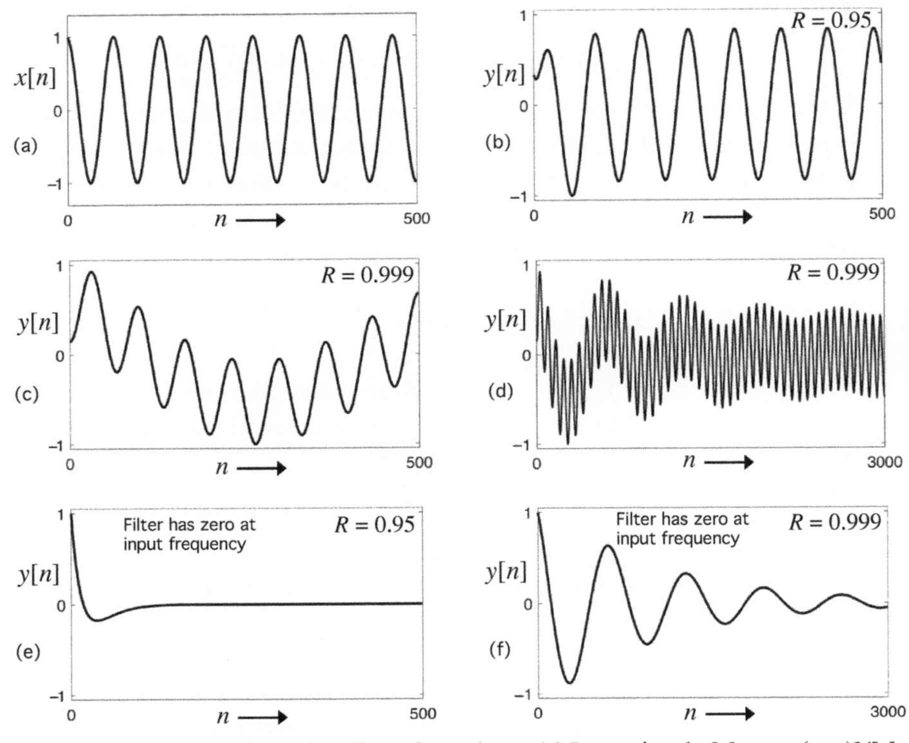

Figure 7.36 Demonstrating the effect of transients. (a) Input signal $x[n] = \cos(\omega_0 n)\mathcal{U}[n]$, (b)–(f) outputs of the second-order IIR filter for different values of pole radius R as indicated. Part (d) is the same as (c), with more samples plotted. For parts (e) and (f), the filter's transmission zero ϕ coincides with the input frequency ω_0. See text for explanation.

It is clear that the transient lingers for a much longer time before the output begins to resemble a sinusoid.

Next, let us change the angle of the filter's zero and make it $\phi = \omega_0$. Thus, the filter has a transmission zero at the input frequency. If the input were $x[n] = \cos(\omega_0 n)$ starting at $n = -\infty$, the output would be zero for all n. But since the input starts only at $n = 0$, the output is as shown in Fig. 7.36(e) for pole radius $R = 0.95$. Thus, the output has an initial transient shaped by the poles $Re^{\pm j\theta}$, and then it approaches zero as expected. If we increase the pole radius to $R = 0.999$, the transient lingers for a much longer time before the output decays to nearly zero, as shown in Fig. 7.36(f).

7.12.3 Transients in the FIR Case

Consider an Nth-order causal FIR filter

$$H(z) = \sum_{n=0}^{N} h[n]z^{-n}. \tag{7.156}$$

If $x[n] = a^n \mathcal{U}[n]$, then

$$y[n] = \sum_{k=0}^{n} h[k]x[n-k] = \sum_{k=0}^{n} h[k]a^{n-k} = a^n \sum_{k=0}^{n} h[k]a^{-k}. \tag{7.157}$$

For $n \geq N$ this becomes

$$y[n] = a^n \sum_{k=0}^{N} h[k]a^{-k} = H(a)a^n, \tag{7.158}$$

which is the steady-state output. Thus, for the FIR case the output reaches the steady state perfectly after the first N samples. By contrast, for IIR filters, the transients have terms of the form p^n, which do not become zero for finite n, although they can be neglected for large enough n under some conditions, as explained earlier. For the FIR case, the transient term lasts only for the first N samples:

$$y[0], y[1], \ldots, y[N-1]. \tag{7.159}$$

In this transient phase, the output is

$$\begin{aligned} y[n] = a^n \sum_{k=0}^{n} h[k]a^{-k} &= a^n \sum_{k=0}^{N} h[k]a^{-k} - a^n \sum_{k=n+1}^{N} h[k]a^{-k} \\ &= \underbrace{H(a)a^n}_{\text{steady state}} \underbrace{-a^n \sum_{k=n+1}^{N} h[k]a^{-k}}_{\text{transient}}. \end{aligned} \tag{7.160}$$

Thus, the first term shows the steady state, and the second term shows the transient.

7.12.4 Transient in Most General Causal Rational LTI System

We know that the most general rational LTI systems have the transfer function

$$H(z) = \frac{\sum_{k=0}^{L} e_k z^{-k}}{1 + \sum_{k=1}^{N} b_k z^{-k}}, \tag{7.161}$$

with numerator order L and denominator order N (assuming $e_L \neq 0$ and $b_N \neq 0$ without loss of generality). If $L \geq N$, we can use long division and rewrite

$$H(z) = \sum_{n=0}^{L-N} c[n]z^{-n} + \frac{\sum_{k=0}^{N-1} a_k z^{-k}}{1 + \sum_{k=1}^{N} b_k z^{-k}} = C(z) + \frac{A(z)}{B(z)}, \tag{7.162}$$

where the FIR part $C(z)$ has the unique inverse transform $c[n]$. So the impulse response has the FIR part $c[n]$, and an IIR part which contains terms of the form

$p_i^n \mathcal{U}[n]$, where the p_i are the poles of $H(z)$, that is, the poles of the second term in Eq. (7.162). Thus, the transient has an FIR part in addition to linear combinations of p_i^n. If some of the poles are not distinct, then we have slight variations of these terms as well. For example, if there is a double pole at p_1, then the transient has terms of the form p_1^n and np_1^n as well, because the impulse response has terms of the form

$$(\alpha_0 + \alpha_1 n)p_1^n \mathcal{U}[n]. \tag{7.163}$$

Similarly, if p_1 is a pole of multiplicity K, then the transient has terms of the form

$$p_1^n,\ np_1^n,\ \ldots,\ n^{K-1}p_1^n, \tag{7.164}$$

in addition to contributions from the other poles.

7.12.5 Transients in Most General Causal LTI System

How do we express the transients for the case of nonrational systems? Consider a causal IIR filter $H(z) = \sum_{n=0}^{\infty} h[n]z^{-n}$, which may not have the form of Eq. (7.161). Such systems cannot be characterized by a set of poles and zeros. Examples of discrete-time irrational systems are easy to generate, for example,

$$e^{z^{-1}} = 1 + z^{-1} + \frac{z^{-2}}{2!} + \frac{z^{-3}}{3!} + \cdots. \tag{7.165}$$

Also see Sec. 6.9.4. The right-hand side follows from the standard power-series expansion for e^w, which converges for all finite w in the complex w-plane (i.e., Eq. (7.165) converges for all $z \neq 0$). It can be shown that the impulse response of Eq. (7.165) is absolutely summable, so the system is stable (Problem 7.15). In this section we show that whether a system is rational or not, it is always possible to get an understanding of steady-state and transient components without bringing the notion of "poles" into the picture.[10] Thus, given the causal LTI system

$$H(z) = \sum_{n=0}^{\infty} h[n]z^{-n}, \tag{7.166}$$

let $x[n] = a^n \mathcal{U}[n]$. Then

$$\begin{aligned} y[n] = \sum_{k=0}^{n} h[k]x[n-k] = \sum_{k=0}^{n} h[k]a^{n-k} &= a^n \sum_{k=0}^{n} h[k]a^{-k} \\ &= a^n \sum_{k=0}^{\infty} h[k]a^{-k} - a^n \sum_{k=n+1}^{\infty} h[k]a^{-k} \\ &= \underbrace{H(a)a^n}_{\text{steady state}} \underbrace{-a^n \sum_{k=n+1}^{\infty} h[k]a^{-k}}_{\text{transient}}. \end{aligned}$$

[10] This discussion is only of theoretical interest because irrational systems cannot be implemented with a finite amount of computation and storage (Sec. 14.4.4).

For example, if $x[n] = e^{j\omega_0 n}\mathcal{U}[n]$, we have

$$y[n] = \underbrace{H(e^{j\omega_0})e^{j\omega_0 n}}_{\text{steady state}} \quad \underbrace{- e^{j\omega_0 n} \sum_{k=n+1}^{\infty} h[k]e^{-j\omega_0 k}}_{\text{transient}} \tag{7.167}$$

for $n \geq 0$. We now show that as long as the system is stable, the transient term will decay to zero. Recall that stability is equivalent to $\sum_{k=0}^{\infty} |h[k]| < \infty$. This means in particular that

$$\sum_{k=n+1}^{\infty} |h[k]| \;\to\; 0 \tag{7.168}$$

as $n \to \infty$, which implies in turn that $\sum_{k=n+1}^{\infty} h[k]e^{-j\omega_0 k} \;\to\; 0$ as $n \to \infty$. This is because

$$\left| \sum_{k=n+1}^{\infty} h[k]e^{-j\omega_0 k} \right| \leq \sum_{k=n+1}^{\infty} |h[k]| \;\to\; 0, \tag{7.169}$$

as $n \to \infty$. Thus, the transient term in Eq. (7.167) indeed decays to zero, and steady state eventually prevails, as long as the causal LTI system is stable.

So, *as far as the mathematics of transients go, this is all there is to it!* For the rational case, we expressed the transient component in terms of poles because it gave us additional practical insight. For example, we learned that the transients involve terms of the form p_i^n, np_i^n, and so on, which take longer to decay if the poles are closer to the unit circle.

7.13 Step Response of a Digital Filter

The output of a filter in response to the unit-step input $x[n] = \mathcal{U}[n]$ is called the step response. For a causal digital filter $H(z) = \sum_{n=0}^{\infty} h[n]z^{-n}$, the step response is

$$s[n] = \sum_{k=-\infty}^{\infty} x[k]h[n-k] = \sum_{k=0}^{n} h[n-k] \tag{7.170}$$

for $n \geq 0$ (and zero for $n < 0$). Clearly, the above can also be written as

$$s[n] = \sum_{k=0}^{n} h[k]. \tag{7.171}$$

This is nothing but the accumulated value of the impulse response from $k = 0$ to $k = n$. For stable filters, since $\sum |h[k]| < \infty$, this converges to a constant as n grows:

$$\lim_{n\to\infty} s[n] = \sum_{k=0}^{\infty} h[k] = H(e^{j0}). \tag{7.172}$$

Thus, the step response eventually settles down to a constant equal to the zero-frequency response. This is not surprising, because the input $x[n] = \mathcal{U}[n]$ is like the zero-frequency input, but starting at zero time. Let us consider some specific examples of stable filters. If $h[n] = cp^n\mathcal{U}[n]$ (first-order filter with pole at p) then

$$s[n] = c\sum_{k=0}^{n} p^k = c\left(\frac{1-p^{n+1}}{1-p}\right) \quad \rightarrow \quad \frac{c}{1-p} \text{ (large } n). \tag{7.173}$$

Similarly, for a second-order filter with distinct poles at p_1, p_2, we have $h[n] = (c_1p_1^n + c_2p_2^n)\mathcal{U}[n]$. Assuming, for example, $c_1 = c_2 = c$, we have

$$h[n] = c(p_1^n + p_2^n)\mathcal{U}[n], \tag{7.174}$$

so that

$$s[n] = c\left(\frac{1-p_1^{n+1}}{1-p_1} + \frac{1-p_2^{n+1}}{1-p_2}\right) \quad \rightarrow \quad \frac{c}{1-p_1} + \frac{c}{1-p_2} \text{ (large } n). \tag{7.175}$$

To plot $s[n]$ we assume the constant c is such that $H(e^{j0}) = 1$, so that the limits in Eqs. (7.173) and (7.175) are unity. For the one-pole case (7.173), the step response is shown in Figs. 7.37(a) and (b) for $p = 0.9$ and $p = 0.97$. We see that it takes much longer to reach the steady state $s[n] \approx 1$ when p is closer to unity. Next consider the two-pole case (7.175) with a complex-conjugate pair of poles at $p_1 = Re^{j\theta}$, $p_2 = Re^{-j\theta}$. For fixed pole angle $\theta = 0.03\pi$ we plot $s[n]$ for two values of the pole radius, $R = 0.9$ and $R = 0.97$. Once again, when the pole is closer to the unit circle, it takes much longer to reach the steady state $s[n] \approx 1$. Furthermore, unlike in the case of the real pole, the step response is **oscillatory** around the steady value of $s[n] = 1$. This is because the impulse response is proportional to

$$p_1^n + p_2^n = R^n\left(e^{j\theta n} + e^{-j\theta n}\right) = 2R^n\cos(\theta n), \tag{7.176}$$

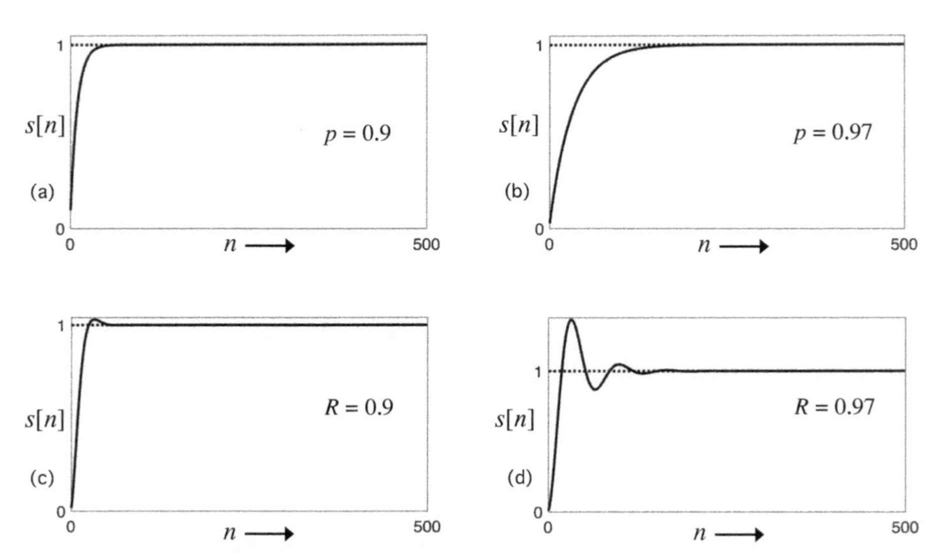

Figure 7.37 Step responses of causal stable filters. (a) and (b) First-order filters with two choices of pole p. (c) and (d) Second-order filters with complex-conjugate poles, with fixed pole angle and two choices of pole radius R.

which oscillates with frequency θ, and a damping factor R^n. As R gets closer to unity, the dampling factor R^n takes more time to decay. This explains the behavior of the plots.

7.14 An Amazing Application of Allpass Filters*

In Sec. 7.5 we showed how a notch filter and an antinotch filter can be designed based on a single allpass filter $G(z)$, as indicated in Eq. (7.43). In this section we will mention a generalization of this idea. This comes from an important result in filter theory, which says that if $H(z)$ is a digital lowpass Butterworth, Chebyshev, or elliptic filter of odd order N, then it can be expressed in the form

$$H(z) = \frac{A_0(z) + A_1(z)}{2}, \tag{7.177}$$

where $A_0(z)$ and $A_1(z)$ are allpass filters with real coefficients. Furthermore, the orders n_0 and n_1 of the filters $A_0(z)$ and $A_1(z)$ are such that

$$N = n_0 + n_1. \tag{7.178}$$

For example, the fifth-order IIR elliptic lowpass filter in Fig. 7.31 can be implemented as a sum of a third-order and a second-order allpass filter. A similar result is true if N is even, except that the allpass filters will have complex coefficients. These results will not be proved here because they are quite involved, and the reader can refer to Saramäki [1985] and Vaidyanathan et al. [1986, 1987] for further details. We will be content with explaining the importance of the results.

First note that $H(z)$ in Eq. (7.43) is a special case of Eq. (7.177) with $A_0(z) = 1$ and $A_1(z) = G(z)$. Inspired by the expression for the second filter $F(z)$ in Eq. (7.43), suppose we define another filter

$$F(z) = \frac{A_0(z) - A_1(z)}{2}. \tag{7.179}$$

Then, using the allpass property $|A_k(e^{j\omega})| = 1$, we can readily show (Problem 7.10) that

$$|H(e^{j\omega})|^2 + |F(e^{j\omega})|^2 = 1, \qquad \forall\omega, \tag{7.180}$$

that is, $H(e^{j\omega})$ and $F(e^{j\omega})$ are *power complementary*. For example, if $|H(e^{j\omega})| \approx 1$ at some frequency ω, then $|F(e^{j\omega})| \approx 0$, and vice versa. What this means is that if $H(z)$ is a good lowpass filter, then $F(z)$ is automatically a good highpass filter, as we shall demonstrate. In fact, it can be shown that if $H(z)$ is a lowpass Butterworth, Chebyshev, or elliptic filter, then $F(z)$ is a highpass Butterworth, Chebyshev, or elliptic filter, respectively.

To demonstrate, Fig. 7.38(a) shows the magnitude response $|H(e^{j\omega})|$ of a fifth-order elliptic filter. In this case, since $N = 5$, this $H(z)$ can be written in the form (7.177), where the allpass filters have orders 3 and 2, and have the form

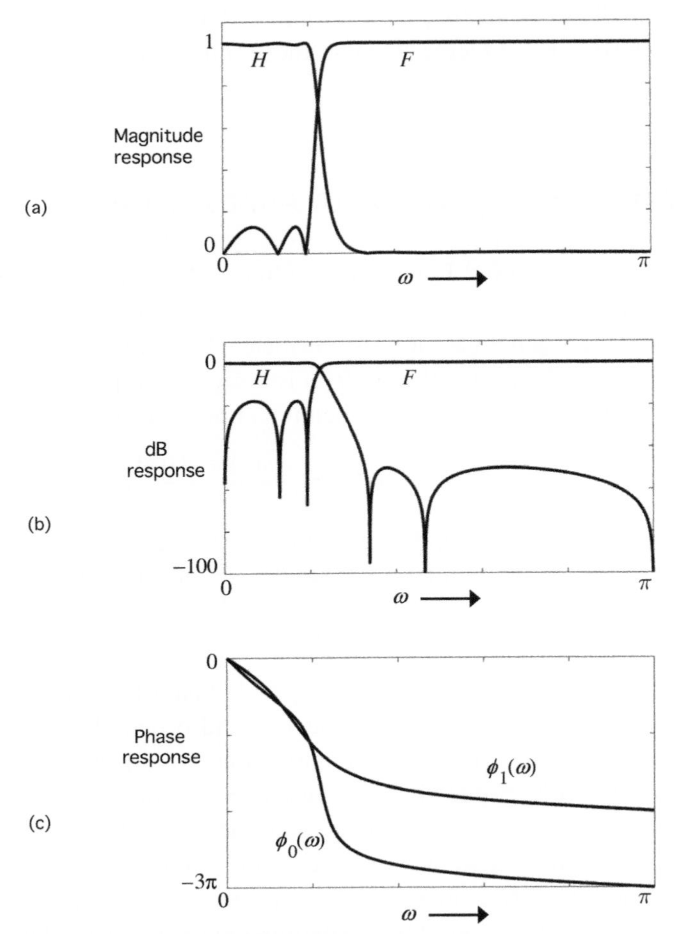

Figure 7.38 Responses of elliptic filters $H(z)$ and $F(z)$ designed from allpass filters $A_0(z)$ and $A_1(z)$ as in Eqs. (7.177) and (7.179). (a) Magnitude responses of $H(e^{j\omega})$ and $F(e^{j\omega})$, (b) dB responses of $H(e^{j\omega})$ and $F(e^{j\omega})$, and (c) phase responses of the allpass filters $A_0(z)$ and $A_1(z)$.

$$A_0(z) = \frac{a_{0,3} + a_{0,2}z^{-1} + a_{0,1}z^{-2} + z^{-3}}{1 + a_{0,1}z^{-1} + a_{0,2}z^{-2} + a_{0,3}z^{-3}},$$

$$A_1(z) = \frac{a_{1,2} + a_{1,1}z^{-1} + z^{-2}}{1 + a_{1,1}z^{-1} + a_{1,2}z^{-2}}. \tag{7.181}$$

If we define $F(z)$ as in Eq. (7.179) and plot $|F(e^{j\omega})|$, we indeed get a highpass elliptic filter, as shown in the same figure. Figure 7.38(b) shows the dB responses of these two filters for further clarity. In this example the coefficients of the allpass filters are

$$a_{0,1} = -2.0822,\ a_{0,2} = 1.7808,\ a_{0,3} = -0.5509,$$
$$a_{1,1} = -1.3293,\ a_{1,2} = 0.5702. \tag{7.182}$$

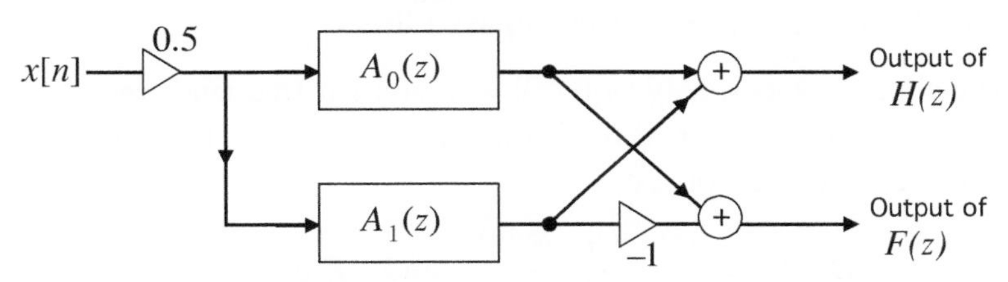

Figure 7.39 Efficient implementation of two elliptic filters $H(z)$ and $F(z)$ using parallel connections of two allpass filters $A_0(z)$ and $A_1(z)$.

The curious student should substitute these coefficient values into Eq. (7.181) and verify that the magnitude responses of $H(z)$ and $F(z)$ are indeed as shown in the figure (Problem 7.18). It is quite interesting that the addition and subtraction of two allpass filters would lead to such selective filters. A few remarks are now in order.

1. *High computational efficiency.* Note that $H(z)$ and $F(z)$ can be implemented together, as in Fig. 7.39. Recall from Sec. 7.4.2 that any real-coefficient mth-order allpass filter can be implemented using m real multipliers. (Also see Sec. 14.6.1 for other ways to achieve this.) Thus, $A_0(z)$ and $A_1(z)$ require only n_0 and n_1 real multipliers, respectively, so that the total number of multipliers in Fig. 7.39 is $n_0 + n_1 = N$. That is, we can implement *two* elliptic filters at the cost of only N multipliers where N is the order of $H(z)$ and $F(z)$. By contrast, if we use the direct-form structure (Fig. 5.7) to implement each of the filters $H(z)$ and $F(z)$ independently, that would take about $4N$ multipliers. Thus, being aware of the allpass decomposition property (7.177) for elliptic filters makes the implementation much more efficient!
2. *Lowpass–highpass pair.* Furthermore, in some applications it is required to have a lowpass–highpass pair of filters to operate simultaneously on an input signal. One example is in digital audio, where one wishes to design *graphic equalizers* by splitting the audio signal into different frequency bands. Another is in *subband coding* of signals for data compression applications. For early work, see the references in Vaidyanathan [1993]. Also see the remarks under "More real-world applications: audio and image signals" in Sec. 4.6.3.
3. *Robustness to quantization.* There is yet another advantage to the implementation given in Fig. 7.39. In any digital implementation, the filter coefficients, and the internal signals in the computational graph, have to be quantized to a finite number of bits (say 8 bits, 16 bits, and so forth). Such quantization affects the filter response $H(e^{j\omega})$, increasing the peak ripple sizes in the passband and stopband. There are a number of other negative effects due to quantization, such as *roundoff noise* (see Sec. 8.5.1) and spurious oscillations called *limit cycles* [Oppenheim and Schafer, 2010]. If the filter is implemented as a sum of two allpass filters, and if each allpass filter is implemented using one of the allpass lattice structures described in Secs. 14.3.2 and 14.6, then these damaging effects due to quantization can be decreased to a significant extent [Gray, 1980]. The details are quite involved, and we do not elaborate on them here.

The Phase Responses of the Allpass Filters

The allpass filters satisfy $|A_i(e^{j\omega})| = 1$, so their frequency responses have the form $A_i(e^{j\omega}) = e^{j\phi_i(\omega)}$, where $\phi_i(\omega)$ is the phase response. Thus, the filters $H(z)$ and $F(z)$ have frequency responses

$$H(e^{j\omega}) = \frac{e^{j\phi_0(\omega)} + e^{j\phi_1(\omega)}}{2} \quad \text{and} \quad F(e^{j\omega}) = \frac{e^{j\phi_0(\omega)} - e^{j\phi_1(\omega)}}{2}. \tag{7.183}$$

So we expect that in the passband of $H(e^{j\omega})$ where $|H(e^{j\omega})| \approx 1$, the phases satisfy $\phi_0(\omega) \approx \phi_1(\omega)$ and in the stopband where $|H(e^{j\omega})| \approx 0$, the phases differ by π. This is indeed the case, as seen from the plots of the phase responses in Fig. 7.38(c).

The careful reader will notice that the phase response of each allpass filter is monotonically decreasing. This is not an accident. The poles of the two allpass filters $A_i(z)$ come from the poles of $H(z)$, which satisfies $|p_i| < 1$. It can be shown that if an allpass filter has all its poles satisfying $|p_i| < 1$, then its phase response is indeed monotonically decreasing in $0 \leq \omega < 2\pi$, and spans a range of $2\pi m$ where m is the order of the allpass filter. A proof can be found on page 76 of Vaidyanathan [1993]. For real coefficient allpass filters this means that the phase response is monotonically decreasing in the range $0 \leq \omega < \pi$, and spans a range of πm. That is why $\phi_0(\omega)$ spans a range of 3π and $\phi_1(\omega)$ spans a range of 2π in Fig. 7.38(c).

7.15 FIR Spectral Factors*

In Sec. 7.8.2 we showed how to find Butterworth transfer functions by finding an appropriate spectral factor of a function. In this section we explain how to find spectral factors in the discrete-time case. We will focus on the FIR case. The discussion generalizes readily to IIR rational transfer functions. Consider an FIR filter $F(z)$ such that

$$F(e^{j\omega}) \geq 0.$$

Thus the frequency response is non-negative everywhere. It is clear that $F(z)$ has linear phase, in fact zero phase. Now, since $F(e^{j\omega}) \geq 0$ we can always write

$$F(e^{j\omega}) = F_1(e^{j\omega})F_1^*(e^{j\omega}), \tag{7.184}$$

for some $F_1(e^{j\omega})$. This $F_1(e^{j\omega})$ is a spectral factor, and is of course not unique. More precisely, its magnitude is unique:

$$|F_1(e^{j\omega})| = \sqrt{F(e^{j\omega})}, \tag{7.185}$$

but its phase is not unique. In fact, if we write

$$F_1(e^{j\omega}) = |F_1(e^{j\omega})|e^{j\phi(\omega)}, \tag{7.186}$$

then Eq. (7.184) is satisfied for any phase response $\phi(\omega)$, so the phase of the spectral factor is completely arbitrary.

7.15.1 Spectral Factors May Not be FIR

Now, even when $F(z)$ is FIR, a spectral factor with arbitrary phase response may not even be FIR (it may even be an irrational function of $e^{j\omega}$). For example, suppose we choose $\phi(\omega) = 0$ so that the spectral factor itself is non-negative (i.e., in particular it has linear phase). Then

$$F_1(e^{j\omega}) = \sqrt{F(e^{j\omega})}. \tag{7.187}$$

To be more specific with an example, let

$$F(z) = 4 + z + z^{-1}, \tag{7.188}$$

so that $F(e^{j\omega}) = 4 + 2\cos\omega > 0$. Then

$$F_1(e^{j\omega}) = \sqrt{4 + 2\cos\omega}. \tag{7.189}$$

There is no FIR $F_1(z)$ such that $F_1(e^{j\omega})$ has the above value (as we shall justify at the end of this section).

7.15.2 FIR Spectral Factors

We now show, by construction, that there exist FIR spectral factors for FIR $F(z)$ when it satisfies $F(e^{j\omega}) \geq 0$. For this, recall that $F(e^{j\omega}) \geq 0$ implies $f[n] = f^*[-n]$ (Sec. 3.13), so $F(z)$ has the form

$$F(z) = \sum_{n=-N}^{N} f[n]z^{-n}. \tag{7.190}$$

Furthermore, its zeros come in reciprocal conjugate pairs $\{z_k, 1/z_k^*\}$. That is, if z_k is a zero then so is $1/z_k^*$ (see Problem 4.17). Moreover, if z_k happens to be on the unit circle, that is, $z_k = e^{j\omega_k}$, then it has to have even multiplicity. Otherwise $F(e^{j\omega})$ will change sign at ω_k, contradicting $F(e^{j\omega}) \geq 0$. Thus, we can always write

$$F(z) = |c|^2 \prod_{k=1}^{N} (1 - z_k z^{-1}) \prod_{k=1}^{N} (1 - z_k^* z),$$

where c is some constant. If we now define

$$F_1(z) = c \prod_{k=1}^{N} (1 - z_k z^{-1}),$$

then the above can be written as

$$F(z) = F_1(z)[F_1(1/z^*)]^*, \tag{7.191}$$

that is,

$$F(z) = F_1(z)\widetilde{F}_1(z), \tag{7.192}$$

where $\widetilde{F}_1(z) = [F_1(1/z^*)]^*$ (Sec. 1.3). Since $[F_1(1/z^*)]^* = [F_1(z)]^*$ on the unit circle, it follows that both $F_1(z)$ and $\widetilde{F}_1(z)$ are FIR spectral factors of $F(z)$, and Eq. (7.191) is called the spectral factorization of $F(z)$.

7.15.3 Number of FIR Spectral Factors

How many FIR spectral factors can there be? Since $F(z)$ is as in Eq. (7.190), it has N pairs of zeros

$$\{z_k,\ 1/z_k^*\}. \tag{7.193}$$

We can assign either z_k or $1/z_k^*$ to construct $F_1(z)$. Since z_k may not be distinct, there are at most 2^N FIR spectral factors. Also, if $z_k = 1/z_k^*$ for some of the k (i.e., z_k is on the unit circle), then the number of distinct FIR spectral factors will be smaller than 2^N. For the special case where $F(z)$ has *real coefficients* $f[n]$, the zeros also occur in conjugate pairs $\{z_k, z_k^*\}$, so that the zeros have the pattern

$$\{z_k,\ z_k^*,\ 1/z_k,\ 1/z_k^*\}. \tag{7.194}$$

In this case we can choose the spectral factor such that it has real coefficients (by choosing both z_k and z_k^* to be in $F_1(z)$), so there always exists a real-coefficient spectral factor. If we don't choose zeros like this, then the spectral factor can turn out to have nonreal coefficients.

7.15.4 Summary and Further Remarks

1. Given an FIR $F(z)$ with $F(e^{j\omega}) \geq 0$, it can be written in the form (7.190) for some integer $N \geq 0$, and there exists an infinite number of filters $F_1(e^{j\omega})$ such that $F(e^{j\omega}) = F_1(e^{j\omega})F_1^*(e^{j\omega})$. Each such $F_1(e^{j\omega})$ is called a spectral factor. There always exists a linear-phase spectral factor as long as it is allowed to be nonrational – just take $F_1(e^{j\omega}) = \sqrt{F(e^{j\omega})}$.
2. If the spectral factor itself is restricted to be FIR, then there are at most 2^N spectral factors, where N is the quantity in Eq. (7.190). In this case we can write $F(z) = F_1(z)[F_1(1/z^*)]^*$. (One can generate an arbitrary number of spectral factors in a degenerate way, because $e^{j\theta}z^{-K}F_1(z)$ is also a spectral factor for arbitrary integer K and arbitrary real θ. These are not counted as distinct spectral factors.)
3. The unique FIR spectral factor with all its zeros in $|z| \leq 1$ is called the *minimum-phase* spectral factor. The unique FIR spectral factor with all its zeros in $|z| \geq 1$ is called the *maximum-phase* spectral factor.
4. If an FIR $F(z)$ satisfying $F(e^{j\omega}) \geq 0$ also has real coefficients (i.e., real impulse response $f[n]$), then there exists an FIR spectral factor $F_1(z)$ with real coefficients $f_1[n]$. Furthermore, the minimum-phase and maximum-phase factors can be guaranteed to have real coefficients.
5. There may not exist a linear-phase FIR spectral factor. For example, consider Eq. (7.189), for which $F(z) = 4 + z + z^{-1}$. This does not have a linear-phase FIR spectral factor. *Proof.* $F(z)$ has two real zeros α and $1/\alpha$ with $0 < |\alpha| < 1$. So the spectral factors are $c(1-\alpha z^{-1})$ and $c^*(1-\alpha z)$ for appropriate c. Neither of these has linear phase because $|\alpha| \neq 1$, and the linear-phase symmetry is not satisfied.
6. Finally, consider the example $F(z) = z^2 + 8z + 18 + 8z^{-1} + z^{-2}$. This satisfies $F(e^{j\omega}) \geq 0$ and does indeed have a linear-phase FIR spectral factor. This is

because $z^2+8z+18+8z^{-1}+z^{-2} = (1+4z^{-1}+z^{-2})(1+4z+z^2)$. So, $(1+4z^{-1}+z^{-2})$ is a linear-phase FIR spectral factor.

7.16 Summary of Chapter 7

This chapter presented a number of methods and examples of digital filter design. We started off with simple examples such as narrowband bandpass filters based on second-order IIR transfer functions. Adjustable notch filters were then presented. Then, allpass filters were introduced and some of their properties developed. It was shown how allpass filters can be used to design a pair of notch and antinotch filters very efficiently. This pair satisfies the power complementary property, which has some advantages.

Then we introduced a systematic method for the design of lowpass filters that satisfy design specifications required by the designer, such as ripple sizes δ_1, δ_2 and band edges ω_p, ω_s. This was based on transforming continuous-time filters such as Butterworth, Chebyshev, and elliptic filters into discrete-time filters, by using a mapping called the bilinear transform. This technique yields stable IIR digital filters. The mathematical properties of the bilinear transform were studied in detail. The Butterworh filter was also studied extensively, including its maximally flat property, its pole locations, and so on.

We then presented a simple but powerful method for the design of linear-phase FIR filters, called the window design method. Examples were presented such as the triangular or Bartlett window, the Hamming window, and the Kaiser window. Lowpass filters based on these windows were presented and compared. We also compared FIR and IIR filter designs and demonstrated that IIR designs require much lower order for the same set of filter specifications. Our example showed that a fifth-order IIR elliptic filter can meet the same specifications that a 47th-order FIR filter satisfies. However, causal IIR filters cannot have linear phase, which can readily be achieved by FIR filters. Nonlinear phase creates phase distortion, which can be avoided by using FIR linear-phase filters. We demonstrated the effect of phase distortion in image filtering. Phase distortion may not be noticeable to the human ear but is quite visible to the eye. So linear-phase filters are important in image processing.

A detailed discussion of transient responses in IIR filters was then given. The concept of frequency response is meaningful only in the steady state and not in the transient state. Unfortunately, transients linger for a long time when poles are close to the unit circle. Since sharp cutoff filters tend to have poles close to the unit circle, they have longer transient duration, which is a price paid for the sharpness of the frequency response. The presence of transients in FIR filters was also analyzed.

Towards the end of the chapter it was explained that IIR digital filters such as Butterworth, Chebyshev, and elliptic filters can be implemented as sums of two allpass filters. It was explained that this leads to great computational effciency. For example, a fifth-order elliptic lowpass filter and highpass filter can both be implemented using a total of only five multipliers. Other advantages of allpass-based implementations, such as robustness to quantization, were also pointed out.

The literature on digital filtering is enormous. We have only given an introductory exposure in this chapter. The idea that filtering can be achieved by computation on a digital machine was one of the most important motivations for the development of digital signal processing in the middle 1960s. Since then, digital filters have impacted a vast area of engineering. Digital filtering was addressed in one of the first books on signal processing by Gold and Rader [1969]. A pioneering signal processing book which included detailed discussions on digital filter design was published by Rabiner and Gold [1975]. The landmark book by Oppenheim and Schafer [1975] also addressed digital filter design. Today, the textbook by the same authors [Oppenheim and Schafer, 2010] continues to serve as a standard reference. The early book by Hamming [1989] provides delicious reading, quickly introducing the ideas to readers with limited signal processing background. Many other books have also dealt with the topic at different levels of detail [Antoniou, 1993, 2006; Diniz et al., 2010; Mitra, 2011; Proakis and Manolakis, 2007]. References for allpass filters include Constantinides [1970], Mitra and Hirano [1974], Regalia et al. [1988], Vaidyanathan [1993] and Vaidyanathan et al. [1986, 1987].

PROBLEMS

Note: If you are using MATLAB for the computing assignments at the end, the "stem" command is convenient for plotting sequences. The "plot" command is useful for plotting real-valued functions such as $|X(e^{j\omega})|$. If you type "help window" you will learn about commands that design various windows for filter design.

7.1 Show that the structure of Fig. 7.3 can be rearranged so that there are only two multipliers instead of three. (Assume $\pm 2^K$ is not counted as a multiplier for integer K.)

7.2 Consider the causal allpass filter $H(z) = (b^* + z^{-1})/(1 + bz^{-1})$, where b^* denotes the complex conjugate of b. Derive an expression for the impulse response $h[n]$. Show that it can be expressed as

$$h[n] = Ap^n \mathcal{U}[n] + B\delta[n],$$

where p is the pole. What are the values of the constants A and B?

7.3 Let $H(z) = (b_2 + b_1 z^{-1} + z^{-2})/(1 + b_1 z^{-1} + b_2 z^{-2})$, where the b_i are real.

(a) Compute the coefficients of the numerator and denominator of the product $H(z)H(1/z)$ and show indeed that $H(z)H(1/z) = 1$ is satisfied for all z.

(b) Using the fact that the b_k are real, show that "$H(z)H(1/z) = 1$ for all z" implies in particular that $|H(e^{j\omega})| = 1$ for all ω, that is, $H(z)$ is indeed allpass.

7.4 Let $H(z) = (b_2^* + b_1^* z^{-1} + z^{-2})/(1 + b_1 z^{-1} + b_2 z^{-2})$, where the b_i are possibly complex.

(a) Compute the coefficients of the numerator and denominator of the product $H(z)\widetilde{H}(z)$ by explicit multiplication, and show indeed that $H(z)\widetilde{H}(z) = 1$ for all z.

(b) Show that this implies in particular that $|H(e^{j\omega})| = 1$ for all ω, that is, $H(z)$ is indeed allpass.

Note: The tilde notation $\widetilde{H}(z)$ is defined in Sec. 7.4.1 and more elaborately in Sec. 1.3.

7.5 Let $H(z) = (-p^* + z^{-1})/(1 - pz^{-1})$, which is allpass. Assume the pole is inside the unit circle, so $p = Re^{j\theta}$, where $0 < R < 1$. Write $z = \rho e^{j\omega}, \rho \geq 0$, so that z is outside the unit circle if and only if $\rho > 1$. With these notations, show that

$$|H(z)|^2 = \frac{1 + R^2\rho^2 - 2R\rho\cos(\omega - \theta)}{R^2 + \rho^2 - 2R\rho\cos(\omega - \theta)}.$$

Hence, show that

$$|H(z)| \begin{cases} > 1 & \text{if } \rho < 1, \\ < 1 & \text{if } \rho > 1, \\ = 1 & \text{if } \rho = 1. \end{cases}$$

This proves Eq. (7.27) for first-order allpass filters with pole magnitude $|p| < 1$.

7.6 Let $H_c(s) = (1 + 2s)/(3 + 4s)$ be a continuous-time transfer function, and let $H(z)$ be the bilinearly transformed version. Find $H(z)$. Where are its poles and zeros?

7.7 Let $G(z) = 1 + 4z^{-1} + z^{-2}$ be an FIR filter. This is an all-zero filter (i.e., there are no nontrivial poles; all poles are at $z = 0$). Now let $G_c(s)$ be a continuous-time filter such that if it is bilinearly transformed, we will get the digital FIR filter $G(z)$. Find $G_c(s)$ and its zeros and poles.

7.8 *Mappings based on first differences.* Suppose we wish to transform an analog filter $H_c(s)$ into a digital filter $H(z)$. Assume that the following transformation is used: $s = 1 - z^{-1}$. This is called the backward difference approach. (The motivation for this substitution is that s represents differentiation and $1 - z^{-1}$ represents a first difference.)

(a) Suppose $H_c(s)$ has all poles in $\text{Re}[s] < 0$. Does it necessarily mean that $H(z)$ has all poles inside the unit circle?

(b) Suppose $H(z)$ has all poles inside the unit circle. Does it mean that $H_c(s)$ has all poles in $\text{Re}[s] < 0$?

(c) Instead of the above mapping, assume that we use the mapping $s = z - 1$ (forward difference approach). Repeat parts (a) and (b).

After doing this problem you should be able to see why the bilinear transformation is more popular!

7.9 In this problem we find the coefficients of a second-order digital lowpass Butterworth filter $H(z)$, with 3 dB point at $\omega_c = 0.2\pi$.

(a) Using the bilinear transform, find the 3 dB point Ω_c of a continuous-time Butterworth filter $H_c(s)$.
(b) For a second-order Butterworth filter $H_c(s)$ with the above Ω_c, find the locations of the poles s_0 and s_1 of the filter using the closed-form expression derived in this chapter.
(c) Compute the coefficients of the all-pole Butterworth transfer function $H_c(s) = K/(s - s_0)(s - s_1)$, where K is chosen such that $H_c(0) = 1$.
(d) Now find the coefficients of the digital Butterworth filter $H(z)$ using bilinear transformation.

7.10 Let $H(z)$ and $F(z)$ be filters defined as in Eqs. (7.177) and (7.179), where $A_0(z)$ and $A_1(z)$ are stable allpass with $|A_i(e^{j\omega})| = 1$ for all ω. Then prove that the power complementary property (7.180) is satisfied.

7.11 *Allpass transformation.* We know that a rational allpass filter with all poles p_k such that $|p_k| < 1$ satisfies the modulus property (7.27). Using this, prove the following: if a causal rational filter $H(z)$ is stable allpass and we replace all appearances of z^{-1} in $H(z)$ with a causal rational stable allpass filter $F(z)$:

$$G(z) = H(z)\Big|_{z^{-1}=F(z)}, \tag{P7.11}$$

then the resulting causal filter $G(z)$ is (a) stable and (b) allpass. Converting a transfer function $H(z)$ into another by using an allpass filter as above is called *allpass transformation* [Constantinides, 1970]. It has been used in the past to perform frequency transformations that are more general than what we studied in Sec. 4.4

7.12 *Minimum-phase filters and energy concentration.* An FIR filter is said to have minimum phase if all its zeros z_k satisfy $|z_k| \leq 1$, that is, they are on or inside the unit circle. In this problem we prove a property called the energy concentration property for minimum-phase filters.
(a) Let $F_1(z) = \sum_{n=0}^{N} f_1[n]z^{-n}$ and $F_2(z) = \sum_{n=0}^{N} f_2[n]z^{-n}$ be two filters such that

$$F_1(z) = \Big(f[0] + f[1]z^{-1} + \cdots + f[N-1]z^{-(N-1)}\Big)(1 - \alpha z^{-1}),$$
$$F_2(z) = \Big(f[0] + f[1]z^{-1} + \cdots + f[N-1]z^{-(N-1)}\Big)(\alpha^* - z^{-1}),$$

where $|\alpha| > 1$. That is, $F_2(z)$ is obtained from $F_1(z)$ by replacing a zero α outside the unit circle with a zero $1/\alpha^*$ inside the unit circle. Show that $|F_1(e^{j\omega})| = |F_2(e^{j\omega})|$, so the filters are two different spectral factors of the same magnitude-squared function.
(b) With $F_1(z)$ and $F_2(z)$ as above, show that

$$\sum_{k=0}^{n} |f_2[k]|^2 \geq \sum_{k=0}^{n} |f_1[k]|^2,$$

for *any* n such that $0 \leq n \leq N$.

(c) Now consider two filters

$$H_m(z) = \sum_{n=0}^{N} h_m[n]z^{-n}, \quad H(z) = \sum_{n=0}^{N} h[n]z^{-n},$$

such that $|H_m(e^{j\omega})|^2 = |H(e^{j\omega})|^2 = P(e^{j\omega})$. So $H_m(z)$ and $H(z)$ are two spectral factors of a filter $P(z)$ satisfying $P(e^{j\omega}) \geq 0$. Assume that $H_m(z)$ has all zeros satisfying $|z_k| \leq 1$ (i.e., $H_m(z)$ has "minimum phase"). Show then that

$$\sum_{k=0}^{n} |h_m[k]|^2 \geq \sum_{k=0}^{n} |h[k]|^2,$$

for *any* n such that $0 \leq n \leq N$.

Thus, minimum-phase filters concentrate the energy near zero time. Linear-phase FIR filters, which are useful to avoid phase distortion, do not have this property because their zeros come in pairs like $(z_k, 1/z_k^*)$ (Sec. 4.6). This can also be qualitatively understood from the observation that linear-phase lowpass filters usually have large coefficients in the middle and small coefficients at the ends (Fig. 4.21).

7.13 *Reactances and allpass filters.* In the continuous-time world a rational transfer function $G(s)$ of degree $N > 0$ with real coefficients is said to be a reactance if it satisfies the following two properties: (a) $\text{Re}[G(j\Omega)] = 0$ for all frequencies Ω and (b) $\text{Re}[G(s)] > 0$ for all s in the right-half plane, that is, for all s such that $\text{Re}[s] > 0$. Now define a discrete-time transfer function $H(z)$ as follows:

$$H(z) = \left. \frac{1 - G(s)}{1 + G(s)} \right|_{s=(1-z^{-1})/(1+z^{-1})}.$$

Show then that $H(z)$ represents an allpass function with all poles strictly inside the unit circle. (*Note:* It turns out that $G(s)$ is a reactance if and only if it is the input impedance of a lossless electrical network (LC network with positive elements). This establishes the link between digital allpass functions and continuous-time LC network impedances.)

7.14 Let $H(z)$ be a causal stable rational transfer function such that $|H(e^{j\omega})| < 1$ for all ω. Consider the feedback loop shown in Fig. P7.14, where $H(z)$ is such that no delay-free loops are created. Let $F(z) = Y(z)/X(z)$ be the transfer function of the closed-loop system.

(a) Which of the following statements is true? Justify. (i) $F(z)$ is necessarily stable. (ii) $F(z)$ is necessarily unstable. (iii) $F(z)$ could be stable or unstable, depending on what exactly $H(z)$ is.

(b) Repeat for the case where $H(z)$ is as above, except that it only satisfies $|H(e^{j\omega})| \leq 1$ for all ω.

7.15 Show that the impulse response of Eq. (7.165) is absolutely summable, so that the system is stable.

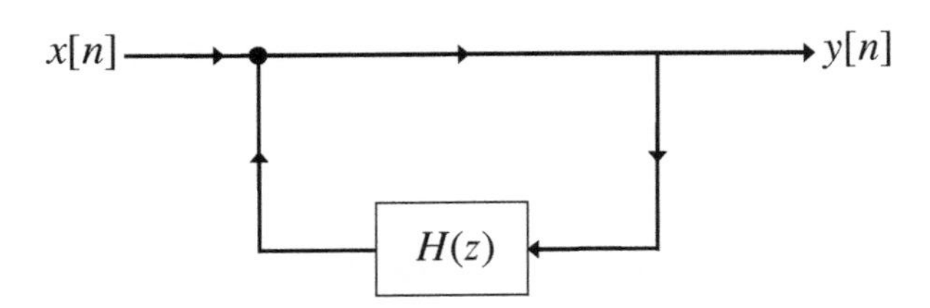

Figure P7.14 A feedback system.

7.16 For the causal LTI system with transfer function

$$H(z) = 1 + 2z^{-1} + 3z^{-2} + \frac{5}{1 - \frac{1}{2}z^{-1}},$$

suppose the input is $x[n] = e^{j\omega_0 n}\mathcal{U}[n]$. Find expressions for the transient and steady-state terms in the output.

7.17 (Computing assignment) *Frequency transformations.* For the digital Butterworth filter $H(z)$ designed in Problem 7.9, plot the magnitude response $|H(e^{j\omega})|$. Also plot the magnitude responses of the frequency-transformed versions $H(-z), H(z^2)$, and $H(-z^2)$.

7.18 (Computing assignment) Substitute the coefficient values (7.182) into Eq. (7.181) and verify that the magnitude responses of $H(z)$ and $F(z)$ in Eqs. (7.177) and (7.179) are indeed as shown in Fig. 7.38(a).

7.19 (Computing assignment) *IIR digital Butterworth filters.* In this problem you will plot the responses of some IIR digital Butterworth filters. This is just to demonstrate that you can get great approximations of ideal filters simply by designing a ratio of two polynomials in $e^{-j\omega}$. We will design digital Butterworth filters with order N, and passband cutoff ω_c (3 dB point) equal to $\alpha\pi$. So $\alpha = 0.3$ corresponds to $\omega_c = 0.3\pi$. We will plot the magnitude of $H(e^{j\omega})$ at M points in the range $0 \leq \omega \leq \pi$.

(a) *Instructions if you use MATLAB.* The command $[B, A] = \mathit{butter}(N, \alpha)$ designs digital Butterworth lowpass filters with order N, and passband cutoff $\alpha\pi$. Here, B and A are the coefficients of the numerator and denominator of the transfer function $H(z) = B(z)/A(z)$. The MATLAB command $[H, W] = \mathit{freqz}(B, A, M)$ computes $H(e^{j\omega})$ at M points of frequency in the range $0 \leq \omega \leq \pi$.

(b) In this problem we take $\alpha = 0.3$ and $M = 2^9$. Plot $|H(e^{j\omega})|$ (magnitude plot and dB plot) in the range $0 \leq \omega \leq \pi$ for the following values of filter order: $N = 11$, $N = 21$, and $N = 31$. Notice how the filter response gets sharper and sharper as the order increases.

7.20 (Computing assignment) *FIR frequency response.* Consider an FIR digital filter with impulse response $h[n]$ of length 11, with $h[0]$ through $h[10]$ given below:

$$-0.0872, 0.0370, 0.0984, 0.1755, 0.2388, 0.2632,$$
$$0.2388, 0.1755, 0.0984, 0.0370, -0.0872.$$

(a) Compute and plot the magnitude response of $H(e^{j\omega})$ in the range $0 \leq \omega \leq \pi$. You will find that this is a lowpass filter. Thus a simple polynomial of degree 9 can produce a beautiful lowpass filter if the coefficients are chosen appropriately.

(b) Also plot the magnitude in dB. When $H(e^{j\omega}) = 0$ at some point in the stopband this corresponds to $-\infty$ dB. If the computer complains, just add a tiny value like 10^{-10} to $|H(e^{j\omega})|$ before plotting the dB response.

(c) Also plot the magnitude responses of the frequency-transformed versions $H(-z), H(z^2)$, and $H(-z^2)$.

7.21 (Computing assignment) *IIR frequency response*. Compute and plot the magnitude of the frequency response $H(e^{j\omega})$ in the range $0 \leq \omega \leq \pi$ for the LTI system described by the linear constant coefficient difference equation given below:

$$\begin{aligned} y[n] = {} & 4.401595y[n-1] - 7.883242y[n-2] \\ & +7.1701575y[n-3] - 3.308688y[n-4] + .619276y[n-5] \\ & +0.001502x[n] - 0.002742x[n-1] + 0.001691x[n-2] \\ & +0.001691x[n-3] - 0.002742x[n-4] + 0.001502x[n-5]. \end{aligned}$$

In addition to the plot of $|H(e^{j\omega})|$, please also include a dB plot for the above example. If you are using MATLAB, the commands "freqz" and "abs" are useful for this. You will find that the above system is an amazingly good lowpass filter.

7.22 (Computing assignment) *Transients*. Consider the causal LTI system

$$H(z) = \frac{2 - (5/6)z^{-1}}{1 - (5/6)z^{-1} + (1/6)z^{-2}} .$$

Suppose we apply the one-sided "single frequency" signal $x[n] = e^{j\omega_0 n}\mathcal{U}[n]$ as the input to this system (with zero initial conditions). Then we know that the output has the form

$$y[n] = Ae^{j\omega_0 n} + y_{tr}[n], \quad n \geq 0.$$

The first term is the steady-state term and the second term is the transient (whose shape depends on the poles).

(a) Find the value of the constant A when $\omega_0 = 0.1\pi$.

(b) Find a closed-form expression for the transient $y_{tr}[n]$. Find its magnitudes for $n = 3$ and $n = 10$ when $\omega_0 = 0.1\pi$.

7.23 (Computing assignment) *Notch filters in action*. Consider again the examples of the notch filters designed in Sec. 7.3, with various plots shown in Fig. 7.8. For the same input signal, and with all other parameters remaining unchanged, repeat this experiment for two new values of R, namely $R = 0.7$ and $R = 0.995$. Plot the filter magnitude responses $|H(e^{j\omega})|$ and the filter outputs, similar to Fig. 7.8. Provide an explanation as to why the outputs behave like this for these values of R.

7.24 (Computing assignment) *Step responses.* In Sec. 7.13 we discussed step responses and plotted examples of the same for second-order digital filters in Figs. 7.37(c) and (d). With all quantities as in that example, just change the pole radius to $R = 0.7$ and plot the step response. Plot it also for $R = 0.995$. Provide an explanation as to why the step responses behave like this for these values of R.

7.25 (Computing assignment) *Trick for narrowband filters.* In this problem $H(z)$ refers to the filter in Problem 7.20. In that problem you would have noticed that $H(z^2)$ has two passbands, one at $\omega = 0$ and one at $\omega = \pi$. Furthermore, since $H(e^{j2\omega})$ is the squeezed version of $H(e^{j\omega})$, each passband of $H(z^2)$ is twice as narrow as the passband of $H(z)$. Now define the product filter

$$F_2(z) = H(z)H(z^2).$$

This is nothing but the cascade of $H(z)$ with $H(z^2)$. Since $H(z)$ has only one passband (which is at $\omega = 0$), the second passband of $H(z^2)$ gets attenuated by $H(z)$. So the product $F_2(z) = H(z)H(z^2)$ has only one passband, and it is a lowpass filter again. But the difference is that this passband is narrower than that of $H(z)$, because the passbands of $H(z^2)$ are twice as narrow as the passband of $H(z)$. In short, the simple trick of cascading $H(z)$ with $H(z^2)$ can produce a sharper lowpass filter.

(a) Plot the magnitude responses of $H(z)$, $H(z^2)$, and $F_2(z) = H(z)H(z^2)$.

(b) We can push the cascading trick further. Plot the magnitude responses of $H(z^3)$ and $F_3(z) = H(z)H(z^3)$. Notice that $F_3(z)$ is an even sharper lowpass filter than $F_2(z)$.

(c) Repeat for $H(z^4)$ and $F_4(z) = H(z)H(z^4)$.

This provides a simple way to design very narrowband lowpass filters starting from wider-band filters. If L is the length of $H(z)$ (assumed FIR), then $F_M(z) = H(z)H(z^M)$ has length $(L-1)M + L$, but this does not mean it takes that many multipliers to implement it. The impulse response of $G(z) \stackrel{\Delta}{=} H(z^M)$ is nothing but $h[n]$ with $M - 1$ zeros inserted between samples:

$$g[n] = \begin{cases} h[n/M] & \text{if } n \text{ is a multiple of} M, \\ 0 & \text{otherwise.} \end{cases}$$

Thus, with $M = 2$, we get Eq. (4.55). Because of the above, $H(z^M)$ still requires only L multipliers, so $F_M(z) = H(z)H(z^M)$ requires only $2L$ multipliers, regardless of what M is! In short, the product $F_M(z) = H(z)H(z^M)$ offers a very efficient way to implement narrowband filters. This is called the interpolated FIR, or **IFIR** method. This idea is described in a number of references, and the pioneering reference is Neuvo et al. [1984]. Notice, however, that we cannot increase M arbitrarily because if M is too large, more than one passband of $H(z^M)$ will fall within the passband of $H(z)$.

7.26 (Computing assignment) *Window design of FIR filters.* In this problem we design window-based lowpass filters with filter length $L = 51$ and cutoff $\omega_c = 0.3\pi$. Consider the following three designs: (a) rectangular window based, (b) Hamming window based, and (c) Blackman window based, where the **Blackman window** is defined as

$$w[n] = \begin{cases} 0.42 + 0.5\cos\left(\frac{2\pi}{N}n\right) + 0.08\cos\left(\frac{4\pi}{N}n\right) & |n| \leq M, \\ 0 & \text{otherwise.} \end{cases}$$

(Recall our notation in Sec. 7.9.3: filter length $L = 2M + 1$ and filter order $N = 2M$. Also, in some books you will see $v[n] = w[n - M]$ listed, which is nonzero in $0 \leq n \leq N$. Then the second term becomes $0.5\cos(2\pi(n-M)/N) = -0.5\cos(2\pi n/N)$, so it acquires a minus sign.) For each of these, plot $|W(e^{j\omega})|$ and $|H(e^{j\omega})|$ in dB. For ease of comparison, normalize the plots so that the value at zero frequency is unity (0 dB). What are the approximate attenuations of these filters at the largest side lobe in the stopband? Which of the three designs has the largest transition bandwidth?

7.27 (Computing assignment) *Kaiser window-based FIR filters.* In this problem we design Kaiser window lowpass filters with filter length $L = 51$ and cutoff $\omega_c = 0.3\pi$. Design and plot $|W(e^{j\omega})|$ and $|H(e^{j\omega})|$ in dB, for $\beta = 4$ and $\beta = 8$.

8 Difference Equations with Nonzero IC

8.1 Introduction

This chapter introduces recursive difference equations (d.e.) where the initial conditions are nonzero. The output $y[n]$ of such a system is studied in detail. The input $x[n]$ itself may or may not be zero, and applications of both types are demonstrated. One application is in the design of digital waveform generators such as oscillators. Another application is in the computation of the monthly payment on a loan. These are included to demonstrate that the ground covered by recursive difference equations is quite broad.

The chapter begins with homogeneous difference equations (i.e., equations with zero input) with nonzero initial conditions. We analyze the types of nonzero outputs that such systems can produce. These outputs are called homogeneous solutions. It will be seen that the general form of these solutions can be expressed in terms of the poles of the system in an elegant closed form. More general systems with nonzero inputs and initial conditions are then considered. As in the previous chapters, poles play a crucial role in the system response. We conclude the chapter with a summary of five different meanings of a "pole" that have been encountered so far in this book, including in this chapter.

8.2 Recursive Difference Equations

In Chapter 5 we introduced recursive difference equations of the form

$$y[n] = -\sum_{k=1}^{N} b_k y[n-k] + \sum_{k=0}^{N} a_k x[n-k]. \tag{8.1}$$

These are called linear constant coefficient difference equations. When the initial conditions (IC) are chosen to be zero, that is,

$$y[-1] = y[-2] = \cdots = y[-N] = 0, \tag{8.2}$$

it was shown that Eq. (8.1) behaves like an **LTI system** with transfer function

$$H(z) = \frac{A(z)}{B(z)} = \frac{\sum_{k=0}^{N} a_k z^{-k}}{1 + \sum_{k=1}^{N} b_k z^{-k}}. \tag{8.3}$$

We will assume there are no common zeros between $A(z)$ and $B(z)$. With p_i denoting the poles of the system, the denominator can be factored into

$$B(z) = 1 + \sum_{k=1}^{N} b_k z^{-k} = \prod_{i=1}^{N} (1 - p_i z^{-1}). \tag{8.4}$$

To avoid trivialities, we also assume

$$p_i \neq 0 \tag{8.5}$$

in Eq. (8.4). Equivalently, we assume $b_N \neq 0$. Notice in particular that Eq. (8.4) implies

$$B(p_i) = 1 + \sum_{k=1}^{N} b_k p_i^{-k} = 0, \quad 1 \leq i \leq N. \tag{8.6}$$

8.3 The Homogeneous Difference Equation

If the input is identically zero, that is, if

$$x[n] = 0, \quad \forall n, \tag{8.7}$$

then Eq. (8.1) becomes

$$y[n] = -\sum_{k=1}^{N} b_k y[n-k]. \tag{8.8}$$

This is called the "homogeneous" difference equation. Even though $x[n] = 0$ for all n, Eq. (8.8) can have nonzero solutions $y[n]$ provided the *initial conditions are nonzero, that is, one or more of* $y[-1], y[-2], \ldots, y[-N]$ *is nonzero*. Such solutions of Eq. (8.8) are called **homogeneous** solutions. To generate such a solution, we start from some nonzero initial condition $y[-1], y[-2], \ldots, y[-N]$ and compute the samples

$$y[0], y[1], y[2], \ldots \tag{8.9}$$

and

$$y[-N-1], y[-N-2], y[-N-3], \ldots. \tag{8.10}$$

The samples (8.9) are computed by running Eq. (8.8) forward in time, that is, for $n = 0, 1, 2, \ldots$, in that order. The samples (8.10) are computed by running Eq. (8.8) backwards in time as in Sec. 6.7, that is, by rewriting it as

$$y[n-N] = \frac{1}{b_N}\Big(-y[n] - \sum_{k=1}^{N-1} b_k y[n-k]\Big), \tag{8.11}$$

and computing this for $n = -1, -2, -3, \ldots$, in that order. This requires the assumption $b_N \neq 0$, which just means that Eq. (8.8) truly has order N. Homogeneous solutions are useful for generating specific waveforms such as sinusoids, radar pulses, and baseband pulses in digital communications, and we will discuss some of these in Sec. 8.5.

8.3.1 Example of a Homogeneous Solution

We now show that

$$y[n] = p^n, \quad p \neq 0 \tag{8.12}$$

is a solution of Eq. (8.8) for certain values of p. To see this, we simply substitute $y[n] = p^n$ into Eq. (8.8) and examine the condition under which the equation

$$p^n = -\sum_{k=1}^{N} b_k p^{n-k} \tag{8.13}$$

holds. By factoring out p^n, this can be rearranged as

$$1 + \sum_{k=1}^{N} b_k p^{-k} = 0, \tag{8.14}$$

that is, $B(p) = 0$, where $B(z)$ is the denominator in Eq. (8.3). Summarizing, the exponential $y[n] = p^n$ is a solution to the homogeneous equation (8.8) *if and only if* p is a zero of $B(z)$, that is, p **is a pole** of $H(z)$.

8.3.2 A More General Homogeneous Solution

Now consider a linear combination of the form

$$y[n] = c_i p_i^n + c_l p_l^n, \tag{8.15}$$

where p_i and p_l are distinct poles. We can readily show that this is also a solution to Eq. (8.8). To see this, again substitute Eq. (8.15) into Eq. (8.8) and rearrange as follows:

$$c_i p_i^n \left(1 + \sum_{k=1}^{N} b_k p_i^{-k}\right) + c_l p_l^n \left(1 + \sum_{k=1}^{N} b_k p_l^{-k}\right) = 0. \tag{8.16}$$

To show that this equation is satisfied, simply note that the quantities within the large round brackets are zero, in view of Eq. (8.6). This proves the claim that Eq. (8.15) is a solution to Eq. (8.8). Similarly, we can show that

$$y[n] = \sum_{i=1}^{N} c_i p_i^n \tag{8.17}$$

is a homogeneous solution, that is, a solution to Eq. (8.8), for *any* choice of the constants c_i. For clarity, we emphasize a couple of points here.

1. The expression (8.17) satisfies Eq. (8.8) for all n. In particular, therefore, if we first compute

$$y[-1], y[-2], \ldots, y[-N] \tag{8.18}$$

from Eq. (8.17) for some fixed set of coefficients c_i, and then use Eq. (8.8) to compute $y[n]$ for other values of n (i.e., for all $n > -1$ and all $n < -N$), the result will be the same as computing all $y[n]$ from Eq. (8.17). Indeed, this is what it means to say that Eq. (8.17) satisfies Eq. (8.8).
2. So, we can generate the solution Eq. (8.17) either by calculating it for all n from Eq. (8.17) for the given set of c_i, or by calculating only the samples (8.18) from Eq. (8.17) and then using Eq. (8.8) to find $y[n]$ for all other n.

It is clear that the N quantities $c_1, c_2, \ldots, c_N$ and the N initial conditions (8.18) are related by the equation

$$\begin{bmatrix} y[-1] \\ y[-2] \\ y[-3] \\ \vdots \\ y[-N] \end{bmatrix} = \underbrace{\begin{bmatrix} p_1^{-1} & p_2^{-1} & \cdots & p_N^{-1} \\ p_1^{-2} & p_2^{-2} & \cdots & p_N^{-2} \\ p_1^{-3} & p_2^{-3} & \cdots & p_N^{-3} \\ \vdots & \vdots & \vdots & \vdots \\ p_1^{-N} & p_2^{-N} & \cdots & p_N^{-N} \end{bmatrix}}_{\text{call this } \mathbf{P}} \begin{bmatrix} c_1 \\ c_2 \\ c_3 \\ \vdots \\ c_N \end{bmatrix}, \tag{8.19}$$

where $\mathbf{P}$ is an $N \times N$ matrix. Note that the assumption (8.5) is used in writing Eq. (8.19). The significance of this matrix-vector equation will be clear in Sec. 8.3.3. For now, observe that we can rewrite the matrix $\mathbf{P}$ as

$$\mathbf{P} = \underbrace{\begin{bmatrix} 1 & 1 & \cdots & 1 \\ \alpha_1 & \alpha_2 & \cdots & \alpha_N \\ \alpha_1^2 & \alpha_2^2 & \cdots & \alpha_N^2 \\ \vdots & \vdots & \vdots & \vdots \\ \alpha_1^{N-1} & \alpha_2^{N-1} & \cdots & \alpha_N^{N-1} \end{bmatrix}}_{\text{call this } \mathbf{V}} \begin{bmatrix} \alpha_1 & 0 & 0 & \cdots & 0 \\ 0 & \alpha_2 & 0 & \cdots & 0 \\ 0 & 0 & \alpha_3 & \cdots & 0 \\ \vdots & \vdots & \vdots & \ddots & \vdots \\ 0 & 0 & 0 & \cdots & \alpha_N \end{bmatrix}, \tag{8.20}$$

where $\alpha_i = 1/p_i$. The matrix $\mathbf{V}$ is a **Vandermonde matrix**, and it is nonsingular (invertible) if and only if the α_i are distinct (Sec. 5.16), that is,

$$p_i \neq p_m, \quad i \neq m. \tag{8.21}$$

So, $\mathbf{P}$ is nonsingular if and only if the poles are distinct. The usefulness of this observation will soon be clear.

8.3.3 The Most General Homogeneous Solution

It was just shown that any expression of the form (8.17) is a homogeneous solution (i.e., a solution to Eq. (8.8)). We now show that when the poles are distinct (i.e.,

Eq. (8.21) holds), *any* homogeneous solution has the form (8.17). Say, $y[n]$ is an arbitrary homogeneous solution. Let us take the N samples

$$y[-1], y[-2], \ldots, y[-N] \tag{8.22}$$

from this given solution and compute N constants c_i satisfying Eq. (8.19). Even though we have not yet shown that $y[n]$ has the form (8.17), this calculation can always be done by inverting the matrix $\mathbf{P}$ because it is nonsingular when the poles are distinct. Once the c_i are thus computed, suppose we define

$$s[n] = \sum_{i=1}^{N} c_i p_i^n. \tag{8.23}$$

It was already shown that this is necessarily a homogeneous solution. It will now be proved that this $s[n]$ is exactly the same as the homogeneous solution $y[n]$ we started with!

Details of the proof. First observe that Eq. (8.23) implies

$$\begin{bmatrix} s[-1] \\ s[-2] \\ s[-3] \\ \vdots \\ s[-N] \end{bmatrix} = \underbrace{\begin{bmatrix} p_1^{-1} & p_2^{-1} & \cdots & p_N^{-1} \\ p_1^{-2} & p_2^{-2} & \cdots & p_N^{-2} \\ p_1^{-3} & p_2^{-3} & \cdots & p_N^{-3} \\ \vdots & \vdots & \vdots & \vdots \\ p_1^{-N} & p_2^{-N} & \cdots & p_N^{-N} \end{bmatrix}}_{\mathbf{P}} \begin{bmatrix} c_1 \\ c_2 \\ c_3 \\ \vdots \\ c_N \end{bmatrix}. \tag{8.24}$$

But c_i in the above equation was defined such that Eq. (8.19) is satisfied. By comparing Eqs. (8.24) and (8.19), we therefore see that

$$y[-1] = s[-1],\; y[-2] = s[-2],\; \ldots,\; y[-N] = s[-N]. \tag{8.25}$$

Thus, the homogeneous solution $y[n]$ and the homogeneous solution $s[n]$ have the same set of values for $n = -1, -2, \ldots, -N$. But two solutions to Eq. (8.8) with identical initial conditions as in Eq. (8.25) must obviously be identical. This proves that

$$y[n] = s[n], \quad \forall n. \tag{8.26}$$

This means in particular that the given homogeneous solution $y[n]$ has the form shown in the right-hand side of Eq. (8.23). ▽ ▽ ▽

Summarizing, in this section we have proved the following:

1. Equation (8.17) is always a homogeneous solution (i.e., a solution to Eq. (8.8)).
2. All homogeneous solutions have the form (8.17), when the poles p_i are distinct.

8.4 What if the Poles are Not Distinct?

Whether the poles are distinct or not, Eq. (8.17) is always a solution for the homogeneous equation (8.8), although the number of terms in Eq. (8.17) can be reduced by combining the terms for which the p_i have the same value. The main point however is that when the p_i are not distinct, Eq. (8.17) is **not** the most general form of the homogeneous solution: the matrix $\mathbf{P}$ is singular (because $\mathbf{V}$ is singular), and the earlier proof that Eq. (8.17) is the general solution is not valid. The general solution in this case takes a different form and we shall mention it here without proof.

First consider an example. Suppose $N = 3$, and assume that the poles are p_1, p_1, p_2. That is, two of the poles are identical. We say that p_1 has multiplicity 2 (i.e., it is a double pole). In this case, p_1^n and p_2^n are still homogeneous solutions (as shown earlier). But it can be shown that there is also a homogeneous solution of the form

$$np_1^n, \tag{8.27}$$

when p_1 is a double pole. It can in fact be shown that the most general homogeneous solution in this case is

$$c_1p_1^n + d_1np_1^n + c_2p_2^n, \tag{8.28}$$

where c_1, d_1, and c_2 are arbitrary constants. More generally, if p is a pole with multiplicity M (i.e., this pole occurs M times), then it contributes to the homogeneous solution terms of the form

$$p^n,\ np^n, \ldots,\ n^{M-1}p^n. \tag{8.29}$$

So, any linear combination of these M forms is also a solution. Now consider the most general situation: let $p_1, p_2, \ldots, p_L$ be L distinct poles, and let p_k have multiplicity M_k. Then it can be shown that the most general form of the homogeneous solution is

$$y[n] = \sum_{k=1}^{L} \sum_{m_k=0}^{M_k-1} c_{k,m_k} n^{m_k} p_k^n. \tag{8.30}$$

This is similar to the expression taken by the impulse response of $H(z)$ when there are multiple poles (Sec. 6.6.2).

8.5 The Oscillator Example

An oscillator is a waveform generator which generates a sinusoidal signal. These waveform generators are very useful in real-world applications (Sec. 8.5.2). We now show how an oscillator can be designed based on the theory we have learned. Consider the second-order difference equation

$$y[n] = -b_1y[n-1] - b_2y[n-2] + x[n]. \tag{8.31}$$

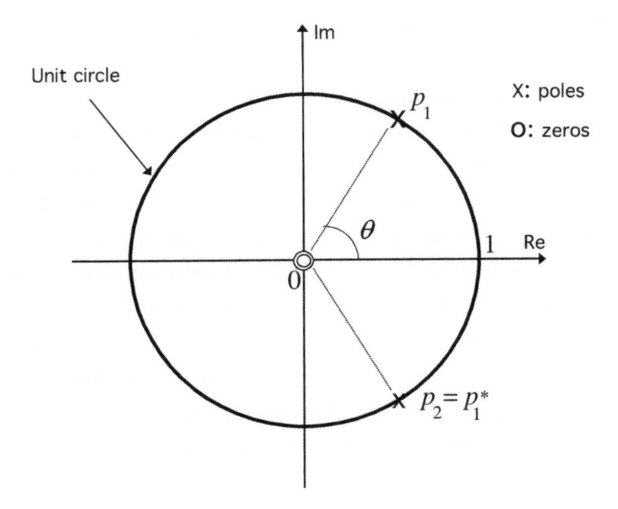

Figure 8.1 The poles and zeros of the second-order example with transfer function (8.34). The poles are on the unit circle, whereas the zeros are both at the origin.

The transfer function of the LTI system (system with zero IC) is

$$H(z) = \frac{1}{1 + b_1 z^{-1} + b_2 z^{-2}} = \frac{1}{(1 - p_1 z^{-1})(1 - p_2 z^{-1})}.$$

We have $N = 2$. Consider the specific case where the poles are

$$p_1 = e^{j\theta}, \quad p_2 = p_1^* = e^{-j\theta}, \tag{8.32}$$

where p_1^* is the complex conjugate of p_1. So the *poles are on the unit circle* of the z-plane (Fig. 8.1). Since

$$\left(1 - e^{j\theta} z^{-1}\right)\left(1 - e^{-j\theta} z^{-1}\right) = 1 - 2\cos(\theta) z^{-1} + z^{-2}, \tag{8.33}$$

it follows that

$$H(z) = \frac{1}{1 - 2\cos(\theta) z^{-1} + z^{-2}}, \tag{8.34}$$

and the homogeneous difference equation simplifies to

$$y[n] = 2\cos\theta\, y[n-1] - y[n-2]. \tag{8.35}$$

The general form of the homogeneous solution (8.17) reduces to

$$y[n] = c_1 e^{j\theta n} + c_2 e^{-j\theta n}. \tag{8.36}$$

If the constants are chosen as $c_1 = c_2 = 0.5$, then

$$y[n] = \cos(\theta n). \tag{8.37}$$

This is a cosine signal with frequency θ. This means that if we start the recursion (8.35) with appropriate initial conditions $y[-1]$ and $y[-2]$, the recursion will generate the cosine waveform $\cos(\theta n)$ for all n. We can find the appopriate IC

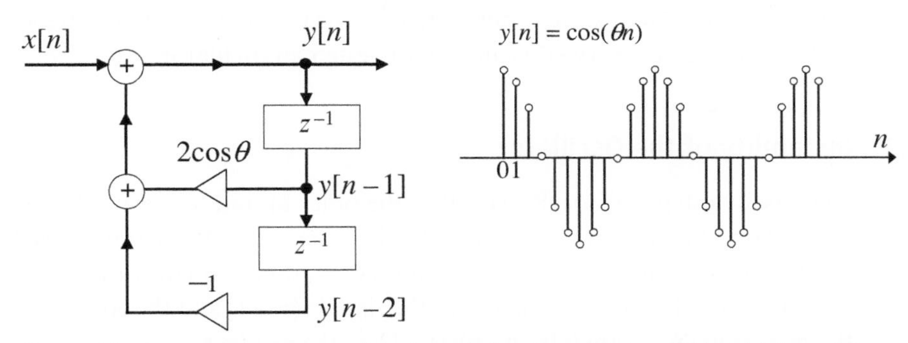

Figure 8.2 Structure or computational graph for the oscillator. With initial conditions $y[-1] = \cos\theta$ and $y[-2] = \cos 2\theta$, and input $x[n] = 0$, the output is $y[n] = \cos(\theta n)$, $n \geq 0$.

$(y[-1],\ y[-2])$ by substituting the values of c_1, c_2 into Eq. (8.19). But some thinking shows that a simpler approach would be to just take

$$y[-1] = \cos(-\theta) = \cos(\theta), \quad y[-2] = \cos(-2\theta) = \cos(2\theta). \tag{8.38}$$

Figure 8.2 shows the direct-form structure implementing this, with zero input. As another example, if we choose $c_1 = 1/2j, c_2 = -1/2j$, then Eq. (8.36) becomes

$$y[n] = \sin(\theta n). \tag{8.39}$$

The correct initial conditions for this are simply

$$y[-1] = \sin(-\theta) = -\sin(\theta), \quad y[-2] = \sin(-2\theta) = -\sin(2\theta). \tag{8.40}$$

Such a waveform generator is called an **oscillator** or a sinewave generator (whether it generates a sine or cosine). In fact, we can run two identical recursions like (8.35) in parallel, with these two sets of initial conditions, to generate $\cos(\theta n)$ and $\sin(\theta n)$ simultaneously, synchronized in time. Thus, the cosine and sine waveforms of frequencies θ, which are in perfect quadrature (i.e., 90° out of phase), can be generated in parallel. Such waveform pairs find applications in digital communication systems.

Advantage of recursive computation. The advantage of using Eq. (8.35) to generate cosines and sines is that the computational complexity is very low. Namely, it takes just **one multiplication** and one addition to generate each output sample $y[n]$. This is useful in dedicated hardware implementations. By contrast, if we use standard software, then $\cos(\theta n)$ and $\sin(\theta n)$ are typically generated by using a power series (Taylor series [Churchill and Brown, 1984]), and truncating the series after many terms. For example,

$$\cos(\theta n) = 1 - \frac{(\theta n)^2}{2!} + \frac{(\theta n)^4}{4!} - \frac{(\theta n)^6}{6!} + \cdots. \tag{8.41}$$

So, for each time instant n, there are many multiplications and additions (depending on how many terms we retain in Eq. (8.41), unlike the simple recursion (8.35). When we use dedicated hardware to generate waveforms, the recursive difference equation (8.35) is therefore a very efficient way to design an oscillator. ▽ ▽ ▽

8.5.1 Instability of the Oscillator

One disadvantage of Eq. (8.35) is that the poles p_1 and p_2 are on the unit circle of the z-plane (Fig. 8.1). That is, $|p_i| = 1$. This means that the system is **unstable**. This has the consequence that any small noise (e.g., digital quantization noise) can get amplified. To elaborate, assume that the signal $y[n-1]$ and the coefficient $2\cos\theta$ in the recursion (8.35) are b-bit numbers. Then the product

$$(2\cos\theta)y[n-1] \tag{8.42}$$

is a $2b$-bit number, and we have to **quantize** it back to b bits before repeating the recursion (8.35). Otherwise, the number of bits needed to represent $y[n]$ will keep growing as time n advances. Thus, in any digital implementation, the homogeneous equation (8.35) is replaced with[1]

$$y[n] = \mathcal{Q}\Big(2\cos\theta y[n-1]\Big) - y[n-2], \tag{8.43}$$

where $\mathcal{Q}(\cdot)$ represents quantization to b bits. If the number of bits b is large enough (say $b \geq 8$), then the effect of quantization can be approximated as follows:

$$\mathcal{Q}\Big(2\cos\theta y[n-1]\Big) \approx 2\cos\theta y[n-1] + e[n], \tag{8.44}$$

where $e[n]$ is random noise representing quantization error. Thus, the homogeneous equation (8.35) is replaced with

$$y[n] = 2\cos\theta y[n-1] - y[n-2] + e[n]. \tag{8.45}$$

Thus, instead of $x[n] = 0$ as in the homogeneous equation (8.35), we now have random noise in place of $x[n]$. In any practical digital implementation with large enough b, the noise $e[n]$ is very small indeed. However, it gets **amplified** because of the recursion (8.45) and the amplification factor is very large for poles very close to the unit circle [Oppenheim and Schafer, 2010]. So, when the poles are **on the unit circle**, as in the case of our oscillator, the noise amplification is severe.

To demonstrate this we now consider an example. Figure 8.3(a) shows the output of the ideal oscillator with no quantization noise, for the case where $\theta = 0.003\pi$ and intial conditions as in Eq. (8.38). This output $y[n]$ is indeed a perfect cosine, $\cos(\theta n)$. Now suppose we add the noise waveform $e[n]$ shown in Fig. 8.3(b), to convert the oscillator equation to Eq. (8.45). This noise is very small indeed, with the random samples $e[n]$ uniformly distributed in

$$-0.005 \leq e[n] \leq 0.005. \tag{8.46}$$

[1] Strictly speaking, the addition with $-y[n-2]$ can also increase the number of bits by one, but we ignore this for simplicity.

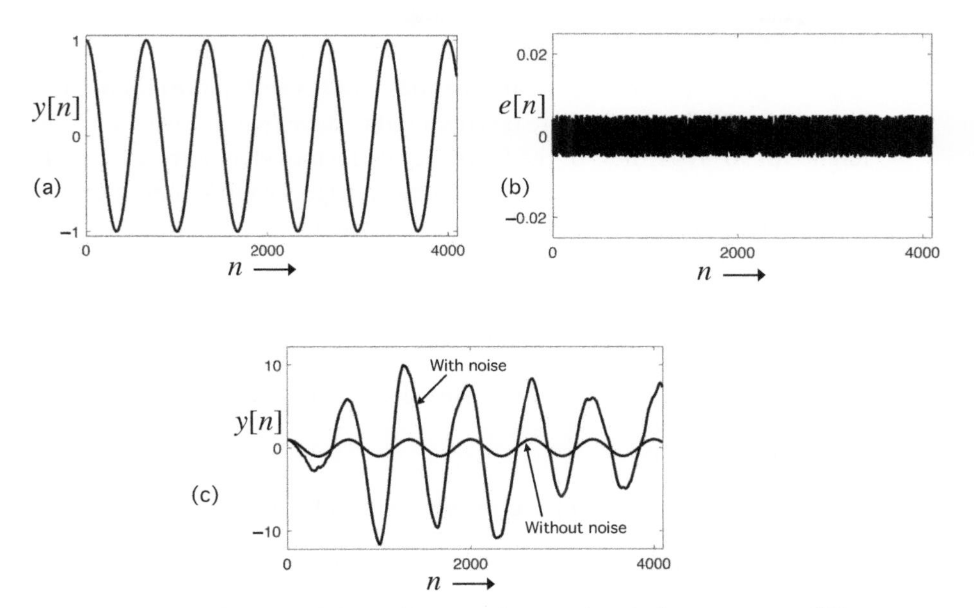

Figure 8.3 (a) The output of the oscillator designed using the homogeneous difference equation (8.35), (b) the noise waveform, and (c) the oscillator output when there is noise.

The output of the oscillator in the presence of this tiny noise is shown in Fig. 8.3(c), and is completely different from the ideal output, which is also shown for reference. Thus, the unstable poles amplify the noise dramatically, and the oscillator output is not useful.

In practice, there are some tricks one can use to stabilize the oscillator. One is to run the recursion for a few hundred samples, and then re-initialize the recursion. Another is to move the poles a little bit to place them inside the unit circle, thus producing a slightly damped sinusoid. We shall not go into these details here.

8.5.2 More General Waveform Generators

More general waveforms, for example, for radar applications or for digital communications, can be obtained by generating sines and cosines at many frequencies $\theta_1, \theta_2, \theta_3, \ldots$, and taking linear combinations. Thus, consider a waveform defined as

$$C_0 + \sum_{k=1}^{M} C_k \cos(\theta_k n) + \sum_{k=1}^{M} S_k \sin(\theta_k n), \tag{8.47}$$

where C_k and S_k are real, and $\theta_k = k\theta_1$. This is nothing but a Fourier series summation for real signals (Sec. 11.5). Even for small M, such a summation can approximate an arbitrary smooth shape reasonably well in an interval. Thus, we can generate each $\cos(\theta_k n)$ and $\sin(\theta_k n)$ using a difference equation like Eq. (8.35) in parallel, and combine the ouputs of these oscillators by using Eq. (8.47) to get the arbitrary waveform.

8.5.3 The Coupled-Form Oscillator

We now present a structure that generates $\cos(\theta n)$ and $\sin(\theta n)$ simultaneously. This is useful in applications where it is required to have two sine waves with a phase difference of 90° regardless of the frequency θ. To introduce a fresh style of reasoning, we derive this structure starting from basic trigonometric identities. Observe that the $(n+1)$th sample of the cosine can be written as

$$\cos(\theta(n+1)) = \cos\theta\cos(\theta n) - \sin\theta\sin(\theta n). \tag{8.48}$$

Similarly, $\sin(\theta(n+1))$ can be expressed as

$$\sin(\theta(n+1)) = \sin\theta\cos(\theta n) + \cos\theta\sin(\theta n). \tag{8.49}$$

Using the notations

$$x_1[n] = \cos(\theta n), \qquad x_2[n] = \sin(\theta n), \tag{8.50}$$

the preceding two equations can be rewritten as

$$\begin{aligned} x_1[n+1] &= \cos\theta\, x_1[n] - \sin\theta\, x_2[n], \\ x_2[n+1] &= \sin\theta\, x_1[n] + \cos\theta\, x_2[n]. \end{aligned} \tag{8.51}$$

Thus, the next value $x_1[n+1]$ is computed from the present values of both signals $x_1[n]$ and $x_2[n]$. The same holds for the computation of $x_2[n+1]$. That is, the two difference equations are **coupled** in the sense that each affects the other. Figure 8.4 shows the structure for implementing this system. Again, we set the input $x[n] = 0$, so Eqs. (8.51) are homogeneous equations. If we start these equations with the initial conditions

$$x_1[0] = 1, \quad x_2[0] = 0, \tag{8.52}$$

we can generate the signals $x_1[n] = \cos(\theta n)$ and $x_2[n] = \sin(\theta n)$ simultaneously for all $n \geq 0$. This system is called the **coupled-form oscillator.** One can think of Fig. 8.4 as a system with two outputs $x_1[n]$ and $x_2[n]$. Or one can regard $x_1[n]$ and $x_2[n]$ as internal state variables and think of Eq. (8.51) as state recursions. State-space descriptions are discussed in detail in Chapter 14.

Matrix formulation. Equation (8.51) can be written in matrix form as

$$\underbrace{\begin{pmatrix} x_1[n+1] \\ x_2[n+1] \end{pmatrix}}_{\mathbf{x}[n+1]} = \underbrace{\begin{pmatrix} \cos\theta & -\sin\theta \\ \sin\theta & \cos\theta \end{pmatrix}}_{\boldsymbol{\Theta}} \underbrace{\begin{pmatrix} x_1[n] \\ x_2[n] \end{pmatrix}}_{\mathbf{x}[n]}. \tag{8.53}$$

Thus, for each instant of time n, the next vector $\mathbf{x}[n+1]$ is obtained by taking the present vector $\mathbf{x}[n]$ and performing a multiplication with the matrix $\boldsymbol{\Theta}$. The matrix $\boldsymbol{\Theta}$ depends entirely on the frequency θ.

Some readers will recognize that this is a unitary or orthogonal matrix, that is, it satisfies $\boldsymbol{\Theta}^T\boldsymbol{\Theta} = \mathbf{I}$. In fact, it can be regarded as a **rotation** operator. That is, the vector $\mathbf{x}[n]$ is rotated counterclockwise by the amount θ, in order to obtain the

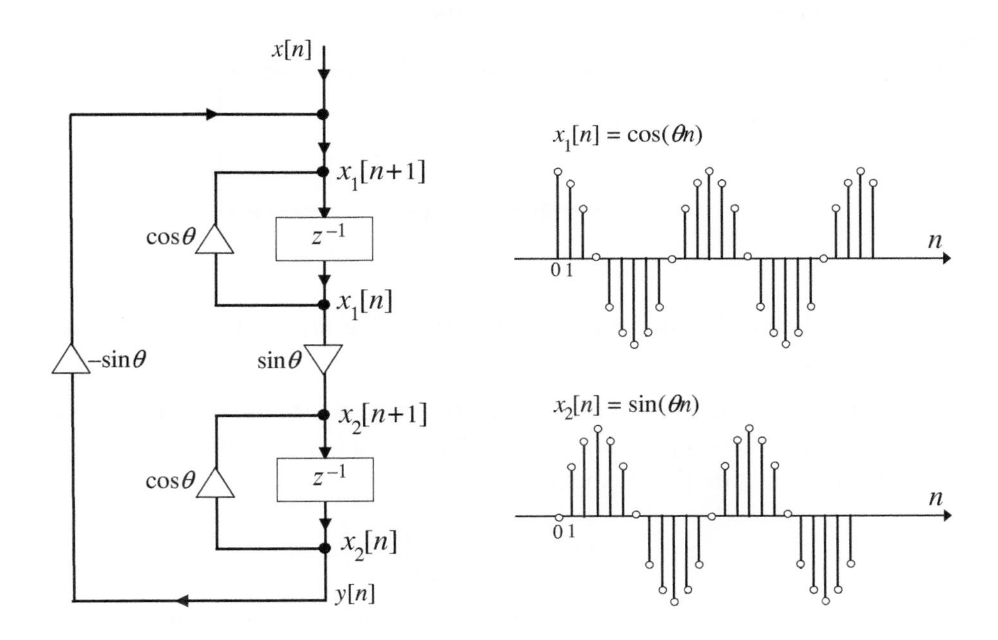

Figure 8.4 Structure for the coupled-form oscillator. With initial conditions $x_1[0] = 1, x_2[0] = 0$ and input $x[n] = 0$, we get $x_1[n] = \cos(\theta n)$ and $x_2[n] = \sin(\theta n), n \geq 0$.

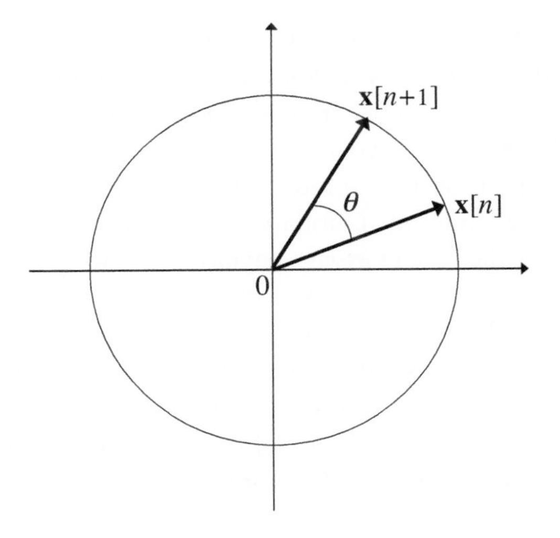

Figure 8.5 Counterclockwise rotation of $\mathbf{x}[n]$ to obtain $\mathbf{x}[n+1]$.

vector $\mathbf{x}[n+1]$ as shown in Fig. 8.5 (see Problem 8.9). So we can generate the pair of waveforms $\cos(\theta n)$ and $\sin(\theta n)$ together simply by using a rotation operator! Figure 8.6 shows the coupled-form structure using matrix notation. Here

$$z^{-1}\mathbf{I} = \begin{bmatrix} z^{-1} & 0 \\ 0 & z^{-1} \end{bmatrix} \tag{8.54}$$

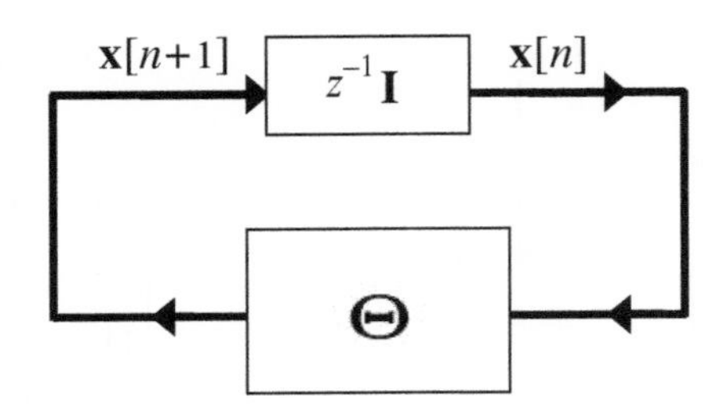

Figure 8.6 The coupled-form oscillator using matrix notation. See text.

is the diagonal matrix of delay elements. The matrix $\boldsymbol{\Theta}$ is also called a **Givens rotation**. ▽ ▽ ▽

Instead of using Fig. 8.4 as an oscillator, we can also use it as an LTI system with input $x[n]$ and output $y[n] = x_2[n]$ if we set the initial conditions $x_i[0] = 0$. The transfer function (Problem 8.6) is

$$H(z) = \frac{\sin\theta\, z^{-2}}{1 - 2\cos\theta\, z^{-1} + z^{-2}}. \tag{8.55}$$

Similar to the transfer function (8.34), this has poles on the unit circle (Fig. 8.1), and is therefore unstable.

8.6 Mortgage: A Real-World Example

Suppose you have borrowed L dollars from a bank, and want to repay it in I installments. For example, you could make a monthly payment of M dollars for Y years. This means that there are $I = 12Y$ installments. The first thing the bank will tell you is that you will have to pay interest — *money is not free.* For example, they could specify that the annual rate of interest is 10%.

The borrowed amount, L dollars, is also called the **principal**. The monthly payment M, also called the **mortgage payment**, has two parts to it. The first is a *portion of the principal* L, and the second is the *interest* on the "yet unpaid part" of the loan. Given (a) the borrowed amount L, (b) the interest rate, and (c) the number of installments I, the bank will calculate the amount M which you have to pay every month. How do they make this calculation? Usually the formula is built into their business software, and the beauty of the simple math that goes with it is lost in the heat of negotiations. We will therefore concentrate on this interesting aspect.

8.6.1 Difference Equation for the Mortgage Problem

Let $U[n]$ denote the amount of **unpaid balance** after the nth monthly payment has been made. Let $p[n]$ denote the portion of the principal being repaid during the nth monthly payment. We then have

$$U[n] = U[n-1] - p[n]. \tag{8.56}$$

During the nth installment, the fixed monthly payment M is split into two parts, namely, the principal part $p[n]$ and the interest on the unpaid part $U[n-1]$ at the end of the $(n-1)$th installment. Let the annual interest rate be equal to P %. Then the interest paid during the nth installment is given by $\alpha U[n-1]$, where

$$\alpha = \frac{P}{12 \times 100}. \tag{8.57}$$

Thus, the monthly payment M has the form

$$M = p[n] + \alpha U[n-1]. \tag{8.58}$$

We can eliminate $p[n]$ from Eqs. (8.56) and (8.58) to arrive at

$$U[n] = (1+\alpha)U[n-1] - M, \qquad n \geq 1. \tag{8.59}$$

This is a recursive difference equation of the form

$$y[n] = \beta y[n-1] + x[n], \tag{8.60}$$

where $\beta = 1 + \alpha$. The input term is

$$x[n] = -M, \quad \forall\, n \geq 1, \tag{8.61}$$

and the initial condition is

$$U[0] = L, \tag{8.62}$$

which is the unpaid balance at the end of the 0th month, that is, at the (un)fortunate moment when the loan got approved. So the mortgage problem is described by the recursive difference equation (8.59) with *nonzero input* (8.61) *and nonzero initial condition* (8.62).

8.6.2 Solution to the Mortgage Equation

For a given initial condition[2] $U[0]$, we can find $U[n]$ for any $n \geq 1$ by repeated application of Eq. (8.59). Thus,

$$\begin{aligned} U[n] &= \Big(1+\alpha\Big)^n U[0] - \left(\frac{(1+\alpha)^n - 1}{\alpha}\right) M \\ &= \Big(1+\alpha\Big)^n \Big(U[0] - \frac{M}{\alpha}\Big) + \frac{M}{\alpha}. \end{aligned} \tag{8.63}$$

If you have agreed to repay the entire loan in I installments, then at the end of the Ith month you would have

$$U[I] = 0. \tag{8.64}$$

[2] For convenience, we have taken $U[0]$ rather than $U[-1]$ as the initial condition.

Substituting $U[0] = L$ and $U[I] = 0$ into Eq. (8.63), one obtains an expression for the monthly payment:

$$M = \left(\frac{\alpha(1+\alpha)^I}{(1+\alpha)^I - 1} \right) L. \tag{8.65}$$

This is an important expression. Observe some interesting facts.

1. The monthly payment is proportional to the loan amount L. The proportionality factor depends on the interest rate (through α) and on the number of installments I.
2. Since $(1 + \alpha) > 1$, we have $(1 + \alpha)^I >> 1$ for sufficiently large I. So we can rewrite the monthly payment as

$$M \approx \alpha L \qquad \text{(large } I\text{)}, \tag{8.66}$$

   so it is proportional to α as well. The quantity I has disappeared from this expression! So, the monthly payment becomes insensitive to the number of installments I, if I is very large. That is, we cannot decrease the monthly payments arbitrarily by increasing the **loan period**, that is, the *duration of the loan*!
3. The causal LTI system described by Eq. (8.59) has its pole at $z = (1 + \alpha)$. This is outside the unit circle, because $\alpha > 0$ unless the bank offered you a negative interest rate, which is rare! The mortgage equation is therefore an **unstable** system, with pole $1 + \alpha > 1$. However, the computation of $U[n]$ is done only for $n \leq I$, and $U[n]$ decays to zero at the end of this period. The instability property is of no concern in this application.

8.6.3 Interest Part and Principal Part

Recall that in each monthly payment there is an interest part $\alpha U[n-1]$ and a principal part $p[n]$. These can be calculated immediately, once $U[n]$ is calculated as described above. From Eq. (8.56), the principal part is calculated as

$$p[n] = U[n-1] - U[n]. \tag{8.67}$$

The total amount you pay at the end of the loan period is MI dollars. From Eq. (8.65) we have

$$\text{Total payment } = MI = \underbrace{\left(\frac{\alpha(1+\alpha)^I I}{(1+\alpha)^I - 1} \right)}_{\text{amplification factor } A} \times L. \tag{8.68}$$

That is, if you borrow L dollars you end up paying AL dollars, where A is the amplification factor. The difference

$$AL - L \tag{8.69}$$

is the **total interest paid** during the loan period. It depends on the interest rate (which governs α) and the number of installments I. All of these quantities can be calculated from the three numbers L (the loan amount), I (the number of installments), and α (which is determined by the interest rate).

8.6.4 Example: Mortgage Calculation

Suppose you borrow \$200,000 for a period of 30 years, at an annual interest rate of 10%. Then

$$\begin{aligned} L &= 200,000 \quad \text{(loan amount)}, \\ I &= 30 \times 12 = 360 \quad \text{(number of installments)}, \\ \alpha &= 10/(12 \times 100) = 0.0083333\ldots. \end{aligned} \tag{8.70}$$

If we use the linear approximation $M \approx \alpha L$ (Eq. (8.66)), then the monthly payment would be $M \approx 1,667$. This, of course, is not used in practice, but adds insight, and is useful for a quick estimation. More accurately, using the correct formulas developed above, we see that the payments are (after slight rounding)

$$\begin{aligned} M &= \$1,755.10 \quad \text{(monthly payment)}, \\ M \times 360 &= \$631,850 \quad \text{(total payment)}, \\ \$631,850 - \$200,000 &= \$431,850 \quad \text{(total interest payment)}. \end{aligned} \tag{8.71}$$

So, if you borrow \$200,000 for 30 years at 10% interest rate, you pay a total interest of \$431,850 in those 30 years, which is more than twice the loan amount! When you start borrowing money, that is what you are getting into!

Interest part and principal part. It is interesting to see what happens during the first monthly payment. With $n = 1$ the interest part is $\alpha U[0] = \alpha L \approx \$1,667$, whereas the principal part is $M - 1,667 \approx \$88$ only. Thus, the first monthly payment goes mostly towards the interest payment. The principal part $p[n]$ and the interest part $\alpha U[n-1]$ are plotted in Fig. 8.7 as a function of n. For small n (i.e., for the first few months), the monthly payment is almost entirely dominated by the interest part. During the last few months of the loan period, the monthly payment is mostly the prinicipal part. ▽ ▽ ▽

To demonstrate another interesting feature of these calculations, suppose that your loan period is 20 years rather than 30 years. Then the number of installments is

$$I = 20 \times 12 = 240. \tag{8.72}$$

We now recalcuate the above quantities and find

$$\begin{aligned} M &= \$1,930 \quad \text{(monthly payment)}, \\ M \times 240 &= \$463,210 \quad \text{(total payment)}, \\ \$463,210 - \$200,000 &= \$263,210 \quad \text{(total interest payment)}. \end{aligned} \tag{8.73}$$

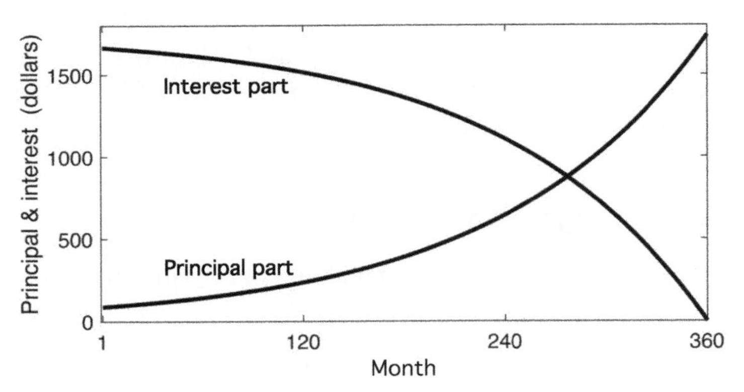

Figure 8.7 The interest part and principal part of the monthly payment. See text.

So the monthly payment increases only by a small amount, but the total interest paid has decreased dramatically! Similarly, if we change the loan period to 15 years we get

$$I = 15 \times 12 = 180. \tag{8.74}$$

Recalculating the above quantities we find

$$\begin{aligned} M &= \$2,149.20 \quad \text{(monthly payment)}, \\ M \times 180 &= \$386,860 \quad \text{(total payment)}, \\ \$386,860 - \$200,000 &= \$186,860 \quad \text{(total interest payment)}. \end{aligned}$$

It is clearly wise to choose a relatively short loan period!

8.7 Meaning of Poles: A Summary

In all our discussions in this chapter, poles have played an important role. This was the case in earlier chapters as well. Let us summarize a bit here. By definition (Sec. 5.8.1), a pole of $H(z)$ is a point p in the z-plane where the denominator of $H(z)$ equals zero. We showed that poles determine the various possible regions of convergence of the z-transform (Sec. 6.3.2, Fig. 6.3). They are also responsible for the recursive terms in the difference equations like Eq. (8.1), and give rise to feedback loops in the computational graphs of $H(z)$ (Fig. 5.7). Poles also determine the interplay between causality and stability (Sec. 6.4.3), and help to shape frequency responses of filters. We now summarize some specific manifestations of poles which give insight into the time-domain or "dynamical" meaning of poles. For simplicity of notation, we assume all poles have multiplicity one.

1. *Impulse response*. Each pole p_i contributes an exponential term p_i^n to the impulse response $h[n]$, as we show in Sec. 5.10 (see Eq. (5.94)).

2. *Transient response.* One consequence of the above is that each pole p_i contributes an exponential term p_i^n to the transient response of the filter (e.g., see Eq. (7.145)).
3. *Homogeneous solution.* Each pole p_i contributes an exponential term p_i^n to the homogeneous solution, as we show in Sec. 8.3 (see Eq. (8.17)). The homogeneous solution can be made exactly equal to p_i^n by appropriate choice of initial conditions.
4. *State transition matrix.* In Chapter 14 on state-space descriptions, we show that the eigenvalues of the state transition matrix $\mathbf{A}$ of any minimal implementation of $H(z)$ are precisely the set of poles of $H(z)$ (Sec. 14.5.2).
5. *Watching the poles.* There is one more interesting time-domain meaning of a pole (Problem 8.10). Let $H(z)$ be a causal stable rational system. If p_i is a pole, then there exists a finite-duration causal input $x[n]$ such that the output is exactly p_i^n for all $n > L$, for some finite integer L. That is, if we wait long enough, then the input is identically zero and the output is the exponential p_i^n representing the pole. In other words, we can "watch the pole" at the system output simply by applying a finite-duration signal at the input. The converse of this is also true (Problem 8.10).

All of these can be qualitatively summarized by saying that if p_i is a pole then p_i^n is a "natural response" or **natural mode** of the system. The quantitative meaning of this is of course as explained above. For systems with multiple inputs and outputs (MIMO systems), the meaning of poles is somewhat similar, although there are some subtleties; see Sec. 13.6 of Vaidyanathan [1993]. For inspiring early work on the time-domain meaning of poles, see Desoer and Schulman [1974]. It is also possible to give dynamical meanings for zeros, similar to what we have done for poles. See Problems 8.12 and 8.13.

8.8 Summary of Chapter 8

This chapter focused on recursive difference equations with nonzero initial conditions. Such a system can have a nonzero output $y[n]$ even when the input $x[n]$ is zero for all n. This is because the initial conditions can trigger an output which sustains itself because of the feedback portion of the recursive equation. Recursive difference equations with zero inputs are called homogeneous equations, and the nonzero outputs they produce are called homogeneous solutions. Given an Nth-order system, if the N poles of the system are p_i, then any expression of the form

$$y[n] = \sum_{i=1}^{N} c_i p_i^n \tag{8.75}$$

is a homogeneous solution for arbitrary constants c_i. This is, in fact, the most general solution when the poles are distinct. Given an arbitrary set of constants $\{c_i\}$, the corresponding homogeneous solution can be obtained simply by choosing the initial conditions $y[-1], y[-2], \ldots, y[-N]$ appropriately. As an application, by appropriate choice of the poles, we can generate sinusoidal waveforms using such equations. These are called oscillators and they are computationally more efficient than generating sines and cosines using power series. The chapter also presented the coupled-form oscillator, which can generate a sine and cosine for the same frequency with the 90° phase shift maintained perfectly. Oscillators have poles on the unit circle, which makes them unstable, but there are methods to stabilize these oscillators.

The chapter also introduced another application of recursive difference equations, namely the computation of mortgages. It was shown that the monthly payment on a loan can be computed using a first-order recursive difference equation. The equation also allows one to calculate the interest and principal parts of the payment every month.

Poles play a crucial role in the behavior of recursive difference equations with zero or nonzero initial conditions. There are many different manifestations of the effects of a pole. These are summarized in Sec. 8.7, and include time-domain meanings of poles.

PROBLEMS

8.1 Consider the causal difference equation corresponding to

$$H(z) = \frac{1 - z^{-1} + z^{-2}}{1 - (3/4)z^{-1} + (1/8)z^{-2}}.$$

(a) Assume the input $x[n] = 0$ for all n. Suppose the initial conditions are $y[-1] = 1$ and $y[-2] = -1$. Then the resulting output $y[n]$ is a homogeneous solution for the system. What are the values of this solution for $n = 0, 1, 2, 3$?

(b) We know the most general form of the homogeneous solution in this case is $c_1 p_1^n + c_2 p_2^n$, where p_i are the two poles. What are the values of c_1 and c_2 which generate the same solution as in part (a)?

(c) What choice of the initial conditions $y[-1]$ and $y[-2]$ will achieve the homogeneous solution $p_1^n + p_2^n$ for all $n \geq 0$?

8.2 In Problem 8.1(a), what are the values of the homogeneous solution for $n = -3, -4$? *Note:* You have to run the difference equation backwards to find this.

8.3 Consider the difference equation $y[n] = (2\cos\theta)y[n-1] - y[n-2] + x[n]$. With zero input, find the initial conditions $y[-1]$ and $y[-2]$ such that the output is $y[n] = 4\cos(\theta n) - \sin(\theta n), n \geq 0$.

8.4 Find a recursive, linear, constant-coefficient difference equation such that the period-2 signal $y[n], n \geq 0$, shown in Fig. P8.4, is a homogeneous solution. For your system, what is the initial condition that generates this solution?

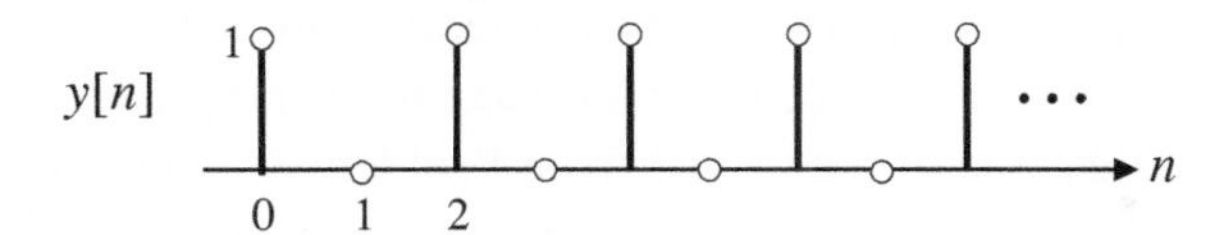

Figure P8.4 The homogeneous solution we want.

8.5 Show how you would modify your answer to Problem 8.4 so that the period-2 signal shown in Fig. P8.5 is a homogeneous solution.

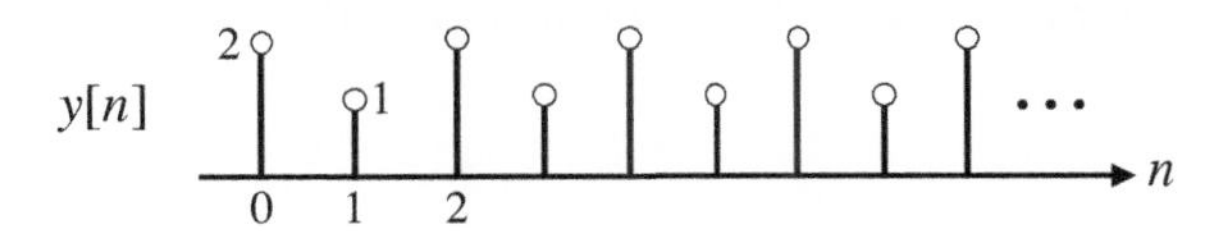

Figure P8.5 The new homogeneous solution we want.

8.6 Show that the transfer function $H(z) = Y(z)/X(z)$ in Fig. 8.4 is given by Eq. (8.55).

8.7 In Sec. 8.5 we showed how the difference equation

$$y[n] = (2\cos\theta)y[n-1] - y[n-2] + x[n] \tag{P8.7a}$$

can be used to generate the cosine output $y[n] = \cos(\theta n)\mathcal{U}[n]$ by setting initial conditions (IC) to be as in Eq. (8.38), and taking the input to be zero (i.e., using Eq. (P8.7a) in the homogeneous mode). We now show that $y[n] = \cos(\theta n)\mathcal{U}[n]$ can also be produced by taking the *initial conditions to be zero*, and applying an appropriate input $x[n]$ with two nonzero samples. That is, we can use Eq. (P8.7a) in its LTI mode instead of nonzero-IC mode.

(a) With Eq. (P8.7a) used as an LTI system (i.e., with zero IC) we know that the transfer function is $H(z) = 1/(1 - 2\cos\theta z^{-1} + z^{-2})$. Show that this has impulse response

$$h[n] = \frac{\sin(n+1)\theta}{\sin\theta}\mathcal{U}[n]. \tag{P8.7b}$$

Thus, $x[n] = \delta[n]$ produces $y[n] = A\cos(\theta n + \psi)\mathcal{U}[n]$ for some A, ψ.

(b) Now assume that instead of applying $x[n] = \delta[n]$, we apply an input with two nonzero samples, that is,

$$x[n] = x[0]\delta[n] + x[1]\delta[n-1]. \tag{P8.7c}$$

Find $x[0]$ and $x[1]$ such that $y[n] = \cos(\theta n)\mathcal{U}[n]$.

(c) More generally, suppose we want the output $y[n] = D\cos(\theta n + \phi)\mathcal{U}[n]$ for some arbitrary real D and ϕ. What are the values of $x[0]$ and $x[1]$ to achieve this (with zero IC)?

8.8 Let $G(z) = 1/(1 - 2R\cos\theta z^{-1} + R^2 z^{-2})$ be a causal system. Show that its impulse response is

$$g[n] = \frac{R^n \sin(n+1)\theta}{\sin\theta}\mathcal{U}[n]. \tag{P8.8}$$

With $R < 1$ this can be used to generate damped oscillations whose amplitude slowly decays to zero. *Hint:* You can use a simple property of the z-transform to modify one of the results from Problem 8.7 to get Eq. (P8.8).

8.9 *The Givens rotation matrix.* The vector $\mathbf{v}$ in Fig. P8.9 is obtained by rotating the vector $\mathbf{u}$ counterclockwise by θ. We can write the vectors in the form

$$\mathbf{u} = \rho \begin{bmatrix} \cos\phi \\ \sin\phi \end{bmatrix}, \quad \mathbf{v} = \rho \begin{bmatrix} \cos(\theta+\phi) \\ \sin(\theta+\phi) \end{bmatrix}.$$

Show that $\mathbf{v} = \mathbf{\Theta} u$, where $\mathbf{\Theta}$ is as in Eq. (8.53).

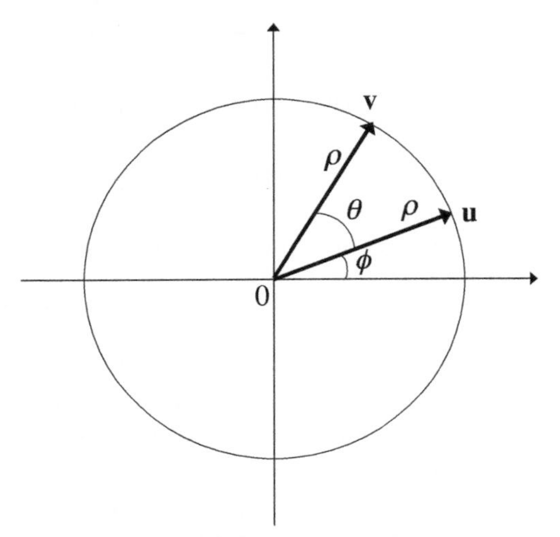

Figure P8.9 Rotation of a vector $\mathbf{u}$ by an angle θ, to produce $\mathbf{v}$.

8.10 *Watching the poles.* Let $H(z) = \sum_{n=0}^{\infty} h[n]z^{-n}, h[0] \neq 0$, be a rational transfer function representing a causal, stable LTI system. So we can write $H(z) = A(z)/B(z)$, where $A(z)$ and $B(z)$ are relatively prime polynomials in z^{-1} (i.e., $A(z)/B(z)$ is irreducible). This problem develops yet another time-domain or "dynamical" interpretation for poles. In this problem we assume all initial conditions to be zero.

(a) Suppose $p \neq 0$ is a pole of $H(z)$. Show that there exists a causal finite-length input $x[n]$ and a finite integer L such that the output $y[n]$ has the form p^n for $n > L$.

(b) Conversely, suppose there exists a causal finite-length input $x[n]$ and a finite integer L such that the output $y[n]$ has the form p^n for $n > L$. Show that p is a pole of $H(z)$.

8.11 This is based on Problem 8.10. Consider a causal, stable LTI system with transfer function

$$H(z) = \frac{2 - (5/6)z^{-1}}{1 - (5/6)z^{-1} + (1/6)z^{-2}}.$$

Let p_1 and p_2 be the poles with $|p_1| \geq |p_2|$. Find a finite-duration input $x[0], x[1], \ldots, x[M]$ such that the output is of the form $y[n] = p_1^n$ for $n > L$ for some integer L. What are the values of M and L in your construction?

8.12 *Dynamical meaning of zeros.* We know that if z_k is a zero of $H(z)$ then the output in response to the input z_k^n is zero for all n. This input is noncausal. In this and the next problems, we develop meanings based on causal inputs. Assume $H(z)$ is in irreducible rational form as in Eq. (8.3). Also, assume that it is causal so that it can be implemented with the difference equation (8.1). Assume $a_0 \neq 0$ and $b_N \neq 0$ to avoid trivialities.

(a) Consider the first-order case ($N = 1$). We know the system has a zero at $z_1 = -a_1/a_0$. Suppose we apply an input of the form $z_1^n \mathcal{U}[n]$, where $\mathcal{U}[n]$ is the unit step. Find an initial value $y[-1]$ such that $y[n] = 0$ for all $n \geq 0$.

(b) More generally, for the Nth-order system, suppose z_1 is a zero and we apply the causal input $z_1^n \mathcal{U}[n]$. Show how you can find an initial state $y[-1], y[-2], \ldots, y[-N]$ such that $y[n] = 0$ for all $n \geq 0$.

(c) Conversely, suppose there exists an initial state $y[-1], y[-2], \ldots, y[-N]$ such that the input $z_1^n \mathcal{U}[n]$ produces zero output for all $n \geq 0$. Show that z_1 is a zero of $H(z)$.

8.13 Let $H(z) = \sum_{n=0}^{\infty} h[n] z^{-n}, h[0] \neq 0$, be a rational transfer function representing a causal, stable LTI system. So, we can write $H(z) = A(z)/B(z)$, where $A(z)$ and $B(z)$ are relatively prime polynomials in z^{-1}. In this problem we assume all initial conditions to be zero.

(a) Let z_0 be a *zero* of the system. Then show that there exists a causal finite-length sequence $s[n]$ such that the input

$$x[n] = z_0^n \mathcal{U}[n] + s[n] \tag{P8.13}$$

produces a causal, finite-length output.

(b) Conversely, let there exist an input of the form (P8.13), where $s[n]$ is causal and finite length, such that the output is of finite length. Then show that z_0 is indeed a zero, assuming $z_0 \neq 0$.

This gives the following engineering interpretation of a *zero*: there exists a causal input such that, if you wait for finite time after applying the input, the input will look like an exponential z_0^n, whereas the output will become zero and stay zero!

8.14 (Computing assignment) *Mortgage calculation.* In this chapter we derived the mortgage equation[3]

$$U[n] = (1 + \alpha)U[n - 1] - M,$$

where $U[n]$ is the unpaid balance at the end of the nth payment, M is the monthly payment, and $\alpha = P/(12 * 100)$ (the annual interest rate being P %). This is a recursive difference equation with *constant* input $-M$ for $n \geq 1$, and

[3] $U[n]$ should not be confused with the unit step $\mathcal{U}[n]$.

initial condition $U[0] = L$ (loan amount). If we pay back the loan after I installments, then $U[I] = 0$ and the monthly payment M can be calculated from this information. In this problem we assume the loan amount $L =$ \$500K and interest rate $P = 12\%$.

(a) Assume a 30-year loan ($I = 30 * 12$). What is the monthly payment? What is the total interest paid ($MI - L$)?

(b) For a 20-year loan, what is the monthly payment? What is the total interest paid?

(c) For a 10-year loan, what is the monthly payment? What is the total interest paid?

Feel free to use a calculator or computer for doing the calculations. But you are not allowed to use any software for interest or mortgage calculations.

8.15 (Computing assignment) In Problem 8.14, what is the monthly payment if the bank lets you use an infinite number of years to pay?

8.16 (Computing assignment) In Problem 8.14, suppose the interest P is only 3%. Then repeat part (a).

8.17 (Computing assignment) *Noise in oscillators.* In Sec. 8.5.1 we discussed the instability of the oscillator which generates sinusoidal waveforms, and showed an example in Fig. 8.3 to demonstrate this. With all parameters as mentioned in Sec. 8.5.1, generate the plots given in Figs. 8.3(b) and (c). For this you will need to generate the noise $e[n]$ described in Sec. 8.5.1. If you are using MATLAB, this can be done using the "rand" command. *Note:* Since the noise is random, the error in the oscillator output is also random. So the noisy output in Fig. 8.3(c) will not look the same every time you repeat the experiment.

8.18 (Computing assignment) *Rotation operators.* With reference to Eq. (8.53), we explained how the vector $\mathbf{x}[n+1]$ is the rotated version of $\mathbf{x}[n]$. With initial conditions as in Eq. (8.52), generate and plot the vectors $\mathbf{x}[n], 0 \leq n \leq 20$, in the (x_1, x_2) plane, assuming $\theta = 0.025\pi$. If you are using MATLAB, the command "scatter" is helpful for this. It plots the tips of the vectors, which is a neat way to visualize them.

9 The Continuous-Time Fourier Transform

9.1 Introduction

This chapter introduces the continuous-time Fourier transform (CTFT, or just FT). This is defined by the integral given in Eq. (9.1). There are some similarities to the discrete-time Fourier transform (DTFT) discussed in Chapter 3, but there are also important differences. Unlike the DTFT, which is a summation, the CTFT is an integral. We will see that the frequency variable now takes the infinite range $(-\infty, \infty)$ instead of the finite range $[-\pi, \pi)$ used by the DTFT.

The first four sections of this chapter discuss the CTFT, the inverse CTFT, basic properties of these, and plenty of examples. The Dirac delta train example (Sec. 9.5) will be found to be useful in the development of the sampling theorem, which deals with the conversion of continuous-time signals to discrete-time samples (Chapter 10). Properties which involve the derivatives of the signal or its Fourier transform are then derived. They are useful in greatly simplifying the derivations of Fourier transforms of many signals, as we shall see. Certain types of symmetries in the signal impose certain constraints on the Fourier transform. For example, if a signal is both real and even, then the Fourier transform is also real and even. These and other symmetries are explained in detail.

There are some signals whose Fourier transforms have the same shape as the signal itself. These are called eigenfunctions of the Fourier transform operator. An example is the Gaussian signal. Such signals are also discussed in detail. Signal correlations and their Fourier transforms are discussed, and their application in what is known as "matched filtering" is explained in detail. Matched filters play an important role in digital communications and radar.

Sections 9.8, 9.11, 9.12, and 9.13, which are optional and marked with an asterisk, add considerable insight. For example, Sec. 9.12 presents the fascinating fact that Mother Nature "computes" a Fourier transform when a plane wave is propagating across an aperture and impinging on a distant screen. Section 9.11 shows how to quantify the smoothness of a signal and shows that a higher degree of smoothness of $x(t)$ implies a faster decay of the Fourier transform as the frequency increases.

Many properties of the CTFT are summarized in Sec. 9.16, and examples of CTFT pairs are tabulated in Sec. 9.17. In Chapter 13 we shall briefly discuss the

Laplace transform, which can be regarded as the generalization of the continuous-time Fourier transform. In Chapter 17 we will present some mathematical details relating to Fourier representations. These pertain to the existence of the infinite integrals involved in the definitions, the convergence of the inverse transform formulas and Fourier series formulas, and so on. They are omitted in the main chapters in order to enable the reader to focus on the engineering aspects.

Notations: In this chapter we do not use Ω for the frequency in continuous time, but instead use ω. Similarly, we do not use subscript c as in $x_c(t)$ for continuous time (except for the Dirac delta, so its notation is still $\delta_c(t)$). This is because almost the entire chapter will only have continuous-time quantities.

9.2 Definition, Examples, and Properties

The continuous-time Fourier transform (CTFT or just FT) is defined as the integral

$$X(j\omega) = \int_{-\infty}^{\infty} x(t)e^{-j\omega t}dt \qquad \text{(FT)}. \tag{9.1}$$

This can in general be complex even if $x(t)$ is real, as our examples will show. We will derive the inverse Fourier transform formula later. The notation

$$x(t) \longleftrightarrow X(j\omega) \tag{9.2}$$

will be used to indicate that $x(t)$ and $X(j\omega)$ make a Fourier transform pair (i.e., $X(j\omega)$ is the FT of $x(t)$ and $x(t)$ is the inverse FT of $X(j\omega)$). For a simple example, suppose $x(t)$ is the Dirac delta function: $x(t) = \delta_c(t)$. Then from Eq. (9.1) it follows that $X(j\omega) = 1$ for all ω. So we write

$$x(t) = \delta_c(t) \longleftrightarrow X(j\omega) = 1, \quad \forall\omega. \tag{9.3}$$

This is shown in Fig. 9.1.

9.2.1 Basic Properties of the FT

We begin with a few simple properties of the FT. First, it is evident that Fourier transformation is a **linear operator**, that is, if $x_1(t) \longleftrightarrow X_1(j\omega)$ and $x_2(t) \longleftrightarrow X_2(j\omega)$, then

$$c_1x_1(t) + c_2x_2(t) \longleftrightarrow c_1X_1(j\omega) + c_2X_2(j\omega). \tag{9.4}$$

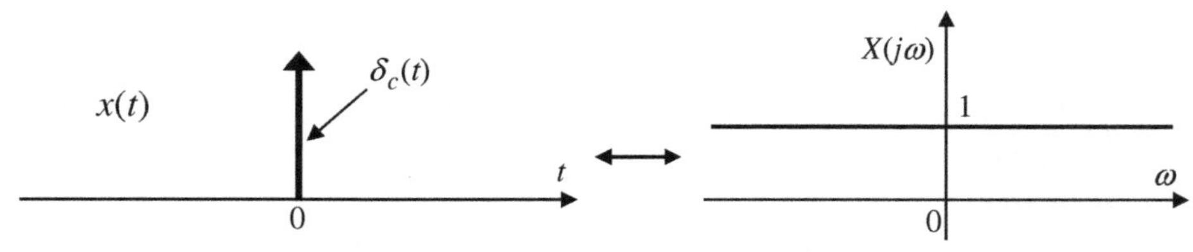

Figure 9.1 The Dirac delta signal $\delta_c(t)$ and its Fourier transform $X(j\omega)$, which is unity for all ω.

From Eq. (9.1) it is clear that the Fourier transform evaluated at zero frequency is

$$X(0) = \int_{-\infty}^{\infty} x(t)dt, \tag{9.5}$$

which can be regarded as the "area under the plot of $x(t)$." Next, notice from Eq. (9.1) that the Fourier transform of the **shifted** version $x(t - t_0)$ is

$$\int_{-\infty}^{\infty} x(t - t_0)e^{-j\omega t}dt = \int_{-\infty}^{\infty} x(t)e^{-j\omega(t+t_0)}dt = e^{-j\omega t_0} \underbrace{\int_{-\infty}^{\infty} x(t)e^{-j\omega t}\, dt}_{X(j\omega)}, \tag{9.6}$$

which proves that

$$x(t - t_0) \longleftrightarrow e^{-j\omega t_0}X(j\omega). \tag{9.7}$$

As an application of this, we readily conclude from Eq. (9.3) that

$$\delta_c(t - t_0) \longleftrightarrow e^{-j\omega t_0}. \tag{9.8}$$

A dual of Eq. (9.7) is the property

$$e^{j\omega_0 t}x(t) \longleftrightarrow X(j(\omega - \omega_0)). \tag{9.9}$$

This is called the **modulation property**, since $e^{j\omega_0 t}x(t)$ is an amplitude-modulated version of $e^{j\omega_0 t}$, the modulating signal being $x(t)$. Equation (9.9) is readily proved as follows:

$$\int_{-\infty}^{\infty} e^{j\omega_0 t}x(t)e^{-j\omega t}dt = \int_{-\infty}^{\infty} x(t)e^{-j(\omega-\omega_0)t}dt = X(j(\omega - \omega_0)). \tag{9.10}$$

We conclude this subsection with the **convolution theorem**. Given two signals $x(t)$ and $h(t)$, their convolution is defined as

$$y(t) = \int_{-\infty}^{\infty} x(\tau)h(t - \tau)d\tau. \tag{9.11}$$

When $x(t)$ is the input to a linear time-invariant (LTI) system with impulse response $h(t)$, then we know that the output is given by the above expression. It will now be shown that

$$Y(j\omega) = H(j\omega)X(j\omega). \tag{9.12}$$

That is,

$$\int_{-\infty}^{\infty} x(\tau)h(t - \tau)d\tau \quad \longleftrightarrow \quad H(j\omega)X(j\omega). \tag{9.13}$$

This is called the convolution theorem. We already proved a similar result for discrete-time systems. The proof in the continuous-time case is similar.

Proof of Eq. (9.12). We have

$$\begin{aligned}
Y(j\omega) &= \int_{-\infty}^{\infty} y(t)e^{-j\omega t}dt \\
&= \int_{t=-\infty}^{\infty}\Big(\int_{\tau=-\infty}^{\infty} x(\tau)h(t-\tau)d\tau\Big)e^{-j\omega t}dt \\
&= \int_{\tau=-\infty}^{\infty} x(\tau)e^{-j\omega\tau}d\tau \int_{t=-\infty}^{\infty} h(t-\tau)e^{-j\omega(t-\tau)}dt \\
&= \underbrace{\int_{\tau=-\infty}^{\infty} x(\tau)e^{-j\omega\tau}d\tau}_{X(j\omega)} \underbrace{\int_{t=-\infty}^{\infty} h(t)e^{-j\omega t}dt}_{H(j\omega)} \\
&= H(j\omega)X(j\omega).
\end{aligned} \tag{9.14}$$

In the third line we have simply inserted $e^{-j\omega\tau}\, e^{j\omega\tau}$ and interchanged the two integrals. The inner integral does not depend on τ because the range of integration is $-\infty < t < \infty$, so this leads to the next line. This completes the proof. ▽ ▽ ▽

Similarly, it can be shown that

$$x_1(t)x_2(t) \longleftrightarrow \frac{1}{2\pi}\int_{-\infty}^{\infty} X_1(ju)X_2(j(\omega-u))du. \tag{9.15}$$

That is, multiplication in time is equivalent to **convolution in frequency** (with appropriate scaling by $1/2\pi$). The proof of Eq. (9.15) becomes easier once we know the inverse Fourier transform relationship (Sec. 9.4).

Comment on notation. The reason for the notation $X(j\omega)$ (rather than just $X(\omega)$) is that, in the engineering literature, $X(s)$ is reserved for the Laplace transform of $x(t)$ (Sec. 3.9 and Chapter 13); when we set $s = j\omega$ in the Laplace transform, it reduces to the Fourier transform $X(j\omega)$. Some books avoid j in the notation by using $\widehat{x}(\omega)$ for the FT of $x(t)$, so that

$$x(t) \longleftrightarrow \widehat{x}(\omega). \tag{9.16}$$

This is common in the mathematics literature. ▽ ▽ ▽

9.2.2 The Pulse Signal

Consider the pulse signal

$$p(t) = \begin{cases} 1 & \text{for } -\frac{1}{2} < t < \frac{1}{2}, \\ 0 & \text{otherwise,} \end{cases} \tag{9.17}$$

shown in Fig. 9.2(a). Its Fourier transform is given by

$$P(j\omega) = \int_{-1/2}^{1/2} e^{-j\omega t}dt = \frac{e^{-j\omega t}}{-j\omega}\Bigg|_{-1/2}^{1/2} = \frac{e^{j\omega/2} - e^{-j\omega/2}}{j\omega} = \frac{\sin(\omega/2)}{\omega/2}. \tag{9.18}$$

Thus

$$p(t) \longleftrightarrow \frac{\sin(\omega/2)}{\omega/2} \tag{9.19}$$

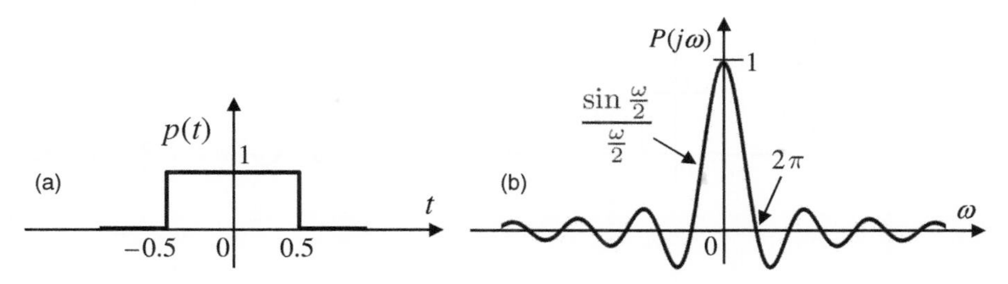

Figure 9.2 The pulse signal $p(t)$ and its Fourier transform, which is the sinc function.

for the pulse $p(t)$ in Eq. (9.17). So $P(j\omega)$ is the sinc function $\sin\theta/\theta$ (Sec. 4.3.4) with θ replaced by $\omega/2$. This is plotted in Fig. 9.2(b). Note that as $\theta \to 0$, we have $\sin\theta \approx \theta$, so that

$$\frac{\sin\theta}{\theta} \to 1, \tag{9.20}$$

as $\theta \to 0$. We therefore have $P(0) = 1$, which is also consistent with Eq. (9.5). Since $\sin(\omega/2) = 0$ for $\omega = 2\pi k$ for any integer k, it follows that

$$P(j\omega) = 0, \quad \omega = 2\pi k, k \neq 0. \tag{9.21}$$

So the zero crossings for $\omega > 0$ are uniformly spaced apart by 2π, and similarly for $\omega < 0$.

9.2.3 The Dilation Property

Consider $x_1(t) = x(at)$, where a is a nonzero real number. We say that $x_1(t)$ is a dilated version of $x(t)$. The FT of this signal is

$$X_1(j\omega) = \int_{-\infty}^{\infty} x_1(t)e^{-j\omega t}dt = \int_{-\infty}^{\infty} x(at)e^{-j\omega t}dt$$

$$= \begin{cases} \dfrac{1}{a}\displaystyle\int_{-\infty}^{\infty} x(\tau)e^{-j(\omega/a)\tau}d\tau & \text{for } a > 0, \\ -\dfrac{1}{a}\displaystyle\int_{-\infty}^{\infty} x(\tau)e^{-j(\omega/a)\tau}d\tau & \text{for } a < 0, \end{cases}$$

where we have made the change of variables $at = \tau$. This shows that $X_1(j\omega) = X(j\omega/a)/|a|$. Thus

$$x(at) \longleftrightarrow \frac{1}{|a|}X\Big(j\frac{\omega}{a}\Big), \tag{9.22}$$

for real nonzero a.

9.2.4 Variations of the Pulse Example

We can derive more examples of FT pairs by starting from Eq. (9.19) and using the properties of the Fourier transform derived earlier. For example, consider the

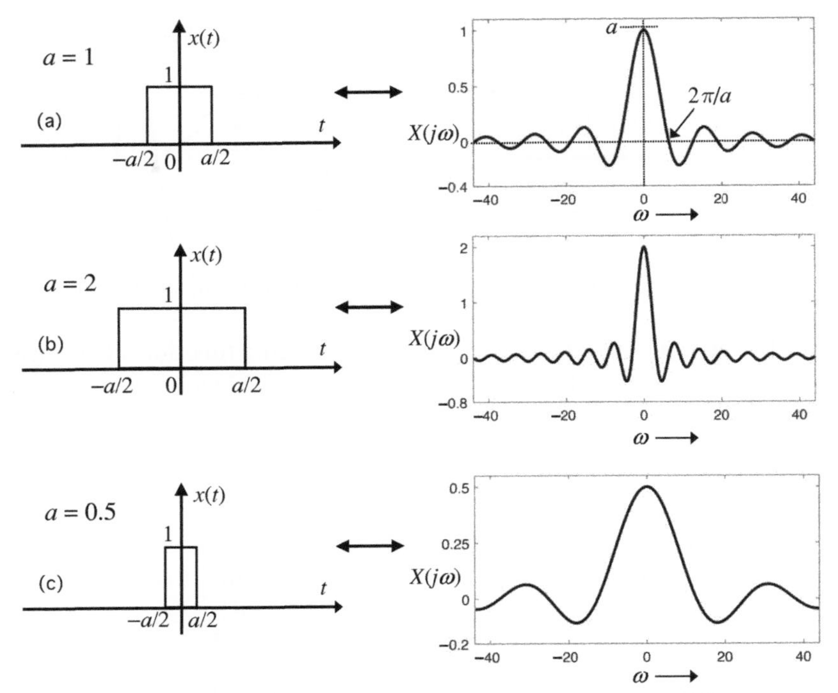

Figure 9.3 (a)–(c) The pulse signal and its Fourier transform for different widths of the pulse. Note that as the pulse width a increases, the main lobe region of the Fourier transform gets narrower.

dilated pulse $x(t) = p(t/a)$. Since Eq. (9.19) has already been proved, it follows from Eq. (9.22) that

$$x(t) = p(t/a) = \begin{cases} 1 & \text{for } -\frac{a}{2} < t < \frac{a}{2} \\ 0 & \text{otherwise} \end{cases} \qquad \longleftrightarrow \qquad a\frac{\sin(a\omega/2)}{a\omega/2} \tag{9.23}$$

for $a > 0$. That is, if we **stretch** the pulse in the time domain by a certain amount, then the sinc in the frequency domain gets **squeezed** by the same amount. See Figs. 9.3(a) and (b). Similarly, confining the pulse to a narrower region in the time domain spreads out the sinc in the frequency domain (Fig. 9.3(c)). Localizing a signal in the time domain delocalizes it in the frequency domain and vice versa, a direct consequence of the dilation property (9.22). An interesting property of the right-hand side of Eq. (9.23) is that

$$\int_{-\infty}^{\infty} a\frac{\sin(a\omega/2)}{a\omega/2}d\omega = 2\pi, \tag{9.24}$$

regardless of the value of a. The importance of this observation will soon become clear.

Proof of Eq. (9.24). We have

$$\int_{-\infty}^{\infty} a\frac{\sin(a\omega/2)}{a\omega/2}d\omega = 2\int_{-\infty}^{\infty} \frac{\sin\theta}{\theta}d\theta = 2\pi. \tag{9.25}$$

Here we have made the change of variables $\theta = a\omega/2$, and then used the well-known result [Haaser and Sullivan, 1991; Kreyszig, 1972] that

$$\int_{-\infty}^{\infty} \frac{\sin\theta}{\theta} d\theta = \pi. \tag{9.26}$$

This completes the proof. ▽ ▽ ▽

9.2.5 The Single-Frequency Signal

By using the property (9.24), a very interesting conclusion can be drawn from the FT pair (9.23). Thus assume $a \to \infty$, that is, the pulse gets longer and longer until it is unity for all t. Then the sinc function on the right-hand side of Eq. (9.23) becomes more and more focused at $\omega = 0$, although its integral is 2π for all a. Thus, as $a \to \infty$, we have

$$p(t/a) \to 1, \ \forall t \quad \text{and} \quad a\frac{\sin(a\omega/2)}{a\omega/2} \to 2\pi\delta_c(\omega). \tag{9.27}$$

This shows that

$$x(t) = 1, \ \forall t \quad \longleftrightarrow \quad X(j\omega) = 2\pi\delta_c(\omega). \tag{9.28}$$

Thus, the FT of the constant or zero-frequency signal (9.28) is a Dirac delta, located at $\omega = 0$. This is demonstrated in Fig. 9.4(a). Using Eq. (9.9) in Eq. (9.28), it follows that

$$e^{j\omega_0 t} \longleftrightarrow 2\pi\delta_c(\omega - \omega_0). \tag{9.29}$$

That is, the single-frequency signal $e^{j\omega_0 t}$ has FT equal to zero everywhere except at $\omega = \omega_0$, where there is a Dirac delta with area 2π. This is demonstrated in Fig. 9.4(b), and is called a **line spectrum**. More generally, a signal whose Fourier

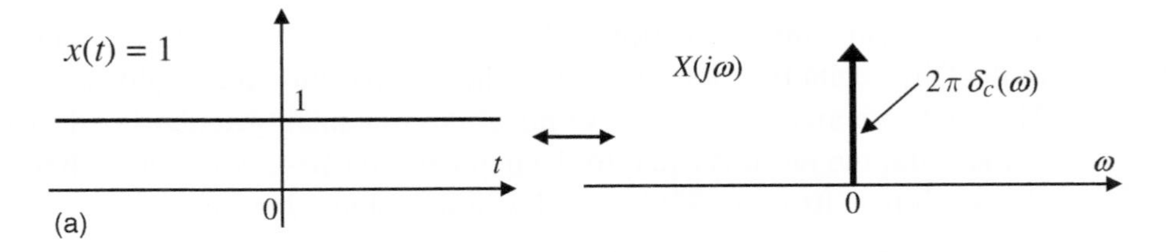

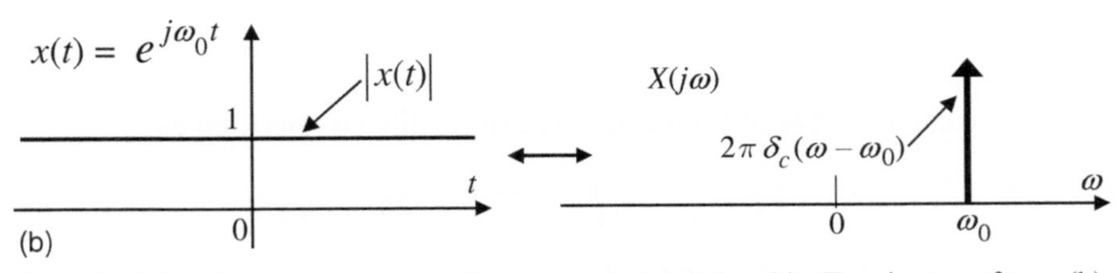

Figure 9.4 (a) The constant or zero-frequency signal $x(t)$ and its Fourier transform. (b) The single-frequency signal with frequency ω_0 and its Fourier transform.

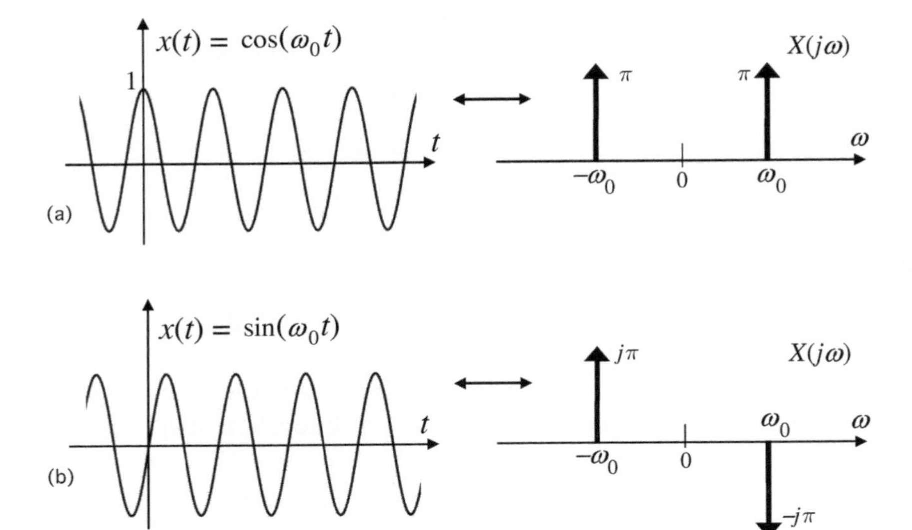

Figure 9.5 (a) The cosine signal and its Fourier transform. (b) The sine signal and its Fourier transform.

transform is a sum of shifted and scaled Dirac delta functions (e.g., as we shall see in Fig. 9.5), is called a line spectral signal.

Next, since

$$\cos(\omega_0 t) = \frac{e^{j\omega_0 t} + e^{-j\omega_0 t}}{2} \quad \text{and} \quad \sin(\omega_0 t) = \frac{e^{j\omega_0 t} - e^{-j\omega_0 t}}{2j}, \tag{9.30}$$

it follows from Eq. (9.29) and the linearity of the FT operator that

$$\cos(\omega_0 t) \longleftrightarrow \pi\delta_c(\omega + \omega_0) + \pi\delta_c(\omega - \omega_0), \tag{9.31}$$

$$\sin(\omega_0 t) \longleftrightarrow j\pi\delta_c(\omega + \omega_0) - j\pi\delta_c(\omega - \omega_0). \tag{9.32}$$

These FT pairs are schematically shown in Fig. 9.5. Note that the numbers next to the Dirac delta functions are not the heights (because the heights are not finite). They indicate area. That is, if we integrate in a small neighborhood around the Dirac delta, the result is equal to the number indicated. Notice also that although $\sin(\omega_0 t)$ is real, its FT (9.32) is complex (it has j and $-j$ in it).

An engineer's identity. Since the Fourier transform for $x(t) = 1$ is by definition $\int_{-\infty}^{\infty} e^{-j\omega t}dt$, it follows from Eq. (9.28) that

$$\int_{-\infty}^{\infty} e^{-j\omega t}dt = 2\pi\delta_c(\omega). \tag{9.33}$$

By using a change of notation, this can readily be rewritten as

$$\int_{-\infty}^{\infty} e^{juv}du = 2\pi\delta_c(v), \tag{9.34}$$

where u and v are real. We will see that this is a useful "mathematical," or rather engineer's, identity for proving some other results about Fourier transforms. ▽▽▽

9.3 More Examples of Fourier Transform Pairs

Figure 9.6(a) shows the pulse signal $x(t) = p(t)$ and its Fourier transform (9.19) again. From this, we will compute the FTs of the signals $x_1(t), x_2(t)$, and $x_3(t)$ in Fig. 9.6, by taking advantage of the properties of the FT discussed so far (which is easier than using the definition (9.1)). First note that $x_1(t)$ in Fig. 9.6(b) is obtained by shifting the pulse to the left by 0.5, that is, $x_1(t) = x(t + 0.5)$. Using Eq. (9.7) it therefore follows that

$$X_1(j\omega) = e^{j\omega/2}X(j\omega) = e^{j\omega/2}\frac{\sin(\omega/2)}{\omega/2}. \tag{9.35}$$

Notice that $X_1(j\omega)$ is complex valued even though $x_1(t)$ is real.[1] Now consider the signal $x_2(t)$ in Fig. 9.6(c), which is a sum of the left-shifted pulse $x_1(t)$ and the right-shifted pulse $-x(t-0.5)$. We will call $x_2(t)$ the **twin-pulse** signal. Since $x(t-0.5)$ has the Fourier transform $e^{-j\omega/2}X(j\omega)$, it follows that

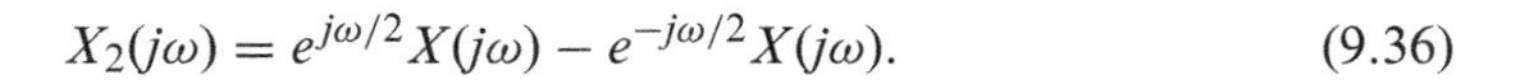

$$X_2(j\omega) = e^{j\omega/2}X(j\omega) - e^{-j\omega/2}X(j\omega). \tag{9.36}$$

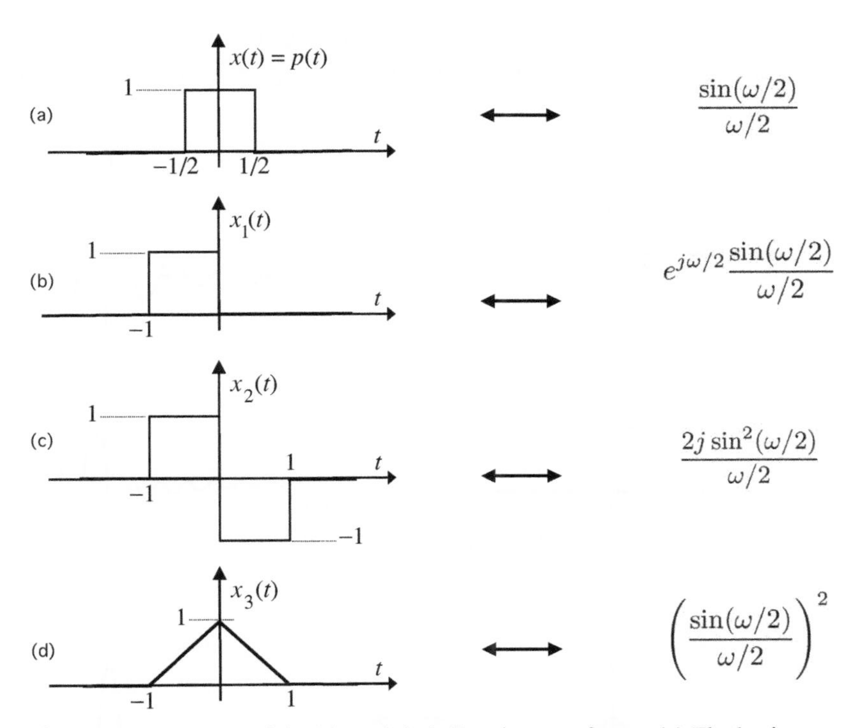

Figure 9.6 Examples of signals and their Fourier transforms. (a) The basic pulse, (b) the shifted pulse, (c) the twin-pulse signal, and (d) the triangular pulse.

[1] It can be shown (Sec. 9.9) that the Fourier transform is real if and only if $x(t)$ satisfies $x(t) = x^*(-t)$. For real $x(t)$ this is equivalent to saying that $x(t)$ is an even function, like the basic pulse in Fig. 9.6(a). Similarly, $X(j\omega)$ is purely imaginary if and only if $x(t) = -x^*(-t)$. This is the case with $x_2(t)$.

This is because of the linearity property (9.4). Thus

$$X_2(j\omega) = \left(e^{j\omega/2} - e^{-j\omega/2}\right) X(j\omega) = \left(e^{j\omega/2} - e^{-j\omega/2}\right) \frac{\sin(\omega/2)}{\omega/2}, \tag{9.37}$$

which simplifies to

$$X_2(j\omega) = \frac{2j \sin^2(\omega/2)}{\omega/2}. \tag{9.38}$$

Note that $X_2(j\omega)$ in Eq. (9.38) is purely imaginary for all ω, although $x_2(t)$ is real. Finally consider the signal $x_3(t)$ in Fig. 9.6(d), which is a triangular pulse. We know that this is nothing but a convolution of the pulse $p(t)$ with itself, that is,

$$x_3(t) = (p * p)(t). \tag{9.39}$$

In view of the convolution theorem (Eq. (9.13)), it then follows that

$$X_3(j\omega) = \left(\frac{\sin(\omega/2)}{\omega/2}\right)^2, \tag{9.40}$$

for all ω. There are other ways to arrive at this result, for example, by noticing that $x_2(t)$ can be regarded as a derivative of $x_3(t)$ (Problem 9.9). The Fourier transforms just derived are summarized in the plots of Fig. 9.7. We have not plotted $X_1(j\omega)$

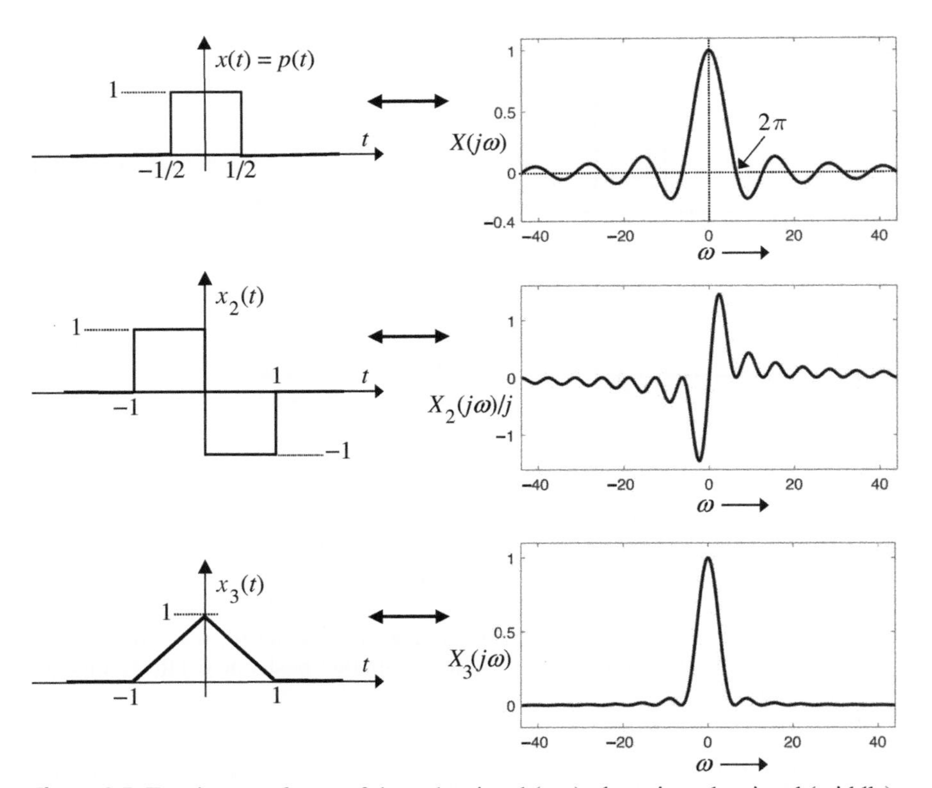

Figure 9.7 Fourier transforms of the pulse signal (top), the twin-pulse signal (middle), and the triangle signal (bottom). The Fourier transform $X_2(j\omega)$ is purely imaginary (see Eq. (9.38)), and we have plotted $X_2(j\omega)/j$ for convenience.

in Eq. (9.35) because it is essentially the sinc function shown at the top of Fig. 9.7, except for the phase factor $e^{j\omega/2}$.

9.4 The Inverse Fourier Transform

Given the continuous-time Fourier transform definition (9.1), it will now be proved that the inverse continuous-time Fourier transform (ICTFT, or IFT) is

$$x(t) = \frac{1}{2\pi}\int_{-\infty}^{\infty} X(j\omega)e^{j\omega t}\,d\omega \qquad \text{(IFT)}. \tag{9.41}$$

Notice right away from Eq. (9.41) that

$$x(0) = \frac{1}{2\pi}\int_{-\infty}^{\infty} X(j\omega)\,d\omega. \tag{9.42}$$

Equation (9.41) is also called the **Fourier integral**.

Proof of Eq. (9.41). The right-hand side of Eq. (9.41) can be rewritten as

$$\begin{aligned}
\frac{1}{2\pi}\int_{-\infty}^{\infty} X(j\omega)e^{j\omega t}\,d\omega &= \frac{1}{2\pi}\int_{-\infty}^{\infty}\Big(\int_{-\infty}^{\infty} x(\tau)e^{-j\omega\tau}\,d\tau\Big)e^{j\omega t}\,d\omega \\
&= \frac{1}{2\pi}\int_{-\infty}^{\infty} x(\tau)\Big(\int_{-\infty}^{\infty} e^{j\omega(t-\tau)}\,d\omega\Big)d\tau \\
&= \int_{-\infty}^{\infty} x(\tau)\delta_c(t-\tau)d\tau = x(t),
\end{aligned} \tag{9.43}$$

which completes the proof. In the first line, we have just substituted the definition of $X(j\omega)$, taking care to use the dummy variable of integration τ so as not to get mixed up with t. In the second line, the integrals have been interchanged. This requires mathematical justification, which we have skipped. In the third line, we have used Eq. (9.34). ▽ ▽ ▽

The double-headed arrow notation $x(t) \longleftrightarrow X(j\omega)$, which we have been using, should not be taken to mean that the FT and IFT are exactly the same operation. Comparing the IFT (9.41) with the FT (9.1), we see that there are two differences: one has $e^{j\omega t}$ while the other has $e^{-j\omega t}$ and (less importantly) one has a $1/2\pi$. In fact, if $f = \omega/2\pi$ is used as the frequency variable instead of ω, this asymmetry of $1/2\pi$ disappears because $df = d\omega/2\pi$. But the use of ω (rads/sec) has some notational advantages over f (Hz), so we shall continue to use ω.

9.4.1 The Ideal Lowpass Filter

Since the FT and IFT are similar integrals, evaluation of the IFT is very similar to that of the FT. For example, let

$$H(j\omega) = \begin{cases} 1 & \text{for} -\omega_c < \omega < \omega_c, \\ 0 & \text{otherwise.} \end{cases} \tag{9.44}$$

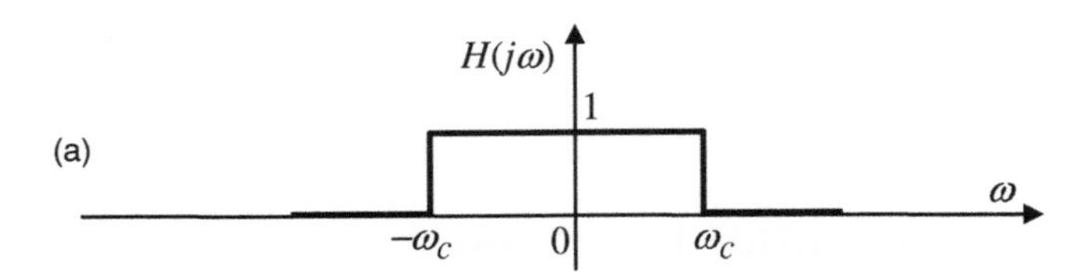

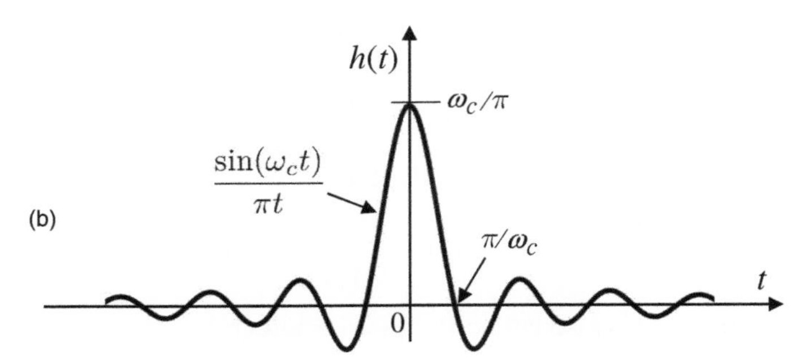

Figure 9.8 (a) Frequency response $H(j\omega)$ of the ideal lowpass filter and (b) the corresponding impulse response $h(t)$.

This is shown in Fig. 9.8(a), and is nothing but the frequency response of an **ideal lowpass filter** with cutoff frequency ω_c. This is also called a brick-wall filter because of the abrupt (discontinuous) transition from passband to stopband. Then the inverse FT, which is the impulse response of the filter, is given by

$$h(t) = \frac{1}{2\pi}\int_{-\infty}^{\infty} H(j\omega)e^{j\omega t}\,d\omega = \frac{1}{2\pi}\int_{-\omega_c}^{\omega_c} e^{j\omega t}\,d\omega. \tag{9.45}$$

We evaluated a similar integral in Eq. (9.18). The final answer is readily seen to be

$$h(t) = \frac{\sin(\omega_c t)}{\pi t}, \tag{9.46}$$

which is a sinc function. This is plotted in Fig. 9.8(b). Thus, the filter is noncausal, since $h(t)$ can be nonzero for $t < 0$. It is also unstable because $\int_{-\infty}^{\infty} |h(t)|dt$ is not finite. These results are similar to what we found in the discrete-time case.

9.4.2 Playing with the FT and IFT Operators

For a moment let us resort to the notation $\widehat{x}(\omega)$ for convenience, instead of $X(j\omega)$ (see end of Sec. 9.2.1). Thus

$$\widehat{x}(\omega) = FT(x(t)) = \int_{-\infty}^{\infty} x(t)e^{-j\omega t}\,dt, \tag{9.47}$$

$$x(t) = IFT(\widehat{x}(\omega)) = \frac{1}{2\pi}\int_{-\infty}^{\infty} \widehat{x}(\omega)e^{j\omega t}\,d\omega. \tag{9.48}$$

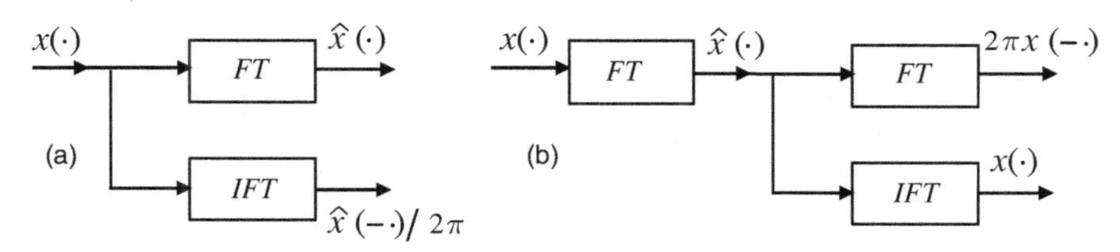

Figure 9.9 (a), (b) A pictorial representation of the notations and properties given in Eqs. (9.49)–(9.51). Here $\widehat{x}(\cdot)$ is the FT of $x(\cdot)$.

Just for this discussion, we will make it even simpler: we will dispense with the use of t and ω and write

$$FT(x(\cdot)) = \widehat{x}(\cdot), \quad IFT(\widehat{x}(\cdot)) = x(\cdot). \tag{9.49}$$

Whether a functional argument is denoted by t or ω or some other letter is not relevant. Comparing Eqs. (9.47) and (9.48), we note two differences between the FT and IFT operators. (a) One has $e^{j\omega t}$ and the other has $e^{-j\omega t}$. (b) One has a $1/2\pi$, while the other does not. Based on these observations, it follows that if we replace FT with IFT and vice versa in Eq. (9.49), we get

$$IFT(x(\cdot)) = \frac{\widehat{x}(-\cdot)}{2\pi}, \qquad FT(\widehat{x}(\cdot)) = 2\pi x(-\cdot). \tag{9.50}$$

What do these equations mean?

1. The first equation in (9.50) says that the IFT of a certain function $x(\cdot)$ (say, a plot) is the FT of the *same function reversed and divided by* 2π. This is sometimes called the **duality** property of Fourier transformation. Thus, if we have already computed the FT of certain shapes, this result is useful to find the inverse FT of similar shapes without elaborate calculations.
2. The second equation in (9.50) says that

$$FT(FT(x(\cdot))) = 2\pi x(-\cdot). \tag{9.51}$$

More verbally, the Fourier transform of the Fourier transform of $x(t)$ is the time-reversed version of $x(t)$ scaled by 2π. There are interesting theoretical implications of this, for example, see Problem 9.19. Figure 9.9 gives a pictorial representation of all these ideas.

9.5 The Dirac Delta Train

Consider a continuous-time signal of the form

$$s(t) = \sum_{n=-\infty}^{\infty} \delta_c(t - nT), \tag{9.52}$$

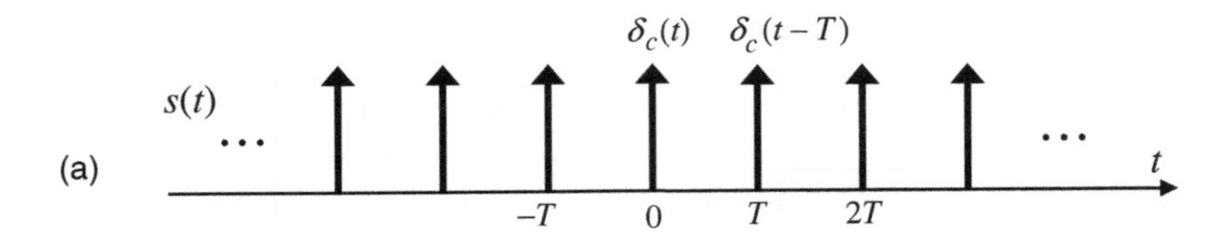

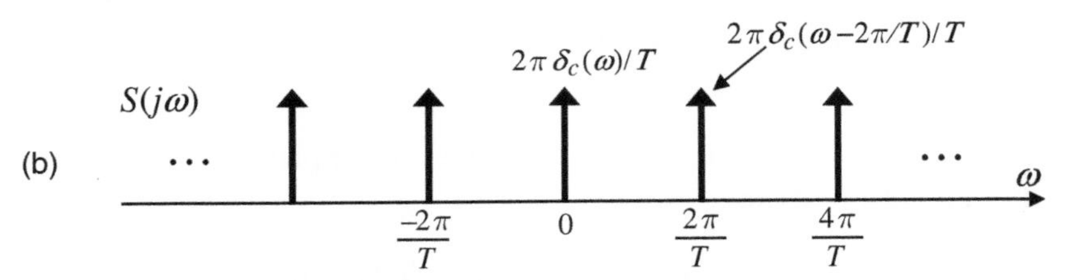

Figure 9.10 (a) The Dirac delta train $s(t) = \sum_{n=-\infty}^{\infty} \delta_c(t - nT)$ and (b) its Fourier transform shown in Eq. (9.56).

where $T > 0$. This is called an impulse train or a Dirac delta train. It is zero everywhere except at the uniformly spaced points $-T, 0, T, 2T, \ldots$, where there is a Dirac delta, as shown in Fig. 9.10(a). We now show that the Fourier transform of this signal is also a Dirac delta train. From the definition of the FT it follows that

$$\begin{aligned} S(j\omega) &= \int_{-\infty}^{\infty} \sum_{n=-\infty}^{\infty} \delta_c(t - nT) e^{-j\omega t} dt \\ &= \sum_{n=-\infty}^{\infty} \int_{-\infty}^{\infty} \delta_c(t - nT) e^{-j\omega t} dt = \sum_{n=-\infty}^{\infty} e^{-j\omega nT}. \end{aligned} \tag{9.53}$$

Now, from our discussions on the DTFT, we know the following:

$$x[n] = 1, \ \forall n \longleftrightarrow X(e^{j\omega}) = 2\pi \sum_{k=-\infty}^{\infty} \delta_c(\omega - 2\pi k). \tag{9.54}$$

Since the DTFT of $x[n]$ is $\sum_{n=-\infty}^{\infty} e^{-j\omega n}$, Eq. (9.54) can be rewritten as

$$\sum_{n=-\infty}^{\infty} e^{-j\omega n} = 2\pi \sum_{k=-\infty}^{\infty} \delta_c(\omega - 2\pi k). \tag{9.55}$$

Substituting this into Eq. (9.53) it follows that

$$S(j\omega) = 2\pi \sum_{k=-\infty}^{\infty} \delta_c(\omega T - 2\pi k) = \frac{2\pi}{T} \sum_{k=-\infty}^{\infty} \delta_c\Big(\omega - \frac{2\pi}{T} k\Big), \tag{9.56}$$

where we have used the property $\delta_c(at) = \delta_c(t)/a$ for $a > 0$ (Sec. 2.12.2). This establishes the Fourier transform pair

$$s(t) = \sum_{n=-\infty}^{\infty} \delta_c(t - nT) \longleftrightarrow \frac{2\pi}{T} \sum_{k=-\infty}^{\infty} \delta_c\Big(\omega - \frac{2\pi}{T}k\Big). \tag{9.57}$$

A plot of this FT pair is shown in Fig. 9.10. This result will be very useful when we discuss the sampling theorem.

9.6 The Derivative Properties and Applications

From the inverse FT relation $x(t) = \int_{-\infty}^{\infty} X(j\omega)e^{j\omega t}\, d\omega/2\pi$, we see that

$$\begin{aligned}\frac{dx(t)}{dt} &= \frac{1}{2\pi}\frac{d}{dt}\int_{-\infty}^{\infty} X(j\omega)e^{j\omega t} d\omega \\ &= \frac{1}{2\pi}\int_{-\infty}^{\infty} X(j\omega)\frac{de^{j\omega t}}{dt} d\omega = \frac{1}{2\pi}\int_{-\infty}^{\infty} j\omega X(j\omega)e^{j\omega t} d\omega,\end{aligned} \tag{9.58}$$

where it has been assumed that the derivative and integral operations are interchangeable. Equation (9.58) shows that

$$\frac{dx(t)}{dt} \longleftrightarrow j\omega X(j\omega). \tag{9.59}$$

Similarly, from $X(j\omega) = \int_{-\infty}^{\infty} x(t)e^{-j\omega t} dt$ we have

$$\frac{dX(j\omega)}{d\omega} = \int_{-\infty}^{\infty} x(t)\frac{de^{-j\omega t}}{d\omega} dt = -\int_{-\infty}^{\infty} jtx(t)e^{-j\omega t} dt, \tag{9.60}$$

which shows that $dX(j\omega)/d\omega \longleftrightarrow -jtx(t)$, that is,

$$tx(t) \longleftrightarrow \frac{jdX(j\omega)}{d\omega}. \tag{9.61}$$

These properties are often useful to simplify the derivation of $X(j\omega)$ for certain signals. For example, Eq. (9.58) is useful when the FT of $dx(t)/dt$ can be evaluated more readily than $X(j\omega)$. In these cases we simply use Eq. (9.59) to obtain

$$X(j\omega) = \frac{\text{FT of } dx(t)/dt}{j\omega}, \quad \omega \neq 0. \tag{9.62}$$

We then take $X(0)$ to be the limiting value of the right-hand side as $\omega \to 0$. If this limit is not easy to find, we can use $X(0) = \int_{-\infty}^{\infty} x(t)dt$, which is often easier. If $X(j\omega)$ is not continuous at $\omega = 0$, other methods have to be used to find out what happens at zero frequency. It can be shown (Sec. 17.4.1) that when $x(t)$ is absolutely integrable then $X(j\omega)$ is bounded and continuous. In these cases $X(0)$ can be found as the limit of Eq. (9.62). But in examples like $x(t) = \mathcal{U}(t)$ and $x(t) = 1$, $x(t)$ is not absolutely integrable, and $X(j\omega)$ is in fact neither continuous nor bounded at $\omega = 0$. So other methods are used to find $X(j\omega)$ (Sec. 9.2.5, Problem 9.5).

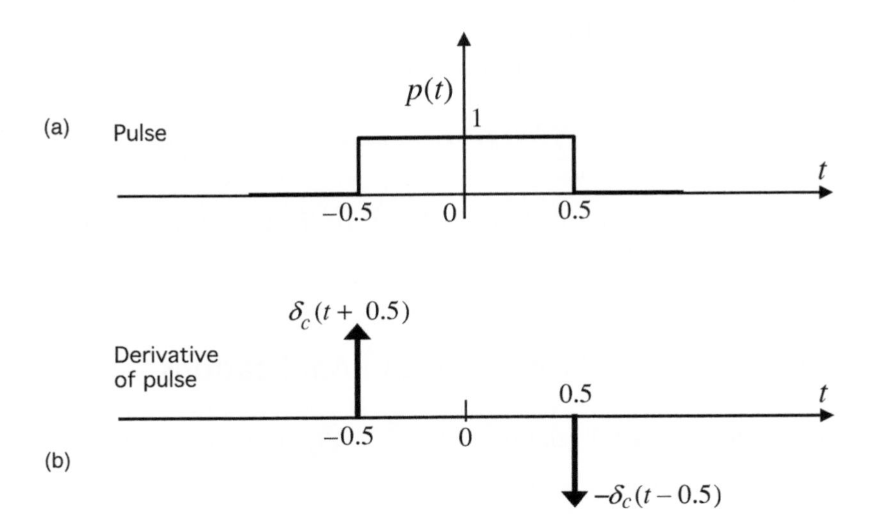

Figure 9.11 (a) The pulse signal and (b) its derivative.

Revisiting the pulse example. Consider again the pulse $p(t)$ in Eq. (9.17). This is discontinuous at $t = \pm 0.5$, and is not differentiable in the strict sense. However, if we allow Dirac delta functions in our answer, we can find an expression for the derivative! Using the results of Sec. 2.12.3, it follows that

$$q(t) \triangleq \frac{dp(t)}{dt} = \delta_c(t + 0.5) - \delta_c(t - 0.5). \tag{9.63}$$

This is depicted in Fig. 9.11. It can easily be deduced that

$$Q(j\omega) = e^{j\omega/2} - e^{-j\omega/2} = 2j\sin(\omega/2), \tag{9.64}$$

where Eq. (9.8) has been used. But from Eq. (9.59) we have $Q(j\omega) = j\omega P(j\omega)$, so that

$$P(j\omega) = \frac{2j\sin(\omega/2)}{j\omega} = \frac{\sin(\omega/2)}{\omega/2}, \quad \omega \neq 0. \tag{9.65}$$

For $\omega = 0$, we readily find $P(0) = \int p(t)dt = 1$, which is also the limit of the above expression as $\omega \to 0$. So $P(j\omega) = \sin(\omega/2)/(\omega/2)$ for all ω. The main point is that we can derive the FT of the pulse by using the derivative property instead of an explicit integration as in Eq. (9.18). While explicit integration as in Eq. (9.18) is quite trivial in this example, there are cases where it can be tedious, and the use of the derivative property simplifies matters (e.g., Problems 9.4 and 9.7). ▽ ▽ ▽

Interestingly, clever use of the derivative properties (9.59) and (9.61) leads to some ingenious examples, as we shall see in Sec. 9.7.

9.7 The Gaussian and its Fourier Transform

Consider the signal

$$x(t) = \frac{e^{-t^2/2}}{\sqrt{2\pi}}, \tag{9.66}$$

which is called the Gaussian signal. This is nothing but the Gaussian probability density function (pdf), with zero mean and a standard deviation of unity. Since this is a pdf, we have $x(t) \geq 0$ for all t, and furthermore

$$\int_{-\infty}^{\infty} x(t)dt = 1. \tag{9.67}$$

A plot of $x(t)$ is shown in Fig. 9.12. To compute the Fourier transform of $x(t)$, we can substitute into Eq. (9.1) and try to simplify it. But we will follow a different procedure, just to demonstrate the power of the derivative properties discussed in Sec. 9.6. For this, first observe from Eq. (9.66) that

$$\frac{dx(t)}{dt} = -\frac{te^{-t^2/2}}{\sqrt{2\pi}}, \tag{9.68}$$

which proves that

$$\frac{dx(t)}{dt} = -tx(t). \tag{9.69}$$

By taking the Fourier transform on both sides and using the properties (9.59) and (9.61), we conclude that

$$j\omega X(j\omega) = \frac{-jdX(j\omega)}{d\omega}. \tag{9.70}$$

Thus, $X(j\omega)$ satisfies a first-order differential equation, and we can find $X(j\omega)$ simply by solving it. For this we only have to rewrite Eq. (9.70) as

$$\frac{dX(j\omega)}{X(j\omega)} = -\omega d\omega, \tag{9.71}$$

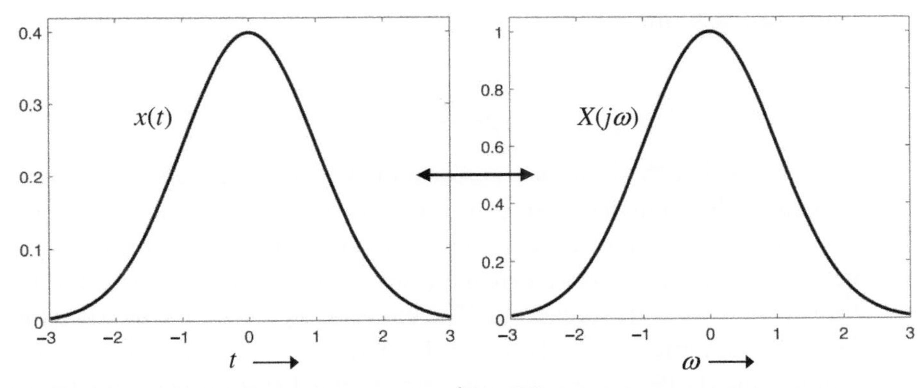

Figure 9.12 The Gaussian signal $x(t) = e^{-t^2/2}/\sqrt{2\pi}$ and its Fourier transform $X(j\omega) = e^{-\omega^2/2}$. Note that $x(t)$ and $X(j\omega)$ have identical shape.

and integrate both sides! The result is

$$\ln X(j\omega) = -\frac{\omega^2}{2} + A, \tag{9.72}$$

where A is the constant of integration (which we will find soon). This can be rewritten as

$$X(j\omega) = Be^{-\omega^2/2}, \tag{9.73}$$

where $B = e^A$. The constant B can readily be evaluated as follows:

$$B = X(0) = \int_{-\infty}^{\infty} x(t)dt = 1. \tag{9.74}$$

The first equality follows from Eq. (9.73), the second from Eq. (9.5), and the third from Eq. (9.67). So we have shown that

$$X(j\omega) = e^{-\omega^2/2}, \tag{9.75}$$

that is,

$$\frac{e^{-t^2/2}}{\sqrt{2\pi}} \longleftrightarrow e^{-\omega^2/2}. \tag{9.76}$$

Thus the Fourier transform of the Gaussian signal (9.66) has exactly the same Gaussian shape, except for a scale factor of $\sqrt{2\pi}$. See Fig. 9.12. An important point here is that we have been able to find the FT of the Gaussian pulse without explicit evaluation of the integral $X(j\omega) = \int_{-\infty}^{\infty} x(t)e^{-j\omega t}dt$.

The dilated Gaussian. Applying the dilation property (9.22), it also follows that

$$\frac{e^{-t^2/2\sigma^2}}{\sqrt{2\pi}} \longleftrightarrow \sigma e^{-\sigma^2\omega^2/2}, \tag{9.77}$$

which can be written as

$$\frac{e^{-t^2/2\sigma^2}}{\sqrt{2\pi\sigma^2}} \longleftrightarrow e^{-\sigma^2\omega^2/2}. \tag{9.78}$$

This shows that the Fourier transform of the Gaussian pulse with standard deviation σ is also Gaussian (except for a scale factor) but it has a different standard deviation, namely, $1/\sigma$. Figure 9.13 shows this Fourier transform pair for two values of σ. Note that as σ gets smaller, the signal $x(t)$ gets squeezed but the Fourier transform $X(j\omega)$ gets spread out. The more we try to confine the signal in the time domain, the more it spreads out in the frequency domain, consistent with the dilation property (9.22). We already demonstrated this property for the case of the pulse-sinc Fourier transform pair in Eq. (9.23). ▽ ▽ ▽

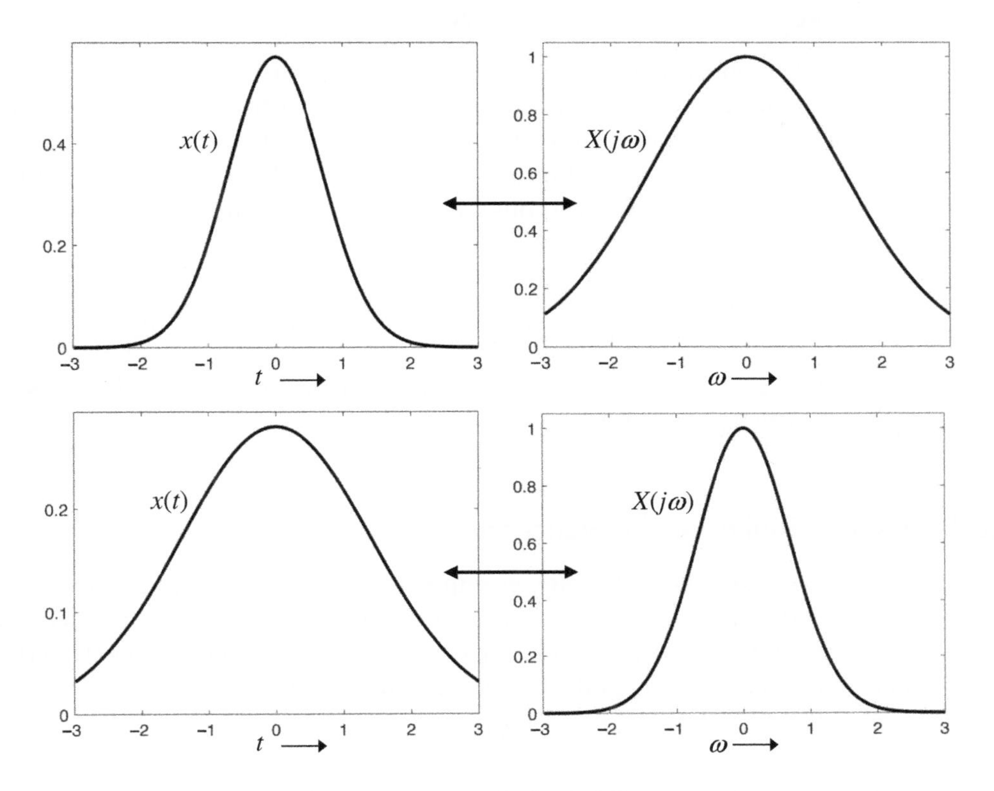

Figure 9.13 The Gaussian signal $x(t) = e^{-t^2/2\sigma^2}/\sqrt{2\pi\sigma^2}$ with arbitrary σ and its Fourier transform $X(j\omega) = e^{-\sigma^2\omega^2/2}$. (a) $\sigma = 0.7$ and (b) $\sigma = 1/0.7$. As the signal $x(t)$ gets squeezed, the Fourier transform $X(j\omega)$ gets spread out.

9.8 Eigenfunctions of the Fourier Transform*

If $x(t)$ and $X(j\omega)$ have the same shape, that is,

$$x(t) = f(t) \quad \text{and} \quad X(j\omega) = \lambda f(\omega) \tag{9.79}$$

for some $f(t)$, we say that $f(t)$ is an **eigenfunction** of the continuous-time Fourier transform operator. In this case the plots of $x(t)$ and $X(j\omega)$ look exactly the same except for the scale factor λ. The scale factor λ is called the **eigenvalue** of the Fourier transform operator corresponding to the eigenfunction $f(t)$. From the example (9.76), we see that the signal with Gaussian shape

$$f(t) = \frac{e^{-t^2/2}}{\sqrt{2\pi}} \tag{9.80}$$

is an eigenfunction, and the eigenvalue is $\lambda = \sqrt{2\pi}$. If $f(t)$ is an eigenfunction then so is $cf(t)$ for any $c \neq 0$. So the quantity $1/\sqrt{2\pi}$ can be dropped in Eq. (9.80). Notice from Eq. (9.78) that the Gaussian signal $f(t) = e^{-t^2/2\sigma^2}/\sqrt{2\pi\sigma^2}$ is not an eigenfunction when $\sigma \neq 1$. In this case, the signal and its Fourier transform (shown in Eq. (9.78)) do not look the same (Fig. 9.13). Only the *normalized Gaussian* ($f(t)$ with $\sigma = 1$) is an eigenfunction of the Fourier transform operator.

An interesting question then is, are there other eigenfunctions of the Fourier transform besides the normalized Gaussian? The answer is yes. One simple example can be derived from the Fourier transform pair (9.57) for the impulse train. If $T = \sqrt{2\pi}$, the two impulse trains will look exactly identical (except for a scale factor):

$$\sum_{n=-\infty}^{\infty} \delta_c\left(t - n\sqrt{2\pi}\right) \longleftrightarrow \sqrt{2\pi} \sum_{k=-\infty}^{\infty} \delta_c\left(\omega - k\sqrt{2\pi}\right). \tag{9.81}$$

Thus, the signal on the left-side is an eignefunction of the Fourier transform operator, with eigenvalue $\sqrt{2\pi}$.

9.8.1 Infinite Families of Eigenfunctions

In fact there exist infinitely many examples of eignefunctions of the FT operator. We conclude this section by mentioning some results in this regard. The details are developed in the Problems section. First, an interesting result (Problem 9.18) is that if $f(t)$ is an eigenfunction of the Fourier transform operator, then

$$\frac{df(t)}{dt} - tf(t) \tag{9.82}$$

is also an eigenfunction (as long as it is not identically zero). Using this result, and starting from Eq. (9.76), the following Fourier transform pairs can be established (Problem 9.18):

$$\begin{aligned} f_1(t) = te^{-t^2/2} &\longleftrightarrow -j\sqrt{2\pi}\,\omega e^{-\omega^2/2}, \\ f_2(t) = (2t^2 - 1)e^{-t^2/2} &\longleftrightarrow -\sqrt{2\pi}(2\omega^2 - 1)e^{-\omega^2/2}, \\ f_3(t) = (2t^3 - 3t)e^{-t^2/2} &\longleftrightarrow j\sqrt{2\pi}(2\omega^3 - 3\omega)e^{-\omega^2/2}. \end{aligned} \tag{9.83}$$

Thus the functions $f_1(t)$, $f_2(t)$, and $f_3(t)$ are eigenfunctions, with eigenvalues

$$-j\sqrt{2\pi}, \ -\sqrt{2\pi}, \text{ and } j\sqrt{2\pi}, \tag{9.84}$$

respectively.

The above examples belong to a family of eigenfunctions of the form

$$f_n(t) = H_n(t)e^{-t^2/2}, \tag{9.85}$$

where $H_n(t)$ are called **Hermite** polynomials and $f_n(t)$ are the **Gauss–Hermite** family of eigenfunctions. For example, $H_0(t) = 1$ in Eq. (9.85) yields the Gaussian $f_0(t)$ and

$$H_1(t) = t, \ H_2(t) = 2t^2 - 1, \text{ and } H_3(t) = 2t^3 - 3t \tag{9.86}$$

yield the examples in Eq. (9.83). The first four eigenfunctions $f_0(t)$, $f_1(t)$, $f_2(t)$, and $f_3(t)$ are plotted in Fig. 9.14. Since $cf_n(t)$ is an eigenfunction for arbitrary $c \neq 0$, the y-axis markings in the figure should not be taken seriously.

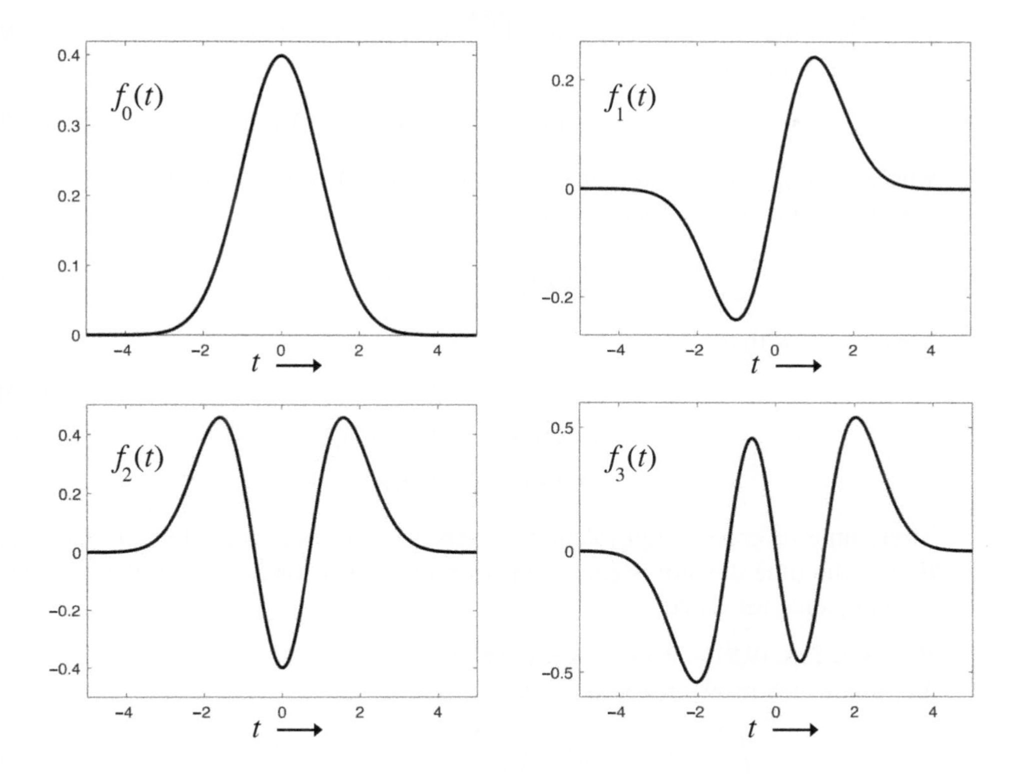

Figure 9.14 The first four Gaussian–Hermite functions. These are eigenfunctions of the Fourier transform operator. Note that $f_0(t)$ is nothing but the Gaussian function.

9.8.2 Only Four Possible Eigenvalues

It turns out that the examples do not end here. It can be shown that there are infinitely many other families of eigenfunctions of the Fourier transform operator (Problem 9.28). Finally, it turns out that although there are infinitely many eigenfunctions, there are only **four** possible eigenvalues no matter what the eigenfunctions are, namely,

$$\lambda = \sqrt{2\pi}, \; -j\sqrt{2\pi}, \; -\sqrt{2\pi}, \text{ and } j\sqrt{2\pi}. \tag{9.87}$$

No other eigenvalues are possible (Problem 9.19).

9.9 Symmetry and Realness

We now mention a number of useful properties that follow readily from the definition of FT reproduced below:

$$X(j\omega) = \int_{-\infty}^{\infty} x(t)e^{-j\omega t}dt. \tag{9.88}$$

First recall that $X(j\omega)$ can in general be complex even if $x(t)$ is real. For complex $X(j\omega)$ we can always write

$$X(j\omega) = |X(j\omega)|e^{j\phi(\omega)} = X_r(\omega) + jX_i(\omega), \tag{9.89}$$

where $|X(j\omega)|$ is the magnitude, $\phi(\omega)$ is the phase, $X_r(\omega)$ is the real part, and $X_i(\omega)$ the imaginary part of $X(j\omega)$. The phase is given by

$$\phi(\omega) = \arg(X(j\omega)) = \tan^{-1}\frac{X_i(\omega)}{X_r(\omega)}. \tag{9.90}$$

We now show that

$$x(-t) \longleftrightarrow X(-j\omega), \tag{9.91}$$

$$x^*(t) \longleftrightarrow X^*(-j\omega), \tag{9.92}$$

$$x^*(-t) \longleftrightarrow X^*(j\omega). \tag{9.93}$$

Thus, time reversal is equivalent to reversal in the frequency domain. Conjugation in the time domain is equivalent to *reversal and conjugation* in the frequency domain, and vice versa.

Proofs of Eqs. (9.91)–(9.93). We have $\int_{-\infty}^{\infty} x(-t)e^{-j\omega t}dt = \int_{-\infty}^{\infty} x(t)e^{j\omega t}dt = X(-j\omega)$, which proves the first property. Next,

$$\int_{-\infty}^{\infty} x^*(t)e^{-j\omega t}dt = \left(\int_{-\infty}^{\infty} x(t)e^{j\omega t}dt\right)^* = X^*(-j\omega), \tag{9.94}$$

proving the second property. The third property follows from the first two. ▽ ▽ ▽

Now assume that $x(t)$ is a real signal, or equivalently, $x(t) = x^*(t)$. From Eq. (9.92) it follows that realness of $x(t)$ is equivalent to the property $X(j\omega) = X^*(-j\omega)$, which is the Hermitian symmetry property. Similarly it follows from Eq. (9.93) that realness of $X(j\omega)$ is equivalent to $x(t) = x^*(-t)$. Thus

$$x(t) \text{ is real} \quad \Longleftrightarrow \quad X(j\omega) = X^*(-j\omega), \tag{9.95}$$

$$X(j\omega) \text{ is real} \quad \Longleftrightarrow \quad x(t) = x^*(-t). \tag{9.96}$$

9.9.1 Types of Symmetry

Recall (Sec. 2.2) that $x(t)$ is **symmetric** or **even** if $x(t) = x(-t)$ (although $x(t)$ may not be real), and similarly $x(t)$ is **antisymmetric** or **odd** if $x(t) = -x(-t)$. From Eq. (9.91) it follows that

$$x(t) = x(-t) \quad \Longleftrightarrow \quad X(j\omega) = X(-j\omega), \tag{9.97}$$

$$x(t) = -x(-t) \quad \Longleftrightarrow \quad X(j\omega) = -X(-j\omega). \tag{9.98}$$

Thus $x(t)$ is even (odd) if and only if $X(j\omega)$ is even (odd), and this statement does not have anything to do with the realness or otherwise of $x(t)$ or $X(j\omega)$.

Next, we say that $x(t)$ is **Hermitian** or Hermitian symmetric if $x(t) = x^*(-t)$, and **skew Hermitian** or Hermitian antisymmetric if $x(t) = -x^*(-t)$. Clearly, for real signals these two properties are equivalent to even and odd properties, respectively. From Eq. (9.96) we see that $x(t)$ is Hermitian if and only if $X(j\omega)$ is real, and similarly from Eq. (9.95) $X(j\omega)$ is Hermitian if and only if $x(t)$ is real.

In short, *realness in one domain* is equivalent to *the Hermitian property in the other domain.*

9.9.2 Real Signals

From Eq. (9.95) it follows that

$$x(t) \text{ is real} \quad \Longleftrightarrow \quad |X(j\omega)| = |X(-j\omega)| \text{ and } \phi(\omega) = -\phi(-\omega), \tag{9.99}$$

where $\phi(\omega) = \arg(X(j\omega))$. With $X(j\omega) = X_r(\omega) + jX_i(\omega)$, where $X_r(\omega)$ and $X_i(\omega)$ are the real and imaginary parts, this can be rewritten as

$$x(t) \text{ is real} \quad \Longleftrightarrow \quad X_r(\omega) = X_r(-\omega) \text{ and } X_i(\omega) = -X_i(-\omega). \tag{9.100}$$

That is, $x(t)$ is real if and only if the real part of $X(j\omega)$ is even and the imaginary part of $X(j\omega)$ is odd. Finally, by combining Eqs. (9.95) and (9.96) it follows that

$$x(t) \text{ is real } \textit{and} \text{ even} \quad \Longleftrightarrow \quad X(j\omega) \text{ is real and even.} \tag{9.101}$$

Extensions for signals with other types of symmetry can be found in Problems 9.13 and 9.14. Here we just mention one property. Suppose $x(t)$ is real, and also odd, that is, $x(t) = -x(-t)$. Then from Eq. (9.98) it follows that $X(j\omega)$ is also odd. That is,

$$X_r(\omega) = -X_r(-\omega) \text{ and } X_i(\omega) = -X_i(-\omega). \tag{9.102}$$

But since Eq. (9.100) must also be true, it follows that $X_r(\omega) = 0$. So $X(j\omega)$ is purely imaginary. Similarly, it can be proved that if $X(j\omega)$ is imaginary and odd, then $x(t)$ is real and odd. Summarizing:

$$x(t) \text{ is real and odd} \quad \Longleftrightarrow \quad X(j\omega) \text{ is imaginary and odd.} \tag{9.103}$$

Revisiting some examples. The pulse $p(t)$ in Eq. (9.17) is real and even, and so indeed is its Fourier transform (9.18), as required by Eq. (9.101). Now consider the shifted pulse

$$p_0(t) = \begin{cases} 1 & \text{for } 0 < t < 1, \\ 0 & \text{otherwise.} \end{cases} \tag{9.104}$$

In view of the shifting property (9.7), it follows that

$$\begin{aligned} P_0(j\omega) &= e^{-j\omega/2} \frac{\sin(\omega/2)}{\omega/2} \\ &= \cos(\omega/2)\Big(\frac{\sin(\omega/2)}{\omega/2}\Big) - j\sin(\omega/2)\Big(\frac{\sin(\omega/2)}{\omega/2}\Big). \end{aligned} \tag{9.105}$$

Thus, although the signal $p_0(t)$ is real, its Fourier transform $P_0(j\omega)$ is complex. It is clear from the above expression that the real part of $P_0(j\omega)$ is even, and the imaginary part is odd. This is indeed in accordance with Eq. (9.100).

Next consider the Gaussian FT pair in Eq. (9.78). Here the signal is real and even, and so is its Fourier transform (Fig. 9.13), consistent with Eq. (9.101). However, consider some of the other eigenfunctions of the FT, say $f_1(t)$ given in Eq. (9.83). We have

$$f_1(t) = te^{-t^2/2} \longleftrightarrow -j\sqrt{2\pi}\,\omega e^{-\omega^2/2}. \tag{9.106}$$

So $f_1(t)$ is real and odd (see Fig. 9.14), and its FT is indeed imaginary and odd, consistent with the property (9.103). ▽ ▽ ▽

9.9.3 The Right-Sided Exponential

Now we consider the right-sided exponential

$$x(t) = e^{-at}\mathcal{U}(t) \tag{9.107}$$

to demonstrate some of the properties mentioned above. Here $\mathcal{U}(t)$ is the unit step. If $Re[a] > 0$, this is a decaying exponential, and in this case the Fourier transform exists. Note that if $Re[a] < 0$, the signal is unbounded and the FT does not exist. For $Re[a] = 0$, $x(t)$ becomes the unit step and the FT exists only if we allow Dirac delta functions into the answer (Problem 9.5). Figure 9.15(a) shows a plot of $x(t)$ for $a = 1/2$. Assuming that $Re[a] > 0$, the FT can be evaluated as follows:

$$X(j\omega) = \int_{-\infty}^{\infty} e^{-at}\mathcal{U}(t)e^{-j\omega t}dt = \int_0^{\infty} e^{-(a+j\omega)t}dt = \frac{-e^{-(a+j\omega)t}}{a+j\omega}\Bigg|_{t=0}^{\infty}.$$

Since $Re[a] > 0$, $e^{-(a+j\omega)t} \to 0$ as $t \to \infty$. So the above reduces to

$$X(j\omega) = \frac{1}{a+j\omega}. \tag{9.108}$$

Since this is complex we can write $X(j\omega) = |X(j\omega)|e^{j\phi(\omega)}$. To simplify further, we will consider a special case.

Case of real *a*. Assume a is real so that $x(t)$ itself is real. Then

$$|X(j\omega)| = \frac{1}{\sqrt{a^2+\omega^2}}, \quad \phi(\omega) = -\tan^{-1}\frac{\omega}{a}. \tag{9.109}$$

Figures 9.15(b) and (c) show plots of these for $a = 1/2$. Since $x(t)$ is real, we see indeed that $|X(j\omega)|$ is even and $\phi(\omega)$ is odd. Note that $|X(j\omega)|$ is $1/a$ for $\omega = 0$ and decreases to 0 as $\omega \to \pm\infty$. We have $\phi(0) = 0$ and $\phi(\omega) \to -\pi/2$ as $\omega \to \infty$. Since a is real we can also write

$$X(j\omega) = \frac{a-j\omega}{a^2+\omega^2} = \underbrace{\frac{a}{a^2+\omega^2}}_{X_r(\omega)} + j\underbrace{\frac{-\omega}{a^2+\omega^2}}_{X_i(\omega)}, \tag{9.110}$$

which shows that $X_r(\omega)$ and $X_i(\omega)$ are even and odd, respectively. ▽ ▽ ▽

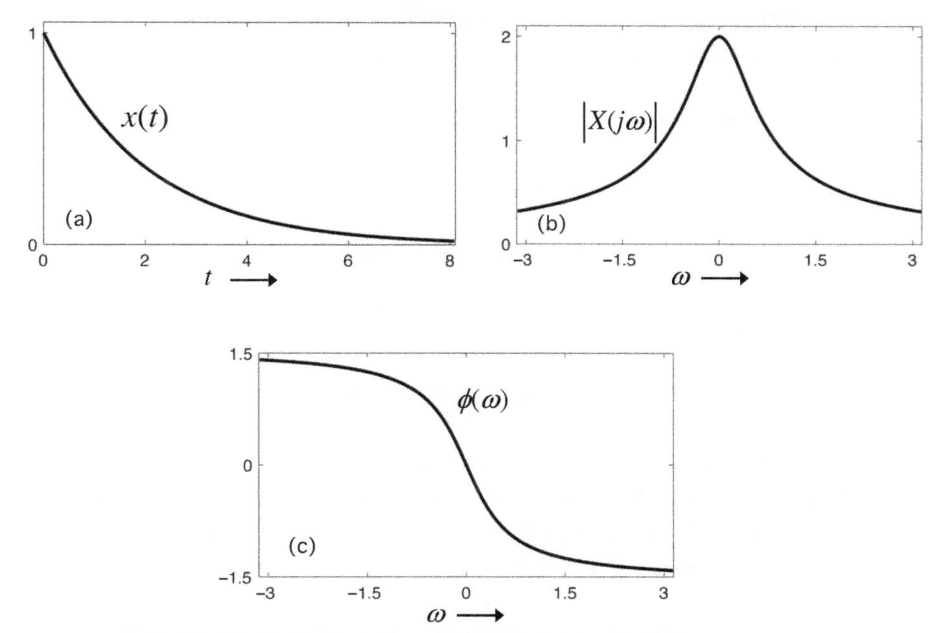

Figure 9.15 (a) The right-sided exponential $x(t) = e^{-at}\mathcal{U}(t)$, (b) the magnitude of the Fourier transform $X(j\omega)$, and (c) the phase $\phi(\omega) = \arg(X(j\omega))$. Here $a = 1/2$.

9.9.4 The Hilbert Transformer

The ideal Hilbert transformer is an LTI system with frequency response

$$H(j\omega) = \begin{cases} -j & \text{for } \omega > 0, \\ j & \text{for } \omega < 0. \end{cases} \tag{9.111}$$

This is discontinuous at $\omega = 0$. But since $|H(j\omega)| = 1$, this can be regarded as an **allpass** filter. It can be shown that the impulse response is (Problem 9.5)

$$h(t) = \frac{1}{\pi t}, \quad t \neq 0. \tag{9.112}$$

This is unbounded as $|t| \to 0$. However, $h(t)$ also changes sign as we cross over from $t < 0$ to $t > 0$. We can take $h(0) = 0$, consistent with the observation that $H(j\omega)$ is odd, so that $h(0) = \int H(j\omega)d\omega/2\pi = 0$ (with the integral interpreted as a Cauchy principal value, see Sec. 17.7). Equation (9.111) is the ideal unrealizable Hilbert transformer, and can only be approximated by realizable filters.

9.9.4.1 What Does the Hilbert Transformer "Do"?

If we apply the filter (9.111) to a signal $x(t)$, then how is it affected in the time domain? The output is clearly the convolution

$$y(t) = \frac{1}{\pi} \int_{-\infty}^{\infty} \frac{x(\tau)}{t - \tau} d\tau. \tag{9.113}$$

But what does this operation "actually do"? It is easier to understand the time-domain effect by considering the case where

$$x(t) = \cos(\omega_0 t + \theta), \quad \omega_0 > 0.$$

Since $\cos(\omega_0 t + \theta) = 0.5[e^{j(\omega_0 t+\theta)} + e^{-j(\omega_0 t+\theta)}]$, it follows that

$$y(t) = 0.5[-je^{j(\omega_0 t+\theta)} + je^{-j(\omega_0 t+\theta)}], \tag{9.114}$$

because the input $e^{j\omega_0 t}$ produces the output $H(j\omega_0)e^{j\omega_0 t}$. So

$$y(t) = \cos(\omega_0 t + \theta - \pi/2) = \sin(\omega_0 t + \theta). \tag{9.115}$$

Thus, a cosine is converted to a sine, regardless of its frequency. More specifically, letting $\theta = 0$ it follows that $x(t) = \cos(\omega_0 t)$ produces $y(t) = \sin(\omega_0 t)$. Similarly, setting $\theta = \pi/2$ we have $x(t) = -\sin(\omega_0 t)$ and $y(t) = \cos(\omega_0 t)$. Thus a cosine is turned into a sine and vice versa (except for a sign flip in one case):

$$\cos(\omega_0 t) \longmapsto \sin(\omega_0 t), \quad \sin(\omega_0 t) \longmapsto -\cos(\omega_0 t). \tag{9.116}$$

Next, if the input $x(t)$ is an arbitrary signal, then what can we say about the output of the Hilbert transformer? For the special case where $x(t)$ is a sum of sines and cosines: $x(t) = \sum_{k=1}^{N} A_k \cos(\omega_k t) + \sum_{k=1}^{N} B_k \sin(\omega_k t)$, we have

$$y(t) = \sum_{k=1}^{N} A_k \sin(\omega_k t) - \sum_{k=1}^{N} B_k \cos(\omega_k t), \tag{9.117}$$

by linearity of the system. So the cosines and sines get interchanged (with a sign flip as shown). This can readily be generalized to arbitrary $x(t)$. Thus, starting from the inverse FT formula (9.41), we have

$$x(t) = \frac{1}{2\pi}\int_{-\infty}^{\infty} X(j\omega)\cos(\omega t)d\omega + \frac{j}{2\pi}\int_{-\infty}^{\infty} X(j\omega)\sin(\omega t)d\omega. \tag{9.118}$$

It should be clear that the Hilbert transformer converts $x(t)$ into

$$y(t) = \frac{1}{2\pi}\int_{-\infty}^{\infty} X(j\omega)\sin(\omega t)d\omega - \frac{j}{2\pi}\int_{-\infty}^{\infty} X(j\omega)\cos(\omega t)d\omega. \tag{9.119}$$

The Hilbert transformer finds application in communication systems. More specifically, in systems that use amplitude modulation (AM), there is a special subclass called single side band (SSB) modulation. The SSB-modulated signals are produced at the radio transmitter with the help of a Hilbert transformer. Details can be found in books on communication systems. For example, see Lathi [1989]. A detailed discussion of Hilbert transformers and their approximations in discrete time can be found in Oppenheim and Schafer [2010].

9.10 Convolution, Correlation, and Energy

The **cross-correlation** of a signal $x(t)$ with the signal $y(t)$ is defined by the integral

$$r_{xy}(\tau) \stackrel{\Delta}{=} \int_{-\infty}^{\infty} x(t)y^*(t-\tau)dt. \tag{9.120}$$

This is nothing but the inner product (Sec. 18.3.2) between $x(t)$ and $y(t-\tau)$, and τ is called the **lag** (amount of shift in one signal before the inner product is taken). Note that Eq. (9.120) can be regarded as the convolution of $x(t)$ with the reverse-conjugated version

$$y_r(t) \stackrel{\Delta}{=} y^*(-t). \tag{9.121}$$

But since $Y_r(j\omega) = Y^*(j\omega)$ according to Eq. (9.93), it follows from the convolution theorem that the Fourier transform of $r_{xy}(\tau)$ is

$$S_{xy}(j\omega) = X(j\omega)Y_r(j\omega) = X(j\omega)Y^*(j\omega). \tag{9.122}$$

Thus

$$r_{xy}(\tau) = \int_{-\infty}^{\infty} x(t)y^*(t-\tau)dt \quad \longleftrightarrow \quad X(j\omega)Y^*(j\omega). \tag{9.123}$$

By interchanging $x(t)$ with $y(t)$ we see that this implies $r_{yx}(\tau) \longleftrightarrow Y(j\omega)X^*(j\omega)$ which proves, in view of Eq. (9.93), that

$$r_{yx}(\tau) = r_{xy}^*(-\tau). \tag{9.124}$$

This can also be shown directly by starting from the definition (9.120) (Problem 9.23). Next, from the definition (9.120), we have

$$r_{xy}(0) = \int_{-\infty}^{\infty} x(t)y^*(t)dt. \tag{9.125}$$

But since we also have $r_{xy}(\tau) \longleftrightarrow X(j\omega)Y^*(j\omega)$, it follows that $r_{xy}(0) = \int_{-\infty}^{\infty} X(j\omega)Y^*(j\omega)d\omega/2\pi$. This proves that

$$\int_{-\infty}^{\infty} x(t)y^*(t)dt = \frac{1}{2\pi}\int_{-\infty}^{\infty} X(j\omega)Y^*(j\omega)d\omega. \tag{9.126}$$

This is called **Parseval's** relation. It says that the inner product between $x(t)$ and $y(t)$ is equal to the inner product between their Fourier transforms, scaled by $1/2\pi$. So we say that the Fourier transform preserves the inner product.

9.10.1 Autocorrelations and Energy

In particular, when $x(t) = y(t)$, the function $r_{xx}(\tau)$ is called the **autocorrelation** of $x(t)$:

$$r_{xx}(\tau) = \int_{-\infty}^{\infty} x(t)x^*(t-\tau)dt. \tag{9.127}$$

From Eq. (9.123) it follows that the Fourier transform of this is $|X(j\omega)|^2$. Thus

$$r_{xx}(\tau) \longleftrightarrow S_{xx}(j\omega) = |X(j\omega)|^2 \geq 0. \tag{9.128}$$

Since Eq. (9.124) is true, we have in particular

$$r_{xx}(\tau) = r^*_{xx}(-\tau). \tag{9.129}$$

That is, the autocorrelation is Hermitian symmetric. When $x(t) = y(t)$, Parseval's relation (9.126) becomes

$$\int_{-\infty}^{\infty} |x(t)|^2 dt = \frac{1}{2\pi} \int_{-\infty}^{\infty} |X(j\omega)|^2 d\omega. \tag{9.130}$$

The left-hand side, which is also $r_{xx}(0)$, represents the **energy** of the signal $x(t)$. Equation (9.130) says that the energy of $x(t)$ is equal to the energy in its Fourier transform $X(j\omega)$ (scaled by $1/2\pi$). Finally, it is readily shown (Problem 9.23) that

$$r_{xx}(0) \geq |r_{xx}(\tau)|, \quad \forall \tau. \tag{9.131}$$

Summarizing, we have

$$\text{Energy of } x(t) = \int_{-\infty}^{\infty} |x(t)|^2 dt = r_{xx}(0) \geq |r_{xy}(\tau)|, \quad \forall \tau. \tag{9.132}$$

The chirp: a real-world signal. Consider Fig. 9.16(a), which shows a **chirp** signal $x(t)$. A chirp signal is like a sinusoid whose frequency itself increases (or decreases) with time. For example, the chirp in Fig. 9.16(a) has the form

$$x(t) = \cos((\omega_1 + \Delta_1 t)t)\, \mathcal{U}(t), \quad \Delta_1 > 0, \tag{9.133}$$

so that the frequency increases from the initial value ω_1 in a linear fashion, with slope Δ_1. So this is also called a *linear frequency-modulated* or **LFM** signal.

Chirp signals are real-world signals. We hear them all the time in busy areas of town. The ambulance or fire truck rushing past you generates a siren which sounds like a chirp because its frequency increases as it approaches and decreases as it recedes (due to the Doppler effect). The chirp is also used in other applications including radar signal processing and digital communications, because of some good properties of its correlation, as we shall see. We can regard $\omega_1 + \Delta_1 t$ as the *instantaneous frequency* of the chirp signal. Figure 9.16(b) shows another example $y(t)$ of such a signal, with a different starting frequency ω_2. This plot appears imperfect towards the end because the sampling density for plotting is not large enough for the very high frequencies near the end!

Figure 9.17 shows the autocorrelation $r_{xx}(\tau)$ of the signal $x(t)$, and the cross-correlation $r_{xy}(\tau)$. From this plot we clearly see that the symmetry property (9.129) and the inequality $r_{xx}(0) \geq |r_{xx}(\tau)|$ in Eq. (9.131) are satisfied. Since the signals are real valued in the example, all conjugates can be dropped. The autocorrelation shows a **sharp peak** at $\tau = 0$, whereas the cross correlation is very small for most τ. Similar remarks hold for $r_{yy}(\tau)$, which is also shown in the figure. We have not shown $r_{yx}(\tau)$ because $r_{xy}(\tau) = r_{yx}(-\tau)$ from Eq. (9.124). $\triangledown \triangledown \triangledown$

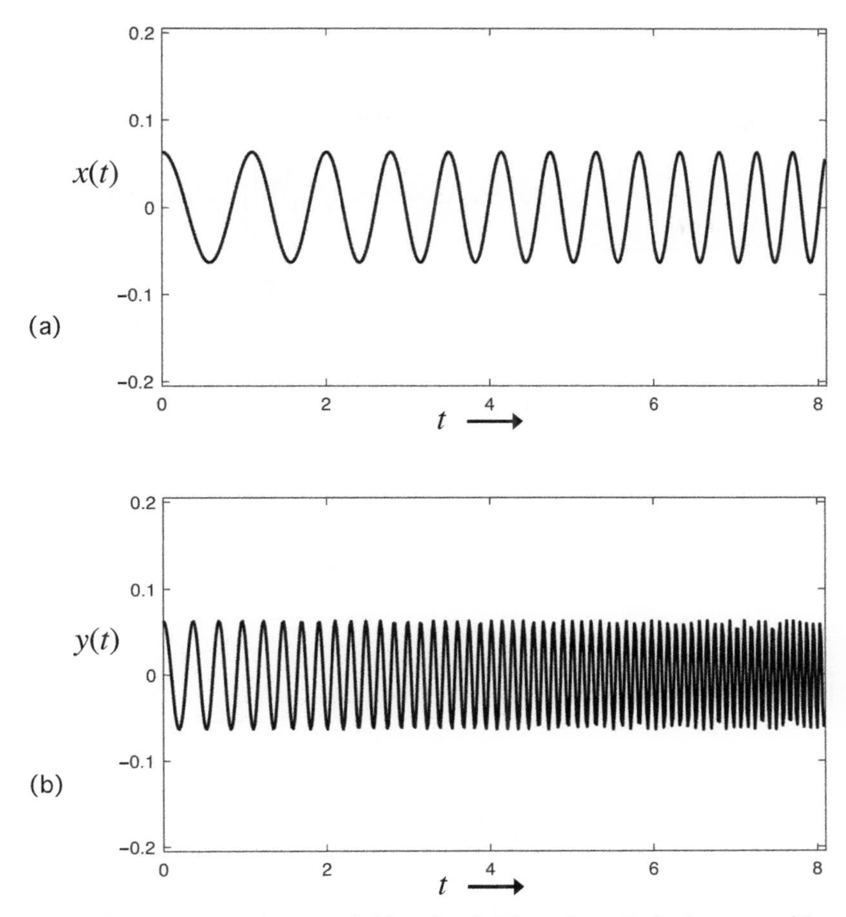

Figure 9.16 (a), (b) Examples of chirp signals whose frequencies increase with time. The second plot appears imperfect towards the end because the sampling density for plotting is not large enough as the frequency increases.

From an engineering viewpoint, we shall see that the most important properties of chirps in the above example are the following:

1. The autocorrelations have very sharp peaks at $\tau = 0$.
2. The cross correlations are small for all τ, compared to autocorrelation peaks.

9.10.2 Matched Filtering

To explain how the above properties might be useful in practical applications, imagine a *digital communications system* where the transmitter sends one of the two known waveforms, $x(t)$ or $y(t)$, to a distant receiver. For example, $x(t)$ could represent a "**0**" and $y(t)$ represent a "**1**" in binary (one-bit) communication. The task of the receiver is to figure out which of the two signals was transmitted. Given the received signal $s(t)$, which is one of $x(t)$ or $y(t)$ in the absence of noise, the receiver

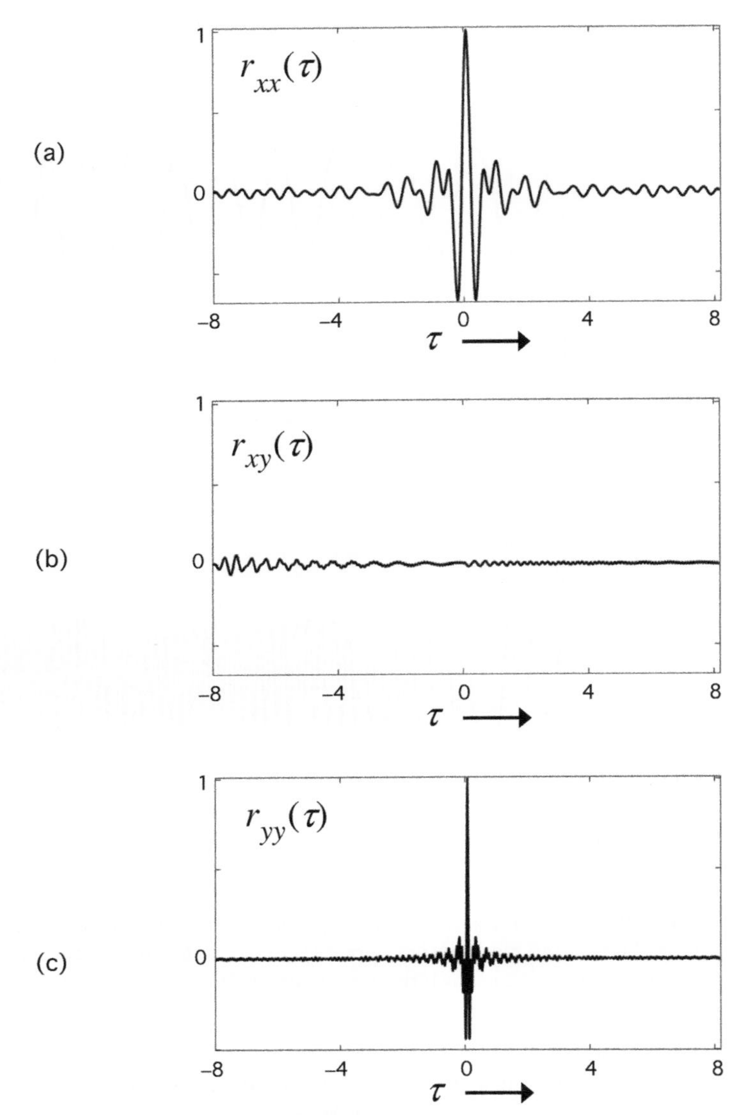

Figure 9.17 Correlations for the chirp signals in Fig. 9.16. (a) Autocorrelation $r_{xx}(\tau)$, (b) cross-correlation $r_{xy}(\tau)$, and (c) autocorrelation $r_{yy}(\tau)$. Since $r_{yx}(\tau) = r_{xy}(-\tau)$, it is not shown.

simply computes the correlations $r_{sx}(\tau)$ and $r_{sy}(\tau)$ by using locally available copies of the known shapes $x(t)$ and $y(t)$. If $r_{sx}(\tau)$ has a sharp peak and $r_{sy}(\tau)$ is uniformly small everywhere (as in Fig. 9.17), then the receiver concludes that $s(t) = x(t)$, that is, a "**0**" was transmitted. If it is the other way around, then the receiver concludes that $s(t) = y(t)$, that is, a "**1**" was transmitted. So we see that the choice of chirp signals is very convenient in such applications, although there are other possible choices for the waveforms.

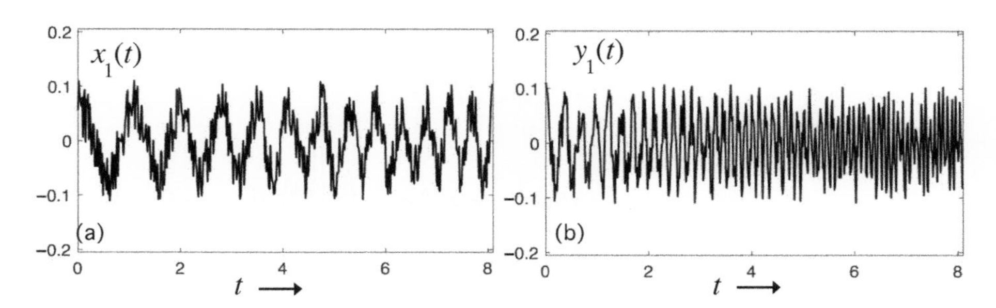

Figure 9.18 (a), (b) The chirp signals in Fig. 9.16 with noise added.

In practice, the received signal $s(t)$ is the transmitted signal contaminated with **additive noise**. For example, Figs. 9.18(a) and (b) show noisy versions

$$x_1(t) = x(t) + e(t), \quad y_1(t) = y(t) + f(t), \tag{9.134}$$

of $x(t)$ and $y(t)$ shown earlier in Fig. 9.16. Here, $e(t)$ and $f(t)$ are random noise sources that are unavoidable in communication systems. Thus, depending on whether $x(t)$ or $y(t)$ was transmitted, the received signal $s(t)$ is either $x_1(t)$ or $y_1(t)$. If we compute the correlations of the noisy signals in Figs. 9.18(a) and (b) with the noise-free signals $x(t)$ and $y(t)$, then we will get the plots in Fig. 9.19. Note that in spite of the strong noise, the peak in the plot $r_{x_1x}(\tau)$ is still strong compared to the entire plot $r_{x_1y}(\tau)$. Similarly, the peak in the plot $r_{y_1y}(\tau)$ is still very strong compared to the plot $r_{y_1x}(\tau)$ (although we have not shown the latter to save space).

So, all that we have to do at the receiver is this: given the received noisy signal $s(t)$, which is either $x_1(t)$ or $x_2(t)$, compute and plot the correlations $r_{sx}(\tau)$ and $r_{sy}(\tau)$, using the noise-free copies $x(t)$ and $y(t)$ available locally to the receiver. If $r_{sx}(\tau)$ has a sharp peak and $r_{sy}(\tau)$ is uniformly small everywhere (as in Fig. 9.19), then the receiver concludes that $x(t)$ (a "**0**") was transmitted. If it is the other way around, then $y(t)$ (a "**1**") was transmitted. Some further points should be noted in conclusion:

1. *Matched filters.* The correlation $r_{sx}(\tau)$ can also be regarded as the *convolution* of $s(t)$ with $g_1(t) = x^*(-t)$. Similarly, $r_{sy}(\tau)$ is the convolution of $s(t)$ with $g_2(t) = y^*(-t)$. Thus, the receiver is actually implementing two LTI systems (filters) with impulse responses $g_1(t)$ and $g_2(t)$. See Fig. 9.20. This is called a **filter bank** with two filters. Depending on which of the two filters produces a sharp peak at the output, the receiver can tell which signal was transmitted. We say that the filter $g_1(t)$ is **matched** to $x(t)$ because it produces a large peak when $x(t)$ is the input. Similarly, $g_2(t)$ is matched to $y(t)$. In practice, the filtering operations are typically done using **digital filters** – discrete-time versions of the filters described here. Matched filters are widely used in waveform detection in communication as well as in **radar** signal processing. In radar, a transmitted pulse shape is reflected by an object and returns to the receiver with some distortions and noise. Matched filtering is used at the receiver to decide whether there was indeed a target in a specified direction or not.

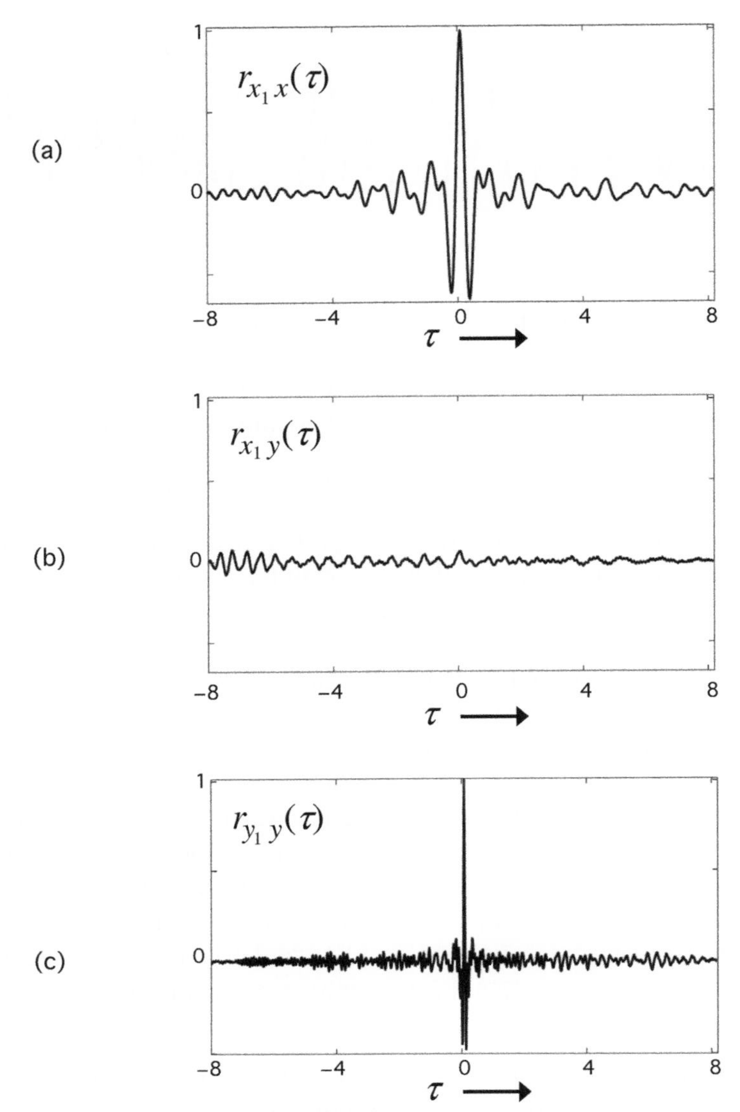

Figure 9.19 (a), (c) Correlations between the various noisy and noise-free signals. Note that the correlation peaks in $r_{x_1x}(\tau)$ and $r_{y_1y}(\tau)$ are very clear in spite of the noise.

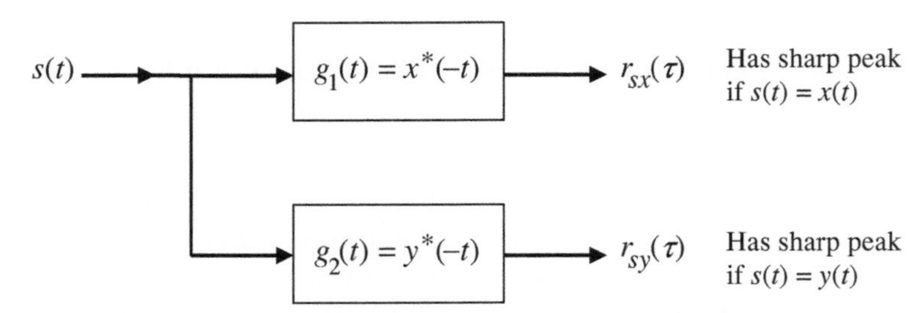

Figure 9.20 The matched filter bank. The outputs of the filters are the convolutions of $s(t)$ with the impulse responses $g_k(t)$. The outputs are therefore the correlations $r_{sx}(\tau)$ and $r_{sy}(\tau)$.

2. *Intuition about chirps.* Why do chirp waveforms work well? The intuitive explanation comes from the fact that finite-duration sinusoids with different frequencies tend to be **almost orthogonal** (i.e., their inner products are close to zero) if the durations are long enough. Since the different segments of the chirp have different frequencies, $x(t)$ and $x(t-\tau)$ tend to be orthogonal unless τ is very small. This gives a qualitative explanation for the sharp peak of $r_{xx}(\tau)$ at $\tau = 0$. For similar reasons, if the initial frequencies and the durations are properly adjusted for the two chirps, they will have very small cross correlation $r_{xy}(\tau)$ for all τ.
3. *Optimizing the signals.* While the chirp example works well, there are systematic ways to optimize the waveforms $x(t)$ and $y(t)$. Ideally, one would like the autocorrelations $r_{xx}(\tau)$ and $r_{yy}(\tau)$ to have sharp peaks, and $r_{xy}(\tau)$ to be as small as possible for all τ. Can we make $r_{xy}(\tau) = 0$ for all τ? In view of Eq. (9.123) this is possible only when $X(j\omega)Y^*(j\omega) = 0$ for all ω, that is, the nonzero portions of $X(j\omega)$ and $Y(j\omega)$ should have no overlap at all in the frequency domain. In practice, we can only approximate this condition with finite-duration signals. The problem of matched filter design becomes an **optimization problem**, which should take into account the filter complexity, the bandwidth allocated to the signals $x(t)$ and $y(t)$, and the time-domain durations of signals among other things.
4. *Generalizations.* The idea of matched filtering can readily be generalized to the case where the transmitter sends one of M signal shapes $\{s_k(t)\}$ – ***M*-ary communication** as opposed to binary. Again, a noisy copy $s(t)$ of the transmitted signal is received at the receiver. The receiver uses a bank of M filters, one matched to each possible waveform. Thus, the kth receiving filter has impulse response $g_k(t) = s_k^*(-t)$. If the lth filter produces the most prominent peak at its output, then the conclusion is that $s_l(t)$ was transmitted.
5. *Gravity wave detection.* Matched filtering played a key role in the detection of gravity waves in the **LIGO** project.[2] It turns out that some of the gravity waves received from collapsing twins (e.g., two black holes, two neutron stars, or one of each) are approximately chirp waveforms. Their shapes can be predicted from the physics of the collapsing objects. This allows one to design filters at the receiver, matched to specific sources of gravity waves. Because of the large distances from such objects to the earth, the received LIGO signals are extremely weak, and the signal-to-noise ratio is very small. For example, the first gravity waves (GW 150914, detected in Hanford and Livingston in 2015) were from two merging black holes, each appproximately 30 times heavier than our sun, and nearly 1.3 billion light years away. It is almost a miracle that sensitive detectors could be built to detect waves from so far away! Matched filters played a role in this milestone.

[2] Laser Interferometer Gravitational-wave Observatory (www.ligo.caltech.edu/).

9.11 Smoothness, Decay, and Convergence*

We saw many different examples of Fourier transforms in this chapter. In some cases the Fourier transform is zero for all ω outside a finite range, like

$$-\sigma < \omega < \sigma. \tag{9.135}$$

Such signals are called **σ-bandlimited** (σ-BL) signals, and will be discussed in detail in Chapter 10. One bandlimited example is the sinc function $h(t)$ in Fig. 9.8(b), whose FT is the ideal lowpass filter $H(j\omega)$. Other examples include sinusoids, for example, those in Fig. 9.5. We have also seen some examples that are not bandlimited. Thus, the signals in Fig. 9.7 have FTs which decay to zero only asymptotically as $|\omega| \to \infty$. Even for these signals, the decay rates are not the same – the decay of $X(j\omega)$ in Fig. 9.7(c) is faster than the other two. Finally, there are examples like the Gaussian signal, whose FT is Gaussian (Eq. (9.76)) and decays exponentially with ω^2, that is, it decays very rapidly as $|\omega| \to \infty$.

The rate of decay of the FT is an important property. According to the **sampling theorem** (Chapter 10), if a signal $x(t)$ is bandlimited then it can be recovered from its samples $x(nT)$ perfectly as long as the sample spacing T is small enough. Even though no finite-duration signal can be exactly bandlimited (Sec. 10.9), many finite-duration signals can be approximated well by bandlimited signals. This approximation works better for signals whose Fourier transform decays faster as $|\omega| \to \infty$. So the rate of decay of the FT plays an important role in digital signal processing. As another example of its importance, the rate of decay also governs the *convergence rate* of the inverse FT formula, as we shall explain in Sec. 9.11.3.

In the theory of Fourier transforms, there are some elegant results which tell us that the rate of decay of $X(j\omega)$ is related to the **smoothness** of the signal $x(t)$ in the time domain. There are ways to quantify smoothness. For example, a continuous signal is smoother than a discontinuous signal, and a differentiable signal is even smoother. A signal that is differentiable K times everywhere can be regarded as "smoother" than a signal which is only $K-1$ times differentiable everywhere. It has been shown by mathematicians that a signal which is more smooth in this sense has a more rapidly decaying Fourier transform as $|\omega| \to \infty$. These ideas will be explained in this section. To describe the behavior of $X(j\omega)$ for large ω, it is convenient to introduce a notation [Apostol, 1974]:

Definition 9.1 *The big Oh notation.* We say that a function $X(j\omega)$ is $O(1/\omega^n)$ for large ω if $\omega^n X(j\omega)$ is bounded for large ω, that is, there exists an $\omega_m > 0$ and a finite $c > 0$ such that

$$|X(j\omega)| \le \frac{c}{|\omega|^n}, \quad \forall\ |\omega| > \omega_m. \tag{9.136}$$

Most importantly, as $|\omega| \to \infty$, the function $X(j\omega)$ *decays at least as fast* as $1/\omega^n$. Note that if $X(j\omega)$ is $O(1/\omega^n)$ then it is also $O(1/\omega^{n-1})$. ◇

The "big Oh" notation is different from the so-called "small oh" notation (which we shall not require): $X(j\omega)$ is $o(1/\omega^n)$ if $\omega^n X(j\omega) \to 0$ as $|\omega| \to \infty$.

9.11.1 Convolving the Pulse with Itself: Splines

Before presenting the main result, we will present an example which demonstrates clearly the connection between smoothness in the time domain and decay rate in the frequency domain. Recall that the FT $P(j\omega)$ of the pulse $p(t)$ in Fig. 9.6(a) is the sinc function $\sin(\omega/2)/(\omega/2)$. Since $\sin(\omega/2)$ is oscillatory and does not decay, the decay rate for the sinc function is only $O(1/\omega)$. Thus the FT of the pulse has decay rate $O(1/\omega)$. Next, the triangle in Fig. 9.6(d) is the convolution $(p * p)(t)$ and therefore has FT $= P^2(j\omega)$, and it decays like $O(1/\omega^2)$, which is faster. Suppose we convolve the pulse with itself one more time to obtain $(p * p * p)(t)$. Then the FT becomes $P^3(j\omega)$ (sinc cubed), which decays like $O(1/\omega^3)$ which is even faster. More generally, let us define $p_N(t)$ for any integer $N > 0$ as follows:

$$p_N(t) = \Big(\underbrace{p * p * \cdots * p}_{p \text{ occurs } N+1 \text{ times}} \Big)(t), \tag{9.137}$$

where the subscript N is the number of convolutions ("$*$" signs) involved (i.e., p occurs $N+1$ times on the right-hand side). Thus

$$\begin{aligned}
\text{pulse } p_0(t) = p(t) &\longleftrightarrow \frac{\sin(\omega/2)}{\omega/2}, \\
\text{triangle } p_1(t) = (p * p)(t) &\longleftrightarrow \left(\frac{\sin(\omega/2)}{\omega/2}\right)^2, \\
p_2(t) = (p * p * p)(t) &\longleftrightarrow \left(\frac{\sin(\omega/2)}{\omega/2}\right)^3, \\
&\vdots
\end{aligned} \tag{9.138}$$

We therefore have

$$p_N(t) = \Big(p * p * \cdots * p \Big)(t) \longleftrightarrow \left(\frac{\sin(\omega/2)}{\omega/2}\right)^{N+1}. \tag{9.139}$$

The function $p_N(t)$ is called an Nth-order **spline** [Schoenberg, 1973]. For example, $p_2(t)$ is the **quadratic spline**, $p_3(t)$ is the **cubic spline**, and so forth. Note that $p_N(t)$ has duration $N+1$. Except for a time shift, these functions are similar to the functions $b_N(t)$ in Sec. 3.10.1, that is,

$$b_N(t) = p_N\left(t - \frac{N+1}{2}\right), \tag{9.140}$$

which is nonzero only in $0 < t < N+1$. Thus, $b_1(t)$ is the triangle which is nonzero in $0 < t < 2$ and $b_2(t)$ has the form (3.162), reproduced below:

$$b_2(t) = \begin{cases} t^2/2 & \text{for } 0 \le t < 1, \\ 0.75 - (t-1.5)^2 & \text{for } 1 \le t < 2, \\ (t-3)^2/2 & \text{for } 2 \le t < 3, \\ 0 & \text{otherwise.} \end{cases} \tag{9.141}$$

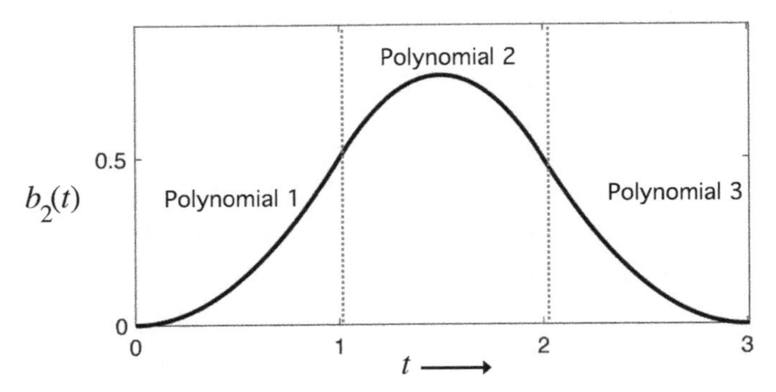

Figure 9.21 Plot of the quadratic spline function $b_2(t)$. This is obtained by glueing together three polynomial pieces, as shown in Eq. (9.141). More generally, the Nth-order spline has $N+1$ polynomial pieces of degree N each. See text.

This is plotted in Fig. 9.21. Notice that $b_2(t)$ is made up of three polynomial pieces and each polynomial is a quadratic.

We shall refer to both $b_N(t)$ and $p_N(t)$ as splines, and sometimes refer to $b_N(t)$ as ***B*-splines**. It can be shown that $b_N(t)$ is a polynomial of degree N in each of the regions $n \le t < n+1$, where $n = 0, 1, \ldots, N$ (as demonstrated in Eq. (9.141)). In the interior of these regions it is therefore *infinitely differentiable*. What is more beautiful is that even at the boundaries (junction points) of these regions, that is, at the points

$$t = 0, 1, 2, \ldots, N+1, \tag{9.142}$$

the spline $b_N(t)$ is differentiable $N-1$ times. These boundaries (9.142) are called **knots** of the spline. Differentiability at a knot means that the derivatives (slopes) of the polynomial pieces to the left and right of each knot match at the knots, so that the derivative has a unique value there. A similar meaning applies for higher derivatives. Thus, the Nth-order spline $b_N(t)$ (hence $p_N(t)$) is **everywhere differentiable $N-1$ times**. It turns out that the $(N-1)$th derivative is continuous everywhere, so we say that the Nth-order spline is *continuously differentiable $N-1$ times*.

For example, $b_2(t)$, which is called the **quadratic spline** (because it has t^2), is continuously differentiable once everywhere (Problem 9.26). Similarly, $b_3(t)$, which is called the **cubic spline**, is continuously differentiable twice everywhere. Thus, splines are polynomial pieces glued together at the knots in such a way that they are not only continuous, but also differentiable $N-1$ times even at the knots. The knots of $p_N(t)$ are merely shifted versions of (9.142).

Figure 9.22 shows the first few splines, and Fig. 9.23 shows the magnitudes of their Fourier transforms. Notice how quickly the Fourier transform decays as $\omega \to \infty$, for larger and larger N. The main point of this example is to indicate that the increased smoothness of splines in the time domain as N increases is related to the fact that the Fourier transform decays faster as N increases (see Eq. (9.139)). This is the connection between **smoothness in time** and **decay in frequency**. Because of their smoothness, splines are very useful in signal processing theory and applications,

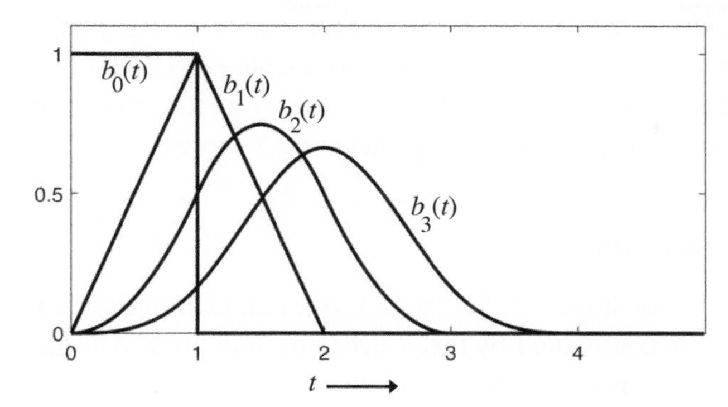

Figure 9.22 Plots of the spline functions $b_N(t)$ of orders $N = 0, 1, 2, 3$.

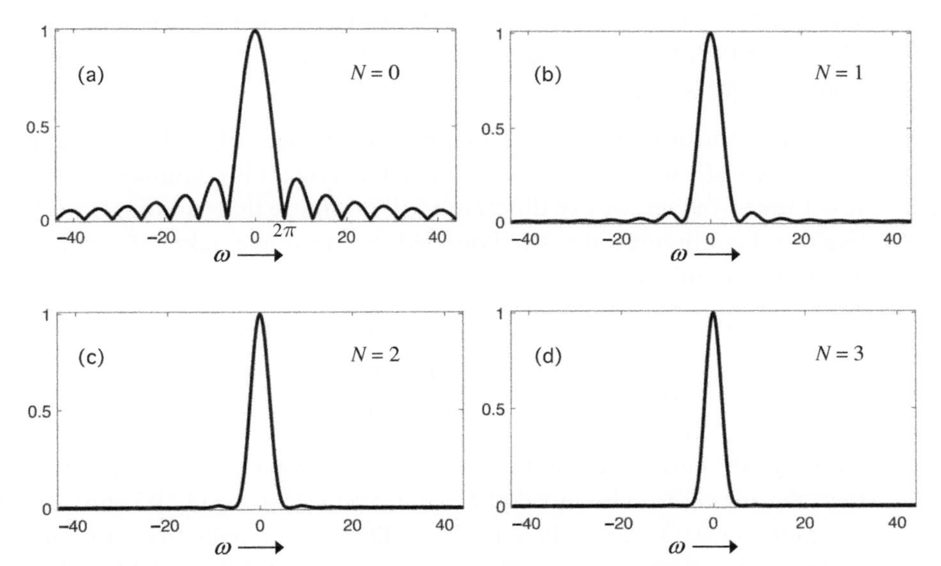

Figure 9.23 (a), (d) Plots of the Fourier transform magnitudes $|B_N(j\omega)|$ of spline functions of orders $N = 0, 1, 2, 3$. These decay like $1/\omega^{N+1}$.

for example, in the **interpolation** of signals [Unser et al., 1993; Vaidyanathan, 2001]. Spline interpolation of images is a standard operation in many image processing platforms, because of the smoothness of the resulting interpolated images.

Short Summary of Splines

1. The Nth-order spline $p_N(t)$ is defined as in Eq. (9.137). The shifted spline $b_N(t)$ is also referred to as a B-spline.
2. $p_N(t)$ has the Fourier transform shown in Eq. (9.139).
3. $p_N(t)$ has duration $N + 1$.
4. $p_N(t)$ is obtained by glueing together $N + 1$ polynomial pieces, each of degree N and duration 1. At the junction of the polynomial pieces (called knots), the glueing is so smooth that the function is differentiable there $N - 1$ times.

5. The $(N-1)$th derivative of $p_N(t)$ is continuous everywhere. So we say that $p_N(t)$ (hence $b_N(t)$) is **continuously differentiable** *everywhere* $N-1$ times.
6. While the smoothness (differentiability) of the spline increases as N increases, the Fourier transform decays faster as N increases.

9.11.2 Generalization

The family of splines in Sec. 9.11.1 gives an example where increased smoothness in time is accompanied by faster decay in frequency. Another extreme example is a bandlimited signal which is infinitely smooth, and its Fourier transform has the best possible decay: it equals zero for $|\omega| > \sigma$ for some finite σ. The preceding examples are special cases of a general result, which can be stated as follows:

Theorem 9.1 *Smoothness and decay.* If $x(t)$ is differentiable N times with bounded Nth derivative, then $X(j\omega)$ is $O(1/\omega^{N+1})$. That is, $X(j\omega)$ decays at least as fast as $1/\omega^{N+1}$ as $|\omega| \to \infty$. ◇

This will be justified in Sec. 9.11.3 (specifically in Sec. 9.11.3.3). Now let us examine the spline $p_N(t)$ in the light of the above theorem. It is continuously differentiable $N-1$ times. In fact, we can differentiate it one more time: although the Nth derivative is not continuous, it is still bounded. For example, take $p_1(t)$. It is continuous and differentiable once:

$$\frac{dp_1(t)}{dt} = \begin{cases} 1 & \text{for } -1 < t < 0, \\ -1 & \text{for } 0 < t < 1, \\ 0 & \text{for} |t| > 1. \end{cases} \tag{9.143}$$

Although the derivative is not continuous, it is bounded. If we differentiate this further, the answer is unbounded at $t = 0, 1$, and -1, so we say it is not differentiable any more (unless we are willing to accept Dirac delta functions in the answer). Similarly, $p_2(t)$ is continuously differentiable once, and differentiable twice where the second derivative is bounded but not continuous (Problem 9.26). More generally, $p_N(t)$ has bounded (although not continuous) Nth derivative. So, according to the above theorem, its FT should decay like $O(1/\omega^{N+1})$. This is indeed the case, as seen from Eq. (9.139).

Fourier series and decay rate. In Chapter 11 we will discuss the Fourier series. Since the Fourier series coefficients c_n of a periodic signal $x(t)$ are proportional to the sampled versions of the Fourier transform $X_T(j\omega)$ of one period $x_T(t)$ (Sec. 11.3.1), the coefficients c_n decay with increasing n in the same manner that $X_T(j\omega)$ decays as a function of ω. For example, if $x_T(t)$ is thrice differentiable with bounded third derivative, then we can claim that c_n is $O(1/n^4)$. ▽ ▽ ▽

Recall here that the Gaussian signal has a Gaussian Fourier transform (see Eq. (9.76)). So, both $x(t)$ and $X(j\omega)$ decay very rapidly, which is quite a unique property! The signal is continuous and infinitely differentiable in both domains. Also see Problem 9.15 for some optimality properties of the Gaussian.

9.11.3 Further Details

We now state some detailed results that relate the time-domain properties of $x(t)$ to the decay rate of the Fourier transform $X(j\omega)$. A justification of Theorem 9.1 will also follow from this discussion. There will be no loss of continuity for the reader who skips this section during first reading.

9.11.3.1 FT of L_1 and L_2 signals

First, assume that $x(t)$ is an L_1 function, that is, it is absolutely integrable (i.e., $|x(t)|$ is Lebesgue integrable). Then it can be shown that the Fourier transform integral exists as a Lebesgue integral, and furthermore satisfies the following properties (for proofs, see Sec. 11 in Chapter 11 of Haaser and Sullivan [1991]):

1. $X(j\omega)$ is a *continuous function* of ω. Thus, if $X(j\omega)$ is discontinuous for some signal, then we can conclude that $x(t)$ is not absolutely integrable. For example, if $X(j\omega)$ is the rectangular pulse (which is discontinuous), then $x(t)$ is the sinc function which is not absolutely integrable.
2. $X(j\omega)$ is *bounded*. In fact, $|X(j\omega)| \leq \int_{-\infty}^{\infty} |x(t)|dt$ because

$$|X(j\omega)| = \left|\int_{-\infty}^{\infty} x(t)e^{-j\omega t}dt\right| \leq \int_{-\infty}^{\infty} |x(t)e^{-j\omega t}|dt = \int_{-\infty}^{\infty} |x(t)|dt.$$

3. $X(j\omega) \to 0$ as $\omega \to \pm\infty$. This is called the **Riemann–Lebesgue** lemma.

If $x(t)$ is not absolutely integrable, but has finite energy (i.e., it is an L_2 function), then it can be shown that the energy of $X(j\omega)$ in the high-frequency region $\omega_m \leq |\omega| \leq \infty$ decays to zero as $\omega_m \to \infty$. That is,

$$\frac{1}{2\pi}\int_{-\infty}^{-\omega_m} |X(j\omega)|^2 d\omega + \frac{1}{2\pi}\int_{\omega_m}^{\infty} |X(j\omega)|^2 d\omega \quad \longrightarrow 0 \tag{9.144}$$

as $\omega_m \to \infty$. In this sense we can say that $X(j\omega)$ decays to zero as $\omega \to \infty$. So for signals which are L_1 or L_2 (which covers most signals we encounter in practice), the FT decays to zero as $\omega \to \infty$. This result has nothing to do with the degree of smoothness of $x(t)$ – the only assumption so far is the L_1 or L_2 property of $x(t)$.

9.11.3.2 FT of Bounded Signals

For L_1 and L_2 signals we saw that $X(j\omega) \to 0$ as $|\omega| \to \infty$, but this does not specify the rate of decay. Assume now that $x(t)$ is a bounded signal, that is, $|x(t)| \leq B$ for some finite $B > 0$. Then we can say something stronger about the decay of the FT. Thus, consider the partial integral

$$X_\tau(j\omega) = \int_{-\tau}^{\tau} x(t)e^{-j\omega t}dt. \tag{9.145}$$

It is clear that

$$|X_\tau(j\omega)| \leq B\left|\int_{-\tau}^{\tau} e^{-j\omega t}dt\right| = B\left|\frac{2\sin(\tau\omega)}{\omega}\right| \leq \frac{2B}{|\omega|}. \tag{9.146}$$

Since the bound on the right-hand side is independent of τ, we conclude that

For any **bounded signal** $x(t)$, *the FT* $X(j\omega)$ *is* $O(1/\omega)$. (9.147)

The decay can, of course, be faster than $O(1/\omega)$ depending on other properties in the time domain. It is interesting to reconcile this result with some examples mentioned elsewhere in the book:

1. The rectangular pulse $p(t)$ is bounded, and has Fourier transform equal to the sinc function, which is indeed $O(1/\omega)$ (since $\sin(\omega/2)$ oscillates for ever). In fact, practically all examples in this book have bounded $x(t)$ so that the FT decays at least as fast as $O(1/\omega)$.
2. For unbounded signals, $X(j\omega)$ (if it exists) may decay more slowly. For example, the signal

$$x(t) = \begin{cases} \dfrac{1}{\sqrt{t}} & t > 0, \\ 0 & t \leq 0, \end{cases} \tag{9.148}$$

is unbounded as $t \to 0^+$. It has the Fourier transform (Problem 9.25)

$$X(j\omega) = \begin{cases} e^{-j\pi/4}\sqrt{\dfrac{\pi}{|\omega|}} & \text{for } \omega > 0, \\ e^{j\pi/4}\sqrt{\dfrac{\pi}{|\omega|}} & \text{for } \omega < 0. \end{cases} \tag{9.149}$$

Clearly $X(j\omega)$ decays only like $O(1/\omega^{0.5})$, rather than $O(1/\omega)$. Note that the Fouier transform is unbounded as $|\omega| \to 0$ and $X(0)$ is undefined, consistent with the fact that $\int x(t)dt$ does not exist. In examples like this where neither $\int x(t)dt$ nor $\int x^2(t)dt$ exists, the Fourier transform is not an L_1 or L_2 FT, and the CTFT integral is usually interpreted either as an improper Riemann integral or a Cauchy principal value (Sec. 17.7).
3. The Dirac delta $x(t) = \delta_c(t)$, which is unbounded at $t = 0$, has FT $X(j\omega) = 1$ for all ω, which does not decay at all. In this example $X(j\omega)$ is $O(1)$. This might be confusing to some readers: since $\delta_c(t) \geq 0$ and $\int \delta_c(t)dt = 1$, can we not regard it as an L_1 function? In that case, the Riemann–Lebesgue lemma mentioned above says that $X(j\omega) \to 0$ as $|\omega| \to \infty$. So, is this not a contradition? The explanation is that the integral in $\int \delta_c(t)dt = 1$ is not a rigorously defined Lebesgue integral. Rather, it is obtained as the limiting case of the integral of a pulse which gets narrower and narrower under the unit area constraint (Sec. 2.12).

9.11.3.3 FT of Signals with Bounded Derivatives

Now assume that $x(t)$ is N times differentiable with bounded (though possibly not continuous) Nth derivative. We will show that $X(j\omega)$ is $O(1/\omega^{N+1})$. For this, recall

(Sec. 9.6) that the Fourier transform of the derivative $x^{(N)}(t)$ is $(j\omega)^N X(j\omega)$. Since this derivative is bounded, it follows (see Eq. (9.147)) that

$$(j\omega)^N X(j\omega) \quad \text{is} \quad O(1/\omega). \tag{9.150}$$

We therefore conclude that

$$X(j\omega) \quad \text{is} \quad O(1/\omega^{N+1}), \tag{9.151}$$

whenever $x(t)$ is N times differentiable with bounded (though not continuous) Nth derivative. This justifies Theorem 9.1.

9.11.3.4 Decay Rate and Rate of Convergence

If a Fourier transform is $O(1/\omega^{n+1})$, we say that it has a faster decay than a Fourier transform which is only $O(1/\omega^n)$. Why is it good to have a faster decay? The rate at which the Fourier transform decays determines the accuracy with which truncated versions of the Fourier integral and the Fourier series represent a signal. More specificially, the partial integral

$$x_\alpha(t) = \frac{1}{2\pi}\int_{-\alpha}^{\alpha} X(j\omega)e^{j\omega t}d\omega \tag{9.152}$$

converges to $x(t)$ more quickly (i.e., for smaller values of α) if the decay rate of $X(j\omega)$ is higher. Similarly, in the case of the Fourier series (Chapter 11), the partial sums

$$x_K(t) = \sum_{k=-K}^{K} c_k e^{j\frac{2\pi k}{T}t} \tag{9.153}$$

converge to $x(t)$ faster if the decay rate of Fourier series coefficients c_k is higher. Thus the convergence rates of the Fourier integral and the Fourier series depend directly on the decay rates in the Fourier domain, which in turn depend on the smoothness of the functions $x(t)$ in the time domain.

9.11.4 Closing Remarks

In this section we discussed the rate of decay of a Fourier transform for large ω, and showed that the decay is related to the smoothness of the signal $x(t)$. For further details and early references on this topic, the reader should see Carslaw [1950] and Appendix 3-B in Chapter 3 of Papoulis [1977a]. In the mathematics literature the smoothness of a function is sometimes characterized in terms of what is called the Hölder **regularity** index [Daubechies, 1992]. The regularity index has the form $\alpha = n+\beta$, where n is a non-negative integer and $0 < \beta \leq 1$. If $x(t)$ has the regularity index $\alpha = n+\beta$, then it is differentiable n times and the nth derivative $y(t)$ is continuous with a Hölder continuity index β.[3] If the regularity index of a function is α,

[3] What this means is that the nth derivative $y(t)$ satisfies $|y(t_0) - y(t_1)| \leq c|t_0 - t_1|^\beta$ for some $c > 0$. Here, β is also called the Lipschitz constant of $y(t)$.

then the Fourier transform is $O(1/\omega^{1+\alpha})$. In the theory of wavelet representations, the design of smooth wavelets is therefore related to the design of functions with rapidly decaying Fourier transforms. A brief discussion of wavelets is given in Sec. 11.10.

9.12 Fourier Transforms in Nature*

We will now argue that the Fourier transform arises automatically in Nature. In some sense, Nature has already been computing the Fourier transform even before we humans invented or discovered it! Consider Fig. 9.24, where a monochromatic plane electromagnetic (EM) wave, say visible light, is incident from the left and passes through a **transmission function** or mask $A(y)$. Here, y is the vertical spatial variable as indicated. The transmission function $A(y)$ is simply a multiplicative factor for the field. For example, if

$$A(y) = \begin{cases} 1 & \text{for } |y| \leq a/2, \\ 0 & \text{otherwise,} \end{cases} \tag{9.154}$$

this corresponds to placing a single opening or **slit** of width a in the path of the light. More generally, $A(y)$ could be a transparency (viewgraph or glass slide) representing a nice image pattern; it would be a 2D function $A(y_1, y_2)$ but for simplicity we discuss the 1D version here.

Now consider a screen far away from the mask, and let Q be some point on the screen. Light reaches Q from all points of the mask $A(y)$. We have shown rays from three such points on $A(y)$ at the vertical coordinate locations $y = 0, d, -d$. Assuming $d << R$, where R is the distance shown, these three rays are almost parallel. For fixed R notice also that the location of Q on the screen is fully defined by the angle θ shown in the figure.

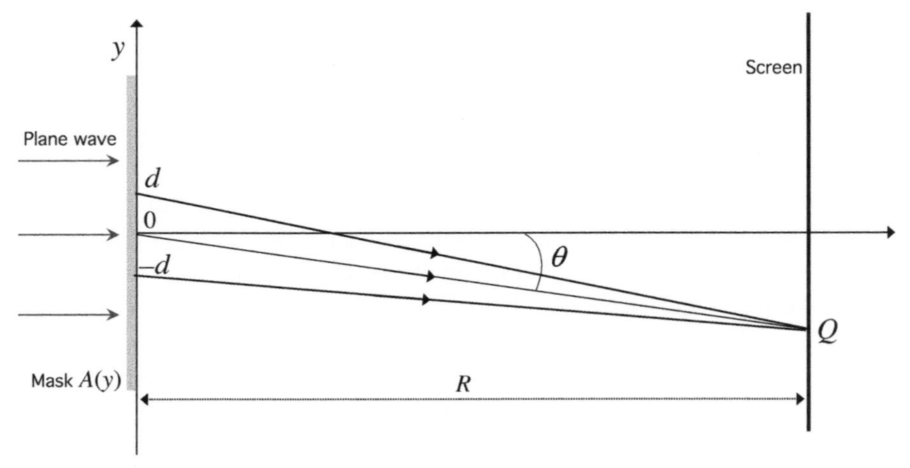

Figure 9.24 A plane wave passing through a transparency with mask function $A(y)$, and hitting a screen far away.

We will show that the image seen on the screen is related to the Fourier transform of the transmission function $A(y)$. Nature therefore automatically computes the Fourier transform. Truly this is one of the most "real-world" signatures of the Fourier transform!

9.12.1 Electric Field at a Distant Point Q

It is clear that the path from $y = d$ to the point Q is longer than the path from $y = 0$, the excess being $d \sin\theta$ (see Fig. 9.25). Similarly, the path from $y = -d$ to the point Q is shorter than the path from $y = 0$, by the amount $d \sin\theta$. Now, a monochromatic plane EM wave propagating along some direction z is described by $e^{j(\omega_c t - \kappa z)}$, where ω_c is the frequency and κ is the wavenumber. Thus, an excess path length Δz gives the additional phase factor $e^{-j\kappa\Delta z}$. Because of this the electric field at Q reaching from $y = d$ can be described by

$$E_Q(d) = A(d)E_Q(0)e^{-j\kappa d \sin\theta}. \tag{9.155}$$

The factor $A(d)$ arises because the mask attenuates the field by $A(d)$. Similarly, the electric field at Q reaching from $y = -d$ can be described by

$$E_Q(-d) = A(-d)E_Q(0)e^{j\kappa d \sin\theta}. \tag{9.156}$$

Thus, the total electric field at Q (i.e., at the angle θ) due to all points on the mask $A(y)$ can be expressed as

$$E_{total}(\theta) = E_Q(0)\int_{-a/2}^{a/2} A(y)e^{-j\kappa y \sin\theta}dy, \tag{9.157}$$

where the mask exists in the range $-a/2 \leq y \leq a/2$ and the transmission is assumed to be blocked beyond that. The above derivation is valid when $a << R$ and the screen is far away from the mask $A(y)$. These are called **far-field** approximations.

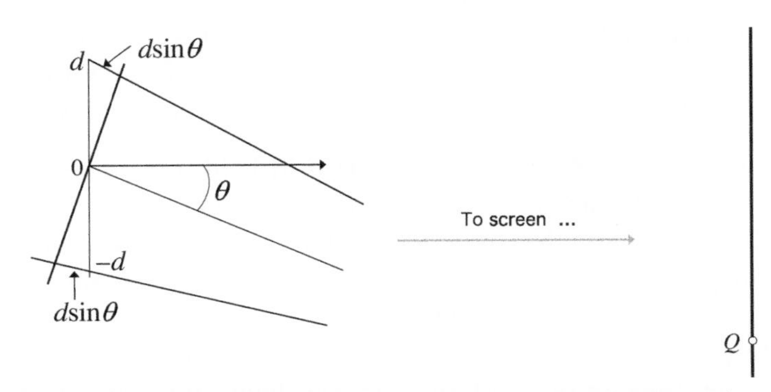

Figure 9.25 A magnified view of the details in Fig. 9.24. From the vertical line on the left to the distant point Q on the screen, the ray from d travels an extra distance $d \sin\theta$ compared to the ray from 0, and the ray from $-d$ travels a correspondingly smaller distance.

9.12.2 Connection to the Fourier Transform

We now rewrite Eq. (9.157) in a more convenient form by defining the variable $\omega = \kappa \sin\theta$. Since the wavenumber κ is related to the wavelength λ by $\kappa = 2\pi/\lambda$, we have

$$\omega = \frac{2\pi \sin\theta}{\lambda}. \tag{9.158}$$

Notice that θ and ω are in the following ranges:

$$-\frac{\pi}{2} \leq \theta \leq \frac{\pi}{2} \quad \text{so that} \quad \frac{-2\pi}{\lambda} \leq \omega \leq \frac{2\pi}{\lambda}. \tag{9.159}$$

For every θ we can find a unique ω in this range, and vice versa. So ω is simply a different representation of the angle θ. We therefore refer to ω as **angular frequency**, or just "angle" when there is no room for confusion.[4] Using the definition of ω, we can rewrite Eq. (9.157) as

$$\frac{E_{total}}{E_Q(0)} = \int_{-a/2}^{a/2} A(y)e^{-j\omega y}\,dy. \tag{9.160}$$

For a given mask $A(y)$, the right-hand side is completely determined by ω (i.e., the angle θ). So we use the notation $B(j\omega) = E_{total}/E_Q(0)$ and rewrite Eq. (9.160) as

$$B(j\omega) = \int_{-a/2}^{a/2} A(y)e^{-j\omega y}\,dy, \tag{9.161}$$

which is the Fourier transform of $A(y)$. Summarizing, the total electric field $E_{total} = B(j\omega)E_Q(0)$ at the angle ω on the distant screen is proportional to the Fourier transform of the transmission function $A(y)$, evaluated at ω.

Remarks

Let us reflect a bit. The above result says that if we shine light on a transparency $A(y)$ and let the light propagate to a distant screen, then on the screen we "see" the Fourier transform of the transmission function automatically! There is no sophisticated equipment anywhere – no optical gadgets (like lenses), no computers, no circuits. Mother Nature simply "evaluates" the Fourier transform and makes it available on the screen. That is all. A number of remarks are now in order.

1. Since EM waves travel at the speed of light, one might say that the Fourier transformation above occurs *at the speed of light*.
2. The assumption that the screen is "far away" is called the far-field or the **Fraunhofer** approximation. For the near-field scenario a more accurate theory called the *Fresnel* diffraction theory is used.
3. In practice it is not so easy to observe the Fourier transform of $A(y)$ on the screen because of interfering ambient light, other objects in the environment, and so forth. However, in a carefully controlled laboratory atmosphere we can

[4] This ω should not be confused with ω_c in the expression $e^{j(\omega_c t - \kappa z)}$ for the wave.

observe this. In fact, this can be done easily by using a convex lens. Because of its focusing property, the lens allows us to reduce the distance between the mask $A(y)$ and the screen (see below). The use of a lens creates other errors and approximations, such as finite aperture error and so forth, which we do not elaborate here.

4. Since $B(j\omega)$ can be complex valued, typically, only the magnitude-square $|B(j\omega)|^2$ of the complex field is recorded (as an intensity pattern) on the screen.

9.12.3 Using Lenses to Get the Fourier Transform

For readers familiar with the principles of geometric optics, we mention that if we place the mask $A(y)$ on the *left focal plane* of a **convex** lens and put the screen on the *right focal plane*, the Fourier transform of $A(y)$ appears on the screen. This can be proved under some mild assumptions using a similar approach as above.

The idea is demonstrated in Fig. 9.26(a) with an example. The mask shown here is called a **single slit**. It has a simple opening that allows light to pass; light is blocked everywhere else. So $A(y)$ is as in Fig. 9.26(b). Since its Fourier transform is the sinc function, a plot of $|B(j\omega)|^2$ is as shown in Fig. 9.26(c). We have shown both the ω axis and the $\sin\theta$ axis for clarity. The sinc function has uniform zero crossings when the horizontal axis is $\sin\theta$ (and not θ). Notice that the wavelength λ affects the width of the main lobe in Fig. 9.26(c). Thus, if we do not use monochromatic light, then the image spreads out differently for the different wavelength components. The superposition of these multiple images on the screen will not be a clear image.

You must have learned in basic physics that there are other masks, such as the **double slit**, multiple slits, and so forth. The intensity pattern produced at the second focal plane by any of these can readily be computed by using the basic examples in Fourier transform theory presented in this chapter. Figure 9.27(a) shows one final example. Here the mask is a sequence of uniformly spaced slits, sometimes called a

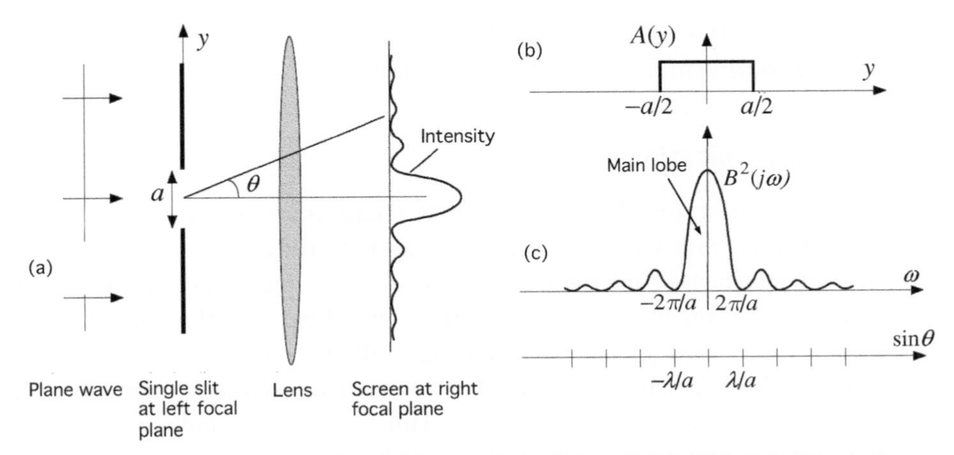

Figure 9.26 (a) A single slit at the left focal plane of a convex lens and the corresponding intensity pattern at the right focal plane, (b) the mask function $A(y)$, and (c) the square of its Fourier transform.

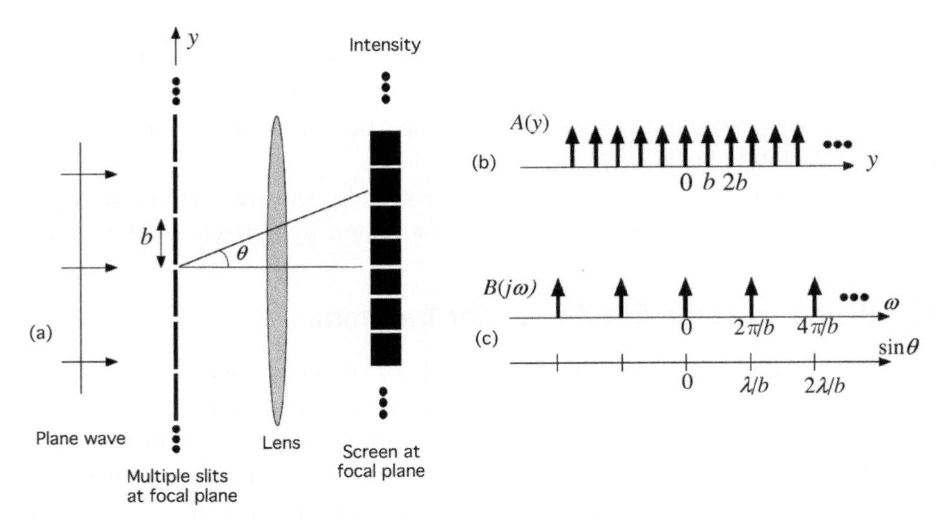

Figure 9.27 (a) A diffraction grating (infinite number of uniformly spaced slits) located at the left focal plane and the corresponding intensity pattern at the right focal plane, (b) the mask pattern $A(y)$, and (c) its Fourier transform.

diffraction **grating**. Assuming that the slit widths are infinitesimally small and that there are an infinite number of slits, the transmission function $A(y)$ is an impulse train (Fig. 9.27(b)). From Sec. 9.5 we know that its Fourier transform $B(j\omega)$ is also an impulse train, as shown in Fig. 9.27(c). In practice, we have a finite number of slits and each slit has nonzero width. So $B(j\omega)$ is an appropriately modified version of what is shown in Fig. 9.27(c); namely, each impulse in Fig. 9.27(c) spreads out slightly, and the heights vary slightly as well. This can readily be derived by using basic Fourier transform properties presented in this chapter. The intensity pattern on the screen at the right focal plane is qualitatively as shown in Fig. 9.27(a).

The above results are sometimes expressed by saying, "The lens computes the Fourier transform." But in reality, the Fourier transform operation was already there even without the lens, as explained in Sec. 9.12.2. The lens merely moves the Fourier transform from infinity to your desk.

9.12.4 Further Implications

The observation that Fourier-transform relations are implicitly there in Nature is far reaching. Fourier theory has been used in studying the resolution of optical instruments such as **microscopes**. If two distinct objects have to be resolvable on a screen, then the minimum separation required between the objects can be found using Fourier theory. Smaller wavelengths allow us to resolve more closely spaced objects. This is because an object of extent a spreads out to λ/a on the screen (when plotted against $\sin\theta$, see Fig. 9.26(c)). Thus, a smaller λ (higher frequency) results in a smaller spread, and enables objects to be resolved better.

Referring to Fig. 9.26, we know that the intensity on the screen is the magnitude-squared of the Fourier transform of the transmisison function $A(y)$. This remains nearly true if we remove the lens and move the screen to a point "far away." What is fascinating is that this continues to be true if the plane EM wave on the left is replaced with a stream of electrons or even neutrons. This property is at the very heart of quantum behavior, as explained beautifully in Feynman et al. [1963] (see vol. 3). Thus, Fourier relations continue to arise naturally in the theory of **quantum behavior** of matter.

The reader may remember from early physics classes the **uncertainty principle** in quantum mechanics. It says that if Δx and Δp are the uncertainties in the simultaneous determination of the position and momentum of a particle, then

$$\Delta x \Delta p \geq h, \tag{9.162}$$

where h is Planck's constant [Halliday and Resnick, 1978]. Compare this with the uncertainty principle in Fourier transform theory (Problem 9.15), which says that if a signal is more confined in the time domain, then it is more spread out in the frequency domain. More precisely, if the root-mean-square (rms) durations in time and frequency are D_t and D_f, then

$$D_t D_f \geq 0.5. \tag{9.163}$$

The similarity between Eq. (9.162) and Eq. (9.163) is not a coincidence. We know that when the slit gets narrower in Fig. 9.26, the main lobe of the sinc on the screen spreads out. This Fourier-transform property is connected to the quantum idea that when the slit tries to confine a particle more tightly vertically in space, the vertical momentum of the particle spreads out more (which corresponds to the spreading out of the sinc function).

Another major application of the Fourier relations is in **crystallography**. Crystals are regular arrangements of atoms (or rather ions) in the form of periodic structures or lattices [Feynman et al., 1963]. X-rays are used for imaging a crystal because the interatomic distances are too small to be resolved with optical wavelengths. The idea of using X-ray reflection patterns to identify crystal structures originated in the work of William Henry Bragg and William Lawrence Bragg (father and son, who shared the 1915 Nobel Prize in Physics).

Later, in 1929, Lawrence Bragg suggested, perhaps for the first time, the use of Fourier transform theory to identify the structure of complex crystals. The X-ray diffraction pattern obtained from a crystal lattice is related to its 3D Fourier transform. Based on a measurement of this, it is possible to find out what the crystal structure itself is. The structures of many crystals have been determined using this method, including proteins (which are very complex macromolecules). Oftentimes, only partial information of the 3D Fourier transform is available, such as a 2D projection or a magnitude-only recording. There are methods to recover the information about the crystal structure from such partial information, by combining with other basic rules from physical chemistry.

Fourier transform theory has also been crucial in the identification of the structure of crystalline DNA (deoxyribo nucleic acid) molecules. According to James Watson (who discovered the DNA double-helical structure along with Francis

Crick), Crick's expertise with the Fourier transform greatly helped them in finding the structure of DNA (see p. 113 of Watson [1968]).

You can read more about Fourier transforms in optics from Goodman [2017], Guenther [1990], and Papoulis [1968]. An interesting paper on the early history of Fourier transforms in crystallography was written by Beevers and Lipson [1985]. Other references include Coppens [1997] and Stout and Jensen [1989].

9.13 Hilbert Transform Relations*

We now discuss the relation between the real part $X_r(\omega)$ and the imaginary part $X_i(\omega)$ of the the Fourier transform $X(j\omega) = X_r(\omega) + jX_i(\omega)$. For arbitrary $x(t)$, there is no relation between these. Even for real-valued $x(t)$, the components $X_r(\omega)$ and $X_i(\omega)$ can be totally unrelated, the only constraint being that $X_r(\omega)$ is even and $X_i(\omega)$ is odd.

9.13.1 Causal $x(t)$ and Hilbert Transform Relations

Now assume that $x(t)$ is possibly complex but causal, or more strictly,

$$x(t) = 0,\ t \leq 0 \qquad \text{(causality)}. \tag{9.164}$$

We claim that when Eq. (9.164) holds, $X_r(\omega)$ and $X_i(\omega)$ are related as follows:

$$X_i(\omega) = -\frac{1}{\pi}\int_{-\infty}^{\infty} \frac{X_r(\nu)}{\omega - \nu} d\nu, \qquad X_r(\omega) = \frac{1}{\pi}\int_{-\infty}^{\infty} \frac{X_i(\nu)}{\omega - \nu} d\nu. \tag{9.165}$$

Thus, $X_r(\omega)$ can be found from $X_i(\omega)$ (and vice versa) by performing a convolution in the frequency domain with $1/\omega$ (and multiplying by $1/\pi$ or $-1/\pi$). The relations (9.165) between the real and imaginary parts of the FT of a causal signal are called *Hilbert transform relations* [Papoulis, 1977a]. The reason for this name is that the above integrals look like the time-domain convolution in Eq. (9.113), which represented a Hilbert transformer filter. The difference is that the convolutions above are in the frequency domain, unlike Eq. (9.113).

The relations (9.165) are also variously known as *Poisson transforms* or *Kramers–Kronig transforms* because these relations were discovered in different scientific communities independently.

Proof of Eq. (9.165). Given the causal signal $x(t)$, define

$$x_h(t) = \frac{x(t) + x^*(-t)}{2}, \quad x_{sh}(t) = \frac{x(t) - x^*(-t)}{2}, \tag{9.166}$$

so that $x(t) = x_h(t) + x_{sh}(t)$. Clearly, $x_h(t) = x_h^*(-t)$ and $x_{sh}(t) = -x_{sh}^*(-t)$, so these are the Hermitian and skew-Hermitian components of $x(t)$. With $x(t) \longleftrightarrow X(j\omega) = X_r(\omega) + jX_i(\omega)$, we know $x^*(-t) \longleftrightarrow X^*(j\omega) = X_r(\omega) - jX_i(\omega)$. So

$$x_h(t) \longleftrightarrow X_h(j\omega) = X_r(\omega), \quad x_{sh}(t) \longleftrightarrow X_{sh}(j\omega) = jX_i(\omega). \tag{9.167}$$

Since $x(t)$ satisfies Eq. (9.164), there is no overlap between $x(t)$ and $x(-t)$. So we have

$$x_h(t) = \begin{cases} 0.5x(t) & t > 0, \\ 0.5x^*(-t) & t < 0 \end{cases} \qquad x_{sh}(t) = \begin{cases} 0.5x(t) & t > 0, \\ -0.5x^*(-t) & t < 0, \end{cases} \tag{9.168}$$

which shows that

$$x_{sh}(t) = s(t)x_h(t), \tag{9.169}$$

where

$$s(t) = \begin{cases} 1 & t > 0, \\ -1 & t < 0, \end{cases} \tag{9.170}$$

and $s(0) = 0$. So the FTs of $x_{sh}(t)$ and $x_h(t)$ are related by the convolution

$$X_{sh}(j\omega) = \frac{1}{2\pi}\int_{-\infty}^{\infty} X_h(jv)S(j(\omega - v))dv. \tag{9.171}$$

It is easily shown (Problem 9.5) that the FT of $s(t)$ is

$$S(j\omega) = \frac{2}{j\omega}, \quad \omega \neq 0, \tag{9.172}$$

and $S(j0) = 0$. So, using Eq. (9.167), Eq. (9.171) reduces to

$$jX_i(\omega) = \frac{1}{2\pi j}\int_{-\infty}^{\infty} X_r(v)\frac{2}{\omega - v}dv, \quad \text{that is,} \quad X_i(\omega) = -\frac{1}{\pi}\int_{-\infty}^{\infty} \frac{X_r(v)}{\omega - v}dv, \tag{9.173}$$

which is the first relation in (9.165). Similarly, we can also write $x_h(t) = s(t)x_{sh}(t)$ instead of Eq. (9.169), and proceed as above to obtain the second Hilbert transform relation in (9.165). ▽ ▽ ▽

One subtlety here is that the integrands in (9.165) are unbounded as $v \to \omega$. So we have to interpret them carefully. For example, the first integral should be interpreted as

$$\int_{-\infty}^{\infty} \frac{X_r(v)}{\omega - v}dv = \lim_{\epsilon \to 0}\left(\int_{-\infty}^{\omega-\epsilon} \frac{X_r(v)}{\omega - v}dv + \int_{\omega+\epsilon}^{\infty} \frac{X_r(v)}{\omega - v}dv\right). \tag{9.174}$$

An integral interpreted in this way is sometimes referred to as the Cauchy principal value (also see Sec. 17.7).

9.13.2 Causal $x[n]$ and Hilbert Transform Relations

We now consider the discrete-time case, which is similar with minor differences [Oppenheim and Schafer, 1975]. Further insights can also be found in Vaidyanathan and Mitra [1982]. Let $x[n]$ be causal (and possibly complex) with Fourier transform

$$X(e^{j\omega}) = X_r(e^{j\omega}) + jX_i(e^{j\omega}). \tag{9.175}$$

Then the real part $X_r(e^{j\omega})$ and imaginary part $X_i(e^{j\omega})$ are related as follows:

$$X_i(e^{j\omega}) = -\frac{1}{2\pi}\int_{-\pi}^{\pi} X_r(e^{j\nu})\cot\Big(\frac{\omega-\nu}{2}\Big)d\nu, \tag{9.176}$$

$$X_r(e^{j\omega}) = \frac{1}{2\pi}\int_{-\pi}^{\pi} X_i(e^{j\nu})\cot\Big(\frac{\omega-\nu}{2}\Big)d\nu \quad + \text{Re}\,(x[0])\,. \tag{9.177}$$

Since the integrand is unbounded as $\nu \to \omega$, we have to interpret the integral as a limit, similar to Eq. (9.174).

Proof of Eqs. (9.176) and (9.177). Define

$$x_h[n] = \frac{x[n] + x^*[-n]}{2}, \quad x_{sh}[n] = \frac{x[n] - x^*[-n]}{2}, \tag{9.178}$$

so that $x[n] = x_h[n] + x_{sh}[n]$. Since $x_h[n] = x_h^*[-n]$ and $x_{sh}[n] = -x_{sh}^*[-n]$, we say $x_h[n]$ is the Hermitian part of $x[n]$ and $x_{sh}[n]$ the skew-Hermitian part. Since

$$x^*[-n] \longleftrightarrow X^*(e^{j\omega}) = X_r(e^{j\omega}) - jX_i(e^{j\omega}), \tag{9.179}$$

it follows that

$$x_h[n] \longleftrightarrow X_r(e^{j\omega}), \quad x_{sh}[n] \longleftrightarrow jX_i(e^{j\omega}). \tag{9.180}$$

Now define

$$s[n] = \begin{cases} 1 & n > 0, \\ -1 & n < 0, \\ 0 & n = 0. \end{cases} \tag{9.181}$$

This sequence has the Fourier transform (Sec. 3.8.3.1)

$$S(e^{j\omega}) = \frac{1 + e^{-j\omega}}{1 - e^{-j\omega}} = \frac{\cot(\omega/2)}{j}, \quad \omega \neq 0, \tag{9.182}$$

and $S(e^{j0}) = 0$. Since $x[n]$ is causal, it then follows that

$$x_{sh}[n] = s[n]x_h[n]. \tag{9.183}$$

So $X_{sh}(e^{j\omega})$ (i.e., $jX_i(e^{j\omega})$) can be expressed using a convolution of $S(e^{j\omega})$ and $X_h(e^{j\omega})$ (i.e., $X_r(e^{j\omega})$):

$$jX_i(e^{j\omega}) = \frac{1}{2\pi j}\int_{-\pi}^{\pi} X_r(e^{j\nu})\cot\Big(\frac{\omega-\nu}{2}\Big)d\nu, \tag{9.184}$$

which proves Eq. (9.176). Similarly, we have

$$x_h[n] = s[n]x_{sh}[n] + \text{Re}\,(x[0])\,\delta[n], \tag{9.185}$$

from which we get Eq. (9.177). $\triangledown \triangledown \triangledown$

9.14 Some Reflections on Fourier the Genius

The great mathematician Joseph Fourier (1768–1830) came up with the idea that signals can be represented in the form of infinite series of sines and cosines. Although there are many other representations of signals in terms of different types of basis functions, Fourier's sine and cosine series continues to be with us, and has touched every corner of science and engineering. One reason for this is that the sine and cosine basis functions (or more accurately, functions like $e^{j\omega t}$) are eigenfunctions of linear time-invariant systems, and Fourier representations arise naturally, as we saw in Chapter 3. Another reason is that the concept of frequency, which is so basic to science and engineering, is already at the heart of Fourier's representations.

What we have just studied in this chapter is the Fourier *integral* representation for *nonperiodic* continuous-time signals (CTFT theory). The discrete-time version was discussed in Chapter 3 (DTFT theory). Fourier's original idea of a *series* representation for *periodic* (or finite-duration) signals in continuous time will be elaborated in Chapter 11 (Fourier series theory). In Chapter 12 we will study yet another type of representation. It is the only version which can be implemented by digital computation, and is called the DFT. It admits a fast algorithm for its computation called the FFT. The connection between the above four types of Fourier representations is discussed and summarized in Sec. 12.7.

The reader should certainly learn more about Fourier's interesting life history from books and online sources. In addition to his scientific achievements, Fourier also had an interesting political life. It is said that his name is one of the 72 names inscribed on the Eiffel Tower. Even though Fourier proposed his series representation about 200 years ago, a rigorous mathematical proof of the convergence of the series was to come much later, after the Lebesgue integral was established. In the last 100 years or so, many mathematicians have contributed to results relating to the Fourier series and the Fourier integral. This includes results on pointwise convergence of the infinite sum, existence of the inverse transform, and so forth. All our discussions so far have emphasized engineering intuition and applications, rather than mathematical rigor. However, Chapter 17 is dedicated to a discussion of mathematical intricacies involved in Fourier representations, such as conditions for the convergence of infinite sums, conditions for existence of inverses, and so forth.

In the signals and systems community, many books have been writtten on Fourier representations. Besides signal processing classics such as Oppenheim and Willsky [1997], early books by Papoulis [1962, 1968, 1977a, 1980], Bracewell [1986], and Churchill and Brown [1987] have been popular for many decades. Some of the excellent books for mathematically rigorous proofs of many results include Apostol [1974], Haaser and Sullivan [1991], Rudin [1974], and Zygmund [1988].

9.15 Summary of Chapter 9

This chapter gave a detailed exposition of the continuous-time Fourier transform (CTFT). Many properties of the CTFT were derived and illuminating examples were given throughout the chapter. This includes the pulse signal, the Dirac delta train, the one-sided exponential, the Gaussian signal, the Hilber transformer, and so on. The inverse CTFT was derived and it was used to find the impulse response of the ideal lowpass filter in continuous time. The CTFT and its inverse are linear operators. Section 9.16 summarizes the key properties of the CTFT. Many examples of continuous-time Fourier transform pairs were presented, some of which are summarized in Sec. 9.17.

The derivative properties of the CTFT are especially useful to derive the Fourier transforms of many signals without explicit integration, as demonstrated by several examples in the chapter. CTFTs of real-valued signals have some special properties, and these were studied in detail. Next, autocorrelations and their Fourier transforms were studied, and the application of autocorrelations in match filtering was demonstrated. This application has been important in digital communications, in radar systems, and in the LIGO project.

It was shown that the normalized Gaussian signal is its own Fourier transform except for a scale factor. So it is an example of an eigenfunction of the CTFT operator. An entire family of such eigenfunctions was presented, called the Gauss–Hermite family. It was pointed out that although there are infinitely many eigenfunctions, the eigenvalues can only have four possible values, namely, $\sqrt{2\pi}, -\sqrt{2\pi}, j\sqrt{2\pi}, -j\sqrt{2\pi}$.

The relation between the smoothness of a signal in the time domain and its decay rate in the frequency domain was studied in detail. Smooth signals have rapidly decaying Fourier transforms. Signals called splines were introduced, which have provable smoothness properties in the time domain. Splines have important applications in signal processing, such as in image interpolation.

For causal signals we proved that the real and imaginary parts of the CTFT are related to each other, and satisfy a relation called the Hilbert transform relation. This is also known as the Poisson transform or the Kramers–Kronig transform.

Another fascinating fact was explained, namely that Mother Nature "computes" a Fourier transform when a plane wave is propagating across an aperture and impinging on a distant screen. This observation has far-reaching impact in optics, crystallography, and quantum physics.

9.16 Table of CTFT Properties

We summarize here some of the important properties of the continuous-time Fourier transform (CTFT, or just FT) discussed in this chapter. Recall that the CTFT is defined as

$$X(j\omega) = \int_{-\infty}^{\infty} x(t)e^{-j\omega t}dt,$$

and that the inverse Fourier transform (IFT, or ICTFT) is given by

$$x(t) = \int_{-\infty}^{\infty} X(j\omega)e^{j\omega t}\frac{d\omega}{2\pi}.$$

The notation

$$x(t) \longleftrightarrow X(j\omega)$$

means that $x(t)$ and $X(j\omega)$ constitute a Fourier transform pair, that is, they satisfy the above two equations.

1. Fourier transformation is a *linear* operator, that is, if $x_1(t) \longleftrightarrow X_1(j\omega)$ and $x_2(t) \longleftrightarrow X_2(j\omega)$, then $c_1x_1(t) + c_2x_2(t) \longleftrightarrow c_1X_1(j\omega) + c_2X_2(j\omega)$.
2. $X(0) = \int_{-\infty}^{\infty} x(t)dt$ and $x(0) = \int_{-\infty}^{\infty} X(j\omega)d\omega/2\pi$.
3. $x(t-t_0) \longleftrightarrow e^{-j\omega t_0}X(j\omega)$; $x(t-t_0)$ is called the *shifted* version of $x(t)$.
4. $e^{j\omega_0 t}x(t) \longleftrightarrow X(j(\omega-\omega_0))$; $e^{j\omega_0 t}x(t)$ is called the *amplitude-modulated* version of $x(t)$.
5. $dx(t)/dt \longleftrightarrow j\omega X(j\omega)$.
6. $tx(t) \longleftrightarrow jdX(j\omega)/d\omega$.
7. $y(t) \triangleq \int_{-\infty}^{\infty} x(\tau)h(t-\tau)d\tau \longleftrightarrow X(j\omega)H(j\omega)$ (convolution in time).
8. $x_1(t)x_2(t) \longleftrightarrow \int_{-\infty}^{\infty} X_1(ju)X_2(j(\omega-u))du/2\pi$ (convolution in frequency).
9. $x(-t) \longleftrightarrow X(-j\omega)$; time reversal is equivalent to frequency reversal.
10. $x(at) \longleftrightarrow X(j\omega/a)/|a|$ for real $a \neq 0$ (dilation property).
11. $x^*(-t) \longleftrightarrow X^*(j\omega)$.
12. $x^*(t) \longleftrightarrow X^*(-j\omega)$.
13. $x(t)$ real $\Longleftrightarrow$ $X(j\omega) = X^*(-j\omega)$. So $|X(j\omega)| = |X(-j\omega)|$ and $\arg(X(j\omega)) = -\arg(X(-j\omega))$.
14. $x(t)$ real and even $\Longleftrightarrow$ $X(j\omega)$ real and even ($x(t)$ even means $x(t) = x(-t)$).
15. $FT(FT(x(t))) = 2\pi x(-t)$; $FT(FT(FT(FT(x(t))))) = 4\pi^2 x(t)$ (Sec. 9.4.2).
16. $r_{xy}(\tau) \triangleq \int_{-\infty}^{\infty} x(t)y^*(t-\tau)dt \longleftrightarrow X(j\omega)Y^*(j\omega)$; $r_{xy}(\tau)$: cross-correlation between $x(t)$ and $y(t)$.
17. $r_{xx}(\tau) \triangleq \int_{-\infty}^{\infty} x(t)x^*(t-\tau)dt \longleftrightarrow |X(j\omega)|^2$; $r_{xx}(\tau)$: autocorrelation of $x(t)$.

18. $\int_{-\infty}^{\infty} x(t)y^*(t)dt = \int_{-\infty}^{\infty} X(j\omega)Y^*(j\omega)d\omega/2\pi$; Parseval's relation: inner product preserved.

19. $\int_{-\infty}^{\infty} |x(t)|^2 dt = \int_{-\infty}^{\infty} |X(j\omega)|^2 d\omega/2\pi$; Parseval's relation: energy preserved.

20. $x(t)$ differentiable N times $\forall t$ (with bounded Nth derivative) $\implies$ $X(j\omega)$ is $O(1/\omega^{N+1})$, so that $|X(j\omega)| \leq c/|\omega|^{N+1}$ for large ω.

21. Example of eigenfunction of the FT operator: $e^{-t^2/2} \longleftrightarrow \sqrt{2\pi}e^{-\omega^2/2}$; there are infinitely many examples.

22. The only possible eigenvalues of the FT operator: $\sqrt{2\pi}, -\sqrt{2\pi}, j\sqrt{2\pi}, -j\sqrt{2\pi}$.

23. For any $T > 0$, $\sum_{n=-\infty}^{\infty} x(nT)\delta_c(t-nT) \longleftrightarrow \frac{1}{T}\sum_{k=-\infty}^{\infty} X\big(j(\omega - k\omega_s)\big)$, $\omega_s = 2\pi/T$ (Chapter 10).

24. If $x(t)$ is time limited then it cannot be bandlimited, and vice versa (unless $x(t) = 0, \ \forall t$) (Chapter 10).

25. If $x(t)$ (is possibly complex and) satisfies $x(t) = 0, t \leq 0$, then the real and imaginary parts of its CTFT are related as

$$X_i(\omega) = -\frac{1}{\pi}\int_{-\infty}^{\infty} \frac{X_r(v)}{\omega - v}dv, \qquad X_r(\omega) = \frac{1}{\pi}\int_{-\infty}^{\infty} \frac{X_i(v)}{\omega - v}dv.$$

This is called the Hilbert transform relation. It is also known as the Poisson transform or the Kramers–Kronig transform.

9.17 Table of CTFT Pairs

Signal	Fourier Transform (CTFT)
$x(t) = \delta_c(t)$	$X(j\omega) = 1$
$x(t) = \delta_c(t - t_0)$	$X(j\omega) = e^{-j\omega t_0}$
$p(t) = \begin{cases} 1 & \text{for } -\frac{1}{2} < t < \frac{1}{2} \\ 0 & \text{otherwise} \end{cases}$	$P(j\omega) = \dfrac{\sin(\omega/2)}{\omega/2}$ (sinc)
$x(t) = 1 \ \forall t$	$X(j\omega) = 2\pi\,\delta_c(\omega)$
$e^{j\omega_0 t}$	$2\pi\,\delta_c(\omega - \omega_0)$
$\cos(\omega_0 t)$	$\pi\,\delta_c(\omega + \omega_0) + \pi\,\delta_c(\omega - \omega_0)$
$\sin(\omega_0 t)$	$j\pi\,\delta_c(\omega + \omega_0) - j\pi\,\delta_c(\omega - \omega_0)$
$x_2(t) = \begin{cases} 1 & -1 < t < 0 \\ -1 & 0 < t < 1 \end{cases}$	$\dfrac{2j\sin^2(\omega/2)}{\omega/2}$ (twin pulse, Fig. 9.6(c))
$x_3(t) = \begin{cases} t+1 & -1 < t < 0 \\ 1-t & 0 < t < 1 \end{cases}$	$\left(\dfrac{\sin(\omega/2)}{\omega/2}\right)^2$ (triangular pulse, Fig. 9.6(d))
$h(t) = \dfrac{\sin(\omega_c t)}{\pi t}$ (ideal LPF, Fig. 9.8)	$H(j\omega) = \begin{cases} 1 & \text{for } -\omega_c < \omega < \omega_c \\ 0 & \text{otherwise} \end{cases}$
$s(t) = \sum_{n=-\infty}^{\infty} \delta_c(t - nT)$	$S(j\omega) = \dfrac{2\pi}{T} \sum_{k=-\infty}^{\infty} \delta_c\left(\omega - \dfrac{2\pi}{T}k\right)$ (Fig. 9.10)
$\dfrac{e^{-t^2/2}}{\sqrt{2\pi}}$	$e^{-\omega^2/2}$ (Gaussian; Fig. 9.12)
$x(t) = e^{-at}\mathcal{U}(t),\ Re[a] > 0$	$X(j\omega) = \dfrac{1}{a + j\omega}$ (Fig. 9.15)
$h(t) = \dfrac{1}{\pi t},\quad t \neq 0$ (Hilbert transformer, Sec. 9.9.4)	$H(j\omega) = \begin{cases} -j & \text{for } \omega > 0 \\ j & \text{for } \omega < 0 \end{cases}$
$x(t) = \begin{cases} \dfrac{1}{\sqrt{t}} & t > 0 \\ 0 & t \leq 0 \end{cases}$ (Sec. 9.11.3.2)	$X(j\omega) = \begin{cases} e^{-j\pi/4}\sqrt{\dfrac{\pi}{\lvert\omega\rvert}} & \text{for } \omega > 0 \\ e^{j\pi/4}\sqrt{\dfrac{\pi}{\lvert\omega\rvert}} & \text{for } \omega < 0 \end{cases}$

PROBLEMS

9.1 For the signals shown in Fig. P9.1, find the Fourier transforms. For this you can use the defining integral of the CTFT if you like. But you can also combine the properties of the Fourier transform with examples of known CTFT pairs worked out in this chapter, and get the answers more easily.

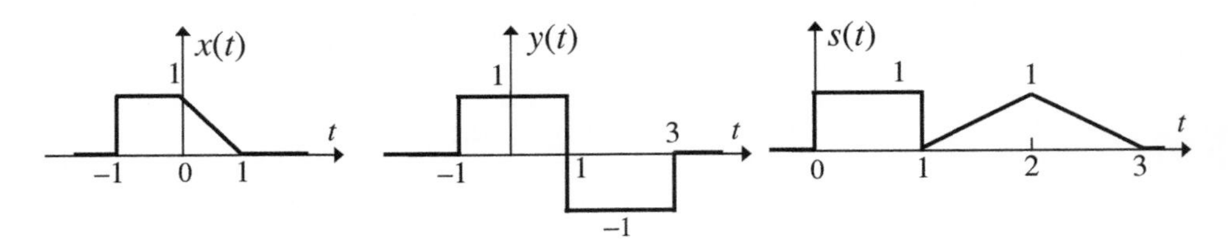

Figure P9.1 Signals for the problem.

9.2 Figure P9.2 shows a truncated Dirac delta train $x(t) = \sum_{n=-3}^{3} \delta_c(t - nT)$.

(a) Compute the Fourier transform by using the integral (9.1).

(b) Now observe that $x(t)$ can be regarded as a product of the infinite train $s(t) = \sum_{n=-\infty}^{\infty} \delta_c(t - nT)$ with a rectangular pulse, so that its CTFT is (proportional to) the convolution of Eq. (9.56) with a sinc function in the frequency domain. Using this, obtain another expression for $X(j\omega)$. This expression may or may not "look like" the answer you got in part (a), but it is equivalent.

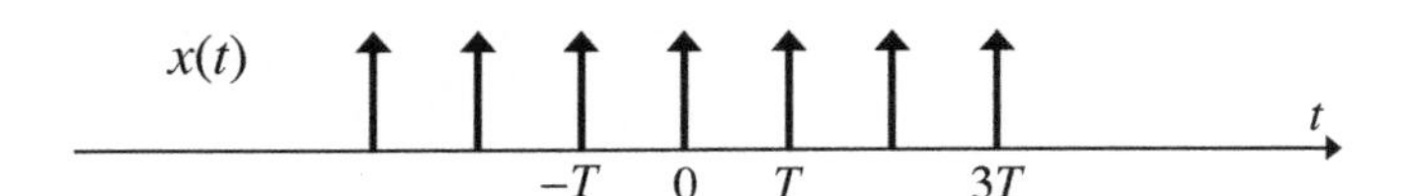

Figure P9.2 Truncated Dirac delta train.

9.3 Find the Fourier transform of the truncated pulse train $y(t)$ shown in Fig. P9.3. *Hint:* This signal is closely related to the truncated impulse train in Problem 9.2. Using this relation, you can get the result readily from the answer to Problem 9.2.

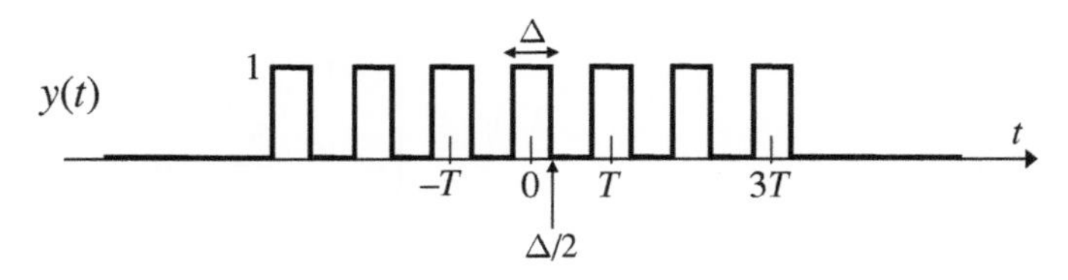

Figure P9.3 Truncated pulse train.

9.4 For the signal in Fig. P9.4, compute the Fourier transform $X(j\omega)$ for all $\omega \neq 0$ without evaluating the defining integral for $X(j\omega)$, but only using the properties of the Fourier transform (e.g., derivative property, shift property, and so forth) and the fact that the Fourier transform of $\delta_c(t)$ is unity for all ω. What is $X(j\omega)$ at $\omega = 0$?

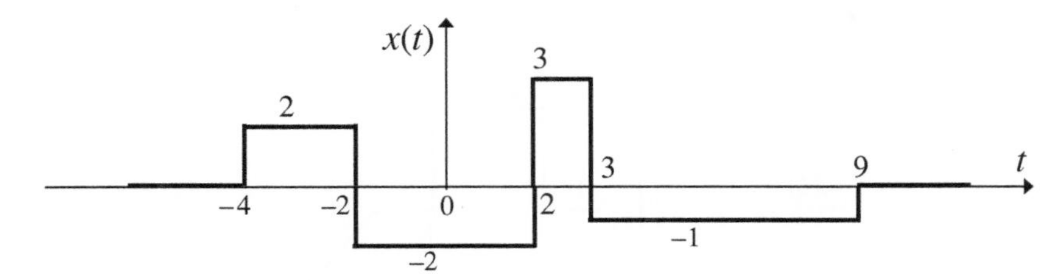

Figure P9.4 A piecewise constant signal.

9.5 This problem has three parts.

(a) For the signal $x(t)$ in Fig. P9.5, find the Fourier transform by using the appropriate derivative propery, and show that

$$X(j\omega) = \frac{1}{j\omega}, \quad \omega \neq 0.$$

Handsketch $jX(j\omega)$. Also plot the phase of $X(j\omega)$. *Note:* $X(j\omega)$ is unbounded as $|\omega| \to 0$. However, $jX(j\omega)$ changes sign when zero frequency is crossed. We can take $X(j0) = 0$, consistent with the observation that $x(t)$ is odd, so that $X(j0) = \int x(t)dt = 0$.

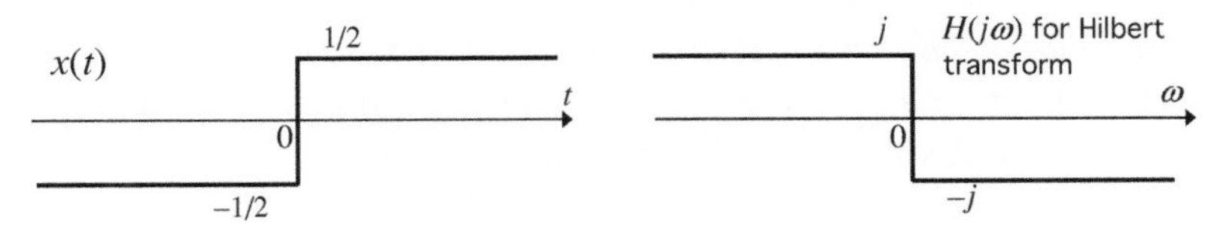

Figure P9.5 Signal and filter for the problem.

(b) Using the above result, show that

$$\mathcal{U}(t) \longleftrightarrow \frac{1}{j\omega} + \pi\delta_c(\omega), \quad \text{(P9.5)}$$

where $\mathcal{U}(t)$ is the unit step. *Note:* This should be interpreted carefully. While the first term is unbounded as $|\omega| \to 0$, its value *at* $\omega = 0$ is taken to be zero, as in part (a). The integral of the first term is zero, whereas the second term makes sure that the integral of the Fourier transform is π.

(c) Consider the Hilbert transformer with frequency response $H(j\omega)$ as in Fig. P9.5. By using the appropriate derivative propery of Fourier transforms, show that the impulse response is as in Eq. (9.112).

9.6 Show that the CTFT of $\sin(\omega_0 t)$ given in Eq. (9.32) can be derived from Eq. (9.31) by using the derivative property (9.59).

9.7 For the signal shown in Fig. P9.7, compute the Fourier transform *without evaluating* the defining integral for $X(j\omega)$. You can do this by first computing the derivative of $x(t)$ and freely using (a) properties of the FT such as the derivative property, shifting property, and so forth and (b) known answers for standard Fourier transform examples like the pulse, Dirac delta, and so forth, given in this chapter.

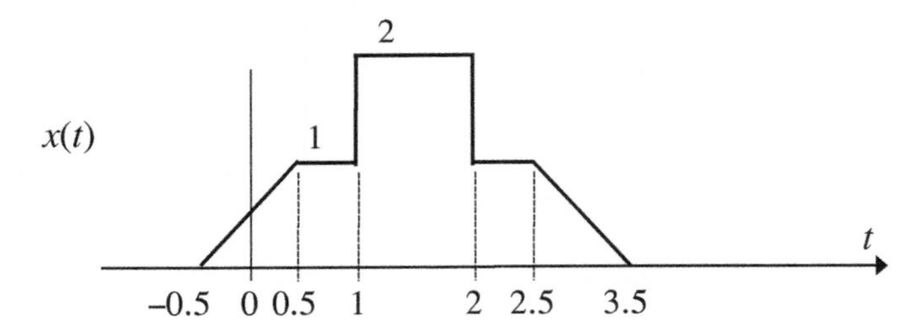

Figure P9.7 A piecewise linear signal.

9.8 Consider the truncated cosine

$$x(t) = \begin{cases} \cos(\omega_0 t) & -\tau/2 < t < \tau/2, \\ 0 & \text{otherwise.} \end{cases}$$

Express $X(j\omega)$ as a superposition of shifted sinc functions.

9.9 Consider Fig. 9.6. By using the observation that $x_2(t)$ can be regarded as the derivative of $x_3(t)$, and the fact that $X_2(j\omega)$ is as in Eq. (9.38), prove that $X_3(j\omega)$ is as in Eq. (9.40).

9.10 *Dilation and shift*. Let $d(t) = x(2 - 3t)$. Express $D(j\omega)$ in terms of $X(j\omega)$.

9.11 Let $x(t) = e^{-a|t|}$, where $a > 0$. Find an expression for $X(j\omega)$ and handsketch $|X(j\omega)|$. Explain what happens when $a < 0$.

9.12 Prove the "convolution in frequency" property (9.15).

9.13 *Odd symmetry*. In Sec. 9.9 we proved some symmetry properties for real signals. For example, if $x(t)$ is real and even, then $X(j\omega)$ is also real and even. In this problem we explore symmetry properties further.

(a) If $x(t)$ is real and odd ($x(t) = -x(-t)$), then show that $X(j\omega)$ is imaginary and odd.

(b) Hence show that if $x(t)$ is imaginary and odd, then $X(j\omega)$ is real and odd.

(c) Hence show that if $X(j\omega)$ is imaginary and odd, then $x(t)$ is real and odd. This is the converse of part (a).

It follows that $x(t)$ is real and odd if and only if $X(j\omega)$ is imaginary and odd.

9.14 This problem develops some details that were skipped in Sec. 9.9.

(a) Show that $x(t)$ is skew Hermitian ($x(t) = -x^*(-t)$) if and only if $X(j\omega)$ is imaginary (i.e., $X(j\omega) = -X^*(j\omega)$).

(b) Show that $x(t)$ is imaginary if and only if $X(j\omega)$ is skew Hermitian.

9.15 For a real signal $x(t)$, consider the two non-negative quantities D_t and D_f defined by

$$D_t^2 = \frac{1}{E}\int_{-\infty}^{\infty} t^2 x^2(t)dt, \quad D_f^2 = \frac{1}{2\pi E}\int_{-\infty}^{\infty} \omega^2 |X(j\omega)|^2 d\omega, \tag{P9.15a}$$

where E is the energy, that is, $E = \int x^2(t)dt$. We say that D_t is the root-mean-square (rms) time-domain duration and D_f the rms frequency-domain duration of $x(t)$. In this problem we show

$$D_t D_f \geq 0.5. \tag{P9.15b}$$

This is called the uncertainty principle: if we make the time-domain duration small, then the frequency-domain duration has to increase such that the above inequality is preserved. In what follows, all integrals are in the range $-\infty$ to ∞. Assume that all relevant integrals exist, and that $x(t)$ has "sufficient decay" so that $tx^2(t) \to 0$ as $t \to \pm\infty$.

(a) Prove the inequality

$$\left(\int tx(t)\frac{dx(t)}{dt}dt\right)^2 \leq \int t^2x^2(t)dt \int \left(\frac{dx(t)}{dt}\right)^2 dt.$$

(b) Prove that the right-hand side of the above inequality equals $E^2 D_t^2 D_f^2$.

(c) Show that $\int tx(t)\frac{dx(t)}{dt}dt = -0.5E$. (Use integration by parts, i.e., $\int udv = uv - \int vdu$.)

(d) Combine these results to prove $D_t D_f \geq 0.5$.

(e) Finally, show from the above that $D_t D_f = 0.5$ if and only if $x(t) = Ae^{-\alpha t^2}$ for some real A and $\alpha > 0$. In other words, $x(t)$ is a Gaussian pulse. So the Gaussian minimizes the time-frequency duration product. For the Gaussian, both $x(t)$ and $X(j\omega)$ are Gaussian and decay very rapidly, which is quite a unique property!

Hint: Use the Cauchy–Schwarz inequality.

9.16 In Problem 3.27 we showed that

$$f(\omega) = \frac{\sin(\omega L/2)}{\sin(\omega/2)}, \quad L \text{ even}, \tag{P9.16a}$$

is not a valid Fourier transform for any discrete-time signal $x[n]$ because when L is even, $f(\omega)$ has period 4π rather than 2π. Now consider the following continuous-time signal:

$$x(t) = \sum_{n=-(K+1)}^{K} \delta_c(t - 0.5 - n). \tag{P9.16b}$$

This is similar to the Dirac delta train in Fig. 9.10 with $T = 1$, except that all Dirac deltas are uniformly shifted by $1/2$, and the signal truncated to a finite duration.

(a) Plot $x(t)$ for $K = 5$. Note that $x(t)$ is even, that is, $x(t) = x(-t)$.

(b) Show that the CTFT of $x(t)$ is precisely $f(\omega)$ for appropriate even L. What is this L?

Note: Since $x(t)$ is real and even, its FT is also real and even (Eq. (9.101)).

9.17 Let

$$x(t) = \frac{\alpha}{\pi}\frac{\sin(\alpha t)}{\alpha t}, \quad \alpha > 0,$$

and define $y(t) = (x * x)(t)$ (i.e., the convolution of $x(t)$ with itself). Give a closed-form expression for $y(t)$. Also let $z(t) = (y * y)(t)$. Find a closed-form expression for $z(t)$.

9.18 Suppose $f(t)$ is an eigenfunction of the CTFT operator. Then it turns out that

$$\frac{df(t)}{dt} - tf(t)$$

is also an eigenfunction (as long as it is not identically zero $\forall t$).

(a) Prove the above claim.

(b) Similarly, show that $df(t)/dt + tf(t)$ is an eigenfunction (as long as it is not identically zero $\forall t$).

(c) We know that $e^{-t^2/2}$ is an eigenfunction of the CTFT operator. Starting from this and using one or both of the above properties, prove that $f_1(t)$, $f_2(t)$, and $f_3(t)$ defined in Eq. (9.83) are also eigenfunctions.

9.19 Let $FT(\cdot)$ be the continuous-time Fourier transform operator (see notation in Sec. 9.4.2). Let

$$y(t) = FT(FT(FT(FT(x(\cdot))))).$$

That is, we take the Fourier transform of $x(t)$ four times to obtain $y(t)$.

(a) Express $y(t)$ in terms of $x(t)$ directly (i.e., without using integral notations). For example, if

$$x(t) = \begin{cases} te^{-t^2/2} & -1 < t < 1, \\ 0 & \text{otherwise}, \end{cases}$$

then what is $y(t)$?

(b) Suppose $x(t)$ is an eigenfunction of the $FT(\cdot)$ operator, with eigenvalue λ. Using the result in part (a), prove that $\lambda^4 = (2\pi)^2$. This proves that the eigenvalue λ can only have four possible values:

$$\lambda = \sqrt{2\pi}, \quad -j\sqrt{2\pi}, \quad -\sqrt{2\pi}, \quad j\sqrt{2\pi}.$$

Hint: Review Secs. 9.4.2 and 9.8.

9.20 *Convolving the Gaussian with itself.* Let $x(t) = e^{-t^2/2}/\sqrt{2\pi}$. What is the convolution of $x(t)$ with itself?

9.21 For the signals shown in Fig. P9.21, consider the integrals $\int_{-\infty}^{\infty} X(j\omega) Y_k^*(j\omega)d\omega$ for $k = 1, 2$. Which of these integrals is zero?

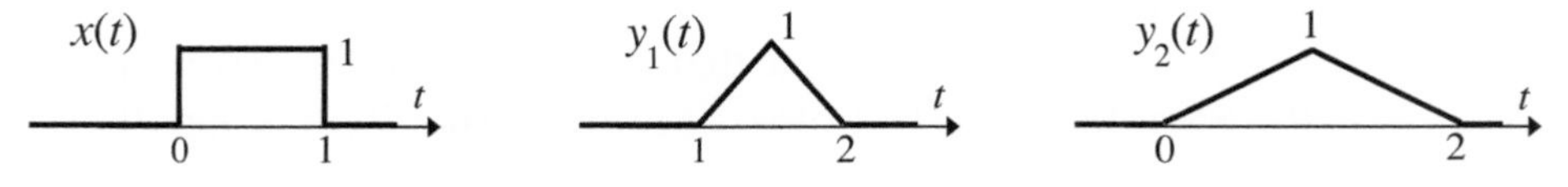

Figure P9.21 Signals for Problem 9.21.

9.22 What is the energy of the sinc signal $x(t) = \sin t/t$?

9.23 We now consider some properties of correlations.

(a) Using only the definition (9.120), show that the cross correlation $r_{xy}(\tau)$ satisfies

$$r_{xy}^*(-\tau) = r_{yx}(\tau). \tag{P9.23}$$

In particular, therefore, $r_{xx}^*(-\tau) = r_{xx}(\tau)$, so the autocorrelation is Hermitian symmetric. So for real signals, $r_{xx}(\tau)$ is even.

(b) Using the Cauchy–Schwarz inequality, show that the autocorrelation $r_{xx}(\tau)$ satisfies Eq. (9.131).

9.24 Let $x(t) = e^{-t}\mathcal{U}(t)$ and $y(t) = e^{-|t|}$. In terms of the big Oh notation (Definition 9.1), how would you describe the decay rate of $X(j\omega)$? Is it $O(1/\omega)$ or $O(1/\omega^2)$? How about the decay rate of $Y(j\omega)$? By examining the plots of $x(t)$ and $y(t)$, can you explain why $Y(j\omega)$ decays faster than $X(j\omega)$?

9.25 Consider the chirp signals $c(t) = \cos(\pi t^2/2)\mathcal{U}(t)$ and $s(t) = \sin(\pi t^2/2)\mathcal{U}(t)$ and define the integrals

$$C(v) = \int_0^v c(t)dt, \quad S(v) = \int_0^v s(t)dt.$$

These are called the **Fresnel integrals** and they are plotted in Fig. P9.25. It can be shown that the limits of these quantities exist as $v \to \infty$, and in fact [Abramowitz and Stegun, 1965]

$$C(\infty) = S(\infty) = 0.5.$$

This shows that

$$\int_0^\infty e^{j\pi(t^2/2)}dt = \frac{1+j}{2}.$$

Based on this, it is possible to find a Fourier-transform pair such that both $x(t)$ and $X(j\omega)$ are unbounded at zero, and decay very slowly for large values of the respective arguments. For this, let us make a change of variables in the above integral as follows: $\pi t^2/2 = \omega\tau$, where $\omega > 0$ is a constant. Show then that

$$\int_0^\infty \frac{e^{-j\omega\tau}}{\sqrt{\tau}}d\tau = e^{-j\pi/4}\sqrt{\pi/\omega}, \quad \omega > 0.$$

Based on this result, show that the Fourier transform of Eq. (9.148) is given by Eq. (9.149).

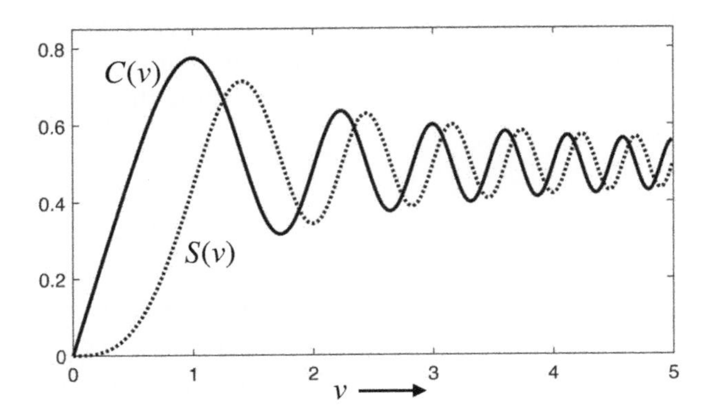

Figure P9.25 The Fresnel integrals.

9.26 *Differentiability of splines.* Consider the spline function $b_2(t)$ in Eq. (9.141). This is a polynomial in each of the three regions shown, and zero everywhere else. So we say this is a piecewise polynomial function. Clearly, this is differentiable everywhere in each open interval of the form $n < t < n + 1$, for $n = 0, 1, 2$. For $t < 0$ and $t > 3$, it is trivially the zero polynomial, hence differentiable there as well. The points $t = 0, 1, 2, 3$, where the individual polynomials meet, are the knots of the spline.

(a) Plot the derivative $db_2(t)/dt$. You will find that it is piecewise linear, and its value on the left of each knot matches the value on the right of the knot. So we say the derivative exists at the knots and is continuous. In short, the derivative exists everywhere including the knots.

(b) Plot the second derivative $d^2b_2(t)/dt^2$. You will find that this is piecewise constant, bounded, and discontinuous at the knots (the left derivative and the right derivative do not match at the knots). So the next derivative $d^3b_2(t)/dt^3$ does not exist, strictly speaking. However, using Dirac delta for the derivatives at discontinuities as in Sec. 2.12.3, we can plot $d^3b_2(t)/dt^3$. Do this, and you will find that it is unbounded.

Summarizing, the first derivative of $b_2(t)$ is bounded and continuous, the second derivative is bounded but discontinuous, and the third derivative is unbounded.

9.27 Find an example of a sequence $x[n]$ such that $X(e^{j\omega})$ is differentiable everywhere with bounded derivative, but the z-transform $X(z)$ does not have a nonempty ROC.

9.28 (Computing assignment) *More general eigenfunctions.* This problem shows how you can generate eigenfunctions of the Fourier transform almost trivially by starting from any real even function, and making a clever construction.

(a) We know that the sinc function $P(j\omega)$ is the Fourier transform of the pulse

$$p(t) = \begin{cases} 1 & -0.5 < t < 0.5, \\ 0 & \text{otherwise.} \end{cases}$$

Find a linear combination of these two functions:

$$x(t) = p(t) + \alpha_1 P(jt)$$

(i.e., find α_1) such that $x(t)$ is an eigenfunction of the Fourier transform, that is, $X(j\omega) = \lambda_1 x(\omega)$. What is the value of λ_1 in your example?

(b) We know that the square of the sinc function $Y(j\omega) = P^2(j\omega)$ is the Fourier transform of the symmetric triangle $y(t)$ formed by convolving $p(t)$ with itself. Find a linear combination of these two functions:

$$z(t) = y(t) + \alpha_2 Y(jt)$$

such that it is an eigenfunction of the Fourier transform, that is, $Z(j\omega) = \lambda_2 z(\omega)$. What is the value of the constant λ_2?

(c) Plot $x(t)$ and $z(t)$ in the range $-20 \leq t \leq 20$ (you may use any plotting software).

The interested reader can start with any real even function, appropriately dilate it to obtain a signal $s(t)$, and combine $s(t)$ and $S(jt)$ to obtain more examples!

9.29 (Computing assignment) *Chirps and correlations*. In Sec. 9.10.1 we discussed the chirp waveform (9.133) and demonstrated that chirps with different sets of parameters (ω_i, Δ_i) typically have small cross correlations, while each chirp has an autocorrelation with a large peak (Fig. 9.17). Play with the parameters (ω_i, Δ_i) to generate an example of two chirps like that. Plot figures similar to Figs. 9.16 and 9.17. Your plots need not look exactly like these; they are just guidelines showing what a pair of good chirps should be like.

9.30 (Computing assignment) *Splines*. In Fig. 9.22 we plotted examples of *B*-splines, and showed their Fourier transform magnitudes in Fig. 9.23. Plot $b_4(t)$ in the region $0 \leq t \leq 5$ and plot $|B_4(j\omega)|$ in $|\omega| \leq 40$. You will see that $b_4(t)$ is very smooth and $|B_4(j\omega)|$ is practically zero for $|\omega|$ exceeding some small threshold. This is because of the $O(1/\omega^5)$ property of $B_4(j\omega)$.

9.31 (Computing assignment) *Eigenfunctions of the FT operator*. In Problem 9.18 we proved that the $f_k(t)$ in Eq. (9.83) are eigenfunctions of the FT operator. Using the same technique, prove that

$$f_4(t) = (4t^4 - 12t^2 + 3)e^{-t^2/2}$$

is also an eigenfunction of the FT operator. Plot this for $-5 \leq t \leq 5$.

10 Sampling a Continuous-Time Signal

10.1 Introduction

In order to process a continuous-time signal $x(t)$ digitally, it is first sampled in time as shown in Fig. 10.1(a). The figure shows the case of **uniform** sampling, where adjacent samples are separated by a fixed amount T. Thus, the samples form a sequence in time as follows:

$$\cdots x(-T), x(0), x(T), x(2T) \cdots \tag{10.1}$$

where T is the sample spacing in seconds. These samples $x(nT)$ are then digitized, that is, converted into binary form with finite number of bits. At first it appears that the sampling process results in loss of information because the values of $x(t)$ between samples are lost. But if the signal $x(t)$ has a property called the **bandlimited** property, then it is actually possible to reconstruct it perfectly for all t from the samples $x(nT)$, as long as T is smaller than a certain threshold.

In this chapter we will discuss bandlimited signals, sampling theory, and the method of reconstruction from samples. Our focus will be on uniform sampling. The chapter discusses sampling theorems for bandlimited signals of the lowpass type as well as the bandpass type. The reconstruction from samples is based on the use of a linear filter as an **interpolator**. Fourier transforms will play a crucial role in our derivations and our discussions throughout the chapter. When the sampling rate is not sufficiently large, the sampling process leads to a phenomenon called **aliasing**. This is discussed in detail and several real-world manifestations of aliasing are also discussed (Sec. 10.11).

In practice, the sampled signal is typically processed by a digital signal processing device, before it is converted back into a continuous-time signal. The building blocks in such a digital signal processor are discussed in Sec. 10.7. It turns out that signals that are bandlimited cannot also be of finite duration (Sec. 10.9). This has not stopped digital signal processing from evolving into a major field of research touching our lives in many ways. This is because time-limited signals can often be satisfactorily approximated by bandlimited signals, and vice versa.

More detailed discussions on bandlimited signals can be found in Chapter 15. A brief overview of sampling techniques based on sparsity (rather than on the bandlimited property) is given in Chapter 16.

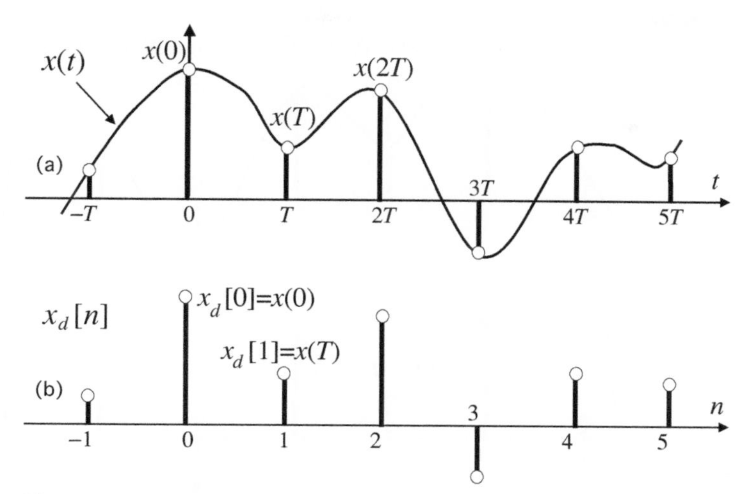

Figure 10.1 (a) A continuous-time signal $x(t)$ and its uniformly spaced samples. (b) The sampled version $x_d[n]$, viewed as a discrete-time signal with dimensionless integer time argument n.

Notations. In this chapter we do not use the special notation Ω for the frequency in continuous time, but instead use ω. Similarly, we do not use subscript c as in $x_c(t)$ for continuous time (except for the Dirac delta, so its notation is still $\delta_c(t)$). This is because almost every page will have continuous-time quantities! However, whenever we need to distinguish between CTFT and DTFT, we use a subscript d for discrete time as in $x_d[n] = x(nT)$, instead of using $x[n] = x(nT)$. This way, the frequency domain has $X_d(e^{j\omega})$ and $X(j\omega)$, avoiding a clash of notations such as $X(e^{j\omega})$ and $X(j\omega)$.

10.2 Representing a Sampled Version of a Signal

The uniformly sampled version of $x(t)$ can simply be represented as a sequence

$$x_d[n] = x(nT), \quad -\infty < n < \infty, \tag{10.2}$$

as shown in Fig. 10.1(b). Notice carefully that $x_d[n]$ (with square brackets) is a discrete-time signal with its argument n defined only for integers. On the other hand, $x(nT)$ (with round brackets) comes from a continuous-time signal $x(t)$ with its argument defined for all real t. For the purpose of analysis, it is often convenient to define the sampled version as a product of $x(t)$ with the Dirac delta train (impulse train)

$$s(t) = \sum_{n=-\infty}^{\infty} \delta_c(t - nT). \tag{10.3}$$

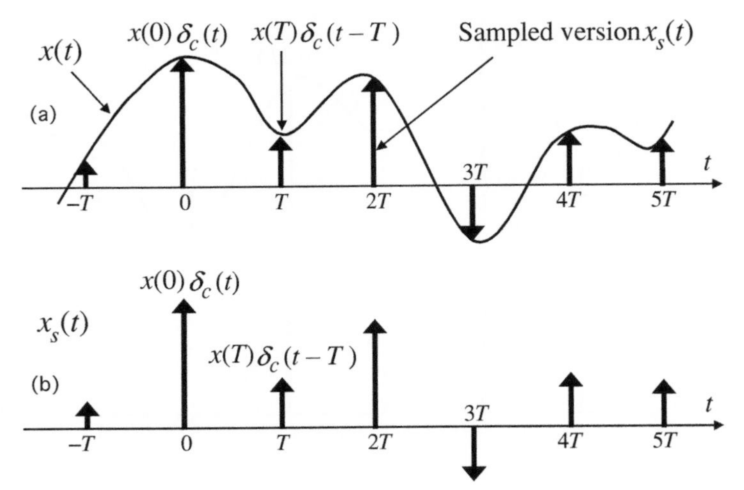

Figure 10.2 (a), (b) Another way to represent the sampling process. The sampled version $x_s(t)$ is here viewed as a train of Dirac delta functions. Note that $x_s(t)$ is a function of the continuous-time argument t, and is zero for all t except at the sample locations.

Thus, the uniformly sampled signal is

$$x_s(t) = x(t) \sum_{n=-\infty}^{\infty} \delta_c(t - nT) \tag{10.4}$$

$$= \sum_{n=-\infty}^{\infty} x(nT)\delta_c(t - nT). \tag{10.5}$$

This is demonstrated in Fig. 10.2(a). In this representation the sampled version is a train of Dirac delta functions, as shown in Fig. 10.2(b). The advantage of this representation is that it allows us to find an expression for the Fourier transform $X_s(j\omega)$ of the sampled version easily. In the representation of Fig. 10.1(b), we think of the *sampled version as a discrete-time sequence* $x_d[n]$ with dimensionless, integer, time argument n. In the representation of Fig. 10.2(b), we think of the *sampled version as a continuous-time signal* $x_s(t)$ with argument t in seconds. Note that $x_s(t)$ is zero for all t except at the sample locations.

Terminology. At this point it is convenient to introduce some notations and terminology. T is called the sample spacing. The **sampling rate** or sampling frequency is defined to be its reciprocal, and there are two conventions for this:

$$T = \text{sample spacing in secs},$$
$$f_s = \frac{1}{T} = \text{sampling rate in Hz},$$
$$\omega_s = \frac{2\pi}{T} = 2\pi f_s = \text{sampling rate in rads/sec}. \tag{10.6}$$

Note that f_s represents the number of samples per second. For example, if $T = 0.01$ secs, then $f_s = 100$ Hz = 100 samples per second and $\omega_s = 2\pi * 100$ rads/sec. We will see that the notation ω_s is useful as it repeatedly appears in all discussions involving the Fourier transform of the sampled signal. ▽ ▽ ▽

10.2.1 Fourier Transform of the Sampled Version

We now find an expression for the Fourier transform (FT) of the sampled signal $x_s(t)$. For this, recall that the Fourier transform of the Dirac delta train (10.3) is given by

$$S(j\omega) = \frac{2\pi}{T} \sum_{k=-\infty}^{\infty} \delta_c \left(\omega - \frac{2\pi}{T}k \right). \tag{10.7}$$

Since the sampled version $x_s(t)$ is the product of $x(t)$ with $s(t)$, its FT $X_s(j\omega)$ is obtained by a **convolution** in the frequency domain. More precisely:

$$X_s(j\omega) = \frac{1}{2\pi} (X * S)(j\omega), \tag{10.8}$$

so that

$$X_s(j\omega) = \frac{1}{T} \sum_{k=-\infty}^{\infty} X\left(j\left(\omega - \frac{2\pi}{T}k \right) \right), \tag{10.9}$$

where we have used the fact that convolution of $X(j\omega)$ with $\delta_c(\omega - \omega_0)$ produces the shifted version $X(j(\omega - \omega_0))$ (Sec. 3.10.3). For greater clarity, here is the explicit derivation of Eq. (10.9):

$$\begin{aligned} X_s(j\omega) &= \frac{1}{2\pi} \int_{-\infty}^{\infty} X(ju)S(j(\omega - u))du \\ &= \frac{1}{T} \int_{-\infty}^{\infty} X(ju) \sum_{k=-\infty}^{\infty} \delta_c \left(\omega - u - \frac{2\pi}{T}k \right) du \\ &= \frac{1}{T} \sum_{k=-\infty}^{\infty} \int_{-\infty}^{\infty} X(ju)\delta_c \left(\omega - u - \frac{2\pi}{T}k \right) du \\ &= \frac{1}{T} \sum_{k=-\infty}^{\infty} X\left(j\left(\omega - \frac{2\pi}{T}k \right) \right). \end{aligned} \tag{10.10}$$

Thus, the FT $X_s(j\omega)$ of the sampled version of $x(t)$ is a *sum of an infinite number of* **uniformly shifted** *copies* of the original FT $X(j\omega)$, scaled by $1/T$. Figures 10.3(a) and (b) explain this pictorially. The meaning of Fig. 10.3(c) will be explained in the following subsection. The term $k = 0$ in Eq. (10.9) is just $X(j\omega)$ scaled by $1/T$, as shown in the figure. The other terms, namely

$$\frac{1}{T} X\left(j\left(\omega - \frac{2\pi}{T}k \right) \right), \tag{10.11}$$

are uniformly shifted versions of this term, where the shifts are in integer multiples of the sampling rate $\omega_s = 2\pi/T$. Two important remarks are in order:

1. It is clear that $X_s(j\omega)$ is a **periodic function** of ω with period ω_s (the sampling rate). This is because, if we replace ω with $\omega + \omega_s$ in Eq. (10.9) and simplify it, we get the same expression back.

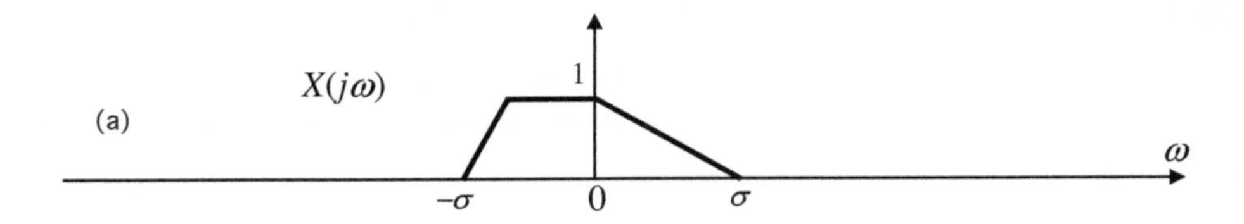

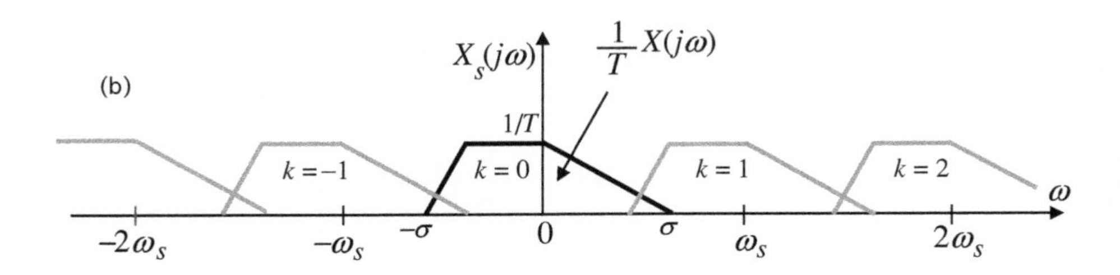

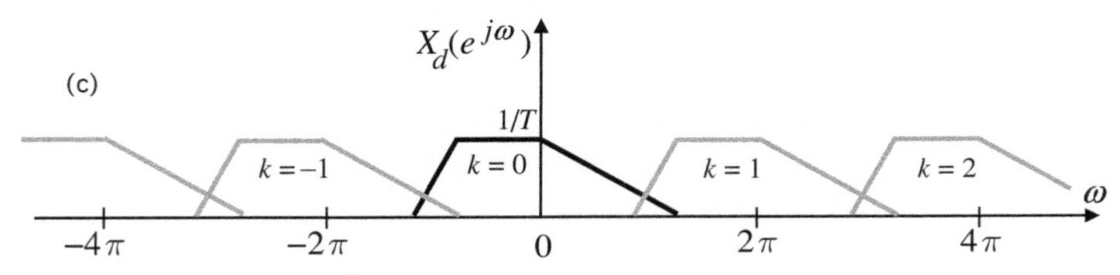

Figure 10.3 (a) Fourier transform $X(j\omega)$ of the original signal $x(t)$. (b) Fourier transform $X_s(j\omega)$ of the sampled version of $x(t)$, where $\omega_s = 2\pi/T$ is the sampling rate. (c) Fourier transform $X_d(e^{j\omega})$ of the sampled version, if the latter is viewed as a discrete-time signal $x_d[n] = x(nT)$.

2. If T is decreased (samples are more closely spaced), then the copies of Eq. (10.9) become more widely separated, and vice versa. Denser sampling in the time domain implies that the copies in the frequency domain are sparsely packed, and vice versa.

10.2.2 The DTFT and CTFT of the Sampled Version

With $X_d(e^{j\omega})$ denoting the discrete-time Fourier transform (DTFT) of the sampled version $x_d[n] = x(nT)$, we have

$$X_d(e^{j\omega}) = \sum_{n=-\infty}^{\infty} x_d[n]e^{-j\omega n} = \sum_{n=-\infty}^{\infty} x(nT)e^{-j\omega n}. \tag{10.12}$$

On the other hand, from Eq. (10.5), it follows that the continuous-time Fourier transform (CTFT) of the sampled version $x_s(t)$ can also be written as

$$X_s(j\omega) = \sum_{n=-\infty}^{\infty} x(nT)e^{-j\omega nT}, \tag{10.13}$$

where we have used the fact that the CTFT of $\delta_c(t - nT)$ is $e^{-j\omega nT}$. Comparing the above two expressions, it follows that the DTFT and the CTFT of the sampled version are related as follows:

$$X_d(e^{j\widehat{\omega}}) = X_s(j\omega)\Big|_{\omega T=\widehat{\omega}} = X_s(j\omega)\Big|_{\omega=\widehat{\omega}/T}. \tag{10.14}$$

That is, $X_d(e^{j\widehat{\omega}})$ at any frequency $\widehat{\omega}$ is obtained by replacing ω with $\widehat{\omega}/T$ in $X_s(j\omega)$. So it follows from Eq. (10.9) that

$$X_d(e^{j\widehat{\omega}}) = \frac{1}{T}\sum_{k=-\infty}^{\infty} X\left(j\left(\frac{\widehat{\omega} - 2\pi k}{T}\right)\right). \tag{10.15}$$

Dropping the hat on the ω, which was only for temporary convenience, the Fourier transform of the uniformly sampled version of $x(t)$ with sample spacing T can be written in two ways:

$$x_s(t) \longleftrightarrow X_s(j\omega) = \frac{1}{T}\sum_{k=-\infty}^{\infty} X\left(j\left(\omega - \frac{2\pi}{T}k\right)\right) \quad \text{(CTFT)}, \tag{10.16}$$

$$x_d[n] \longleftrightarrow X_d(e^{j\omega}) = \frac{1}{T}\sum_{k=-\infty}^{\infty} X\left(j\left(\frac{\omega - 2\pi k}{T}\right)\right) \quad \text{(DTFT)}, \tag{10.17}$$

where $x_s(t) = \sum_{n=-\infty}^{\infty} x(nT)\delta_c(t-nT)$ and $x_d[n] = x(nT)$. It is clear from the above that

$$X_s(j\omega) = X_d(e^{j\omega T}). \tag{10.18}$$

Figures 10.3(b) and (c) show the relation between $X_s(j\omega)$ and $X_d(e^{j\omega})$. Note that $X_s(j\omega)$ is periodic in ω with period ω_s, whereas $X_d(e^{j\omega})$ is periodic in ω with period 2π. Also, ω in $X_s(j\omega)$ is in rads/sec, whereas ω in $X_d(e^{j\omega})$ is in rads.

10.3 Sampling a Bandlimited Signal

We say that a signal $x(t)$ is **bandlimited** if its Fourier transform satisfies the following:

$$X(j\omega) = 0 \quad \text{for} \quad |\omega| \geq \sigma, \tag{10.19}$$

for some positive frequency $\sigma < \infty$. That is, the energy of the signal is limited to the band

$$-\sigma < \omega < \sigma. \tag{10.20}$$

The frequency σ is called the *band limit*. A signal satisfying Eq. (10.19) is also called a σ-**BL** signal, pronounced "sigma-bandlimited." The signal with FT as in Fig. 10.3(a) is an example of a σ-BL signal. Note that 2σ is the total **bandwidth** of $x(t)$. Sometimes we use the notation

$$f_m = \frac{\sigma}{2\pi}. \quad (10.21)$$

Here, f_m is the band limit in Hz and σ is in rads/sec, analogous to Eq. (10.6).

10.3.1 The Sampling Theorem

Consider again Fig. 10.3(b), which shows the Fourier transform of the uniformly sampled version of $x(t)$. This is a superposition of uniformly spaced copies of $X(j\omega)$. The adjacent copies are spaced apart by ω_s, and each copy has a width of 2σ (which is the bandwidth of $x(t)$). So there is an **overlap** between adjacent copies because

$$\omega_s < 2\sigma, \quad (10.22)$$

that is, the spacing ω_s between copies is smaller than the width 2σ of each copy. The overlap phenomenon is also called **aliasing**. It is clear that there is no aliasing as long as

$$\omega_s \geq 2\sigma, \quad (10.23)$$

that is, if the sampling rate is at least twice the band limit σ. Aliasing and absence of aliasing are both demonstrated in Fig. 10.4. When $\omega_s \geq 2\sigma$, we can therefore recover $X(j\omega)$ of Fig. 10.5(a) from $X_s(j\omega)$ of Fig. 10.5(b) by multiplying the latter with the function $H(j\omega)$, shown in Fig. 10.5(c), because it is obvious from the figure that

$$X(j\omega) = X_s(j\omega)H(j\omega), \quad (10.24)$$

when there is no aliasing. This multiplication in the frequency domain corresponds to lowpass filtering with a frequency response

$$H(j\omega) = \begin{cases} T & -\sigma < \omega < \sigma, \\ 0 & \text{otherwise.} \end{cases} \quad (10.25)$$

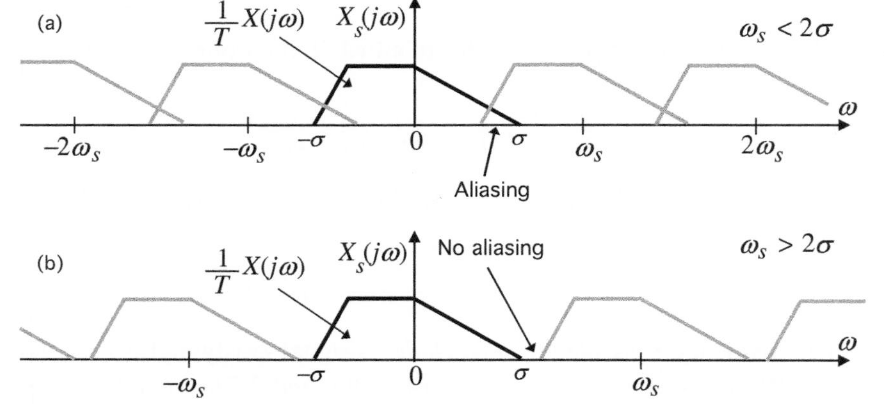

Figure 10.4 Sampling rate and aliasing. (a) Aliasing occurs when $\omega_s < 2\sigma$; (b) there is no aliasing when $\omega_s \geq 2\sigma$.

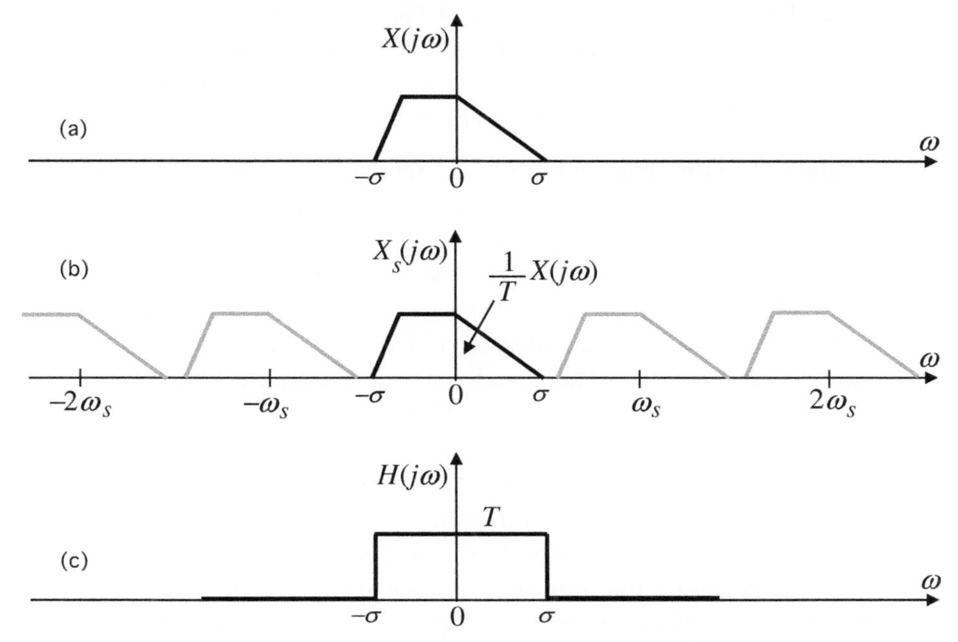

Figure 10.5 (a) Fourier transform of a σ-BL signal $x(t)$, (b) FT of its sampled version $x_s(t)$, where $\omega_s > 2\sigma$, and (c) frequency response of a filter to recover $x(t)$ from $x_s(t)$.

Note that since $X_s(j\omega)$ has the original copy $X(j\omega)$ scaled by $1/T$, the filter $H(j\omega)$ has to have passband response T rather than unity. The main result proved above can be stated as follows.

Theorem 10.1 *The uniform sampling theorem.* Assume $x(t)$ is σ-BL, as demonstrated in Fig. 10.5(a), and let $x_s(t)$ be the uniformly sampled version with sample spacing T, so that the sampling rate is $\omega_s = 2\pi/T$. If $\omega_s \geq 2\sigma$, then $x(t)$ can be reconstructed from the sampled version $x_s(t)$ by filtering it with a lowpass reconstruction filter with frequency response (10.25). ◇

Two comments are in order:

1. In order to reconstruct a signal from its samples based on the sampling theorem, **we need to know the band limit** σ. This is because the sampling rate to be chosen, and the reconstruction filter to be used, both depend on this knowledge.
2. We can rewrite $\omega_s \geq 2\sigma$ in terms of f_s (the sampling rate $1/T$ in Hz) and f_m (the band limit in Hz). For this, recall that $f_s = \omega_s/2\pi$ and $f_m = \sigma/2\pi$. So the sampling theorem says that if

$$\text{sampling rate } f_s \geq 2f_m, \quad \text{that is,} \quad \text{sample spacing } T \leq \frac{1}{2f_m}, \tag{10.26}$$

then $x(t)$ can be reconstructed from samples using the lowpass filter. Equation (10.26) is equivalent to saying that the sampling rate is at least as large as the **bandwidth** 2σ of the signal.

The above theorem was proved by **Claude Shannon** in 1949 and in a slightly different form by **Harry Nyquist** in 1928. So it is called Shannon's sampling theorem or Nyquist's sampling theorem. It was also noticed by **Whittaker** in 1929, and is sometimes referred to as Whittaker's theorem [Whittaker, 1929].

Some real-world sampling rates. The **speech** signal generated by humans typically has highest frequency of interest $f_m \approx 4$ kHz. Any energy remaining beyond this frequency is of relatively less importance. So, the speech signal $m(t)$ at the microphone output is first passed through a good analog lowpass filter $G(j\omega)$ with cutoff at 4 kHz, so that any incidental energy beyond 4 kHz (including noise) is attentuated. $G(j\omega)$ is called an **antialias filter**. The filter output $x(t)$ is therefore practically bandlimited to 4 kHz. This signal $x(t)$ is then sampled at a rate $f_s \geq 2f_m = 8$ kHz. See Fig. 10.6, which shows a typical filter frequency response and the magnitudes of the Fourier transforms of $m(t)$ and $x(t)$. For convenience, we plot them as functions of f in Hz (instead of ω in rads/sec). We show only $f \geq 0$ because, for real signals, the FT magnitude is an even function.

In practice, we always sample at a frequency slightly higher than the minimum rate $2f_m$, because antialising filters are not ideal filters, they only approximate them (see Fig. 10.6). So it is common to use $f_s = 10$ kHz for speech. This is sufficient for telephone-quality speech.

For **music** waveforms the story is different. Music (Sec. 11.8) is often rich in harmonics (integer multiples of the pitch frequency). There is considerable information beyond 4 kHz, generated by the human voice and more importantly, by musical instruments. The highest frequency of interest is typically taken to be $f_m = 20$ kHz, so that the sampling rate is $f_s \geq 2f_m = 40$ kHz. Again, in practice, the analog music waveform from a microphone is filtered by an antialiasing filter $G(j\omega)$ with cutoff at 20 kHz, and then sampled. A sampling rate higher than the minimum rate of 40 kHz is always prefered in practice. Typical examples of sampling rates are 44.1 kHz (for CD mastering), 48 kHz (for studio recording), and so forth. Sometimes significantly higher sampling rates than this are used to achieve very high

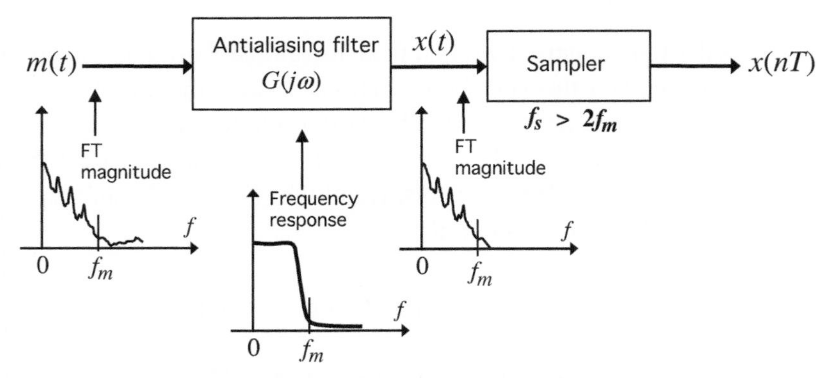

Figure 10.6 A lowpass speech signal $m(t)$ from a microphone is filtered through an antialiasing filter to minimize the energy in $f \geq f_m$, before it is sampled. The filter output $x(t)$ is then sampled at a rate $f_s > 2f_m$.

fidelity. In practice, sampling is implemented by an analog to digital converter (or **A/D converter**), which samples the signal as well as digitizing the samples into a finite number of bits. ▽ ▽ ▽

Sampling rates for boring lectures. Students are always inventing ingeneous sampling strategies, especially when the lectures are boring. Say a class meets on Mondays, Wednesdays, and Fridays. Attending only the Monday lectures would be like *uniform downsampling* by a factor of 3, and attending Mondays and Wednesdays would be *nonuniform downsampling*. Some students are of course very dedicated. They attend all lectures, and in addition watch the lecture recordings again – clearly a case of oversampling, but you learn better that way: if there was an error in your understanding of a concept, it gets cleared up when you listen again. In any case, these are some typical sampling rates every professor witnesses in college.

10.3.2 The Reconstruction Formula

Equation (10.24), which is valid when $\omega_s \geq 2\sigma$, implies that we can recover the original σ-BL signal $x(t)$ from the sampled version $x_s(t)$ by convolving it with the impulse response $h(t)$ of the filter $H(j\omega)$:

$$x(t) = (h * x_s)(t), \tag{10.27}$$

where $x_s(t)$ is the sampled version

$$x_s(t) = \sum_{n=-\infty}^{\infty} x(nT)\delta_c(t - nT). \tag{10.28}$$

The filter impulse response $h(t)$, which is the inverse Fourier transform of $H(j\omega)$ in Fig. 10.5(c), is given by

$$h(t) = \frac{T}{2\pi}\int_{-\sigma}^{\sigma} e^{j\omega t} d\omega = T\,\frac{\sin \sigma t}{\pi t}. \tag{10.29}$$

This is a sinc function, and is plotted in Fig. 10.7. Note that for $t \to 0$ we have $h(t) \to T\sigma t/\pi t = T\sigma/\pi$, so that $h(0) = T\sigma/\pi$, as indicated in the figure. The zero crossings of $h(t)$ occur when $\sigma t = k\pi$ for nonzero integer k, that is, at

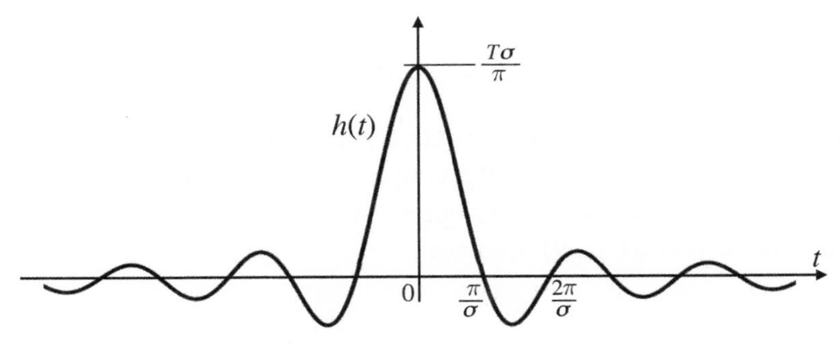

Figure 10.7 Impulse response $h(t) = T \sin \sigma t/\pi t$ of the lowpass reconstruction filter $H(j\omega)$.

$$t_k = \frac{k\pi}{\sigma}, \quad k \neq 0 \qquad \text{(zero crossings of } h(t)), \tag{10.30}$$

as shown in the figure. It is insightful to write the filtering or convolution operation (10.27) more explicitly. From Eqs. (10.27) and (10.28) we have

$$\begin{aligned} x(t) = (h * x_s)(t) &= h(t) * \left(\sum_{n=-\infty}^{\infty} x(nT)\delta_c(t - nT) \right) \\ &= \sum_{n=-\infty}^{\infty} x(nT) \left(h(t) * \delta_c(t - nT) \right). \end{aligned} \tag{10.31}$$

Since the convolution of $h(t)$ with $\delta_c(t-nT)$ is the shifted impulse response $h(t-nT)$, we finally have

$$x(t) = \sum_{n=-\infty}^{\infty} x(nT)h(t - nT). \tag{10.32}$$

This reconstruction formula is more convenient than Eq. (10.27) as it explicitly shows how to reconstruct $x(t)$ from the samples $x(nT)$. Equation (10.32) is also called an **interpolation formula**, as it recovers $x(t)$ for all values of t between any pair of adjacent samples. It is also called the **sinc interpolation** formula, since $h(t)$ is a sinc function.

10.4 The Nyquist Rate

The sampling theorem assures us that the sampling rate $\omega_s \geq 2\sigma$ is sufficient for recovering a σ-BL signal $x(t)$ from its samples. Now consider the special case where

$$\omega_s = 2\sigma. \tag{10.33}$$

This is called the **Nyquist rate**. The case where $\omega_s > 2\sigma$ is said to be *oversampling*, and when $\omega_s < 2\sigma$ we call it *undersampling*. For Nyquist-rate sampling, clearly Eq. (10.24) and the reconstruction formula (10.32) continue to be true. Since $\omega_s = 2\pi/T$, it follows from Eq. (10.33) that

$$\sigma = \frac{\pi}{T}, \tag{10.34}$$

when we sample at the Nyquist rate. Figures 10.8(a)–(c) show the FT of the unsampled signal $X(j\omega)$, the FT of the sampled signal $X_s(j\omega)$, and the reconstruction filter response $H(j\omega)$ for Nyquist sampling. At the Nyquist rate, the reconstruction filter impulse response (10.29) becomes

$$h(t) = \frac{\sin \frac{\pi t}{T}}{\frac{\pi t}{T}}. \tag{10.35}$$

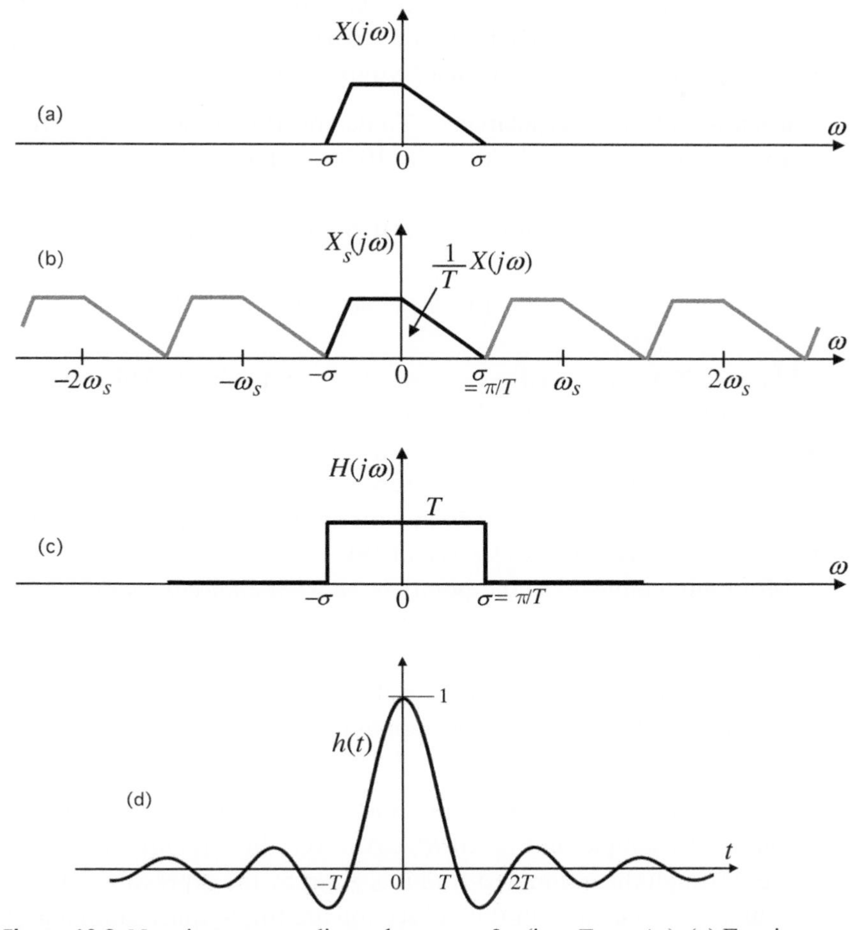

Figure 10.8 Nyquist-rate sampling, where $\omega_s = 2\sigma$ (i.e., $T = \pi/\sigma$). (a) Fourier transform of a σ-BL signal $x(t)$, (b) FT of its sampled version $x_s(t)$, (c) frequency response of the filter to reconstruct $x(t)$ from $x_s(t)$, and (d) impulse response $h(t) = \sin(\pi t/T)/(\pi t/T)$ of the filter $H(j\omega)$.

The reconstruction formula (10.32) is therefore

$$x(t) = \sum_{n=-\infty}^{\infty} x(nT) \frac{\sin \dfrac{\pi(t-nT)}{T}}{\dfrac{\pi(t-nT)}{T}}. \tag{10.36}$$

From Eq. (10.35) we see that

$$h(t) = \begin{cases} 1 & \text{for } t = 0, \\ 0 & \text{for } t = kT, \text{where } k = \text{nonzero integer}, \end{cases} \tag{10.37}$$

or in short,

$$h(kT) = \delta[k]. \tag{10.38}$$

Thus, $h(t)$ has uniformly spaced zero crossings, separated exactly by the sample spacing T. See Fig. 10.8(d). Equation (10.38) is called the Nyquist property and the filter $H(j\omega)$ satisfying it is called a Nyquist(T) filter.

The miracle of sinc interpolation. To demonstrate how the beautiful sinc interpolation formula works, consider Fig. 10.9(a). Here each sinc plot shows one term of Eq. (10.36). The height is exactly $x(nT)$, and the zero crossings are at integer multiples of T (because of Eq. (10.37)). So the set of functions

$$\cdots \;\; h(t+T), \;\; h(t), \;\; h(t-T), \;\; h(t-2T), \;\; \cdots \tag{10.39}$$

all have uniformly spaced zero crossings separated by T. This means that, at $t = 0$ where $x(0)h(t) = x(0)$, *all other sinc functions* are zero. The sum (10.36) therefore adds up to $x(0)$ at the point $t = 0$, as expected. Similarly, at each sample point $t = kT$, only the kth term in Eq. (10.36) is nonzero, and yields the answer $x(kT)$. In short, the terms in Eq. (10.36) do not interfere with each other at sample locations – there is no *intersample interference* or **ISI**.

More miraculously, at the points *in between adjacent samples*, the infinite sum (10.36) adds up to $x(t)$ exactly. So the shifted and scaled sinc functions in Fig. 10.9(a) add up to the original bandlimited signal $x(t)$, as shown in Fig. 10.9(b). It is very difficult to be convinced of this perfect reconstruction property by thinking in the time domain. The proof comes from the frequency-domain picture shown in Fig. 10.8. ▽ ▽ ▽

Notice that we can conceive of infinitely many continuous-time signals that "pass through" the samples in Fig. 10.9(b). And yet, there is *only one σ-BL function* that passes through these samples! And it is given by the expression (10.36).

While the discussions in this subsection are for Nyquist sampling, the preceding remark can be generalized: suppose $x(t)$ and $y(t)$ are two σ-BL signals, and let $x(nT)$ and $y(nT)$ be the samples, with sampling rate $\omega_s \geq 2\sigma$. Then we can never have $x(nT) = y(nT)$ for all n unless $x(t) = y(t)$ for all t. In short, *two distinct σ-BL signals sampled at $\omega_s \geq 2\sigma$ can never yield the same set of samples.* There are many ways to see this. For example, consider the reconstruction formulas

$$x(t) = \sum_n x(nT)h(t-nT) \quad \text{and} \quad y(t) = \sum_n y(nT)h(t-nT), \tag{10.40}$$

which are both valid because $\omega_s \geq 2\sigma$. Then

$$x(t) - y(t) = \sum_{n=-\infty}^{\infty} \left(x(nT) - y(nT)\right) h(t-nT). \tag{10.41}$$

If $x(nT) = y(nT)$ for all n, then the above difference is zero, which proves that $x(t) = y(t)$ for all t indeed. The reasoning is even simpler in the frequency domain, as some readers would have realized.

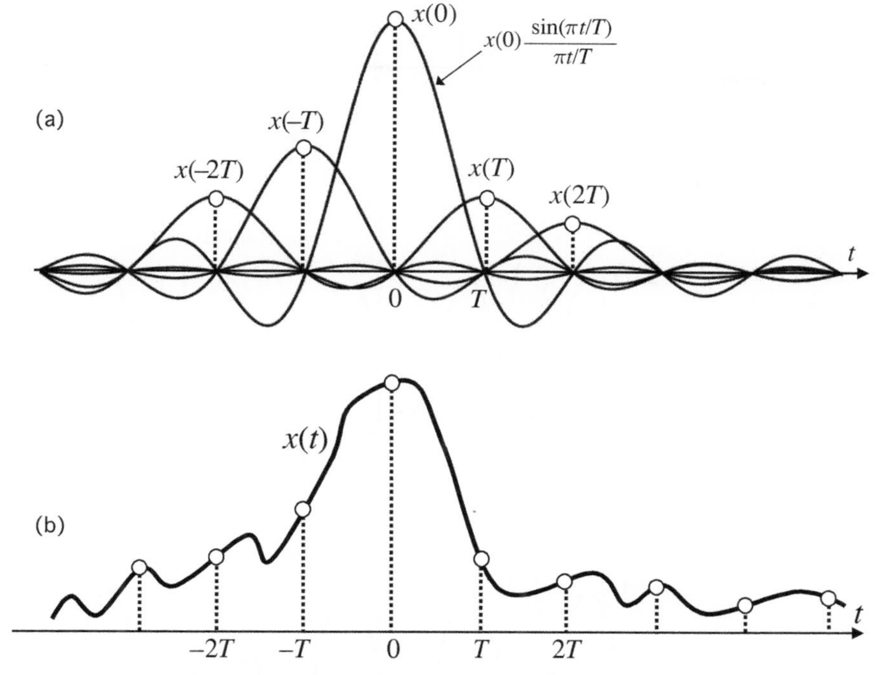

Figure 10.9 (a) Shifted and scaled sinc functions $x(nT)h(t - nT)$ for $n = -2, -1, 0, 1, 2$ for Nyquist sampling rate $\omega_s = 2\sigma$ (i.e., $T = \pi/\sigma$). (b) The reconstructed σ-BL signal obtained by adding all the sinc functions in (a).

10.5 Sampling a Sinusoid

Now let us consider the sinusoidal signal $x(t) = \sin \omega_0 t$. How many samples per second are needed, in order to reconstruct $x(t)$ from $x(nT)$? The answer depends on what is known about the signal a priori. For example, if we know the frequency ω_0, then of course we don't need any samples, because we can "reconstruct" $x(t)$ for any t just by using the formula $x(t) = \sin \omega_0 t$. So let us suppose that ω_0 is not known. Assume it is only known that it is in the range

$$0 \leq \omega_0 < \sigma \tag{10.42}$$

for some known σ. We know the Fourier transform of $x(t) = \sin \omega_0 t$ is a superposition of Dirac deltas at $\pm\omega_0$, as shown in Fig. 10.10(a). This shows that $x(t)$ is a σ-BL signal. From the sampling theorem it follows that if the sampling rate is $\omega_s \geq 2\sigma > 2\omega_0$, we can use a lowpass filter to recover $x(t)$ from $x(nT)$. Instead of $\omega_s > 2\omega_0$, suppose we choose

$$\omega_s = 2\omega_0. \tag{10.43}$$

Then what happens? In view of Eq. (10.42), we have $\omega_s = 2\omega_0 < 2\sigma$, so the condition required by the sampling theorem is not quite satisfied. Figure 10.10(b) shows

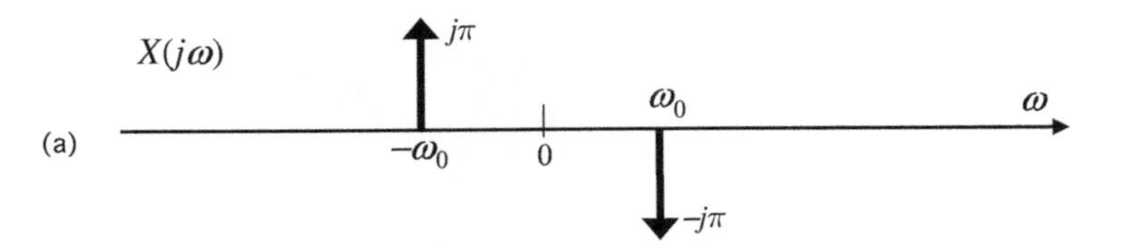

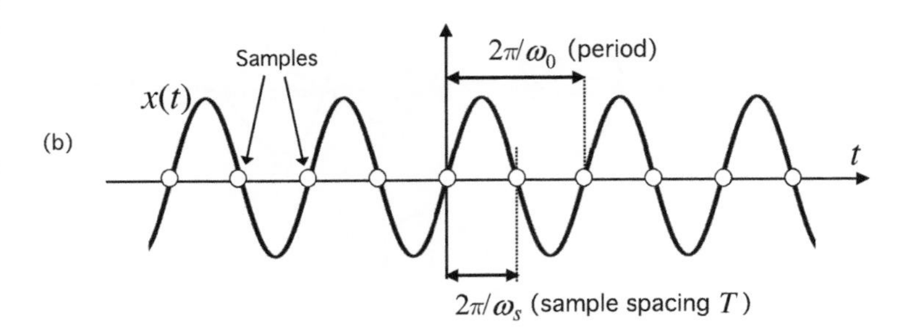

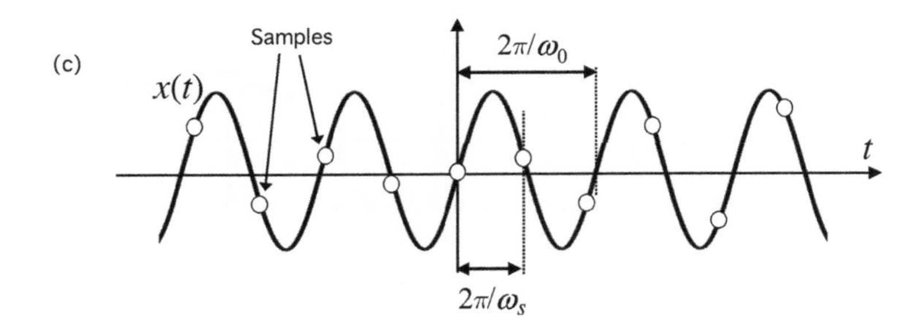

Figure 10.10 Sampling the sinusoid $x(t) = \sin \omega_0 t$. (a) Fourier transform of $x(t)$, (b) samples of $x(t)$ when the sampling rate is $\omega_s = 2\omega_0$, and (c) samples of $x(t)$ when the sampling rate is slightly higher than $2\omega_0$.

what happens in the time domain. The signal $x(t) = \sin \omega_0 t$ has period $2\pi/\omega_0$. The sample spacing T is exactly half of this period, because

$$T = \frac{2\pi}{\omega_s} = \frac{1}{2}\frac{2\pi}{\omega_0}. \qquad (10.44)$$

So, as shown in Fig. 10.10(b), all the samples fall at the zero crossings! In short,

$$x(nT) = 0, \quad \text{for all } n, \qquad (10.45)$$

which shows that the samples have no information, and we cannot recover $x(t)$. To see it in another way (Problem 10.6), the reader can use the formula (10.16) and verify that in the plot of $X_s(j\omega)$, there is overlap of Dirac delta functions $\delta_c(\cdot)$ and $-\delta_c(\cdot)$, leading to a complete cancellation, resulting in $X_s(j\omega) = 0$.

For the above sinusoid, if the sampling rate is $\omega_s = 2\omega_0 + \epsilon$ for some $\epsilon > 0$, then no matter how small ϵ is, the sampling theorem guarantees that $x(t) = \sin \omega_0 t$ can be recovered from the samples $x(nT)$. In this case the samples are not all zero any more, as demonstrated in Fig. 10.10(c). Again, the fact that we can take these samples $x(nT)$ and use them in the formula

$$x(t) = \sum_{n=-\infty}^{\infty} x(nT)h(t - nT) \tag{10.46}$$

to recover the sinusoid perfectly for all t is not easy to see, just by staring at the formulas in the time domain. The proof was based on the frequency-domain argument of Sec. 10.3.1.

Suppose we have $x(t) = \cos \omega_0 t$ instead of $\sin \omega_0 t$. If we use the sampling rate $\omega_s = 2\omega_0$, then what are the sample values? Can $x(t)$ be reconstructed from these samples? Think about it!

Can sampling theorem go wrong? It should finally be mentioned that if we *know* that the signal is a sinusoid, or more generally, a sum of sinusoids like

$$\sum_{k=1}^{D} a_k e^{j\omega_k t}, \tag{10.47}$$

there are ways to find the frequencies ω_k and amplitudes a_k from a **finite number of samples,**[1] provided we know some σ such that $|\omega_k| < \sigma$ for all k. Once ω_k and a_k are found, the entire signal can be reconstructed for any t. Since we only require a finite number of samples, the number of samples per second, averaged over the range $-\infty < t < \infty$, is zero, that is, the average sampling rate is zero! Since the example $x(t) = \sin \omega_0 t$ is a special case of Eq. (10.47), the same remark also holds for this case. Does this mean that the sampling theorem is contradicted?

Not to worry. The sampling theorem merely says that if we have the a priori information that a signal is σ-BL, and if $\omega_s > 2\sigma$, then we can reconstruct the signal from uniform samples by filtering. The theorem does *not* say that this is the only way to reconstruct signals from samples. In fact, it is sometimes possible to reconstruct signals from uniform samples even if $\omega_s < 2\sigma$, or even if a signal is not σ-BL at all. In such cases, the signals satisfy some other property, and we know a priori what that property is. For example, see Problem 10.16. Also see Chapter 16, which explains sampling methods based on the a priori knowledge that the signal has some "sparsity" property. ▽ ▽ ▽

10.6 Non-uniqueness of the Reconstruction Filter

Let $x(nT)$ be samples of a σ-BL signal $x(t)$ with sampling rate $\omega_s > 2\sigma$, that is, $T < \pi/\sigma$. Then there is an entire range of filters $H(j\omega)$, not just one, which will reconstruct $x(t)$ from $x(nT)$. To see this, consider Fig. 10.11(a), which shows the FT $X_s(j\omega)$ of the sampled version, with multiple copies of $X(j\omega)/T$, as usual. Since $\omega_s > 2\sigma$, there is a gap between these copies. In these gaps, the response of the filter $H(j\omega)$ is immaterial. The only condition that the filter has to satisfy is as follows:

$$H(j\omega) = \begin{cases} T & \text{for } -\sigma < \omega < \sigma, \\ 0 & \text{for } k\omega_s - \sigma < \omega < k\omega_s + \sigma, k \neq 0. \end{cases} \tag{10.48}$$

[1] 2D samples, to be precise.

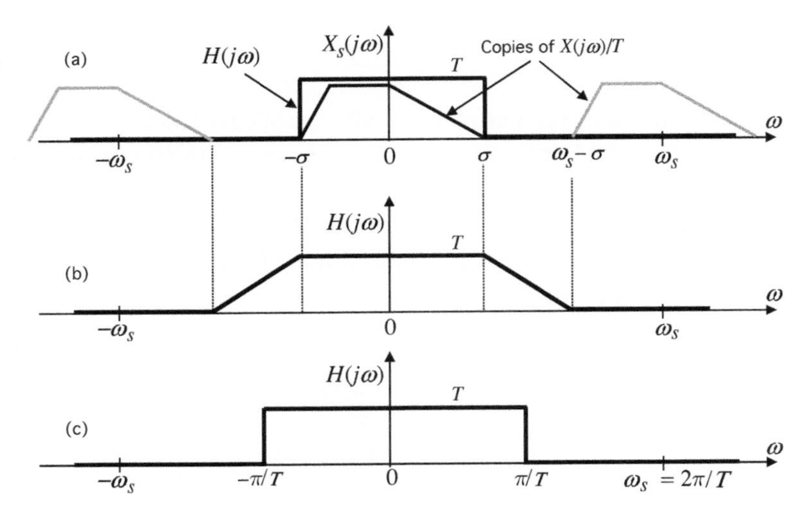

Figure 10.11 Non-uniqueness of the reconstruction filter $H(j\omega)$ when $\omega_s > 2\sigma$. Plots in (a)–(c) show examples of $H(j\omega)$ which reconstruct $x(t)$ from the samples $x(nT)$ perfectly.

That is, the gain should be T where the original copy $X(j\omega)/T$ resides, and zero where the shifted copies $X(j(\omega - k\omega_s))/T, k \neq 0$ reside. As long as the filter satisfies this, we have $H(j\omega)X_s(j\omega) = X(j\omega)$, and the filter output is the original signal $x(t)$. Figure 10.11(a) shows the traditional choice of $H(j\omega)$, and Figs. 10.11(b) and (c) show two other choices satisfying Eq. (10.48). It is clear that the reconstruction or interpolation equation

$$x(t) = \sum_{n=-\infty}^{\infty} x(nT)h(t - nT) \tag{10.49}$$

is valid for all three filters, in fact for any filter which satisfies Eq. (10.48). Notice finally that for $\omega_s = 2\sigma$ (Nyquist rate) there is no gap between copies, and the only reconstruction filter is the ideal lowpass filter shown in Fig. 10.11(a); in this case the three filters shown in the figure reduce to the same filter. So, *for Nyquist rate sampling, the filter for perfect reconstruction is unique.*

Comments on the three filters. Recall that the ideal lowpass filter in Fig. 10.11(a) has a sinc impulse response which is **unstable**. Thus, if we are doing Nyquist-rate sampling, the only filter available for reconstruction is unstable. The advantage of the filter in Fig. 10.11(b) is that it has a transition band

$$\sigma < \omega < \omega_s - \sigma \tag{10.50}$$

between the passband and stopband. Its impulse response is not the sinc function, in fact it decays like $1/t^2$ for large t. So it is a **stable** filter (Problem 10.13). While this filter is also unrealizable, like the one in part (a), it is easier to find practical designs that approximate it. This is one of the reasons why we **always oversample** in practice. Finally, the filter in Fig. 10.11(c) has frequency response

$$H(j\omega) = \begin{cases} T & \text{for } -\dfrac{\pi}{T} < \omega < \dfrac{\pi}{T}, \\ 0 & \text{otherwise.} \end{cases} \tag{10.51}$$

This is an ideal lowpass filter with passband edge equal to half the sampling frequency $\omega_s = 2\pi/T$. It has the familiar impulse response

$$h(t) = \frac{\sin \pi t/T}{\pi t/T}. \tag{10.52}$$

It is also unstable, like the filter in Fig. 10.11(a). However, the filter (10.52) has a unique property not enjoyed by the other filters, the so-called **orthogonal basis** property (15.9), as we shall explain in Sec. 15.2.2. ▽ ▽ ▽

Another insight here is that, whenever the reconstruction filter has **excess bandwidth**, it is capable of representing a larger class of bandlimited signals. Thus, in Fig. 10.11, although the signal of interest is σ-BL, the filter in part (c) is capable of representing (π/T)-BL signals, where $\pi/T > \sigma$. One downside of this is that we can never guarantee that the reconstructed signal is even a σ-BL signal, when samples $x(nT)$ are contaminated with additive noise (e.g., noise due to digitization). To elaborate, let us say we only have available a noisy version

$$x(nT) + e(nT) \tag{10.53}$$

of the samples $x(nT)$. The signal reconstructed from this is

$$\underbrace{\sum_{n=-\infty}^{\infty} x(nT)h(t-nT)}_{x(t)} \quad + \quad \underbrace{\sum_{n=-\infty}^{\infty} e(nT)h(t-nT)}_{e(t)}. \tag{10.54}$$

The first term is the ideal reconstructed signal and the second term is noise. While the first term is confined to the band $|\omega| < \sigma$, the second term represents a signal with larger bandwidth because $h(t)$ (and hence $h(t-nT)$) occupies a larger bandwidth. The energy in the second term gets larger as the excess bandwidth of the filter gets larger. Thus, although excess bandwidth makes it easier to design filters, it also allows a larger portion of the noise $e(nT)$ to make its way into the reconstructed signal. This is explained more quantitatively in Sec. 15.4.4.

10.7 Digital Signal Processing

Figure 10.12 shows how a continuous-time signal $m(t)$ is digitized and processed by a digital signal processor, and converted back to a continuous-time signal $y(t)$.

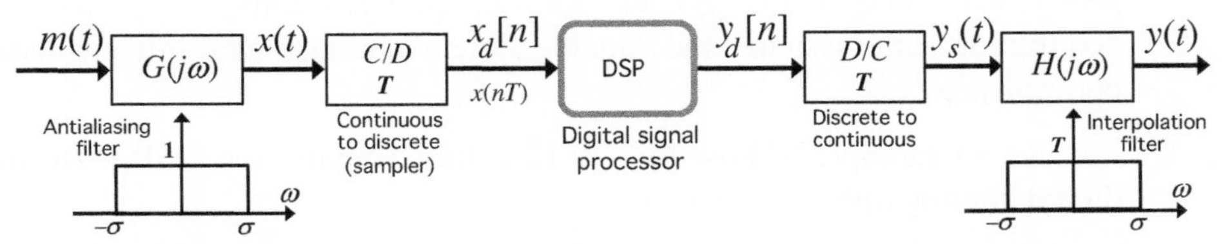

Figure 10.12 The various stages involved in digital signal processing. The message $m(t)$ is filtered by an antialias filter, sampled, and then processed digitally. The result $y_d[n]$ is then converted into a continuous-time signal $y(t)$ by interpolation (linear filtering).

Here, $m(t)$ represents a message, say a speech signal. First this is passed through an antialiasing filter $G(j\omega)$ to bandlimit it (similar to Fig. 10.6). The output $x(t)$ is a σ-BL signal, and is sampled to obtain

$$x_d[n] = x(nT) \qquad (C/D \text{ converter}). \tag{10.55}$$

Here, "C/D converter" stands for "continuous-time to discrete-time converter." You can think of it as another name for the sampler. The signal $x_d[n]$ is then processed digitally. This digital signal processing (DSP) could be a simple digital filtering operation, a sophisticated noise reduction algorithm, an edge enhancer, or some other processing. The box labeled D/C, which follows the DSP, is the "discrete-time to continuous-time converter." It is a conceptual device which converts $y_d[n]$ into an impulse train with samples separated by T secs:

$$y_s(t) = \sum_{n=-\infty}^{\infty} y_d[n]\delta_c(t - nT) \qquad (D/C \text{ converter}). \tag{10.56}$$

Note that $y_s(t)$ should be regarded as a function of continuous time. This is then filtered by the interpolation filter to produce

$$y(t) = (y_s * h)(t) = \sum_{n=-\infty}^{\infty} y_d[n]h(t - nT). \tag{10.57}$$

Note that the C/D and D/C converters use the same T (sample spacing) as indicated in the boxes.

A/D and D/A converters. In practice, we use an analog to digital or *A/D converter* instead of the conceptual C/D converter. The A/D converter not only performs sampling, it also converts the samples $x(nT)$ into a finite-precision or b-bit **binary** representation. Thus, $x_d[n] = x(nT)$ should be replaced everywhere with $x_d[n]+e[n]$, where $e[n]$ is the quantization error. Also, in practice, the conceptual D/C converter is replaced with a digital to analog or *D/A converter* which converts the binary representation of $y_d[n]$ into analog samples $y_d[n]$ before producing $y_s(t)$. Once this is done, analog filtering with $H(j\omega)$ is performed on these samples to obtain the final continuous-time output (10.57).

The effects of quantization can be ignored if the number of bits can be assumed to be large ("high bit rate" assumption). In this case the terms "discrete-time system" and "digital system" are used interchangeably. Note that the terms "continuous-time system" and "analog system" are also used interchangeably for convenience. ▽ ▽ ▽

To give an example of how Fig. 10.12 works, suppose the DSP performs some digital filtering operation so that

$$y_d[n] = \sum_{k=0}^{N} h_d[k]x_d[n - k] \tag{10.58}$$

and

$$Y_d(e^{j\omega}) = H_d(e^{j\omega})X_d(e^{j\omega}), \tag{10.59}$$

where $H_d(e^{j\omega}) = \sum_n h_d[n]e^{-j\omega n}$ is the digital filter frequency response. Then what is the FT of the final continuous-time output $y(t)$? We have

$$\begin{aligned}
Y(j\omega) &= \sum_{n=-\infty}^{\infty} y_d[n] \int_{-\infty}^{\infty} h(t-nT)e^{-j\omega t}dt \quad \text{(from Eq. (10.57))} \\
&= \underbrace{\sum_{n=-\infty}^{\infty} y_d[n]e^{-j\omega nT}}_{Y_d(e^{j\omega T})} H(j\omega) \\
&= H(j\omega)H_d(e^{j\omega T})X_d(e^{j\omega T}) \quad \text{(from Eq. (10.59)).}
\end{aligned} \tag{10.60}$$

When $\omega_s > 2\sigma$, the overlap or aliasing indicated in Fig. 10.4(a) is absent so that

$$X_d(e^{j\omega T}) = X_s(j\omega) = \frac{X(j\omega)}{T}, \quad |\omega| < \sigma \tag{10.61}$$

(see Eq. (10.18)). So, Eq. (10.60) implies

$$Y(j\omega) = \frac{H(j\omega)H_d(e^{j\omega T})}{T}X(j\omega), \quad |\omega| < \sigma. \tag{10.62}$$

Assuming that the interpolation filter $H(j\omega)$ is an ideal lowpass filter as shown in Fig. 10.12, this can be rewritten as

$$Y(j\omega) = H_d(e^{j\omega T})X(j\omega), \quad \text{for all } \omega. \tag{10.63}$$

The reason why this holds for all ω is that $x(t)$ and $y(t)$ are both σ-BL because of $G(j\omega)$ and $H(j\omega)$. So the entire system between $x(t)$ and $y(t)$ behaves like a continuous-time or analog filter

$$H_c(j\omega) \approx \begin{cases} H_d(e^{j\omega T}) & |\omega| < \sigma, \\ 0 & \text{otherwise.} \end{cases} \tag{10.64}$$

Summarizing, this is what has been shown: in the system shown in Fig. 10.12, assume the DSP block is a digital filter $H_d(z)$ as shown in Fig. 10.13(a). Then the equivalent analog system from input $m(t)$ to output $y(t)$ is an analog filter (Fig. 10.13(b)) and its frequency response is given by Eq. (10.64). Figure 10.13(a) shows an example plot of $H_d(e^{j\omega})$ and Fig. 10.13(c) shows the equivalent analog filter $H_c(j\omega)$ for the cases $\sigma = \pi/T$ and $\sigma < \pi/T$. The equivalent analog filter $H_c(j\omega)$ is nothing but the digital filter $H_d(e^{j\omega})$ with the frequency axis rescaled so that $-\pi < \omega < \pi$ is mapped to

$$-\pi/T < \omega < \pi/T,$$

and then truncated to $|\omega| < \sigma$ because of the interpolation filter $H(j\omega)$. Notice that the usual periodic repetitions in the digital filter response $H_d(e^{j\omega})$ are eliminated because of this truncation.

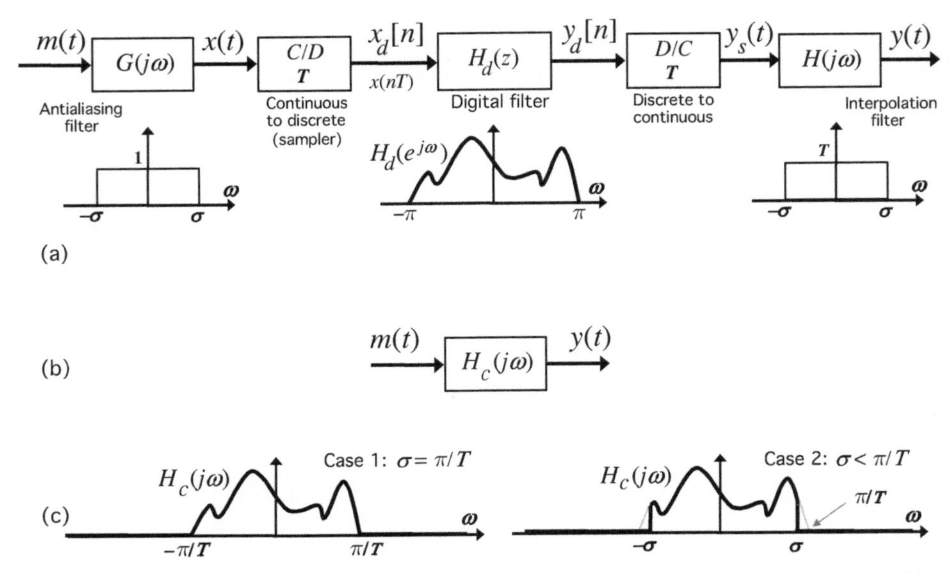

Figure 10.13 (a) The system of Fig. 10.12 where the DSP block is a digital filter $H_d(z)$. (b) The equivalent analog system from input $m(t)$ to output $y(t)$ is an analog filter $H_c(j\omega)$, with frequency response $H_c(j\omega) = H_d(e^{j\omega T})$ for $|\omega| < \sigma$, and zero outside. (c) Plot of $H_c(j\omega)$ for two cases.

Why implement analog filters digitally? The fact that we can implement analog filters $H_c(j\omega)$ by implementing digital filters $H_d(e^{j\omega})$ is very useful in practice. The reader might wonder where the advantage is, because $G(j\omega)$ and $H(j\omega)$ still need to be implemented as analog filters. The fact is that lowpass filters like $G(j\omega)$ and $H(j\omega)$ are quite standard, and approximations are easy to design and implement. However, in some applications we require more arbitrary filtering functions, for example, like the shape $H_c(j\omega)$ shown in Fig. 10.13(c). Such shapes are much easier to realize as **digital** filters. Digital signal processing gives a great deal of flexibility and diversity in terms of what types of filters can be designed and implemented. So, in some applications, it is indeed preferable to realize an analog LTI system $H_c(j\omega)$ by first realizing the digital version $H_d(e^{j\omega})$ and then using the system of Fig. 10.13(a) to interface with continuous time.

Another point to be noticed is that in practice one chooses ω_s to be significantly larger than 2σ (i.e., $\sigma < \pi/T$ significantly). Then the analog filters $G(j\omega)$ and $H(j\omega)$ can be more relaxed filters like the one in Fig. 10.11(b), and are especially easy to design and implement. So the only nontrivial part is the design of the arbitrary-shaped digital filter $H_d(e^{j\omega})$, for which there exist many good algorithms. ▽ ▽ ▽

10.8 Pulse Sampling

The uniform sampling described in Sec. 10.3.1 is called impulse sampling because we multiply a signal $x(t)$ with an impulse train (Dirac delta train) to get the samples

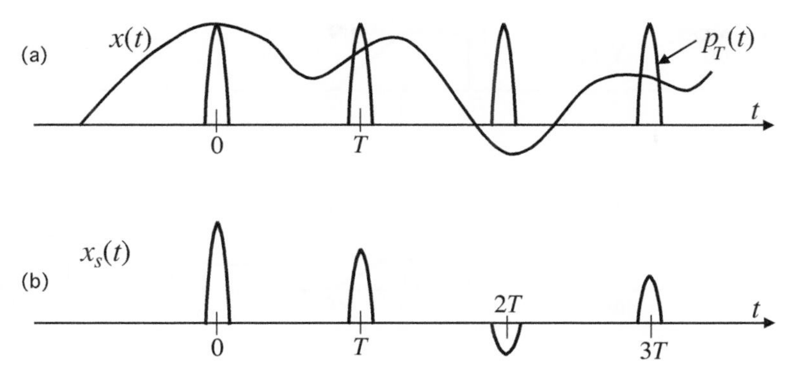

Figure 10.14 (a) A signal $x(t)$ and (b) the pulse-sampled version $x_s(t) = x(t)p_T(t)$. Here $p_T(t)$ is a periodic pulse train, with period T.

at isolated instances of time $t = 0, \pm T, \pm 2T, \ldots$ (Fig. 10.2). Consider now Fig. 10.14(a), where $x(t)$ is multiplied by a periodic pulse train $p_T(t)$ which satisfies

$$p_T(t) = p_T(t + T). \tag{10.65}$$

Thus the sampling pulse train has period T just like the impulse train used earlier. But now the pulses are allowed to have nonzero width. The pulse train can be regarded as a practical approximation to the ideal impulse train, which is unrealizable. We define the pulse-sampled signal to be

$$x_s(t) = x(t)p_T(t). \tag{10.66}$$

This is demonstrated in Fig. 10.14(b). Suppose $x(t)$ is σ-BL and we define the sampling rate to be $\omega_s = 2\pi/T$ as before. If

$$\omega_s \geq 2\sigma \tag{10.67}$$

as in the sampling theorem of Sec. 10.3.1, can we still recover $x(t)$ from the "sampled" version $x_s(t)$? Rather surprisingly, the answer is yes.

The original signal $x(t)$ can be reconstructed from the pulse-sampled version $x_s(t)$ by lowpass filtering, as long as $p_T(t)$ has nonzero average value over a period, that is,

$$c_0 \triangleq \frac{1}{T}\int_{-T/2}^{T/2} p_T(t)dt \neq 0. \tag{10.68}$$

This is proved next.

Fourier transform of the pulse-sampled signal. We now use a basic idea which will be introduced in Chapter 11. Namely, $p_T(t)$ can be expressed as a Fourier series because it is periodic. The Fourier series has the form

$$p_T(t) = \sum_{k=-\infty}^{\infty} c_k e^{jk\omega_s t}, \tag{10.69}$$

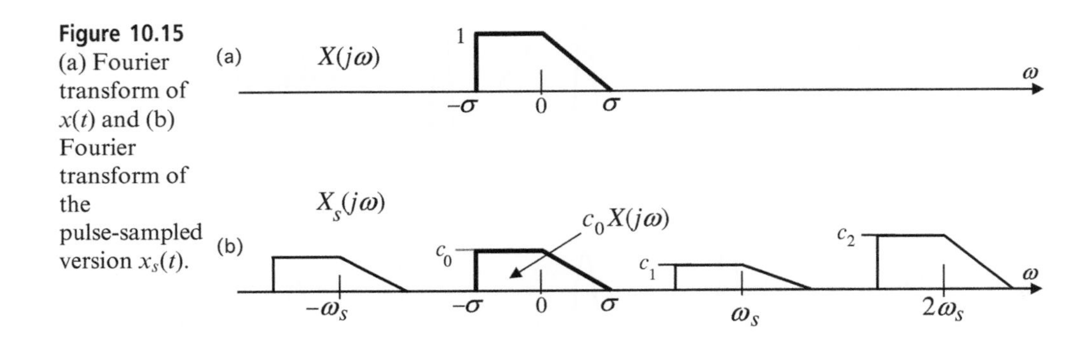

Figure 10.15 (a) Fourier transform of $x(t)$ and (b) Fourier transform of the pulse-sampled version $x_s(t)$.

where $\omega_s = 2\pi/T$. We then have $x_s(t) = x(t)p_T(t) = x(t)\sum_k c_k e^{jk\omega_s t}$, which can be written as

$$x_s(t) = \sum_{k=-\infty}^{\infty} c_k \left(x(t)e^{jk\omega_s t}\right). \tag{10.70}$$

Taking the Fourier transform of both sides, and using the fact that the Fourier transform of $x(t)e^{jk\omega_s t}$ is $X(j(\omega - k\omega_s))$, we get

$$X_s(j\omega) = \sum_{k=-\infty}^{\infty} c_k X(j(\omega - k\omega_s)). \tag{10.71}$$

So the Fourier transform of the sampled version is a linear combination of uniformly shifted versions of the original Fourier transform $X(j\omega)$, the shifts being integer multiples of the sampling rate ω_s. This is demonstrated in Fig. 10.15. There is no overlap between any two shifted copies, because of the assumption $\omega_s \geq 2\sigma$. $\bigtriangledown \bigtriangledown \bigtriangledown$

So we see that there is no difference from the impulse sampling case (Fig. 10.5) except for the fact that the copies in Fig. 10.15(b) do not have identical heights. It is clear that as long as $c_0 \neq 0$, the original signal $x(t)$ can be recovered by lowpass filtering as before. Notice that the argument works regardless of the values of the other Fourier coefficients $c_k, k \neq 0$. The expression for c_0 in Eq. (10.68) follows by integrating both sides of Eq. (10.69) and observing that $\int_{-T/2}^{T/2} e^{jk\omega_s t} dt = T\delta[k]$ (since $\omega_s = 2\pi/T$).

10.9 Time-Limited versus Bandlimited Signals

In this section we show that if a signal $x(t)$ is bandlimited, it cannot also be time-limited (i.e., have finite duration), unless it is zero for all time. A bandlimited signal is therefore constrained to be not time-limited. In fact, a bandlimited signal is constrained in many other ways in the time domain. For example, it has to be differentiable for all t, and it cannot be identically constant in any region of time and so forth, as we shall explain later (Sec. 15.5).

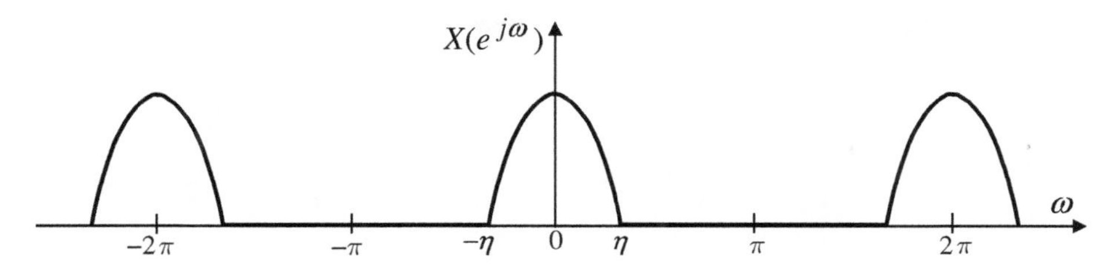

Figure 10.16 Fourier transform of a discrete-time bandlimited signal $x[n]$.

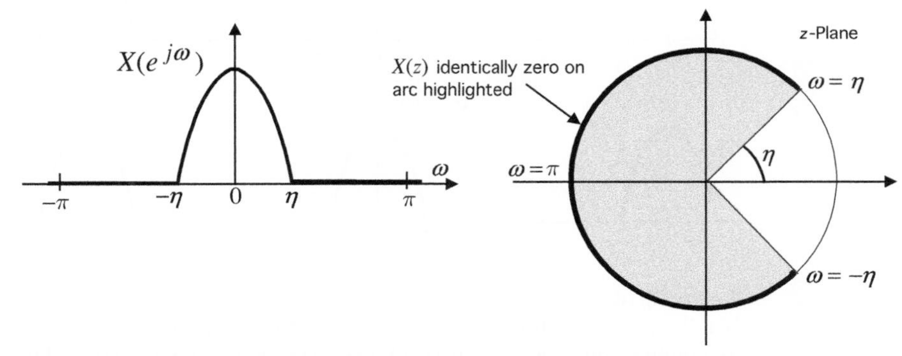

Figure 10.17 If $x[n]$ is η-BL and of finite duration, the polynomial $X(z)$ has to be identically zero on a portion of the unit circle in the z-plane, which is not possible.

10.9.1 Discrete-Time Bandlimited Signals

We begin with an observation about discrete-time signals $x[n]$. Since the FT $X(e^{j\omega})$ is periodic in this case, it is not possible to have $X(e^{j\omega}) = 0$ for all $|\omega| \geq \eta$ (unless $X(e^{j\omega}) = 0$ for all ω, that is, $x[n] = 0$ for all n). So we have to be careful with our definition of a bandlimited signal in discrete time. We say that $x[n]$ is η-BL for some $\eta \in (0, \pi)$ if

$$X(e^{j\omega}) = 0, \qquad \eta \leq |\omega| \leq \pi, \tag{10.72}$$

as demonstrated in Fig. 10.16. Now imagine that $x[n]$ is η-BL with $\eta < \pi$ and also has finite duration, say it is restricted to be nonzero only in $0 \leq n \leq N-1$. Then the z-transform is

$$X(z) = \sum_{n=0}^{N-1} x[n]z^{-n}, \tag{10.73}$$

which is a polynomial in z^{-1}. Equation (10.72) implies that $X(z)$ is identically zero at all points on an arc on the unit circle, as demonstrated in Fig. 10.17. But since $X(z)$ is a polynomial in z^{-1}, it only has $N-1$ zeros in the z-plane, and it cannot be identically zero on any contour unless $X(z)$ is zero everywhere (i.e., $x[n] = 0$ for all n). This shows that a discrete-time signal $x[n]$ cannot be both time-limited and bandlimited, unless it is zero for all n.

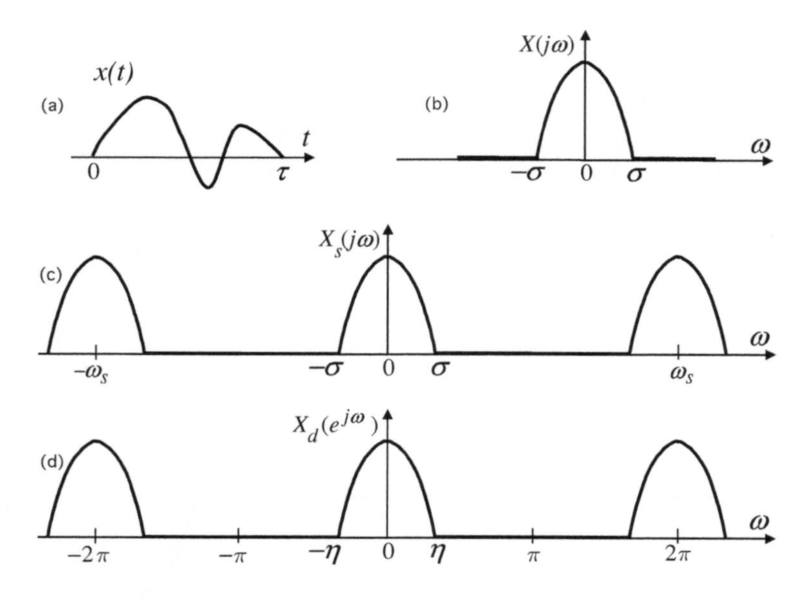

Figure 10.18 Proving that a nonzero time-limited signal cannot be bandlimited. (a) A time-limited signal $x(t)$, (b) Fourier transform of $x(t)$ assuming it *could* be σ-BL, (c) CTFT of the sampled signal $x_s(t)$, and (d) DTFT of the sampled signal $x_d[n] = x(nT)$, viewed as a discrete-time signal. In the demonstration, the sampling rate ω_s is significantly higher than the Nyquist rate 2σ. See text.

10.9.2 Continuous-Time Bandlimited Signals

Suppose $x(t)$ is a σ-BL signal with finite duration, with nonzero values occurring only in $0 \leq t \leq \tau$, where $\tau < \infty$ (Fig. 10.18(a)). For the sake of argument, assume this can also be bandlimited as in Fig. 10.18(b). If we use a sufficiently large sampling rate $\omega_s > 2\sigma$ then in the FT $X_s(j\omega)$, there is plenty of gap between copies of $X(j\omega)/T$ as demonstrated in Fig. 10.18(c). This means that the sampled version $x_d[n] = x(nT)$ has a bandlimited Fourier transform as shown in Fig. 10.18(d), where

$$\eta = \sigma T = \sigma \frac{2\pi}{\omega_s}. \tag{10.74}$$

This is because $X_s(j\omega) = X_d(e^{j\omega T})$ (see Eq. (10.18)). But since $x_d[n] = x(nT)$ and $x(t)$ has finite duration, $x_d[n]$ also has finite duration. In short, $x_d[n]$ is both an η-BL signal with $\eta < \pi$ (since $\omega_s > 2\sigma$) and a finite-duration signal. This is not possible unless $x_d[n] = x(nT) = 0$ for all n (as proved in Sec. 10.9.1). This of course means that

$$x(t) = \sum_{n=-\infty}^{\infty} x(nT)h(t - nT) = 0 \quad \text{for all } t, \tag{10.75}$$

where $h(t)$ is the sinc reconstruction filter as usual. This proves that $x(t)$ cannot be both time-limited and bandlimited, unless it is zero for all t.

The practical side of it. In practice, we often have finite-duration segments of bandlimited signals which we would like to sample. These truncated versions are not bandlimited, as they are time-limited. However, if the signal to be sampled is sufficiently smooth, then the Fourier transform of the time-limited version typically decays rapidy for high frequencies (Sec. 9.11). So the aliasing created by sampling can be kept within acceptable levels. Furthermore, the truncated signals are almost always oversampled (i.e., sampled at a rate significantly higher than the Nyquist rate of the untruncated signal). This also helps to keep aliasing under control.

10.10 Bandpass Sampling Theorems

The sampling theorem for σ-BL signals introduced in Sec. 10.3 is useful if $x(t)$ is known to be a lowpass signal whose FT can be any nonzero function in the region $|\omega| < \sigma$. Now consider Fig. 10.19(a), which is the FT of a bandpass signal $x(t)$. The FT is nonzero only in $\omega_1 < \omega < \omega_2$. This is an example of a complex signal $x(t)$ because a real signal would have to satisfy $X(-j\omega) = X^*(j\omega)$, which is evidently not true here. Notice that $x(t)$ can be regarded as a σ-BL signal with $\sigma = \omega_2$. The sampling theorem of Sec. 10.3 tells us that if we sample this at a rate $\omega_s \geq 2\omega_2$, then $x(t)$ can be reconstructed from the samples by lowpass filtering. While this is

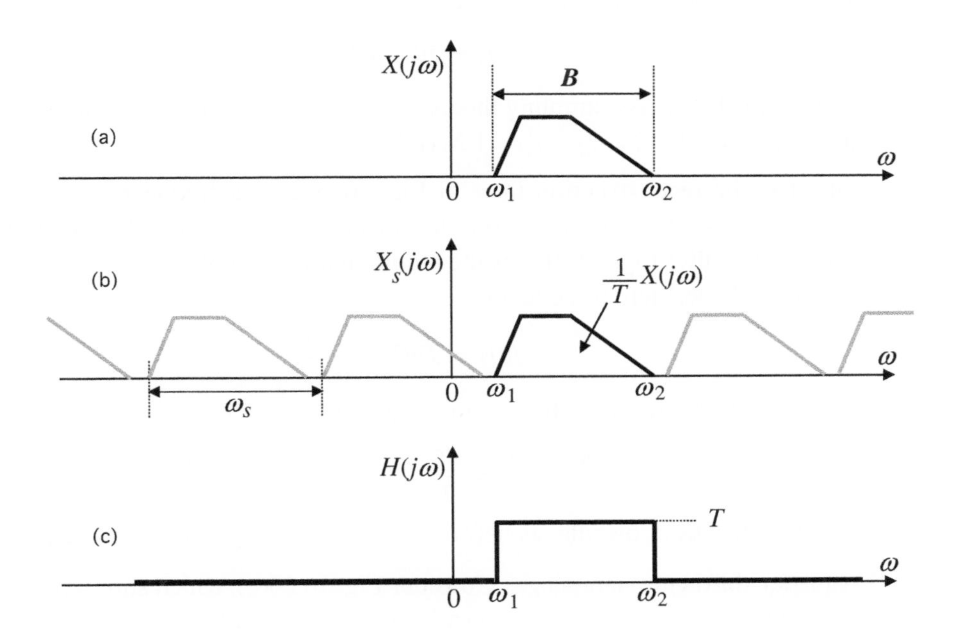

Figure 10.19 (a) Fourier transform of a complex bandpass signal $x(t)$, (b) FT of the sampled version $x_s(t)$ with sampling rate $\omega_s \geq$ bandwidth B of $X(j\omega)$, and (c) frequency response of a filter which reconstructs $x(t)$ from samples $x_s(t)$.

certainly correct, it is not the most efficient way to sample this signal, if we **know a priori** that it is bandlimited to $\omega_1 < \omega < \omega_2$. Notice that the total bandwidth of the signal is only

$$B = \omega_2 - \omega_1, \tag{10.76}$$

which can be much smaller than $2\omega_2$. Intuition tells us that the required sampling rate ought to be this bandwidth. Indeed, it can be shown that if $x(t)$ is uniformly sampled at the rate

$$\omega_s \geq B, \tag{10.77}$$

then it can be reconstructed from the samples $x(nT)$. To see this, simply apply the formula (10.9) for the FT $X_s(j\omega)$ of the sampled version. This yields the plot shown in Fig. 10.19(b). As before, multiple copies of $X(j\omega)/T$ are created by sampling, and the copies are separated by ω_s. So, as long as the condition (10.77) is satisfied, there is **no aliasing**, that is, no overlap between copies. We can therefore recover $x(t)$ from the sampled version $x_s(t)$ simply by using the bandpass filter shown in Fig. 10.19(c). In short, the sampling rate need only be as large as the bandwidth (Eq. (10.77)) – this is called the **bandpass sampling theorem** for signals with one frequency band, as in Fig. 10.19(a).

An example. To give an example, let $\omega_i = 2\pi f_i$ where $f_1 = 100$ MHz and $f_2 = 101$ MHz. Then a "brute-force" approach using the lowpass sampling theorem of Sec. 10.3 would suggest a minimum sampling rate

$$f_s = 202 \ \text{MHz}.$$

If we use the bandpass sampling theorem, on the other hand, the required sampling rate would only be $f_s = f_2 - f_1 = 1$ MHz! ▽ ▽ ▽

Note that the reconstruction filter in Fig. 10.19(c) is a frequency-shifted version of the lowpass filter (Fig. 10.5(c)) used in traditional sampling. So the impulse response of the filter Fig. 10.19(c) is the modulated version of the sinc function in Fig. 10.7. It can be written in the form

$$h(t) = Te^{j\omega_a t}\frac{\sin \sigma t}{\pi t}, \tag{10.78}$$

where $T = 2\pi/\omega_s$ is the uniform sample spacing as before, and

$$\sigma = \frac{\omega_2 - \omega_1}{2} = \frac{B}{2} \quad \text{and} \quad \omega_a = \frac{\omega_2 + \omega_1}{2}. \tag{10.79}$$

The reconstruction formula, as before, is $x(t) = \sum_{n=-\infty}^{\infty} x(nT)h(t - nT)$.

Frequency shifting by sampling. Consider Fig. 10.20(a), which shows the CTFT of a narrowband bandpass signal $x(t)$ with bandwidth B and center frequency ω_0. This can be regarded as a signal of the form $x(t) = m(t)e^{j\omega_0 t}$, where $m(t)$ is a lowpass signal. Thus $X(j\omega) = M(j(\omega - \omega_0))$ so that the lowpass message $M(j\omega)$ gets translated to high frequency ω_0. Such translation to high frequency is often done in radio communication systems, where ω_0 is the carrier frequency and B is the bandwidth of the

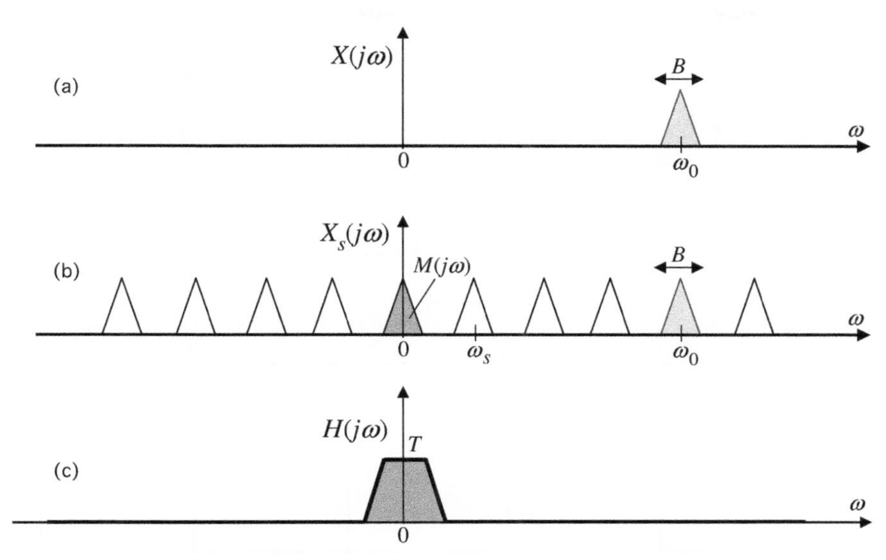

Figure 10.20 (a) Fourier transform of a narrowband complex bandpass signal $x(t)$, (b) FT of the sampled version $x_s(t)$ with sampling rate $\omega_s >$ bandwidth B of $X(j\omega)$, and (c) frequency response of a *lowpass* reconstruction filter. The reconstructed signal is the original $x(t)$, but downshifted in frequency, to become a lowpass version $m(t)$. Here, ω_s is assumed to be a submultiple of ω_0.

message $m(t)$ we wish to transmit. Typically $\omega_0 >> B$. The high-frequency signal $x(t) = m(t)e^{j\omega_0 t}$, called an amplitude-modulated (AM) signal, is easier to transmit than $m(t)$ itself.

When $x(t)$ is received, we demodulate it to recover the message $m(t) = x(t)e^{-j\omega_0 t}$. Another way to achieve this is to sample the high-frequency signal $x(t)$ at a rate $\omega_s > B$. This results in the signal $x_s(t)$ with Fourier transform shown in Fig. 10.20(b). There is no overlap between copies because $\omega_s > B$. Furthermore, if ω_0 is an integer multiple of ω_s, then one of the shifted copies of $X(j\omega)$ is located exactly at zero frequency as shown in the figure. By using a lowpass filter (Fig. 10.20(c)) we can recover the lowpass message signal $m(t)$ (with $M(j\omega)$ as in Fig. 10.20(b)) which was intended to be transmitted. Even if $\omega_0 \neq n\omega_s$ for integer n, $X_s(j\omega)$ still contains a slightly shifted copy of $M(j\omega)$ somewhere near $\omega = 0$, and it can be extracted by lowpass filtering. The main point of this discussion, in short, is that a narrowband high-frequency signal can be downconverted into a low-frequency signal simply by using sampling at a very low rate $\omega_s > B$ compared to ω_0, and then using lowpass filtering. ▽ ▽ ▽

10.10.1 Signals with Positive and Negative Frequency Bands

We know that if $x(t)$ is a real signal then $X(-j\omega) = X^*(j\omega)$, so that $|X(-j\omega)| = |X(j\omega)|$. So a real bandpass signal with one frequency band for $\omega > 0$ also has a corresponding band for $\omega < 0$. The left and right bands are indicated as $X_L(j\omega)$ and $X_R(j\omega)$ in Fig. 10.21(a), where it is assumed that the band edge $\omega_1 = B$ for

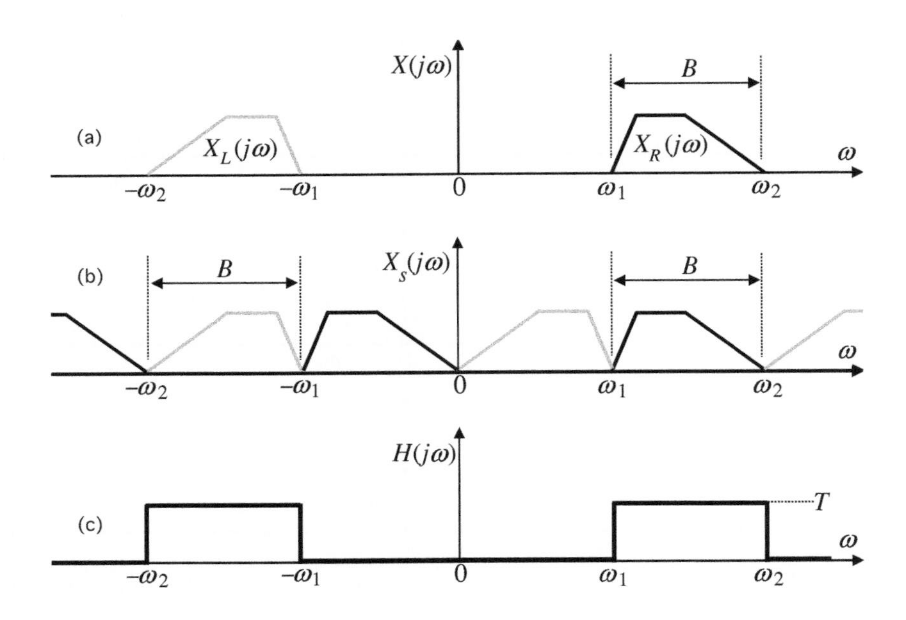

Figure 10.21 (a) Fourier transform of a real bandpass signal $x(t)$, (b) FT of the sampled version $x_s(t)$ with sampling rate $\omega_s = 2B$, and (c) frequency response of a filter which reconstructs $x(t)$ from $x_s(t)$. Notice that there is *no* aliasing in this case.

simplicity. The **total bandwidth** of the positive and negative frequency bands is now $2B$. Does this mean that we can use a sampling rate of

$$\omega_s \geq 2B \tag{10.80}$$

without causing aliasing? With $\omega_s = 2B$, Fig. 10.21(b) shows the FT of the sampled version. Indeed, there is no overlap between copies in this example, and we can reconstruct $x(t)$ from the samples $x(nT)$ by using the two-sided bandpass filter shown in Fig. 10.21(c). Thus, sampling this $x(t)$ at a rate equal to the total bandwidth $2B$ is indeed sufficient.

However, we have to be careful before we turn this example into a theorem. Thus, consider the example in Fig. 10.22(a), where the bands are slightly shifted, so that $B < \omega_1 < 2B$. If the sampling rate is $\omega_s = 2B$, then since $\omega_s > B$, the copies of $X_R(j\omega)$ do not overlap with each other, and similarly copies of $X_L(j\omega)$ do not overlap with each other. However, as shown in Fig. 10.22(b), copies of $X_R(j\omega)$ overlap with copies of $X_L(j\omega)$. There *is* aliasing, and we cannot use a bandpass filter to recover $x(t)$ from samples.

The point therefore is that the sampling rate $\omega_s = 2B$ works in some examples and does not work in others. Copies of $X_R(j\omega)$ do not overlap with those of $X_L(j\omega)$, only if the band edges are appropriately located. The bandpass sampling theorem for real signals, or more generally for signals with two bands located symmetrically with respect to the origin as in Fig. 10.23(a), is as follows.

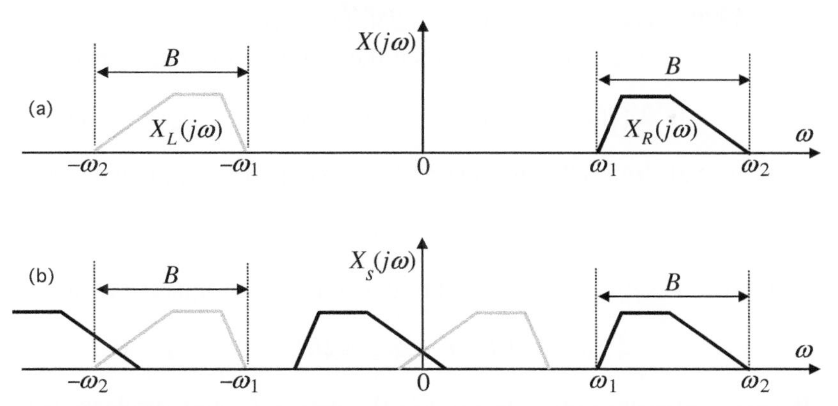

Figure 10.22 (a) Fourier transform of a real bandpass signal $x(t)$ and (b) FT of the sampled version $x_s(t)$ with sampling rate $\omega_s =$ total bandwidth $2B$ of $X(j\omega)$. Notice that there *is* aliasing in this case. See text.

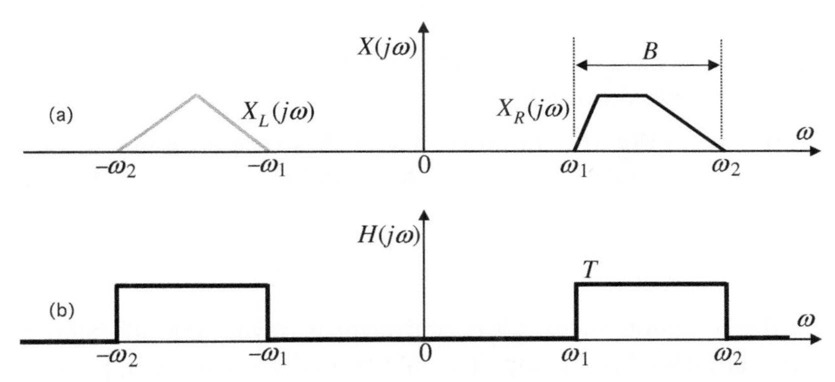

Figure 10.23 (a) Fourier transform of a two-sided bandpass signal $x(t)$ and (b) filter for reconstruction from samples, when the bandpass sampling condition (Theorem 10.2) is satisfied.

Theorem 10.2 *The bandpass sampling theorem.* Assume $x(t)$ is a bandpass signal with two bands located symmetrically with respect to the origin as in Fig. 10.23(a), where the shapes of the two bands are not necessarily related. Assume the sampling rate is $\omega_s = 2B$. Then there is no aliasing (no overlap between copies of $X(j\omega)$ in the FT of the sampled version) if and only if

$$\omega_1 = kB,$$

for some integer k. In this case $x(t)$ can be recovered from the samples using the bandpass filter shown in Fig. 10.23(b). ◇

Under the conditions of the theorem, there is no aliasing if $\omega_s = 2B$. But unfortunately, this does not necessarily mean that there is no aliasing if $\omega_s > 2B$ (Problem 10.12). This is unlike the sampling results presented earlier in this chapter.

Proof of Theorem 10.2. In Fig. 10.23(a) we have $X(j\omega) = X_L(j\omega) + X_R(j\omega)$. If we sample $x(t)$ at the rate $\omega_s = 2B$, then the resulting Fourier transform has the copies

$$X\big(j(\omega - 2kB)\big) = X_L\big(j(\omega - 2kB)\big) + X_R\big(j(\omega - 2kB)\big). \tag{10.81}$$

The set of all copies due to the first term occupy the frequency regions

$$-\omega_2 + 2mB < \omega < -\omega_1 + 2mB, \quad -\infty < m < \infty, \tag{10.82}$$

and the copies due to the second term occupy the frequency regions

$$\omega_1 + 2lB < \omega < \omega_2 + 2lB, \quad -\infty < l < \infty. \tag{10.83}$$

Between adjacent copies in (10.82) the gap is B, and similarly between adjacent copies in (10.83) the gap is B. So the two sets of copies do not overlap if and only if they interleave perfectly. That is, for every l, there should be an m such that

$$-\omega_1 + 2mB = \omega_1 + 2lB, \tag{10.84}$$

that is, $\omega_1 = (m - l)B$. This will happen if and only if ω_1 is an integer multiple of B. This completes the proof. $\nabla\nabla\nabla$

Returning to Fig. 10.21(a), we had $\omega_1 = B$, so this condition was satisfied. In Fig. 10.22(a) we had $B < \omega_1 < 2B$, so the condition was violated. Note that if $\omega_1 = kB$ then

$$\omega_2 = (k + 1)B,$$

so the condition $\omega_1 = kB$ is equivalent to saying that all band edges are integer multiples of B. Finally, it should be noticed that the impulse response of the reconstruction filter in Fig. 10.23(b) is a superposition of two terms like (10.78), namely

$$h(t) = Te^{-j\omega_a t}\frac{\sin\sigma t}{\pi t} + Te^{j\omega_a t}\frac{\sin\sigma t}{\pi t} = 2T\cos(\omega_a t)\frac{\sin\sigma t}{\pi t}. \tag{10.85}$$

10.10.2 Sampling Multiband Signals

What should be the sampling rate for a signal $x(t)$ which has multiple frequency bands, say K bands with bandwidths $B_1, B_2, \ldots, B_K$? There is no simple theorem for this because copies of one band created by sampling can overlap other bands unless the $2K$ band edges are chosen very carefully. As this can get complicated, a simpler method is often used in practice. Namely we use a bank of K bandpass filters to split the signal into K signals. The output of each filter has only one band (as in Fig. 10.19(a)), and it can be sampled at a rate equal to its bandwidth B_k. Thus, each one of the K signals can be reconstructed from its sampled version, and these can be combined to obtain the original signal $x(t)$. Note that the total number of samples per second in this scheme is the sum of the individual sampling rates B_k, that is, $\sum_k B_k$. This is indeed the total bandwidth of the multi-band signal $x(t)$.

10.10.3 Sub-Nyquist Sampling

For a signal bandlimited to $-\sigma < \omega < \sigma$ the minimum sampling frequency is $\omega_s = 2\sigma$, which is called the Nyquist rate. This is twice the maximum frequency σ. For a complex signal bandlimited to $\omega_1 < \omega < \omega_2$ (Fig. 10.19(a)) it might be tempting to call $2\omega_2$ the Nyquist rate, because ω_2 is the highest frequency. While it is true that sampling at this frequency will avoid aliasing, we have seen above that this is not necessary. It is sufficient to sample at a rate $\omega_s \geq B = \omega_2 - \omega_1$; no aliasing is created by such sampling.

The term "sub-Nyquist sampling" has sometimes been used to describe the idea that we can sample at a rate smaller that $2\omega_2$, and still avoid aliasing. This term is applicable only when the Nyquist rate is defined as twice the highest frequency. But this definition of the Nyquist rate is meaningful only in the case of lowpass signals bandlimited to $|\omega| < \sigma$. For signals with more general band locations in frequency (such as Fig. 10.19(a)), one can define the Nyquist rate ω_N to be the frequency such that there is no aliasing if the sampling rate $\omega_s \geq \omega_N$. That is, as long as $\omega_s \geq \omega_N$, there is no overlap of the copies $X(j(\omega - k\omega_s))$ for different integers k. Mathematically, this absence of aliasing is equivalent to

$$X(j(\omega - k\omega_s)) \cdot X(j(\omega - m\omega_s)) \equiv 0 \quad \text{for all } \omega \text{, when } k \neq m. \tag{10.86}$$

With this definition, the signal in Fig. 10.19(a) has Nyquist rate B, and not $2\omega_2$. Interestingly, there exist some situations where even when Eq. (10.86) is not satisfied (i.e., even when there is aliasing), it is possible to recover $x(t)$ from its samples! Such a possibility arises when we have some other a priori information about the signal. In Chapter 16 it will be seen that such a situation arises when a signal is known to have a sparse representation in some basis. Other examples include signals with specific models, such as in Problem 10.16. Situations like this, where $x(t)$ can be recovered from its samples in spite of aliasing, can be regarded as truly sub-Nyquist sampling.

10.11 Insights on Aliasing

Consider again Fig. 10.3(b), which shows the Fourier transform

$$X_s(j\omega) = \frac{1}{T} \sum_{k=-\infty}^{\infty} X\left(j\left(\omega - k\omega_s\right)\right) \tag{10.87}$$

of the uniformly sampled version of $x(t)$. This is a superposition of uniformly spaced copies of $X(j\omega)$. The adjacent copies are spaced apart by the sampling rate $\omega_s = 2\pi/T$, and if ω_s is not large enough, there is overlap between adjacent copies. This overlap was earlier referred to as aliasing. There are other interesting effects, arising again from the expression (10.87), which are also referred to as aliasing in a general sense. We now mention some of these.

Consider a single-frequency signal $x(t) = e^{j\omega_1 t}$, whose FT is the Dirac delta $2\pi\delta_c(\omega - \omega_1)$. Figure 10.24(a) shows the FT of the sampled version (10.87). The

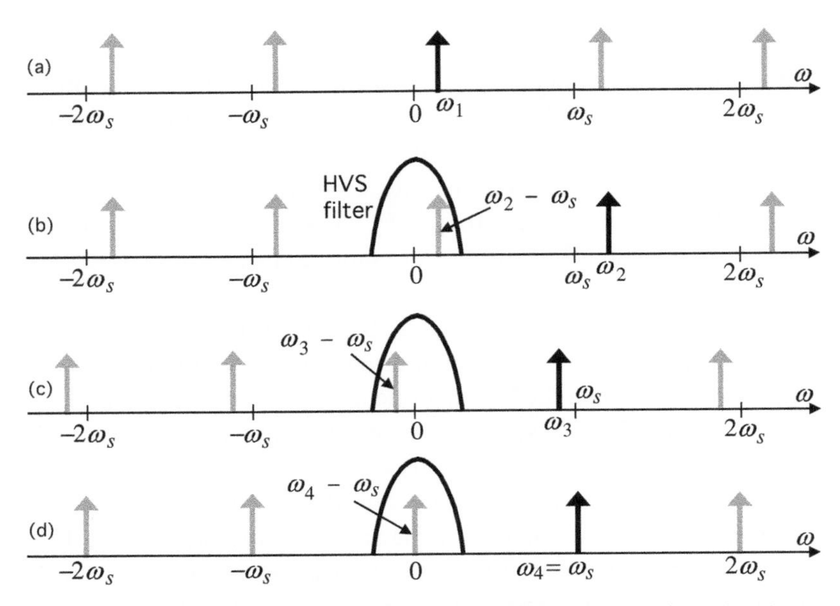

Figure 10.24 Fourier transforms of the sampled versions of various signals, with sampling frequency ω_s. (a) FT of the sampled version of $e^{j\omega_1 t}$, (b) FT of the sampled version of $e^{j\omega_2 t}$, where $\omega_2 = \omega_1 + \omega_s$, (c) FT of the sampled version of $e^{j\omega_3 t}$, where $\omega_3 < \omega_s$, and (d) FT of the sampled version of $e^{j\omega_4 t}$, where $\omega_4 = \omega_s$. See text.

original term $k = 0$ is shown in black, and the copies corresponding to $k \neq 0$ are shown in light gray. Now consider a new signal $y(t) = e^{j\omega_2 t}$, which has a higher frequency

$$\omega_2 = \omega_1 + \omega_s. \tag{10.88}$$

If this is sampled at same rate ω_s as before, then the sampled version $Y_s(j\omega)$ is exactly as before, as shown in Fig. 10.24(b). Thus, there is no difference between the sampled versions of $x(t)$ and $y(t)$ (i.e., $y(nT) = x(nT)$). This phenomenon, which makes **two frequencies appear the same** after sampling, is also referred to as aliasing. We will explain the rest of the figure in a moment.

10.11.1 Watching a Car Chase on TV: A Real-World Example

One practical manifestation of this is the following. Recall that the signal $e^{j\omega_2 t}$ can be interpreted as a vector which rotates counterclockwise ω_2 radians per second, that is, $f_2 = \omega_2/2\pi$ **rotations per second**. Imagine you are watching TV, and there is a speeding car moving on the screen. The rotating wheel is like a rotating vector with frequency ω_2 that depends on the speed of the car. Now, the video you are watching has been made by a camera which produces a certain number of still images per second, say $f_s = 60$ images per second. It is this sequence of images that creates the illusion of a motion picture. Thus we have a situation where the rotating vector $e^{j\omega_2 t}$ is sampled by the camera at some rate $\omega_s = 2\pi f_s$. If the rotating wheel has a high frequency ω_2 as represented by the black line in Fig. 10.24(b), then the sampling process creates multiple frequencies as shown by all the lines in the figure. The human visual system or **HVS**, which interprets the image on TV, is approximately like a lowpass filter (more accurately, a bandpass filter). This is indicated as "HVS filter" in Fig. 10.24(b). Thus, the high frequency ω_2 gives the visual impression of a low frequency ω_1. That is, a car moving at a high speed (corresponding to ω_2) appears to have its wheel rotating much slower (corresponding to ω_1). No doubt, many of you have observed this phenomenon when watching movies.

Now consider another example where the rotating wheel of the car has frequency ω_3 as shown in Fig. 10.24(c), which is slightly *smaller* than ω_s. As usual, the sampled version has all the frequencies shown in light gray, and the HVS filter captures only the frequency $\omega_3 - \omega_s$, which is negative. What this means is that even though the car on the TV is moving in a certain direction, the wheel appears to be rotating in the opposite direction!

Finally, Fig. 10.24(d) shows the case where the wheel is rotating at a speed ω_4 which matches the sampling rate. In this case the signal captured by the HVS filter has zero frequency, that is the wheel appears to be still, even though the car is moving fast! All of these are different manifestations of sampling, and can be referred to as aliasing. This is analogous to our earlier observation that narrowband signals with a large center frequency can be downconverted into low frequency by sampling followed by lowpass filtering (see discussions before Fig. 10.20).

10.11.2 Fluorescent Lighting: Another Real-World Example

We don't have to be movie watchers in order to observe aliasing phenomena in real life. It happens all the time in rooms with *fluorescent lighting*. A fluorescent light bulb has ionized gas in it. The fluctuating electric field created by the 60 Hz voltage at the terminals causes the electrons in the outer orbits of the atoms to absorb and release energy many times per second. The voltage has the form $\cos(\omega_0 t)$ and the intensity of light is proportional to the square $\cos^2 \omega_0 t = (1 + \cos 2\omega_0 t)/2$. So the intensity contains a component at double the frequency, as demonstrated in Fig. 10.25. If we have 60 Hz AC voltage, this fluctuating illumination is approximately like "sampling" the objects in the room with a sampling rate of 120 Hz. If an object in the room, say a fan, rotates at a certain speed, say 130 times per second, it may appear to be rotating only 10 times per second (because $130-120 = 10$). Similarly, a rotation of 110 cycles per second will look like 10 rotations per second in the *reverse direction* (because $110 - 120 = -10$). This is a phenomenon that we have observed in our daily lives, and it is called the **strobe effect**. In machine shops where there are large rotating machines (Fig. 10.25), this can be a safety issue, because a rotating machine may look still, inviting someone to touch it! So in such environments, **incandescent bulbs** are safer. In such bulbs a tungsten filament is heated to produce light, and the resulting light intensity does not follow the rapid fluctuations of the 60 Hz current as well as fluorescent bulbs do.

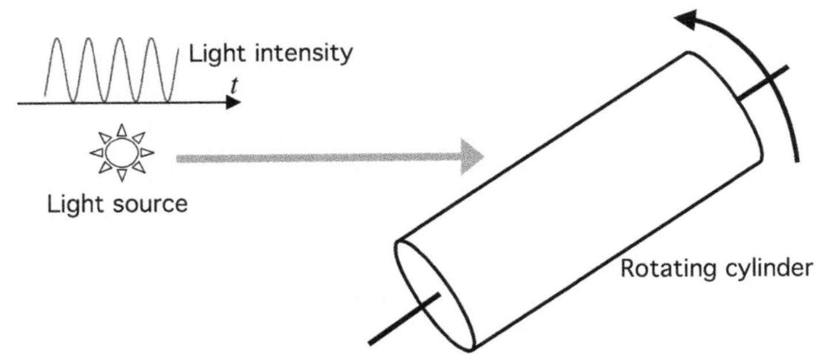

Figure 10.25 A source of light illuminating a rotating cylinder. The fluctuation of light intensity creates the effect of sampling the rotation, and the perceived speed of rotation can be very different from the actual speed. This is called the strobe effect. See text.

10.12 Aliasing Demonstration in Images

In Sec. 2.13, 2D sinusoids were introduced, which the reader may want to review at this time. In this section we will sample such sinusoids and demonstrate the visual effects of aliasing in the 2D images representing these sinusoids. Consider the 2D signal

$$x(t_1, t_2) = \frac{1 + \sin(\omega_1 t_1 + \omega_2 t_2)}{2}, \tag{10.89}$$

where $\omega_1 = 0.2\pi,\ \omega_2 = 0.15\pi$. This is essentially a 2D sinusoid with a constant added to make it non-negative, so that it can be displayed as an intensity plot. Figure 10.26(a) shows the uniformly sampled version

$$x[n_1, n_2] = \frac{1 + \sin(\omega_1 n_1 + \omega_2 n_2)}{2}, \tag{10.90}$$

where the sample spacings are $T_1 = T_2 = 1$. This implies a sampling rate of $\omega_s = 2\pi/T_i = 2\pi$ in each direction. Since $|\omega_1|$ and $|\omega_2|$ are less than π, this sample spacing is sufficient, and there is no aliasing. In Fig. 10.26(a) we have $0 \leq n_1, n_2 \leq 511$, so it is a 512×512 image. The dark lines we see in the figure are perpendicular to the frequency vector $[\omega_1 \quad \omega_2]^T$ as explained in Sec. 2.13. Next, Fig. 10.26(b) shows the sampled version

$$x[2n_1, 2n_2] = \frac{1 + \sin(2\omega_1 n_1 + 2\omega_2 n_2)}{2}, \tag{10.91}$$

with $T_1 = T_2 = 2$, that is, a smaller sampling rate $\omega_s = \pi$ in each direction. This is a 256×256 image. Since $|\omega_1|$ and $|\omega_2|$ are less than $\pi/2$ as well, there is still **no aliasing**. We see the same set of parallel lines, they are just squeezed more tightly because of the downsampling. In this example, downsampling does not imply undersampling.

As a second example, let $\omega_1 = 0.4\pi,\ \omega_2 = 0.6\pi$. We still have no aliasing when $T_1 = T_2 = 1$ and the plot of Eq. (10.90) is as in Fig. 10.27(a), and the dark lines in the figure are perpendicular to the frequency vector $[\omega_1 \quad \omega_2]^T$ as expected. Now consider the downsampled version where $T_1 = T_2 = 2$, so the sampling rate is

(a)

(b)

Figure 10.26 (a) Sinusoidal image (10.89) with $\omega_1 = 0.2\pi$, $\omega_2 = 0.15\pi$, sampled with $T_1 = T_2 = 1$. (b) The same image downsampled by two ($T_1 = T_2 = 2$). Downsampling does *not* create aliasing.

$$\omega_s = \frac{2\pi}{2} = \pi \tag{10.92}$$

in each direction. Since $\omega_2 = 0.6\pi$, we have $2\omega_2 = 1.2\pi$, that is,

$$\omega_s < 2\omega_2. \tag{10.93}$$

So the sampling rate is not large enough in the t_2 direction, and this creates aliasing. In this example, downsampling does imply undersampling. You can see from Eq. (10.91) that the frequencies of the sampled version are now $2\omega_1 = 0.8\pi$ and $2\omega_2 = 1.2\pi$. Since $2\omega_2 = 1.2\pi > \pi$, it aliases to the equivalent frequency

Figure 10.27 (a) Sinusoidal image (10.89) with $\omega_1 = 0.4\pi$, $\omega_2 = 0.6\pi$, sampled with $T_1 = T_2 = 1$. (b) The same image downsampled by two ($T_1 = T_2 = 2$). Downsampling *does* create aliasing in this case because ω_2 is too large for the smaller sampling rate.

$$2\omega_2 = 1.2\pi \equiv 1.2\pi - 2\pi = -0.8\pi. \tag{10.94}$$

So the frequency vector is

$$[2\omega_1 \quad 2\omega_2]^T = [0.8\pi \quad -0.8\pi]^T, \tag{10.95}$$

whose orientation is very different from the original frequency vector

$$[\omega_1 \quad \omega_2]^T = [0.4\pi \quad 0.6\pi]^T. \tag{10.96}$$

Thus the downsampled version has the form shown in Fig. 10.27(b). The parallel dark lines, orthogonal to the frequency vector (10.95), have a completely different orientation than the image before downsampling! This is one way in which aliasing manifests in images: the orientations of certain high-frequency patterns change. In Fig. 10.27 we have shown only small portions of the images, so that the lines can be clearly seen. A rendition of the 512×512 image on paper would make the lines too tightly packed (in this example) to be comfortably distinguished without rendition artifacts.

10.13 Are there Other Ways to Sample Signals?

Information sciences have revolutionized our lives today. Shannon was the father of information theory. Nyquist was one of the fathers of digital communications, and we live in a world of digital communications today. The sampling theorem, attributed to Shannon, Nyquist, and Whittaker, is the backbone of digital signal processing and digital communication. The revolution of digital signal processing which happened in the twentieth century would not have been possible without the sampling theorem.

We have not discussed extensions of sampling techniques such as *derivative sampling* theorems and *nonuniform* sampling theorems. The derivative sampling theorem for σ-BL signals says that instead of sampling $x(t)$ at the rate $\omega_s \geq 2\sigma$, we can sample $x(t)$ and its derivative $dx(t)/dt$ simultaneously at half the rate $\omega_s \geq \sigma$ each, uniformly. From these two sets of samples it is possible to reconstruct $x(t)$. See Problem 5.13 of Vaidyanathan [1993] and references therein. Extensions to higher derivatives are also possible.

Nonuniform sampling theorems allow one to sample $x(t)$ at points t_n which are not uniformly spaced, but satisfy certain other conditions. One of these is that the average sampling rate should be large enough (e.g., $\geq 2\sigma$ for the σ-BL case). For example, suppose we sample a σ-BL signal only in the region $t \leq 0$, but at a rate $> 4\sigma$ (i.e., faster than twice the Nyquist rate). Then the sampling rate averaged in $-\infty < t < \infty$ is still $> 2\sigma$. So, can we recover $x(t)$ for all t (including $t > 0$) from these samples taken only from $t \leq 0$? Intuition seems to say that this should be possible in theory. But this would mean that $x(t), t > 0$, is completely predictable from $x(t), t \leq 0$, for a bandlimited signal. Is this so? See Papoulis [1975], Vaidyanathan [1987c], and references therein for more discussions. Also see remarks at the end of Sec. 15.5.2. These nonuniform sampling methods are not as commonly used as uniform sampling because of numerical instabilities in the reconstruction. There have been many other generalizations of sampling. The interested reader can refer to Papoulis [1977b], Vaidyanathan and Liu [1988, 1990], Unser [2000], Vaidyanathan [2001], and Chapter 10 of Vaidyanathan [1993].

Another type of unconventional sampling is the sampling of signals that are not bandlimited at all. An example can be found in Problem 10.16, and arises from wavelet theory [Daubechies, 1992]. Sampling based on the so-called “finite rate of innovation" was introduced by Vetterli et al. [2002]. Last but not least, the paradigm of compressive sensing is an entirely new way to look at sampling theory and has revolutionized sampling in the last two decades [Baraniuk, 2007; Candès and Wakin, 2008]. We will discuss this briefly in Chapter 16.

10.14 Transmitting Over a Bandlimited Channel

We now mention a very different context in which the bandlimited property arises, namely in digital communications [Lathi, 1989; Proakis and Salehi, 2008]. In Sec. 5.11 we briefly mentioned digital communication systems, which the student may want to review at this time. Thus, consider Fig. 10.28 where a symbol stream $s[n]$ is to be transmitted over a channel. The channel is often modeled as an LTI system with a frequency response $F(j\omega)$, followed by an additive noise source $e_c(t)$. (The notation used here is a bit different from Fig. 5.14, which considered only the discrete-time equivalent.) In Fig. 10.28 the symbol stream $s[n]$, represented by $\sum_n s[n]\delta_c(t - nT)$, is transmitted over the channel and the output of the channel is sampled at the receiver. These samples are digitally processed to recover the symbol stream $s[n]$. The goal is to ensure perfect recovery, that is, achieve $r[n] = s[n]$.

An important question that arises in this context is, how many symbols can we transmit over the channel per second, so that it is still possible to perfectly recover the symbols $s[n]$? The answer depends on the channel bandwidth. Any physical channel is a bandlimited system, that is, its frequency response $F(j\omega)$ is as indicated in the figure with some finite bandwidth $2B$ (indicated in Hz). Assuming $F(j2\pi f) \neq 0$ in $-B < f < B$, it can be shown that perfect symbol recovery is possible as long as

$$\frac{1}{T} < 2B. \tag{10.97}$$

Note that $1/T$ is the symbol rate (number of symbols per second). So the preceding says that if the symbol rate is less than the bandwidth, we can achieve perfect recovery. Thus, the larger the channel bandwidth, the larger the symbol rate that can be used successfully. This explains why the communications industry always wants to open up more bandwidth for future generations. For example, the 6G network

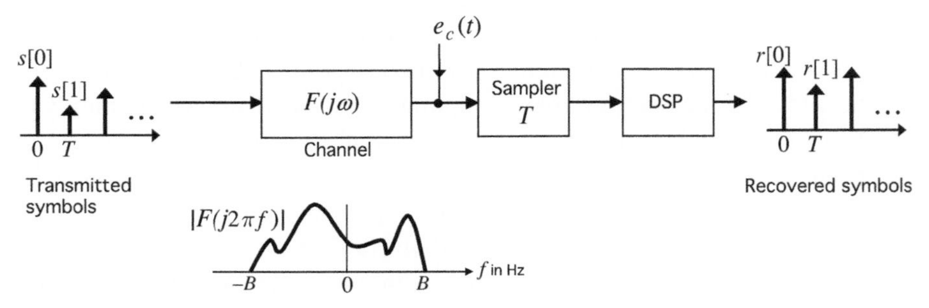

Figure 10.28 A symbol sequence $s[n]$ transmitted over a bandlimited channel and the recovered symbol sequence $r[n]$ at the receiver end.

(which hopes to use the region between 90 GHz and 300 GHz) will be able to offer much more bandwidth per user than 5G (which uses mmWave bands between 24.25 GHz and 52.6 GHz) [Poulakis, 2022].

Contrast with sampling theory. Recall here that if a signal $x(t)$ is bandlimited to $2B$ Hz, then the sampling rate should be

$$\frac{1}{T} \geq 2B, \tag{10.98}$$

so that we can recover $x(t)$ from the samples $x(nT)$. This condition is the exact opposite of (10.97). While this is amusing, it should not be confusing to the reader, as the contexts are different. Equation (10.97) arises because if we squeeze in too many symbols per second, then the channel with limited bandwidth may distort the symbols beyond recovery. On the other hand, Eq. (10.98) arises because if we do not collect enough samples per second from $x(t)$ then we do not have enough information to recover $x(t)$ by linear filtering. ▽ ▽ ▽

It should be mentioned that there is another important limiting factor in communication systems, namely noise. Any channel introduces a certain amount of random noise ($e_c(t)$ in Fig. 10.28) in addition to the bandlimiting effect. This leads to errors in symbol recovery. In practice, each symbol is quantized to a discrete set of levels, and represents a certain number of bits of information. What really matters is the number of bits per second (bit rate) that can be transmitted with acceptably low error rate. For a desired error rate and fixed signal power, the achievable bit rate increases as the channel becomes less noisy. There are great formulas that connect the allowed bit rate to the channel bandwith, transmitted signal power, and channel noise power [Lathi, 1989; Proakis and Salehi, 2008]. We do not go into these details here. Suffice it to mention that the foundation for such formulas comes from information theory.

10.15 Summary of Chapter 10

In this chapter we introduced sampling. The uniformly sampled version of a continuous-time signal $x(t)$ can be expressed in two ways:

(a) as an impulse train $x_s(t) = \sum_{n=-\infty}^{\infty} x(nT)\delta_c(t - nT)$, or

(b) as a discrete-time sequence $x_d[n] = x(nT)$.

Here, T is the sample spacing. The sampling rate is

$$f_s = \frac{1}{T} \text{ Hz} \quad \text{or equivalently} \quad \omega_s = \frac{2\pi}{T} \text{ rads/sec.}$$

The Fourier transform of the sampled version can be expressed in two ways:

$$x_s(t) \longleftrightarrow X_s(j\omega) = \frac{1}{T}\sum_{k=-\infty}^{\infty} X\left(j\left(\omega - \frac{2\pi}{T}k\right)\right) \quad \text{(CTFT)},$$

$$x_d[n] \longleftrightarrow X_d(e^{j\omega}) = \frac{1}{T}\sum_{k=-\infty}^{\infty} X\left(j\left(\frac{\omega - 2\pi k}{T}\right)\right) \quad \text{(DTFT)}.$$

If any two terms in the above sum overlap, we say there is aliasing. A σ-BL signal $x(t)$ satisfies $X(j\omega) = 0$ for $|\omega| \geq \sigma$. This can be sampled without aliasing if $\omega_s \geq 2\sigma$. In this case we can reconstruct $x(t)$ from $x[nT]$ using lowpass filtering:

$$x(t) = \sum_{n=-\infty}^{\infty} x(nT)h(t - nT),$$

and this result is called the sampling theorem for lowpass signals. The sampling theorem is at the center of digital signal processing, as explained in Sec. 10.7. Typical sampling rates for speech and music signals are 10 kHz and 44.1 kHz, respectively. In the above,

$$h(t) = T\frac{\sin(\sigma t)}{\pi t}$$

is the impulse response of an ideal lowpass filter with cutoff σ and passband height T. In fact, the reconstruction filter $h(t)$ is nonunique when $\omega_s > 2\sigma$. The case where $\omega_s = 2\sigma$ is called Nyquist sampling. In this case $h(t)$ is unique, and is given by

$$h(t) = \sin(\pi t/T)/(\pi t/T) \qquad \text{(Nyquist sampling)}.$$

This has zero crossings at kT for integers $k \neq 0$. If $\omega_s < 2\sigma$ there is aliasing. There are many manifestations of aliasing and these were discussed in detail. For example, the wheel of a moving car in a movie appears to be rotating too slowly, or even in the wrong direction, because of sampling (in movies, only a finite number of images are shown per second). The same happens with a rotating fan in fluorescent lighting.

Extensions of the lowpass sampling theorem to bandpass signals were also presented in this chapter. Also proved was the pulse sampling theorem where the sampling pulse is spread out over a short duration, unlike the Dirac delta train. Even in this case it is possible to recover a σ-BL $x(t)$ by using a lowpass filter.

A discrete-time signal is bandlimited or η-BL if $X(e^{j\omega}) = 0$ for $\eta \leq |\omega| \leq \pi$ for some $\eta < \pi$. We further showed that a bandlimited signal cannot also be time-limited unless it is identically zero for all time. This result holds both for continuous-time and discrete-time signals.

PROBLEMS

Notes: In all problems, "sampling" means "uniform sampling," unless mentioned otherwise. Also, "rad" stands for radians as in rads/sec (radians per second) and krads/sec (i.e., 10^3 rads/sec).

If you are using MATLAB for the computing assignments at the end, the "stem" command is convenient for plotting sequences. The "plot" command is useful for plotting real-valued functions such as $|X(e^{j\omega})|$.

10.1 Figure P10.1 shows the Fourier transforms of three bandlimited signals, $x(t), y(t)$, and $s(t)$, where ω is in krads/sec.

(a) Suppose we sample the signals at the rate $\omega_s = 10$ krads/sec. Sketch the Fourier transforms of the sampled versions $x_s(t), y_s(t)$, and $s_s(t)$. Which of these suffers from aliasing?

(b) What is the minimum sampling rate ω_s to avoid aliasing, in each of the three cases?

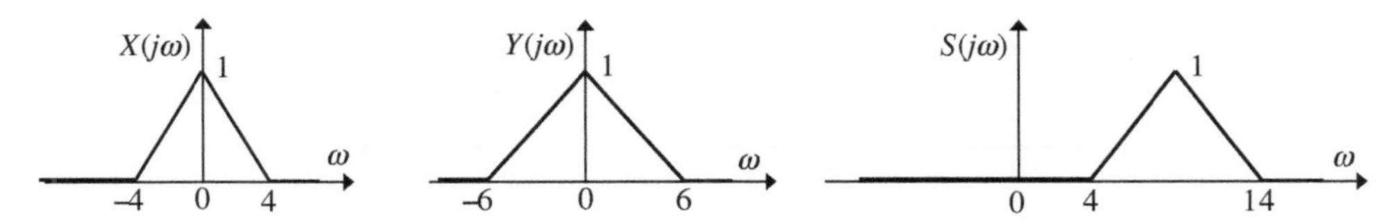

Figure P10.1 Bandlimited signals for Problem 10.1.

10.2 In Problem 10.1(a), define $x_d[n] = x(nT), y_d[n] = y(nT)$, and $s_d[n] = s(nT)$. For these discrete-time signals, plot $X_d(e^{j\omega})$, $Y_d(e^{j\omega})$, and $S_d(e^{j\omega})$ (i.e., the respective DTFTs), in the range $-\pi \leq \omega \leq \pi$.

10.3 Suppose we uniformly sample the signal $x(t)$ in Fig. P10.1 at the rate $\omega_s = 10$ krads/sec. Show plots of three possible reconstruction filters $H(j\omega)$ which will give perfect recovery of $x(t)$ from samples. Can you find a stable reconstruction filter?

10.4 Suppose we wish to sample $x(t) = e^{j\omega_0 t}$ and filter the sampled version using the lowpass filter $H(j\omega)$ shown in Fig. P10.4, where all frequencies are in krads/sec. Assume $\omega_0 = 10$ krads/sec so that $X(j\omega)$ is as shown. We consider four possible sampling rates: $\omega_s = 9, 10, 11, 14$ krads/sec. In the figure, $T = 2\pi/\omega_s$.

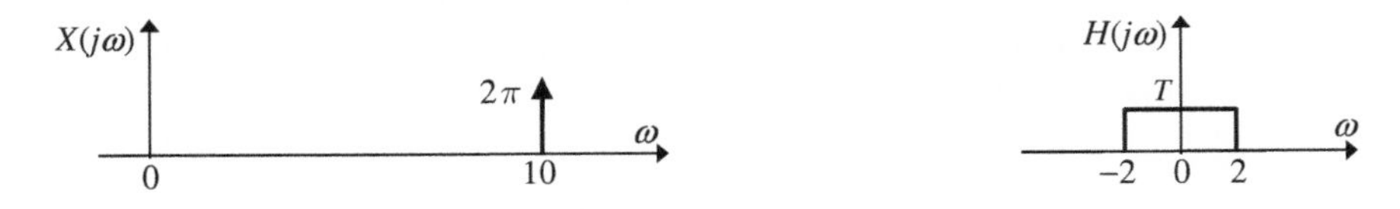

Figure P10.4 Signal and filter for Problem 10.4.

(a) For each of the sampling rates, plot the Fourier transform of the output $y(t)$ of the filter $H(j\omega)$.

(b) For each of the sampling rates, write down the output $y(t)$ in the time domain.

(c) Is the output zero for any of the sampling rates? Is the output a negative frequency signal for any of the sampling rates?

10.5 Figure P10.5 shows the Fourier transforms of four signals $x(t), y(t), p(t)$, and $q(t)$. Here, $Y(j\omega) = X(j(\omega - \omega_0))$ and $k \geq 0$ is an integer. We now consider uniform sampling of these signals. We know that $x(t)$ can be sampled at the rate $\omega_s = 2\omega_1$ without causing aliasing.

(a) What is the minimum sampling rate for $y(t)$, in order to avoid aliasing?

(b) What is the minimum sampling rate for $p(t)$, in order to avoid aliasing? With this sampling rate, what is the reconstruction filter $H(j\omega)$ which will reconstruct $p(t)$ from the samples?

(c) If we sample $q(t)$ at the rate $\omega_s = 2\sigma$ (which is the total bandwidth), will it cause aliasing?

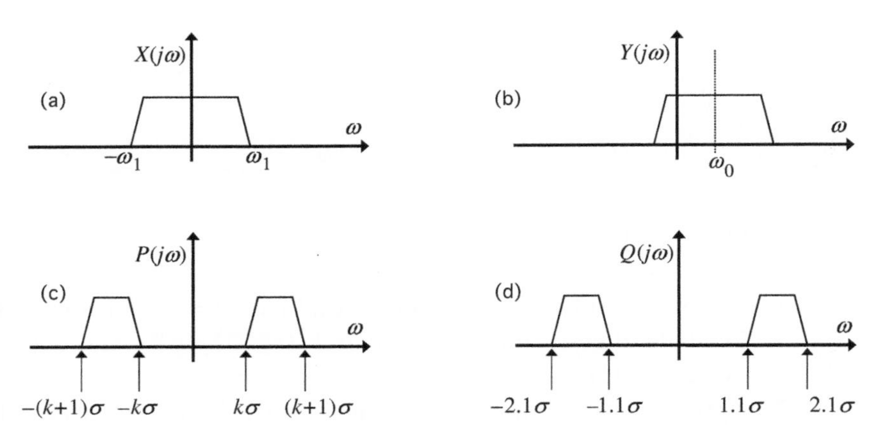

Figure P10.5 Pertaining to the sampling problem.

10.6 For $x(t) = \sin \omega_0 t$ we know that the Fourier transform has two Dirac delta functions. Now consider the sampled version $x_s(t)$ with sampling rate $\omega_s = 2\omega_0$. Show that the copies of Dirac deltas in successive shifted copies $X(j(\omega - k\omega_s))$ cancel perfectly and therefore $X_s(j\omega) = 0$ for all ω.

10.7 Let $x(t) = [(\sin t)/t]^6$. For uniform sampling of this signal, find the minimum sampling rate ω_s (Nyquist rate) to avoid aliasing.

10.8 Consider the bandpass signal $x(t)$ with Fourier transform as shown in Fig. P10.8. Let $x_s(t) = \sum_n x(nT)\delta_c(t - nT)$ be the sampled signal with sampling rate $\omega_s = 2B$.

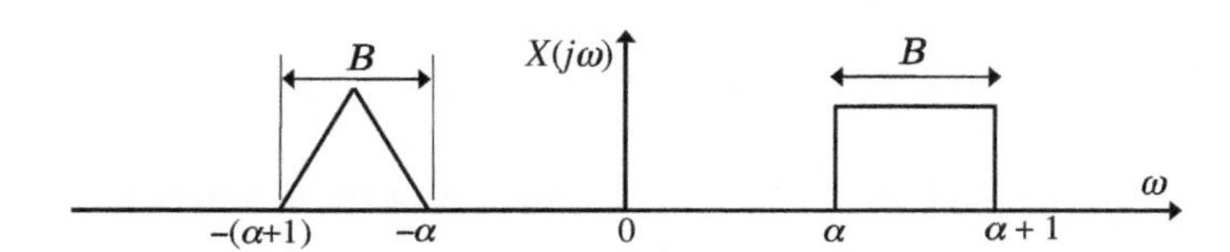

Figure P10.8 Bandpass sampling, bands located symmetrically.

(a) Plot the Fourier transform of $x_s(t)$ for two cases: $\alpha = 1$ and $\alpha = 0.2$.

(b) In which of the above two cases is aliasing avoided in sampling?

10.9 Suppose the signal $x(t)$ with $X(j\omega)$ as in Fig. P10.9 is sampled at frequency $\omega_s = 2\sigma$, and the sampled version filtered with the bandpass filter $G(j\omega)$ shown in the figure to obtain the output $y(t)$ (where $T = 2\pi/\omega_s$).

(a) Express $y(t)$ directly in terms of $x(t)$ in the time domain.

(b) Define $s(t) = \sum_{n=-\infty}^{\infty} y(nT)h(t - nT)$ where $h(t) = \sin(\sigma t)/(\sigma t)$. Express $s(t)$ directly in terms of $x(t)$.

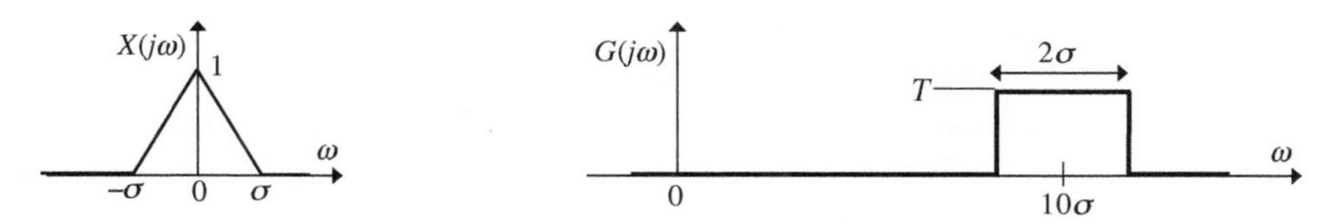

Figure P10.9 Signal and filter for Problem 10.9.

10.10 Consider a signal $x(t)$ with Fourier transform as shown in Fig. P10.10, where ω is in rads/sec (although the ω-axis is not drawn to scale). Suppose we sample $x(t)$ at $\omega_s = 2$ rads/sec and filter the result using the lowpass filter $H(j\omega)$ shown, to obtain $y(t)$. Plot $Y(j\omega)$. What should the height c of the filter be, so that $Y(j\omega)$ has peak value of unity?

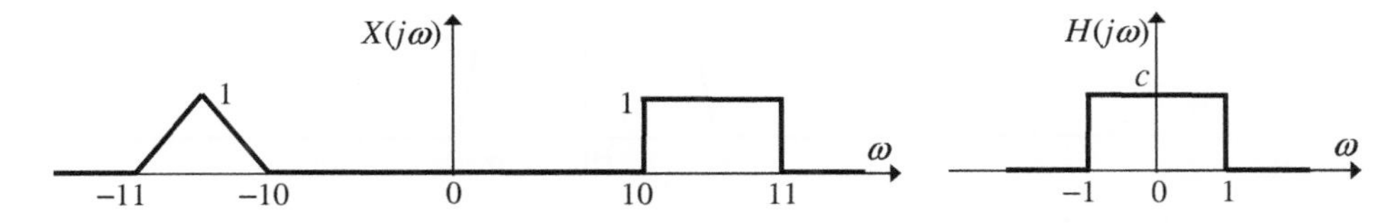

Figure P10.10 Signal and filter for Problem 10.10.

10.11 Let $x(t)$ be σ-BL. Show that $|x(t)|^2$ is 2σ-BL. Also, find a specific example of σ-BL $x(t)$ such $|x(t)|^2$ is *not* σ-BL.

10.12 *Aliasing due to increased sampling rate.* Figure P10.12 shows a bandpass signal with the two bands positioned such that the bandpass sampling condition of Theorem 10.2 is satisfied. (Here, k is an integer.) So we can sample at the rate $\omega_s = 2B$ without causing aliasing. Does this also mean that we can sample at any rate $\omega_s > 2B$ without aliasing? The answer is in general no, as we shall see in this problem. This is unlike the sampling of a lowpass signal (Theorem 10.1) or a one-band bandpass signal, where sampling at any rate higher than the Nyquist rate would guarantee freedom from aliasing.

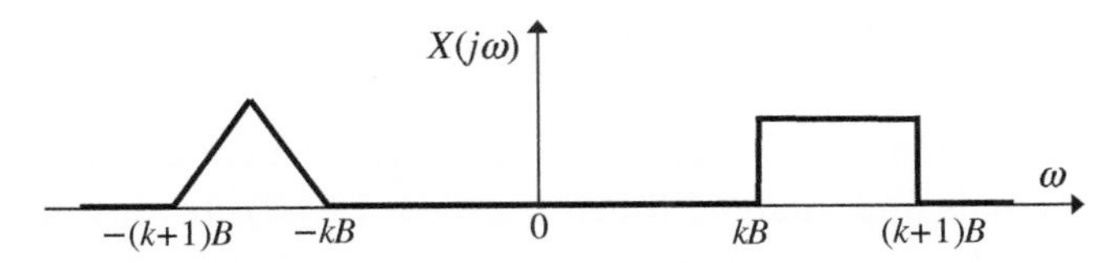

Figure P10.12 Signal for Problem 10.12.

(a) For our bandpass signal, consider the example where $B = 1$ and $k = 10$. Then we know that the sampling rate $\omega_s = 2$ does not create aliasing. Show, however, that if $\omega_s = 3$, then there is aliasing.

(b) More generally, suppose the sampling rate is $\omega_s = LB$, where $L \geq 2$ is an integer. Show that aliasing happens if and only if L is a divisor of $2k+1$. In particular, therefore, if $L \geq 2$ is even, there is no aliasing.

10.13 *Relaxed ideal filter.* Consider the relaxed ideal lowpass filter $H(j\omega)$ shown in Fig. P10.13, where $0 < \alpha < \beta < \infty$ and the response is symmetric with respect to zero frequency. The transition from the passband to the stopband is gradual (and not discontinuous as in a brick-wall lowpass filter, which has $\alpha = \beta$). Derive a closed-form expression for the impulse response $h(t)$ of this filter. Show that

$$|h(t)| \leq \frac{c}{t^2}$$

for all t, for some finite $c > 0$. Because of this, it can be shown that it is a stable filter (i.e., $h(t)$ is absolutely integrable). (The discrete-time version of this problem was considered in Problem 4.25.)

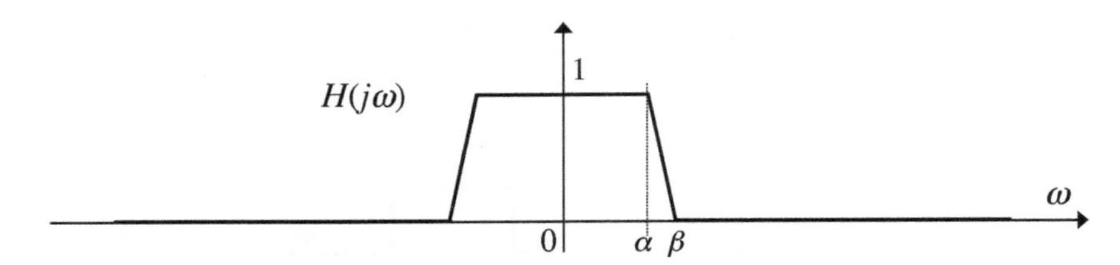

Figure P10.13 A relaxed lowpass filter.

10.14 *Causality and bandlimitedness (discrete time).* We know that a discrete-time signal $x[n]$ is said to be σ-BL if the Fourier transform $X(e^{j\omega})$ is zero for $\sigma \leq |\omega| \leq \pi$, for some $\sigma > 0$. This is demonstrated in Fig. P10.14 in two equivalent ways.

(a) Suppose $x[n]$ is σ-BL with $\sigma < \pi$ and furthermore $x[n]$ is causal, that is, $x[n] = 0$ for $n < 0$. Show that $x[n] = 0$ for all n. In short, a causal sequence cannot be bandlimited unless it is the zero sequence.

(b) As a generalization of part (a), show that if $x[n]$ is a left-sided or right-sided sequence, it cannot be σ-BL for any $\sigma < \pi$ unless it is zero for all n. (A sequence is right-sided if it is zero for all $n < n_0$ where n_0 is some finite integer. A sequence $x[n]$ is left-sided if $x[-n]$ is right-sided.)

10.15 *Causality and bandlimitedness (continuous time).* Let $x(t)$ be a continuous-time signal such that $x(t) = 0$ for $t \leq 0$. Show that $x(t)$ cannot be bandlimited unless it is zero everywhere.

Note: One way to prove this is to use the connection between bandlimited signals and analytic functions (Sec. 15.5.2). In this problem we are asking you to give a different proof, by taking advantage of Problem 10.14 and the sampling theory presented in this chapter.

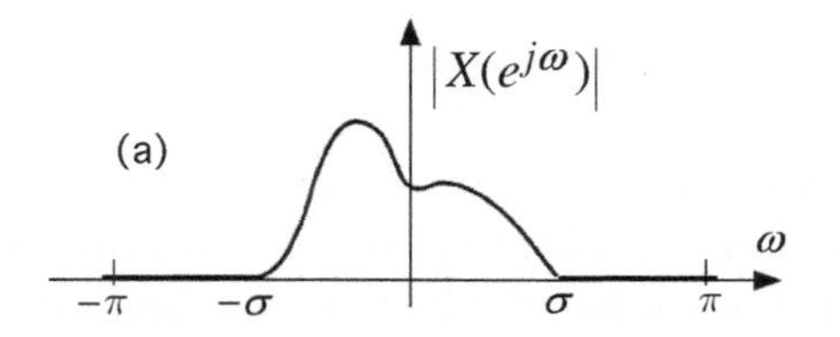

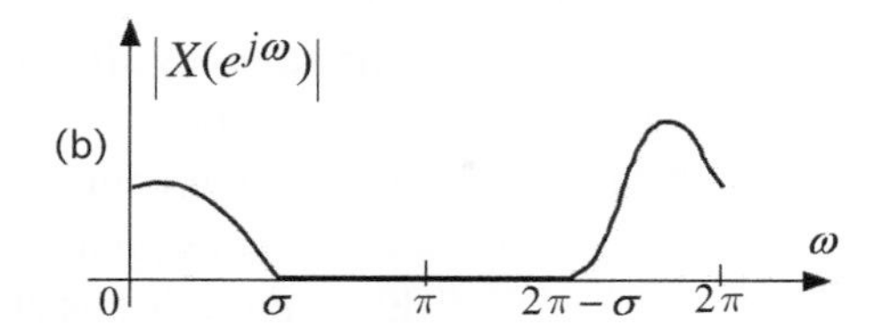

Figure P10.14 Causality and bandlimitedness.

10.16 *Sampling non-bandlimited signals.* All the sampling theorems in this chapter assumed the signal to be bandlimited one way or another (lowpass, bandpass, two-sided bandpass, and so forth). It turns out that there exist sampling theorems for signals that are not bandlimited at all. We now present such an example. This can be regarded as truly *sub-Nyquist* sampling (see Sec. 10.10.3). Thus, consider a signal which can be modeled as

$$x(t) = \sum_{k=-\infty}^{\infty} c_k \phi(t - kT). \tag{P10.16}$$

For example, if $\phi(t)$ is the sinc function, then the above model reduces to the bandlimited model. Notice also that if $\phi(t)$ is not bandlimited, then $x(t)$ is also not bandlimited. Now, assume *we know* that the signal has the above model for some known $\phi(t)$. We will see that, under some conditions on $\phi(t)$, we can recover $x(t)$ from uniformly spaced samples $x[n] = x(nT)$, where the sample spacing T is the same T that appears in Eq. (P10.16).

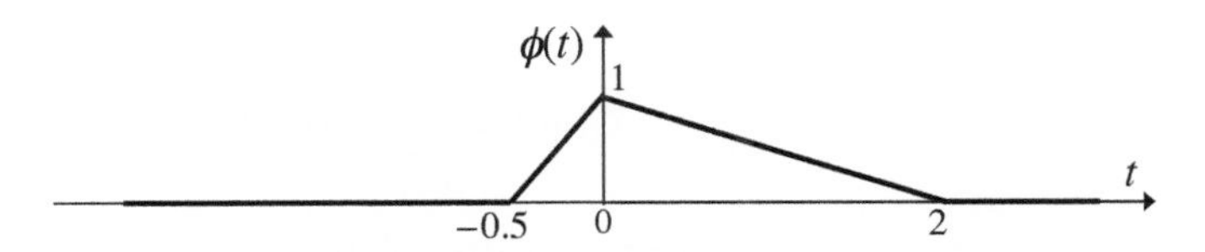

Figure P10.16 Pertaining to non-bandlimited sampling.

(a) Just to develop a feeling for Eq. (P10.16), give a handsketch of $x(t)$ when $\phi(t)$ is as in Fig. P10.16. Assume $T = 1$, $c_0 = 1, c_1 = c_{-1} = 0.5$, and $c_k = 0$ otherwise.

(b) More generally, for any $\phi(t)$, show that the samples $x[n] = x(nT)$ can be written as

$$x[n] = \sum_{k=-\infty}^{\infty} c_k \phi_d[n - k],$$

where $\phi_d[n] = \phi(nT)$ (sampled version of $\phi(t)$).

(c) Define $\Phi_d(z)$ to be the z-transform of $\phi_d[n]$, and assume $G(z) = 1/\Phi_d(z)$ is a stable digital filter. Show that we can recover c_k from the samples $x[n]$ by convolution, that is, $c_k = \sum_m x[m]g[k - m]$.

(d) For the special case where $T = 1$ and $\phi(t)$ is as in Fig. P10.16, what is $G(z)$, and what is its causal impulse response? Is it stable?

Thus, as long as $G(z) = 1/\Phi_d(z)$ is stable and realizable, we can recover c_k from the samples $x[n]$. Then $x(t)$ can be reconstructed for all t using the model of Eq. (P10.16).

10.17 (Computing assignment) *Sampling a Gaussian pulse.* From Sec. 9.7 we know that the Gaussian signal $x(t) = e^{-t^2/2}/\sqrt{2\pi}$ has Fourier transform $X(j\omega) = e^{-\omega^2/2}$. Even though this signal is not bandlimited, we see from the plot of $X(j\omega)$ in Fig. 9.12 that it is almost bandlimited. For example, it has very little energy for $|\omega| > 3$. Since $x(t)$ is also small for $|t| > 3$, we will ignore that part and sample the portion from $-3 \le t \le 3$.

(a) Suppose we accept that $x(t)$ is σ-BL with $\sigma = 3$ rad/sec, and sample it at $\omega_s = 2\sigma = 6$ rad/sec in the range $-3 \le t \le 3$. We will obtain only five samples $x(nT)$. With

$$X_s(j\omega) = \sum_{n=-2}^{2} x(nT)e^{-j\omega nT}$$

denoting the Fourier transform of the sampled version, plot $|X_s(j\omega)|$ in $-10 \le \omega \le 10$. This is periodic with period $\omega_s = 6$ rad/sec. Also plot $|X_s(j\omega)|$ and $|X(j\omega)|$ in the range $-3 \le \omega \le 3$ and you will see that they resemble each other reasonably well (except for a scale factor $1/T$ due to sampling). So there is negligible aliasing.

(b) Instead of the above, suppose we pretend that $x(t)$ is σ-BL with $\sigma = 2$ (which is not a good assumption) and sample it at $\omega_s = 2\sigma = 4$ rad/sec in the range $-3 \le t \le 3$. Now we will obtain only three samples $x(nT)$. With

$$X_s(j\omega) = \sum_{n=-1}^{1} x(nT)e^{-j\omega nT}$$

denoting the Fourier transform of the sampled version, plot $|X_s(j\omega)|$ in $-10 \le \omega \le 10$. This is periodic with period $\omega_s = 4$ rad/sec. Also plot $|X_s(j\omega)|$ and $|X(j\omega)|$ in the range $-2 \le \omega \le 2$ and you will see that they do not resemble each other well because of aliasing. Furthermore, the information in $X(j\omega)$ in the region $2 \le |\omega| \le 3$ is not available, since the portion of $X_s(j\omega)$ in $-2 \le \omega \le 2$ simply keeps repeating.

10.18 (Computing assignment) *Sampling a line spectral signal.* We now consider a sum of sinusoids of the form

$$x(t) = \sum_{k=1}^{D} a_k \sin(\omega_k t),$$

where $D = 5$. The frequencies in rads/sec are

$$\omega_1 = 0.1\pi, \quad \omega_2 = 0.3\pi, \quad \omega_3 = 0.7\pi, \quad \omega_4 = 1.1\pi, \quad \omega_5 = 2.1\pi,$$

and the amplitudes are $a_1 = 1, a_2 = 0.8, a_3 = 0.6, a_4 = 0.4, a_5 = 0.2$. Since the Fourier transform of a sine is a line spectrum (Sec. 9.2.5), the FT of $x(t)$

is also a line spectrum. Since $x(t)$ is bandlimited with highest frequency ω_5, we can sample it at a rate $\omega_s > 2\omega_5 = 4.2\pi$ without any aliasing.

(a) Define the sampled version $x[n] = x(nT)$, where $T = 2\pi/\omega_s$ and $\omega_s = 5\pi$ rads/sec. This ω_s is large enough to avoid aliasing. Clearly $x[n]$ is infinitely long, so let us define the truncated version

$$x_{trunc}[n] = \begin{cases} x[n] & \text{for } 0 \le n \le 300, \\ 0 & \text{otherwise,} \end{cases}$$

for convenience. Let

$$x_{s,trunc}(t) = \sum_n x_{trunc}[n]\delta_c(t - nT)$$

be the sampled and truncated version written in terms of Dirac deltas. With $X_{s,trunc}(j\omega)$ denoting the Fourier transform of $x_{s,trunc}(t)$, plot the magnitude $|X_{s,trunc}(j\omega)|$ in the range $-\omega_s/2 \le \omega \le \omega_s/2$. You will see a beautiful line spectrum, except that the lines are a little bit spread out because of the above truncation. Also, the heights decrease progressively as we move from $\omega = 0$ to $\omega = \omega_s/2$ because of the a_k. The lines and their amplitudes can be seen so clearly because there is no aliasing (since ω_s is large enough). The locations of the lines are at $\pm\omega_k$ because each $\sin(\omega_k t)$ contributes to lines at $\pm\omega_k$.

(b) Now repeat the above with $\omega_s = 1.5\pi$ rads/sec. Since this is less than $2\omega_4$ and $2\omega_5$, there will be aliasing. In the plot of $|X_{s,trunc}(j\omega)|$ in the range $-\omega_s/2 \le \omega \le \omega_s/2$ you will still see 10 lines (five on the positive side) but the heights are not monotonically decreasing as we move from $\omega = 0$ to $\omega = \omega_s/2$. The lines are permuted because of aliasing. Thus $\pm\omega_4$, which were originally at $\pm 1.1\pi$, get aliased to $\pm 0.4\pi$ and $\pm\omega_5$, which were originally at $\pm 2.1\pi$, get aliased to $\pm 0.6\pi$. (This can be verified by adding appropriate integer multiples of ω_s to $\pm\omega_4$ and $\pm\omega_5$.) Summarizing, if a line spectral signal is sampled at a rate that is not large enough, then the high-frequency lines get aliased and move to points in between some low-frequency lines (or may even overlap exactly with a low-frequency line, and possibly cancel it).

10.19 (Computing assignment) *Generating bandlimited examples.* Let $x(t) = \sum_n c_n h(t - nT)$, where $h(t)$ is the sinc function

$$h(t) = \frac{\sin(\pi t/T)}{\pi t/T}.$$

We know this $x(t)$ is (π/T)-BL, which is very useful to generate examples of bandlimited signals, by choosing c_n appropriately. We now plot an example of such a bandlimited signal. Assume $T = 1$ for simplicity. Let $c_0 = c_1 = c_2 = 1$, and $c_n = 0$ otherwise. Plot $x(t)$ in the range $0 \le t \le 10$. (Plot about 1000 points so the grid size for plotting is $\Delta = 0.01$.) You will find a nice smooth plot. You can play with c_n and generate more examples like this.

11 The Fourier Series

11.1 Introduction

In Chapter 3, where we introduced discrete-time LTI systems, we found that the discrete-time Fourier transform (DTFT) arose naturally, and therefore studied it in some detail. This was our first exposure to Fourier representations in this book because of our focus on discrete-time systems and digital signal processing (DSP). Then, in order to study the theory of sampling (an integral part of DSP courses), we had to introduce the student to the continuous-time Fourier transform (CTFT), which we did in Chapter 9 in considerable detail.

In this chapter we will discuss yet another type of Fourier representation called the Fourier series. This is for continuous-time signals which are either periodic or have a finite duration. We abbreviate this as **CTFS** (continuous-time Fourier series) or just **FS**. The purpose of including the chapter here is for reference value and completeness. To make the reading easier, mathematical details regarding convergence of the Fourier series are not given in this chapter; they can be found in Chapter 17. There are three optional sections at the end of the chapter, marked with asterisks. It is our hope that the reader enjoys them. The first one is on the insights we can get about musical signals and musical scales, based on our understanding of Fourier theory (Sec. 11.8). The second one is on the connection between Fourier approximations and the well-known function approximation property of multilayer neural networks (Sec. 11.9). The third one is an overview of wavelet representations which are important in signal processing, and which are very different from Fourier series representations (Sec. 11.10).

In Chapter 12 we will see a fourth type of Fourier representation called the DFT, and in Sec. 12.7 summarize the connection between the four types of Fourier representations. The DFT is of great importance in digital signal processing and data sciences.

Even though our focus on DSP led us to introduce these Fourier representations in a certain order, traditionally it has been more common to introduce the Fourier series first. One reason for doing so is that many students already have an exposure to Fourier series from physics courses, and therefore feel comfortable with it. Another is that historically, the Fourier series was introduced first, and then the other representations evolved. But in this book the DTFT was introduced first.

Having said that, we also like to add that the theory of DTFT can be regarded as the theory of Fourier series in disguise, as we shall explain in Sec. 11.3.2. In that subtle sense we have stayed true to tradition after all!

11.2 Fourier Series Fundamentals

Let $x(t)$ be a (possibly complex) period-T signal (i.e., periodic with period T), so that

$$x(t) = x(t + T), \tag{11.1}$$

as demonstrated in Fig. 11.1.

Then, under some mild conditions (elaborated in Sec. 17.5), $x(t)$ can be expressed as a sum of single-frequency signals as follows:

$$x(t) = \sum_{k=-\infty}^{\infty} a_k e^{jk\omega_0 t}, \tag{11.2}$$

where

$$\omega_0 = 2\pi/T. \tag{11.3}$$

Equation (11.2) is called the **Fourier series** for $x(t)$. The coefficients a_k are called the Fourier series coefficients, or just **Fourier coefficients**. It can be shown (Sec. 11.4.1) that they are given by

$$a_k = \frac{1}{T}\int_0^T x(t)e^{-jk\omega_0 t}dt. \tag{11.4}$$

The frequency ω_0 is called the **fundamental** frequency, and $k\omega_0$ is called the kth **harmonic** (where the integer k can be positive or negative). Thus, any periodic signal $x(t)$ is a superposition of single-frequency signals with frequencies $k\omega_0$ that are uniformly spaced apart by ω_0. These frequencies define a **frequency grid** as shown in Fig. 11.2. A set of frequencies $\{k\omega_0\}$ which are integer multiples of a fixed frequency ω_0 are said to be **harmonically related**. We use the usual notation

$$x(t) \quad \longleftrightarrow \quad a_k \tag{11.5}$$

to denote that a_k are the Fourier coefficients of the signal $x(t)$.

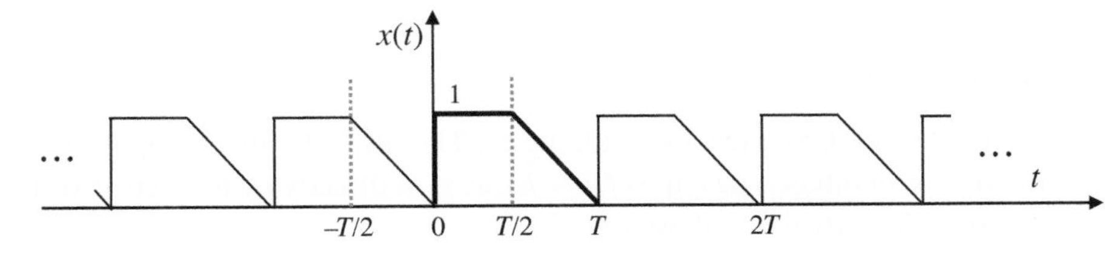

Figure 11.1 A periodic signal with period T.

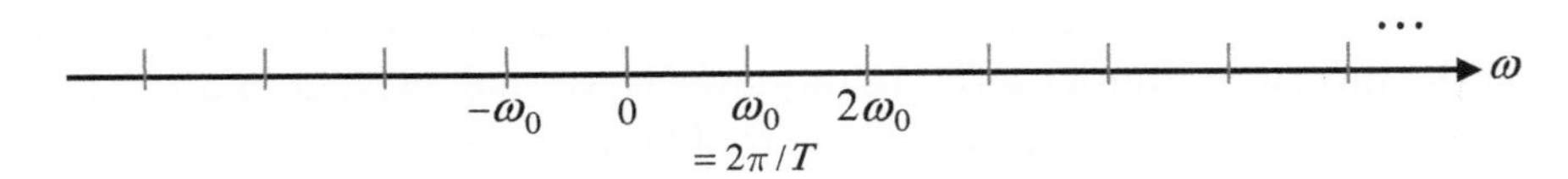

Figure 11.2 The frequency grid $k\omega_0$ used by the Fourier series representation (11.2).

Notice that the single-frequency components in Eq. (11.2) can be written as

$$\eta_k(t) = e^{jk\omega_0 t} = e^{j2\pi kt/T}, \tag{11.6}$$

so that $\eta_k(t) = \eta_k(t + T)$. So these are periodic, and therefore the infinite sum (11.2) is periodic with repetition interval T as expected. Since $x(t) = x(t + T)$, Eq. (11.4) can also be written as

$$a_k = \frac{1}{T}\int_{-T/2}^{T/2} x(t)e^{-jk\omega_0 t}dt. \tag{11.7}$$

More generally, the above integral can be in the range $[\alpha, \alpha + T)$ for any real α. So we sometimes write

$$a_k = \frac{1}{T}\int_T x(t)e^{-jk\omega_0 t}dt, \tag{11.8}$$

where the notation $\int_T$ means that the range of integration is one period, as in $\alpha \le t < \alpha + T$ for arbitrary real α. The region $[0, T)$ (or sometimes $[-T/2, T/2)$, depending on convenience) is said to be the **fundamental interval** of $x(t)$. It is convenient to define a **finite-duration signal**

$$x_T(t) = \begin{cases} x(t) & 0 \le t < T, \\ 0 & \text{otherwise}, \end{cases} \tag{11.9}$$

by restricting $x(t)$ to just one period. The interval $[0, T)$, where $x_T(t)$ can be nonzero, is also called the **support** of $x_T(t)$. It is clear that this finite-duration signal can also be represented using the Fourier series (11.2) as follows:

$$x_T(t) = \begin{cases} \sum_{k=-\infty}^{\infty} a_k e^{jk\omega_0 t} & 0 \le t < T, \\ 0 & \text{otherwise}. \end{cases} \tag{11.10}$$

In short, for a periodic signal $x(t)$ we can use the Fourier series (11.2) to represent $x(t)$ for all t, whereas for a finite-duration signal $x_T(t)$ we use the Fourier series to represent it in its support, as in Eq. (11.10).

11.2.1 Simple Examples

Next, let us look at a few basic examples. The single-frequency signal $x(t) = e^{j\widehat{k}\omega_0 t}$, where $\widehat{k}$ is an integer, has $a_k = \delta[k - \widehat{k}]$, as seen directly by inspection of Eq. (11.2). In particular, when $\widehat{k} = 0$, we get

$$x(t) = 1, \ \forall t \quad \longleftrightarrow \quad a_k = \delta[k]. \tag{11.11}$$

Notice carefully that even though $x(t) = x(t+T)$ in this example, we cannot say that the period is T because $x(t+\tau) = x(t)$ for any τ in this example. Next, the signal $\cos\omega_0 t$ is periodic with period $T = 2\pi/\omega_0$. Since

$$\cos\omega_0 t = \frac{e^{j\omega_0 t} + e^{-j\omega_0 t}}{2}, \tag{11.12}$$

we see that its Fourier series coefficients are $a_1 = a_{-1} = 0.5$, and $a_k = 0$ otherwise. Similarly, since

$$\sin\omega_0 t = \frac{e^{j\omega_0 t} - e^{-j\omega_0 t}}{2j}, \tag{11.13}$$

its Fourier series coefficients are $a_1 = -a_{-1} = -0.5j$, and $a_k = 0$ otherwise. Similarly, we can handle $\cos(\widehat{k}\omega_0 t)$ and so forth. Summarizing, we have

$$\begin{aligned}
x(t) = e^{j\widehat{k}\omega_0 t} &\longleftrightarrow a_k = \delta[k-\widehat{k}], \\
x(t) = \cos(\widehat{k}\omega_0 t) &\longleftrightarrow a_k = 0.5\Big(\delta[k-\widehat{k}] + \delta[k+\widehat{k}]\Big), \\
x(t) = \sin(\widehat{k}\omega_0 t) &\longleftrightarrow a_k = -0.5j\Big(\delta[k-\widehat{k}] - \delta[k+\widehat{k}]\Big).
\end{aligned} \tag{11.14}$$

Historically, when Fourier introduced his series in the eighteenth century, he was interested in the representation of real signals using sums of sines and cosines. While the generalization (11.2) is more convenient, we shall also consider the representation of real signals with cosines and sines in Sec. 11.5.

11.3 Relation to Other Fourier Representations

Before discussing examples and properties, we will explain the connection between the Fourier series and the Fourier representations introduced earlier, such as the DTFT and CTFT. This will help to reduce repetitions, and to obtain a unified perspective.

11.3.1 Connection between Fourier Series and CTFT

Let $X_T(j\omega)$ be the CTFT of $x_T(t)$, that is,

$$X_T(j\omega) = \int_0^T x_T(t)e^{-j\omega t}dt = \int_0^T x(t)e^{-j\omega t}dt. \tag{11.15}$$

Then $x_T(t)$ can be represented by using the inverse CTFT (Sec. 9.4)

$$x_T(t) = \frac{1}{2\pi}\int_{-\infty}^{\infty} X_T(j\omega)e^{j\omega t}\,d\omega, \quad \forall t. \tag{11.16}$$

This is nothing but the *Fourier integral* for $x_T(t)$. Thus, a finite-duration signal can be represented either as the sum in Eq. (11.10) or as the integral (11.16). The Fourier series representation in Eq. (11.10) is valid in the support $[0, T)$, whereas the Fourier integral (or CTFT) representation (11.16) is valid for all t; for t outside $[0, T)$, the right-hand side of Eq. (11.16) reduces to zero as expected.

What is special about the Fourier series representation in Eq. (11.10) is that it says that even though Eq. (11.16) is correct, we *need not* use a continuum of frequencies $-\infty < \omega < \infty$ when $x(t)$ has finite support $[0, T)$. It is *sufficient* to use the discrete grid in Fig. 11.2 with frequency spacing $\omega_0 = 2\pi/T$ inversely proportional to duration T. So, there is more economy in the representation (11.10). As the duration T of the signal $x_T(t)$ increases, the grid gets finer, that is, the spacing ω_0 decreases. In the limit as $T \to \infty$, the Fourier series representation approaches the Fourier integral (11.16). Notice finally that the Fourier series coefficients (11.4) can be written as

$$a_k = \frac{1}{T} X_T(j\omega)\bigg|_{\omega=k\omega_0} = \frac{X_T(jk\omega_0)}{T}. \tag{11.17}$$

That is, the Fourier series coefficients a_k are (proportional to) the samples of the CTFT $X_T(j\omega)$ taken on the uniform frequency grid $k\omega_0$. Figures 11.3(a)–(c) schematically show $x(t), x_T(t)$, and the above relation between a_k and $X_T(j\omega)$. Notice finally that Eq. (11.9) is not the only way to define $x_T(t)$. In some examples it is more convenient to define it as

$$x_T(t) = \begin{cases} x(t) & -T/2 \leq t < T/2, \\ 0 & \text{otherwise.} \end{cases} \tag{11.18}$$

Even in this case, Eq. (11.17) is valid because a_k has the equivalent expression (11.7).

The relation (11.17) means that we can evaluate the Fourier series coefficients by first evaluating the CTFT of one period $x_T(t)$. An example is shown in Fig.

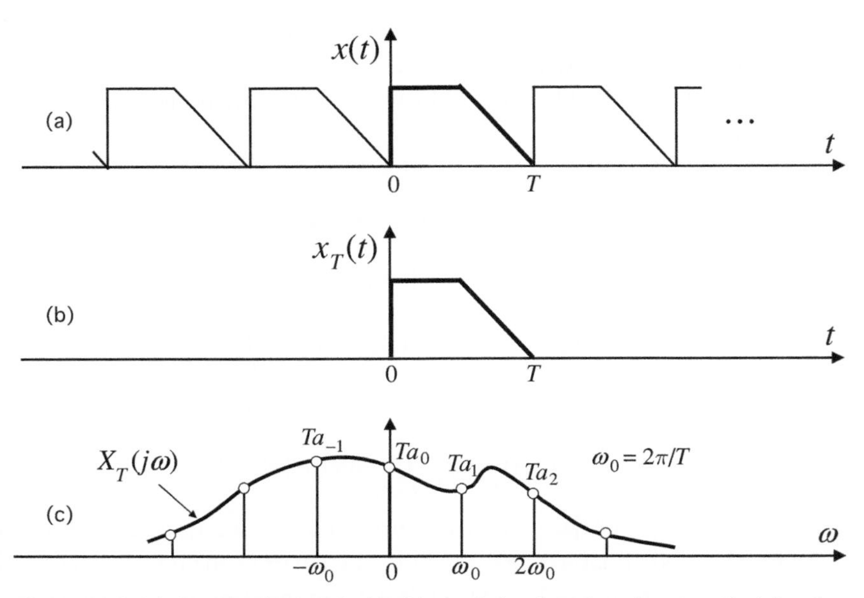

Figure 11.3 (a) A periodic signal $x(t)$, (b) the finite-duration signal $x_T(t)$ defined from one period, and (c) the relation between the Fourier transform of $x_T(t)$ and the Fourier series coefficients a_k of $x(t)$.

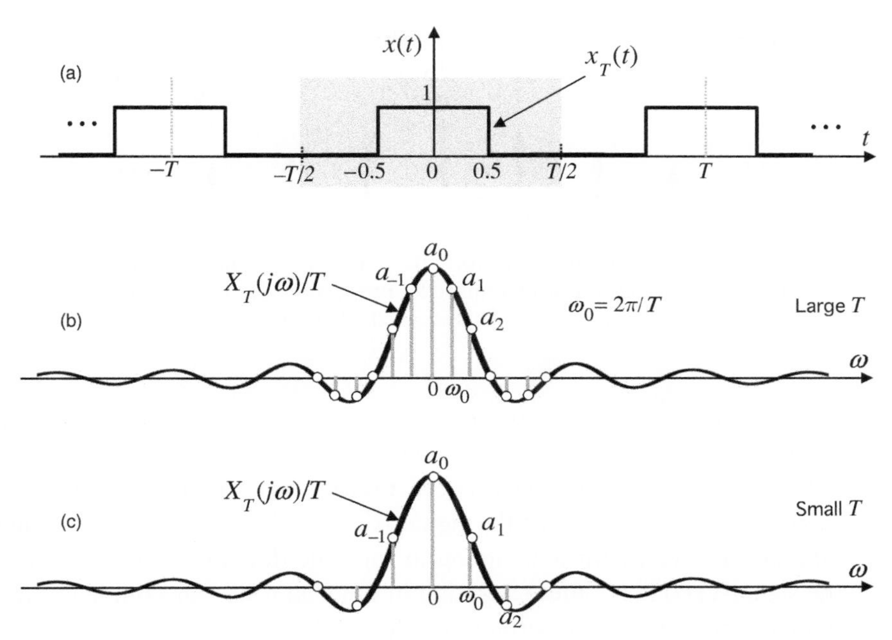

Figure 11.4 (a) A periodic pulse train; (b), (c) the Fourier coefficients a_k for two different values of the period T. For larger T, the sample spacing ω_0 is smaller, that is, the frequency grid is denser.

11.4(a), where $x(t)$ is the periodic pulse train. In this case we can take $x_T(t)$ to be the central pulse as indicated, and we already know (Sec. 9.2.2) that its CTFT is the sinc function

$$X_T(j\omega) = \frac{\sin(\omega/2)}{\omega/2}. \tag{11.19}$$

So Eq. (11.17) yields the Fourier series coefficients

$$a_k = \frac{X_T(jk\omega_0)}{T} = \frac{1}{T}\frac{\sin(k\omega_0/2)}{k\omega_0/2}. \tag{11.20}$$

Since $\omega_0 = 2\pi/T$, this simplifies to

$$a_k = \frac{\sin(k\pi/T)}{k\pi}. \tag{11.21}$$

This is shown in Figs. 11.4(b) and (c) for two values of T. For larger T, the sample spacing ω_0 is smaller, that is, the required frequency grid is denser.

Connection to sampling theory. One beautiful insight that arises from this is the following. In Chapter 10 it was shown that if a signal is **bandlimited**, then it can be represented by a set of uniform samples in the time domain, the required sampling density or rate being proportional to the bandwidth. Similarly, according to Fourier series theory, if a signal $x_T(t)$ is **time-limited** (say, limited to $0 \le t < T$) then we can perform frequency-domain sampling, without losing any information. The required

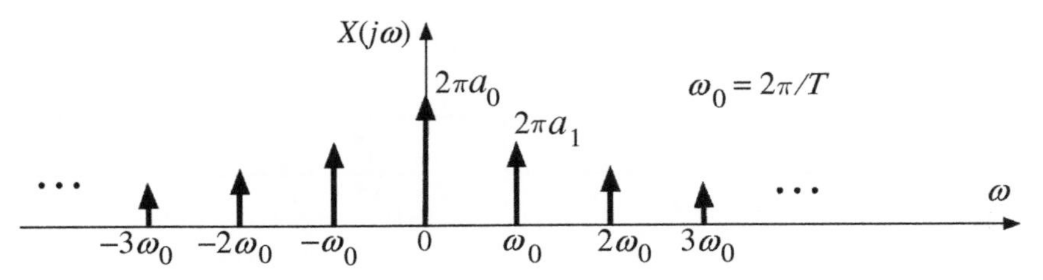

Figure 11.5 The Fourier transform (CTFT) of a periodic signal $x(t)$ consists of Dirac delta functions, with amplitudes proportional to the Fourier series coefficients a_k (Eq. (11.22)). This is called a line spectrum.

sampling density in the frequency domain increases in proportion to the duration T (since the sample spacing is $\omega_0 = 2\pi/T$).

So, this is the connection between Fourier series representation and sampling theory. In sampling theory the signal is of "finite duration in frequency" (band-limited) and we perform sampling in the time domain. In Fourier series theory, the signal $x_T(t)$ is of "finite duration in time" and we can sample in the frequency domain without losing information. That is all! ▽ ▽ ▽

Another simple relation between the CTFT and Fourier series follows from taking the CTFT on both sides of Eq. (11.2). This yields

$$X(j\omega) = 2\pi \sum_{k=-\infty}^{\infty} a_k \delta_c(\omega - k\omega_0) = \frac{2\pi}{T} \sum_{k=-\infty}^{\infty} X_T(jk\omega_0)\delta_c(\omega - k\omega_0), \quad (11.22)$$

for $-\infty < \omega < \infty$. That is, the CTFT of the periodic signal $x(t)$ is obtained by sampling the CTFT of the finite-duration signal $x_T(t)$ at uniformly spaced frequencies $k\omega_0$, and scaling by $2\pi/T$. In particular, therefore, the CTFT $X(j\omega)$ *of a periodic signal* $x(t)$ *is a train of Dirac delta functions* located at frequencies $k\omega_0$, with amplitudes proportional to the Fourier series coefficients a_k as demonstrated in Fig. 11.5. So the CTFT of a periodic signal is called a **line spectrum** because it only has "lines" (Dirac delta functions, to be more precise). The lines are located at harmonically related frequencies.

11.3.2 Connection between Fourier Series and DTFT

Next, what is the connection between Fourier series and DTFT (discrete-time Fourier transform)? For a discrete-time signal $x_d[n]$, the DTFT is given by[1]

$$X_d(e^{j\omega}) = \sum_{n=-\infty}^{\infty} x_d[n]e^{-j\omega n}, \quad (11.23)$$

[1] The subscript d is to avoid any confusion with continuous-time quantities; we don't want to see both $X(j\omega)$ and $X(e^{j\omega})$ on the same page!

and the inverse DTFT (Fourier integral) is

$$x_d[n] = \frac{1}{2\pi} \int_{-\pi}^{\pi} X_d(e^{j\omega}) e^{j\omega n} d\omega. \tag{11.24}$$

Recall that $X_d(e^{j\omega})$ is periodic in ω with period 2π. We can therefore regard $x_d[n]$ as the Fourier series coefficients for $X_d(e^{j\omega})$! Indeed, the integral (11.24) looks exactly like (11.7).

Thus, instead of periodicity in time, we now have periodicity in frequency. So the time-domain coefficients $x_d[n]$ take the role taken in Eq. (11.2) by the Fourier series coefficients a_k. In short, the connection between the Fourier series representation (11.2)–(11.4) and the DTFT representation (11.23) and (11.24) is that time and frequency are interchanged. In Eqs. (11.2)–(11.4) we have periodic time domain and discrete frequency domain, whereas in Eqs. (11.23) and (11.24) we have discrete time domain and periodic frequency domain.

A study of DTFT is therefore equivalent to a study of Fourier series. Since we already studied DTFT in Sec. 3.8, we can be brief in our presentation of the properties of Fourier series. A summary of the connection between different types of Fourier representations can be found in Sec. 12.7.

11.4 Properties of Fourier Series

Many of the properties of the Fourier series are similar to properties of other Fourier representations, which have been discussed in earlier chapters and summarized in tables in Secs. 3.13 and 9.16. We shall therefore discuss only a few selected properties, and summarize the rest in the table of Sec. 11.12.

11.4.1 Orthogonality of Fourier Basis

The Fourier series expression

$$x(t) = \sum_{k=-\infty}^{\infty} a_k e^{jk\omega_0 t}, \quad \omega_0 = 2\pi/T \tag{11.25}$$

is a linear combination of the functions

$$\eta_k(t) = e^{jk\omega_0 t}. \tag{11.26}$$

Each of these functions is periodic, with a common repetition interval of T. We therefore see that as a_k varies over all possible combinations of values, the summation (11.25) *spans a linear space* of signals with repetition interval T. We now claim that the **Fourier basis** functions $\eta_k(t)$ are orthogonal over an interval of duration T. More specifically, $\int_0^T \eta_k(t)\eta_m^*(t)dt = T\delta[k-m]$, or equivalently

$$\int_0^T e^{jk\omega_0 t} e^{-jm\omega_0 t} dt = T\delta[k-m] \quad \text{(orthogonality)}. \tag{11.27}$$

We therefore say that the Fourier series (11.25) represents an expansion of the signal $x(t)$ in terms of the **orthogonal basis** $\{e^{jk\omega_0 t}\}$. For $k = m$, Eq.(11.27) is obvious. For $k \neq m$, Eq. (11.27) follows by setting $l = k - m$ (so $l \neq 0$) and noting that

$$\int_0^T e^{jl\omega_0 t}dt = \frac{e^{jl\omega_0 T} - 1}{jl\omega_0} = \frac{e^{j2\pi l} - 1}{jl\omega_0} = 0. \tag{11.28}$$

The reader can also verify that the limits of the integration can be of the form $\int_\alpha^{\alpha+T}$ for any real α. For example, Eq. (11.27) is equivalent to

$$\int_{-T/2}^{T/2} e^{jk\omega_0 t}e^{-jm\omega_0 t}dt = T\delta[k-m] \tag{11.29}$$

as well.

Proof of Eq. (11.4). The expression (11.4) for the Fourier coefficients follows readily from the above orthogonality. Thus, from Eq. (11.25) we have

$$\begin{aligned}\int_0^T x(t)e^{-jm\omega_0 t}dt &= \sum_{k=-\infty}^{\infty} a_k \int_0^T e^{jk\omega_0 t}e^{-jm\omega_0 t}dt \\ &= T\sum_{k=-\infty}^{\infty} a_k\delta[k-m] = Ta_m,\end{aligned}$$

where Eq. (11.27) is used to get the second equality. This proves Eq. (11.4). ▽ ▽ ▽

It can be shown (Problem 11.4) that the orthogonality property (11.27) also leads to **Parseval's relation**, which states that if $x(t)$ and $y(t)$ are period-T signals with Fourier series

$$x(t) = \sum_{k=-\infty}^{\infty} a_k e^{jk\omega_0 t}, \quad y(t) = \sum_{k=-\infty}^{\infty} b_k e^{jk\omega_0 t}, \tag{11.30}$$

where $\omega_0 = 2\pi/T$ as usual, then

$$\frac{1}{T}\int_0^T x(t)y^*(t)dt = \sum_{k=-\infty}^{\infty} a_k b_k^*. \tag{11.31}$$

This is analogous to Parseval's relation for other types of Fourier transforms we have seen in Chapters 3 and 9. It essentially says that the inner product computed in the time domain is proportional to the inner product in the frequency domain. For the special case where $x(t) = y(t)$, this reduces to

$$\frac{1}{T}\int_0^T |x(t)|^2 dt = \sum_{k=-\infty}^{\infty} |a_k|^2. \tag{11.32}$$

The left-hand side above is the energy in one period of $x(t)$, divided by the period T. It is often called the **power** in the signal. Equation (11.32) says that the power of $x(t)$ is precisely equal to the energy in the Fourier coefficients a_k.

11.4.2 Convolution and Multiplication

Consider two period-T signals $x(t)$ and $y(t)$ with Fourier coefficients a_k and b_k. The periodic convolution of $x(t)$ and $y(t)$ is defined as

$$s(t) = \int_T x(\tau)y(t-\tau)d\tau, \tag{11.33}$$

where the integral extends over one period $\alpha \leq t < \alpha + T$ (where α is immaterial). Then $s(t)$ is periodic-T since

$$s(t+T) = \int_T x(\tau)y(t+T-\tau)d\tau = \int_T x(\tau)y(t-\tau)d\tau = s(t), \tag{11.34}$$

because $y(t)$ is periodic-T. We claim that $s(t)$ has Fourier series Ta_kb_k. That is,

$$\int_T x(\tau)y(t-\tau)d\tau \quad \longleftrightarrow \quad Ta_kb_k. \tag{11.35}$$

This can be proved by using the connection between Fourier series and DTFT explained in Sec. 11.3.2, and the fact that multiplication of two sequences $x[n]$ and $y[n]$ is related to convolution in frequency (Problem 11.7). Here is a direct proof for more clarity.

Direct proof. We have

$$\begin{aligned}
s(t) &= \int_T x(\tau)y(t-\tau)d\tau \\
&= \int_T \sum_{k=-\infty}^{\infty} a_k e^{jk\omega_0\tau} \sum_{m=-\infty}^{\infty} b_m e^{jm\omega_0(t-\tau)}d\tau \\
&= \sum_{k=-\infty}^{\infty}\sum_{m=-\infty}^{\infty} a_kb_m e^{jm\omega_0 t}\int_0^T e^{j(k-m)\omega_0\tau}d\tau \\
&= \sum_{k=-\infty}^{\infty}\sum_{m=-\infty}^{\infty} Ta_kb_m e^{jm\omega_0 t}\delta[k-m] = \sum_{k=-\infty}^{\infty} Ta_kb_k e^{jk\omega_0 t}.
\end{aligned}$$

In the last line we have used orthogonality (11.27). The last summation above has the form of a Fourier series with coefficients Ta_kb_k, so $s(t)$ has Fourier coefficients Ta_kb_k. $\triangledown\triangledown\triangledown$

Next, it can also be shown that the Fourier series c_k of the product $x(t)y(t)$ is the convolution of a_k and b_k, namely, $c_k = \sum_m a_mb_{k-m}$. That is,

$$x(t)y(t) \quad \longleftrightarrow \quad \sum_{m=-\infty}^{\infty} a_mb_{k-m}. \tag{11.36}$$

This can also be proved (Problem 11.9) by using the connection between Fourier series and DTFT (Sec. 11.3.2), and the fact that convolution of two sequences $x[n]$ and $y[n]$ corresponds to multiplication in frequency. A direct proof is requested in Problem 11.8.

11.5 Fourier Series for Real Signals

Suppose $x(t)$ is a real periodic signal with period T, and consider the Fourier series coefficient given by the usual formula

$$a_k = \frac{1}{T}\int_{-T/2}^{T/2} x(t)e^{-jk\omega_0 t}dt, \tag{11.37}$$

where $\omega_0 = 2\pi/T$ as usual. By conjugating both sides and using the realness of $x(t)$, we see immediately that

$$a_k^* = a_{-k}. \tag{11.38}$$

Thus for real signals, even though a_k can be complex, it has Hermitian symmetry with respect to k. Using this, the Fourier series expansion can be rewritten in a form that is more convenient in practice. Thus

$$\begin{aligned} x(t) = \sum_{k=-\infty}^{\infty} a_k e^{jk\omega_0 t} &= a_0 + \sum_{k=1}^{\infty}\Big(a_k e^{jk\omega_0 t} + a_{-k}e^{-jk\omega_0 t}\Big) \\ &= a_0 + \sum_{k=1}^{\infty}\Big(a_k e^{jk\omega_0 t} + a_k^* e^{-jk\omega_0 t}\Big) \\ &= a_0 + \sum_{k=1}^{\infty}|a_k|\Big(e^{j(k\omega_0 t+\phi_k)} + e^{-j(k\omega_0 t+\phi_k)}\Big) \\ &= a_0 + 2\sum_{k=1}^{\infty}|a_k|\cos(k\omega_0 t+\phi_k), \end{aligned}$$

where $a_k = |a_k|e^{j\phi_k}$. By defining $A_k = 2|a_k|$ we can rewrite

$$x(t) = a_0 + \sum_{k=1}^{\infty} A_k\cos(k\omega_0 t + \phi_k), \quad \omega_0 = 2\pi/T. \tag{11.39}$$

Note that a_0 is real and $A_k \geq 0$. This proves that the Fourier series expansion for a *real periodic signal* $x(t)$ is simply a linear combination of cosines with harmonically related frequencies $k\omega_0$ and phases ϕ_k. By writing

$$A_k\cos(k\omega_0 t+\phi_k) = A_k\cos\phi_k\cos(k\omega_0 t) - A_k\sin\phi_k\sin(k\omega_0 t), \tag{11.40}$$

and defining the new coefficients $C_k = A_k\cos\phi_k$ and $S_k = -A_k\sin\phi_k$, we get the popular form of the Fourier series for a real periodic signal $x(t)$:

$$x(t) = a_0 + \sum_{k=1}^{\infty} C_k\cos(k\omega_0 t) + \sum_{k=1}^{\infty} S_k\sin(k\omega_0 t), \tag{11.41}$$

where $\omega_0 = 2\pi/T$ and T is the period of $x(t)$. The coefficients a_0, C_k, and S_k are real valued, as you can readily verify from the above derivation. Clearly they are related to the Fourier series coefficients a_k. We will see that we can also evaluate them directly using formulas (11.53)–(11.55), which follow from the orthogonality of the sine and cosine terms (Sec. 11.5.1).

Even part and odd part. Note that the terms $C_k \cos k(\omega_0 t)$ are even functions of t, and the terms $S_k \sin(k\omega_0 t)$ are odd. Thus a_0 plus the first summation in Eq. (11.41) represents the even part of $x(t)$ and the second sum represents the odd part. If $x(t)$ is *real and even*, its Fourier series therefore reduces to a **cosine series**

$$x(t) = a_0 + \sum_{k=1}^{\infty} C_k \cos(k\omega_0 t), \tag{11.42}$$

whereas for *real and odd* signals the Fourier series reduces to the **sine series**

$$x(t) = \sum_{k=1}^{\infty} S_k \sin(k\omega_0 t). \tag{11.43}$$

More generally, the Fourier series (11.41) of a real periodic signal is a sum of harmonically related cosines representing the even part plus harmonically related sines representing the odd part. ▽ ▽ ▽

11.5.1 Orthogonality of Sines and Cosines

Given a real periodic signal $x(t)$, we can directly identify the coefficients C_k and S_k in its Fourier series (11.41) by noticing that sines and cosines satisfy a wonderful orthogonality property. Assume $k, m > 0$ in the following discussion. We will show that

$$\int_T \cos(k\omega_0 t)\sin(m\omega_0 t)dt = 0, \tag{11.44}$$

$$\int_T \cos(k\omega_0 t)\cos(m\omega_0 t)dt = 0.5T\delta[k-m], \tag{11.45}$$

$$\int_T \sin(k\omega_0 t)\sin(m\omega_0 t)dt = 0.5T\delta[k-m], \tag{11.46}$$

where $\int_T$ means that the range of intergration is one period, as in $\alpha \le t < \alpha + T$ for arbitrary real α. For example, it can be $[-T/2, T/2)$ or $[0, T)$.

Proof. The relations claimed above follow directly from the orthogonality (11.27) of the complex exponentials, which can be rewritten as

$$\int_T e^{jl\omega_0 t}dt = T\delta[l]. \tag{11.47}$$

For example, consider the left-hand side of Eq. (11.45). The integrand can be written as

$$\cos(k\omega_0 t)\cos(m\omega_0 t) = 0.25\left(e^{jk\omega_0 t} + e^{-jk\omega_0 t}\right)\left(e^{jm\omega_0 t} + e^{-jm\omega_0 t}\right). \tag{11.48}$$

Multiplying out the terms in brackets results in

$$e^{j(k+m)\omega_0 t} + e^{j(k-m)\omega_0 t} + e^{j(m-k)\omega_0 t} + e^{-j(k+m)\omega_0 t}. \tag{11.49}$$

Since $k, m > 0$, we have $k + m > 0$. So when $k \neq m$, the integral of each of these terms is zero as seen from Eq. (11.47). This proves Eq. (11.44), and similarly Eqs. (11.45) and (11.46), for $k \neq m$. When $k = m$, the left-hand side of Eq. (11.44) reduces to $0.5 \int_T \sin(2k\omega_0 t)dt = 0$ (use $\omega_0 T = 2\pi$). For $k = m$, the left-hand side of Eq. (11.45) is seen to be

$$\int_T \cos^2(k\omega_0 t)dt = 0.5 \int_T \Big(1 + \cos(2k\omega_0 t)\Big)dt = 0.5T, \tag{11.50}$$

since $\int_T \cos(2k\omega_0 t)dt = 0$ for $k \neq 0$. Similarly, for $k = m$ the left-hand side of Eq. (11.46) is

$$\int_T \sin^2(k\omega_0 t)dt = 0.5 \int_T \Big(1 - \cos(2k\omega_0 t)\Big)dt = 0.5T. \tag{11.51}$$

This completes the proof of Eqs. (11.44)–(11.46). $\nabla\nabla\nabla$

Using the relations (11.44)–(11.46) and the facts that

$$\int_T \cos(k\omega_0 t)dt = \int_T \sin(k\omega_0 t)dt = 0 \quad \text{for } k \geq 1 \tag{11.52}$$

(which follows from $\omega_0 T = 2\pi$), we can readily evaluate (Problem 11.10) the Fourier series coefficients a_0, C_k, and S_k in Eq. (11.41). Thus the constant term is given by

$$a_0 = \frac{1}{T} \int_T x(t)dt, \tag{11.53}$$

whereas the cosine coefficients are

$$C_k = \frac{2}{T} \int_T x(t) \cos(k\omega_0 t)dt, \quad k \geq 1, \tag{11.54}$$

and the sine coefficients are

$$S_k = \frac{2}{T} \int_T x(t) \sin(k\omega_0 t)dt, \quad k \geq 1. \tag{11.55}$$

11.5.2 Fourier Series for Rectified Cosines

The 60 Hz AC voltage that we get from power outlets (50 Hz in some countries, e.g., India) can be written in the form of a cosine

$$x(t) = \cos\omega_0 t, \tag{11.56}$$

where $\omega_0 = 2\pi/T$ with $1/T = 60$ Hz, that is, $T = 1/60$ sec. The signal $x_{hw}(t)$, obtained by retaining the positive half and rejecting the negative half, is called the **half-wave** rectified signal. Thus

$$x_{hw}(t) = \begin{cases} \cos\omega_0 t & \text{when this is non-negative,} \\ 0 & \text{otherwise.} \end{cases} \tag{11.57}$$

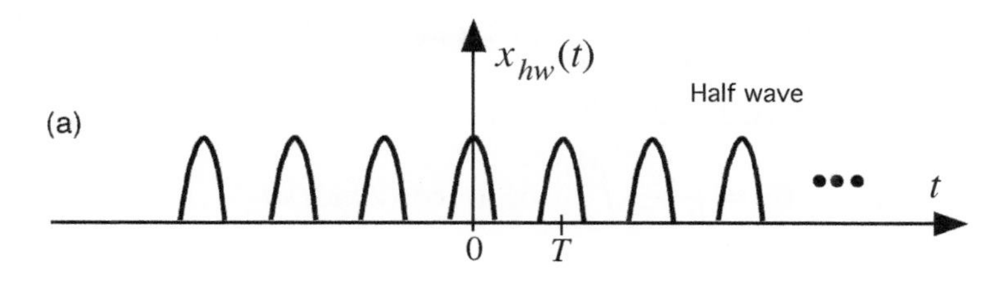

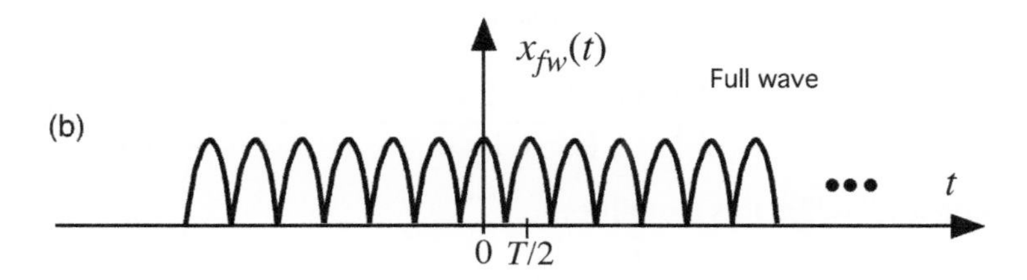

Figure 11.6 (a) The half-wave rectified cosine and (b) the full-wave rectified cosine.

This is shown in Fig. 11.6(a). The signal which is obtained just by taking the magnitude of the cosine, namely

$$x_{fw}(t) = |\cos \omega_0 t|, \tag{11.58}$$

is called the **full-wave** rectified signal, and is shown in Fig. 11.6(b). Rectified cosines can be generated from the AC line voltage by using simple circuits made from diodes and resistors. The rectified signal is often lowpass filtered to extract the "DC component" (the coefficient a_0 in the Fourier series), which is used in power supplies for electronic equipment.[2]

In this section we derive the Fourier series for rectified cosine waveforms. Notice first that the half-wave rectifier still has period T, whereas the full-wave rectifier has period $T/2$ and a fundamental frequency of $2\omega_0$.

11.5.2.1 Full-Wave Rectifier

Since $x_{fw}(t)$ is an even function of t, its Fourier series is simply the cosine series

$$x_{fw}(t) = a_0 + \sum_{k=1}^{\infty} C_k \cos(2k\omega_0 t). \tag{11.59}$$

We can evaluate a_0 and C_k by modifying Eqs. (11.53) and (11.54) to accommodate the fact that the fundamental frequency is $2\omega_0$. Thus

[2] Here, "DC" stands for *direct current*. It just refers to the constant or zero-frequency component of electrical current (or possibly voltage). In our context, this is nothing but a_0, which is the average of $x(t)$ in one period.

$$a_0 = \frac{1}{0.5T}\int_{-T/4}^{T/4} x_{fw}(t)dt = \frac{1}{0.5T}\int_{-T/4}^{T/4}\cos(\omega_0 t)dt = \frac{2}{\pi}, \tag{11.60}$$

where we have used $\omega_0 T = 2\pi$. Next

$$\begin{aligned} C_k &= \frac{2}{0.5T}\int_{-T/4}^{T/4} x_{fw}(t)\cos(2k\omega_0 t)dt \\ &= \frac{4}{T}\int_{-T/4}^{T/4}\cos(\omega_0 t)\cos(2k\omega_0 t)dt \\ &= \frac{2}{T}\int_{-T/4}^{T/4}\Big(\cos(2k+1)\omega_0 t + \cos(2k-1)\omega_0 t\Big)dt \\ &= \frac{4}{T}\left(\frac{\sin[(2k+1)\omega_0 T/4]}{(2k+1)\omega_0} + \frac{\sin[(2k-1)\omega_0 T/4]}{(2k-1)\omega_0}\right). \end{aligned}$$

Using $\omega_0 T = 2\pi$ this simplifies to

$$C_k = \frac{2}{\pi}\left(\frac{\sin[(2k+1)\pi/2]}{2k+1} + \frac{\sin[(2k-1)\pi/2]}{2k-1}\right). \tag{11.61}$$

Since $\sin[(2k+1)\pi/2] = (-1)^k$ and $\sin[(2k-1)\pi/2] = -(-1)^k$, we finally get

$$C_k = \frac{4(-1)^k}{\pi(1-4k^2)}\cdot \tag{11.62}$$

Summarizing, the Fourier series for the full-wave rectified signal is

$$x_{fw}(t) = \frac{2}{\pi} + \sum_{k=1}^{\infty}\frac{4(-1)^k}{\pi(1-4k^2)}\cos(2k\omega_0 t), \tag{11.63}$$

or more explicitly

$$\begin{aligned} x_{fw}(t) = |\cos(\omega_0 t)| &= a_0 + \sum_{k=1}^{\infty} C_k\cos(2k\omega_0 t) \\ &= \frac{4}{\pi}\left(\frac{1}{2} + \frac{1}{3}\cos(2\omega_0 t) - \frac{1}{15}\cos(4\omega_0 t) + \frac{1}{35}\cos(6\omega_0 t) - \cdots\right). \end{aligned} \tag{11.64}$$

Figure 11.7 plots the above Fourier cosine series coefficients a_0 and C_k. In this example, note that C_k are the coefficients of $\cos(2k\omega_0 t)$ and not $\cos(k\omega_0 t)$. From the figure we see that C_k decays rapidly. Only the first three or four coefficients are significant. The Fourier transform of the full-wave rectified signal (11.64) can clearly be written in terms of Dirac delta functions as follows:

$$X_{fw}(j\omega) = \frac{4}{\pi}\left(\pi\delta_c(\omega) + \frac{\pi}{3}\Big(\delta_c(\omega - 2\omega_0) + \delta_c(\omega + 2\omega_0)\Big) - \cdots\right). \tag{11.65}$$

Figure 11.8(a) shows this qualitatively (the heights of the lines are not precise). Notice that instead of showing ω (rads/sec) we have shown $f = \omega/2\pi$ (Hz). Thus $2\omega_0$ corresponds to $2f_0 = 120$ Hz, for 60 Hz domestic electricity. The lines in the plot are at multiples of 120 Hz only.

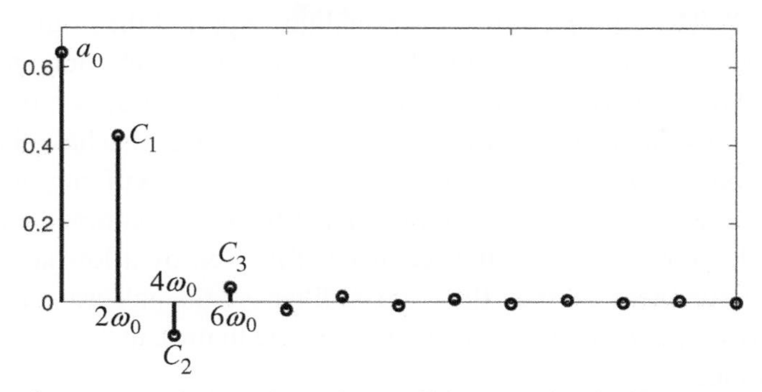

Figure 11.7 The Fourier series coefficients C_k for the full-wave rectified cosine $x_{fw}(t) = |\cos(\omega_0 t)| = a_0 + \sum_{k=1}^{\infty} C_k \cos(2k\omega_0 t)$.

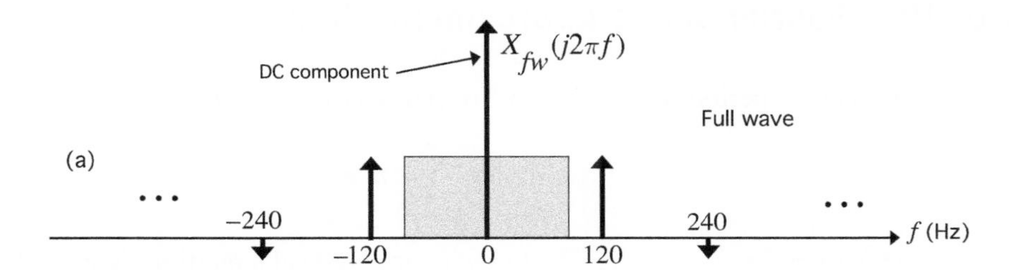

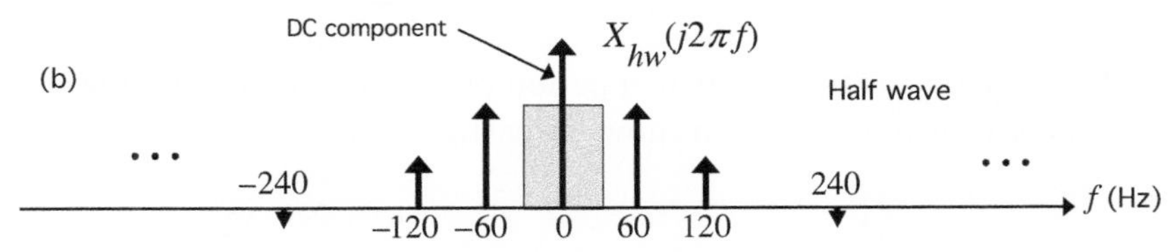

Figure 11.8 Qualitative plots of Fourier transforms of the rectified cosines. (a) Full-wave rectified cosine and (b) half-wave rectified cosine. Shaded areas denote typical passbands of filters to retain DC component.

11.5.2.2 Half-Wave Rectifier

In order to derive the Fourier series expansion for the half-wave rectified cosine $x_{hw}(t)$, it is convenient to notice that it can be written as

$$x_{hw}(t) = \frac{x_{fw}(t) + \cos \omega_0 t}{2}. \tag{11.66}$$

This is readily seen from Fig. 11.6. By using Eq. (11.64) it therefore follows that $x_{hw}(t)$

$$= \frac{2}{\pi}\left(\frac{1}{2} + \frac{\pi}{4}\cos(\omega_0 t) + \frac{1}{3}\cos(2\omega_0 t) - \frac{1}{15}\cos(4\omega_0 t) + \frac{1}{35}\cos(6\omega_0 t) - \cdots\right). \tag{11.67}$$

See Fig. 11.8(b), which shows the qualitative plot of the Fourier transform $X_{hw}(j\omega)$. Unlike the full-wave rectified cosine (11.64) which only has even harmonics $2\omega_0, 4\omega_0$, and so on, the half-wave rectified cosine does have the ω_0 component (since its fundamental period is still $T = 2\pi/\omega_0$). It also has all the even harmonics, but does not have any odd harmonics. If we are rectifying the 60 Hz voltage, the full-wave rectified cosine does not have the 60 Hz component, but the half-wave rectified cosine does. If the rectifier is followed by a lowpass filter to extract the DC component a_0, then the lowpass filter for the half-wave rectifier has to have a narrower passband (see shaded areas in the figure), in order to eliminate the 60 Hz component.

11.6 How Fourier Series Approximates Signals

Let $x(t)$ be a period-T signal with Fourier series expansion

$$x(t) = \sum_{k=-\infty}^{\infty} a_k e^{jk\omega_0 t}, \tag{11.68}$$

where $\omega_0 = 2\pi/T$ as usual. We know from Parseval's relation (Sec. 11.4.1) that

$$\frac{1}{T}\int_0^T |x(t)|^2 dt = \sum_{k=-\infty}^{\infty} |a_k|^2. \tag{11.69}$$

Thus, as long as the energy in a period of $x(t)$ is finite, the infinite sum on the right is a convergent sum, which shows in particular that

$$a_k \to 0 \quad \text{as} \quad |k| \to \infty. \tag{11.70}$$

Thus, although $|a_k|$ may not be monotonically decreasing with increasing k, it eventually decays to zero, as demonstrated for the Fourier series of the rectangular pulse in Fig. 11.4, and the coefficients of the Fourier cosine series (Fig. 11.7) for the full-wave rectified signal.

11.6.1 Truncating the Fourier Series

The above observation suggests that if we truncate the Fourier series to a finite number of terms, the result will be a good approximation of $x(t)$ as long as we keep sufficiently large number of harmonics M. The trucated Fourier series

$$x_M(t) = \sum_{k=-M}^{M} a_k e^{jk\omega_0 t} \tag{11.71}$$

is still a period-T signal because the terms $e^{jk\omega_0 t}$ are periodic. It is an approximation of $x(t)$ which gets closer to $x(t)$ as the number of terms

$$N = 2M + 1 \tag{11.72}$$

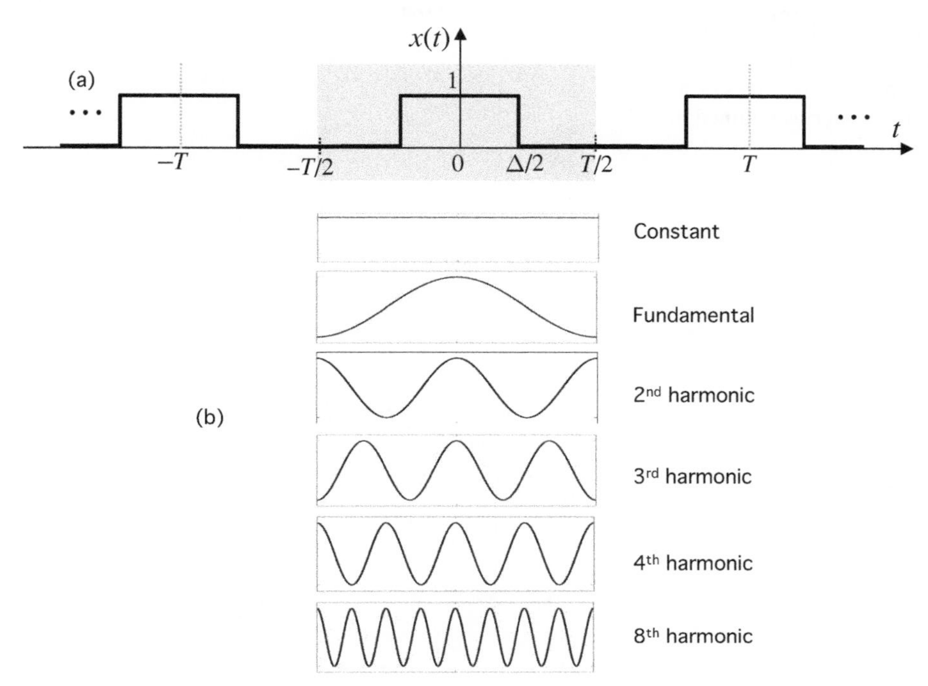

Figure 11.9 (a) The periodic rectangular pulse and (b) examples of basis functions in the Fourier cosine series (11.73), shown for $-T/2 \le t < T/2$.

increases. To demonstrate this, consider again the rectangular pulse train reproduced in Fig. 11.9(a). Since this is a real and even signal, its Fourier series reduces to the cosine series (11.42), and its truncated version has the form

$$x_M(t) = a_0 + 2\sum_{k=1}^{M} a_k \cos k\omega_0 t. \tag{11.73}$$

Figure 11.9(b) shows the cosine basis functions $\cos(k\omega_0 t)$ for several values of k, in the range $-T/2 \le t < T/2$. An appropriate linear combination of such terms results in the rectangular pulse in Fig. 11.9(a). As $M \to \infty$, it seems a miracle that these functions would add up to exactly zero everywhere in $\Delta/2 \le t \le T/2$.

To see how this works, Fig. 11.10 shows the truncated Fourier series of the periodic pulse train for three values of M (number of cosines retained in Eq. (11.73)). In this example, $T = 1$ and $\Delta = 0.5$. Notice how the truncated series gets closer and closer to the ideal pulse train as M increases. For a fixed M, the highest-frequency term included in the truncated series is $\cos(M\omega_0 t)$. As M increases, we therefore see more high-frequency components in the approximation. For any fixed M, the error grows as we get closer to the discontinuity at $t = \Delta/2$. The *maximum error* (which is near the discontinuity) *does not decrease* as M increases. This behavior is referred to as the **Gibbs phenomenon**. This occurs because the truncated series approximates unity to the left of $\Delta/2$ and approximates zero to its right. Because of this conflict, there is larger error there.

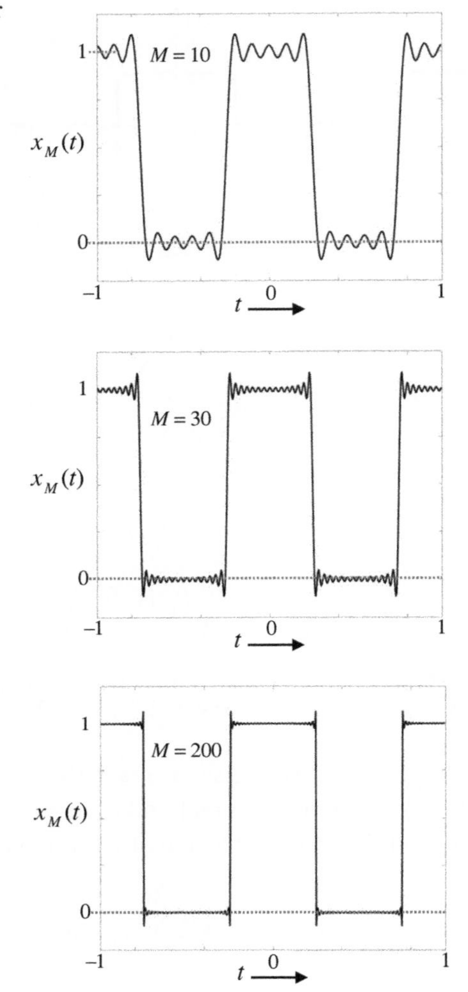

Figure 11.10 Truncated Fourier series for the rectangular pulse train for various values of M (number of terms retained in Eq. (11.73)).

Rate of decay depends on smoothness. With $X_T(j\omega)$ denoting the CTFT of $x_T(t)$ (which is $x(t)$ restricted to one period), we know from Sec. 9.11 that the rate of decay of $X_T(j\omega)$ as $\omega \to \infty$ is larger if $x_T(t)$ is smoother (i.e., more differentiable). Thus, according to Theorem 9.1, if $x_T(t)$ is differentiable K times then $X_T(j\omega)$ decays at least as fast as

$$1/\omega^{K+1}, \tag{11.74}$$

as $\omega \to \infty$. Since $a_k = X_T(jk\omega_0)/T$ (Sec. 11.3), it follows that a_k also decays faster for smoother signals $x_T(t)$. Thus, for a certain accuracy of approximation, the number of terms M required in Eq. (11.71) is smaller for smoother signals. $\triangledown\triangledown\triangledown$

11.6.2 Expression for the Truncation Error

We now derive an expression for the truncation error in Fourier series approximations. The Fourier series for $x(t)$ is shown in Eq. (11.68), and the truncated version is

$$x_M(t) = \sum_{k=-M}^{M} a_k e^{jk\omega_0 t}, \tag{11.75}$$

where $\omega_0 = 2\pi/T$. The truncation error is given by

$$e_M(t) = x(t) - x_M(t) = \sum_{|k|>M}^{\infty} a_k e^{jk\omega_0 t}, \tag{11.76}$$

so that

$$|e_M(t)|^2 = \sum_{|k|>M}^{\infty} \sum_{|m|>M}^{\infty} a_k a_m^* e^{j(k-m)\omega_0 t}. \tag{11.77}$$

The mean square error over one period $0 \le t < T$ is therefore

$$\mathcal{E}_M \triangleq \frac{1}{T}\int_0^T |e_M(t)|^2 dt = \frac{1}{T}\sum_{|k|>M}^{\infty} \sum_{|m|>M}^{\infty} a_k a_m^* \int_0^T e^{j(k-m)\omega_0 t} dt. \tag{11.78}$$

Using the orthogonality property (11.27), this reduces to

$$\mathcal{E}_M = \sum_{|k|>M}^{\infty} |a_k|^2. \tag{11.79}$$

So the mean square error due to truncation is exactly equal to the energy in the dropped portion of the Fourier series coefficients a_k. Thus, once we compute the coefficients a_k, the above expression can be used as a guideline for choosing the number of terms in the approximation: we just have to choose M large enough so that Eq. (11.79) is smaller than whatever error threshold has to be satisfied for the application in hand.

Assuming that $x(t)$ has finite energy in $[0, T]$, we see from Parseval's relation (11.69) that $\sum_k |a_k|^2$ is finite, which means in particular that Eq. (11.79) goes to zero as $M \to \infty$. That is,

$$\begin{aligned}\mathcal{E}_M &= \frac{1}{T}\int_0^T |x(t) - x_M(t)|^2 dt \\ &= \frac{1}{T}\int_0^T \left| x(t) - \sum_{k=-M}^{M} a_k e^{jk\omega_0 t}\right|^2 dt \longrightarrow 0,\end{aligned} \tag{11.80}$$

as $M \to \infty$. Since the above $x(t)$ restricted to $[0, T]$ is an $L_2[0, T]$ signal, we say that the Fourier series for an L_2 signal converges to that signal in the L_2 **sense** or in the **mean square sense**. This should not be confused with pointwise convergence, which will be discussed in Chapter 17.

Case of Real Signals

For a real period-T signal $x(t)$, the Fourier series (11.68) can be rewritten as in Eq. (11.41), where $\omega_0 = 2\pi/T$ and a_0, C_k, and S_k are real. Suppose we approximate this infinite series with

$$x_M(t) = a_0 + \sum_{k=1}^{M} C_k \cos(k\omega_0 t) + \sum_{k=1}^{M} S_k \sin(k\omega_0 t). \tag{11.81}$$

Then the error is

$$e_M(t) = \sum_{k=M+1}^{\infty} C_k \cos(k\omega_0 t) + \sum_{k=M+1}^{\infty} S_k \sin(k\omega_0 t). \tag{11.82}$$

So the square error is

$$\begin{aligned} e_M^2(t) &= \textstyle\sum_{k=M+1}^{\infty} C_k \cos(k\omega_0 t) \sum_{l=M+1}^{\infty} C_l \cos(l\omega_0 t) \\ &\quad + \textstyle\sum_{k=M+1}^{\infty} S_k \sin(k\omega_0 t) \sum_{l=M+1}^{\infty} S_l \sin(l\omega_0 t) \\ &\quad + 2\textstyle\sum_{k=M+1}^{\infty} C_k \cos(k\omega_0 t) \sum_{l=M+1}^{\infty} S_l \sin(l\omega_0 t), \end{aligned} \tag{11.83}$$

and the mean square error of truncation is

$$\begin{aligned} \mathcal{E}_{M,real} \triangleq \frac{1}{T}\int_0^T e_M^2(t)dt &= \tfrac{1}{T} \textstyle\sum_{k=M+1}^{\infty} \sum_{l=M+1}^{\infty} C_k C_l \int_0^T \cos(k\omega_0 t)\cos(l\omega_0 t)\, dt \\ &\quad + \tfrac{1}{T} \textstyle\sum_{k=M+1}^{\infty} \sum_{l=M+1}^{\infty} S_k S_l \int_0^T \sin(k\omega_0 t)\sin(l\omega_0 t)\, dt \\ &\quad + \tfrac{2}{T} \textstyle\sum_{k=M+1}^{\infty} \sum_{l=M+1}^{\infty} C_k S_l \int_0^T \cos(k\omega_0 t)\sin(l\omega_0 t)\, dt. \end{aligned}$$

We showed earlier that the Fourier basis functions $\cos(k\omega_0 t)$ and $\sin(k\omega_0 t)$ have the orthogonality properties (11.44)–(11.46). It therefore follows that

$$\mathcal{E}_{M,real} = \frac{1}{T}\int_0^T e_M^2(t)dt = \frac{1}{2}\sum_{k=M+1}^{\infty}\left(C_k^2 + S_k^2\right). \tag{11.84}$$

This is exactly equal to the average energy in the dropped portions of the coefficients C_k and S_k. Thus, once the coefficients C_k, S_k have been computed, we can use the above expression as a guideline for choosing M.

11.6.3 Optimality of Truncated Fourier Series

Given a signal $x(t), 0 \leq t < T$, suppose we wish to approximate it with a finite number of the Fourier basis functions $e^{jk\omega_0 t}$, $-M \leq k \leq M$, as follows:

$$y_M(t) = \sum_{k=-M}^{M} b_k e^{jk\omega_0 t}, \tag{11.85}$$

where $\omega_0 = 2\pi/T$ as usual. One way to do this is to start with the Fourier series expansion $x(t) = \sum_k a_k e^{jk\omega_0 t}$, and truncate it:

$$x_M(t) = \sum_{k=-M}^{M} a_k e^{jk\omega_0 t}. \tag{11.86}$$

We now claim that this truncated Fourier series $x_M(t)$ is a better approximation of $x(t)$ than *any* other linear combination of the form (11.85), in the sense that the mean square error is minimized. That is, if

$$e_M(t) = x(t) - x_M(t) \quad \text{and} \quad \widehat{e}_M(t) = x(t) - y_M(t), \tag{11.87}$$

then

$$\frac{1}{T}\int_0^T |\widehat{e}_M(t)|^2 dt \geq \frac{1}{T}\int_0^T |e_M(t)|^2 dt, \tag{11.88}$$

with equality if and only if $b_k = a_k$ for all k (i.e., if and only if $y_M(t) = x_M(t)$). Thus, if the basis functions to be used are given by $e^{jk\omega_0 t}, -M \leq k \leq M$, then the **truncated Fourier series is the optimal approximation** of $x(t)$.

Proof of Eq. (11.88). Simply observe that

$$\widehat{e}_M(t) = \sum_{|k|\leq M} (a_k - b_k)e^{jk\omega_0 t} + \sum_{|k|>M} a_k e^{jk\omega_0 t}. \tag{11.89}$$

Using the orthogonality property (11.27), it follows as before that

$$\widehat{\mathcal{E}}_M \triangleq \frac{1}{T}\int_0^T |\widehat{e}_M(t)|^2 dt = \sum_{|k|\leq M} |a_k - b_k|^2 + \sum_{|k|>M} |a_k|^2. \tag{11.90}$$

But we already showed that $\mathcal{E}_M$ is given by Eq. (11.79). So

$$\widehat{\mathcal{E}}_M = \sum_{|k|\leq M} |a_k - b_k|^2 + \mathcal{E}_M, \tag{11.91}$$

proving that $\widehat{\mathcal{E}}_M \geq \mathcal{E}_M$ with equality if and only if $a_k = b_k$ for $|k| \leq M$. $\nabla\nabla\nabla$

11.6.4 Truncating to Principal Components

It should be mentioned that for a fixed number of terms in the Fourier series approximation, there is a different way to perform the truncation, namely, to retain the M coefficients a_k with the largest magnitudes, instead of the first M coefficients. For example, consider the truncated cosine series (11.73) for real, even signals. Compare it with

$$x_{M,pr}(t) = a_0 + 2\sum_{k=1}^{M} \widehat{a}_k \cos(\widehat{\omega}_k t), \tag{11.92}$$

where $\widehat{a}_k$ are the Fourier coefficients with the largest M magnitudes, and $\widehat{\omega}_k$ are the corresponding frequencies.[3] The above approximation is called the **principal component** approximation (which justifies *pr* in the subscript in Eq. (11.92)). It is easy to show (Problem 11.17) that the principal component approximation will yield a mean square error

$$\mathcal{E}_{M,pr} \leq \mathcal{E}_M. \tag{11.93}$$

[3] That is, $\widehat{a}_k = a_{l_k}$ and $\widehat{\omega}_k = l_k\omega_0$, for an appropriate permutation $\{l_k\}$ of $\{1, 2, 3, \ldots\}$.

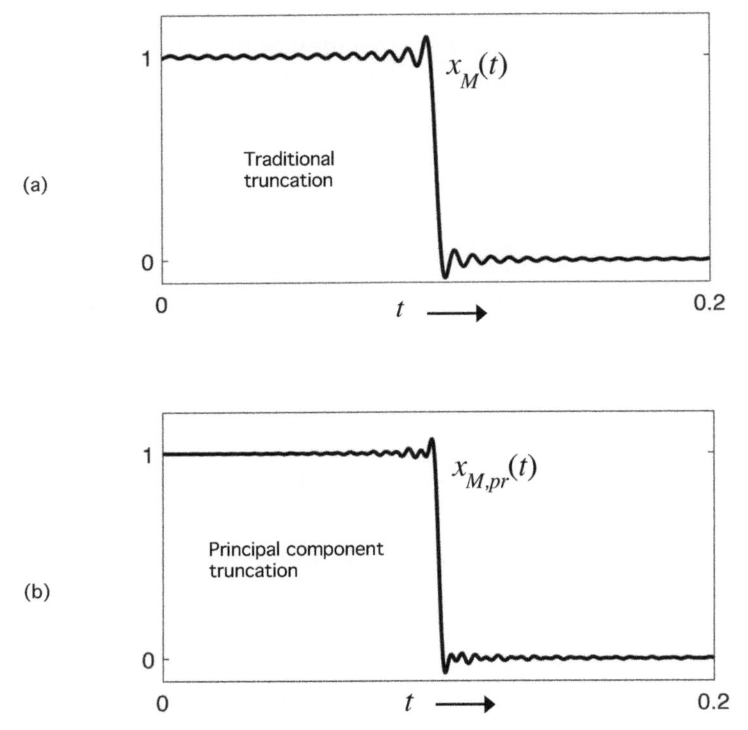

Figure 11.11 The Fourier series approximations obtained using (a) traditional truncation and (b) principal component approach.

So the principal component approximation is more accurate than the traditional approximation $x_M(t) = a_0 + 2\sum_{k=1}^{M} a_k \cos k\omega_0 t$. To illustrate, consider again the pulse train of Fig. 11.9(a). Figure 11.11 shows plots of the approximations $x_M(t)$ and $x_{M,pr}(t)$. Here, $\Delta = 0.2$, $T = 1$, and $M = 150$. It is clear from the plots that the principal component approximation is more accurate than the traditional one. In this example, the mean square errors are

$$\mathcal{E}_M \approx 6.65 \times 10^{-4}, \quad \mathcal{E}_{M,pr} \approx 5 \times 10^{-4}, \tag{11.94}$$

so that the improvement factor is about 1.33.

11.7 Sum of Periodic Signals

Suppose $x_1(t)$ has period T_1, so that $x_1(t) = x_1(t + T_1)$ and there is no smaller repetition interval than T_1. Similarly, let $x_2(t)$ have period T_2. If T_1/T_2 is rational, that is,

$$\frac{T_1}{T_2} = \frac{P}{Q} \tag{11.95}$$

for positive integers P and Q, then $x(t) = x_1(t) + x_2(t)$ is also periodic. The proof is as follows:

$$\begin{aligned} x(t + QT_1) &= x_1(t + QT_1) + x_2(t + QT_1) \\ &= x_1(t + QT_1) + x_2(t + PT_2) \\ &= x_1(t) + x_2(t) = x(t). \end{aligned} \tag{11.96}$$

Here we have used $QT_1 = PT_2$ from Eq. (11.95), and the fact that $x_i(t)$ has period T_i. Thus, $x(t)$ is periodic and has repetition interval

$$T = QT_1 = PT_2, \tag{11.97}$$

which is a common integer multiple of T_1 and T_2. Thus the period is a common integer multiple of the individual periods. Now we can assume without loss of generality that P and Q in Eq. (11.95) are coprime, that is, they do not have any common factor > 1. In this case T is the least common integer multiple or **lcm** of T_1 and T_2, that is,[4]

$$T = \text{lcm}\,(T_1, T_2). \tag{11.98}$$

The reason is that, if $\widehat{Q}T_1 = \widehat{P}T_2$ is a smaller common multiple, then $T_1/T_2 = P/Q = \widehat{P}/\widehat{Q}$ for $\widehat{P} < P$, which contradicts the fact that P and Q are coprime.

Note that the period of the sum $x(t) = x_1(t) + x_2(t)$ can be either T or T/K for some integer $K > 1$, depending on the details of the example. So the period in general can be smaller than $\text{lcm}\,(T_1, T_2)$. It can even be smaller than both T_1 and T_2. See Problem 11.18 for examples covering different possibilities. In the above, we assumed that T_1/T_2 is rational. If this is not the case, we cannot say that $x(t) = x_1(t) + x_2(t)$ is periodic. See Problem 11.19.

11.7.1 What is so Fundamental about the Fundamental?

If a signal $x(t)$ has period T, then we know T is the *smallest* repetition interval. The Fourier series is a superposition of terms of the form $e^{jk\omega_0 t}$, where all frequencies are integer multiples of ω_0 defined by

$$\omega_0 = \frac{2\pi}{T}.$$

Thus ω_0 (which we call the fundamental frequency) is the *largest* real number such that all frequencies $k\omega_0$ are integer multiples of it. Now consider an example of a superposition of two frequencies:

$$x(t) = e^{jk\omega_0 t} + e^{jm\omega_0 t}, \tag{11.99}$$

where k and m are integers. If k and m are not coprime, then we can write $k = g\widehat{k}$ and $m = g\widehat{m}$, where the integer $g > 1$ is the gcd of k and m, and $\widehat{k}$ and $\widehat{m}$ are coprime.

[4] Note that T and T_i are in general not integers, but we can still call T a common integer multiple of T_i because P and Q in Eq. (11.97) are integers. The concept of lcm generalizes to non-integer T_i in this way.

Since $k\omega_0 = \widehat{k}(g\omega_0)$ and $m\omega_0 = \widehat{m}(g\omega_0)$, this shows that $g\omega_0 > \omega_0$ is a larger frequency such that the two frequencies in Eq. (11.99) are integer multiples of it. So the fundamental is $g\omega_0 > \omega_0$ and not ω_0. Correspondingly, the period is the smaller quantity $2\pi/g\omega_0$ rather than $2\pi/\omega_0$. More generally, if $x(t)$ is a superposition of

$$e^{jk_1\omega_0 t}, e^{jk_2\omega_0 t}, \ldots, e^{jk_M\omega_0 t}, \tag{11.100}$$

for integer k_i, and if the set $(k_1, k_2, \ldots, k_M)$ is coprime (i.e., there is no integer factor > 1 common to all k_i), then ω_0 is the largest frequency such that all frequencies can be expressed as its integer multiples. So we say that ω_0 is the fundmental. On the other hand, if the set $(k_1, k_2, \ldots, k_M)$ has gcd $g > 1$ then $g\omega_0$ is the largest frequency such that all frequencies in Eq. (11.100) are integer multiplies of it. So $g\omega_0$ is the fundamental in this case. And the period of the signal is the smaller quantity $2\pi/g\omega_0 < 2\pi/\omega_0$ because the period is always the reciprocal of the fundamental frequency times 2π.

Summarizing, a superposition of signals of the form (11.100), where the k_i are nonzero integers, has fundamental frequency ω_0 if and only if the set

$$(k_1, k_2, \ldots, k_M) \tag{11.101}$$

is **coprime**. Otherwise the fundamental frequency is $g\omega_0$, where g is the gcd of the integers $(k_1, k_2, \ldots, k_M)$.

Examples. If $x(t) = e^{j2\omega_0 t} + e^{j4\omega_0 t} + e^{j6\omega_0 t}$, then all frequencies are common multiples of $2\omega_0$ so the fundamental is $2\omega_0$ and the period is $2\pi/(2\omega_0) = \pi/\omega_0$. If $x(t) = e^{j2\omega_0 t} + e^{j15\omega_0 t}$, then the fundamental is ω_0 because 2 and 15 are coprime. Note that if we compare this with the Fourier series expression $\sum_k a_k e^{jk\omega_0 t}$, we find that the fundamental component $e^{j\omega_0 t}$ is missing. However, the fundamental frequency for $x(t)$ is still ω_0. The **missing fundamental** at ω_0 by itself does not imply that the fundamental frequency is larger than ω_0. Next consider $x(t) = e^{j6\omega_0 t} + e^{j15\omega_0 t}$. Since the gcd of 6 and 15 is 3, the two frequencies can be written as $6\omega_0 = 2(3\omega_0)$ and $15\omega_0 = 5(3\omega_0)$, where 2 and 5 are coprime. So the fundamental is actually $3\omega_0$ and the period is $2\pi/(3\omega_0)$. ▽ ▽ ▽

11.8 Music Signals*

Many of us love music. Historically, music evolved over centuries, mostly guided by culture, aesthetics, and emotions. And yet there is so much additional beauty we can see if we look at it scientifically – either as a physicist or as a signal processing person. There are some aspects to music which are quite universal, just like basic science is.

In Sec. 2.3.2 we briefly touched upon music signals and mentioned that musical notes are periodic signals. Each musical note is therefore associated with a definite pitch. For example, in the middle (fourth) octave of the modern piano, the note A

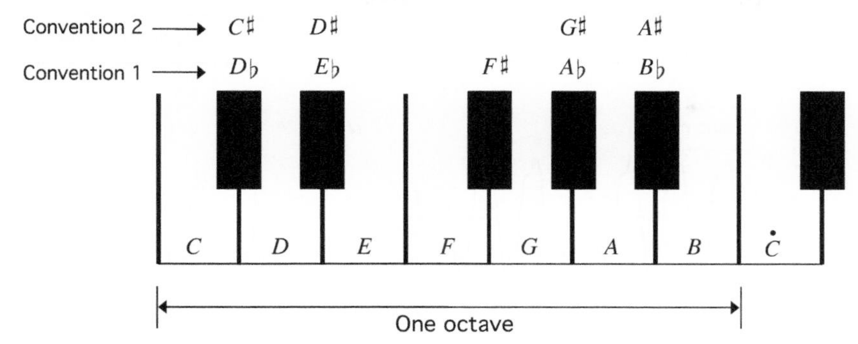

Figure 11.12 One octave of a musical keyboard. The seven notes are indicated on the white keys, and the finer distinctions represented by the black keys are also indicated. Notice the two naming conventions for the black keys. For example, $D\flat$ (D-flat) and $C\sharp$ (C-sharp) indicate the same key.

is nearly 440 Hz (444 Hz in some cases) and the note C is 264 Hz. In this section we will see that a knowledge of Fourier representations adds great insight into music signals and musical scales. For convenience, the musical keyboard is shown again in Fig. 11.12, with the twelve notes in the octave clearly marked. The note $\dot{C}$ in the figure is C of the next octave.

The octave has a very special significance, which should be pointed out right away. It turns out that the note C in the fifth octave has pitch exactly twice that of C in the fourth octave. More generally, any note in the $(n+1)$th octave has pitch $f_2 = 2f_1$, where f_1 is its pitch in the nth octave. Thus, in Fig. 11.12 $\dot{C} = 2C$. The pitches of a note, say C, in successive octaves have the form

$$f_1, 2f_1, 4f_1, 8f_1, \ldots. \tag{11.102}$$

All these notes, separated exactly by one or more octaves, "sound the same" to the human ear. For example, they all sound like C if $f_1 = 264$ Hz. This perceptual property is called **octave equivalence**.

11.8.1 Fourier Series and Tonal Quality

A musical note in its steady state is periodic with some period T secs and pitch $f_1 = 1/T$ Hz. It can therefore be represented by a Fourier series. Music signals from different instruments have different Fourier series coefficients, even if they have identical pitch. Thus, let $x_{piano}(t)$ and $x_{violin}(t)$ represent the middle octave C (or C_4) generated by a piano and a violin with common pitch $f_1 = 264$ Hz. The Fourier series representations are

$$x_{piano}(t) = \sum_{n=0}^{\infty} a_n \cos(n\omega_1 t + \phi_n),$$

$$x_{violin}(t) = \sum_{n=0}^{\infty} b_n \cos(n\omega_1 t + \theta_n),$$

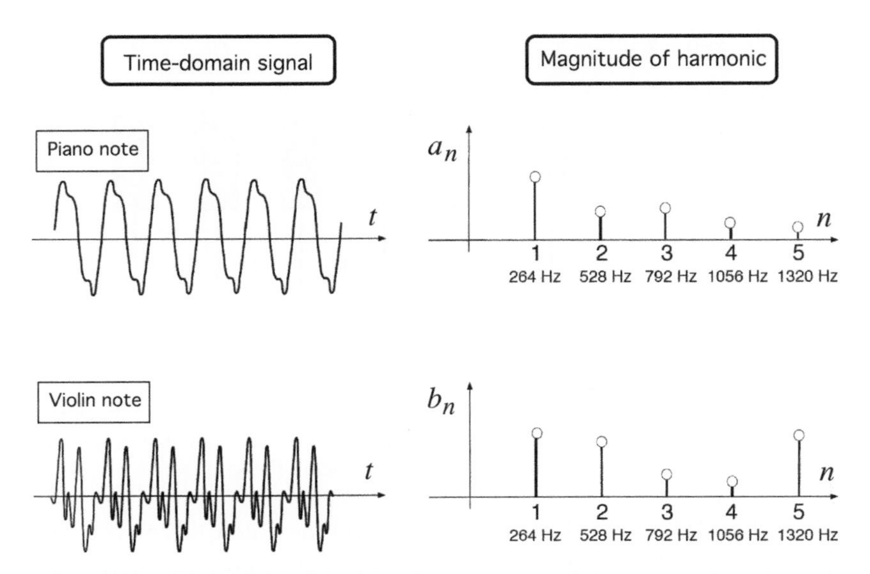

Figure 11.13 The musical waveforms generated by the piano and the violin for the note C_4 (middle-octave C). Both have fundamental frequency 264 Hz, that is, period $T = 3.79$ msecs. The approximate magnitudes of the first few Fourier coefficients are also shown.

where $\omega_1 = 2\pi f_1$ and $a_n, b_n \geq 0$ without loss of generality. It is common experience that these two notes sound very different. So do the appearances of the signals $x_{piano}(t)$ and $x_{violin}(t)$, as shown on the left in Fig. 11.13. It turns out that the phase angles ϕ_n and θ_n are not of primary importance as far as the human ear is concerned (e.g., see Chapter 6 of Backus [1977]).[5] The most significant difference between these two signals is that the coefficients a_n are different from b_n. Figure 11.13 gives a sketch of these coefficients. The figure also indicates in small letters the frequencies of the harmonics or overtones (multiples of 264 Hz).

The plots convey the main idea, but are qualitative (handsketched); actual plots can vary depending on various parameters. As shown in the plot, it is generally true that b_5 is quite large compared to a_5, so the violin signal has a stronger fifth-harmonic content than the piano [Halliday and Resnick, 1978]. The violin signal has strong high-frequency harmonics, consistent with the fact that the time-domain plot $x_{violin}(t)$ shows fast variations at many places. Thus, signals generated by different musical instruments have differernt **timbre** or **tonal quality**, because their harmonic contents (i.e., Fourier series coefficients) are different [Backus, 1977; Fletcher and Rossing, 1998; Olson, 1967].

[5] This was first conjectured by Helmholtz (a nineteenth-century German physicist) and is also referred to as Ohm's law of acoustics. It is not a universally accepted theory [Risset and Wessel, 1999].

11.8.2 Consonance of Musical Notes

Two musical notes sounded together may or may not be pleasant to the ear. When they sound pleasant the notes are said to be **consonant**, otherwise dissonant. Many great minds have delved into the question of consonance from a scientific viewpoint, including Pythagoras (a Greek mathematician of the sixth century BC), Galileo (a sixteenth to seventeenth-century Italian scientist), Helmholtz (a nineteenth-century German physicist), and later musicians [Gamow, 1961].

Pythagoras knew that if the ratio of the frequencies (pitches) f_1 and f_2 of two musical notes is such that

$$\frac{f_2}{f_1} = \frac{n_2}{n_1}, \tag{11.103}$$

where n_1 and n_2 are "small" integers, then the result is pleasant to the ear. This led to the idea that for consonance, the pitch ratio must be a rational number with small n_1 and n_2. Since f_2/f_1 is the ratio of the two pitches, it defines the **musical interval** between the two notes. Notice that perception of musical intervals is based on the *ratio* and not the *difference* in pitch.

1. For example, if $f_2 = 2f_1$ the interval is called an **octave**. This is the most consonant interval. As mentioned earlier, these two notes sound like the same note because of octave equivalence. When sounded together, we just perceive a single note with fundamental frequency f_1.
2. After the octave, the **perfect fifth** interval

   $$f_2 = (3/2)f_1 \tag{11.104}$$

   is considered to be the most consonant. Similarly, the **fourth** ($f_2 = (4/3)f_1$) and the **major third** ($f_2 = (5/4)f_1$) are considered to be very consonant.

An explanation for such consonant behavior was proposed by Helmholtz based on the theory of **beats**. We know that a musical note with pitch f_1 contains the fundamental frequency f_1 as well as a significant number of **harmonics** with frequencies nf_1 ($n = 2, 3, 4\ldots$). Similarly, the note with pitch f_2 has harmonics mf_2. Helmholtz suggested (see Chapter 8 of Backus [1977]) that if some of the harmonics of f_1 are very close (but not equal) to those of f_2, then the beat frequency, proportional to the difference

$$mf_2 - nf_1, \tag{11.105}$$

can be "heard" and creates dissonance.[6] While there is some truth to this, it has been found in later years that beats do not necessarily create dissonance. For example, certain pairs of keys in the piano, sounded together, are expected to produce a certain number of beats in a "properly tuned" western piano (e.g., see Chapter 13 in Backus [1977], or Oster [1973]). Listening tests with pure tones (sinusoids) have

[6] For example, if $mf_2 - nf_1 = 3$ Hz then we hear three beats per second, that is, the sound intensity increases and decreases three times per second – a very noticeable phenomenon. You can readily find out more about beats from physics books and online encyclopedias.

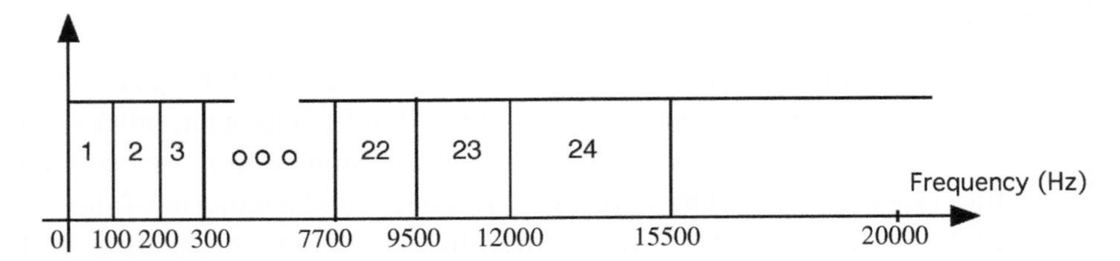

Figure 11.14 Sketch of how the audio-frequency region is divided into 24 critical bands. The 25th band can be considered to be the audible region beyond 15,500 Hz.

also indicated that beats are not necessarily regarded as dissonance (see Sec. III of Rasch and Plomp [1999]).

Critical bands and dissonance. It turns out that the so-called critical frequency bands of the human ear [Zwicker and Fastl, 1990] play an important role in consonance and dissonance. Briefly the story is this: the human audible range (20 Hz to 20 kHz) can be conceptually divided into contiguous frequency subbands called critical bands, such that the ear processes two signals within a band *differently* from two signals in two separate bands. The critical bands get wider as the frequency increases, though they have a uniform bandwidth of about 100 Hz for the low-frequency range (20 to 500 Hz). See Fig. 11.14. This has given rise to the idea that the human ear behaves like a filter bank, that is, a collection of non-overlapping bandpass filters. If two or more pure tones (sinusoids) are placed close enough to be within one critical band of the ear, then they produce some dissonance. If two notes with pitches f_1 and f_2 are sounded together, then even if f_1 and f_2 are reasonably well separated, it is possible that some of the harmonics of f_1 are so close to some of the harmonics of f_2 that they fall within one critical band. This can cause dissonance. It turns out (as we will see in Sec. 11.8.3) that if the frequency ratio f_2/f_1 is equal to the ratio of small integers, this is less likely to happen, which explains why small-integer ratios are preferred. Musical scales which have evolved over centuries in different parts of the world tend to satisfy this criterion to different extents, as we shall see. ▽ ▽ ▽

11.8.3 Importance of Small-Integer Ratios

Assume now that the pitches f_i of two notes satisfy the integer ratio condition (11.103). Without loss of generality, assume n_1 and n_2 are coprime integers (i.e., have no common factors other than unity). We will argue that the closeness of the harmonics is directly related to how large the integers n_1 and n_2 are. More precisely, we claim that the spacing between the harmonics nf_1 and mf_2 can take *any value* of the form

$$k\frac{f_1}{n_1}, \tag{11.106}$$

or equivalently kf_2/n_2, for integer k. Thus, large n_1 or n_2 implies that there can exist closely spaced harmonics. That is why it is desirable to have small values for the integers n_1 and n_2. Musicians often relate the degree of consonance to how small the integers are in the ratio n_1/n_2 (see the plot on page 107 of Deutsch [1999]).

Proof of the claim. We have

$$mf_2 - nf_1 = \frac{mn_2 f_1}{n_1} - nf_1 = \Big(mn_2 - nn_1\Big)\frac{f_1}{n_1}. \tag{11.107}$$

Since n_1 and n_2 are coprime, it follows (from Euclid's theorem [McClellan and Rader, 1979]) that there exist integers m and n such that $mn_2 - nn_1$ is equal to unity, or in fact, any desired integer. This means that $mf_2 - nf_1$ can equal any integer multiple of f_1/n_1 for an appropriate choice of the two harmonic numbers m and n. $\bigtriangledown\bigtriangledown\bigtriangledown$

For example, compare the perfect fifth interval (i.e., C:G) which has the pitch ratio $G/C = 3/2$, with a slightly imperfect fifth interval $301/201$. When the pitch ratio is $3/2$, we have

$$mf_2 - nf_1 = \Big(3m - 2n\Big)\frac{f_1}{2}, \tag{11.108}$$

and the harmonics of C and G can get no closer than $f_1/2$. But if the pitch ratio is $301/201 \approx 1.4975$, then

$$mf_2 - nf_1 = \Big(301m - 201n\Big)\frac{f_1}{201}\cdot \tag{11.109}$$

This means that there can exist harmonics with spacing $f_1/201$. So, there can be some very closely spaced harmonics.

11.8.4 Musical Scales

The music-scale system prevalent in western countries is called the **equitempered** system, and the one prevalent in the east is called the **just-intonation system** [Backus, 1977; Olson, 1967]. For example, the South Indian music system called **Carnatic music** is based on just-intonation [Sambamurthy, 1963]. We now make some remarks about musical scales, in light of the Fourier theory we have learned.

11.8.4.1 The Equitempered System

In this system the twelve notes of the octave have exponentially increasing pitch. More precisely:

Note:	C	$D\flat$	D	$E\flat$	E	F	$F^\sharp$	G	$A\flat$	A	$B\flat$	B	$\dot{C}$
Pitch:	1	α	α^2	α^3	α^4	α^5	α^6	α^7	α^8	α^9	α^{10}	α^{11}	α^{12}
				minor 3rd	major 3rd	4th		5th					

(11.110)

The pitch shown is **normalized pitch**, that is, it assumes that the pitch of C is one unit. Thus, if the actual pitch of C is 264 Hz, then the pitch of $D\flat$ is $\alpha \times 264$ Hz. Similarly the pitch of G is $\alpha^7 \times 264$ Hz, and so on. Since the interval C:G is called the **fifth**, we refer to the note G as the fifth. Similarly E is referred to as the *major third*, F as the *fourth*, and $E\flat$ as the *minor third*.

Irrationality of α. What is the value of α in the above system? Since $\dot{C} = 2C$, it follows from above that $\alpha^{12} = 2$, so that

$$\alpha = 2^{1/12}, \tag{11.111}$$

that is, α should be a 12th root of 2. Since the pitch has positive values, α should be taken as the unique positive 12th root of 2, namely

$$\alpha = 1.059463\ldots \approx 1.06. \tag{11.112}$$

Using this we can numerically calculate all the pitches in the octave. For example, with $C = 264$ Hz (fourth-octave C) we have $D\flat = \alpha C = 279.7$ Hz, $D = \alpha^2 C = 296.3$ Hz, and so on, approximately.

It is easy to show that α is an **irrational** number, that is, it cannot be written in the form $\alpha = p/q$, where p and q are integers. The proof is by contradiction. Thus, if possible, let $2^{1/12} = p/q$. Without loss of generality, assume p and q are coprime. Since $2^{1/12} = p/q$ we have $p^{12} = 2q^{12}$. This shows that q^{12} is a common factor of p^{12} and q^{12}. But since p and q are coprime, p^{12} and q^{12} are also coprime. That is, the common factor $q^{12} = 1$, which implies $q = 1$. This of course implies $2^{1/12} = p$, where p is an integer. This contradicts the fact that $1 < 2^{1/12} < 2$. $\bigtriangledown\bigtriangledown\bigtriangledown$

Similarly, it can be verified that α^k is irrational for all integers k in $1 < k < 12$. In any case, with α as in Eq. (11.111), one can work out the pitch for any note in the octave. For example, to find the pitch of G, note that

$$\alpha^7 = 2^{7/12} = 1.49830707\ldots \approx \frac{3}{2}, \tag{11.113}$$

so that

$$G = \alpha^7 C \approx \frac{3}{2}C. \tag{11.114}$$

Since α is the ratio of pitches (interval) between adjacent notes, it is called a **semitone.** Similarly, α^2 is said to be a **full tone** or a whole tone.

11.8.4.2 Examples of Beats when Notes are Sounded Together

In the middle (fourth) octave we have $C = 264$ Hz, so that the adjacent note is

$$D\flat = \alpha C = 2^{1/12} C \approx 279.7. \tag{11.115}$$

The difference $D\flat - C \approx 15.7$ Hz is large enough, and the notes are perceived as two separate notes. However, in a lower octave, these differences get smaller. For example, in the second octave, $D\flat - C \approx 15.7/4 = 3.925$ Hz. So if C and $D\flat$ are played together in the second octave (just as an experiment) we can perceive approximately four beats per second. That is, there is waxing and waning of loudness about four times in one second. So at least in principle we can hear beats between adjacent notes in the lower octaves.

Now, even in the fourth octave it is possible to hear beats between notes for a different reason: some harmonic of a note can get very close to some harmonic of another note. Consider an example. We have $C = 264$ Hz in the fourth octave. So

$$G = \alpha^7 C = 395.6 \text{ Hz}. \tag{11.116}$$

Since $\alpha^7 \approx 3/2$ (Eq. (11.113)), we have $G \approx 3C/2$. So we expect $3C - 2G \approx 0$. But more accurately,

$$3C - 2G \approx 3 * 264 - 2 * 395.6 = 0.8 \text{ Hz}. \tag{11.117}$$

In short, *when C and G are sounded together, beats can be heard* between the third harmonic of C and the second harmonic of G, and the beat frequency is (11.117). We hear a waxing and waning of loudness once every $1/0.8 = 1.25$ secs. As another example, consider the major third $E = \alpha^4 C$. We have

$$\alpha^4 = 2^{4/12} = 2^{1/3} \approx 1.2599 \cdots \approx \frac{5}{4}, \tag{11.118}$$

so that $E \approx 5C/4$, that is, $4E - 5C \approx 0$. We can calculate $4E - 5C$ more accurately. In the fourth octave $C = 264$ Hz, so that

$$4E - 5C = 4\alpha^4 C - 5C \approx 10.4 \text{ Hz}. \tag{11.119}$$

Since the ear cannot perceive more than about six to eight beats per second, this is not easily perceivable as beats for everyone; it is just perceived as some roughness. But if C and E are played together in the third octave, all the frequencies are divided by two, so $4E - 5C \approx 5.2$ Hz, that is, we can clearly perceive about five beats per second.

11.8.4.3 The Just-Intonation System

In the equitempered system we hear beats between (harmonics of) notes because the harmonics of two notes can never exactly coincide. That is, $nX - mY$ can never be zero for two notes X and Y in an octave for any choice of integers m, n. This is because the ratio $X/Y = \alpha^i$ of any pair of notes within an octave is irrational in the equitempered system, as we saw earlier. The just-intonation system mentioned next is such that the ratios X/Y are rational numbers. This is because the normalized pitches of the notes are themselves rational numbers approximating the pitches α^n of the equitempered system given in (11.110). For the just-intonation system we have

C	$D\flat$	D	$E\flat$	E	F	$F\sharp$	G	$A\flat$	A	$B\flat$	B	$\dot{C}$
Sa	Ri_1	Ri_2	Ga_1	Ga_2	Ma_1	Ma_2	Pa	Da_1	Da_2	Ni_1	Ni_2	$\dot{S}a$
1	$\frac{16}{15}$	$\frac{9}{8}$	$\frac{6}{5}$	$\frac{5}{4}$	$\frac{4}{3}$	$\frac{45}{32}$	$\frac{3}{2}$	$\frac{8}{5}$	$\frac{5}{3}$	$\frac{16}{9}$	$\frac{15}{8}$	2
			minor **3**rd	major **3**rd	**4**th		**5**th					

(11.120)

The second row shows alternate names for the notes used in the Indian music system, which follows just intonation. Because the musical intervals are now exact integer ratios, the major third, the fourth, and the fifth in (11.120) are sometimes referred to as the *perfect major third*, the *perfect fourth*, and the *perfect fifth*. Notice that their pitches are ratios of **small integers**, namely, $5/4$, $4/3$, and $3/2$. Also, the ratio of any two pitches, say G/C, is a rational number now.[7]

To examine the situation of beats, let us consider some examples. First notice that $G = 3C/2$, which means that

$$3C - 2G = 0 \tag{11.121}$$

exactly! This means that the third harmonic of C and the second harmonic of G agree exactly, and there are **no beats** if they are sounded together. This is unlike the equitempered system which satisfied Eq. (11.117), resulting in beats. Similarly, since $E = 5C/4$ from (11.120), it follows that

$$4E - 5C = 0, \tag{11.122}$$

and no beats are heard when C and E are sounded together. Next, since $F = 4C/3$ from (11.120), it follows that

$$4C - 3F = 0, \tag{11.123}$$

and there are no beats between the third harmonic of F and the fourth harmonic of C. The fact that certain low-order harmonics of two notes agree exactly is thought to make these musical intervals more **consonant** or pleasant. So, as mentioned before, the musical intervals defined by ratios of small integers (like E, F, G, and A) are often regarded as very consonant – an idea which prevailed even during the time of Pythagoras [Gamow, 1961].

[7] The ratios in (11.120) can be derived from calculations called the "cycle of fifths" and the "cycle of fourths" [Backus, 1977]. These yield two possible alternative values for some of the notes (e.g., D can be 10/9 or 9/8). These microtonal differences (known as "shrutis" in Carnatic music) have given rise to the idea that there are more than 12 "acceptable" notes in an octave, 22 to be precise [Sambamurthy, 1963]. This is a topic with multiple opinions. See Krishnaswamy [2004] for good insights.

Large-integer ratios. There are some notes which are ratios of integers that are not so small. Thus, consider $F\sharp$. From (11.120) we have

$$F\sharp = \frac{45}{32}C, \tag{11.124}$$

which shows that $45C = 32F\sharp$. That is, the 45th harmonic of C creates a consonance with the 32nd harmonic of $F\sharp$. Since harmonics of such high order are not strong, this is only a weak consonance. This comment applies whenever the pitch is a ratio of large integers. $\triangledown\ \triangledown\ \triangledown$

In just intonation, when tuning a musical instrument, one simply has to adjust the tuning until *all beats disappear*. This is unlike in the equitempered system, where the tuner looks for a certain number of beats between pairs of notes as demonstrated with examples earlier. Also see Chapter 13 of Backus [1977].

11.8.4.4 Power-of-Two Denominators: A Cute Property

Observe from (11.120) that, for notes such as G, E, and D in the just-intonation system, the denominator is a power of two:

$$G = \frac{3}{2}C, \quad E = \frac{5}{4}C, \quad D = \frac{9}{8}C. \tag{11.125}$$

The first equation implies

$$3C = 2G = \dot{G}, \tag{11.126}$$

which is G in the next higher octave. Similarly,

$$5C = 4E = \ddot{E}, \tag{11.127}$$

which is E itself, two octaves higher! Finally,

$$9C = 8D = 2^3 D = D \text{ three octaves above.} \tag{11.128}$$

Equation (11.126) shows that if we play C using an instrument that has a good third harmonic, then this **harmonic of C itself sounds like** G in the next octave. So we get the illusion that both C and G are being played together, although only C is played. Similarly, (11.127) shows that when C is played, it also gives the illusion that both C and E are being played. Finally, (11.128) shows that when C is played, it also gives the illusion that both C and D are being played, provided the ninth harmonic of C is strong enough. All of this contributes to musical aesthetics. This also gives a hint as to why C goes so well with G and E in the so-called **triad**

$$C : E : G, \tag{11.129}$$

which is known to be a pleasant combination of notes. Summarizing, C played with plenty of good harmonics already "contains" G (because of (11.126)), E (because of (11.127)), and D (because of (11.128)). Putting it another way, it is as though the D, E, and G "arise from" C, and everything blends well. This is the beauty of the power-of-two denominators!

11.8.4.5 Can Harmonics Hurt?

It is generally held in most cultures that harmonics are important in music. They contribute to richness, consonance, and aesthetics (even though they create beats). But there are certain specific harmonics which necessarily create dissonance with other notes. Consider, for example, the seventh harmonic of C. Assuming $C = 1$, this harmonic has pitch equal to 7. Since $4 < 7 < 8$, this falls two octaves above. You can actually verify that this falls between the notes A and $A\sharp$ in that octave. Thus, $7C$ is separated by about *half a semitone* from both A and $A\sharp$ and this creates dissonance. If we play C and A together, and if these are rich in harmonics, then the human ear perceives the dissonance or unpleasantness created by hearing $7C$ and $4A$ together in the higher octave. This fact is expressed by the statement that "the seventh harmonic of C is out of tune." When designing musical instruments it is therefore considered desirable to suppress the seventh harmonic as much as possible. Similarly, the eleventh and thirteenth harmonics are undesirable. It seems that the existence of such harmonics was noticed more than 100 years ago by Helmholtz.

Now let us assume that we play a single note, say C. Assume it is rich in harmonics. Is there any way that these harmonics can create dissonance even though no other notes are played? That would seem quite mysterious, but it can happen! To see this, observe that the harmonics have the form

$$C, 2C, 3C, 4C, \ldots$$

and are separated by C. But the octaves are marked by power-of-two multiples of C:

$$C, 2C, 4C, 8C, \ldots,$$

which increases exponentially. Thus, the spacings between adjacent notes (say between C and $D\flat$) are larger and larger in higher octaves – they increase exponentially. For example, $D\flat - C$ in the seventh octave is 16 times $D\flat - C$ in the third octave. Similarly, the critical bands of the human ear get wider and wider as the frequency increases (Fig. 11.14). So it is quite possible that if a note has lots of high harmonics, then two or more harmonics fall within a critical band at high frequency. This is called **harmonic crowding**. It has been claimed that this creates dissonance.

11.8.5 Further Reading

It is indeed fascinating to think about music in the light of Fourier representations. However, the story is much larger than what we have portrayed. The physiology of hearing and even psychology play an important role in the perception and appreciation of music. For further study the reader can refer to some of the excellent books available on these topics, such as Backus [1977], Pickles [1988], Fletcher and Rossing [1998], and Deutsch [1999]. One of the earliest papers on the overtones produced by the musical drum was written by Sir C. V. Raman, Nobel Laureate in physics [Raman, 1934]. Even today it is a joy to read that paper.

11.9 Connection to Machine Learning*

We know that for real $x(t)$ we can express the Fourier series as in Eq. (11.39), where $A_k \geq 0$. This can be rewritten as

$$x(t) = \sum_{k=0}^{\infty} A_k \cos(k\omega_0 t + \phi_k) = \sum_{k=0}^{\infty} A_k \cos(\omega_0(kt + b_k)) , \tag{11.130}$$

where $A_k \geq 0$, $b_k = \phi_k/\omega_0$, and a_0 is absorbed into A_0. The truncated approximation is therefore

$$x_M(t) = \sum_{k=0}^{M} A_k \cos(\omega_0(kt + b_k)) . \tag{11.131}$$

Thus, any function $x(t)$ under some mild conditions can be approximated by this sum of cosines. We now make a slight change of notation to remind us that the independent variable is not necessarily time. Thus we replace t with the independent variable x and replace $x_M(t)$ with $f(x)$:

$$f(x) = \sum_{k=0}^{M} A_k \cos(\omega_0(kx + b_k)) . \tag{11.132}$$

Figure 11.15(a) shows a computational graph or structure for this equation, using the standard notations of Secs. 2.8 and 4.2. The only new element is the building block labeled σ, which is defined as

$$\sigma(u) = \cos(\omega_0 u), \tag{11.133}$$

(a)

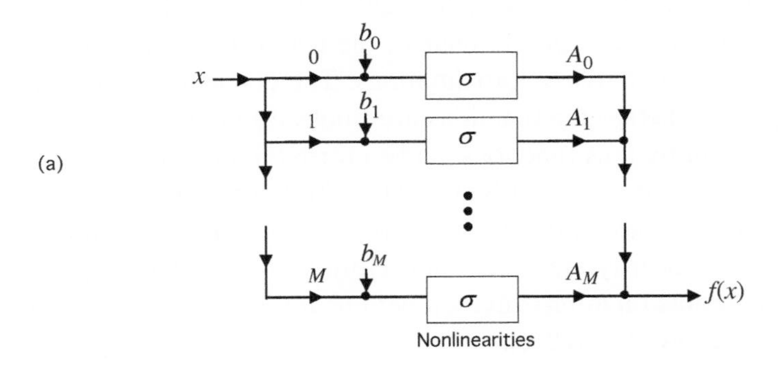

(b)

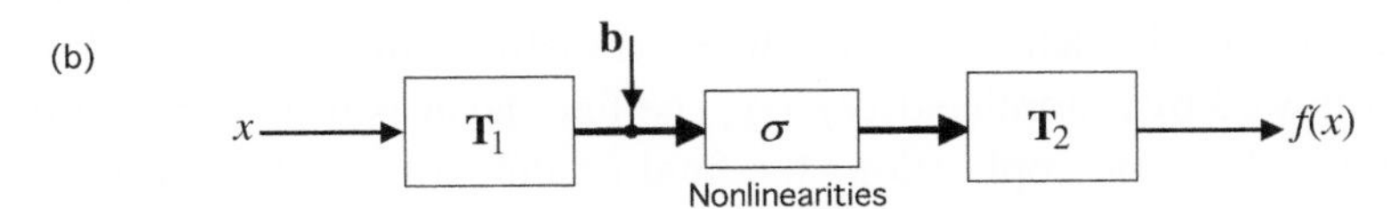

Figure 11.15 (a) The Fourier approximation (11.132), depicted in terms of the computational flowgraph notations of Secs. 2.8 and 4.2. (b) A redrawing of the system using matrix-vector notation. This resembles a two-layer perceptron.

for any real argument u. This $\sigma(u)$ is clearly a nonlinear system. The structure of Fig. 11.15(a) is further redrawn in Fig. 11.15(b) using matrix-vector notation, where

$$\mathbf{T}_1 = \begin{bmatrix} 0 & 1 & 2 & \cdots & M \end{bmatrix}^T, \mathbf{T}_2 = \begin{bmatrix} A_0 & A_1 & A_2 & \cdots & A_M \end{bmatrix}, \tag{11.134}$$

and $\mathbf{b} = \begin{bmatrix} b_0 & b_1 & \cdots & b_M \end{bmatrix}^T$. Now σ takes a vector input, and applies the scalar nonlinearity (11.133) componentwise. For example,

$$\sigma(\mathbf{u}) = \begin{bmatrix} \sigma(u_1) & 0 \\ 0 & \sigma(u_2) \end{bmatrix} \tag{11.135}$$

if $\mathbf{u} = \begin{bmatrix} u_1 & u_2 \end{bmatrix}^T$. Readers familiar with machine learning will recognize that Fig. 11.15(b) represents a **neural network**. More specifically, it is a two-layer feedforward perceptron [Goodfellow et al., 2016; Hertz et al., 1991; Shalev-Shwartz and Ben-David, 2014].

One difference, however, is that the nonlinearity commonly used in modern machine learning practice is not the cosine. Other elements, such as the logistic-sigmoid, the tanh-sigmoid, and ReLU (the rectified linear unit), are commonly used. However, the cosine was also mentioned in the earlier days [Hornik et al., 1989]. The insight we get from this discussion is therefore that the approximation of a function using a Fourier series can be interpreted as the approximation of a function using a two-layer feedforward perceptron! The network shown in Fig. 11.15 is said to be a **two-layer** network because of the use of two transformations $\mathbf{T}_1, \mathbf{T}_2$.

The fact that almost any function can be approximated with a Fourier series has been well known for centuries, and a more rigorous statement of the results, along with some history, can be found in Chapter 17. One of the fundamental results in machine learning theory is that almost any function $f(x)$ can be approximated well using a two-layer network [Cybenko, 1989; Hornik, 1991; Hornik et al., 1989]. While these two are closely related, the neural network results are more general than the Fourier-style approximations. The details of the neural network results depend on whether the functions are Boolean functions, or functions of real variables, or continuous functions, and so forth [Goodfellow et al., 2016; Hertz et al., 1991; Shalev-Shwartz and Ben-David, 2014]. To some extent the results also depend on the choice of the nonlinearity $\sigma(u)$, although the conditions to be satisfied by the nonlinearity are quite mild [Hornik et al., 1989]. A good overview of the various results on approximation with neural networks can be found in Sec. 6.4.1 of Goodfellow et al. [2016].

Figure 11.16 shows a generalization of Fig. 11.15(b) for the case of three layers. Each layer by definition has a linear transformation $\mathbf{T}_i$ followed by a bias vector $\mathbf{b}_i$, followed by a nonlinearity $\sigma(\cdot)$. The final layer is usually referred to as the **output layer**. In some applications the final nonlinearity at the output is omitted, but we have shown it for generality. All layers except the output layer are called **hidden layers** [Shalev-Shwartz and Ben-David, 2014]. So there are two hidden layers in Fig. 11.16, and one hidden layer in Fig. 11.15. For economy, the multilayer neural network is often described in equations as follows:

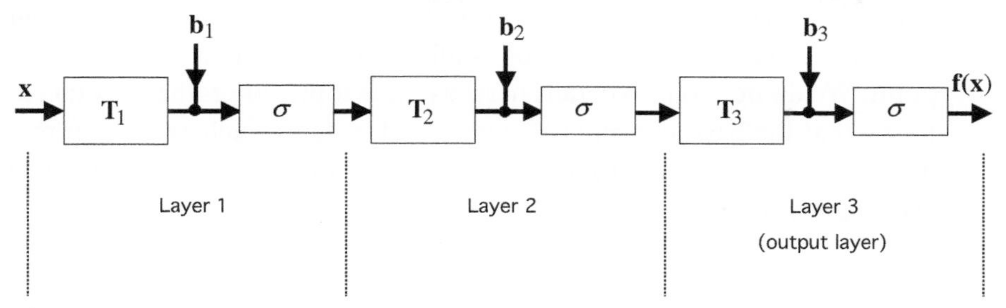

Figure 11.16 Generalization of Fig. 11.15 to multiple layers. This has three layers, and can be regarded as a deep neural network.

$$\mathbf{f}(\mathbf{x}) = \sigma\left(\mathbf{T}_3\left(\sigma\left(\mathbf{T}_2(\sigma(\mathbf{T}_1\mathbf{x} + \mathbf{b}_1)) + \mathbf{b}_2\right)\right) + \mathbf{b}_3\right). \tag{11.136}$$

Even though a two-layer network (i.e., a network with one hidden layer) is sufficient for the approximation of functions [Cybenko, 1989; Hornik, 1991; Hornik et al., 1989], it is usually advantageous to have mutliple layers for various practical reasons, including trainability. Networks with more than two layers (i.e., more than one hidden layer) are usually referred to as **deep** neural networks (DNN). Networks with many hidden layers are very powerful. They have created a revolution in computer vision, generative artificial intelligence, and in many areas of science and engineering.

Other Relevance in Machine Learning

Even though the above connection between a neural network and the Fourier basis is quite straightforward, the influence of signal processing methods in machine learning has gone much further. In Graves et al. [2013], the authors develop an interesting deep neural network or **DNN** for speech recognition. An important preprocessor in this network is the short-time Fourier transformer or **STFT**, which produces the discrete-time FT of short, overlapping, segments of speech to create a time-frequency plot, showing how various frequency components change with time. The STFT has been an integral part of many signal processing algorithms over the decades, and has certainly also been used in important machine learning algorithms.

Another, altogether different, example is the design of functions similar to digital filter frequency responses, using a powerful tool called the **autoencoder** [Liu et al., 2018]. The main goal in that work is to separate out waves arriving from different spatial directions by using spatial filters, before using a machine learning algorithm to find the directions of arrival, or **DOAs**, more accurately. Unlike digital filters, these "autoencoder filters" can be nonlinear, and they are designed by training directly on data (using supervised learning methods). Autoencoders of course have a much broader scope than this [Goodfellow et al., 2016], but it is intriguing that one of the applications is to develop a nonlinear version of digital filtering functions.

Finally, very interesting insights into the operations of deep **convolutional** neural networks have been developed by Mallat [2016], taking a signal processing viewpoint. We do not go into details here, as these topics are rather advanced. Suffice it to say that the basic concepts of signals and systems, such as Fourier methods and filtering, have been very useful in the design and understanding of some machine learning methods.

11.10 Wavelet Representations*

Returning again to Fig. 11.9 where a pulse is represented by a sum of cosines, notice that each cosine is uniformly spread out in $[-T/2, T/2]$. That is, the functions in the Fourier basis are not "localized" or confined to small subregions in the region $[-T/2, T/2]$ of interest. In the region $[\Delta/2, T/2]$ where the pulse is zero, the infinite linear combination of these unlocalized functions adds up to zero. This same linear combination also represents the nonzero value of the pulse in $[-\Delta/2, \Delta/2]$. The abrupt change from 1 to 0 at $t = \pm\Delta/2$ is also approximated by this same linear combination.

Now consider the signal in Fig. 11.17, which is more complicated than the pulse. This signal has slow variations (low frequencies) localized in some regions, and fast variations (high frequencies) localized in other regions as indicated. Again the Fourier series can represent this using the infinite sum of sines and cosines (11.41). In the Fourier basis, a high-frequency basis function, say $\cos(K\omega_0 t)$, which may be required in a certain region (like the "very fast" region in the figure), may not be needed in other regions. But since $\cos(K\omega_0 t)$ is not localized, its effect has to be miraculously cancelled in these other regions by other basis components. This inefficiency arises because of the nonlocal nature of the Fourier basis.

So it appears, at least intuitively, that if we have a different type of basis, where the basis functions are **localized in time** (i.e., have finite durations), then we will have a more efficient representation. There are a number of bases that provide such localized representations. The short-time Fourier transform, mentioned in Sec. 11.9, is a well-known example. Another is the wavelet basis, which has been highly successful in many applications. It is easiest to explain the idea with an example. Thus consider Fig. 11.18, which shows the **Haar wavelet** basis [Haar, 1910]. Assume that we wish to represent a signal $x(t)$ in the region $0 \leq t \leq 1$. The basis

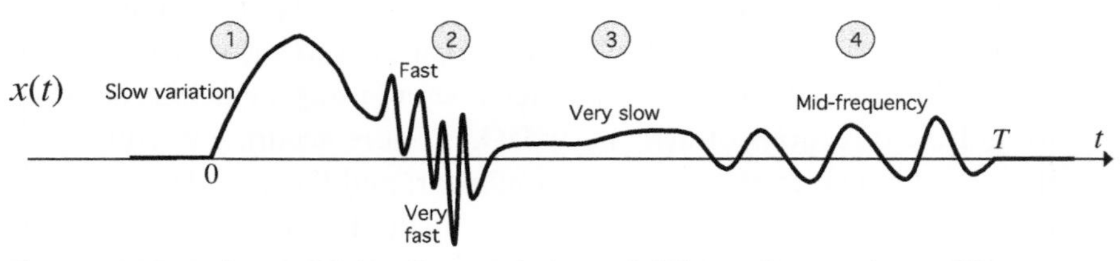

Figure 11.17 A signal with localized variations of different frequencies at different regions of time.

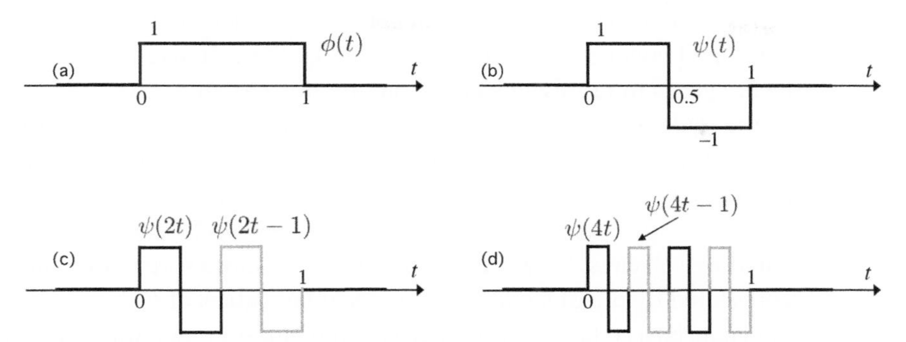

Figure 11.18 Basis functions for the Haar wavelet representation. (a) Scaling function $\phi(t)$, (b) mother wavelet $\psi(t)$, and (c), (d) dilated wavelets and their shifted versions.

function $\phi(t)$ (a constant) is useful to represent the average or DC value of the signal. The function $\psi(t)$ represents a relatively higher rate of change, although it is not a sinusoid. Notice that $\phi(t)$ and $\psi(t)$ are finite-duration functions, restricted to $[0, 1]$. To represent faster variations, we use the dilated version $\psi(2t)$. Notice that $\psi(2t)$ not only changes faster, it is also more localized. Since it can cover the signal behavior only in the region $[0, 0.5]$, a shifted copy $\psi(2t-1)$ is also used in the basis, as shown in the figure. Since

$$\psi(2t-1) = \psi(2(t-0.5)), \tag{11.137}$$

the amount of shift is exactly 0.5. To represent even faster variation, the dilated version $\psi(4t)$ can be used. Since this is even more localized, four copies are used ($\psi(4t)$, $\psi(4t-1)$, and so on), as shown in the figure.
Continuing like this we get a basis in which the functions are $\phi(t)$ and

$$\psi(2^k t - m), \qquad 0 \le m \le 2^k - 1, \quad 0 \le k \le \infty. \tag{11.138}$$

Summarizing, the representation of $x(t)$ takes the form

$$x(t) = \sum_{k=0}^{\infty} \sum_{m=0}^{2^k-1} d_{km}\psi(2^k t - m) \quad + \quad c_0\phi(t). \tag{11.139}$$

This is an example of a wavelet representation called the Haar wavelet representation, and c_0 and d_{km} are the wavelet coefficients in the representation. Notice that the wavelet representation is a **double summation**, unlike the Fourier representation (11.41) which is a single summation. In wavelet representations, the function $\phi(t)$ is usually called the **scaling** function, and $\psi(t)$ is called the **mother wavelet**, or just the wavelet function. A number of remarks are now in order.

1. For larger and larger k, the functions $\psi(2^k t - m)$ capture details of the signal at finer and finer scales. So the variable k is called the **scale**. When the bases were sines and cosines, we used the term "frequency" to represent the amount of variation per second. In the wavelet world, the word "scale" is used. The shift index m makes sure that the basis functions cover the entire time range of

interest. It is also possible to use negative k, in order to cover coarser and coarser scales, and cover a larger region than $[0, 1]$. We can also use negative values of the integer m to represent signals that are nonzero in $t < 0$.

2. As the scale gets finer (large k), the functions $\psi(2^k t - m)$ are more and more **localized**, unlike in the Fourier basis. This is a very useful feature when a high-frequency variation is confined to a very small region (as in the region labeled "very fast" in Fig. 11.17). Such signals are therefore very efficiently represented by wavelets. Slow variations tend to be more spread out and can be well represented by use of small (or even negative) values of k.
3. The Haar basis [Haar, 1910] is perhaps the oldest known wavelet basis, and was known many decades before the major developments of the 1980s [Daubechies, 1988; Mallat, 1989]. One problem with the Haar basis is that the function $\psi(t)$ is discontinuous, and so are all the basis functions $\psi(2^k t - m)$. Thus, when Eq. (11.139) is truncated to a finite number of terms, the truncated version looks very nonsmooth. This is a major disadvantage in signal representation. For example, in image compression, the roughness of the image created due to this truncation can readily be perceived by the eye. One of the landmark contributions by Daubechies [1988] showed that orthogonal wavelet bases (for the L_2 space, Sec. 17.3) can be constructed with much smoother functions $\psi(t)$ than the Haar wavelet. Wavelets with finite duration, which are not only continuous, but also differentiable a certain number of times, are possible.
4. The student can readily verify that the Haar wavelet basis is orthonormal; since $\psi(t)$ has finite duration, it is said to be a **compactly supported** wavelet. Orthonormal wavelet bases with compact support can be constructed with a degree of smoothness that can be chosen by the designer.
5. There is a close connection between wavelets and multirate digital filter banks [Burrus et al., 1998; Rioul and Vetterli, 1991; Strang and Nguyen, 1996; Vaidyanathan, 1993; Vetterli and Herley, 1992]. Orthonormal wavelets, for example, are related to orthonormal or **paraunitary** filter banks [Vaidyanathan, 1987a, 1987b]. Compactly supported wavelets are related to FIR filter banks.
6. Wavelets have also been designed based on spline functions, which we know to be very smooth (Sec. 9.11.1). They are called **spline wavelets** [Chui, 1992; Unser et al., 1993].
7. Given two representations for a signal in terms of two different basis functions, one of them might require fewer basis functions (fewer nonzero coefficients) than the other, in order to achieve a certain accuracy of representation. Bases which can represent a signal accurately with the help of very few functions from the basis lead to **sparse representations**. It turns out that for natural images and other images that arise in many applications, wavelet representations provide very sparse representations, and are of great value in image compression [Lewis and Knowles, 1992; Malvar, 1992; Sayood, 2000]. For example, see the striking image compression example given in Candès and Wakin [2008].

In this section we have made some brief remarks about wavelets, so that the interested student can pursue this topic in depth. The phenomenal developments on wavelets which took place in the 1980s started with the landmark papers by Daubechies [1988] and by Mallat [1989]. Since then there have been many papers. Introductory books and tutorials include Rioul and Vetterli [1991], Akansu and Haddad [1992], Strang and Nguyen [1996], and Burrus et al. [1998]. Other major books in this area include Chui [1992], Daubechies [1992], Vetterli and Kovačević [1995], and Mallat [1998].

11.11 Summary of Chapter 11

A period-T signal $x(t)$ can be expressed as a sum $x(t) = \sum_{k=-\infty}^{\infty} a_k e^{jk\omega_0 t}$ called the Fourier series, where $\omega_0 = 2\pi/T$ is the fundamental frequency and a_k are the Fourier series coefficients. We studied several properties of Fourier series, and these are summarized in the table in Sec. 11.12. With $X(j\omega)$ denoting the traditional CTFT of $x(t)$ and $X_T(j\omega)$ the traditional CTFT of $x_T(t)$ ($x(t)$ truncated to $[0, T)$ or $[-T/2, T/2)$), we have

$$X(j\omega) = 2\pi \sum_k a_k \delta_c(\omega - k\omega_0) \quad \text{and} \quad a_k = X_T(jk\omega_0)/T.$$

This shows the connection between the CTFT and Fourier series. Next, recall that the discrete-time Fourier transform (DTFT, Sec. 3.4.1) is applicable when time is discrete so that the frequency domain has periodicity. The Fourier series is applicable when the opposite is true: there is periodicity in time, and the frequency domain is discrete. This can be regarded as a connection between DTFT and Fourier series. The above remarks therefore show the conceptual connections between all three representations: CTFT, DTFT, and the Fourier series. The connection between the Fourier series and sampling theory was mentioned in Sec. 11.3.1 as well.

The Fourier basis $\{e^{jk\omega_0 t}\}$ has the orthogonality property, which makes the truncated version of the Fourier representation optimal in a certain sense (Sec. 11.6). We also showed that the principal component approximation derived from the Fourier series has advantages over truncation. For real-valued signals, the Fourier series can be rewritten as a sum of a cosine series representing the even part of $x(t)$ and a sine series representing the odd part of $x(t)$. This representation also enjoys an orthogonality property (Sec. 11.5.1).

We also had a detailed discussion of the properties of musical signals, musical scales, consonance, and dissonance. A theoretical appreciation of music is gained through such an analysis, as explained in detail. It was also argued that the Fourier approximation can be regarded as a special case of function approximation done by multilayer neural networks. In this sense the function approximation properties of neural networks are not altogether surprising.

Towards the end of the chapter we gave an overview of wavelet representations and contrasted them with Fourier series representations. Wavelet representations allow time localization of various frequency components and are well suited in many practical situations.

11.12 Table of Fourier Series Properties

A signal $x(t)$ with period T (more generally, repetition interval T) has the Fourier series

$$x(t) = \sum_{k=-\infty}^{\infty} a_k e^{jk\omega_0 t}, \quad \omega_0 = 2\pi/T. \tag{11.140}$$

Here, ω_0 is the fundamental frequency of $x(t)$. The Fourier series coefficients a_k are

$$a_k = \frac{1}{T}\int_T x(t)e^{-jk\omega_0 t}dt.$$

The notation $\int_T$ means integration over one period, that is, over the interval $\alpha \le t < \alpha + T$ for arbitrary α. Choice of α does not affect any results. The notation

$$x(t) \quad \longleftrightarrow \quad a_k \tag{11.141}$$

means that $x(t)$ and a_k constitute a Fourier series pair, that is, they satisfy Eq. (11.140). We summarize the properties of Fourier series below, where $x(t)$ and $y(t)$ are periodic-T, with Fourier coefficients a_k and b_k.

1. *Linearity.* $\alpha x(t) + \beta y(t) \longleftrightarrow \alpha a_k + \beta b_k$.
2. *Reversal.* $x(-t) \longleftrightarrow a_{-k}$ (reversal in time is equivalent to that in frequency).
3. *Evenness.* $x(t)$ is even ($x(t) = x(-t)$) if and only if a_k is even ($a_k = a_{-k}$).
4. *Conjugation.* $x^*(t) \longleftrightarrow a^*_{-k}$ and similarly $x^*(-t) \longleftrightarrow a^*_k$.
5. *Realness.* $x(t)$ is real if and only if $a_k = a^*_{-k}$. Similarly, a_k is real if and only if $x(t) = x^*(-t)$.
6. *Fourier series for real signal.* $x(t) = a_0 + \sum_{k=1}^{\infty} C_k \cos(k\omega_0 t) + \sum_{k=1}^{\infty} S_k \sin(k\omega_0 t)$; $a_0 = \int_T x(t)dt/T,\ C_k = 2\int_T x(t)\cos(k\omega_0 t)dt/T,\ S_k = 2\int_T x(t)\sin(k\omega_0 t)dt/T$.
7. *Real even signals.* $x(t)$ is both real and even if and only if a_k is real and even.
8. *Shifting.* $x(t - t_0) \longleftrightarrow e^{-jk\omega_0 t_0} a_k$.
9. *Modulation.* $x(t)e^{jk_0\omega_0 t} \longleftrightarrow a_{k-k_0}$ for integer k_0.
10. *Differentiation.* $dx(t)/dt \longleftrightarrow jk\omega_0 a_k$.
11. *Periodic convolution.* $\int_T x(\tau)y(t-\tau)d\tau \longleftrightarrow Ta_k b_k$.
12. *Multiplication.* $x(t)y(t) \longleftrightarrow \sum_{m=-\infty}^{\infty} a_m b_{k-m}$ (convolution of a_k and b_k).
13. *Orthogonality of Fourier basis.* $\int_T e^{jk\omega_0 t}e^{-jm\omega_0 t}dt = T\delta[k-m]$.
14. *Parseval's relation.* $\int_T x(t)y^*(t)dt = T\sum_k a_k b_k^*$; $\quad \int_T |x(t)|^2 dt = T\sum_k |a_k|^2$.
15. *Optimal approximation property of truncated Fourier series.* Let $x_M(t) = \sum_{k=-M}^{M} a_k e^{jk\omega_0 t}$, $\widehat{x}_M(t) = \sum_{k=-M}^{M} \widehat{a}_k e^{jk\omega_0 t}$. Then $x_M(t) - x(t)$ has smaller mean square value than $\widehat{x}_M(t) - x(t)$, unless $\widehat{a}_k = a_k$ for all k.

11.13 Table of Fourier Series Examples

Signal in $-T/2 \leq t < T/2$ *Note:* $\omega_0 = 2\pi/T$, and $\widehat{k}$ below is an integer	**FS coefficients** a_k, **or FS representation**
$x(t) = 1$	$a_k = \delta[k]$
$x(t) = e^{j\widehat{k}\omega_0 t}$	$a_k = \delta[k - \widehat{k}]$
$x(t) = \cos(\widehat{k}\omega_0 t)$	$a_k = 0.5\Big(\delta[k - \widehat{k}] + \delta[k + \widehat{k}]\Big)$
$x(t) = \sin(\widehat{k}\omega_0 t)$	$a_k = -0.5j\Big(\delta[k - \widehat{k}] - \delta[k + \widehat{k}]\Big)$
$x_{fw}(t) = \lvert\cos(\omega_0 t)\rvert$ (full-wave rectifier output)	$\frac{2}{\pi} + \sum_{k=1}^{\infty} \frac{4(-1)^k}{\pi(1-4k^2)} \cos(2k\omega_0 t)$ (Fourier cosine series, Sec. 11.5.2)
$x_{hw}(t) = \frac{x_{fw}(t) + \cos\omega_0 t}{2}$ (half-wave rectifier output)	$\frac{1}{\pi} + \frac{\cos\omega_0 t}{2} + \sum_{k=1}^{\infty} \frac{2(-1)^k}{\pi(1-4k^2)} \cos(2k\omega_0 t)$ (Fourier cosine series, Sec. 11.5.2)
$x(t) = e^{j\beta\sin(\omega_0 t)}, \ \beta \geq 0$ (Problem 11.25)	$\sum_{k=-\infty}^{\infty} J_k(\beta) e^{jk\omega_0 t}$ $J_k(\beta) =$ kth-order Bessel function of first kind.
$-\ln\lvert 2\sin(t/2)\rvert$ (unbounded signal, Sec. 17.6.4)	$\cos t + \frac{\cos 2t}{2} + \frac{\cos 3t}{3} + \cdots$

PROBLEMS

11.1 For the signal $x(t)$ in Fig. 11.1, find an expression for the Fourier series coefficients a_k. Also find the Fourier transform (CTFT) $X(j\omega)$.

11.2 Figure P11.2 shows a periodic signal $x(t)$. The signal $x_T(t)$, which is $x(t)$ truncated to one period, is highlighted using heavy lines.

(a) Find the Fourier transform of $x_T(t)$.

(b) Using the result for $X_T(j\omega)$, find the Fourier series coefficients a_k for the periodic signal $x(t)$.

(c) Also find the Fourier transform (CTFT) $X(j\omega)$.

Note: One way to do (a) is to use the derivative property, and use the known expressions for the CTFT of the pulse, the Dirac delta, and so forth.

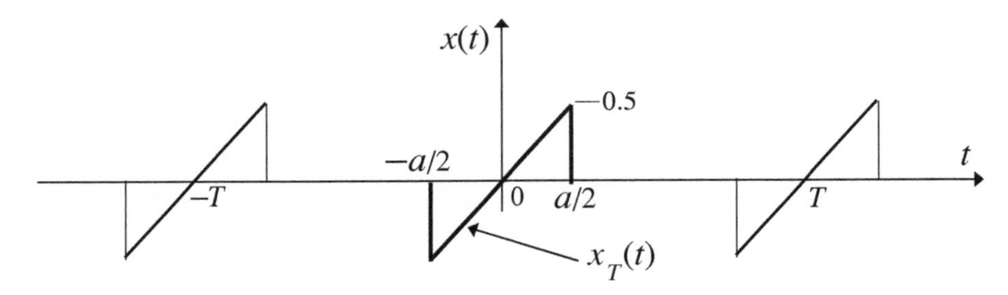

Figure P11.2 Antisymmetric triangular periodic signal.

11.3 Find the Fourier series coefficients a_k for the periodic signal $x(t)$ in Fig. P11.3. *Hint*: Take advantage of the information in Fig. 9.6.

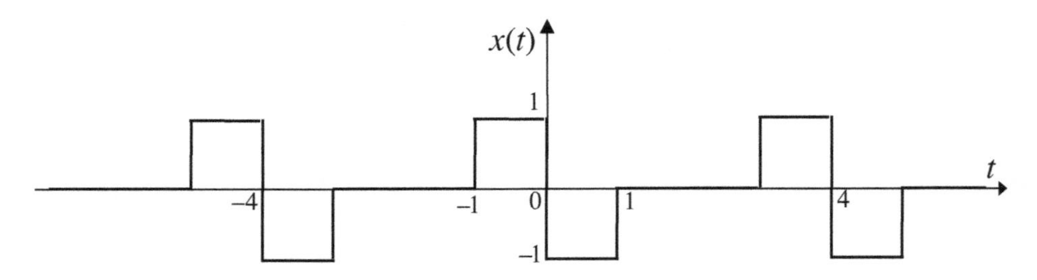

Figure P11.3 Periodic signal for Problem 11.3.

11.4 Let $x(t)$ and $y(t)$ be period-T signals with Fourier series as in Eq. (11.30). Then prove Parseval's relation (11.31) using the orthogonality property (11.27).

11.5 Consider the Dirac delta train $s(t) = \sum_{n=-\infty}^{\infty} \delta_c(t - nT)$ discussed in Sec. 9.5. This is a periodic signal with period T. Find its Fourier series coefficients a_k using the formula (11.7). Based on your answer, give a second proof that the Fourier transform $S(j\omega)$ is indeed as in Eq. (9.57).

11.6 For the signal in Problem 11.1, find the energy in the Fourier series coeffcients a_k, that is, $\sum_k |a_k|^2$.

11.7 Let $x(t)$ and $y(t)$ be period-T signals with Fourier series as in Eq. (11.30). Prove the convolution theorem (11.35) for Fourier series by using the connection between Fourier series and DTFT explained in Sec. 11.3.2, and the fact (Sec. 3.8.1) that multiplication of two sequences $x_1[n]$ and $x_2[n]$ is related to convolution in frequency. *Note:* This may not be simpler than the direct proof in Sec. 11.4.2, but it adds insight.

11.8 Let $x(t)$ and $y(t)$ be period-T signals with Fourier series as in Eq. (11.30). Then prove Eq. (11.36) directly by using

$$c_k = \frac{1}{T}\int_T x(t)y(t)e^{-jk\omega_0 t}dt$$

for the Fourier coefficients of $x(t)y(t)$, and substituting from Eq. (11.30) and simplifying.

11.9 Let $x(t)$ and $y(t)$ be period-T signals with Fourier series as in Eq. (11.30). Then prove Eq. (11.36) using the connection between Fourier series and DTFT (Sec. 11.3.2), and the fact (Sec. 3.4.3) that convolution of two sequences $x[n]$ and $y[n]$ corresponds to multiplication in frequency.

11.10 Using the relations (11.44)–(11.46) and Eq. (11.52), show that the Fourier series coefficients a_0, C_k, and S_k in Eq. (11.41) are indeed given by Eqs. (11.53), (11.54), and (11.55).

11.11 *Parseval's relation for real periodic signal.* For a real periodic signal we know that $x(t)$ can be expressed as in Eq. (11.41). Show that

$$\frac{1}{T}\int_0^T x^2(t)dt = a_0^2 \; + \; 0.5\sum_{k=1}^{\infty} C_k^2 \; + \; 0.5\sum_{k=1}^{\infty} S_k^2.$$

This is Parseval's relation rewritten in terms of a_0, C_k, and S_k.

11.12 In Problem 11.2, since $x(t)$ is real and odd, we know that it can also be written as a Fourier sine series, $x(t) = \sum_{k=1}^{\infty} S_k \sin(k\omega_0 t)$, where $\omega_0 = 2\pi/T$ and S_k are real. Find S_k.

11.13 In Problem 11.3, since $x(t)$ is real and odd, we know that it can also be written as a Fourier sine series, $x(t) = \sum_{k=1}^{\infty} S_k \sin(k\omega_0 t)$, where $\omega_0 = 2\pi/T$ and S_k are real. Find S_k.

11.14 Let $f(t)$ be a possibly infinite-duration signal with CTFT $F(j\omega)$, and let $x(t) = \sum_{k=-\infty}^{\infty} f(t-kT)$. Assume this summation is finite for all t.

(a) Show that $x(t) = x(t-T)$. Thus $x(t)$ is periodic even though it may not be equal to $f(t)$ in the fundamental range $0 \le t < T$, because of overlaps between shifted copies of $f(t)$.

(b) With a_k denoting the Fourier coefficients of $x(t)$, show that $a_k = F(jk\omega_0)/T$, where $\omega_0 = 2\pi/T$. This is similar to Eq. (11.17) with the difference that $x_T(t)$ represents one period of $x(t)$, which $f(t)$ may not.

11.15 Assume $x(t)$ has period T (so that $x(t) = x(t + T)$), with the usual Fourier series $x(t) = \sum_{k=-\infty}^{\infty} a_k e^{jk\omega_0 t}$, where $\omega_0 = 2\pi/T$. Since $x(t) = x(t + 2T)$ as well, suppose someone mistakenly thinks $x(t)$ has period $2T$, and represents it as a Fourier series $x(t) = \sum_{k=-\infty}^{\infty} b_k e^{jk\omega_1 t}$, where $\omega_1 = 2\pi/2T = \pi/T = \omega_0/2$. What is the relation between a_k and b_k?

11.16 Figure P11.16 shows a periodic signal $x(t)$, and a signal $y(t)$ in which every other period in $x(t)$ is removed. Thus, $x(t)$ has period T with Fourier coefficients a_k, and $y(t)$ has period $2T$ with Fourier coefficients b_k. Express a_k in terms of b_k. For example, what are a_0, a_1, and a_2 in terms of b_k?

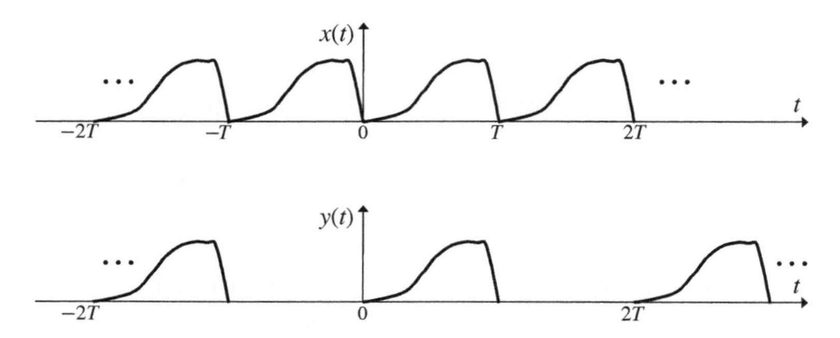

Figure P11.16 Signals for Problem 11.16.

11.17 Referring to Sec. 11.6.4, show that the principal component approximation (11.92) yields a mean square error $\mathcal{E}_{M,pr} \leq \mathcal{E}_M$, where $\mathcal{E}_M$ is the mean square error for $x_M(t) = a_0 + 2\sum_{k=1}^{M} a_k \cos k\omega_0 t$, which is the traditional truncation of the Fourier series.

11.18 Figure P11.18 shows three periodic signals $x_1(t), x_2(t)$, and $x_3(t)$ with periods 2, 2, and 3, respectively. (a) What is the period of the signal $x(t) = x_1(t) + x_2(t)$? (b) What is the period of the signal $y(t) = x_1(t) + x_3(t)$? This problem shows that when we add two periodic signals, the resulting signal can have a smaller or a larger period.

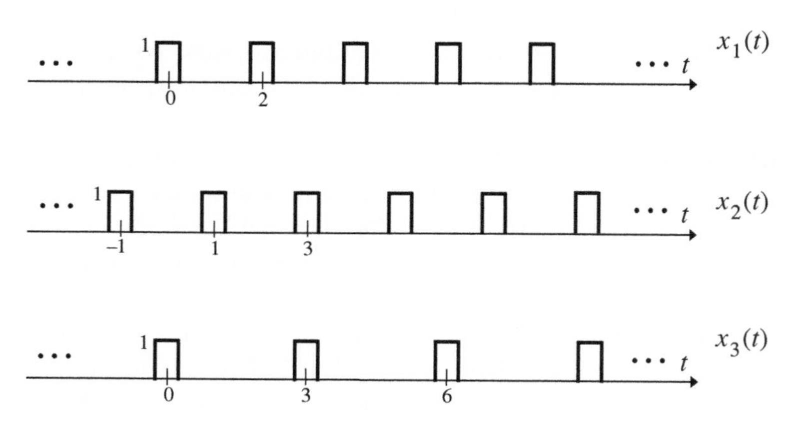

Figure P11.18 Signals for Problem 11.18.

11.19 Continuing with the line of thought in Problem 11.18, the sum of two periodic signals may not be periodic at all. For example, let $s_1(t) = e^{jt}$ and $s_2(t) = e^{j\pi t}$.

(a) What are the periods of $s_1(t)$ and $s_2(t)$?

(b) Let $s(t) = s_1(t) + s_2(t)$. Show that $s(t)$ is not periodic.

We can of course regard $s(t)$ (or any nonperiodic signal) as periodic with infinite period, but that is not especially useful.

11.20 Let $x(t) = 1 + \sum_{k=1}^{3} e^{jk\omega_0 t}$, so that it has period $T = 2\pi/\omega_0$. Suppose we remove the fundamental, and define

$$y(t) = 1 + \sum_{k=2}^{3} e^{jk\omega_0 t}.$$

Then what is the fundamental frequency of $y(t)$, and what is its period?

11.21 What are the fundamental frequencies and periods of the following signals? Assume $\omega_0 > 0$. (a) $x(t) = e^{j4\omega_0 t} + e^{j9\omega_0 t}$, (b) $y(t) = e^{j8\omega_0 t} + e^{j12\omega_0 t}$, and (c) $s(t) = e^{-j8\omega_0 t} + e^{j12\omega_0 t} + e^{j25\omega_0 t}$.

11.22 Let $x(t)$ have period T_x. Let $y(t)$ be the output of an LTI system in response to $x(t)$. Show that $y(t)$ is periodic with period

$$T_y \leq T_x.$$

Thus, an LTI system can never increase the fundamental period although it may be able to decrease it in some cases.

11.23 This problem has two parts.

(a) Find a non-LTI system such that its output $y_1(t)$ in response to any periodic input $x(t)$ is periodic, with period *larger* than the input period.

(b) Find a non-LTI system such that its output $y_2(t)$ in response to any periodic input $x(t)$ is periodic, with period *smaller* than the input period.

11.24 In Problem 11.23(a), let a_k and b_k be the Fourier series coefficients of the input and output. How are these related in the example you came up with? Repeat for part (b).

11.25 *Angle modulation.* Consider a signal of the form

$$x(t) = e^{j\left(\omega_c t + \beta \sin(\omega_m t)\right)} = e^{j\omega_c t} e^{j\beta \sin(\omega_m t)}. \tag{P11.25a}$$

This arises in communication systems which use a method called angle modulation. The familiar frequency modulation used by **FM radio** is a special case of this. The signal $e^{j\omega_c t}$ is called a *carrier* with frequency ω_c. The signal $x(t)$ is an *angle-modulated* version of the carrier because the angle $\omega_c t$ of the carrier is modulated (changed) by the sinusoid $\beta \sin(\omega_m t)$, which is called the *message* signal. The modulation index $\beta \geq 0$ determines how strong the modulation is. We now perform a Fourier analysis of $x(t)$. Let

$$x_m(t) = e^{j\beta \sin(\omega_m t)}.$$

(a) Since $\sin(\omega_m t)$ is periodic with period $T = 2\pi/\omega_m$, we have $x_m(t) = x_m(t + T)$, so $x_m(t)$ can be expressed as a Fourier series. Show that the

Fourier coefficients are $a_n = J_n(\beta)$, where

$$J_n(\beta) = \frac{1}{2\pi} \int_{-\pi}^{\pi} e^{j(\beta \sin u - nu)} du. \qquad \text{(P11.25b)}$$

The function $J_n(\beta)$ is called the nth-order **Bessel function** of the first kind. The modulated signal $x(t)$ can therefore be written

$$x(t) = \sum_{n=-\infty}^{\infty} J_n(\beta) e^{j(\omega_c + n\omega_m)t}. \qquad \text{(P11.25c)}$$

This shows that if a single-frequency signal $e^{j\omega_c t}$ is angle modulated with a sinusoid $\sin(\omega_m t)$, the resulting signal is a superposition of an *infinite* number of single-frequency signals with frequencies

$$\omega_c, \quad \omega_c \pm \omega_m, \quad \omega_c \pm 2\omega_m, \quad \omega_c \pm 3\omega_m, \quad \ldots.$$

The coefficient $J_n(\beta)$ above determines the amplitude of the newly generated frequency $\omega_c + n\omega_m$. Angle modulation should be compared with amplitude modulation (AM, Sec. 3.8.1), which shifts the frequency band, but does not create infinitely many new frequencies.

(b) Show that we can also write $J_n(\beta)$ in the form

$$J_n(\beta) = \frac{1}{\pi} \int_{0}^{\pi} \cos(\beta \sin u - nu) du, \qquad \text{(P11.25d)}$$

which shows that it is real for real β. Figure P11.25 shows plots of $J_n(\beta)$ for three different values of the modulation index β. Notice how the number of significant harmonics increases with β.

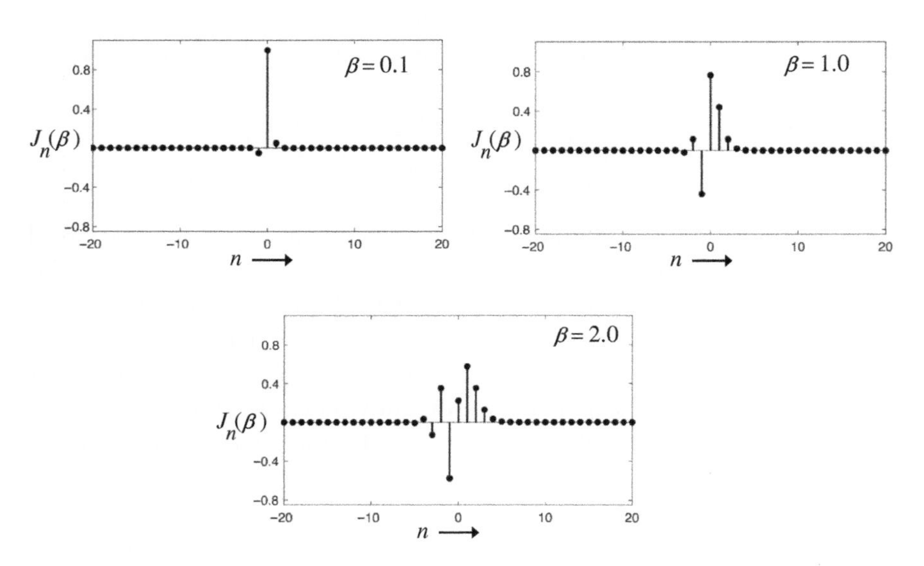

Figure P11.25 Plots of $J_n(\beta)$ for various values of the modulation index β.

11.26 *Angle modulation, more general.* A generalization of the angle-modulated signal in Eq. (P11.25a) is

$$y(t) = e^{j\left(\omega_c t + s(t)\right)},$$

where $s(t)$ is a message signal. Using the fact that

$$e^z = 1 + z + \frac{z^2}{2!} + \frac{z^3}{3!} \cdots$$

for any z such that $|z| < \infty$, show that $y(t)$ in general has infinite bandwidth even if $s(t)$ is σ-BL.

11.27 (Computing assignment) In Problem 11.14, let $f(t) = e^{-t^2/2}$ and $T = 2$. Plot $x(t)$ for $-5 \leq t \leq 5$, using a step size $\Delta t = 0.01$ for the plot. Also compute and plot a_k for $-3 \leq k \leq 3$.

11.28 (Computing assignment) *Truncation error in Fourier series.* We defined the Fourier series truncation errors $\mathcal{E}_M$ (direct truncation) and $\mathcal{E}_{M,pr}$ (principal component truncation) in Secs. 11.6.2 and 11.6.4. For the signal $x(t)$ given in Problem 11.2 with $T = 1$ and $a = 0.3$, plot these errors for $1 \leq M \leq 50$. Plot the two truncated signals $x_M(t)$ and $x_{M,pr}(t)$ (direct truncation and principal component truncation) for some typical values of M (say $M = 10, 50$, and 100).

11.29 (Computing assignment) *Angle-modulated signals.* In Problem 11.25 we found that the Bessel functions $J_n(\beta)$ play a role in the representation of angle-modulated signals. We also plotted $J_n(\beta)$ for some $\beta \leq 2$. Plot $J_n(\beta)$ for $\beta = 3$ and $\beta = 6$. Notice that the number of significant harmonics is even larger now. If you are using MATLAB, the command "besselj" is useful here.

12 The DFT and the FFT

12.1 Introduction

In this chapter we introduce the discrete Fourier transform (DFT), which should not be confused with the discrete-time Fourier transform (DTFT) introduced in Chapter 3. Among all the types of FT discussed in this book, the DFT is the only FT that can be computed with a finite number of operations. In short, it is the only *computable* Fourier transform (Sec. 12.7.1). The fast Fourier transform or FFT, introduced by Cooley and Tukey [1965], is a fast algorithm to compute the DFT. Without the help of the FFT, computation of the Fourier transforms of large sets of data would be impractical. The digital signal processing revolution of the 1960s was possible mainly because of the introduction of the FFT. In the last 60 years there have been many variations of the FFT, but the simplest version, called the radix-2 FFT, is still an integral part of many signal processing platforms.

This chapter will introduce the DFT, the radix-2 FFT, and a number of their properties. We will also introduce circular convolution, which has a special place in DFT theory, and explain the connection to ordinary convolution. Circular convolution paves the way for fast (FFT-like) algorithms for ordinary convolution. Using the FFT, the Fourier transform of a finite-duration signal can be computed efficiently at various desired resolutions in frequency, regardless of the data length, as we shall explain. The connections between the different types of Fourier transforms introduced in this book will be summarized in Sec. 12.7. A summary of properties of the DFT is given in Sec. 12.9 for quick reference.

12.2 Definition and Basic Properties

Given a sequence of N numbers (Fig. 12.1(a))

$$x[n], \qquad 0 \leq n \leq N-1, \tag{12.1}$$

its discrete Fourier transform or **DFT** is defined as

$$X[k] = \sum_{n=0}^{N-1} x[n]e^{-j2\pi kn/N}, \qquad 0 \leq k \leq N-1. \tag{12.2}$$

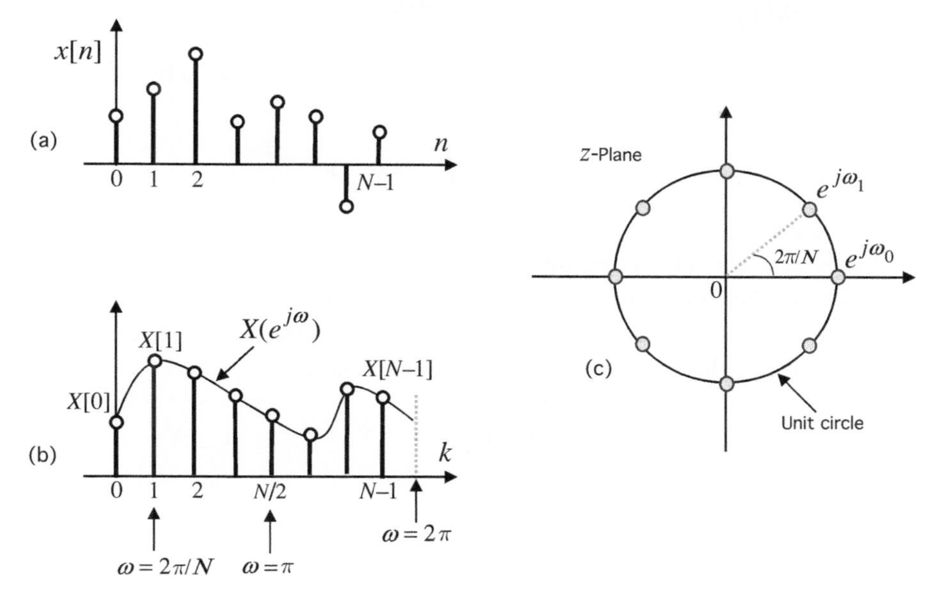

Figure 12.1 (a) A finite-duration signal $x[n]$, (b) its DFT $X[k]$, and (c) the N unit-circle points in the z-plane, representing the frequencies where the Fourier transform is evaluated by the DFT. Here $N = 8$.

Thus the DFT is also an N-point sequence. It is readily verified that the DFT is a *linear transform*. If we consider $x[n]$ to be an infinite sequence which is zero for all n outside the interval $0 \leq n \leq N-1$, then we know that its DTFT (Sec. 3.4.1) is given by

$$X(e^{j\omega}) = \sum_{n=0}^{N-1} x[n]e^{-j\omega n}. \tag{12.3}$$

Comparing with Eq. (12.2) it follows that

$$X[k] = X\left(e^{j2\pi k/N}\right), \tag{12.4}$$

that is, the DFT coefficients $X[k]$ are samples of the DTFT $X(e^{j\omega})$, evaluated at the uniformly spaced frequencies

$$\omega_k = \frac{2\pi k}{N}, \quad 0 \leq k \leq N-1, \tag{12.5}$$

as shown in Fig. 12.1(b). The discrete set of frequencies (12.5) is called the **DFT grid**. In the z-plane, these are uniformly spaced points on the unit circle as demonstrated in Fig. 12.1(c) for $N = 8$.

Comment on notation. Even though the letter X is used both in $X[k]$ and $X(e^{j\omega})$, these are two different functions. The **square brackets** are a reminder that k in $X[k]$ is an integer. The round brackets in $X(e^{j\omega})$ allow the argument to be a continuous (complex) quantity. The relation between these two functions is given by Eq. (12.4).

12.2.1 The Inverse DFT

We know (Sec. 3.5) that $x[n]$ can be recovered from $X(e^{j\omega})$ by using the inverse FT relation

$$x[n] = \int_0^{2\pi} X(e^{j\omega})e^{j\omega n}\frac{d\omega}{2\pi}. \qquad (12.6)$$

The use of this integral requires a knowledge of the Fourier transform for all ω in $0 \leq \omega < 2\pi$. However, since $x[n]$ has finite duration N, it should be intuitive that we can recover it just by knowing the N samples $X[k]$ (instead of having to know $X(e^{j\omega})$ for all ω in $[0, 2\pi)$). This intuition is indeed correct. In fact, we will show in Sec. 12.2.2 that the inverse DFT or **IDFT** relation is

$$x[n] = \frac{1}{N}\sum_{k=0}^{N-1} X[k]e^{j2\pi kn/N}, \qquad 0 \leq n \leq N-1. \qquad (12.7)$$

The notation

$$W_N \triangleq e^{-j2\pi/N} \qquad (12.8)$$

will be used for simplicity, with the subscript N deleted when it is clear from the context. Thus the DFT and the IDFT can be written in the form

$$X[k] = \sum_{n=0}^{N-1} x[n]W^{kn} \qquad \text{(DFT)}, \qquad (12.9)$$

$$x[n] = \frac{1}{N}\sum_{k=0}^{N-1} X[k]W^{-nk} \qquad \text{(IDFT)}. \qquad (12.10)$$

The IDFT is similar to the DFT except for the scale factor $1/N$ and the sign on the exponent. Note that

$$W^N = 1, \qquad (12.11)$$

that is, W is an Nth **root of unity**. Many of the properties of the DFT follow from the properties of W, and we will elaborate on some of these below. As in earlier chapters, the notation

$$x[n] \longleftrightarrow X[k] \qquad (12.12)$$

is used to indicate that $x[n]$ and $X[k]$ form a DFT pair. The integer k is regarded as the DFT frequency and has N distinct values. As we know, the actual frequencies are $\omega_k = 2\pi k/N$.

Notice that the DFT and IDFT are both finite summations. The DFT is a **linear transformation** from a set of N numbers $x[n]$ to a set of N numbers $X[k]$. The IDFT is the inverse of this transformation. From Eqs. (12.9) and (12.10) it is clear that the DFT and the IDFT can be computed with a finite number of multiplications and additions. Notice that the other types of Fourier transforms and inverses we have seen in this book involve either integrals or infinite summations and cannot be computed with a finite number of computations. So the DFT–IDFT pair is the only **computable** Fourier transform pair we have seen so far in this book! There are other computable transforms, such as the DCT (discrete cosine transform) and WHT (Walsh–Hadamard transform), that are commonly used in digital signal processing.

12.2.2 Properties of *W*

Since $W^m = e^{-j2\pi m/N}$, the following N powers of W:

$$1, W, W^2, \ldots, W^{N-1}, \tag{12.13}$$

are distinct, and are in fact the N points on the unit circle of the z-plane representing the DFT grid (Fig. 12.1(c)). These are precisely the N distinct Nth roots of unity. Observe now that for any integer m, the quantity W^m can be rewritten as

$$W^m = W^k, \tag{12.14}$$

where $0 \le k \le N-1$, just by making repeated use of $W^N = 1$. For example, if $N = 4$, then since $W^4 = 1$, we have $W^{10} = W^{(2+2*4)} = W^2$; similarly, $W^{-1} = W^{-1+4} = W^3$. Thus, W^m can be rewritten as one of the N roots of unity in Eq. (12.13). In particular, therefore, the quantities W^{kn} and W^{-kn} in Eqs. (12.9) and (12.10) are all Nth roots of unity. Since $W^m = e^{-j2\pi m/N}$, it follows that

$$W^m = 1 \iff m = \text{multiple of } N. \tag{12.15}$$

Another important property of W is that

$$\sum_{k=0}^{N-1} W^{mk} = \begin{cases} 0 & \text{for } m \text{ not a multiple of } N, \\ N & \text{for } m \text{ a multiple of } N. \end{cases} \tag{12.16}$$

This is obvious when m is a multiple of N, since $W^{mk} = 1$ in that case. To prove this when m is not a multiple of N, note that since $W^m \neq 1$ in this case, we have

$$\sum_{k=0}^{N-1} W^{mk} = \frac{1 - W^{mN}}{1 - W^m} = 0, \tag{12.17}$$

since $W^N = 1$. This proves the claim. Equation (12.16) implies in particular that, when $0 \le n, l \le N-1$,

$$\sum_{k=0}^{N-1} W^{(\ell-n)k} = \begin{cases} 0 & \text{for } \ell \neq n, \\ N & \text{for } \ell = n. \end{cases} \tag{12.18}$$

By using this, the IDFT formula (12.10) is readily proved.

Proof of the IDFT formula. The sum on the right-hand side of Eq. (12.10) can be rewritten as

$$\begin{aligned} \sum_{k=0}^{N-1} X[k] W^{-nk} &= \sum_{k=0}^{N-1} W^{-nk} \sum_{\ell=0}^{N-1} x[\ell] W^{\ell k} \quad \text{(from Eq. (12.9))} \\ &= \sum_{\ell=0}^{N-1} x[\ell] \sum_{k=0}^{N-1} W^{(\ell-n)k} = N x[n] \quad \text{(from Eq. (12.18))}. \end{aligned}$$

This proves Eq. (12.10). ▽ ▽ ▽

Equation (12.18) is referred to as the **orthogonality** property. To see why, define the column vectors

$$\mathbf{w}_m = \begin{bmatrix} 1 \\ W^m \\ W^{2m} \\ \vdots \\ W^{(N-1)m} \end{bmatrix}, \quad 0 \leq m \leq N-1. \tag{12.19}$$

Then Eq. (12.18) is equivalent to

$$\mathbf{w}_n^H \mathbf{w}_\ell = N\delta[n-\ell], \tag{12.20}$$

that is, the vectors $\mathbf{w}_n$ and $\mathbf{w}_\ell$ are orthogonal unless $n = \ell$.

12.2.3 Some Basic Examples

The **unit pulse** $x[n] = \delta[n]$ has the DFT

$$X[k] = \sum_{n=0}^{N-1} x[n]W^{nk} = 1, \quad 0 \leq k \leq N-1. \tag{12.21}$$

Thus the DFT is a constant, as shown in Fig. 12.2(a). Conversely, the **constant signal** $x[n] = 1, \quad 0 \leq n \leq N-1$, has the DFT shown in Fig. 12.2(b), which is the unit pulse scaled by N (Problem 12.2). Thus

$$x[n] = \delta[n] \longleftrightarrow X[k] = 1 \ \forall k \quad \text{and} \quad x[n] = 1 \ \forall n \longleftrightarrow X[k] = N\delta[k]. \tag{12.22}$$

Remember that in all examples on DFT, the arguments are always in the finite range $0 \leq n, k \leq N-1$. From the two examples in Eq. (12.22) it follows that there is a duality between time and frequency domains, which is not surprising because Eqs. (12.9) and (12.10) are so similar. Next consider the complex sinusoid

$$x[n] = e^{j\frac{2\pi k_0}{N}n}, \quad 0 \leq n \leq N-1, \tag{12.23}$$

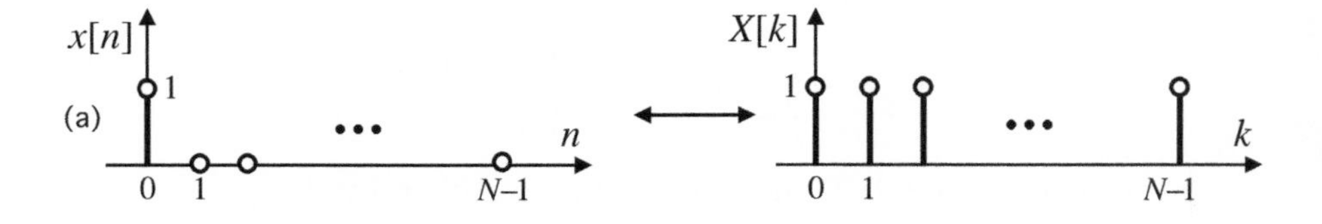

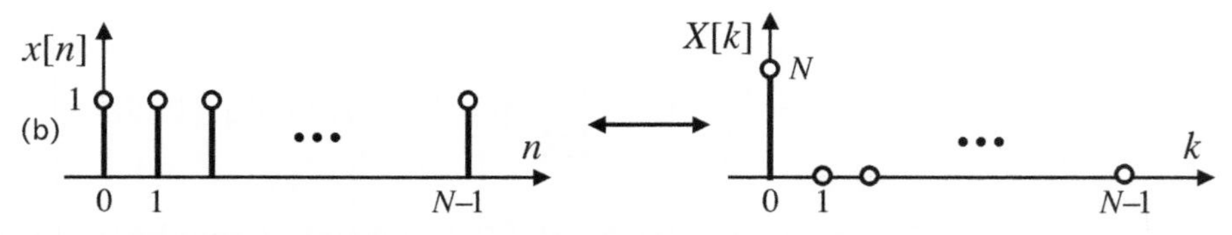

Figure 12.2 Examples of DFT pairs. (a) The unit pulse and its DFT; (b) the constant signal and its DFT. Note that both $x[n]$ and $X[k]$ are defined in the range $0 \leq n, k \leq N-1$.

where k_0 is an integer with $0 \le k_0 \le N-1$. This is a single-frequency signal with frequency $2\pi k_0/N$ on the DFT grid. Then we can rewrite $x[n] = W^{-k_0 n}$, and we have

$$X[k] = \sum_{n=0}^{N-1} W^{-k_0 n} W^{kn} = \sum_{n=0}^{N-1} W^{(k-k_0)n} = N\delta[k-k_0], \tag{12.24}$$

where we have used Eq. (12.16). Thus

$$e^{j\frac{2\pi k_0}{N}n} \quad \longleftrightarrow \quad N\delta[k-k_0]. \tag{12.25}$$

Similarly it can be shown that

$$\delta[n-n_0] \longleftrightarrow e^{-j\frac{2\pi k}{N}n_0}. \tag{12.26}$$

By using Eq. (12.25) we can readily derive the DFT for the cosine signal and sine signal with frequency $2\pi k_0/N$ (for integer k_0):

$$\begin{aligned} \cos\left(\frac{2\pi k_0}{N}n\right) &\longleftrightarrow 0.5N\Big(\delta[k-k_0] + \delta[k-(N-k_0)]\Big), \\ \sin\left(\frac{2\pi k_0}{N}n\right) &\longleftrightarrow -0.5Nj\Big(\delta[k-k_0] - \delta[k-(N-k_0)]\Big). \end{aligned} \tag{12.27}$$

To see how the first property comes about, just observe that

$$\cos\left(\frac{2\pi k_0}{N}n\right) = 0.5\left(e^{j\frac{2\pi k_0}{N}n} + e^{-j\frac{2\pi k_0}{N}n}\right) = 0.5\left(e^{j\frac{2\pi k_0}{N}n} + e^{j\frac{2\pi(N-k_0)}{N}n}\right),$$

where we have used the fact that $e^{j2\pi} = 1$. The second property follows similarly.

12.2.3.1 The Leakage Phenomenon

Next consider the single-frequency signal

$$x[n] = e^{j\widehat{\omega}n}, \quad 0 \le n \le N-1, \tag{12.28}$$

truncated to N samples. We know from Chapter 3 that this has DTFT

$$X(e^{j\omega}) = e^{j(N-1)(\widehat{\omega}-\omega)/2}\left(\frac{\sin(N(\widehat{\omega}-\omega)/2)}{\sin((\widehat{\omega}-\omega)/2)}\right). \tag{12.29}$$

So we can readily evaluate the DFT $X[k]$ using Eq. (12.4), that is,

$$X[k] = e^{j(N-1)(\widehat{\omega}-(2\pi k/N))/2}\left(\frac{\sin(N(\widehat{\omega}-(2\pi k/N))/2)}{\sin((\widehat{\omega}-(2\pi k/N))/2)}\right). \tag{12.30}$$

Since $\sin(N\theta/2)$ has zero crossings at $\theta_m = 2\pi m/N$, the zero crossings of the plot of $X(e^{j\omega})$ are located uniformly at frequencies which are separated from $\widehat{\omega}$ by $2\pi m/N$ for integer m. Figure 12.3(a) shows a plot of $|X(e^{j\omega})|$. The DFT grid $\omega_k = 2\pi k/N$ is also shown by the vertical set of lines for reference. In this figure it is assumed that $\widehat{\omega}$ coincides with a DFT frequency, say $2\pi k_0/N$, as in Eq. (12.23). So, all the zero crossings of Eq. (12.29) are exactly at the DFT frequencies, and therefore the DFT

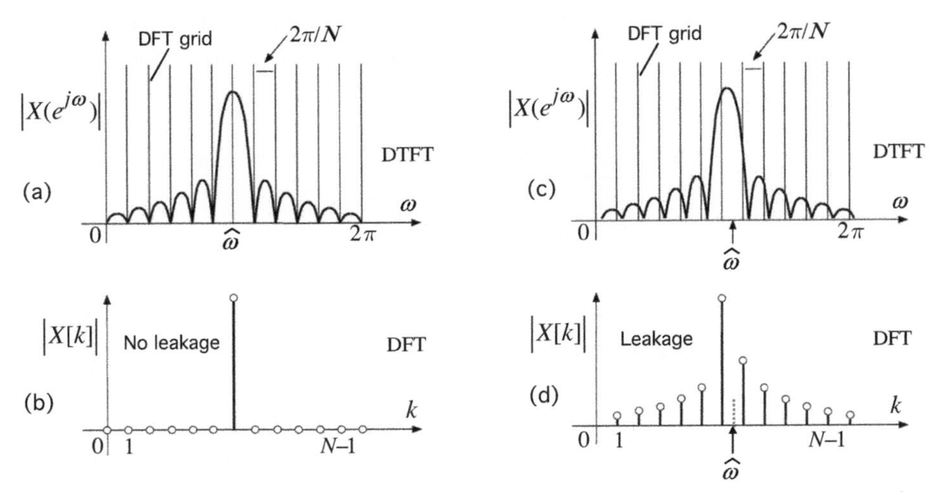

Figure 12.3 The leakage phenomenon. (a) The DTFT magnitude $|X(e^{j\omega})|$ of a signal $e^{j\widehat{\omega}n}$, when the frequency $\widehat{\omega}$ is aligned to a DFT frequency; (b) the corresponding DFT magnitude. (c) The DTFT magnitude of $e^{j\widehat{\omega}n}$, when the frequency $\widehat{\omega}$ is *not* aligned to any DFT frequency; (d) the corresponding DFT magnitude.

$X[k]$ has a single line as shown in Fig. 12.3(b). The frequency $\widehat{\omega}$ comes through as a clean line if we do a DFT analysis of $x[n]$.

Next consider the situation shown in Fig. 12.3(c). Here $\widehat{\omega}$ falls **in between** two DFT frequencies. Now the zero crossings of Eq. (12.29) do not coincide with the DFT frequencies. So when we sample $X(e^{j\omega})$ to obtain $X[k]$, many nonzero samples are produced, as shown in Fig. 12.3(d). Thus, a single frequency $\widehat{\omega}$ does not come through as a clean line when we do a DFT analysis. Instead, the energy in the frequency $\widehat{\omega}$ "leaks" into a number of lines close to $\widehat{\omega}$. This is called the **leakage phenomenon** in DFT analysis. In short, when a frequency $\widehat{\omega}$ does not coincide with a point on the DFT grid, the energy in that frequency spreads into multiple lines in the DFT plot. See Problems 12.32 and 12.33 for more demonstrations.

12.3 The DFT Matrix

The DFT is a linear transform that produces N numbers $X[k]$ from the N numbers $x[n]$, according to Eq. (12.9). We can express the transformation in the form of a matrix-vector multiplication.[1] For example, when $N = 4$ we have

$$\underbrace{\begin{bmatrix} X[0] \\ X[1] \\ X[2] \\ X[3] \end{bmatrix}}_{\mathbf{X}} = \underbrace{\begin{bmatrix} 1 & 1 & 1 & 1 \\ 1 & W & W^2 & W^3 \\ 1 & W^2 & W^4 & W^6 \\ 1 & W^3 & W^6 & W^9 \end{bmatrix}}_{\mathbf{W}} \underbrace{\begin{bmatrix} x[0] \\ x[1] \\ x[2] \\ x[3] \end{bmatrix}}_{\mathbf{x}}. \tag{12.31}$$

[1] A review of matrix notations from Sec. 1.3 will be helpful here. In particular $\mathbf{A}^T$, $\mathbf{A}^*$, and $\mathbf{A}^H$ represent, respectively, the transpose, conjugate, and transpose-conjugate of $\mathbf{A}$.

More generally, the N-point DFT is the matrix transformation given by

$$\mathbf{X} = \mathbf{W}\mathbf{x}, \tag{12.32}$$

where $\mathbf{x}$ and $\mathbf{X}$ are column vectors defined as

$$\mathbf{x} = \begin{bmatrix} x[0] \\ x[1] \\ \vdots \\ x[N-1] \end{bmatrix}, \quad \mathbf{X} = \begin{bmatrix} X[0] \\ X[1] \\ \vdots \\ X[N-1] \end{bmatrix}, \tag{12.33}$$

and $\mathbf{W}$ is an $N \times N$ matrix called the **DFT matrix**. We now discuss some properties of this matrix.

1. All the elements of $\mathbf{W}$ are powers of $W = e^{-j2\pi/N}$. More precisely, the (k, n)th element of $\mathbf{W}$ is

$$[\mathbf{W}]_{kn} = W^{kn} = e^{-j2\pi kn/N}. \tag{12.34}$$

   The elements of $\mathbf{W}$ are in general complex.
2. *Symmetry*. Since $[\mathbf{W}]_{kn} = [\mathbf{W}]_{nk}$ the matrix $\mathbf{W}$ is symmetric, that is, $\mathbf{W}^T = \mathbf{W}$. However, in general it is not Hermitian, that is, $\mathbf{W}^H \neq \mathbf{W}$. This is because $\mathbf{W}^H = (\mathbf{W}^T)^* = \mathbf{W}^* \neq \mathbf{W}$ in general.
3. The mth column of $\mathbf{W}$ is of the form $\mathbf{w}_m$ defined in Eq. (12.19), that is, we can write $\mathbf{W}$ in the form

$$\mathbf{W} = \begin{bmatrix} \mathbf{w}_0 & \mathbf{w}_1 & \cdots & \mathbf{w}_{N-1} \end{bmatrix}. \tag{12.35}$$

4. *Vandermonde property*. Note that each column $\mathbf{w}_m$ of the DFT matrix is a **Vandermonde** vector, that is, it has the form $\begin{bmatrix} 1 & \alpha & \alpha^2 & \cdots & \alpha^{N-1} \end{bmatrix}^T$. A matrix, all of whose columns are Vandermonde vectors, is said to be a Vandermonde matrix (Sec. 5.16). So the DFT matrix $\mathbf{W}$ is a Vandermonde matrix. For example, inspect $\mathbf{W}$ in Eq. (12.31).
5. *Reduced form*. As mentioned in Sec. 12.2.2, any power W^m of W can be rewritten as $W^m = W^k$, where $0 \leq k \leq N-1$. So, the matrix $\mathbf{W}$ can always be reduced to a form where the elements are the Nth roots of unity $1, W, W^2, \ldots, W^{N-1}$. For example, $\mathbf{W}$ in Eq. (12.31) can be rewritten as

$$\mathbf{W} = \begin{bmatrix} 1 & 1 & 1 & 1 \\ 1 & W & W^2 & W^3 \\ 1 & W^2 & W^4 & W^6 \\ 1 & W^3 & W^6 & W^9 \end{bmatrix} = \begin{bmatrix} 1 & 1 & 1 & 1 \\ 1 & W & W^2 & W^3 \\ 1 & W^2 & 1 & W^2 \\ 1 & W^3 & W^2 & W \end{bmatrix}. \tag{12.36}$$

   From the reduced form on the right side it is not obvious that $\mathbf{W}$ is Vandermonde, but the left-hand side clearly shows that it is.
6. *Unitary property*. We now claim that

$$\frac{1}{N}\mathbf{W}^H\mathbf{W} = \mathbf{I}_N. \tag{12.37}$$

That is, $\mathbf{W}/\sqrt{N}$ is a unitary matrix (see Chapter 18). To prove this, just observe that the columns of $\mathbf{W}$ are pairwise orthogonal, as shown by Eq. (12.20). So the (n, ℓ) element of $\mathbf{W}^H\mathbf{W}$ is

$$[\mathbf{W}^H\mathbf{W}]_{n,\ell} = \mathbf{w}_n^H\mathbf{w}_\ell = N\delta[n-\ell], \tag{12.38}$$

which proves Eq. (12.37).

12.3.1 Inverse of the DFT Matrix

The unitary property (12.37) implies in particular that $\mathbf{W}$ is invertible. So, using Eq. (12.32) we can express the inverse DFT operation in the form

$$\mathbf{x} = \mathbf{W}^{-1}\mathbf{X}. \tag{12.39}$$

The inversion of $\mathbf{W}$ does not require any computation. From the unitary property (12.37) it readily follows that $\mathbf{W}^{-1}$ has the closed form

$$\mathbf{W}^{-1} = \frac{\mathbf{W}^H}{N} = \frac{\mathbf{W}^*}{N}. \tag{12.40}$$

For example, if $N = 4$ we have $W = e^{-j2\pi/4} = -j$. So the DFT matrix and its inverse (the IDFT matrix) are

$$\mathbf{W} = \begin{bmatrix} 1 & 1 & 1 & 1 \\ 1 & -j & -1 & j \\ 1 & -1 & 1 & -1 \\ 1 & j & -1 & -j \end{bmatrix}, \quad \mathbf{W}^{-1} = \frac{1}{4}\begin{bmatrix} 1 & 1 & 1 & 1 \\ 1 & j & -1 & -j \\ 1 & -1 & 1 & -1 \\ 1 & -j & -1 & j \end{bmatrix}. \tag{12.41}$$

Since $\mathbf{W} = \mathbf{W}^T$, we have $\mathbf{W}^H = \mathbf{W}^*$, so the inverse transform (12.39) can be written in a number of ways:

$$\mathbf{x} = \mathbf{W}^{-1}\mathbf{X} = \frac{1}{N}\mathbf{W}^H\mathbf{X} = \frac{1}{N}\mathbf{W}^*\mathbf{X}. \tag{12.42}$$

The reader can easily verify that this expression is equivalent to Eq. (12.10) (Problem 12.5).

12.3.2 Parseval's Relation for DFT

We conclude this section by deriving Parseval's relation for the DFT. The reader will recall that Parseval's relation has been derived earlier for other types of Fourier transforms, and it expresses a relation between *inner products* computed in the time and frequency domains. For the case of the DFT, this relation is

$$\sum_{k=0}^{N-1} X[k]Y^*[k] = N\sum_{n=0}^{N-1} x[n]y^*[n]. \tag{12.43}$$

This implies in particular that

$$\sum_{k=0}^{N-1} |X[k]|^2 = N\sum_{n=0}^{N-1} |x[n]|^2. \tag{12.44}$$

That is, the energy in the finite-duration sequence is equal to the energy in the DFT sequence, except for the scale factor N. This energy conservation property, and more generally the inner product conservation (12.43), follow directly from the unitary property (12.37). To see this, write the DFT relations in vector form:

$$\mathbf{Y} = \mathbf{W}\mathbf{y}, \quad \mathbf{X} = \mathbf{W}\mathbf{x}. \tag{12.45}$$

Then

$$\mathbf{Y}^H\mathbf{X} = \mathbf{y}^H\mathbf{W}^H\mathbf{W}\mathbf{x} = N\mathbf{y}^H\mathbf{x}, \tag{12.46}$$

where we have used $\mathbf{W}^H\mathbf{W} = N\mathbf{I}$. Written in terms of the components of the vectors, this is precisely Eq. (12.43).

12.4 The Fast Fourier Transform

The fast Fourier transform or FFT is a fast tool to compute the DFT. As mentioned earlier, its development has created a revolution in the history of signal processing. It was introduced by Cooley and Tukey [1965], although its history can be traced back to the great mathematician and scientist Carl F. Gauss [Heideman et al., 1984]. A fascinating historical account of the FFT has been given by Cooley et al. [1969]. Recall that the DFT of an N-point sequence is given by

$$X[k] = \sum_{n=0}^{N-1} x[n]W_N^{kn}, \quad 0 \leq k \leq N-1. \tag{12.47}$$

In the discussions so far, we replaced W_N by the simpler notation W. But for development of the FFT, it is convenient to keep the subscript N.

For any fixed k, it is clear from Eq. (12.47) that there are N complex multiplications[2] and $N-1$ additions in the computation of $X[k]$. The computation of $X[k]$ for all k therefore involves N^2 multiplications and $N(N-1)$ additions. So we say that the computational complexity is $O(N^2)$. The fast Fourier transform (FFT) algorithm exploits the properties of the quantity

$$W_N \stackrel{\Delta}{=} e^{-j2\pi/N}, \tag{12.48}$$

to reduce the complexity. For the case where N is a power of two, the complexity becomes approximately $0.5N\log_2 N$. For example, if $N = 2^{10} = 1024$, we have

$$N^2 \approx 10^6, \qquad \text{whereas} \qquad 0.5N\log_2 N = 512 \times 10 = 5120, \tag{12.49}$$

which shows that the complexity is reduced by two orders of magnitude! Even when N is not a power of two, the savings can still be substantial if not dramatic (Problem 12.21).

[2] It is true that some of these multipliers are trivial, for example, $W^0 = 1$, $W^{N/2} = -1$, and so forth, and they need not be counted. But these contribute only to minor savings; we shall pay attention to them later.

12.4.1 The Basic Idea Behind FFT

We explain the basic idea by assuming N is a power of two, that is,

$$N = 2^M \tag{12.50}$$

for some integer $M > 0$. The corresponding FFT algorithm is called the **radix-2 FFT** algorithm. In the development we will keep track of multiplications, and worry about additions later. Direct computation of the N-point DFT (12.47) requires N^2 multiplications. Instead of direct computation, suppose we rewrite

$$X[k] = \sum_{n=0}^{N/2-1} x[2n]W_N^{2kn} + \sum_{n=0}^{N/2-1} x[2n+1]W_N^{(2n+1)k}, \quad 0 \le k \le N-1, \tag{12.51}$$

where all the even samples $x[2n]$ have been grouped under one summation and the odd samples $x[2n+1]$ under the other. In the above, $N/2$ is an integer because N is even. Now, from the definition of W_N it follows that

$$W_N^2 = e^{-j2*2\pi/N} = e^{-j2\pi/(N/2)} = W_{N/2}. \tag{12.52}$$

We can therefore further rewrite

$$X[k] = \underbrace{\sum_{n=0}^{N/2-1} x[2n]W_{N/2}^{kn}}_{\text{call this } X_1[k]} + W_N^k \underbrace{\sum_{n=0}^{N/2-1} x[2n+1]W_{N/2}^{nk}}_{\text{call this } X_2[k]}, \quad 0 \le k \le N-1.$$

That is,

$$X[k] = X_1[k] \ + \ W_N^k X_2[k], \qquad 0 \le k \le N-1. \tag{12.53}$$

This leads to the following important observations:

1. $X_1[k]$ can be regarded as the $N/2$-point DFT of the $N/2$-point sequence $x[2n]$ of even samples. Similarly, $X_2[k]$ is the $N/2$-point DFT of the odd samples $x[2n+1]$. As k increases from 0 to $N/2-1$, we obtain

$$X_1[0],\ X_1[1],\ \ldots,\ X_1[N/2-1], \tag{12.54}$$

   but after this, the same values of $X_1[k]$ repeat for $k = N/2, \ldots, N-1$. This is because $W_{N/2}^{N/2} = 1$. The same is true for $X_2[k]$ as well.
2. The factor W_N^k in Eq. (12.53) *does* have N distinct values as k increases from 0 to $N-1$. This is what makes it possible for $X[k]$ to have N distinct values, even though $X_1[k]$ and $X_2[k]$ can only have $N/2$ distinct values each.
3. The factors W_N^k in Eq. (12.53), which are used to combine the two smaller DFTs into the original DFT, are said to be **twiddle factors** in the FFT algorithm.

In view of Eq. (12.53), we can first compute the $N/2$-point DFTs $X_1[k]$ and $X_2[k]$ and then compute $X[k]$ using Eq. (12.53). The most important point here is that the $N/2$-point DFTs $X_1[k]$ and $X_2[k]$ require only about $(N/2)^2$ multiplications each. After this, the use of Eq. (12.53) requires about N multiplications due to the twiddle factors W_N^k. In fact, this can be reduced to $N/2$ multiplications by observing that for $k \geq N/2$ we can write

$$W_N^k = W_N^{N/2} W_N{}^{\ell} = -W_N^{\ell}, \tag{12.55}$$

for some integer ℓ in $0 \leq \ell \leq N/2 - 1$. Thus, the computation (12.53) can be rearranged as

$$\begin{aligned} X[k] &= X_1[k] + W_N^k X_2[k], \quad 0 \leq k \leq N/2 - 1, \\ X[k + N/2] &= X_1[k] - W_N^k X_2[k], \quad 0 \leq k \leq N/2 - 1. \end{aligned} \tag{12.56}$$

Figure 12.4 demonstrates this idea for $N = 8$. Here the eight-point DFT has been rewritten as two four-point DFTs and four twiddle factors. The total number of multiplications in the computation of $X[k]$ from Eq. (12.56) would be

$$\left(\frac{N}{2}\right)^2 + \left(\frac{N}{2}\right)^2 + \frac{N}{2} = \frac{N^2}{2} + \frac{N}{2}, \tag{12.57}$$

which is nearly $N^2/2$ for large N. This is already a significant saving compared to N^2 multiplications required in the direct evaluation of the DFT. For example, if

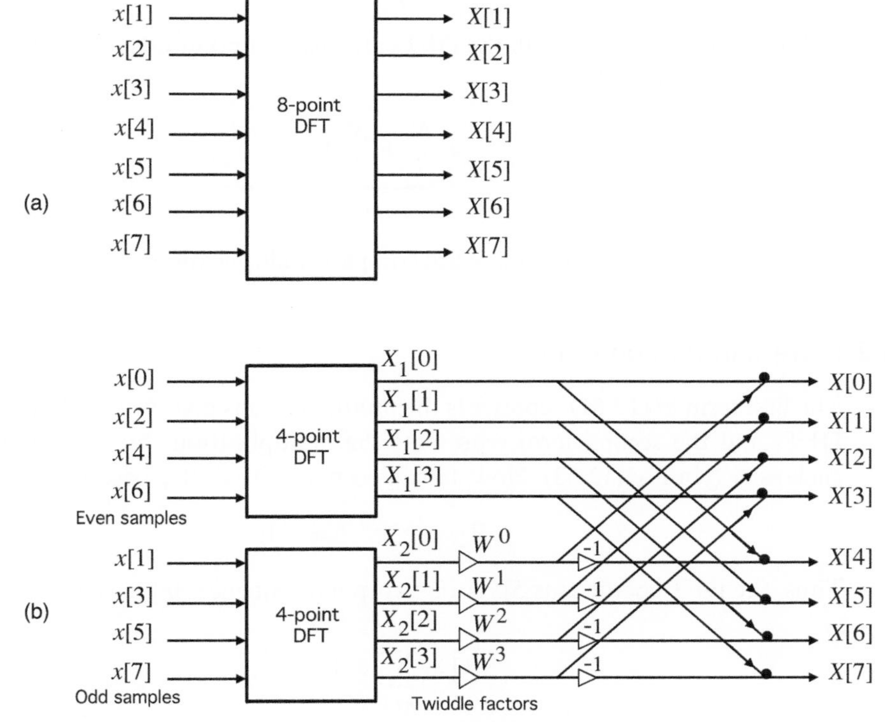

Figure 12.4 (a) Schematic of an eight-point DFT implementation and (b) faster computation of an eight-point DFT using two four-point DFTs and twiddle factors. Here, $W = W_8$.

$N = 2^{10} = 1024$, then the complexity has been reduced from a million to *half* a million multiplications!

Continuing the Split into Smaller DFTs

This idea of splitting a large DFT into smaller DFTs is the key to the success of the FFT algorithm. We can now repeat the same idea on the smaller $(N/2)$-point DFT since $N/2$ is still even (unless it is unity), because $N = 2^M$. Thus, $X_1[k]$ can be written in terms of two $(N/4)$-point DFTs, and the $(N/2)^2$ multiplications reduce to

$$\left(\frac{N}{4}\right)^2 + \left(\frac{N}{4}\right)^2 + \frac{N}{4} = \frac{N^2}{8} + \frac{N}{4}. \tag{12.58}$$

The total number of multiplications for $X[k]$ is therefore

$$2 * \left(\frac{N^2}{8} + \frac{N}{4}\right) + \frac{N}{2} = \frac{N^2}{4} + \frac{N}{2} + \frac{N}{2}. \tag{12.59}$$

If $N/4$ is not unity, it is still a power of two, and we can repeat this again, and all the DFTs involved become $N/8$-point DFTs. The number of multiplications for $X[k]$ would be further reduced to

$$\frac{N^2}{8} + \frac{N}{2} + \frac{N}{2} + \frac{N}{2}. \tag{12.60}$$

This can be repeated until all the DFTs are just two-point DFTs. The final number of multiplications will be

$$\frac{N^2}{2^{M-1}} + \underbrace{\frac{N}{2} + \frac{N}{2} + \cdots + \frac{N}{2}}_{M-1 \text{ times}}, \tag{12.61}$$

where $N = 2^M$. In general, these are complex multiplications.

12.4.2 More Careful Counting

The first term in (12.61) represents the multiplications involved in all the two-point DFTs and the second term represents the multiplications by the twiddle factors such as W_N^k in Eq. (12.53). Now, for a two-point DFT, W_N becomes

$$W_2 = e^{-j2\pi/2} = -1. \tag{12.62}$$

Thus, the DFT coefficients $S[k]$ of a two-point sequence $s[n]$ have the form

$$S[k] = \sum_{n=0}^{1} s[n](-1)^{nk}, \tag{12.63}$$

so that

$$S[0] = s[0] + s[1] \quad \text{and} \quad S[1] = s[0] - s[1], \tag{12.64}$$

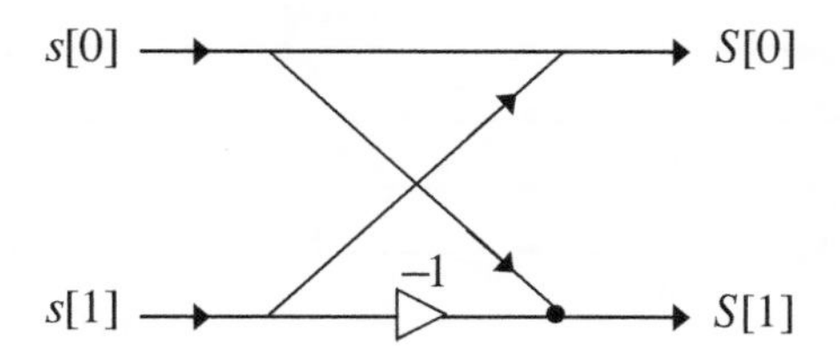

Figure 12.5 The two-point DFT flowgraph. Only two additions, and no multiplications, are required for this computation.

which shows that no multiplications are required. The flowgraph for the two-point DFT is shown in Fig. 12.5. The first term in (12.61) is therefore an overestimate and is a consequence of counting trivial multiplers like -1 as serious multiplications. Ignoring this term, and using the relation $N = 2^M$, we see that the number of multiplications is

$$(M-1)N/2 = \frac{N}{2}\log_2 N - \frac{N}{2}. \tag{12.65}$$

For $N = 2$ we do indeed get zero from here. Now consider the example of $N = 4 = 2^2$. In this case $M = 2$ and the preceding formula yields two multiplications. It turns out, however, that a four-point DFT can be done without any multiplications at all! The reason is that

$$W_4 = e^{-j2\pi/4} = e^{-j\pi/2} = -j, \tag{12.66}$$

which shows that the four powers of W are

$$1, -j, -1, j, \tag{12.67}$$

all of which are trivial multiplications. For arbitrary $N = 2^M$ a careful counting of all such trivial multiplications $(1, -1, j, -j)$ can be done, and it has been shown [Rabiner and Gold, 1975] that the number of **nontrivial complex multiplications** for the radix-2 FFT is

$$\frac{N}{2}\log_2 N - \frac{3N}{2} + 2. \tag{12.68}$$

Setting $N = 2$ or $N = 4$, we see indeed that the preceding formula yields zero. Notice that for $N = 8$, the above formula yields 2. So an eight-point FFT takes only two multiplications, and not $8^2 = 64$ multiplications!

12.4.3 The Complete FFT Flowgraph

Figure 12.6 shows the complete flowgraph for the eight-point FFT. The W in the figure represents W_8. There are three stages or **passes** in the figure. We first compute four two-point DFTs, then compute the two four-point DFTs, and finally compute the eight-point DFT. For arbitrary power-of-two N, there are $\log_2 N$ passes. Each criss-cross in the figure with the twiddle factor preceding it is referred to as a **butterfly**. Thus, each butterfly represents one multiplication and two additions, unless the multiplier is trivial like $1, -1, j$, or $-j$.

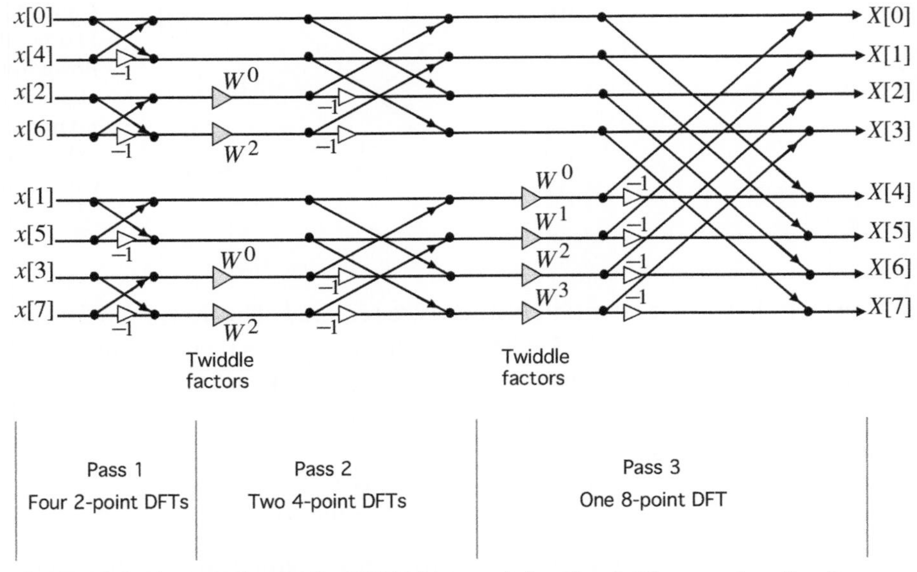

Figure 12.6 The complete radix-2 FFT flowgraph for $N = 8$. There are $\log_2 8 = 3$ passes. In the figure, $W = W_8$.

The FFT flowgraph makes it particularly easy to estimate the number of additions. In each pass we have N complex additions, so the total number of complex additions is clearly $N \log_2 N$. Summarizing, the computational complexity of the N-point radix-2 FFT is

$$\left(\frac{N}{2}\log_2 N - \frac{3N}{2} + 2\right) \text{ complex multiplications,} \quad N\log_2 N \text{ complex additions.} \tag{12.69}$$

For the example of $N = 8$ in the figure we see that there are eight twiddle factors. But since $W_8 = e^{-j2\pi/8}$, the only nontrivial multipliers are W and W^3. That is why substituting $N = 8$ into Eq. (12.68) yields two nontrivial multipliers. Here are more examples comparing the brute-force complexity N^2 with the FFT complexity (12.68):

N	N^2 (brute force)	Eq. (12.68) (FFT)
2	4	0
4	16	0
8	64	2
16	256	10
32	1024	34
64	4096	98

As N increases, the savings due to FFT are impressive indeed!

12.4.4 FFT and Bit Reversal

Notice in Fig. 12.6 that the outputs $X[k]$ are in natural order $X[0], X[1], X[2], \ldots,$ whereas the inputs are arranged in a permuted order:

$$x[0], x[4], x[2], \ldots. \tag{12.70}$$

This ordering, demonstrated in the figure for $N = 8$, is called bit-reversed ordering. This is because, if we represent the numbers $0 \leq n \leq N - 1$ in binary format and reverse the bits, the resulting decimal numbers have precisely this ordering, as demonstrated below.

Index n in decimal	Index n in binary	Reversed binary	New decimal number
0	000	000	0
1	001	100	4
2	010	010	2
3	011	110	6
4	100	001	1
5	101	101	5
6	110	011	3
7	111	111	7

Thus, the FFT algorithm described in this section begins by bit-reversing the input data and then performing pass 1, pass 2, etc.

12.4.5 Concluding Remarks on FFT

The radix-2 FFT algorithm described in this section is called the **decimation in time** algorithm because the DFT of $x[n]$ is computed by working with $x[2n]$ and $x[2n+1]$, and these smaller DFTs are computed in turn by further decimations in time, and so forth. There is a dual algorithm called the **decimation in frequency** algorithm. In this variation, bit reversal takes place at the output rather than the input. In fact, it is also possible to avoid bit reversal altogether by using more complicated flowgraphs, but we shall not go into details here. Chapter 10 of Rabiner and Gold [1975] provides an excellent account of these variations, as well as other generalizations such as the radix-4 FFT. An early book by Brigham [1988], dedicated to the FFT, has a wealth of information on variations of the FFT.

When the number of data samples N is not a power of two, the radix-2 FFT algorithm described here is not applicable. The simplest way to handle this situation would be to append a sufficient number of zeros to the data and make the length a power of two. There are many other methods. One of these, called the **mixed radix FFT algorithm**, is described in Problem 12.21, and works for the case where N is a composite number (i.e., not a prime number). Several other algorithms exist for the case of composite N which is not a power of two. These are based on clever use of deep ideas from number theory [Cooley et al., 1969; Kolba and Parks, 1977]. For the case where N is a prime number, an elegant method for fast computation of the DFT was proposed by Rader [1968].

The history of fast algorithms for DSP is indeed a fascinating one, and many books have been dedicated to the subject [Blahut, 1985; Brigham, 1988]. A revolution in the theory of Fourier transform computation is the early work of Winograd, which established a lower bound on the achievable complexity [Winograd, 1978]. The use of number theory has had a far-reaching influence on fast algorithms for digital signal processing [Blahut, 1985; McClellan and Rader, 1979]. One of these is the invention of the **number-theoretic transform** which works in finite fields, and leads to a fast convolution algorithm free from computational roundoff errors [Agarwal and Burrus, 1975].

12.5 Circular or Cyclic Operations on Signals

From the DFT formula $X[k] = \sum_{n=0}^{N-1} x[n]W^{kn}$, it follows that

$$X[k + iN] = X[k] \quad \text{for any integer } i, \tag{12.71}$$

where $0 \leq k \leq N - 1$ as before. Thus, even though the definition of DFT specifies $X[k]$ only in the range $0 \leq k \leq N - 1$, we can think of $X[k]$ as a *periodic sequence* satisfying Eq. (12.71). Mathematically this is because

$$W^{(k+iN)n} = W^{kn}W^{iNn} = W^{kn}. \tag{12.72}$$

Physically this is not surprising because the $X[k]$ are uniformly spaced samples of $X(e^{j\omega})$, that is,

$$X[k] = X\left(e^{j2\pi k/N}\right), \tag{12.73}$$

and $X(e^{j\omega})$ itself is periodic with period 2π. Similarly, from the inverse DFT formula

$$x[n] = \frac{1}{N}\sum_{k=0}^{N-1} X[k]W^{-nk}, \tag{12.74}$$

we find that if n is replaced with $n + iN$ on the right-hand side, the summation does not change. Thus, the right-hand side repeats periodically as a function of n. Since we started with the assumption that $x[n]$ has finite duration, Eq. (12.74) is therefore valid only in $0 \leq n \leq N - 1$. So we write

$$x[n] = \begin{cases} \dfrac{1}{N}\displaystyle\sum_{k=0}^{N-1} X[k]W^{-nk} & \text{if } 0 \leq n \leq N-1, \\ 0 & \text{otherwise.} \end{cases} \tag{12.75}$$

Since the right-hand side of Eq. (12.74) represents a periodic sequence in n that is different from $x[n]$ in Eq. (12.75), we use the notation $x[[n]]$ to denote this periodic sequence:

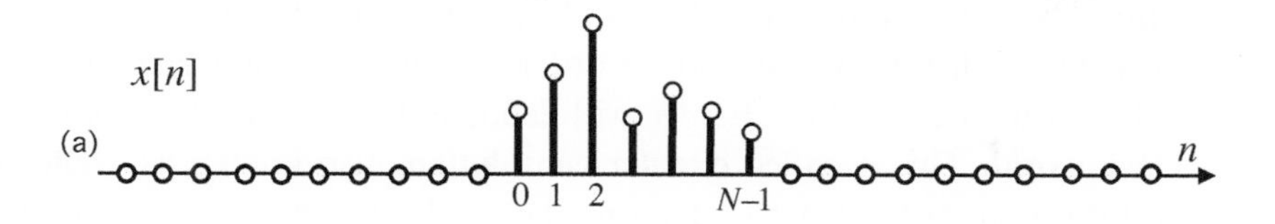

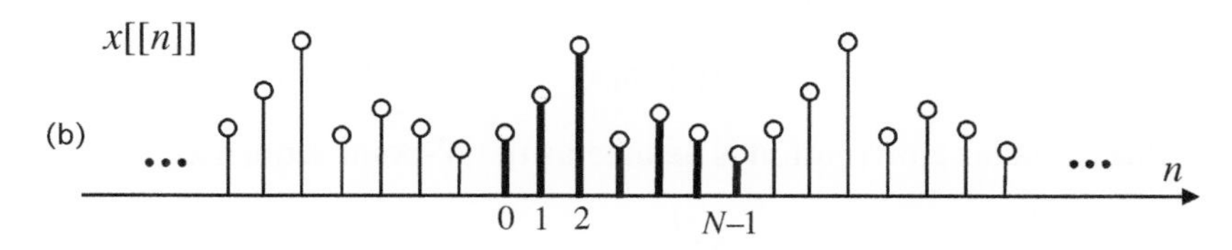

Figure 12.7 (a) A finite-duration signal $x[n]$ and (b) its periodic extension $x[[n]]$.

$$x[[n]] = \frac{1}{N} \sum_{k=0}^{N-1} X[k] W^{-nk}, \quad \forall n. \tag{12.76}$$

We say that $x[[n]]$ is the **periodic extension** of $x[n]$ because, firstly,

$$x[[n]] = x[[n + N]], \quad \forall n, \tag{12.77}$$

and furthermore, $x[[n]] = x[n]$ in $0 \leq n \leq N - 1$. Figure 12.7 shows the relation between $x[n]$ and the periodic extension $x[[n]]$.

The modulo notation. Recall that any integer n can always be written as

$$n = n_0 + Nn_1, \tag{12.78}$$

for a unique integer n_0 in $0 \leq n_0 \leq N - 1$ and a unique integer n_1. This n_0 is nothing but the remainder when n is divided by N, and n_1 is the quotient. This unique remainder n_0 is called "n modulo N" and is, by definition, in the range $0 \leq n_0 \leq N - 1$. In all our discussions, the argument $[n]$ in $x[[n]]$ denotes this remainder n_0. Thus

$$[n] = n \text{ modulo } N, \tag{12.79}$$

so that $0 \leq [n] \leq N - 1$. ▽ ▽ ▽

12.5.1 Circular or Cyclic Convolution

Now consider the convolution $y[n] = (x * h)[n]$ of two sequences $x[n]$ and $h[n]$. If these are N-point sequences, assumed to be zero outside the range $0 \leq n \leq N - 1$, then the convolution is possibly nonzero in

$$0 \leq n \leq 2N - 2, \tag{12.80}$$

and zero outside. So, we can no longer talk about the N-point DFT of $y[n]$, we can only talk about its $2N - 1$ point DFT. In particular, a result like $Y[k] = H[k]X[k]$

cannot be obtained because $Y[k]$ can have $2N-1$ distinct values whereas $H[k]X[k]$ only has N distinct values! So, in order to obtain a result analogous to the convolution theorem of Sec. 3.2, we will define a different kind of convolution in the DFT world. This is called **circular convolution**, also known as **cyclic convolution**, and it preserves the lengths. That is, the result of circular convolution of N-point sequences $x[n]$ and $h[n]$ still has length N, and it has an N-point DFT.

Definition 12.1 *Circular convolution.* Given the N-point sequences

$$x[n], \; h[n], \quad 0 \le n \le N-1, \tag{12.81}$$

their circular convolution is defined as the N-point sequence

$$y[n] = \sum_{l=0}^{N-1} x[l]h[[n-l]], \quad 0 \le n \le N-1. \tag{12.82}$$

Here, $[n-l]$ is $n-l$ modulo N, and satisfies $0 \le [n-l] \le N-1$. ◇

We use the notation

$$y[n] = (x \otimes h)[n] \tag{12.83}$$

for the circular convolution (12.82). This is to distinguish it from ordinary convolution (Sec. 3.2), which is denoted as $(x * h)[n]$. To distinguish it from circular convolution, ordinary convolution is sometimes referred to as **linear convolution**. The term linear just means that the time axis is not interpreted modulo n. It has nothing to do with the term linear as in "linear systems."[3]

The beauty of the definition (12.82) is that if $X[k], H[k]$, and $Y[k]$ are the N-point DFTs of $x[n], h[n]$, and $y[n]$, then their DFTs are related as

$$Y[k] = H[k]X[k], \quad 0 \le k \le N-1. \tag{12.84}$$

This is called the **circular convolution theorem**. The quantity $h[[-l]]$, which is the flipped (time-reversed) version of $h[[l]]$, is called the circularly flipped version of $h[l]$. Similarly, $h[[n-l]]$ is called the circularly shifted version (by n samples) of the circularly flipped version $h[[-l]]$. We explain such circular operations more clearly with examples in Sec. 12.5.2, and provide examples and applications of circular convolution later. This discussion concludes with a proof of the above theorem.

Proof of the circular convolution theorem. From Eq. (12.82) we have

$$\begin{aligned} Y[k] = \sum_{n=0}^{N-1} y[n]W^{nk} &= \sum_{n=0}^{N-1}\sum_{l=0}^{N-1} W^{nk}x[l]h[[n-l]] \\ &= \sum_{l=0}^{N-1} x[l]W^{lk} \sum_{n=0}^{N-1} h[[n-l]]W^{(n-l)k} \end{aligned}$$

[3] If we think of $h[n]$ as a "system" with input $x[n]$ and output $y[n]$, then it can be shown that circular convolution is a linear operator just like linear convolution (Problem 12.14).

$$= \sum_{l=0}^{N-1} x[l]W^{lk} \sum_{m=0}^{N-1} h[[m]]W^{mk}$$
$$= \sum_{l=0}^{N-1} x[l]W^{lk} \sum_{m=0}^{N-1} h[m]W^{mk} = X[k]H[k],$$

which proves the claim. The third line was obtained by making the change of variables $n - l = m$. The summation range remains $0 \le m \le N-1$, because $h[[m]]W^{mk}$ is periodic in m anyways, with repetition index N. In the fourth line we use the fact that $[m] = m$ when $0 \le m \le N-1$. $\nabla\nabla\nabla$

Since $Y[k] = H[k]X[k] = X[k]H[k]$, it follows that circular convolution is commutative, just like ordinary convolution:

$$y[n] = (x \otimes h)[n] = (h \otimes x)[n]. \tag{12.85}$$

This can also be proved directly in the time domain by starting from Eq. (12.82) and making an appropriate change of variables (Problem 12.13).

12.5.2 Circular Shift and Time Reversal

To get further insights into circular convolution, it is useful to define some basic circular operations. First, the **circular** or **cyclic** shift of $x[n]$ is obtained by performing an ordinary shift of $x[[n]]$ and retaining only the samples in the fundamental region $0, 1, 2, \ldots, N-1$. To give a feeling for this, assume $N = 8$, so that $x[n]$ is an eight-point sequence. The ordinary shifted version $x[n-1]$, and the quantity $x[[n-1]]$, are both shown below. The samples in the region $0 \le n \le N-1$ are highlighted in bold.

n:	−2	−1	0	1	2	3	4	5	6	7	8
$x[n]$:	0	0	$x[0]$	$x[1]$	$x[2]$	$x[3]$	$x[4]$	$x[5]$	$x[6]$	$x[7]$	0
$x[n-1]$:	0	0	0	$x[0]$	$x[1]$	$x[2]$	$x[3]$	$x[4]$	$x[5]$	$x[6]$	$x[7]$
$x[[n]]$:	$x[6]$	$x[7]$	**x[0]**	**x[1]**	**x[2]**	**x[3]**	**x[4]**	**x[5]**	**x[6]**	**x[7]**	$x[0]$
$x[[n-1]]$:	$x[5]$	$x[6]$	**x[7]**	**x[0]**	**x[1]**	**x[2]**	**x[3]**	**x[4]**	**x[5]**	**x[6]**	$x[7]$

Similarly, the **circular** or **cyclic** flip (time reversal) of $x[n]$ is obtained by performing an ordinary flip of $x[[n]]$ and retaining only the samples in the fundamental region $0, 1, 2, \ldots, N-1$. Examples of ordinary flip $x[-n]$ and circular flip $x[[-n]]$ are shown below with the samples in $0 \le n \le N-1$ highlighted in bold:

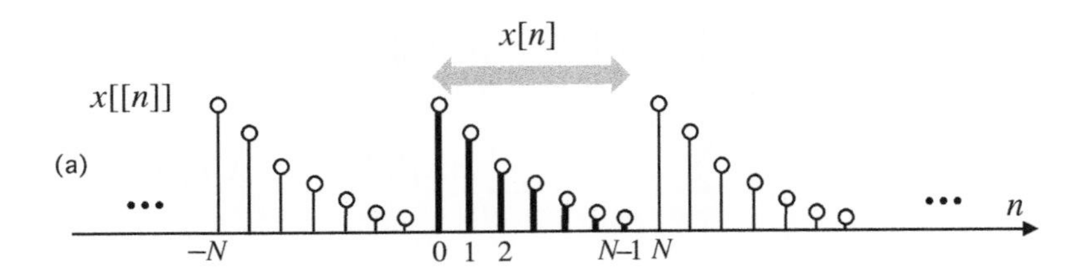

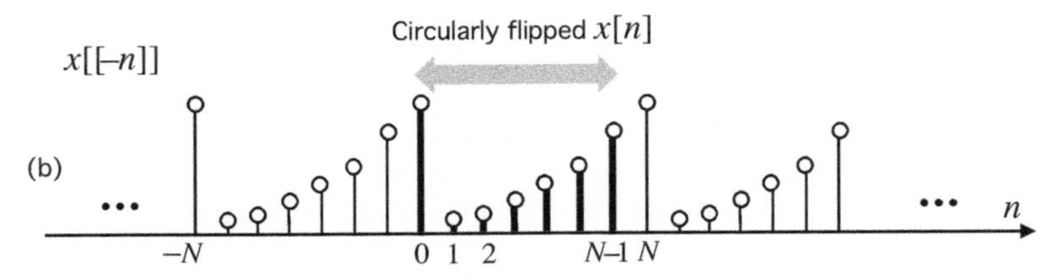

Figure 12.8 (a) The periodic extension $x[[n]]$ of $x[n]$ and (b) its time-reversed (or flipped) version $x[[-n]]$. The portion of $x[[-n]]$ in $0 \leq n \leq N-1$ is called the circularly flipped version of $x[n]$.

n:	-2	-1	0	1	2	3	4	5	6	7	8
$x[n]$:	0	0	$x[0]$	$x[1]$	$x[2]$	$x[3]$	$x[4]$	$x[5]$	$x[6]$	$x[7]$	0
$x[-n]$:	$x[2]$	$x[1]$	$x[0]$	0	0	0	0	0	0	0	0
$x[[n]]$:	$x[6]$	$x[7]$	**x[0]**	**x[1]**	**x[2]**	**x[3]**	**x[4]**	**x[5]**	**x[6]**	**x[7]**	$x[0]$
$x[[-n]]$:	$x[2]$	$x[1]$	**x[0]**	**x[7]**	**x[6]**	**x[5]**	**x[4]**	**x[3]**	**x[2]**	**x[1]**	$x[0]$

This is also demonstrated in Fig. 12.8, where part (a) shows $x[[n]]$ and part (b) shows $x[[-n]]$. The circularly flipped version of $x[n]$ is the portion of $x[[-n]]$ from $0 \leq n \leq N-1$. Note that the sample located at $n = 0$ does not move in a circular flip. Only the **samples from** $n = 1$ **to** $n = N - 1$ **are reversed**.

Even and odd sequences. We now point out an interesting consequence of the definition of cyclic time reversal. In the DFT world, an N-point sequence $x[n]$ defined in $0 \leq n \leq N-1$ is said to be even (or symmetric) if $x[[n]] = x[[-n]]$ and odd (or antisymmetric) if $x[[n]] = -x[[-n]]$. Thus,

$$x[[n]] = \begin{cases} x[[-n]] & \text{even}, \\ -x[[-n]] & \text{odd}. \end{cases} \tag{12.86}$$

So, for an even sequence the sample $x[0]$ is unconstrained, and the samples in $1 \leq n \leq N-1$ are constrained to be symmetric: $x[1] = x[N-1], x[2] = x[N-2]$, and so forth. This is demonstrated in Fig. 12.9(a). For an odd sequence, on the other hand, $x[0] = 0$ and the samples in $1 \leq n \leq N-1$ are constrained to be antisymmetric:

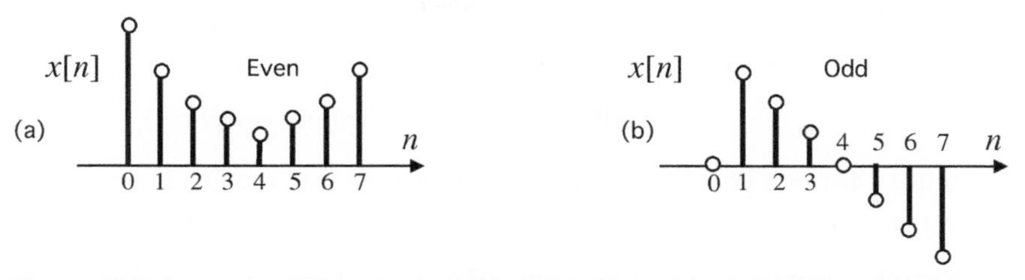

Figure 12.9 Examples of (a) even and (b) odd sequences in the DFT world. Here $N = 8$. See text.

$x[1] = -x[N-1], x[2] = -x[N-2]$, and so forth (Fig. 12.9(b)). More generally, a Hermitian symmetric sequence is defined by the property $x[[n]] = x^*[[-n]]$, and skew-Hermitian by the property $x[[n]] = -x^*[[-n]]$. The properties of the DFTs of such sequences are summarized in Sec. 12.9. ▽▽▽

A summary of circular shift and flip operations is given below, with only the region $0 \leq n \leq N-1$ shown. The quantity $x[[1-n]]$, which is the circularly right-shifted version of the circularly flipped version $x[[-n]]$, is also shown. We will find these circular operations to be very useful later, when we present examples of circular convolution.

n:	0	1	2	3	4	5	6	7
$x[n]$:	$x[0]$	$x[1]$	$x[2]$	$x[3]$	$x[4]$	$x[5]$	$x[6]$	$x[7]$
$x[[n-1]]$:	$x[7]$	$x[0]$	$x[1]$	$x[2]$	$x[3]$	$x[4]$	$x[5]$	$x[6]$
$x[[-n]]$:	$x[0]$	$x[7]$	$x[6]$	$x[5]$	$x[4]$	$x[3]$	$x[2]$	$x[1]$
$x[[1-n]]$:	$x[1]$	$x[0]$	$x[7]$	$x[6]$	$x[5]$	$x[4]$	$x[3]$	$x[2]$

12.5.3 DFT and Circular Operations on Signals

Let $x[n]$ be an N-point sequence and $W = e^{-j2\pi/N}$ as usual. We will show that if $x[n] \longleftrightarrow X[k]$ then the DFT of the circularly shifted version $x[[n-L]]$ is $W^{Lk}X[k]$, that is,

$$x[[n-L]] \longleftrightarrow W^{Lk}X[k] \quad \text{(circular shift property)}, \tag{12.87}$$

for any integer L. The right-hand side above can be rewritten as

$$e^{-j2\pi kL/N}X[k] = e^{-j\omega_k L}X[k], \tag{12.88}$$

where $\omega_k = 2\pi k/N$. This shows the analogy with the shift property in traditional Fourier transform theory. Similarly,

$$W^{-Ln}x[n] \longleftrightarrow X[k-L] \quad \text{(modulation property)}. \tag{12.89}$$

Thus, the time-modulated version $e^{j(2\pi L/N)n}x[n]$ corresponds to a shift in the DFT domain by L samples. Next, it can also be shown that

$$x[[-n]] \longleftrightarrow X[-k] \quad \text{(flipping property)}. \tag{12.90}$$

Since $X[k] = X[k+N]$ anyway, the above says that the DFT of the circularly flipped version $x[[-n]]$ is itself a circularly flipped version of $X[k]$.

Proofs. First consider the claim (12.87). We have

$$x[[n-L]] \longleftrightarrow \sum_{n=0}^{N-1} x[[n-L]]W^{nk} = \sum_{n=-L}^{-L+N-1} x[[n]]W^{(n+L)k}, \tag{12.91}$$

where the equality follows from an obvious change of variables. Since $x[[n]] = x[[n+N]]$, the product $x[[n]]W^{(n+L)k}$ is periodic in n with repetition interval N, so the summation range in the above right-hand side can be taken as any consecutive set of N samples. Thus

$$x[[n-L]] \longleftrightarrow \sum_{n=0}^{N-1} x[[n]]W^{(n+L)k} = W^{Lk}\sum_{n=0}^{N-1} x[[n]]W^{nk} = W^{Lk}X[k], \tag{12.92}$$

which proves the claim. The proof of (12.89) is left as an exercise. Finally, the DFT of $x[[-n]]$ is

$$\begin{aligned} \sum_{n=0}^{N-1} x[[-n]]W^{nk} &= \sum_{n=-(N-1)}^{0} x[[n]]W^{-nk} \\ &= \sum_{n=0}^{N-1} x[[n]]W^{-nk} \\ &= \sum_{n=0}^{N-1} x[n]W^{-nk} = X[-k]. \end{aligned} \tag{12.93}$$

The second equality follows from the fact that $x[[n]]W^{-nk}$ is periodic in n with repetition index N, so the summation can be done over any N successive values of n. The third equality follows because $x[[n]] = x[n]$ in $0 \le n \le N-1$. The last equality follows from the definition of the DFT $X[k]$. ▽ ▽ ▽

Besides the three properties above, a number of other properties of the DFT are summarized in Sec. 12.9. The proofs of those are similar to the proofs of the above properties, and are covered in Problem 12.23.

12.5.4 Circular Convolution Example

Assuming $N = 4$, we now demonstrate four-point circular convolution. Let the two sequences be

n:	0	1	2	3
$x[n]$:	0	1	2	3
$h[n]$:	4	5	6	7

The first step is to circularly flip $h[n]$:

$$\begin{array}{rcccc} n: & 0 & 1 & 2 & 3 \\ x[n]: & 0 & 1 & 2 & 3 \\ h[[-n]]: & 4 & 7 & 6 & 5 \end{array}$$

If we multiply the sequences $x[n]$ and $h[[-n]]$ pointwise, and add the samples, the result is

$$y[0] = 0 + 7 + 12 + 15 = 34. \tag{12.94}$$

The next step is to circularly shift $h[[-n]]$ to the right by one:

$$\begin{array}{rcccc} n: & 0 & 1 & 2 & 3 \\ x[n] & 0 & 1 & 2 & 3 \\ h[[1-n]]: & 5 & 4 & 7 & 6 \end{array}$$

Multiplying $x[n]$ and $h[[1-n]]$ pointwise and adding, we get

$$y[1] = 0 + 4 + 14 + 18 = 36. \tag{12.95}$$

We again circularly shift $h[[1-n]]$ to the right by one:

$$\begin{array}{rcccc} n: & 0 & 1 & 2 & 3 \\ x[n]: & 0 & 1 & 2 & 3 \\ h[[2-n]]: & 6 & 5 & 4 & 7 \end{array}$$

and multiply $x[n]$ and $h[[2-n]]$ pointwise, and add the samples, to get

$$y[2] = 0 + 5 + 8 + 21 = 34. \tag{12.96}$$

One final circular right shift of $h[[2-n]]$ yields

$$\begin{array}{rcccc} n: & 0 & 1 & 2 & 3 \\ x[n]: & 0 & 1 & 2 & 3 \\ h[[3-n]]: & 7 & 6 & 5 & 4 \end{array}$$

so that

$$y[3] = 0 + 6 + 10 + 12 = 28. \tag{12.97}$$

Further right circular shift of $h[[3-n]]$ yields $h[[4-n]] = h[[-n]]$ again, so we can stop. Summarizing, the sequences $x[n], h[n]$, and the circular convolution $y[n]$ are given by

$$\begin{array}{rcccc} n: & 0 & 1 & 2 & 3 \\ x[n]: & 0 & 1 & 2 & 3 \\ h[n]: & 4 & 5 & 6 & 7 \\ y[n]: & 34 & 36 & 34 & 28 \end{array}$$

12.5.5 Linear Convolution Using Circular Convolution

Linear convolution (i.e., the ordinary convolution described in Sec. 3.2) arises naturally in digital filtering. For example, the output of an FIR filter in response to a finite-length input can be evaluated by computing the linear convolution of the input with the impulse response. Circular convolution, on the other hand, is a theoretical byproduct arising from the definition of the DFT. We now show how to perform the practically useful linear convolution operation using circular convolution. The advantage is that circular convolutions can be performed using the fast Fourier transform (Sec. 12.5.6), offering a fast method for linear convolution.

Let $x_1[n]$ and $h_1[n]$ be two finite-duration sequences that are zero outside the range $0 \leq n \leq M-1$. If we perform linear convolution, the result $y_1[n]$ could possibly be nonzero in $0 \leq n \leq 2M-2$. That is, convolution of two M-point sequences in general results in an N-point sequence where

$$N = 2M - 1. \tag{12.98}$$

So, by performing a circular convolution of $x_1[n]$ and $h_1[n]$, we cannot obtain these N samples. To obtain all N samples of the linear convolution of $x_1[n]$ and $h_1[n]$, let us define two N-point sequences $x[n]$ and $h[n]$ by zero-padding the original sequences:

$$x[n] = \begin{cases} x_1[n] & 0 \leq n \leq M-1, \\ 0 & M \leq n \leq N-1, \end{cases} \tag{12.99}$$

and

$$h[n] = \begin{cases} h_1[n] & 0 \leq n \leq M-1, \\ 0 & M \leq n \leq N-1. \end{cases} \tag{12.100}$$

If we were to perform a circular convolution of $x[n]$ and $h[n]$, then the result $y[n]$ would be equal to the correct linear-convolution answer $y_1[n]$. The idea is best demonstrated with an example. Let $M = 3$ and consider the three-point sequences

$$\begin{array}{rccc} x_1[n]: & 1 & 2 & 3 \\ h_1[n]: & 4 & 5 & 6 \end{array}$$

Since $N = 2M - 1 = 5$, we define the five-point sequences $x[n]$ and $h[n]$ by padding two zeros:

$$\begin{array}{rccccc} x[n]: & 1 & 2 & 3 & 0 & 0 \\ h[n]: & 4 & 5 & 6 & 0 & 0 \end{array}$$

To compute the five-point circular convolution, we first circularly flip $h[n]$:

$$\begin{array}{rccccc} x[n]: & 1 & 2 & 3 & 0 & 0 \\ h[[-n]]: & 4 & 0 & 0 & 6 & 5 \end{array}$$

so that $y[0] = 4$. To compute $y[1]$, we circularly shift $h[[-n]]$ to the right by one sample:

$$\begin{array}{rccccc} x[n]: & 1 & 2 & 3 & 0 & 0 \\ h[[1-n]]: & 5 & 4 & 0 & 0 & 6 \end{array}$$

and obtain $y[1] = 1 * 5 + 2 * 4 = 13$. Continuing this process, we get

$$\begin{array}{rccccc} x[n]: & 1 & 2 & 3 & 0 & 0 \\ h[[2-n]]: & 6 & 5 & 4 & 0 & 0 \end{array}$$

so that $y[2] = 1 * 6 + 2 * 5 + 3 * 4 = 28$. Next we look at $h[[3-n]]$:

$$\begin{array}{rccccc} x[n]: & 1 & 2 & 3 & 0 & 0 \\ h[[3-n]]: & 0 & 6 & 5 & 4 & 0 \end{array}$$

and obtain $y[3] = 12 + 15 = 27$. Finally,

$$\begin{array}{rccccc} x[n]: & 1 & 2 & 3 & 0 & 0 \\ h[[4-n]]: & 0 & 0 & 6 & 5 & 4 \end{array}$$

so that $y[4] = 18$. So the padded zeros ensure that each step in the circular convolution performs the same calculation as would a linear convolution of $x_1[n]$ and $y_1[n]$. So the final result

$$y[n] = \{4, 13, 28, 27, 18\} \tag{12.101}$$

is indeed equal to the linear convolution $y_1[n]$ of the three-point sequences $x_1[n]$ and $h_1[n]$.

12.5.6 Fast Convolution Using the FFT

Consider two M-point sequences $x_1[n]$ and $h_1[n]$, $0 \leq n \leq M-1$. The linear convolution of these sequences is given by

$$y_1[n] = \sum_{m=0}^{M-1} x_1[m]h_1[n-m], \quad 0 \leq n \leq 2M-2. \tag{12.102}$$

The number of multiplications required for this is exactly M^2. To verify this, note that it takes $n+1$ multiplications to compute $y_1[n]$ for each n in $0 \leq n \leq M-1$. For larger n, the number of multiplications decreases linearly so that $y_1[2M-2]$ takes only one multiplication. Adding these, we obtain

$$2(1 + 2 + \cdots + M - 1) + M = M^2. \tag{12.103}$$

We now show how the FFT can be used to perform linear convolution with much fewer multiplications. From Sec. 12.5.5, recall that the linear convolution can be performed by using a circular convolution on the extended sequences $x[n]$ and $h[n]$ of length $N = 2M-1$, obtained by zero-padding. The circular convolution theorem (Sec. 12.5.1) shows that this can be performed by computing the N-point DFTs of $x[n]$ and $h[n]$, multiplying the results pointwise, and then performing the inverse DFT. If N is not a power of two, we can pad additional zeros to make it so. Here is a summary of the procedure:

1. Given the M-point sequences $x_1[n]$ and $h_1[n]$, define the zero-padded sequences $x[n]$ and $h[n]$ as in Eqs. (12.99) and (12.100), where N is a power of two such that $N \geq 2M-1$.
2. Compute the N-point DFTs $X[k]$ and $H[k]$ using the FFT.
3. Compute the N products $Y[k] = X[k]H[k]$, $0 \leq k \leq N-1$.
4. Finally, compute the inverse DFT $y[n]$. Then the result of linear convolution (12.102) can be identified as $y_1[n] = y[n], 0 \leq n \leq 2M-2$.

Since the FFT takes about $0.5N \log_2 N$ complex multiplications, the total number of multiplications in the above algorithm is approximately

$$1.5N \log_2 N + N, \tag{12.104}$$

where the extra N multiplications are required in the computation of the products $X[k]H[k]$. The number of multiplications (12.104) is typically much smaller than the number M^2 required in the direct computation of the convolution. Thus, let $M = 500$. Then the number of multiplications in direct convolution is

$$M^2 = 250,000. \tag{12.105}$$

For the FFT-based convolution (with $N = 2^{10} = 1024$) the number of multiplications, is

$$1.5N \log_2 N + N = 16,384, \tag{12.106}$$

which shows savings by a factor of 15. The savings in the number of additions can similarly be estimated.

For real signals, direct convolution uses M^2 real multiplications, whereas the FFT-based method uses (12.104) complex multiplications. The preceding example might therefore appear to be unfair, since each complex multiplication is usually equivalent to four real multiplications. However, for real signals, the FFT computation can be made more efficient by taking advantage of realness (Problems 12.25 and 12.26). Finally, it should be mentioned that there are some simple tricks which allow complex multiplications to be performed with only three real multiplications, at the expense of a slightly increased number of additions (Problem 12.16).

12.6 Computing the FT at Different Resolutions

Given an N-point sequence $x[n]$, the DFT calculates the Fourier transform $X(e^{j\omega})$ at N uniformly spaced points on the grid $\omega_k = 2\pi k/N, 0 \leq k \leq N-1$. In some situations, one may wish to compute $X(e^{j\omega})$ at a lower or higher resolution. We now explain how to do this efficiently. Fig. 12.10 demonstrates these situations.

12.6.1 Computing the FT at Higher Resolution

Suppose we wish to compute $X(e^{j\omega})$ on the **denser grid**

$$\omega_k = \frac{2\pi k}{M}, \; 0 \leq k \leq M-1, \quad M > N. \tag{12.107}$$

Can we still do this by using a DFT-style algorithm? This is easily done by appending or "padding" zeros to $x[n]$ to increase its length artificially:

$$x_a[n] = \begin{cases} x[n] & \text{for } 0 \leq n \leq N-1, \\ 0 & \text{for } N \leq n \leq M-1, \end{cases} \tag{12.108}$$

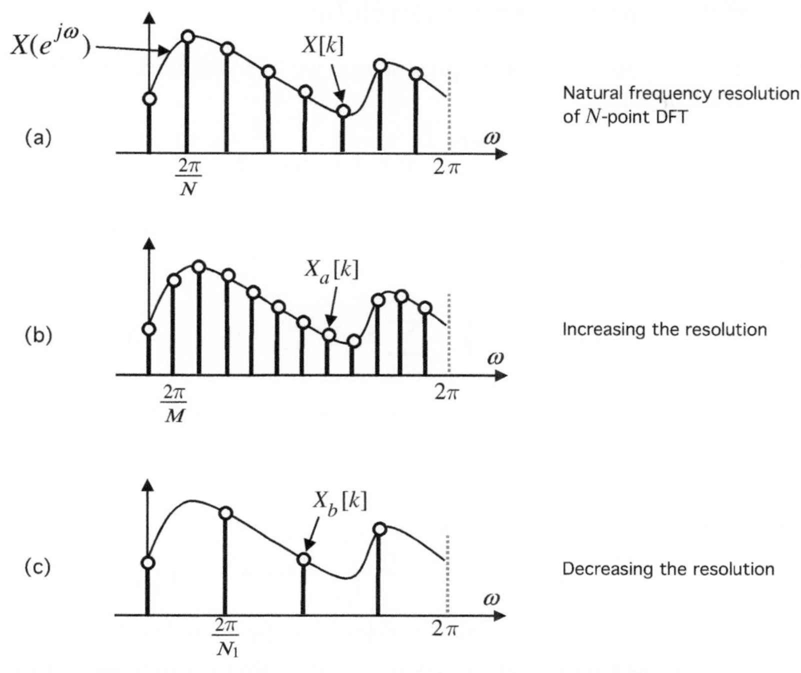

Figure 12.10 (a) The Fourier transform $X(e^{j\omega})$ and DFT $X[k]$ of an N-point sequence $x[n]$, (b) samples of $X(e^{j\omega})$ at higher resolution, and (c) samples of $X(e^{j\omega})$ at lower resolution.

and computing the M-point DFT

$$X_a[k] = \sum_{n=0}^{M-1} x_a[n] W_M^{kn}, \quad 0 \le k \le M-1. \tag{12.109}$$

Since $W_M = e^{-j2\pi/M}$ it follows that

$$X_a[k] = \sum_{n=0}^{M-1} x_a[n] e^{-j\frac{2\pi k}{M}n} = \sum_{n=0}^{N-1} x[n] e^{-j\omega_k n}, \quad 0 \le k \le M-1, \tag{12.110}$$

which yields the desired samples on the denser grid (12.107). Thus, zero-padding in the time domain increases resolution in the frequency domain. When M is composite (e.g., a power of two), the above method can be made computationally efficient using the M-point FFT. Note that there is still some lack of efficiency because the zeros inserted artifically in the time domain result in some multiplications which are unnecessary. This can be avoided for the case where M is a multiple of N (see Problem 12.28). In the above we have not assumed M to be a multiple of N. In this case the increase in frequency resolution M/N is a *rational* number. This is sometimes referred to as *fractional* increase of resolution.

12.6.2 Computing the FT at Lower Resolution

Given an N-point sequence $x[n]$, assume we wish to compute $X(e^{j\omega})$ on a coarser or **sparser** grid like

$$\omega_k = \frac{2\pi k}{N_1}, \quad 0 \le k \le N_1 - 1, \quad N_1 < N. \tag{12.111}$$

This situation may arise when N is very large (large amount of data). The computation of interest now is

$$X\left(e^{j2\pi k/N_1}\right) = \sum_{n=0}^{N-1} x[n] W_{N_1}^{nk}, \quad 0 \le k \le N_1 - 1, \tag{12.112}$$

where $W_{N_1} = e^{-j2\pi/N_1}$ as usual. For simplicity, assume N_1 is a factor of N, that is,

$$N = N_1 N_2 \tag{12.113}$$

for some integer N_2. In this case

$$X\left(e^{j2\pi k/N_1}\right) = X\left(e^{j2\pi N_2 k/N}\right) = X[N_2 k], \tag{12.114}$$

that is, these are downsampled versions of the N-point DFT $X[k]$, obtained by retaining the samples at multiples of N_2. It is obvious that we can first compute all the N-points of the DFT $X[k]$ (say, using an FFT) and then downsample it by N_2 to obtain the desired answer (12.114). But this would not be efficient, because we are interested only in $N_1 < N$ samples of $X[k]$, and many of the computed samples are discarded. Is there a way to compute **only** the desired samples $X[N_2 k]$ instead of all the samples? Indeed there is, as we explain next.

The summation (12.112) looks like an N_1-point DFT because of the $W_{N_1}^{nk}$, but the summation index n ranges over the larger interval $0 \le n \le N - 1$. We will show how to rearrange this so that it becomes an N_1-point DFT, which can then be computed using an N_1-point FFT (if N_1 is composite). For this we simply express the summation index n in the form

$$n = \ell + N_1 m, \quad 0 \le \ell \le N_1 - 1, \ 0 \le m \le N_2 - 1, \tag{12.115}$$

so that

$$X\left(e^{j2\pi k/N_1}\right) = \sum_{\ell=0}^{N_1-1} \sum_{m=0}^{N_2-1} x[\ell + N_1 m] W_{N_1}^{(\ell+N_1 m)k} = \sum_{\ell=0}^{N_1-1} \sum_{m=0}^{N_2-1} x[\ell + N_1 m] W_{N_1}^{\ell k},$$

where we have used the fact that $[W_{N_1}]^{N_1} = 1$. Defining the N_1-point sequence

$$x_b[\ell] = \sum_{m=0}^{N_2-1} x[\ell + N_1 m], \quad 0 \le \ell \le N_1 - 1, \tag{12.116}$$

we then have

$$X\left(e^{j2\pi k/N_1}\right) = \sum_{\ell=0}^{N_1-1} x_b[\ell] W_{N_1}^{k\ell} = X_b[k], \quad 0 \le k \le N_1 - 1, \tag{12.117}$$

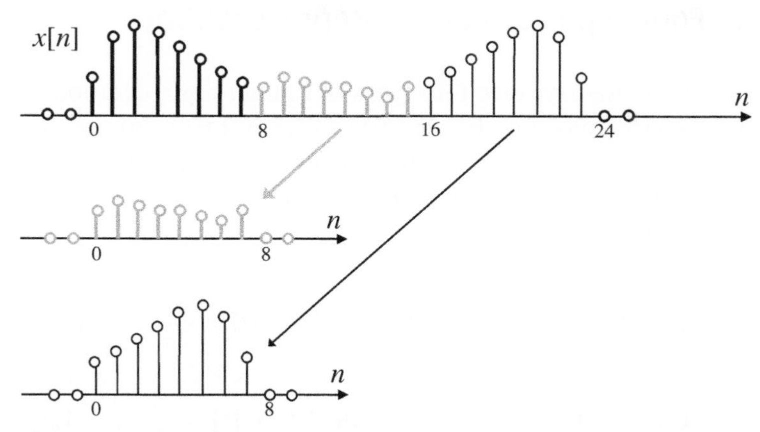

Figure 12.11 The time-domain folding operation involved in computing DFT at a lower resolution.

where $X_b[k]$ is the N_1-point DFT of $x_b[\ell]$. Summarizing, we start from the original N-point sequence $x[n]$ and construct the shorter N_1-point sequence $x_b[n]$ according to Eq. (12.116). The N_1-point DFT of $x_b[n]$ is the downsampled version $X[N_2k]$ we wanted to compute.

This construction of the smaller sequence $x_b[n]$ can be visualized as demonstrated in Fig. 12.11, where we have $N_1 = 8$ and $N_2 = 3$, so that $N = 24$. The set of 24 samples in $x[n]$ is divided into $N_2 = 3$ adjacent groups, shown in different shades. Then all three groups of eight samples are aligned and added together. The result is the eight-point sequence $x_b[n]$. This operation is called *time-domain folding* and is analogous to frequency-domain folding or *aliasing*, which results from sampling a signal (Sec. 10.2). The time-domain folding is a consequence of the fact that we have *undersampled* in the frequency domain. Namely, we have undersampled the DFT $X[k]$ to retain only a subset $X[N_2k]$.

Fractional Reduction of Frequency Resolution

It was assumed above that N_1 is a factor of N. If this is not the case, for example if $N = 300$ and $N_1 = 200$, we have a fractional change of resolution. The resolution is decreased by the factor $N/N_1 = 3/2$, which is a non-integer. For such fractional reduction of frequency resolution, the above method can readily be modified as follows:

1. First define a longer sequence $x_1[n]$ by padding N zeros to $x[n]$. The $2N$ point DFT $X_1[k]$ corresponds to $X(e^{j\omega})$ at the denser grid $\omega_k = 2\pi k/2N$.
2. Now observe that 3 is a divisor of $2N$ (since $2N = 3N_1$). So we can compute the DFT of the $2N$-point sequence $x_1[n]$ at three times lower resolution by using the efficient method described above. This yields $X(e^{j\omega})$ on the grid $\omega_k = 2\pi k/(2N/3) = 2\pi k/N_1$ as desired, without any wasted computation.

12.7 The Four Types of Fourier Representation

In this book we have seen four types of Fourier representation. These are based on (a) the continuous-time Fourier transform (CTFT), (b) the discrete-time Fourier transform (DTFT), (c) the continuous-time Fourier series (CTFS), and (d) the discrete Fourier transform (DFT). For each of these, the transform and its inverse are summarized below:

$$\text{CTFT:}\ \ X(j\omega)=\int_{-\infty}^{\infty}x(t)e^{-j\omega t}dt;\ \ x(t)=\frac{1}{2\pi}\int_{-\infty}^{\infty}X(j\omega)e^{j\omega t}\,d\omega, \quad (12.118)$$

$$\text{DTFT:}\ \ X_d(e^{j\omega})=\sum_{n=-\infty}^{\infty}x_d[n]e^{-j\omega n};\ x_d[n]=\frac{1}{2\pi}\int_{-\pi}^{\pi}X_d(e^{j\omega})e^{j\omega n}d\omega, \quad (12.119)$$

$$\text{CTFS:}\ \ a_k=\frac{1}{T}\int_{-T/2}^{T/2}x(t)e^{-jk\omega_0 t}dt;\ \ x(t)=\sum_{k=-\infty}^{\infty}a_k e^{jk\omega_0 t}, \quad (12.120)$$

$$\text{DFT:}\ \ X_d[k]=\sum_{n=0}^{N-1}x_d[n]W^{kn};\ \ x_d[n]=\frac{1}{N}\sum_{k=0}^{N-1}X_d[k]W^{-nk}, \quad (12.121)$$

where

$$\omega_0=\frac{2\pi}{T}\ \text{in Eq. (12.120)} \quad \text{and} \quad W=e^{-j2\pi/N}\ \text{in Eq. (12.121).}$$

We have used the subscript d for "discrete time" in Eq. (12.119) to avoid confusion between $X(e^{j\omega})$ and $X(j\omega)$. The same is done in Eq. (12.121) for consistency with Eq. (12.119). Figure 12.12 gives a schematic summary of the four types of Fourier representation. For part (c) notice that in Eq. (12.120), the a_k are the Fourier series coefficients of $x(t)$, and as shown in Sec. 11.3.1, the CTFT $X(j\omega)$ of $x(t)$ in Eq. (12.120) is readily deduced:

$$x(t)=\sum_{k=-\infty}^{\infty}a_k e^{jk\omega_0 t} \quad \longleftrightarrow \quad X(j\omega)=2\pi\sum_k a_k\delta_c(\omega-k\omega_0), \quad (12.122)$$

where $\delta_c(\cdot)$ is the Dirac delta. The $X(j\omega)$ in Eq. (12.122) is what we have shown in Fig. 12.12(c).

12.7.1 Differences between Types of FT

The differences between the four types of FT are clear from Fig. 12.12. The first observation is that the domains of values for time and frequency are different in the four cases.

1. For the CTFT, time t and frequency ω are both continuous, with $-\infty < t < \infty$ and $-\infty < \omega < \infty$.

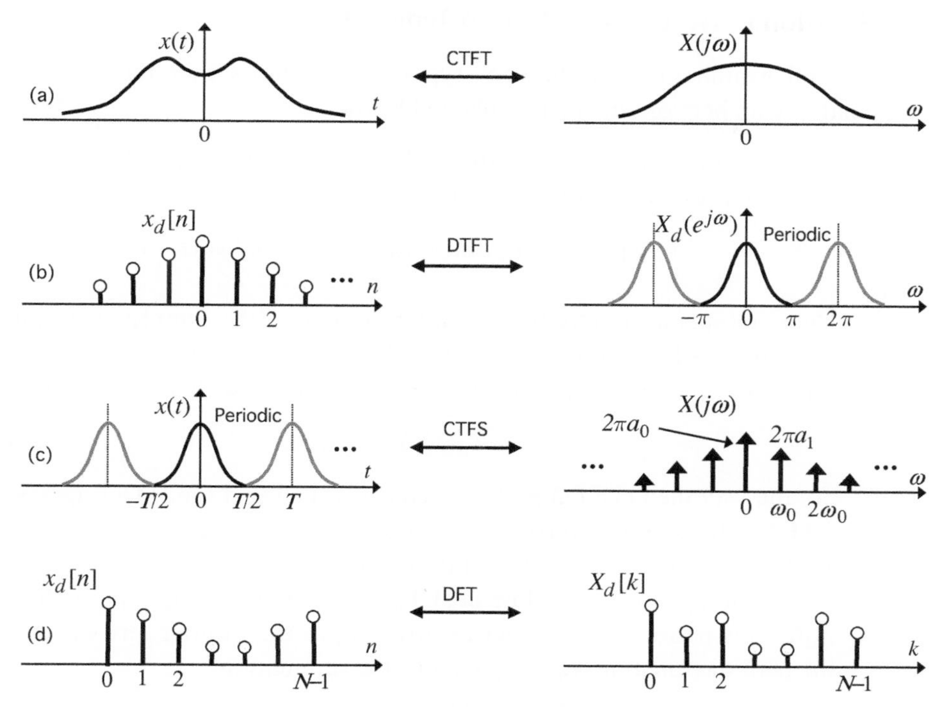

Figure 12.12 (a)–(d) The four types of Fourier representation. The shapes of the plots should not be taken seriously, they are only cartoons.

2. For the DTFT, time n is discrete and frequency ω is continuous, with $-\infty < n < \infty$ and $-\pi < \omega < \pi$. Since time is discrete, $X_d(e^{j\omega})$ is periodic in ω and repeats at intervals of 2π.
3. For the CTFS, time t is continuous and frequency $k\omega_0$ is discrete, with $-T/2 \leq t < T/2$ and $-\infty < k < \infty$. Here, $x(t)$ can be regarded as a periodic signal with $x(t) = x(t + T)$, or as a signal which is zero outside $-T/2 \leq t < T/2$ (or equivalently, $0 \leq t < T$).
4. Finally, for the DFT, time and frequency are both discrete, with $0 \leq n, k \leq N - 1$, and the frequencies $\omega_k = 2\pi k/N$. Here, $X_d[k]$ is periodic with $X_d[k] = X_d[k + N]$ and $x_d[n]$ can be regarded as a periodic signal with $x_d[n] = x_d[n + N]$, or as a signal which is zero outside $0 \leq n \leq N - 1$.

Another interesting point is that all the expressions in Eqs. (12.118)–(12.120) are either integrals or infinite summations. Neither of these can be computed exactly, they can only be approximated on a computer. On the other hand, the DFT and the IDFT in Eq. (12.121) are both finite summations and can be computed exactly with a finite amount of computation with complexity $O(N^2)$. In fact, the complexity can be reduced by using the FFT (Sec. 12.4).

So we see that among all the Fourier representations we have discussed, *only the DFT/IDFT pair is* **computable** *with a finite number of arithmetic operations.*

12.7.2 Relation between the Different Types of FT

Careful examination of the four types of FT shows that there is some close relationship between them, as explained below.

1. *CTFT and DTFT.* We know from Chapter 10 that the sampled version of $x(t)$ can be expressed using two notations:
$$x_d[n] = x(nT) \quad \text{and} \quad x_s(t) = \sum_n x(nT)\delta_c(t - nT). \tag{12.123}$$
Here, $x_d[n]$ is a discrete-time signal representing the samples, whereas $x_s(t)$ is defined for all t (continuous time), with the samples represented by Dirac delta functions at $t = nT$. Then, as seen from Eq. (10.18) with $T = 1$, we have
$$X_d(e^{j\omega}) = X_s(j\omega). \tag{12.124}$$
So, the way the DTFT $X_d(e^{j\omega})$ is defined is such that the **DTFT agrees with the CTFT** of the sampled version $x_s(t)$ when the sample spacing is unity.
2. *DTFT and CTFS.* For the continuous-time Fourier series (CTFS), $x(t)$ is periodic and a_k is discrete. For DTFT, on the other hand, $x_d[n]$ is discrete and $X_d(e^{j\omega})$ is periodic. So we can regard $x_d[n]$ as the Fourier series coefficients of the periodic function $X_d(e^{j\omega})$. That is, the summation
$$\sum_{n=-\infty}^{\infty} x_d[n]e^{-j\omega n} \tag{12.125}$$
is the Fourier series representation of the periodic signal $X_d(e^{j\omega})$. So the mathematical properties of CTFS and DTFT are identical, except that the roles played by **time and frequency are interchanged**. Indeed, the expression for $x_d[n]$ in Eq. (12.119) is similar to that of a_k in Eq. (12.120) (except for a minus sign on the exponent).
3. *CTFT and CTFS.* For periodic $x(t)$ described by CTFS, the CTFT $X(j\omega)$ is still well defined and has the form (12.122). This shows the relation between the CTFT and the Fourier series coefficients a_k.
4. *DFT and the other transforms.* We already know from this chapter that we can regard the DFT as the DTFT sampled at $\omega_k = 2\pi k/N$, that is,
$$X_d[k] = X_d\left(e^{j2\pi k/N}\right). \tag{12.126}$$
Another way to look at the DFT is to regard $x_d[n]$ as one period of a periodic sequence $x_d[[n]]$ (which is the periodic extension of $x_d[n]$, see Sec. 12.5). Since periodicity in time implies discreteness in frequency, we can also regard $X_d[k]$ as the **Fourier series** coefficients of the periodic sequence $x_d[[n]]$. Thus, the DFT is related to the other transforms in multiple ways.

Thus, even though we have studied four types of transform, they are all closely related. If we understand one of them, say the CTFT, then the properties of the other types of transform follow from that. Having said this, we would like to add that it is actually convenient to view the four types of transform as separate because notationally each of them is appropriate for a different scenario. In short, while the above connections between the transforms are insightful, it is a matter of convenience to view them as different.

Some of the differences are significant as well. For example, in Chapter 9 we discussed **eigenfunctions** of the CTFT. Such a discussion will not be meaningful for the DTFT because time is discrete and frequency is continuous. For similar reasons it is not meaningful for the CTFS. However, it is still meaningful for the case of DFT where time and frequency-domain signals can both be represented as vectors of size N. So the eigenfunctions are simply the **eigenvectors** of the $N \times N$ DFT matrix $\mathbf{W}$ (Sec. 12.3). However, unlike the examples of eigenfunctions of CTFT such as the Gaussian and Gauss–Hermite functions (Sec. 9.8.1) reproduced below:

$$\frac{e^{-t^2/2}}{\sqrt{2\pi}} \longleftrightarrow e^{-\omega^2/2}, \qquad te^{-t^2/2} \longleftrightarrow -j\sqrt{2\pi}\,\omega e^{-\omega^2/2}, \qquad \ldots \tag{12.127}$$

the eigenvectors of $\mathbf{W}$ do not in general have simple closed-form expressions, although they can readily be computed. The interested reader is referred to the paper by McClellan and Parks [1972].

12.8 Summary of Chapter 12

The DFT (discrete Fourier transform) was introduced in this chapter and discussed in detail. The DFT converts an N-point sequence $x[n]$ into an N-point sequence $X[k]$. We have

$$\text{DFT } = X[k] = X(e^{j2\pi k/N}) = \text{samples of the DTFT } X(e^{j\omega}).$$

The set of N frequencies $\omega_k = 2\pi k/N$ is called the DFT grid. The main properties of the DFT are summarized in the table in Sec. 12.9.

The fast Fourier transform (FFT) computes the N DFT coefficients $X[k]$ in a highly efficient way. When $N = 2^M$ (a power of two), the FFT is called the radix-2 FFT and it takes about $0.5N\log_2 N$ multiplications, whereas direct computation without FFT would take N^2 multiplications. More accurate expressions for complexity were given in Sec. 12.4.2. When N is not a power of two, there exist other types of FFT algorithm which save computations. One of these is the mixed-radix FFT (Problem 12.21).

One of the most important properties of the DFT is the circular convolution property: given two N-point sequences $x[n]$ and $h[n]$, their circular convolution $y[n] = (x \otimes h)[n]$ has DFT $Y[k] = X[k]H[k]$. We can perform ordinary convolution of $x[n]$ and $h[n]$ by padding them with $N-1$ zeros and performing a circular convolution. This gives rise to fast convolution algorithms. The definition of circular convolution involves circular shifting and time reversal, which were developed in Sec. 12.5.

Given an N-point sequence $x[n]$, the DFT computes its DTFT on the grid $\omega_k = 2\pi k/N$. If we wish to compute the DTFT on a denser or sparser grid, we showed that there are ways to do this with high computational efficiency by performing FFT on appropriately modified sequences.

With $\mathbf{x}$ and $\mathbf{X}$ denoting $x[n]$ and $X[k]$ in vector form, the DFT is $\mathbf{X} = \mathbf{W}\mathbf{x}$, where $\mathbf{W}$ is the $N \times N$ DFT matrix with elements

$$[\mathbf{W}]_{kn} = W^{kn} = e^{-j2\pi kn/N},$$

where $W = e^{-j2\pi/N}$. The inverse DFT can be derived from $\mathbf{x} = \mathbf{W}^{-1}\mathbf{X}$. The DFT matrix is symmetric ($\mathbf{W}^T = \mathbf{W}$) and unitary up to scale ($\mathbf{W}^H\mathbf{W} = N\mathbf{I}$). So the inverse is $\mathbf{W}^{-1} = \mathbf{W}^H/N = \mathbf{W}^*/N$.

In this chapter we also summarized the relationships between the four types of Fourier transform studied in this book: CTFT, DTFT, DFT, and Fourier series.

12.9 Table of DFT Properties

In this section a number of properties of the DFT are listed. Proofs of Properties 5–10 are requested in Problem 12.23. The other properties are either proved in this chapter or are straightforward. Here, sequences such as $x[n]$ are N-point sequences defined for $0 \leq n \leq N-1$. Uppercase letters like $X[k]$ denote the N-point DFT, as in

$$X[k] = \sum_{n=0}^{N-1} x[n] W^{kn},$$

where $W \triangleq e^{-j2\pi/N}$. The notation

$$X[k] \longleftrightarrow x[n]$$

means that $X[k]$ is the DFT of $x[n]$. The notation $x[[n]]$ refers to the periodic extension of $x[n]$ (Sec. 12.5). Note that $X[k] = X[k+N]$ by definition of the DFT. The notations $x[[n-L]]$ and $x[[-n]]$ refer to circular shift and circular reversal (or flip), as defined in Sec. 12.5.2.

1. *Linearity.* $c_1 x_1[n] + c_2 x_2[n] \longleftrightarrow c_1 X_1[k] + c_2 X_2[k]$.
2. *Circular time shift.* $x[[n-L]] \longleftrightarrow W^{Lk} X[k]$.
3. *Exponential modulation.* $W^{-Ln} x[n] \longleftrightarrow X[k-L]$.
4. *Circular time reversal.* $x[[-n]] \longleftrightarrow X[-k]$ (equivalently, $X[N-k]$).
5. *Conjugation.* $x^*[n] \longleftrightarrow X^*[-k]$ (equivalently, $X^*[N-k]$).
6. *Real signal.* If $x[n]$ is real, $X[k] = X^*[-k]$ ($X[k]$ is Hermitian symmetric).
7. *Circular time reversal and conjugation.* $x^*[[-n]] \longleftrightarrow X^*[k]$.
8. *Hermitian signal.* If $x[n] = x^*[[-n]]$ (Hermitian symmetric), then $X[k]$ is real.
9. *Real and even signal.* If $x[n] = x[[-n]] =$ real, then $X[k]$ is real and even as well.
10. *Real and odd signal.* If $x[n] = -x[[-n]] =$ real, then $X[k]$ is imaginary and odd.
11. *Circular convolution definition.* $(x \circledast h)[n] \triangleq \sum_{l=0}^{N-1} x[l]h[[n-l]]$.
12. *Circular convolution theorem.* $(x \circledast h)[n] \longleftrightarrow X[k]H[k]$.
13. *Parseval's relation.* $\sum_{k=0}^{N-1} X[k]Y^*[k] = N \sum_{n=0}^{N-1} x[n]y^*[n]$.
14. *Energy balance.* $\sum_{k=0}^{N-1} |X[k]|^2 = N \sum_{n=0}^{N-1} |x[n]|^2$.

12.10 Table of N-Point DFTs

Signal: $x[n], 0 \le n \le N-1$	**DFT**: $X[k], 0 \le k \le N-1$
$x[n] = \delta[n]$	$X[k] = 1$
$x[n] = 1$	$X[k] = N\delta[k]$
$\delta[n - n_0]$	$e^{-j\frac{2\pi k}{N}n_0}\ (= W^{kn_0})$
$e^{j\frac{2\pi k_0}{N}n}\ (= W^{-k_0 n})$ (k_0 integer in $0 \le k_0 \le N-1$)	$N\delta[k - k_0]$
$\cos\left(\frac{2\pi k_0}{N}n\right)$ (k_0 integer in $0 \le k_0 \le N-1$)	$0.5N\left(\delta[k-k_0] + \delta[k-(N-k_0)]\right)$
$\sin\left(\frac{2\pi k_0}{N}n\right)$ (k_0 integer in $0 \le k_0 \le N-1$)	$-0.5Nj\left(\delta[k-k_0] - \delta[k-(N-k_0)]\right)$
$x[n] = e^{j\widehat{\omega}n}$ ($\widehat{\omega}$ not necessarily on DFT grid)	$X[k] = e^{j(N-1)(\widehat{\omega}-(2\pi k/N))/2} \times \left(\frac{\sin(N(\widehat{\omega}-(2\pi k/N))/2)}{\sin((\widehat{\omega}-(2\pi k/N))/2)}\right)$
Picket-fence sequence ($N = ML$) $x[n] = \begin{cases} 1 & \text{for } n = \text{multiple of } M \\ 0 & \text{otherwise} \end{cases}$	DFT is also picket-fence (Problem 12.10) $X[k] = \begin{cases} L & \text{for } k = \text{multiple of } L \\ 0 & \text{otherwise} \end{cases}$

PROBLEMS

Note: If you are using MATLAB for the computing assignments at the end, the "stem" command is convenient for plotting sequences like $x[n]$ and $X[k]$. The "plot" command is useful for plotting real-valued functions such as $|X(e^{j\omega})|$.

12.1 Find a closed-form expression for the DFT $X[k]$ of the 16-point sequence $x[n]$ Fig. P12.1. Then, by using properties of the DFT, find the DFT of the 16-point sequence $y[n]$.

Figure P12.1 Signals for Problem 12.1.

12.2 Show that the constant signal $x[n]$ in Fig. 12.2(b) has the DFT shown in the same figure.

12.3 For the DFT matrix $\mathbf{W}$ with $N = 5$, write out the elements of the matrix explicitly in the form W^m, where $0 \leq m \leq 4$. Is "-1" one of the entries? For what values of N will -1 be an entry of the DFT matrix?

12.4 For the DFT matrix $\mathbf{W}$ with $N = 5$, write out the elements of the IDFT matrix $\mathbf{W}^{-1}$ explicitly in terms of W^m, where $0 \leq m \leq 4$.

12.5 By writing the inverse DFT formula (12.42) explicitly in terms of the components of $\mathbf{x}$ and $\mathbf{X}$, show that it is equivalent to Eq. (12.10).

12.6 When N is a power of two, the FFT computes DFT using only (12.68) nontrivial complex multiplications, whereas a direct computation of DFT requires about N^2 multiplications. Compute these two complexities for $N = 64, N = 1024$, and $N = 65,536$.

12.7 For $N = 8$, we found in Sec. 12.4.4 that the bit-reversed version of $x[n]$ is $x[0], x[4], x[2], x[6], x[1], x[5], x[3], x[7]$. What is the bit-reversed version when $N = 4$? How about when $N = 16$?

12.8 For the eight-point signal $x[n]$ in Fig. P12.8, plot the following circularly modified signals: $x[[n-2]], x[[n+4]], x[[-n]], x[[3-n]], x[[-1-n]]$.

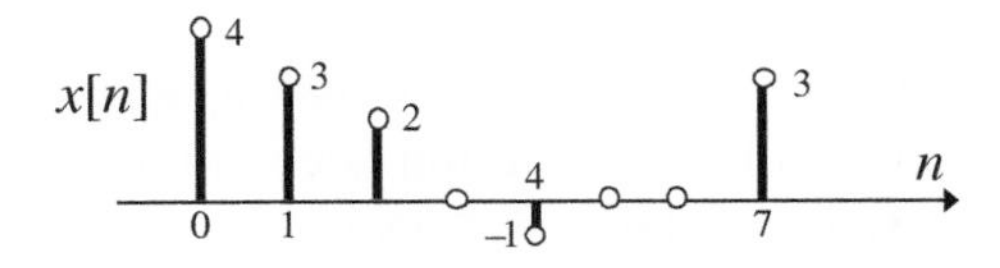

Figure P12.8 Signal for Problem 12.8.

12.9 Consider the 16-point sequences in Fig. P12.1. (a) Find the circular convolution $x_1[n] = (x \otimes x)[n]$ by directly writing all the steps in the time domain. (b) Find the circular convolutions $y_1[n] = (y \otimes y)[n]$ and $s[n] = (x \otimes y)[n]$. For part (b) you may use any method you wish.

12.10 *The picket-fence signal.* Let $N = ML$, where M and L are positive integers, and consider the N-point sequence

$$x[n] = \begin{cases} 1 & \text{for } n = \text{multiple of } M, \\ 0 & \text{otherwise.} \end{cases}$$

Show that this has the DFT given by

$$X[k] = \begin{cases} L & \text{for } k = \text{multiple of } L, \\ 0 & \text{otherwise.} \end{cases}$$

These are shown in Fig. P12.10 for $N = 15, M = 3$, and $L = 5$. This $x[n]$ is sometimes referred to as the "picket-fence" sequence. So, its DFT is also a picket-fence sequence. If the nonzero samples in $x[n]$ are closely spaced, then those in $X[k]$ are spaced far apart, and vice versa. This is analogous to the result that the continuous-time FT of the Dirac delta train is also a Dirac delta train (Sec. 9.5).

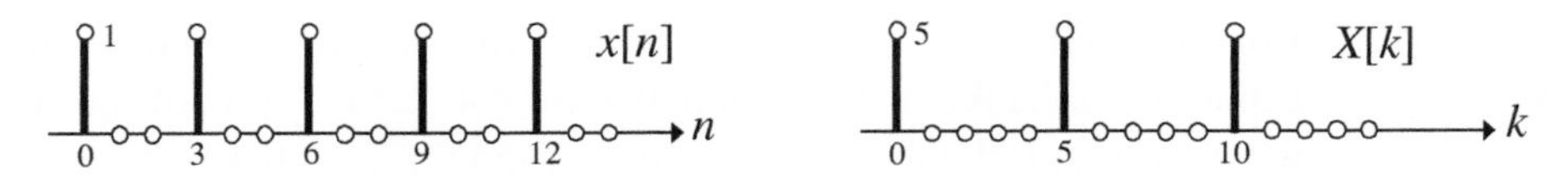

Figure P12.10 The picket-fence signal and its DFT.

12.11 Starting from the definition of the circular convolution sum, find an expression for the circular convolution of the N-point sequences $x[n] = e^{j2\pi n/N}$ and $h[n] = e^{j4\pi n/N}$, where $N > 1$. How do you explain the answer using the circular convolution theorem?

12.12 This is on linear and circular convolutions.

(a) Let $x[n]$ and $h[n]$ be four-point sequences $(0 \leq n \leq 3)$ given by $\{1, 4, -1, 0\}$ and $\{2, 1, -1, 2\}$, respectively. Find the circular convolution $y[n]$.

(b) Let $x[n]$ and $h[n]$ be three-point sequences $(0 \leq n \leq 2)$ given by $\{2, 1, -1\}$ and $\{3, 1, -2\}$, respectively. Compute the linear convolution $y[n]$ by performing a circular convolution of appropriately zero-padded sequences. Verify the correctness of the answer by direct computation of the linear convolution.

12.13 By starting from Eq. (12.82) and making an appropriate change of variables, show that circular convolution is commutative. That is, in the notation of Eq. (12.83), $(x \otimes h)[n] = (h \otimes x)[n]$.

12.14 *Linearity of circular convolution.* For fixed $h[n]$, show that the circular convolution (12.83) is a linear operator. That is, if $x_1[n]$ produces the output $y_1[n]$ and $x_2[n]$ produces the output $y_2[n]$, then $c_1x_1[n] + c_2x_2[n]$ produces the output $c_1y_1[n] + c_2y_2[n]$.

12.15 Find an example of two real sequences $x[n]$ and $h[n]$ of length five such that (a) neither sequence is a constant (in particular, not identically zero for all n)

and (b) the circular convolution of the sequences is zero for all n. Note that this is not possible in the case of linear convolution!

12.16 *Complex multiplication with three real multiplications.* Consider two complex numbers $x = x_r + jx_i$ and $h = h_r + jh_i$, where the subscripts r and i denote real and imaginary parts, respectively. We know the real and imaginary parts of the product $y = xh$ can be computed as

$$y_r = x_r h_r - x_i h_i, \quad y_i = x_r h_i + x_i h_r.$$

This requires *four* real multiplications and two real additions. It turns out that we can reduce this to *three* real multiplications if we use some extra additions. Thus, suppose we first compute the three quantities

$$h_r + h_i, \quad h_r - h_i, \quad \text{and} \quad x_r - x_i.$$

The complexity for this is three real additions (assuming subtraction is equivalent to addition in complexity). By mutiplying these with appropriate quantities and adding, we can get the real and imaginary parts of y. Show how this can be done using three real multiplications and two more additions. So the total complexity is *three* real multiplications and *five* real additions.

Note: While this is admittedly only a simple trick to speed up a complex multiplication, the topic of fast signal processing algorithms is highly sophisticated. Some of it involves deep ideas from discrete mathematics and number theory. For an excellent treatement of early algorithms, see Blahut [1985].

12.17 *Repeated DFTs.* Let $X[k]$ denote the N-point DFT of $x[n]$, and define a new sequence

$$Y[\ell] = \sum_{k=0}^{N-1} X[k] W^{\ell k}, \quad 0 \leq \ell \leq N-1.$$

That is, $Y[\ell]$ is the DFT of the DFT of $x[n]$. Express $Y[\ell]$ directly in terms of $x[n]$. For example, suppose $N = 5$ and $x[n]$ is $\{2, 4, 8, 10, 12\}$ for $0 \leq n \leq 4$. Then what is $Y[\ell]$ for $0 \leq \ell \leq 4$?

12.18 *More on repeated DFTs.* Let $\mathcal{F}$ denote a black box that takes an N-point sequence $x[n], 0 \leq n \leq N-1$, and produces an N-point sequence $y[n], 0 \leq n \leq N-1$, such that the output is the DFT of the input, that is, $y[l] = \sum_{n=0}^{N-1} x[n] W_N^{ln}$ for $0 \leq l \leq N-1$. In short, $\mathcal{F}$ is the DFT operator, and we use the operator notation $y[n] = \mathcal{F}\big(x[n]\big)$ to denote this.

(a) Suppose we use this black box four times in succession:

$$s[n] = \mathcal{F}\Big(\mathcal{F}\Big(\mathcal{F}\Big(\mathcal{F}\big(x[n]\big)\Big)\Big)\Big).$$

That is, we take the DFT of the DFT of the DFT of the DFT of $x[n]$. Show that $s[n] = N^2 x[n]$.

(b) Using the above result show that the $N \times N$ DFT matrix $\mathbf{W}$ has only four possible eigenvalues, $c, -c, jc$, and $-jc$, where $c = \sqrt{N}$.

12.19 Let $x[n]$ be an N-point sequence with DFT $X[k]$. Define the (NM)-point sequence

$$y_I[n] = \begin{cases} x[n/M] & n \text{ a multiple of } M, \\ 0 & \text{otherwise.} \end{cases}$$

Thus, $y_I[n]$ is obtained by inserting zero-valued samples between the samples of $x[n]$. This is called the M-fold expanded version of $x[n]$.

(a) For $N = 8$ and $M = 2$, give an example of the plots of $x[n]$ and the corresponding $y_I[n]$. For reasons of clarity, be sure that the samples of $x[n]$ are chosen to be distinct.

(b) For arbitrary integers $M > 0$ and $N > 0$, express the NM-point DFT $Y_I[k]$ in terms of N-point DFT $X[k]$.

(c) Assume $N = 8$ and $M = 2$, and draw some real-valued $X[k]$ with distinct numbers. Show the values of $Y_I[k]$.

12.20 Let $x[n]$ be an N-point signal ($N > 1$) such that it is its own DFT, that is, $X[n] = x[n]$ for $0 \le n \le N-1$. What can you say about $x[n]$?

12.21 *Mixed-radix FFT.* If N is not a power of two, then we cannot use a radix-2 FFT to compute an N-point DFT. But as long as N is a composite integer (i.e., not a prime number), we can still save computations, (compared to direct computation, which takes N^2 multiplications) by performing a mixed-radix FFT. For example, assume $N = 15 = 3 \times 5$, so that

$$X[k] = \sum_{n=0}^{14} x[n] W_{15}^{nk}, \qquad 0 \le k \le 14. \tag{P12.21a}$$

Express the time index n in the form

$$n = n_0 + 3n_1, \quad 0 \le n_0 \le 2,\ 0 \le n_1 \le 4. \tag{P12.21b}$$

With n_0 and n_1 spanning the indicated ranges, the integer n takes all the values in $0 \le n \le 14$. Similarly, express the frequency index k in the form

$$k = k_0 + 5k_1, \quad 0 \le k_0 \le 4,\ 0 \le k_1 \le 2, \tag{P12.21c}$$

so that all the values in $0 \le k \le 14$ are realized. Now the DFT can be rewritten as

$$X[k] = X[k_0 + 5k_1] = \sum_{n_0=0}^{2} \sum_{n_1=0}^{4} x[n_0 + 3n_1] W_{15}^{(n_0+3n_1)(k_0+5k_1)}. \tag{P12.21d}$$

(a) By using the properties of W, show that this can be rewritten as

$$X[k] = X[k_0 + 5k_1] = \sum_{n_0=0}^{2} W_3^{n_0 k_1} W_{15}^{n_0 k_0} \sum_{n_1=0}^{4} x[n_0 + 3n_1] W_5^{n_1 k_0}. \tag{P12.21e}$$

For each n_0 the inner summation can be regarded as a five-point DFT, mapping from the index n_1 to the index k_0. Similarly, for each k_0, the outer summation can be regarded as a three-point DFT. Thus, the 15-point DFT has been written in terms of several three-point DFTs and five-point DFTs. This is called the mixed-radix FFT algorithm. The multipliers $W_{15}^{n_0 k_0}$ are called the *twiddle factors.*

(b) Draw a flowgraph describing the computations, with black boxes representing three-point DFTs and five-point DFTs, and multipliers representing the twiddle factors (somewhat analogous to Fig. 12.4). Exactly how many five-point DFTs and how many three-point DFTs are involved in the entire computation?

(c) Let us now estimate the complexity. For simplicity, we will not bother to eliminate trivial multiplications. Assume that each three-point DFT requires 3^2 multiplications and each five-point DFT requires 5^2 multiplications. How many muliplications are involved in the compuation of $X[k]$ for all k? How does this compare with the 15^2 or 225 multiplications used for direct brute-force computation?

The mixed-radix idea can readily be generalized for the computation of N-point DFTs with $N = N_1 N_2$, and more generally for composite numbers of the form $N = N_1 N_2 \cdots N_K$.

12.22 In Problem 12.21, suppose we do not count *trivial multipliers*, that is, multiplications by $\pm 2^n$ and $\pm j2^n$ for integer n, which can be implemented with binary shifts. What is the resulting number of multiplications in the mixed-radix 15-point DFT algorithm?

12.23 With $X[k]$ denoting the N-point DFT of the N-point sequence $x[n]$, prove the following properties. It will be useful to review the notations in Sec. 12.5 first.

(a) $x^*[n] \longleftrightarrow X^*[-k]$ (equivalently, $X^*[N-k]$).

(b) If $x[n]$ is real, then $X[k] = X^*[-k]$.

(c) $x^*[[-n]] \longleftrightarrow X^*[k]$.

(d) If $x[n] = x^*[[-n]]$ (Hermitian symmetric), then $X[k]$ is real.

(e) If $x[n]$ is real and even ($x[n] = x[[-n]]$), then $X[k]$ is real and even.

(f) If $x[n]$ is real and odd ($x[n] = -x[[-n]]$), then $X[k]$ is imaginary and odd.

12.24 *Practice with symmetry properties.* Figure P12.24 shows several sequences defined for $0 \leq n \leq 3$. Identify all sequences which have real-valued DFT. Also

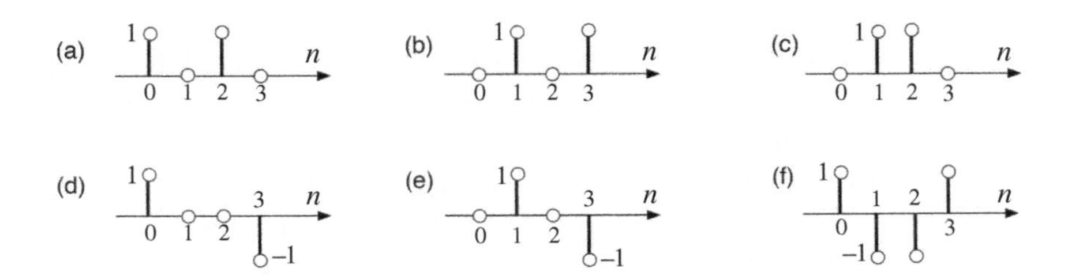

Figure P12.24 Signals for the DFT problem.

identify all sequences which have imaginary-valued DFT. (Use the properties of the DFT (Sec. 12.9) to minimize work. There is no need to evaluate the DFT explicitly.)

12.25 *Multiplexing two real signals.* Suppose we are required to compute the DFT of a real signal using the FFT algorithm. From the FFT flowgraph we know that the "realness" is quickly lost after the first few passes, and we end up doing mostly complex multiplications. Thus, the FFT algorithm does not exploit the realness of data to save computations. If the FFT algorithm already exists in the form of software or hardware, and we do not want to modify it, we have to employ some tricks outside the algorithm to take advantage of the realness of data. For example, it is possible to "multiplex" two real signals and compute their DFTs using a single complex DFT algorithm, as we show in this problem. Let $x_1[n]$ and $x_2[n]$ be two real N-point signals, and define the new complex N-point signal

$$x[n] = x_1[n] + jx_2[n].$$

With $X_1[k]$ and $X_2[k]$ denoting the DFTs of $x_1[n]$ and $x_2[n]$, we can write them in terms of the real and imaginary parts:

$$X_1[k] = X_{1r}[k] + jX_{1i}[k], \quad X_2[k] = X_{2r}[k] + jX_{2i}[k].$$

Then

$$X[k] = \left(X_{1r}[k] - X_{2i}[k]\right) + j\left(X_{1i}[k] + X_{2r}[k]\right).$$

By exploiting the fact that $x_1[n]$ and $x_2[n]$ are real, show how to obtain the four real quantities $X_{1r}[k]$, $X_{1i}[k]$, $X_{2r}[k]$, and $X_{2i}[k]$ from $X[k]$ by using additions, subtractions, and trivial multiplications. This means that the DFTs $X_1[k]$ and $X_2[k]$ can be computed at nearly the cost of one complex DFT. *Note:* The properties listed in Sec. 12.9 can be helpful.

12.26 *Efficient DFT for one real signal.* In Problem 12.25 we showed that a single FFT can be used to compute the DFTs of two real signals. Assume now that we have just one real N-point signal $x[n]$ and wish to compute its DFT $X[k]$. Assuming N to be even, we will show how to compute $X[k]$ using one $N/2$-point DFT and some minor overhead. Define a new $N/2$-point sequence

$$y[n] = x[2n] + jx[2n+1], \quad 0 \le n \le N/2 - 1.$$

(a) Using the technique of Problem 12.25, show how the N-point DFT of $x[n]$ can be recovered from the $N/2$-point DFT of $y[n]$, by using about $N/2$ extra multiplications, and some extra additions and trivial multiplications.

(b) Assuming that $N = 2^{10}$, what is the total number of complex multiplications involved in the computation of $X[k]$? How many complex multiplications have you saved compared to a direct FFT on $x[n]$ which ignores its "realness"?

12.27 *Fourier transform with frequency offset.* We know that the DFT calculates the Fourier transform $X(e^{j\omega})$ of an N-point signal $x[n]$ at the uniformly spaced frequencies $\omega_k = 2\pi k/N, 0 \le k \le N-1$. Suppose instead that we want $X(e^{j\omega})$ to be computed with a frequency offset θ, that is, at the frequencies

$$\omega_k = \frac{2\pi k}{N} - \theta, \quad 0 \le k \le N-1,$$

where $0 < \theta < 2\pi/N$. Given a DFT algorithm (which cannot be changed inside), explain how you would use it to perform the above computation.

12.28 *DFT at higher resolution.* In Sec. 12.6.1 we assumed $M > N$ and computed the Fourier transform of the N-point sequence $x[n]$ on the denser grid $\omega_k = 2\pi k/M$. We did not assume M is a multiple of N, but now let us suppose $M = KN$ for some integer K.

(a) Show that $X(e^{j\omega})$ on the denser grid can be expressed as

$$X(e^{j\omega_k}) = \sum_{n=0}^{N-1} x[n] W_{KN}^{kn}, \quad 0 \le k \le KN-1,$$

where $W_L = e^{-j2\pi/L}$ as usual.

(b) The KN values of k in $0 \le k \le KN-1$ can also be represented as

$$k = k_0 + Kk_1, \quad 0 \le k_0 \le K-1, \; 0 \le k_1 \le N-1.$$

For each k there is a unique k_0 and k_1 above. Show that

$$X(e^{j\omega_k}) = \sum_{n=0}^{N-1} x_{k_0}[n] W_N^{k_1 n}, \qquad (P12.28)$$

where $x_{k_0}[n] = x[n] W_{KN}^{k_0 n}$ for each k_0 in $0 \le k_0 \le K-1$. For fixed k_0, the sum in Eq. (P12.28) is the N-point DFT of $x_{k_0}[n]$. So we can compute $X(e^{j\omega_k})$ for all $M = KN$ values of k by computing K DFTs (one for each k_0), each of size N (original data size).

(c) If K and N are powers of two we can use FFTs, and the above method has complexity $\approx K(N/2)\log_2(N)$. This is better than doing a KN-point FFT directly (complexity $\approx (KN/2)\log_2(KN)$). For $N = 2^8$ and $K = 8$, what are the above two complexities?

12.29 *Nonuniform sampling in frequency.* We know that the DFT calculates the Fourier transform $X(e^{j\omega})$ of an N-point signal $x[n]$ at the uniformly spaced frequencies $\omega_k = 2\pi k/N, 0 \le k \le N-1$. The inverse DFT formula shows that we can recover all N points of $x[n]$ from these samples of $X(e^{j\omega})$. It turns out that in order to recover $x[n]$, the N samples of $X(e^{j\omega})$ need not be uniformly spaced in frequency. Thus, consider N distinct frequencies

$$0 \le \omega_0 < \omega_1 < \cdots < \omega_{N-1} < 2\pi. \qquad (P12.29a)$$

Define the vector of samples of $X(e^{j\omega})$ as

$$\mathbf{X}_{nu} = \left[X(e^{j\omega_0}) \quad X(e^{j\omega_1}) \quad \cdots \quad X(e^{j\omega_{N-1}})\right]^T, \qquad (P12.29b)$$

where the subscript nu is a reminder of "nonuniform." This is sometimes referred to as nonuniform DFT, for lack of a better jargon. We have

$$\mathbf{X}_{nu} = \mathbf{V}\mathbf{x}, \tag{P12.29c}$$

where $\mathbf{V}$ is an $N \times N$ matrix and $\mathbf{x} = \begin{bmatrix} x[0] & x[1] & \cdots & x[N-1] \end{bmatrix}^T$ (as in Eq. (12.31)).

(a) For the case $N = 4$, explicitly write down the elements of $\mathbf{V}$ in terms of the ω_i. You will find that the rows of $\mathbf{V}$ are Vandermonde vectors, but the columns are not (unlike the DFT matrix, whose rows and columns are Vandermonde vectors).

(b) For any N, prove that $\mathbf{V}$ is nonsingular under the assumptions mentioned above. Thus, $\mathbf{x}$ can always be recovered from the N samples $X(e^{j\omega_k})$.

Remarks: Since $\mathbf{x} = \mathbf{V}^{-1}\mathbf{X}_{nu}$, we have

$$x[n] = \sum_{k=0}^{N-1} X(e^{j\omega_k})U_{nk} = \sum_{k=0}^{N-1} X(e^{j\omega_k})f_k[n], \quad 0 \le n \le N-1, \tag{P12.29d}$$

where $\mathbf{U} = \mathbf{V}^{-1}$ and $f_k[n] \triangleq U_{nk}$ is the kth basis function in the representation.

12.30 *Orthogonality and nonuniform sampling.* We know that the DFT matrix is unitary up to scale (Sec. 12.3). Now consider the setup of Problem 12.29, where $\mathbf{V}$ is a generalization of the DFT matrix.

(a) Show that the kth and mth rows of $\mathbf{V}$ are mutually orthogonal if and only if $\omega_k - \omega_m = 2\pi l/N$ for integer l. That is, the frequencies must be separated by integer multiples of $2\pi/N$.

(b) Hence show that $\mathbf{V}$ is unitary up to scale, or more specifically $\mathbf{V}\mathbf{V}^H = N\mathbf{I}$, if and only if the frequencies are uniformly spaced, that is,

$$\omega_k = \frac{2\pi k}{N} + \phi, \quad 0 \le k \le N-1, \tag{P12.30}$$

for some ϕ in $0 \le \phi < 2\pi/N$. This is nothing but the DFT grid shifted by ϕ. (Note that $\mathbf{V}\mathbf{V}^H = N\mathbf{I}$ is equivalent to $\mathbf{V}^H\mathbf{V} = N\mathbf{I}$, since $\mathbf{V}$ is a square matrix.)

(c) Thus, even though we can reconstruct $x[n]$ from any set of N samples of $X(e^{j\omega})$ in $0 \le \omega < 2\pi$, we obtain a unitary transformation from time to frequency if and only if the frequency grid is either the DFT grid or a trivially shifted version of it! Under this condition, show further that the basis functions $f_k[n]$ in Eq. (P12.29d) are orthogonal, that is, $\sum_{n=0}^{N-1} f_k[n]f_m^*[n] = 0$ when $k \ne m$. So (P12.29d) is an orthogonal-basis representation for $x[n]$ when the frequency grid is as in Eq. (P12.30).

Remarks: When $\mathbf{V}$ is unitary up to scale, there are some advantages. Thus, imagine that only a noisy version of the samples of $X(e^{j\omega})$ is available. If $\mathbf{V}$ is unitary, it can be shown that the error in $X(e^{j\omega_k})$ is not severely amplified when we estimate $x[n]$ from the noisy $X(e^{j\omega_k})$. For arbitray nonsingular $\mathbf{V}$, the noise amplification can be severe.

12.31 *Frequency decoupling and nonuniform sampling.* Nonuniform sampling of $X(e^{j\omega})$ as in Problem 12.29 leads to the representation (P12.29d) for $x[n]$. The coefficient $X(e^{j\omega_k})$ of the kth term depends only on ω_k, and not on $\omega_l, l \neq k$. One wonders whether $f_k[n]$ depends only on ω_k or also on ω_l for some $l \neq k$. We explore this now.

(a) When $\mathbf{V}$ is the DFT matrix, show that $f_k[n]$ depends on the frequency ω_k but not on $\omega_l, l \neq k$.

(b) Let $N = 2$ and assume $\mathbf{V}$ is not the DFT matrix. Write the samples $x[0]$ and $x[1]$ explicitly in the form

$$\begin{bmatrix} x[0] \\ x[1] \end{bmatrix} = X(e^{j\omega_0})\mathbf{u}_0 + X(e^{j\omega_1})\mathbf{u}_1,$$

where $\mathbf{u}_0$ and $\mathbf{u}_1$ are the columns of $\mathbf{U} = \mathbf{V}^{-1}$. In general, does $\mathbf{u}_0$ depend on both ω_0 and ω_1? How about $\mathbf{u}_1$? Based on this, argue that $f_k[n]$ can in general depend on ω_k *and* on ω_l, $l \neq k$.

Thus, the basis function $f_k[n]$ in Eq. (P12.29d) may not represent a single frequency ω_k, that is, there is no decoupling of frequencies in this representation.

12.32 (Computing assignment) *Leakage in DFT.* Let $x[n] = \sin(\omega_0 n)$ for $0 \leq n \leq N-1$ and zero otherwise, and let $X[k]$ be the N-point DFT. We will plot $|X[k]|$ for various values of ω_0 and learn some interesting details. Let $N = 32$ and $k_0 = 10$.

(a) Let $\omega_0 = 2\pi k_0/N$. So $\sin(\omega_0 n)$ has the two frequencies $\pm\omega_0$, both of which are on the DFT grid. Plot $|X[k]|$. You will see two nonzero values, corresponding to frequencies $\pm 2\pi k_0/N$. The negative frequency $-2\pi k_0/N$ is equivalent to a frequency in $\pi < \omega < 2\pi$, so it shows up for the DFT index k in the range $N/2 < k < N$.

(b) Now change ω_0 to the value $\omega_0 = 2\pi(k_0 + 0.1)/N$. So ω_0 and $-\omega_0$ are not on the DFT grid any more, and there is leakage as explained in Sec. 12.2.3 in the context of Fig. 12.3. Plot $|X[k]|$, and you will indeed see that ω_0 and $-\omega_0$ spread into multiple lines in the DFT plot.

(c) Repeat part (b) for $\omega_0 = 2\pi(k_0 + 0.5)/N$ and plot $|X[k]|$. You will find that the leakage is more severe. It is not possible to tell whether $x[n]$ contains two frequencies or multiple frequencies.

12.33 (Computing assignment) *Leakage in DFT* (multiple frequencies). Let

$$x[n] = \begin{cases} a_1 \sin(\omega_1 n) + a_2 \sin(\omega_2 n) + a_3 \sin(\omega_3 n) + a_4 \sin(\omega_4 n), & 0 \leq n \leq N-1, \\ 0 & \text{otherwise,} \end{cases}$$

where

$$\omega_1 = 2 \cdot \left(\frac{2\pi}{32}\right), \quad \omega_2 = 5 \cdot \left(\frac{2\pi}{32}\right), \quad \omega_3 = 10 \cdot \left(\frac{2\pi}{32}\right), \quad \omega_4 = 14 \cdot \left(\frac{2\pi}{32}\right)$$

and $a_1 = 0.8, a_2 = 0.8, a_3 = 0.9, a_4 = 0.4$. Let $X[k]$ be the N-point DFT of $x[n]$ as usual.

(a) Let $N = 32$. Plot $|X[k]|$. Since ω_i are on the DFT grid, you will see eight clear lines (two for each ω_i).

(b) Let everything be as in part (a), but let ω_2 and ω_3 be changed to

$$\omega_2 = (5 + 0.3) \cdot \left(\frac{2\pi}{32}\right), \quad \omega_3 = (10 + 0.3) \cdot \left(\frac{2\pi}{32}\right).$$

Plot $|X[k]|$. Since ω_2 and ω_3 are not on the DFT grid any more, they will spread out because of the leakage (Sec. 12.2.3). You will not see eight clear lines any more.

(c) Let everything be as in part (b), but let $N = 256$. Plot $|X[k]|$. Now the frequencies ω_2 and ω_3 are still not on the N-point DFT grid, so there will be leakage. However, the spreading is more controlled and you see eight clear peaks because N is large. Thus, by increasing N the effect of spreading can be controlled.

12.34 (Computing assignment) *Computing FT at lower resolution.* Let $x[n] = \cos(\omega_0 n)$ for $0 \leq n \leq 255$ and zero otherwise, and let $\omega_0 = (40.5\pi/256)$. We would like to compute $X(e^{j\omega})$ only at 64 points, namely, $\omega = 2\pi k/64$, $0 \leq k \leq 63$. We will do this in two ways.

(a) *Method 1.* Just compute the 256-point DFT $X[k]$ of the 256-point sequence $x[n]$ and retain the downsampled version $X[4k]$. Plot $|X[4k]|$, $0 \leq k \leq 63$.

(b) *Method 2.* Use the following two-step procedure:
- Compute the 64-point sequence $x_b[\ell]$ as described in Sec. 12.6.2.
- Now compute the 64-point DFT $X_b[k]$ of $x_b[\ell]$ and plot $|X_b[k]|$. This should be the same as the plot you got in part (a).

For completeness, plot the 64-point sequence $x_b[\ell]$ as well. The second method is computationally more efficient, as explained in Sec. 12.6.2.

12.35 (Computing assignment) *Orthogonality.* Consider the matrix $\mathbf{V}$ in Problem 12.29 again. Let $N = 4$. For the special case where $\mathbf{V}$ is the DFT matrix, compute and print $\mathbf{V}^H\mathbf{V}$. Now let $\omega_2 = 2\pi(2.5/N)$ instead of $2\pi(2/N)$. Again compute and print $\mathbf{V}^H\mathbf{V}$. You will find that this is not a diagonal matrix any more. This is consistent with what is proved in Problem 12.30 regarding orthogonality of columns of $\mathbf{V}$.

12.36 (Computing assignment) *Multiplexing real signals.* Consider two real four-point sequences $x_1[n]=\{0, 2, 3, 1\}$ and $x_2[n]=\{1, 2, 4, 1\}$ (with $0 \leq n \leq 3$).

(a) Find the four-point DFTs $X_1[k]$ and $X_2[k]$ by direct computation.

(b) Instead of direct computation, we now use the trick in Problem 12.25.
- Find the DFT $X[k]$ of the complex sequence $x[n]$ defined in that problem.
- Extract $X_{1r}[k], X_{2i}[k], X_{1i}[k]$, and $X_{2r}[k]$ from the real and imaginary parts of $X[k]$. (All these quantities were defined in Problem 12.25.)

- Find $X_1[k]$ and $X_2[k]$ from these.

These $X_1[k]$ and $X_2[k]$ should agree with what you found in part (a).

12.37 (Computing assignment) *Signal repetition*. Let $x[n]$ be zero everywhere except in $0 \leq n \leq 15$. Choose arbitrary values for $x[0], x[1], \ldots, x[15]$. (To avoid trivialities, make sure these are all nonzero, and not equal to a constant.) Define $y[n]$ as follows:

$$y[n] = \begin{cases} x[n] & \text{for } 0 \leq n \leq 15, \\ x[n-16] & \text{for } 16 \leq n \leq 31, \\ 0 & \text{otherwise.} \end{cases}$$

That is, $x[n]$ is a 16-point sequence and $y[n]$ is a 32-point sequence with the first 16 samples of $x[n]$ repeated again. Let $X[k]$ be the 16-point DFT of $x[n]$, and $Y[k]$ the 32-point DFT of $y[n]$.

(a) Plot $|X[k]|$ and $|Y[k]|$. (If using MATLAB, use the "stem" command.)

(b) You will find that $Y[2k] = 2X[k]$, whereas $Y[2k+1] = 0$. Explain this behavior theoretically.

(c) More generally, let $x[n]$ be an N-point sequence (possibly nonzero in $0 \leq n \leq N-1$ and zero outside). Let $y[n]$ be the MN-point sequence obtained by repeating these N points M times:

$$y[n] = \begin{cases} x[n] & \text{for } 0 \leq n \leq N-1, \\ x[n-N] & \text{for } N \leq n \leq 2N-1, \\ \vdots & \\ x[n-(M-1)N] & \text{for } (M-1)N \leq n \leq MN-1, \\ 0 & \text{otherwise.} \end{cases}$$

Let $X[k]$ be the N-point DFT of $x[n]$, and $Y[k]$ the MN-point DFT of $y[n]$. Show that $Y[Mk] = MX[k]$ and $Y[l] = 0$ otherwise.

12.38 (Computing assignment) *FFT complexity*. For $N = 2^M$ we know that the radix-2 FFT requires

$$\frac{N}{2}\log_2 N - \frac{3N}{2} + 2$$

multiplications (see Eq. (12.68)). Plot the above expression for $1 \leq M \leq 12$. In the same graph, also plot N^2, which is the complexity for computing the DFT by "brute force." You will see that the FFT complexity is negligible compared to brute force. Show the plots for $1 \leq M \leq 5$ separately for added clarity.

13 The Laplace Transform

13.1 Introduction

In Sec. 3.9 an overview of continuous-time LTI systems was presented, which the student should review at this time. In that context, we defined the Laplace transform of a continuous-time signal $x(t)$ as follows:

$$X(s) = \int_{-\infty}^{\infty} x(t)e^{-st}dt, \tag{13.1}$$

where s is a complex variable. With $s = j\omega$, this yields the continuous-time Fourier transform (CTFT). Thus, the Laplace transform is a generalization of CTFT. Its properties are therefore similar to those of the CTFT, which were discussed in detail in Chapter 9 and summarized in Sec. 9.16. For example, if $x(t) = \delta_c(t)$ then $X(s) = \int_{-\infty}^{\infty} \delta_c(t)e^{-st}dt = 1$ for all s and similarly

$$x(t) = \delta_c(t - t_0) \quad \longleftrightarrow \quad X(s) = e^{-st_0} \; \forall s, \tag{13.2}$$

where the notation $\longleftrightarrow$ means "Laplace transform pair." The Laplace transform also has similarities to the z-transform, discussed extensively in Chapter 6. Like the z-transform summation, the integral (13.1) converges only for certain values of s in the complex s-plane, as we shall see. Notice that, for the z-transform, the **unit circle** of the z-plane represents frequencies, whereas for the Laplace transform the **imaginary axis** of the s-plane represents frequencies. In this context, the student may also remember the bilinear transform (Sec. 7.7), which converts a continuous-time transfer function $H_c(s)$ to a discrete-time transfer function $H(z)$. Because of these connections to earlier discussions, only a brief presentation of the Laplace transform is given here.

13.2 A Basic Example

Let us begin with the simple example

$$x(t) = e^{-at}\mathcal{U}(t), \tag{13.3}$$

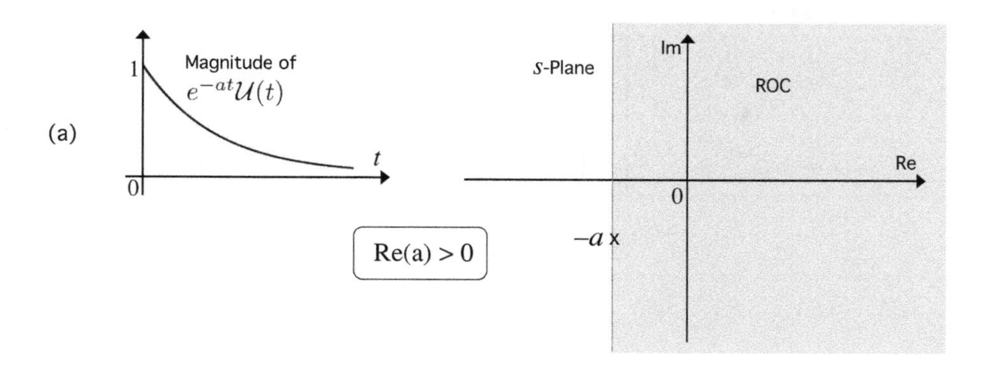

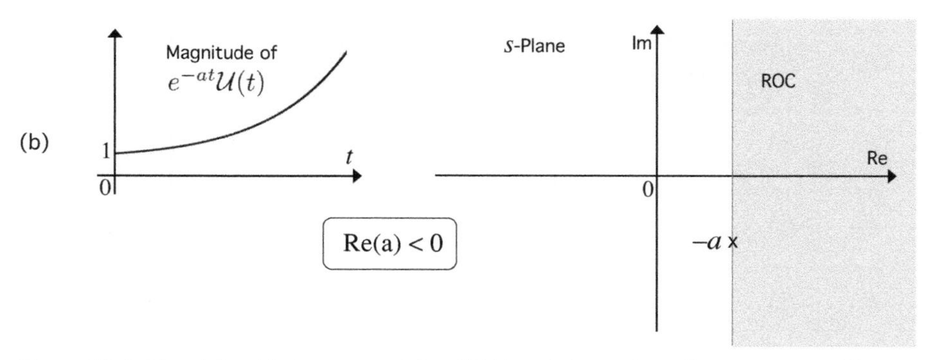

Figure 13.1 The right-sided exponential $e^{-at}\mathcal{U}(t)$ and the region of convergence of its Laplace transform. (a) Re(a) > 0 and (b) Re(a) < 0.

where a is possibly complex. Then

$$X(s) = \int_0^\infty e^{-at}e^{-st}dt = \int_0^\infty e^{-(s+a)t}dt = \frac{1}{-(s+a)}\left[e^{-(s+a)t}\right]_0^\infty. \tag{13.4}$$

If $\text{Re}(s+a) > 0$, then $e^{-(s+a)\infty} = 0$. This shows that

$$X(s) = \frac{1}{s+a}, \quad \text{Re}(s) > -\text{Re}(a). \tag{13.5}$$

If $\text{Re}(s+a) \leq 0$, the integral does not converge. The region $\text{Re}(s) > -\text{Re}(a)$ is called the region of convergence or **ROC** of the Laplace transform in this example. Figure 13.1 summarizes this example for two cases: $\text{Re}(a) > 0$ and $\text{Re}(a) < 0$.

Next, for the *left-sided exponential*

$$x(t) = -e^{-at}\mathcal{U}(-t), \tag{13.6}$$

we can similarly show that the Laplace transform is

$$X(s) = \frac{1}{s+a}, \quad \text{Re}(s) < -\text{Re}(a). \tag{13.7}$$

This example is also summarized in Fig. 13.2.

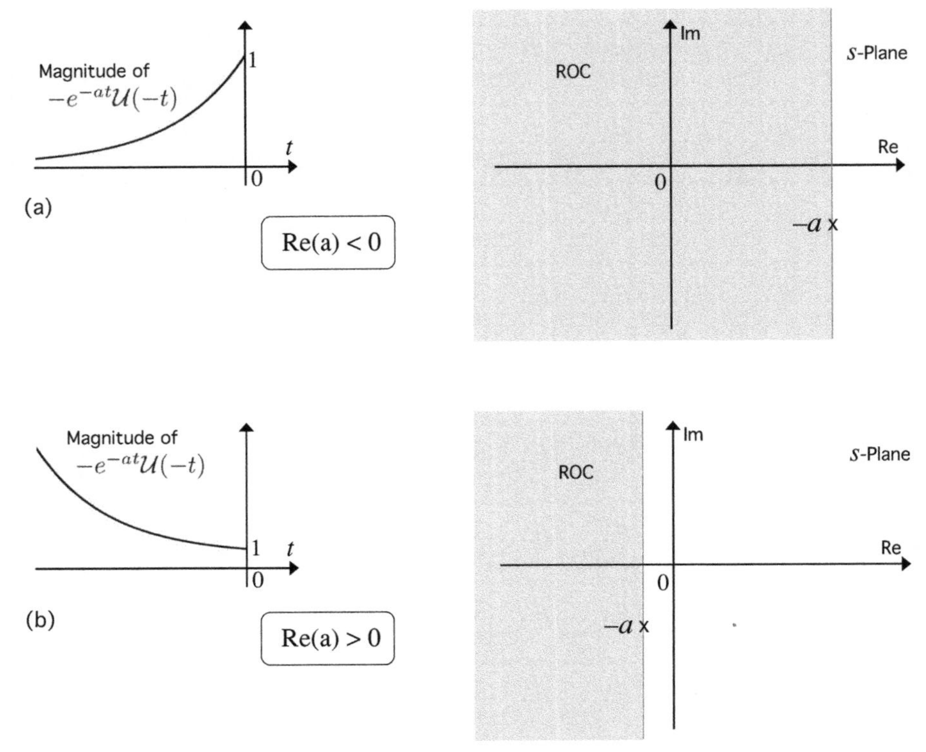

Figure 13.2 The left-sided exponential $-e^{-at}\mathcal{U}(-t)$ and the region of convergence of its Laplace transform. (a) $\mathrm{Re}(a) < 0$ and (b) $\mathrm{Re}(a) > 0$.

Note that in both examples the pole of $X(s)$ is at $s = -a$, and determines the ROC. Thus, $X(s) = 1/(s + a)$ has two possible inverse Laplace transforms depending on the ROC:

$$X(s) = \frac{1}{s+a} \quad \longleftrightarrow \quad x(t) = \begin{cases} e^{-at}\mathcal{U}(t) & \mathrm{Re}(s) > -\mathrm{Re}(a), \\ -e^{-at}\mathcal{U}(-t) & \mathrm{Re}(s) < -\mathrm{Re}(a). \end{cases} \tag{13.8}$$

It is clear that, given an expression for $X(s)$, its ROC has to be known in order to find the inverse transform. When specifying the Laplace transform we should therefore mention the ROC. From the above example we obtain the following conclusions:

1. For the right-sided exponential $x(t) = e^{-at}\mathcal{U}(t)$, the following are true if and only if $\mathrm{Re}(a) > 0$:
   (a) the ROC of $X(s)$ includes the imaginary axis of the s-plane;
   (b) the Fourier transform $X(j\omega)$ therefore exists; and
   (c) $x(t)$ decays to zero as $|t| \to \infty$ (consistent with the fact that $X(j\omega)$ exists).
   On the other hand, for the left-sided exponential $x(t) = -e^{-at}\mathcal{U}(-t)$, the above three are true if and only if $\mathrm{Re}(a) < 0$.

2. Notice in particular that the *Laplace transform can exist even when the Fourier transform does not*. For example, $x(t) = e^t\mathcal{U}(t)$ has $X(s) = 1/(s-1)$ for $\text{Re}(s) > 1$. Since this ROC does not include $s = j\omega$, the FT does not exist.
3. The exponential $x(t) = e^{-at}$ (i.e., the two-sided exponential) does not have a Laplace transform, that is, the ROC of $X(s)$ is empty, no matter what the value of a is. This is because we can write

$$x(t) = e^{-at}\mathcal{U}(t) + e^{-at}\mathcal{U}(-t). \tag{13.9}$$

   The Laplace transforms for the first and second terms have ROCs $\text{Re}(s) > -\text{Re}(a)$ and $\text{Re}(s) < -\text{Re}(a)$, respectively, which do not have any overlap. So the Laplace transform of $x(t) = e^{-at}$ does not exist for any a.
4. For example, $e^{j\omega_0 t}$ does not have a Laplace transform. So it does not have a Fourier transform either. But don't we usually accept $2\pi\delta_c(\omega-\omega_0)$ as its Fourier transform? How do we reconcile these contradictory statements? The fact is that whenever we say that a Laplace transform exists in a certain ROC, we also imply that it is finite everywhere in that ROC. Thus, if the Laplace transform $X(s)$ exists and its ROC includes the imaginary axis, then the Fourier transform $X(j\omega)$ is finite for all ω. But since $2\pi\delta_c(\omega - \omega_0)$ is not finite at $\omega = \omega_0$, it cannot be regarded as a Laplace transform evaluated on $s = j\omega$. The answer $2\pi\delta_c(\omega - \omega_0)$ is accepted because this gives "correct answers" for a number of problems involving the Fourier transform. In addition, it is consistent with the insight that $e^{j\omega_0 t}$ has all its energy concentrated at $\omega = \omega_0$.

From Eq. (13.8) we can readily deduce that

$$\begin{aligned} e^{j\omega_0 t}\mathcal{U}(t) \quad &\longleftrightarrow \quad \frac{1}{s - j\omega_0}, \\ \cos(\omega_0 t)\mathcal{U}(t) \quad &\longleftrightarrow \quad \frac{s}{s^2 + \omega_0^2}, \\ \sin(\omega_0 t)\mathcal{U}(t) \quad &\longleftrightarrow \quad \frac{\omega_0}{s^2 + \omega_0^2}, \end{aligned} \tag{13.10}$$

where the ROC is $\text{Re}(s) > 0$ in each case.

13.3 Inverting the Laplace Transform

Now consider the example

$$H(s) = \frac{s+3}{s^2 + 3s + 2} = \frac{s+3}{(s+1)(s+2)}. \tag{13.11}$$

We show how to find the inverse Laplace transform $h(t)$ of this, based on what has been seen in Sec. 13.2. As you would expect from Eq. (13.8), the answer is not unique and depends on the choice of the ROC. Clearly the poles are at $s = -1$ and $s = -2$, and the zeros at $s = -3$ and $s = \infty$. We can use partial fraction expansion (PFE), similar to what we did in Sec. 6.6, and rewrite

$$H(s) = \frac{2}{s+1} - \frac{1}{s+2}. \tag{13.12}$$

The first term has two possible inverse Laplace transforms, similar to Eq. (13.8), and so does the second term. Thus

$$\frac{2}{s+1} \quad \longleftrightarrow \quad \begin{cases} 2e^{-t}\mathcal{U}(t) & \text{if ROC is Re}(s) > -1, \\ -2e^{-t}\mathcal{U}(-t) & \text{if ROC is Re}(s) < -1, \end{cases} \tag{13.13}$$

$$\frac{-1}{s+2} \quad \longleftrightarrow \quad \begin{cases} -e^{-2t}\mathcal{U}(t) & \text{if ROC is Re}(s) > -2, \\ e^{-2t}\mathcal{U}(-t) & \text{if ROC is Re}(s) < -2. \end{cases} \tag{13.14}$$

This yields three possible inverse transforms for Eq. (13.12), namely

$$h(t) = \begin{cases} 2e^{-t}\mathcal{U}(t) - e^{-2t}\mathcal{U}(t) & \text{if ROC is Re}(s) > -1, \\ -2e^{-t}\mathcal{U}(-t) + e^{-2t}\mathcal{U}(-t) & \text{if ROC is Re}(s) < -2, \\ -2e^{-t}\mathcal{U}(-t) - e^{-2t}\mathcal{U}(t) & \text{if ROC is } -2 < \text{Re}(s) < -1. \end{cases} \tag{13.15}$$

The three regions of convergence which give rise to the three answers are summarized in Fig. 13.3. Notice the following:

1. When the ROC is to the *right* of a vertical line, the inverse transform $h(t)$ is *right* sided.
2. When the ROC is to the *left* of a vertical line, the inverse transform $h(t)$ is *left* sided.
3. When the ROC is between two vertical lines, the inverse transform $h(t)$ is two sided.
4. Only one ROC includes the imaginary axis, and in this case $H(j\omega)$ (Fourier transform) exists and $h(t)$ is absolutely integrable, that is, $H(s)$ represents a stable LTI system.

These results can be generalized for rational $H(s)$ with arbitrary number of poles and zeros, similar to what we did for z-transforms in Chapter 6. Thus, it can be shown that a rational $H(s)$ has a stable inverse $h(t)$ if and only if there are no poles on the imaginary axis; in this case, if the ROC is chosen to include the imaginary axis, then $H(j\omega)$ exists and $h(t)$ represents a stable system. It should also be clear to the student that if $H(s)$ has poles with M distinct values for the real parts, then there are $M + 1$ possible ROCs, hence $M + 1$ possible inverse transforms.

The fourth combination. Returning again to Eq. (13.12), since each term has two answers (as in Eqs. (13.13) and (13.14)), there are four possible combinations of these. Three of these combinations are given in Eq. (13.15), and the fourth is

$$2e^{-t}\mathcal{U}(t) + e^{-2t}\mathcal{U}(-t). \tag{13.16}$$

The first term corresponds to the ROC $\text{Re}(s) > -1$ and the second term to $\text{Re}(s) < -2$. But these two cannot be simultaneously true for any s. That is, the intersection of the regions $\text{Re}(s) > -1$ and $\text{Re}(s) < -2$ is empty. So (13.16) is not a valid answer

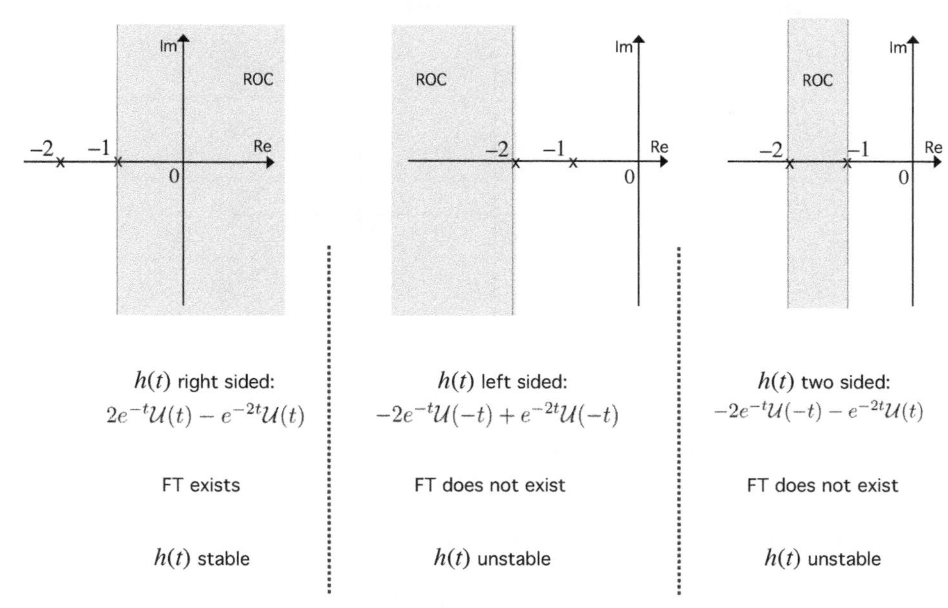

Figure 13.3 Three possible ROCs in the s-plane for the Laplace transform $H(s)$ in Eq. (13.12). The corresponding inverse transforms $h(t)$ are also shown. In this example the right-sided inverse represents a stable system; $H(j\omega)$ exists only in this case.

for the inverse transform of $H(s)$, because the ROC corresponding to it is empty. Another way to look at it is this: we already argued that for two-sided exponentials like e^{-t}, the ROC is empty, so the Laplace transform does not exist. For similar reason, a linear combination like

$$c_0 e^{-t}\mathcal{U}(t) + c_1 e^{-t}\mathcal{U}(-t), \quad c_0, c_1 \neq 0, \tag{13.17}$$

has no ROC. Since the second term in (13.16) grows even faster than the second term in (13.17) as $t \to -\infty$, it follows that (13.16) has no ROC either. $\triangledown \triangledown \triangledown$

13.4 Closed Form for Inverse Transform

An insightful connection between the Laplace transform and the Fourier transform allows us to find a nice formula for the inverse Laplace transform. Let $s = \sigma + j\omega$ for some s in the ROC of $X(s)$, where σ and ω are real. Then $X(s) = \int_{-\infty}^{\infty} x(t)e^{-st}dt$ can be written as

$$X(s) = \int_{-\infty}^{\infty} x(t)e^{-\sigma t}e^{-j\omega t}dt. \tag{13.18}$$

We can regard this as the *Fourier transform* of the modified signal

$$x_1(t) = x(t)e^{-\sigma t}. \tag{13.19}$$

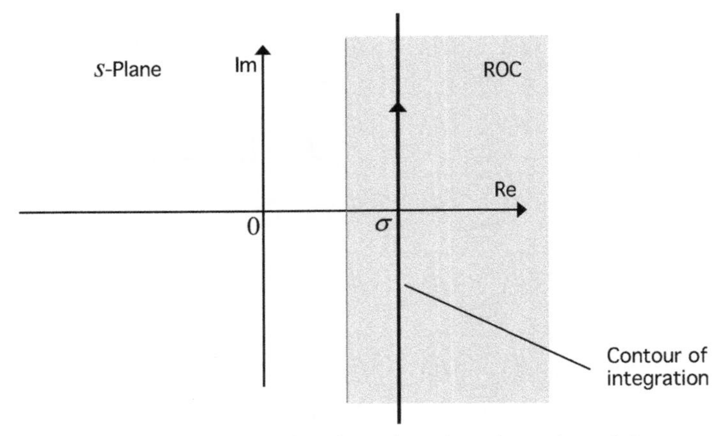

Figure 13.4 Pertaining to the discussion about inversion of the Laplace transform by contour integration in the s-plane.

Using the inverse FT formula (9.41), we therefore have

$$x(t)e^{-\sigma t} = \frac{1}{2\pi}\int_{-\infty}^{\infty} X(s)e^{j\omega t}d\omega. \tag{13.20}$$

Thus, the inverse Laplace transform $x(t)$ can be computed just by computing the inverse Fourier transform (13.20) and multiplying the answer with $e^{\sigma t}$. Now, $x(t)$ can be rewritten as:

$$x(t) = \frac{1}{2\pi}\int_{-\infty}^{\infty} X(s)e^{\sigma t}e^{j\omega t}d\omega, \tag{13.21}$$

where σ is any real number in the ROC. The above integral can be interpreted as a contour integral

$$x(t) = \frac{1}{2\pi j}\int_{C} X(s)e^{st}ds, \tag{13.22}$$

where the contour is any vertical line in the ROC, as shown in Fig. 13.4.[1] Even though this is an elegant closed-form formula, it is often more convenient to find the inverse Laplace transform of rational functions by using partial fractions as we did in the examples (13.11) and (13.12).

13.5 Some Properties and their Usefulness

A list of properties of the Laplace transform is given in the table in Sec. 13.9. These are very similar to the properties of Fourier transforms and z-transforms proved earlier. So the proofs are left as exercises in the problems section. The convolution theorem mentioned in the table is central to LTI system theory, as we saw in Sec. 3.9. We will now use some of the other properties to derive new Laplace transform pairs starting from known examples such as

[1] This can be regarded as a closed contour in the ROC by adding a semicircle of infinite radius to the right of the line.

$$e^{-at}\mathcal{U}(t) \longleftrightarrow \frac{1}{s+a} \qquad (\text{ROC Re}(s) > -\text{Re}(a)), \tag{13.23}$$

which was derived in Sec. 13.2.

13.5.1 Examples Using the Derivative Property

For example, using the derivative property

$$tx(t) \longleftrightarrow -\frac{dX(s)}{ds} \tag{13.24}$$

from the table in Sec. 13.9, it follows from the example (13.23) that

$$te^{-at}\mathcal{U}(t) \longleftrightarrow \frac{1}{(s+a)^2}, \tag{13.25}$$

and the ROC is still $\text{Re}(s) > -\text{Re}(a)$. More generally (Problem 13.4), we can show that

$$t^n e^{-at}\mathcal{U}(t) \longleftrightarrow \frac{n!}{(s+a)^{n+1}} \quad \text{with ROC Re}(s) > -\text{Re}(a), \tag{13.26}$$

for any integer $n \geq 0$. Similarly, by starting from the Laplace transform pair

$$-e^{-at}\mathcal{U}(-t) \longleftrightarrow \frac{1}{s+a} \quad \text{with ROC Re}(s) < -\text{Re}(a), \tag{13.27}$$

we can show that

$$-t^n e^{-at}\mathcal{U}(-t) \longleftrightarrow \frac{n!}{(s+a)^{n+1}} \quad \text{with ROC Re}(s) < -\text{Re}(a), \tag{13.28}$$

for any integer $n \geq 0$.

13.5.2 Examples Based on Other Properties

We now use some of the other properties in Sec. 13.9 to derive more examples from (13.23).

The Pulse Signal

Consider the pulse signal

$$p(t) = \begin{cases} 1 & \text{for } 0 < t < 1, \\ 0 & \text{otherwise.} \end{cases} \tag{13.29}$$

This can be rewritten as $p(t) = \mathcal{U}(t) - \mathcal{U}(t-1)$. From example (13.23) with $a = 0$, we know

$$\mathcal{U}(t) \longleftrightarrow \frac{1}{s}, \quad \text{Re}\, s > 0. \tag{13.30}$$

Using the linearity and shift property of the Laplace transform (see table in Sec. 13.9), it follows that the above pulse has the Laplace transform

$$P(s) = \frac{1}{s} - \frac{e^{-s}}{s} = \frac{1-e^{-s}}{s}. \tag{13.31}$$

Each of the terms above has the ROC $\mathrm{Re}(s) > 0$. So $P(s)$ has an ROC which surely includes $\mathrm{Re}(s) > 0$. But it is more than that. To see this, note that a direct use of the definition would yield

$$P(s) = \int_0^1 e^{-st} dt = \frac{e^{-st}}{-s}\Big|_{t=0}^{1} = \frac{1 - e^{-s}}{s} \tag{13.32}$$

for all s such that $\mathrm{Re}(s) > -\infty$.

The Dilated Pulse

Next consider the dilated pulse

$$p_2(t) \triangleq p(t/2) = \begin{cases} 1 & \text{for } 0 < t < 2, \\ 0 & \text{otherwise.} \end{cases} \tag{13.33}$$

By using the dilation property from Sec. 13.9, we find

$$P_2(s) = \frac{1 - e^{-2s}}{s}. \tag{13.34}$$

The above examples are admittedly simple, and could have been handled directly by using the definition of the Laplace transform. The purpose of the above discussion is to demonstrate that we can often derive the Laplace transform of a signal by starting from known examples and using the properties in the table. This is sometimes simpler.

13.6 LTI Systems Based on Differential Equations

Now consider an equation of the form

$$\sum_{k=0}^{N} a_k \frac{d^k y(t)}{dt^k} = \sum_{k=0}^{N} b_k \frac{d^k x(t)}{dt^k}. \tag{13.35}$$

This is called an Nth-order differential equation. Here, a_k and b_k are constants that do not depend on t or x or y. So this equation is called a *constant-coefficient* differential equation. Furthermore, since $d^k x(t)/dt^k$ and $d^k y(t)/dt^k$ are linear in $x(t)$ and $y(t)$, Eq. (13.35) is *linear* in $x(t)$ and $y(t)$. In short, it is a *linear, constant-coefficient, differential equation.* In many ways, its properties are analogous to those of recursive difference equations discussed in Chapter 5. Thus, if we regard $x(t)$ as the input and $y(t)$ as the output of a system satisfying Eq. (13.35), then the system behaves as an LTI system, under an assumption called the "zero initial condition (IC)" assumption. To be more clear with these ideas, suppose $x(t) = 0$ for $t < 0$, and

$$x(t), \quad t \geq 0, \tag{13.36}$$

is given to us. Then we can find $y(t)$ for all $t \geq 0$ if we know the N "initial conditions" at $t = 0$, namely,

$$\left.\frac{d^k y(t)}{dt^k}\right|_{t=0}, \quad 0 \leq k \leq N-1 \quad \text{(initial conditions)}. \tag{13.37}$$

Zero initial condition just means that the above N quantities are zero:

$$y(0) = 0, \quad \left.\frac{dy(t)}{dt}\right|_{t=0} = 0, \quad \ldots, \quad \left.\frac{d^{N-1}y(t)}{dt^{N-1}}\right|_{t=0} = 0. \tag{13.38}$$

In a way similar to what we did in Sec. 5.7, it can be shown that the mapping $x(t) \longmapsto y(t)$ defined by Eq. (13.35) is linear and time invariant when Eq. (13.38) holds. The system can therefore be described by a transfer function $H(s) = Y(s)/X(s)$. We now show how to express $H(s)$ in terms of the coefficients a_k and b_k in Eq. (13.35). From Sec. 13.9 we know that

$$\frac{d^k x(t)}{dt^k} \longleftrightarrow s^k X(s), \quad \frac{d^k y(t)}{dt^k} \longleftrightarrow s^k Y(s). \tag{13.39}$$

Taking the Laplace transform on both sides of Eq. (13.35) and using linearity, we obtain

$$\sum_{k=0}^{N} a_k s^k Y(s) = \sum_{k=0}^{N} b_k s^k X(s), \tag{13.40}$$

so that the transfer function $H(s) = Y(s)/X(s)$ is

$$H(s) = \frac{\sum_{k=0}^{N} b_k s^k}{\sum_{k=0}^{N} a_k s^k} = \frac{A\prod_{k=1}^{N}(s - z_k)}{\prod_{k=1}^{N}(s - p_k)}. \tag{13.41}$$

This a rational function in s. So we have shown that the transfer function of an LTI system described by the linear constant-coefficient differential equation (13.35) is rational in s. Such an LTI system is referred to as a **rational LTI** system, similar to the discrete-time systems in Chapter 5. In the factored form (13.41), p_k and z_k are the poles and zeros of $H(s)$, respectively. As mentioned in Sec. 7.6.1, the continuous-time causal LTI system with transfer function (13.41) is stable if and only if

$$\text{Re}\,(p_k) < 0 \quad \text{(stability condition)}, \tag{13.42}$$

that is, if and only if all the poles are in the left half of the s-plane. See Problem 13.6 for other insightful ways to derive Eq. (13.41).

13.7 The LCR Circuit

We now present an example based on electrical circuits. Readers not familiar with circuits can skip this section without loss of continuity. Consider the LCR circuit

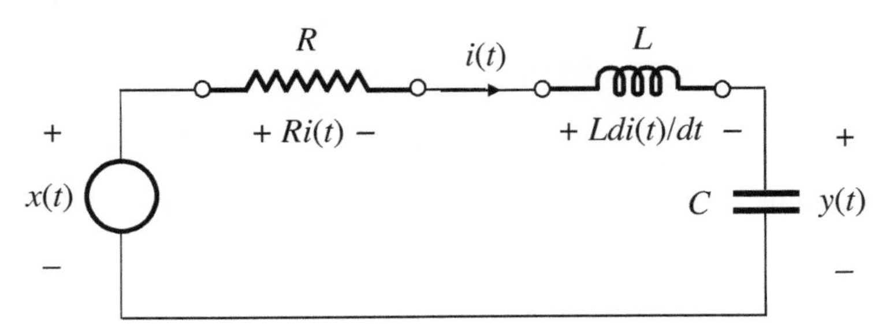

Figure 13.5 An LCR circuit.

shown in Fig. 13.5. Since this is a physical system, it is causal. With capacitor voltage $y(t)$ regarded as the output and voltage source $x(t)$ as the input, we first write the differential equation describing this system. With $i(t)$ denoting the current in the indicated direction, the voltages across the resistor and inductor are as shown in the figure (Sec. 2.11.2). Thus, Kirchhoff's law for voltages yields

$$L\frac{di(t)}{dt} + Ri(t) + y(t) = x(t). \tag{13.43}$$

But since the capacitor voltage $y(t)$ and current $i(t)$ are related as $i(t) = Cdy(t)/dt$, the preceding equation becomes

$$LC\frac{d^2y(t)}{dt^2} + RC\frac{dy(t)}{dt} + y(t) = x(t). \tag{13.44}$$

This is a second-order linear, constant-coefficient, differential equation. With zero initial conditions, it therefore behaves as an LTI system with transfer function of the form (13.41), that is,

$$H(s) = \frac{1}{1 + sRC + s^2LC}. \tag{13.45}$$

Students familiar with circuits can obtain this expression almost by inspection of Fig. 13.5. Thus, the system from $x(t)$ to $y(t)$ is a voltage divider. The impedance of C is $1/sC$ and the total impedance seen by $x(t)$ is $R + sL + 1/sC$. So the voltage across C is, in the s-domain,

$$Y(s) = \left(\frac{1/sC}{R + sL + 1/sC}\right)X(s). \tag{13.46}$$

This yields an expression for $H(s) = Y(s)/X(s)$ that is the same as Eq. (13.45).

The poles of the system (13.45) are the roots of the denominator polynomial, and they are given by

$$p_1 = -\frac{R}{2L} + \sqrt{\left(\frac{R}{2L}\right)^2 - \frac{1}{LC}}, \quad p_2 = -\frac{R}{2L} - \sqrt{\left(\frac{R}{2L}\right)^2 - \frac{1}{LC}}, \tag{13.47}$$

where it is assumed that $L \neq 0$ and $C \neq 0$. For a passive circuit, R, C, and L are non-negative. Furthermore, if the circuit is lossy then $R > 0$. In this case it is readily verified that the poles are in the left-half plane. So the lossy LCR circuit represents a stable system. The poles (13.47) form a complex-conjugate pair if

$$R < \sqrt{\frac{4L}{C}}. \tag{13.48}$$

Otherwise they are real poles. For a "non-passive" circuit (more commonly called an active circuit), some of the circuit elements can be negative, and the circuit may not be stable. In fact, even for a passive circuit with $R = 0$ (i.e., a **lossless** circuit), the poles are on the imaginary axis:

$$p_1 = j\sqrt{\frac{1}{LC}}, \qquad p_2 = -j\sqrt{\frac{1}{LC}}. \tag{13.49}$$

In this case also the circuit is unstable, since Eq. (13.42) is violated.

Zero IC and initial stored energy. The circuit behaves like an LTI system and has the transfer function $H(s)$ only under zero initial conditions, that is,

$$y(0) = 0 \quad \text{and} \quad \left.\frac{dy(t)}{dt}\right|_{t=0} = 0. \tag{13.50}$$

The first condition simply says that the voltage across the capacitor is zero, that is, the capacitor has no stored charge. Since $i(t) = Cdy(t)/dt$, the second condition says $i(0) = 0$, that is, the initial current in the circuit is zero. This means that there is no stored magnetic energy in the inductor. Thus, zero IC means that there is no stored energy in any of the electrical storage elements. ▽▽▽

13.7.1 Impulse Response

Assume $L, C > 0$ and $R \geq 0$, so the circuit is passive. If

$$R > \sqrt{\frac{4L}{C}}, \tag{13.51}$$

the poles are real and the impulse response of Eq. (13.45) is a linear combination of the form

$$h(t) = \left(c_1 e^{p_1 t} + c_2 e^{p_2 t}\right)\mathcal{U}(t), \qquad p_1 < 0,\ p_2 < 0 \tag{13.52}$$

(Problem 13.13), which is a sum of decaying exponentials. But when R is small enough so that condition (13.48) holds, the impulse response is more interesting. The "small resistance" condition (13.48) means that the poles form a complex-conjugate pair (13.47):

$$p_1 = -\alpha + j\beta, \quad p_2 = -\alpha - j\beta, \tag{13.53}$$

where

$$\alpha = \frac{R}{2L} \geq 0 \quad \text{and} \quad \beta = \sqrt{\frac{1}{LC} - \frac{R^2}{4L^2}} > 0. \tag{13.54}$$

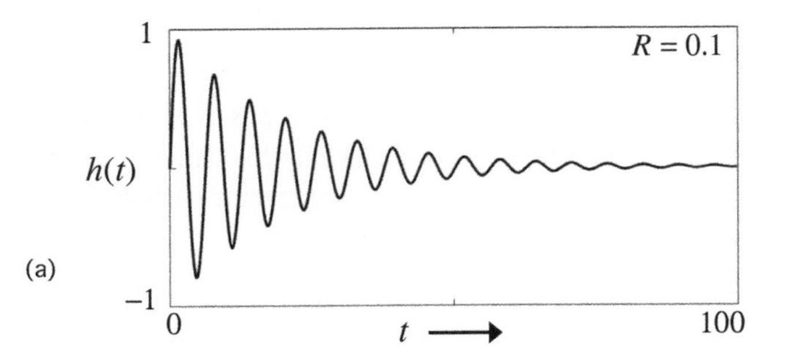

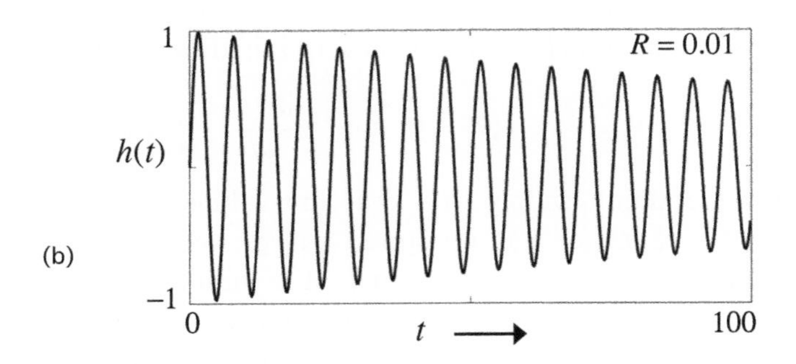

Figure 13.6 Impulse response of the oscillatory LCR circuit for (a) $R = 0.1$ and (b) $R = 0.01$. In both cases, $L = 1$ and $C = 1$.

In this case it can be shown (Problem 13.13) that the impulse response of Eq. (13.45) is

$$h(t) = \frac{1}{\beta LC} e^{-\alpha t} \sin(\beta t)\mathcal{U}(t). \tag{13.55}$$

This is called a **damped sine wave**, the damping factor being the exponential $e^{-\alpha t}$. The frequency of the sine wave β is said to be the **natural frequency** of the LCR circuit. This gets larger and larger as the resistance decreases, until it becomes

$$\beta = \sqrt{\frac{1}{LC}} \tag{13.56}$$

for the **lossless** circuit with $R = 0$. For large R, the damping factor $e^{-\alpha t}$ decays fast and the sine wave is quickly attenuated. For small R, $e^{-\alpha t}$ decays more slowly and the "ringing" of the sinusoid $\sin \beta t$ lasts longer. This is demonstrated in Fig. 13.6. When the circuit is lossless ($R = 0$), there is no damping:

$$h(t) = \frac{1}{\beta LC} \sin(\beta t)\, \mathcal{U}(t). \tag{13.57}$$

So the lossless circuit has a steady, undecaying, sinusoidal impulse response. That is, it behaves as an **oscillator** with oscillation frequency β, analogous to the discrete-time oscillator in Sec. 8.5. Equation (13.57) represents an unstable system, as $h(t)$ is not absolutely integrable. This is not a surprise, because the poles are on the imaginary axis in this case (Eq. (13.49)).

13.7.2 Frequency Response

From the transfer function expression (13.45), we can compute the frequency response by setting $s = j\omega$. The result is

$$H(j\omega) = \frac{1}{1 - \omega^2 LC + j\omega RC}. \tag{13.58}$$

The magnitude-squared response is therefore

$$|H(j\omega)|^2 = \frac{1}{(1 - \omega^2 LC)^2 + \omega^2 R^2 C^2}, \tag{13.59}$$

and the phase response is

$$\phi(\omega) = -\tan^{-1}\left(\frac{\omega RC}{1 - \omega^2 LC}\right). \tag{13.60}$$

The nature of the plot $|H(j\omega)|^2$ depends on the relative values[2] of R and $\sqrt{L/C}$. We now mention some results in this regard, and proofs are requested in Problem 13.14. First, if $R < \sqrt{2L/C}$, then the plot exhibits a peak at the frequency

$$\omega_p = \sqrt{\frac{1}{LC} - \frac{R^2}{2L^2}}. \tag{13.61}$$

This is called the **resonant frequency** of the LCR circuit. (If $R > \sqrt{2L/C}$, then the right-hand side of Eq. (13.61) is imaginary, so the peak frequency is undefined.) Substituting the value of ω_p into Eq. (13.59) and simplifying, we find that the peak value of the magnitude is

$$|H(j\omega_p)| = \frac{1}{\sqrt{\frac{R^2 C}{L}\left(1 - \frac{R^2 C}{4L}\right)}}, \tag{13.62}$$

which for $R << \sqrt{L/C}$ can be approximated as

$$|H(j\omega_p)| \approx \frac{1}{R}\sqrt{\frac{L}{C}}. \tag{13.63}$$

If $R \geq \sqrt{2L/C}$, then there is no peak at any nonzero frequency – the plot is monotone decreasing in $0 \leq \omega \leq \pi$. These behaviors are demonstrated in Figs. 13.7(a) and (b). The condition for existence of a peak in $|H(j\omega)|^2$, namely,

$$R < \sqrt{\frac{2L}{C}} \quad \text{(condition for peak or resonance)}, \tag{13.64}$$

should be compared with the condition for oscillatory behavior of $h(t)$ (Sec. 13.7.1), namely,

$$R < \sqrt{\frac{4L}{C}} \quad \text{(condition for oscillations)}, \tag{13.65}$$

[2] It turns out that $\sqrt{L/C}$ has the same dimensions as R; both are specified in ohms.

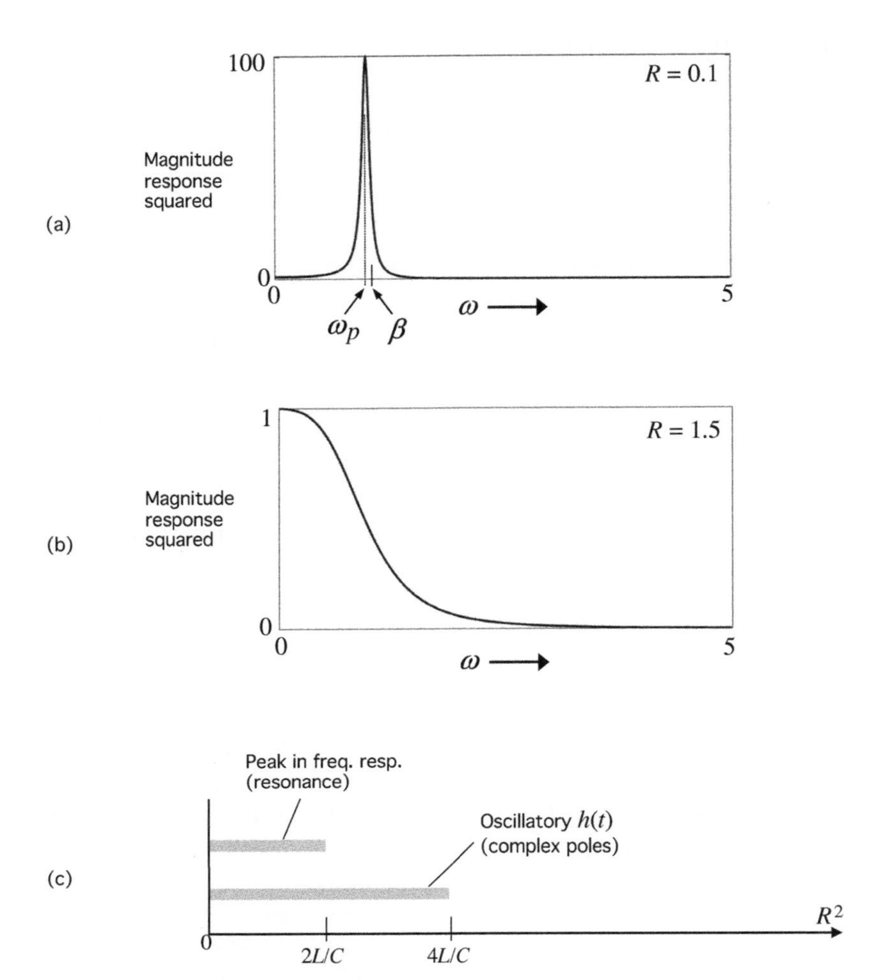

Figure 13.7 (a), (b) Plots of $|H(j\omega)|^2$ for the LCR circuit for two values of the resistance R, with $L = 1$ and $C = 1$. There is a peak in the plot only when $R < \sqrt{2L/C} = \sqrt{2}$. (c) Ranges of R for which different circuit behaviors occur.

which is also the condition for the poles to be complex. (See discussions after Eq. (13.48).) Figure 13.7(c) summarizes the behaviors for different ranges of R. Notice that the peak frequency (resonance frequency) ω_p in Eq. (13.61) and the oscillation frequency

$$\beta = \sqrt{\frac{1}{LC} - \frac{R^2}{4L^2}} \tag{13.66}$$

(from Eq. (13.54)) are such that

$$\omega_p \leq \beta, \tag{13.67}$$

with equality if and only if $R = 0$. In Fig. 13.7(a), where $L = 1$, $C = 1$, and $R = 0.1$, we have $\omega_p = 0.9975$, $\beta = 0.9987$.

13.8 Summary of Chapter 13

In this chapter we had a discussion of the Laplace transform, which is as fundamental to continuous-time systems as the z-transform is to discrete-time systems. Several examples were presented, and the application in circuit analysis was demonstrated with the help of an LCR circuit. Similar to the z-transform, the Laplace transform can be regarded as a generalization of the Fourier transform. Since we have had extensive discussions on these other types of transforms, our presentation of the Laplace transform was brief.

13.9 Table of Laplace Transform Properties

We summarize here some of the important properties of the Laplace transform. Recall that the Laplace transform is defined as

$$X(s) = \int_{-\infty}^{\infty} x(t)e^{-st}dt, \tag{13.68}$$

and has region of convergence (ROC) $\mathcal{R}$ of the form $R_1 < \text{Re}(s) < R_2$. The inverse transform can be written as

$$x(t) = \frac{1}{2\pi}\int_{-\infty}^{\infty} X(s)e^{\sigma t}e^{j\omega t}d\omega, \tag{13.69}$$

where $s = \sigma + j\omega$ is in the ROC, σ is constant, and the integration is over ω. The notation

$$x(t) \longleftrightarrow X(s)$$

means that $x(t)$ and $X(s)$ constitute a Laplace transform pair. In what follows, we also indicate the Laplace transforms of modified versions such as $x(at)$, $x^*(t)$, and so forth. Their ROCs are not indicated as they can readily be found in terms of $\mathcal{R}$.

1. *Linearity*. If $x_1(t) \longleftrightarrow X_1(s)$ and $x_2(t) \longleftrightarrow X_2(s)$, then $c_1x_1(t) + c_2x_2(t) \longleftrightarrow c_1X_1(s) + c_2X_2(s)$.
2. *Time shift*. $x(t - t_0) \longleftrightarrow e^{-st_0}X(s)$.
3. *Modulation*. $e^{s_0 t}x(t) \longleftrightarrow X(s - s_0)$.
4. *Derivative in time*. $d^kx(t)/dt^k \longleftrightarrow s^kX(s)$.
5. *Derivative in s-domain*. $t^nx(t) \longleftrightarrow (-1)^n d^nX(s)/ds^n$.
6. *Convolution theorem*. $y(t) \triangleq \int_{-\infty}^{\infty} x(\tau)h(t-\tau)d\tau \longleftrightarrow X(s)H(s)$.
7. *Time reversal*. $x(-t) \longleftrightarrow X(-s)$.
8. *Dilation*. $x(at) \longleftrightarrow X(s/a)/|a|$ (for real $a \neq 0$).
9. *Conjugation*. $x^*(t) \longleftrightarrow X^*(s^*)$.
10. *Conjugation and reversal*. $x^*(-t) \longleftrightarrow X^*(-s^*)$.

13.10 Table of Laplace Transform Pairs

Signal	**Laplace transform** $X(s) = \int_{-\infty}^{\infty} x(t)e^{-st}dt$
$x(t) = \delta_c(t)$	$X(s) = 1, \ \forall s$
$x(t) = \delta_c(t - t_0)$	$X(s) = e^{-st_0}, \ \forall s$
$e^{-at}\mathcal{U}(t)$	$X(s) = \frac{1}{s+a}$ $\quad \mathrm{Re}(s) > -\mathrm{Re}(a)$
$-e^{-at}\mathcal{U}(-t)$	$X(s) = \frac{1}{s+a}$ $\quad \mathrm{Re}(s) < -\mathrm{Re}(a)$
e^{-at} (e.g., $x(t) = 1, \ \forall t$)	Laplace transform does not exist
$t^n e^{-at}\mathcal{U}(t)$	$\frac{n!}{(s+a)^{n+1}}$ $\quad \mathrm{Re}(s) > -\mathrm{Re}(a)$
$-t^n e^{-at}\mathcal{U}(-t)$	$\frac{n!}{(s+a)^{n+1}}$ $\quad \mathrm{Re}(s) < -\mathrm{Re}(a)$
$e^{j\omega_0 t}\mathcal{U}(t)$	$\frac{1}{s - j\omega_0}$ $\quad \mathrm{Re}(s) > 0$
$\cos(\omega_0 t)\mathcal{U}(t)$	$\frac{s}{s^2 + \omega_0^2}$ $\quad \mathrm{Re}(s) > 0$
$\sin(\omega_0 t)\mathcal{U}(t)$	$\frac{\omega_0}{s^2 + \omega_0^2}$ $\quad \mathrm{Re}(s) > 0$
$p(t) = \begin{cases} 1 & \text{for } 0 < t < 1 \\ 0 & \text{otherwise} \end{cases}$	$P(s) = \frac{1 - e^{-s}}{s}$ $\quad \mathrm{Re}(s) > -\infty$

PROBLEMS

13.1 Prove the two *derivative properties* for Laplace transforms, given in Sec. 13.9.

13.2 By starting from example (13.23) and using the properties in Sec. 13.9, prove the following:

$$\cos(\omega_0 t)\mathcal{U}(t) \longleftrightarrow \frac{s}{s^2+\omega_0^2} \quad \text{with ROC Re}\,(s) > 0,$$

$$\sin(\omega_0 t)\mathcal{U}(t) \longleftrightarrow \frac{\omega_0}{s^2+\omega_0^2} \quad \text{with ROC Re}\,(s) > 0,$$

for real ω_0.

13.3 Prove the *convolution theorem* for Laplace transforms, given in Sec. 13.9.

13.4 By starting from example (13.23) and using the properties in Sec. 13.9, prove (13.26).

13.5 By starting from example (13.27) and using the properties in Sec. 13.9, prove (13.28).

13.6 Assume that the system (13.35) is LTI (i.e., that the initial conditions are zero). Then we know that the exponential input $e^{\beta t}$ produces the exponential output $H(\beta)e^{\beta t}$. Using this "eigenfunction property," prove that the transfer function is indeed as in Eq. (13.41).

13.7 Prove the *time-shift property* and the *modulation property* for Laplace transforms, given in Sec. 13.9.

13.8 Prove the *conjugation property* and the *conjugation and reversal property* for Laplace transforms, given in Sec. 13.9.

13.9 Consider the LTI system

$$H(s) = \frac{2s+7}{s^2+7s+12}.$$

Using partial fractions and the basic results in Sec. 13.2, find all three inverse transforms $h(t)$ and indicate the ROC in each case. Is there an inverse $h(t)$ that represents a stable system? If so, is it causal?

13.10 Consider the LTI system

$$H(s) = \frac{2s-1}{s^2-s-2}.$$

Using partial fractions and the basic results in Sec. 13.2, find all three inverse transforms $h(t)$ and indicate the ROC in each case. Is there an inverse $h(t)$ that represents a stable system? If so, is it causal?

13.11 Consider the LTI system

$$H(s) = \frac{2s}{s^2+1}.$$

Using partial fractions and the basic results in Sec. 13.2, find all the inverse transforms $h(t)$ and indicate the ROC in each case. (a) Is there an inverse $h(t)$ that represents a stable system? (b) Is there an inverse $h(t)$ that is two-sided?

13.12 Prove the *dilation property* for Laplace transforms, given in Sec. 13.9.

13.13 Consider the LCR circuit in Fig. 13.5, with transfer function (13.45). Assume $L, C > 0$, and $R \geq 0$. This system has poles as in (13.47).

(a) For $p_1 \neq p_2$ (i.e., $R \neq \sqrt{4L/C}$), show that

$$h(t) = \frac{1}{LC(p_1 - p_2)}\left(e^{p_1 t} - e^{p_2 t}\right)\mathcal{U}(t).$$

(b) In particular, therefore, when R satisfies Eq. (13.51), p_1 and p_2 are real and negative. So the impulse response of Eq. (13.45) is as in Eq. (13.52), which is a sum of decaying exponentials. What are the values of c_1 and c_2 in terms of L, C, and R?

(c) Suppose $R = \sqrt{4L/C}$ instead of Eq. (13.51). What is the impulse response $h(t)$ in terms of L, C, and R?

(d) Suppose R is small enough so that it satisfies Eq. (13.48). Show that the impulse response $h(t)$ of Eq. (13.45) is as in Eq. (13.55).

13.14 Consider the LCR circuit in Fig. 13.5, with transfer function (13.45). Assume $L, C, R > 0$, and that R is small enough to satisfy Eq. (13.64).

(a) Show that the zeros of the derivative of $|H(j\omega)|^2$ are at $\omega = 0$ and $\omega = \pm\omega_p$, where ω_p is as in Eq. (13.61). So these frequencies are the local extrema of $|H(j\omega)|^2$.

(b) Show that the local extremum of $|H(j\omega)|^2$ at $\omega = \omega_p$ is in fact a local *maximum*.

(c) Show that $|H(j\omega_p)|$ is as in Eq. (13.62).

(d) Show finally that $|H(j\omega_p)| \geq H(0)$, so that ω_p is indeed the frequency where $|H(j\omega)|^2$ is a global maximum.

14 State-Space Descriptions

14.1 Introduction

In this chapter we introduce state-space descriptions for computational graphs (or structures) representing discrete-time LTI systems. State-space descriptions are fundamental. They are not only useful in theoretical analysis, they can also be used to derive alternative structures for a transfer function $H(z)$ starting from a known structure. For example, we can derive the cascade-form structures starting from the direct form (Sec. 5.9) using a state transformation. In fact, all minimal structures for $H(z)$ (i.e., structures which do not have redundant delay elements) can in principle be derived by starting from the direct form and performing a so-called similarity transformation. In addition, state-space descriptions give a different perspective on stability, and on minimality of the implementation, as we shall see. The way in which initial conditions affect the output can be handled especially elegantly using state-space descriptions. Before we begin, recall that the Nth-order rational LTI transfer function

$$H(z) = \frac{\sum_{k=0}^{N} a_k z^{-k}}{1 + \sum_{k=1}^{N} b_k z^{-k}} = \frac{A(z)}{B(z)} \tag{14.1}$$

can be implemented with the causal recursive difference equation

$$y[n] = -\sum_{k=1}^{N} b_k y[n-k] + \sum_{k=0}^{N} a_k x[n-k], \tag{14.2}$$

represented by the computational graph (or structure) shown in Fig. 14.1 for $N = 3$ (Sec. 5.9). We begin by developing the state-space description for this structure. We then explain how to do this for more general structures, including multi-input multi-output (MIMO) systems. Then we derive some of the most important properties of state-space descriptions of arbitrary structures. This includes general expressions for transfer functions and impulse responses for the special case where the initial conditions are zero, and also several deeper properties such as minimality and so forth. Finally, we revisit allpass filters (introduced in Sec. 7.4) and show that

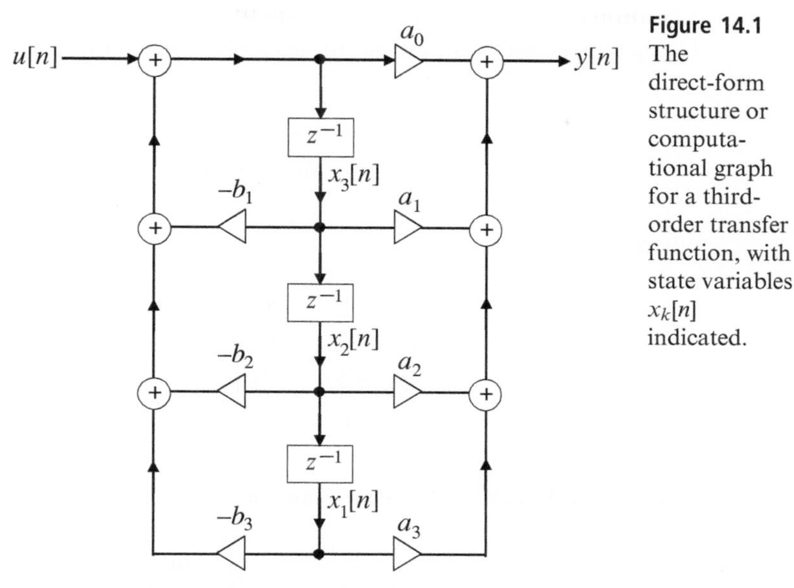

Figure 14.1 The direct-form structure or computational graph for a third-order transfer function, with state variables $x_k[n]$ indicated.

many equivalent structures can be derived for allpass filters using state transformations. These are called lattice structures. One of these is especially attractive for the implementation of notch filters (Sec. 7.5), as we shall see in Sec. 14.6.2.

14.2 State-Space Description for Direct Form

For the direct-form structure in Fig. 14.1, we have labeled the outputs of the delay elements as $x_k[n]$. These are the signals stored internally in the system, and they are required in the computation of $y[n]$ for every n. These internally stored signals $x_k[n]$ are called **state variables**. An Nth-order $H(z)$ requires N delay elements in the structure, and therefore there are N state variables. *Since the notation $x_k[n]$ is used for state variables, we have used $u[n]$ for the input.* This $u[n]$ should not be confused with $\mathcal{U}[n]$, which is the unit step!

In our past discussions we were mostly interested in the input $u[n]$ and the output $y[n]$, and did not write down expressions for the state variables. In this section we first do this. Thus, from Fig. 14.1 it follows that

$$\begin{aligned} x_1[n+1] &= x_2[n], \\ x_2[n+1] &= x_3[n], \\ x_3[n+1] &= -b_3 x_1[n] - b_2 x_2[n] - b_1 x_3[n] + u[n], \end{aligned} \tag{14.3}$$

and

$$\begin{aligned} y[n] &= a_3 x_1[n] + a_2 x_2[n] + a_1 x_3[n] + a_0 x_3[n+1] \\ &= (a_3 - a_0 b_3) x_1[n] + (a_2 - a_0 b_2) x_2[n] + (a_1 - a_0 b_1) x_3[n] + a_0 u[n]. \end{aligned} \tag{14.4}$$

Equations (14.3) are called state equations, and Eq. (14.4) the output equation. Using matrix-vector notation, these can be written more compactly as

$$\mathbf{x}[n+1] = \mathbf{A}\mathbf{x}[n] + \mathbf{B}\mathbf{u}[n], \tag{14.5}$$

$$\mathbf{y}[n] = \mathbf{C}\mathbf{x}[n] + \mathbf{D}\mathbf{u}[n], \tag{14.6}$$

where

$$\mathbf{A} = \begin{bmatrix} 0 & 1 & 0 \\ 0 & 0 & 1 \\ -b_3 & -b_2 & -b_1 \end{bmatrix}, \qquad \mathbf{B} = \begin{bmatrix} 0 \\ 0 \\ 1 \end{bmatrix}, \tag{14.7}$$

$$\mathbf{C} = \begin{bmatrix} a_3 - a_0 b_3 & a_2 - a_0 b_2 & a_1 - a_0 b_1 \end{bmatrix}, \quad \mathbf{D} = a_0. \tag{14.8}$$

The matrix $\mathbf{A}$ is called the **state transition matrix**, and

$$\mathbf{x}[n] = \begin{bmatrix} x_1[n] \\ x_2[n] \\ x_3[n] \end{bmatrix} \tag{14.9}$$

is the **state vector** at time n. Although we have used bold-faced notations for the input $\mathbf{u}[n]$ and the output $\mathbf{y}[n]$ for generality, they are scalars in our example, that is, $\mathbf{u}[n] = u[n]$ and $\mathbf{y}[n] = y[n]$. Figure 14.2 shows a block diagram of the state-space description (14.5) and (14.6). The matrix $\mathbf{A}$ has size $N \times N$. The sizes of the other matrices depend on the size of the input vector $\mathbf{u}[n]$ and the output vector $\mathbf{y}[n]$, as indicated in the caption to Fig. 14.2.

Equation (14.5) is a first-order recursive difference equation, like we have seen in Chapter 5, but since $\mathbf{x}[n]$ is a vector, there are N first-order equations. Furthermore, these are coupled equations (e.g., $x_3[n+1]$ depends not only on $x_3[n]$ but also on $x_1[n]$ and $x_2[n]$ in Eq. (14.3)). Thus, the state-space description converts a single Nth-order difference equation (14.2) to a set of N first-order, coupled, difference equations.

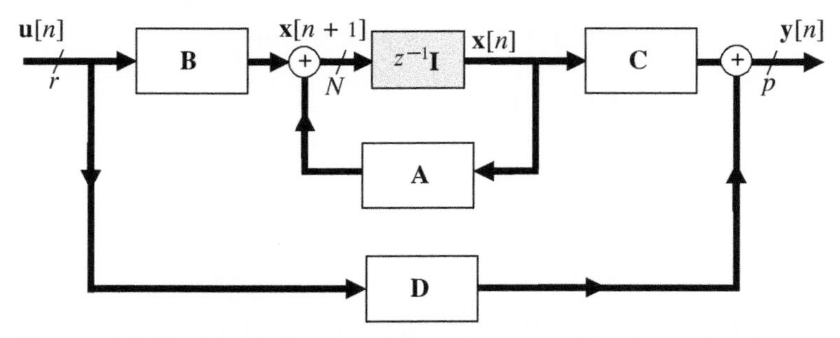

Figure 14.2 The block diagram of the state-space description (14.5) and (14.6). The matrix $\mathbf{A}$ has size $N \times N$. For an r-input, p-output system, $\mathbf{B}$ is $N \times r$, $\mathbf{C}$ is $p \times N$, and $\mathbf{D}$ is $p \times r$.

14.3 State-Space Description for Other Structures

We now show how to obtain the state-space descriptions for some of the other popular structures for LTI systems that arise in signal processing.

14.3.1 State-Space Description for the Cascade Form

The cascade-form structure was introduced in Sec. 5.9 and Fig. 14.3 shows a specific example. The transfer function is clearly

$$H(z) = \left(\frac{a_0 + a_1 z^{-1}}{1 + b_{11} z^{-1}}\right)\left(\frac{1}{1 + b_{21} z^{-1}}\right). \tag{14.10}$$

To write the state-space equations for this system, the state variables, which are the outputs of the delay elements, are first labeled as $x_k[n]$. Note that the inputs of these delay elements are therefore $x_k[n+1]$. By inspection of the figure, here are the state equations:

$$x_1[n+1] = -b_{11}x_1[n] + u[n], \tag{14.11}$$

$$x_2[n+1] = -b_{21}x_2[n] + a_0x_1[n+1] + a_1x_1[n]. \tag{14.12}$$

We should eliminate $x_1[n+1]$ from the right-hand side of the second equation, so that these equations have the form (14.5). This is easily done by substituting from the first equation:

$$\begin{aligned} x_2[n+1] &= -b_{21}x_2[n] + a_0\Big(-b_{11}x_1[n] + u[n]\Big) + a_1x_1[n] \\ &= \big(a_1 - a_0b_{11}\big)x_1[n] - b_{21}x_2[n] + a_0u[n]. \end{aligned} \tag{14.13}$$

Summarizing, the state equations are

$$\begin{aligned} x_1[n+1] &= -b_{11}x_1[n] + u[n], \\ x_2[n+1] &= \big(a_1 - a_0b_{11}\big)x_1[n] - b_{21}x_2[n] + a_0u[n], \end{aligned} \tag{14.14}$$

which can be written in matrix form as follows:

$$\begin{bmatrix} x_1[n+1] \\ x_2[n+1] \end{bmatrix} = \underbrace{\begin{bmatrix} -b_{11} & 0 \\ a_1 - a_0b_{11} & -b_{21} \end{bmatrix}}_{\mathbf{A}} \begin{bmatrix} x_1[n] \\ x_2[n] \end{bmatrix} + \underbrace{\begin{bmatrix} 1 \\ a_0 \end{bmatrix}}_{\mathbf{B}} u[n]. \tag{14.15}$$

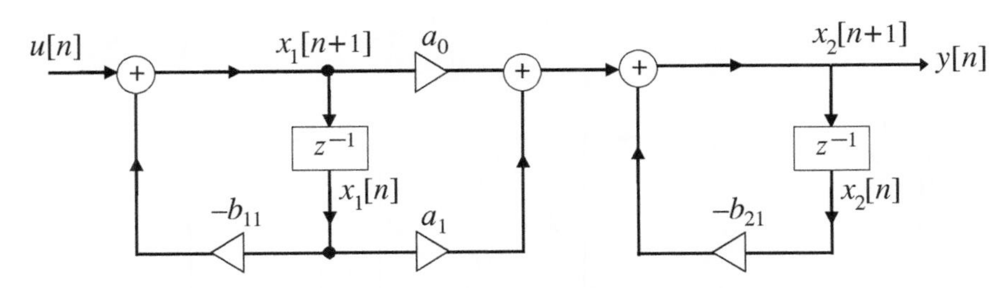

Figure 14.3 A cascade of first-order systems, with the state variables $x_k[n]$ indicated.

Next, the output equation is just $y[n] = x_2[n+1]$, which can be expressed in terms of $x_k[n]$ and $u[n]$ as

$$y[n] = \underbrace{\begin{bmatrix} a_1 - a_0 b_{11} & -b_{21} \end{bmatrix}}_{\mathbf{C}} \begin{bmatrix} x_1[n] \\ x_2[n] \end{bmatrix} + \underbrace{a_0}_{\mathbf{D}} u[n]. \tag{14.16}$$

The matrices $\mathbf{A}, \mathbf{B}, \mathbf{C}, \mathbf{D}$ are indicated, and these equations are clearly in the standard forms (14.5) and (14.6) of the state-space equations.

14.3.2 State-Space Description for the Lattice Structure

Now consider the structure shown in Fig. 14.4. This is a system with one delay and some adders and multipliers, and represents a causal LTI system. In this figure $|k| < 1$ and $\widehat{k}$ is given by

$$\widehat{k} = \sqrt{1 - |k|^2}. \tag{14.17}$$

This structure is quite different from the familiar direct-form structure, and it is called the **lattice** structure, or the **normalized lattice**, to distinguish it from other types (Sec. 14.6). It may not be obvious from inspection what the transfer function is. One way to find it out would be to write the state-space equations. Define the output of the delay element as the state variable $x[n]$. The input $u[n]$ and output $y[n]$ are also indicated. The state equation and output equation are clearly

$$\begin{aligned} x[n+1] &= -kx[n] + \widehat{k}u[n], \\ y[n] &= \widehat{k}x[n] + k^* u[n]. \end{aligned} \tag{14.18}$$

So in this case the matrices $\mathbf{A}, \mathbf{B}, \mathbf{C}, \mathbf{D}$ are all scalar quantities:

$$\mathbf{A} = -k, \quad \mathbf{B} = \widehat{k}, \quad \mathbf{C} = \widehat{k}, \quad \mathbf{D} = k^*. \tag{14.19}$$

To find the transfer function we first take the z-transform of Eq. (14.18):

$$\begin{aligned} zX(z) &= -kX(z) + \widehat{k}U(z), \\ Y(z) &= \widehat{k}X(z) + k^* U(z). \end{aligned} \tag{14.20}$$

From the first equation we find $X(z) = \widehat{k}U(z)/(z+k)$. Substituting into the second equation and simplifying, we get

$$Y(z) = \left(\frac{\widehat{k}^2 + k^* z + |k|^2}{z+k}\right) U(z) = \left(\frac{1 + k^* z}{z+k}\right) U(z) = \left(\frac{k^* + z^{-1}}{1 + kz^{-1}}\right) U(z),$$

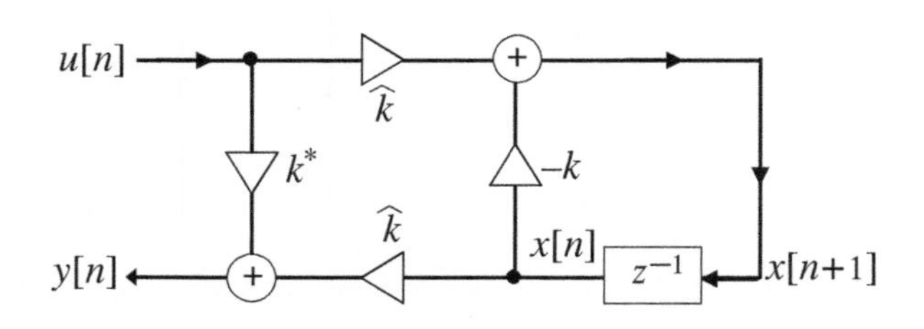

Figure 14.4 The first-order lattice structure. Here $\widehat{k} = \sqrt{1 - |k|^2}$.

where Eq. (14.17) has been used to get the second equality. This shows that the transfer function is

$$H(z) = \frac{k^* + z^{-1}}{1 + kz^{-1}}. \tag{14.21}$$

From Sec. 7.4 we know that this is the familiar allpass function. Thus, the lattice structure of Fig. 14.4 is an allpass filter! This structure is very different from the two-multiplier direct-form structure we saw in Fig. 7.10(a). Even though the structure has four multipliers and is more complicated than the direct form, it has other advantages [Gray and Markel, 1975]. The reason for the assumption $|k| < 1$ made in the beginning is clear from Eq. (14.21): the pole of $H(z)$ is at $z = -k$ and we want it to be inside the unit circle for stability.

14.4 More General LTI Structures

We showed how to write the state-space description for a number of structures like the direct form, the cascade form, and the lattice. More generally, consider an arbitrary implementation of a causal LTI system. Since the system is LTI, the implementation is made of three building blocks: delay elements, multipliers, and adders, as in the above examples. Figure 14.5 shows this general form of any structure implementing a causal LTI system. The interconnections between the three types of building blocks define the flowgraph or computational graph as in the preceding examples. It can contain feedback loops also, as in the case of IIR systems. The building blocks process scalar inputs to produce scalar outputs, but the overall system in general can be a multiple-input, multiple-output or **MIMO** system. That is, it can have a vector input $\mathbf{u}[n]$ and a vector output $\mathbf{y}[n]$:

$$\mathbf{u}[n] = \begin{bmatrix} u_1[n] & u_2[n] & \cdots & u_r[n] \end{bmatrix}^T, \quad \mathbf{y}[n] = \begin{bmatrix} y_1[n] & y_2[n] & \cdots & y_p[n] \end{bmatrix}^T.$$

When $p = r = 1$, we have a single-input, single-output or **SISO** system, also called a **scalar** system.

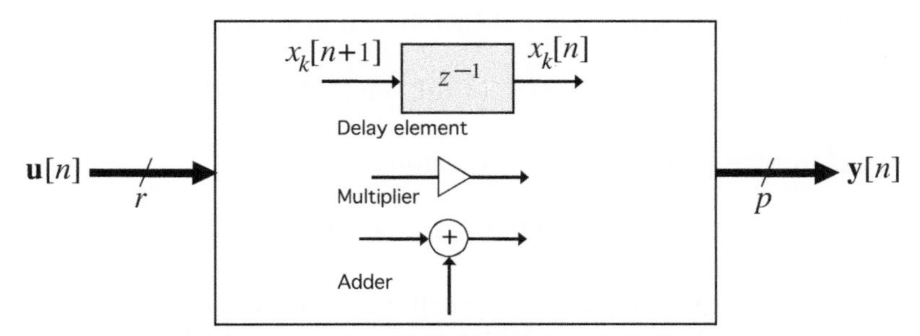

Figure 14.5 A discrete-time causal LTI system is an interconnection or a flowgraph (with forward and feedback connections) made of the three building blocks shown.

As before, the outputs of the delay elements are the state variables, denoted $x_k[n]$, so the inputs of the delay elements are $x_k[n+1]$. By tracing the details of the flow-graph, we can always come up with a state equation of the form (14.5), where the $x_k[n+1]$ are on the left-hand side and the only signals on the right-hand side are $x_i[n]$ and $\mathbf{u}[n]$. Similarly, we can write an output equation like (14.6). These are reproduced below:

$$\mathbf{x}[n+1] = \mathbf{A}\mathbf{x}[n] + \mathbf{B}\mathbf{u}[n], \tag{14.22}$$

$$\mathbf{y}[n] = \mathbf{C}\mathbf{x}[n] + \mathbf{D}\mathbf{u}[n]. \tag{14.23}$$

This was demonstrated for several examples of structures earlier. Here the state vector is a column vector

$$\mathbf{x}[n] = \begin{bmatrix} x_1[n] & x_2[n] & \cdots & x_N[n] \end{bmatrix}^T, \tag{14.24}$$

where N denotes the number of delay elements in the structure. Thus we can always obtain the above state equations, no matter how complex the interconnections are. Once this is done, it is easy to find the impulse response and transfer function of this LTI system, as we shall show next. Thus, state-space descriptions offer a systematic way to describe a complicated LTI structure and derive the transfer function and impulse repsonse. Other benefits of state-space descriptions will be clear later.

14.4.1 Transfer Function

Taking the z-transforms of Eqs. (14.22) and (14.23), it follows that

$$z\mathbf{X}(z) = \mathbf{A}\mathbf{X}(z) + \mathbf{B}\mathbf{U}(z), \tag{14.25}$$

$$\mathbf{Y}(z) = \mathbf{C}\mathbf{X}(z) + \mathbf{D}\mathbf{U}(z), \tag{14.26}$$

where $\mathbf{X}(z) = \sum_n \mathbf{x}[n]z^{-n}$, and so forth. From the first equation we have $(z\mathbf{I} - \mathbf{A})\mathbf{X}(z) = \mathbf{B}\mathbf{U}(z)$, that is,

$$\mathbf{X}(z) = \left(z\mathbf{I} - \mathbf{A}\right)^{-1}\mathbf{B}\mathbf{U}(z). \tag{14.27}$$

Substituting into Eq. (14.26), it therefore follows that

$$\mathbf{Y}(z) = \left(\mathbf{C}\left(z\mathbf{I} - \mathbf{A}\right)^{-1}\mathbf{B} + \mathbf{D}\right)\mathbf{U}(z), \tag{14.28}$$

which proves that the transfer function is

$$\mathbf{H}(z) = \mathbf{D} + \mathbf{C}\left(z\mathbf{I} - \mathbf{A}\right)^{-1}\mathbf{B}. \tag{14.29}$$

This is a $p \times r$ matrix, so we also call it a **transfer matrix**. Each of its entries $H_{li}(z)$ is a rational function. For a SISO system ($p = r = 1$), $\mathbf{D} = D$ is a scalar, $\mathbf{C}$ is a row vector, and $\mathbf{B}$ is a column vector. More explicitly, for a second-order SISO system ($N = 2, p = r = 1$), we can write

$$H(z) = D + \underbrace{\begin{bmatrix} c_{11} & c_{12} \end{bmatrix}}_{\mathbf{C}} \underbrace{\begin{bmatrix} z - a_{11} & -a_{12} \\ -a_{21} & z - a_{22} \end{bmatrix}}_{z\mathbf{I}-\mathbf{A}}^{-1} \underbrace{\begin{bmatrix} b_{11} \\ b_{21} \end{bmatrix}}_{\mathbf{B}}. \tag{14.30}$$

For example, substituting the expressions (14.19) into Eq. (14.29), one immediately obtains the allpass function (14.21). Similarly, substituting the expressions for $\mathbf{A}, \mathbf{B}, \mathbf{C}, \mathbf{D}$ from Eqs. (14.15) and (14.16) into Eq. (14.30), we get the cascade-form transfer function (14.10), although this involves a bit more work (Problem 14.2). Considering that the expression (14.10) is essentially obvious from inspection of the structure in Fig. 14.3, the state-space method might appear to be laborious in the cascade-form example. But the value lies in the fact that it gives a systematic way to analyze *any* LTI structure and find its transfer function and impulse response.

14.4.2 Time Recursion and Impulse Response

The way one runs the state-space equations (14.22) and (14.23) is similar to the recursive difference equations in Chapter 5. We start at some initial time, say $n = 0$. Thus we are given the input $\mathbf{u}[n], n \geq 0$, starting at $n = 0$, and the initial condition or **initial state vector** $\mathbf{x}[0]$. We then recursively compute

$$\begin{aligned}
\mathbf{x}[1] &= \mathbf{A}\mathbf{x}[0] + \mathbf{B}\mathbf{u}[0], \\
\mathbf{x}[2] &= \mathbf{A}\mathbf{x}[1] + \mathbf{B}\mathbf{u}[1] = \mathbf{A}^2\mathbf{x}[0] + \mathbf{A}\mathbf{B}\mathbf{u}[0] + \mathbf{B}\mathbf{u}[1], \\
\mathbf{x}[3] &= \mathbf{A}\mathbf{x}[2] + \mathbf{B}\mathbf{u}[2] = \mathbf{A}^3\mathbf{x}[0] + \mathbf{A}^2\mathbf{B}\mathbf{u}[0] + \mathbf{A}\mathbf{B}\mathbf{u}[1] + \mathbf{B}\mathbf{u}[2], \\
&\vdots
\end{aligned} \tag{14.31}$$

It therefore follows that

$$\mathbf{x}[n] = \mathbf{A}^n\mathbf{x}[0] + \sum_{m=1}^{n} \mathbf{A}^{m-1}\mathbf{B}\mathbf{u}[n-m], \quad n \geq 1. \tag{14.32}$$

Thus the state vector at time n has two contributions: the first term comes from the initial condition $\mathbf{x}[0]$. The second term is a filtered version of the input $\mathbf{u}[n]$. The output equation (14.23) can now be used to find a similar expression for the output:

$$\mathbf{y}[n] = \begin{cases} \mathbf{C}\mathbf{x}[0] + \mathbf{D}\mathbf{u}[0] & n = 0, \\ \mathbf{C}\mathbf{A}^n\mathbf{x}[0] + \sum_{m=1}^{n} \mathbf{C}\mathbf{A}^{m-1}\mathbf{B}\mathbf{u}[n-m] + \mathbf{D}\mathbf{u}[n] & n \geq 1. \end{cases} \tag{14.33}$$

Note that the system is automatically **causal** by virtue of its definition: for each n, the output $\mathbf{y}[n]$ does not depend on future inputs $\mathbf{u}[n+i], i > 0$. The recursive system behaves like an LTI system when the initial condition is zero, that is,

$$\mathbf{x}[0] = \mathbf{0} \quad \text{(zero initial state)}. \tag{14.34}$$

When this is the case, we have

$$\mathbf{y}[n] = \begin{cases} \mathbf{Du}[0] & n = 0, \\ \sum_{m=1}^{n} \mathbf{CA}^{m-1}\mathbf{Bu}[n-m] + \mathbf{Du}[n] & n \geq 1. \end{cases} \quad (14.35)$$

This can be written in the form of a convolution sum

$$\mathbf{y}[n] = \sum_{m=-\infty}^{n} \mathbf{h}[m]\mathbf{u}[n-m], \quad (14.36)$$

where $\mathbf{h}[m]$ is the matrix sequence given by

$$\mathbf{h}[m] = \begin{cases} \mathbf{0} & m < 0, \\ \mathbf{D} & m = 0, \\ \mathbf{CA}^{m-1}\mathbf{B} & m > 0. \end{cases} \quad (14.37)$$

Note that $\mathbf{h}[m]$ is a $p \times r$ matrix, just like $\mathbf{H}(z)$. We call it the impuse response matrix. For a SISO system ($p = r = 1$), $\mathbf{D} = D$ is a scalar, $\mathbf{C}$ is a row vector, and $\mathbf{B}$ is a column vector. More explicitly, for a second-order SISO system ($N = 2, p = r = 1$), $h[n]$ can be written as

$$h[n] = D\delta[n] \; + \; \underbrace{\begin{bmatrix} c_{11} & c_{12} \end{bmatrix}}_{\mathbf{C}} \underbrace{\begin{bmatrix} a_{11} & a_{12} \\ a_{21} & a_{22} \end{bmatrix}^{n-1}}_{\mathbf{A}} \underbrace{\begin{bmatrix} b_{11} \\ b_{21} \end{bmatrix}}_{\mathbf{B}} \mathcal{U}[n-1]. \quad (14.38)$$

In summary, the impulse response and transfer function can both be expressed in closed form in terms of the state-space description $(\mathbf{A}, \mathbf{B}, \mathbf{C}, \mathbf{D})$. With $\mathbf{H}(z)$ given by Eq. (14.29) and $\mathbf{h}[m]$ given by Eq. (14.37), it can indeed be verified that

$$\mathbf{H}(z) = \sum_{m=-\infty}^{\infty} \mathbf{h}[m]z^{-m} \quad (14.39)$$

in the region of convergence of this summation.

Interpreting $H(z)$ and $h[n]$ in the MIMO case. The system $\mathbf{H}(z)$, which is a $p \times r$ transfer matrix, has the input–output relation

$$\mathbf{Y}(z) = \mathbf{H}(z)\mathbf{X}(z), \quad (14.40)$$

where $\mathbf{Y}(z)$ is a $p \times 1$ vector and $\mathbf{U}(z)$ is an $r \times 1$ vector. Thus

$$Y_l(z) = \sum_{i=1}^{r} H_{li}(z)U_i(z). \quad (14.41)$$

So we can regard $H_{li}(z)$ as the transfer function from the ith input to the lth output. Similarly, the (l, i)th element $h_{li}[n]$ of the impulse response matrix $\mathbf{h}[n]$ is the impulse response from the ith input to the lth output. That is, if all the inputs are set to zero except $u_i[n]$, and we set $u_i[n] = \delta[n]$, then $y_l[n] = h_{li}[n]$ for each l, so that the output vector is

$$\mathbf{y}[n] = \begin{bmatrix} h_{1i}[n] \\ h_{2i}[n] \\ \vdots \\ h_{pi}[n] \end{bmatrix}. \tag{14.42}$$

For a SISO system, $h[n]$ is the output in response to the impulse input. But for a MIMO system, $\mathbf{h}[n]$ is itself not the output vector because it is a $p \times r$ matrix by definition. The correct interpretation of $\mathbf{h}[n]$ is as explained above. ▽ ▽ ▽

14.4.3 Homogeneous Solutions and More General Outputs

Now assume that the input $\mathbf{u}[n] = \mathbf{0}$ for all n, and the initial condition $\mathbf{x}[0]$ is not restricted to be zero. Then it follows from Eqs. (14.32) and (14.33) that

$$\mathbf{x}[n] = \mathbf{A}^n\mathbf{x}[0], \quad n \geq 0, \tag{14.43}$$

and

$$\mathbf{y}[n] = \mathbf{C}\mathbf{A}^n\mathbf{x}[0], \quad n \geq 0. \tag{14.44}$$

Since the input is zero, these are called **homogeneous solutions**, similar to what we had in Chapter 8. Thus, all homogeneous solutions for the state and output have the above forms. Finally, when the input and the initial state are both nonzero, we have

$$\mathbf{y}[n] = \mathbf{C}\mathbf{A}^n\mathbf{x}[0] + \sum_{m=0}^{n} \mathbf{h}[m]\mathbf{u}[n-m], \tag{14.45}$$

where the first term is the homogeneous component and the second term is convolution. The convolution sum begins at $m = 0$, because $\mathbf{h}[m]$ is causal, and ends at $m = n$, because the input $\mathbf{u}[\cdot]$ is causal. Equation (14.45) neatly summarizes the most general form of the output. If the system is regarded as the mapping

$$\{\mathbf{x}[0], \mathbf{u}[0], \mathbf{u}[1], \mathbf{u}[2], \ldots\} \longmapsto \{\mathbf{y}[0], \mathbf{y}[1], \mathbf{y}[2], \ldots\}, \tag{14.46}$$

then it is a linear transformation. But if it is regarded as the mapping

$$\{\mathbf{u}[0], \mathbf{u}[1], \mathbf{u}[2], \ldots\} \longmapsto \{\mathbf{y}[0], \mathbf{y}[1], \mathbf{y}[2], \ldots\}, \tag{14.47}$$

then it is linear only when $\mathbf{x}[0] = \mathbf{0}$, and in that case $\mathbf{Y}(z) = \mathbf{H}(z)\mathbf{U}(z)$ is valid, and the transfer function $\mathbf{H}(z)$ characterizes the system. This is similar to what we discussed in Secs. 5.6 and 5.7.1.

Notice that the initial state vector $\mathbf{x}[0]$ is nothing but the set of numbers stored in the delay elements at time zero, when the input signal was started. For example, these are the three quantities $x_k[0]$ in Fig. 14.1. These are analogous to the initial conditions $y[-1], y[-2], \ldots$, in the difference equation (14.2). We only say "analogous" because the quantities $x_k[n]$ in Fig. 14.1 are *not* exactly equal to the delayed output signals $y[n-k]$.

14.4.4 The Class of Realizable LTI Systems

We know that if a discrete-time LTI system is rational, that is, it has a transfer function of the form (14.1), then it can be described by the difference equation (14.2). Conversely, any LTI system described by Eq. (14.2) is rational, because it has the transfer function (14.1). Now, we say that the difference equation is "realizable" because it can be implemented with a **finite** number of multiplications, additions, and delay elements (storage units). In short, therefore, rational LTI systems are realizable. How about the converse? Suppose a (causal) LTI system $H(z)$ is realizable in the above sense. Can we then say that it is necessarily rational? This is indeed the case, as proved next.

Proof. Consider any realization of the realizable function $H(z)$. It has a finite number of multipliers, adders, and N delay elements. The interconnections between these building blocks define a signal flowgraph, as schematically shown in Fig. 14.5. From this graph we can always write state-space equations as in Eqs. (14.22) and (14.23), where $\mathbf{A}$ is $N \times N$ and the other matrices have appropriate sizes. Once we have this description, it readily follows that the transfer function has the form

$$H(z) = D + \mathbf{C}(z\mathbf{I} - \mathbf{A})^{-1}\mathbf{B}. \tag{14.48}$$

Now, the matrix inverse shown above can always be rewritten as

$$(z\mathbf{I} - \mathbf{A})^{-1} = \frac{\text{Adj}\,(z\mathbf{I} - \mathbf{A})}{\det\,(z\mathbf{I} - \mathbf{A})}, \tag{14.49}$$

where $\text{Adj}\,(z\mathbf{I} - \mathbf{A})$ is the adjugate of $(z\mathbf{I} - \mathbf{A})$ (see Chapter 18) and the denominator is the determinant of $(z\mathbf{I} - \mathbf{A})$. The adjugate is an $N \times N$ matrix, whose elements are determinants of submatrices of $(z\mathbf{I} - \mathbf{A})$. These elements are therefore polynomials in z. The denominator $\det\,(z\mathbf{I} - \mathbf{A})$ is also a polynomial in z. So each element of the matrix in Eq. (14.49) is a rational function in z, or equivalently, a rational function in z^{-1}. The transfer function (14.48) is therefore rational as well, and it can be described by a finite-order difference equation in the time domain. This shows that any discrete-time realizable system is rational, and can be associated with a finite-order difference equation. $\triangledown \triangledown \triangledown$

Summarizing, a discrete-time LTI system is realizable (in the sense that it is implementable using a finite number of multipliers, adders, and delay elements) if and only if it is rational. Otherwise, it is unrealizable. Notice that in the continuous-time world, the situation is different. For example, systems made of transmission lines have building blocks that are "distributed" rather than "lumped" circuit elements. LTI systems based on these can have irrational transfer functions $H(s)$, although they are perfectly realizable. The meaning of realizability there is different from how we have defined realizability for discrete-time systems.

14.5 Properties of State-Space Descriptions*

We now mention a number of properties of state-space descriptions briefly. Further details and proofs can be found in many books. For example, see Chapter 13 of Vaidyanathan [1993] and references therein. Needless to say, all discussions here are for causal, rational, LTI systems. Remember throughout that N is the number of state variables or delay elements in the structure (so that $\mathbf{A}$ is $N \times N$), r is the number of inputs, and p the number of outputs.

14.5.1 Degree versus Order

The minimum number of delays required to implement a rational system $\mathbf{H}(z)$ is called the **degree** of $\mathbf{H}(z)$, denoted

$$\mu = \deg\,[\mathbf{H}(z)]. \tag{14.50}$$

The degree is also referred to as the **McMillan** degree. For MIMO systems the degree can be different from the term "order," which we have been using throughout the book. The order is the largest power of z^{-1} anywhere in the expression for $\mathbf{H}(z)$, after cancelling any common factors in the rational expressions. For example, suppose $H(z)$ is a SISO transfer function as in Eq. (14.1), with at least one of a_N, b_N nonzero, and no common factors between numerator and denominator. Since the highest power z^{-N} appears in the transfer function and cannot be cancelled in any way, the system has order N. This transfer function also requires N delays for implementation, so

$$\deg\,[\mathbf{H}(z)] = \text{order } N \qquad \text{(SISO system)}.$$

So, for a causal SISO rational transfer function, the degree equals the order. For the MIMO case the degree can be different from the order. For example, consider

$$\mathbf{H}(z) = \begin{bmatrix} az^{-1} & 0 \\ 0 & bz^{-1} \end{bmatrix}, \tag{14.51}$$

where a and b are nonzero. This has order one (because z^{-1} is the highest power appearing in any of the entries $H_{li}(z)$), but the degree is two (Problem 14.8).

A structure $(\mathbf{A}, \mathbf{B}, \mathbf{C}, \mathbf{D})$ for $\mathbf{H}(z)$ is said to be **minimal** if it contains the minimum number of delay elements necessary for its implementation, namely, the degree. For example, the first-order allpass filter in Fig. 7.10(b) with two delay elements is not minimal, whereas the structure with one delay (Fig. 7.10(a)) is minimal.

14.5.2 Poles, Eigenvalues, and Stability

The poles of $\mathbf{H}(z)$ in Eq. (14.29) can be related to the eigenvalues of the state transition matrix $\mathbf{A}$. Each pole is an eigenvalue of $\mathbf{A}$ (Problem 14.12). Conversely, each eigenvalue of $\mathbf{A}$ is a pole as long as the structure is minimal. Since the poles of a causal stable system are inside the unit circle, it follows that for a stable system all eigenvalues of $\mathbf{A}$ satisfy

$$|\lambda_i| < 1,$$

when $(\mathbf{A}, \mathbf{B}, \mathbf{C}, \mathbf{D})$ is minimal. Because of this, a matrix $\mathbf{A}$ whose eigenvalues satisfy $|\lambda_i| < 1$ is often called a **stable matrix**.

14.5.2.1 FIR Case

Now consider the **FIR** case. For FIR systems where the impulse response is nonzero only in a finite range $0 \leq n \leq L-1$, we know that all poles are at $z = 0$. So we expect all the eigenvalues of $\mathbf{A}$ to be zero (assuming the implementation is minimal). This is indeed true. For example, consider the transfer function (14.1). With $N = 3$, a state-space description with three delays (which is the minimum) is shown in Eqs. (14.7) and (14.8). For the FIR case we have $b_k = 0$ in Eq. (14.1), so that

$$\mathbf{A} = \begin{bmatrix} 0 & 1 & 0 \\ 0 & 0 & 1 \\ -b_3 & -b_2 & -b_1 \end{bmatrix} = \begin{bmatrix} 0 & 1 & 0 \\ 0 & 0 & 1 \\ 0 & 0 & 0 \end{bmatrix}. \tag{14.52}$$

This is an upper-triangular matrix, so its eigenvalues are its diagonal elements (see Chapter 18), all of which are indeed zero. More examples of state-space equations for FIR systems can be found in the problems section.

14.5.2.2 Lyapunov's Equation

In some applications we generate a matrix $\mathbf{A}$ by some means, and want to make sure it is stable without having to compute its eigenvalues. For example, we may just want to prove that matrices $\mathbf{A}$ that arise from some physical process or some iterative computation are always guaranteed to be stable. There are some ways to do this. One of these is based on a result attributed to Lyapunov. Here we mention this result for the interested reader. We state it in two parts for clarity. In the following, $\mathbf{A}$ is some $N \times N$ matrix, and $\mathbf{C}$ is some $p \times N$ matrix.

1. Given $\mathbf{A}$, if (a) there exists $\mathbf{C}$ such that $(\mathbf{C}, \mathbf{A})$ is observable (i.e., the matrix $\mathbf{S}$ introduced later in Eq. (14.57) has rank N) and (b) there exists a positive-definite matrix $\mathbf{P}$ such that

$$\mathbf{A}^H\mathbf{P}\mathbf{A} + \mathbf{C}^H\mathbf{C} = \mathbf{P}, \tag{14.53}$$

   then $\mathbf{A}$ is stable.
2. Conversely, suppose $\mathbf{A}$ is stable, and $\mathbf{C}$ is such that $(\mathbf{C}, \mathbf{A})$ is observable. Then there exists a unique matrix $\mathbf{P}$ satisfying Eq. (14.53) and this $\mathbf{P}$ is positive definite.

The proof can be found in many books [e.g., Vaidyanathan, 1993].

14.5.3 Similarity Transformation

Let $(\mathbf{A}, \mathbf{B}, \mathbf{C}, \mathbf{D})$ be the state-space description of some structure implementing a transfer function $\mathbf{H}(z)$. Now define a new structure by changing the matrices to

$$\widehat{\mathbf{A}} = \mathbf{T}^{-1}\mathbf{A}\mathbf{T}, \quad \widehat{\mathbf{B}} = \mathbf{T}^{-1}\mathbf{B}, \quad \widehat{\mathbf{C}} = \mathbf{C}\mathbf{T}, \quad \widehat{\mathbf{D}} = \mathbf{D}, \tag{14.54}$$

where $\mathbf{T}$ is an arbitrary $N \times N$ nonsingular matrix. Then the new structure has the same transfer function $\mathbf{H}(z)$. This can be verified simply by substituting (14.54) into the right-hand side of Eq. (14.29), and simplifying. The $\mathbf{T}$ matrix cancels off, leaving the right-hand side unchanged (Problem 14.15). The transformation (14.54) is called a **similarity transform**. The new structure has the same impulse response $\mathbf{h}[n]$, as we can verify by substituting (14.54) into the right-hand side of Eq. (14.37). Note that there exists a physically realizable structure with the state-space description (14.54), because we can simply implement it as in Fig. 14.2 with $(\mathbf{A}, \mathbf{B}, \mathbf{C}, \mathbf{D})$ replaced by $(\widehat{\mathbf{A}}, \widehat{\mathbf{B}}, \widehat{\mathbf{C}}, \widehat{\mathbf{D}})$.

Since $\mathbf{T}$ is completely arbitrary (except that it has to be nonsingular), there exist an infinite number of **equivalent structures** for the same transfer function $\mathbf{H}(z)$. The importance of this observation is that some structures are preferred over others, for various reasons. For example, they may have fewer multipliers (see examples in Sec. 14.6), or they may have smaller quantization noise gain. Thus, given a transfer function, it is possible to look for structures that are "optimal" with respect to some objective, simply by searching over all possible $\mathbf{T}$ in a principled way.

It also turns out that given two minimal structures for a transfer function $\mathbf{H}(z)$, their state-space descriptions are related by some similarity transformation matrix $\mathbf{T}$ (although this is more tricky to prove). Thus, searching for an optimal structure by searching over the family of nonsingular $\mathbf{T}$ does not leave out any structure!

14.5.4 Reachability, Observability, and Minimality

A structure $(\mathbf{A}, \mathbf{B}, \mathbf{C}, \mathbf{D})$ is said to be **reachable** if we can start from any arbitrary initial state $\mathbf{x}[0]$ and arrive at any specified final state $\mathbf{x}[M] = \mathbf{x}_f$ in a finite amount of time M by applying an appropriately chosen input

$$\mathbf{u}[0], \mathbf{u}[1], \ldots, \mathbf{u}[M-1]. \tag{14.55}$$

That is, there should exist an integer M and an input sequence $\mathbf{u}[n]$ of duration M such that this is possible. It can be shown that a structure is reachable if and only if the matrix

$$\mathbf{R} \stackrel{\Delta}{=} \begin{bmatrix} \mathbf{B} & \mathbf{A}\mathbf{B} & \mathbf{A}^2\mathbf{B} & \cdots & \mathbf{A}^{N-1}\mathbf{B} \end{bmatrix} \tag{14.56}$$

has rank N. Note that the size of this matrix is $N \times Nr$. A somewhat similar concept called **controllability** is sometimes discussed: controllability means that we can start from arbitrary $\mathbf{x}[0]$ and arrive at the zero state $\mathbf{x}[M] = \mathbf{0}$ in a finite amount of time M by applying an appropriate input. Reachability implies controllability but not vice versa (Problem 14.10).

A structure $(\mathbf{A}, \mathbf{B}, \mathbf{C}, \mathbf{D})$ is said to be **observable** if there exists a finite M such that we can find the state $\mathbf{x}[n]$ at any time n by observing the output $\mathbf{y}[n], \mathbf{y}[n+1], \ldots, \mathbf{y}[n+M-1]$, and also knowing the input in that time interval. It can be shown that a structure is observable if and only if the matrix

$$\mathbf{S} \stackrel{\Delta}{=} \begin{bmatrix} \mathbf{C} \\ \mathbf{CA} \\ \mathbf{CA}^2 \\ \vdots \\ \mathbf{CA}^{N-1} \end{bmatrix} \tag{14.57}$$

has rank N. Note that the size of this matrix is $Np \times N$. The importance of these concepts arises from the following properties:

1. *Structure reduction.* It can be shown that if a structure $(\mathbf{A}, \mathbf{B}, \mathbf{C}, \mathbf{D})$ for $\mathbf{H}(z)$ is not reachable, then there exists an equivalent structure for $\mathbf{H}(z)$ with fewer delay elements, that is, the structure $(\mathbf{A}, \mathbf{B}, \mathbf{C}, \mathbf{D})$ is not minimal. Similarly, if a structure is not observable, then there exists an equivalent structure with fewer delay elements.
2. *Minimality condition.* Conversely, if a structure is both reachable and observable, then it is minimal, that is, there does not exist a structure for this $\mathbf{H}(z)$ with fewer delay elements.

In short, a structure is *minimal if and only if it is both reachable and observable.* Proofs of all the above results can be found in many books [e.g., see Vaidyanathan, 1993 or references therein].

14.5.5 A Curious Equality

Consider any structure $(\mathbf{A}, \mathbf{B}, \mathbf{C}, \mathbf{D})$ for $\mathbf{H}(z)$. By using the fact that $\mathbf{h}[m] = \mathbf{CA}^{m-1}\mathbf{B}$ for $m \geq 1$, it follows that

$$\begin{bmatrix} \mathbf{C} \\ \mathbf{CA} \\ \vdots \\ \mathbf{CA}^{M-1} \end{bmatrix} \begin{bmatrix} \mathbf{B} & \mathbf{AB} & \cdots & \mathbf{A}^{M-1}\mathbf{B} \end{bmatrix} = \underbrace{\begin{bmatrix} \mathbf{h}[1] & \mathbf{h}[2] & \cdots & \mathbf{h}[M] \\ \mathbf{h}[2] & \mathbf{h}[3] & \cdots & \mathbf{h}[M+1] \\ \vdots & \vdots & \cdots & \vdots \\ \mathbf{h}[M] & \mathbf{h}[M+1] & \cdots & \mathbf{h}[2M-1] \end{bmatrix}}_{\text{call this } \mathcal{H}} \tag{14.58}$$

for any $M \geq 1$. Note that the matrix $\mathcal{H}$ on the right-hand side depends only on the system $\mathbf{H}(z)$, whereas the two matrices on the left depend on the specific structure $(\mathbf{A}, \mathbf{B}, \mathbf{C}, \mathbf{D})$.

In particular, if M is the degree of the system then there exists a structure with M delays, and it is a minimal structure. For this structure the two matrices on the left both have rank M and sizes $pM \times M$ and $M \times rM$, respectively. Applying Sylvester's inequality for ranks (reviewed in Sec. 18.6.1), we see that if ρ is the rank of $\mathcal{H}$ then

$$M + M - M \leq \rho \leq M,$$

which proves that $\rho = M$. Summarizing, if M is the degree of $\mathbf{H}(z)$, then the matrix $\mathcal{H}$ has rank M.

14.5.5.1 Case of SISO Systems

The entries $\mathbf{h}[m]$ in Eq. (14.58) are $p \times r$ matrices. For the special case of $p = r = 1$ (SISO case), we saw a matrix similar to this in Eq. (5.119). That matrix is reproduced below with column ordering reversed (the last column appears first and so on, which does not change the rank):

$$\begin{bmatrix} h[1] & h[2] & \cdots & h[N] \\ h[2] & h[3] & \cdots & h[N+1] \\ \vdots & \vdots & \cdots & \vdots \\ h[N] & h[N+1] & \cdots & h[2N-1] \end{bmatrix}. \qquad (14.59)$$

Note that this is precisely the matrix (14.58) (with N instead of M).

14.5.5.2 Equation (14.59) is Nonsingular if N = Filter Order

In Sec. 5.12 we mentioned without proof that Eq. (14.59) is nonsingular provided the transfer function

$$H(z) = \frac{\sum_{k=0}^{N} a_k z^{-k}}{1 + \sum_{k=1}^{N} b_k z^{-k}} \qquad (14.60)$$

has order exactly N (i.e., at least one of a_N, b_N is nonzero, and there are no common factors between the numerator and the denominator). We can now see the reason for this: when the order is N, the degree is also N in the SISO case. So the $N \times N$ matrix (14.59) is nonsingular because of the result that Eq. (14.58) has rank M, when M is the degree of $\mathbf{H}(z)$.

14.6 Allpass Filters and Lattice Structures

Consider again the four-multiplier lattice structure in Fig. 14.4, which has the state-space description (14.19) and the transfer function (14.21) reproduced below:

$$H(z) = \frac{k^* + z^{-1}}{1 + kz^{-1}}. \qquad (14.61)$$

Here $N = 1$ so that $\mathbf{A}$ has size 1×1 (it is a scalar). So the similarity transform matrix $\mathbf{T}$ is also a scalar T. Suppose we choose $T = \widehat{k}$ to get a new state-space description for the same transfer function (14.61):

$$\widehat{\mathbf{A}} = -k, \quad \widehat{\mathbf{B}} = 1, \quad \widehat{\mathbf{C}} = \widehat{k}^2, \quad \widehat{\mathbf{D}} = k^*, \qquad (14.62)$$

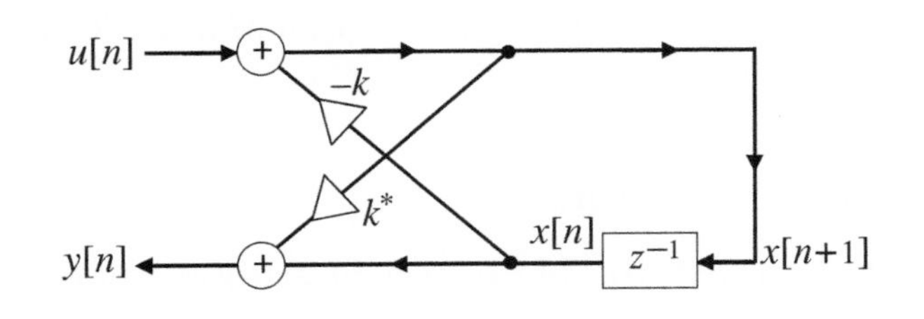

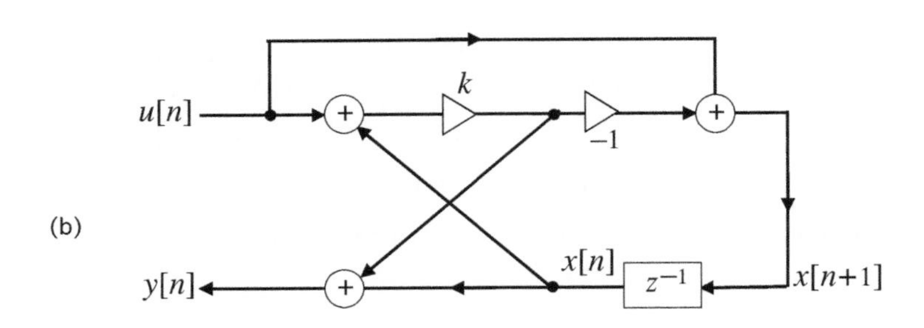

Figure 14.6 Variations of the first-order lattice structure. (a) Two-multiplier version and (b) one-multiplier version for the real-coefficient case.

where $\widehat{k} = \sqrt{1-|k|^2}$. Thus, any structure with state-space parameters given by (14.62) also has the same transfer function (14.61). It can be shown (Problem 14.16) that the structure shown in Fig. 14.6(a) has the state-space description (14.62). This is called a **two-multiplier** lattice structure. Thus, the four-multipler lattice of Fig. 14.4 and the two-multiplier lattice of Fig. 14.6(a) are related by a similarity transform, and therefore they have the same transfer function (14.61).

Returning to the four-multiplier structure of Fig. 14.4 with state-space description (14.19), let us now assume that k is real so that (14.19) becomes

$$\mathbf{A} = -k, \quad \mathbf{B} = \sqrt{1-k^2}, \quad \mathbf{C} = \sqrt{1-k^2}, \quad \mathbf{D} = k. \tag{14.63}$$

Now assume that we apply the similarity transform

$$T = \sqrt{\frac{1+k}{1-k}}. \tag{14.64}$$

Then (14.63) is transformed to

$$\widehat{\mathbf{A}} = -k, \quad \widehat{\mathbf{B}} = 1-k, \quad \widehat{\mathbf{C}} = 1+k, \quad \widehat{\mathbf{D}} = k. \tag{14.65}$$

It can be verified (Problem 14.17) that the structure shown in Fig. 14.6(b) has the state-space description (14.65). This is called a **one-multiplier** lattice structure. Thus, for real k, the four-multipler lattice of Fig. 14.4, the two-multiplier lattice of Fig. 14.6(a), and the one-multiplier lattice of Fig. 14.6(b) are related by similarity transforms, and therefore they have the same allpass transfer function

$$H(z) = \frac{k+z^{-1}}{1+kz^{-1}}. \tag{14.66}$$

The main point of the above examples is to demonstrate the use of similarity transforms in generating equivalent structures for a transfer function. In Chapter 7 we showed that a first-order real allpass filter can be realized with one multiplier, but it required two delays (Fig. 7.10(b)). But the structure of Fig. 14.6(b) requires *only one multiplier and one delay*, which is special. It uses three adders instead of two, but that is usually a small price to pay. While the four-multiplier structure might appear to be unnecessarily complex, it has its own advantages. We shall not go into the details but they can be found in Gray and Markel [1975].

14.6.1 Cascaded Lattice for Higher-Order Allpass

Given a first-order allpass filter

$$\frac{k_2^* + z^{-1}}{1 + k_2 z^{-1}}, \tag{14.67}$$

suppose we replace the delay element with another allpass filter as follows:

$$z^{-1} \longmapsto z^{-1}\left(\frac{k_1^* + z^{-1}}{1 + k_1 z^{-1}}\right). \tag{14.68}$$

Then the result is

$$G(z) = \frac{k_2^* + z^{-1}\dfrac{k_1^* + z^{-1}}{1 + k_1 z^{-1}}}{1 + k_2 z^{-1}\dfrac{k_1^* + z^{-1}}{1 + k_1 z^{-1}}} = \frac{k_2^* + (k_1^* + k_1 k_2^*)z^{-1} + z^{-2}}{1 + (k_1 + k_1^* k_2)z^{-1} + k_2 z^{-2}}. \tag{14.69}$$

For the rational function on the right-hand side, the numerator coefficients are the reversed and conjugated version of the denominator. So, from the results of Sec. 7.4 we conclude that it is an allpass function!

Cascaded lattice. What is less obvious is that if $|k_i| < 1$, that is, if both (14.67) and (14.68) are *stable* allpass filters, then (14.69) is also a stable allpass filter. This is a consequence of a general result, which says: *If a causal rational system $H(z)$ is stable allpass and we replace all appearances of z^{-1} in $H(z)$ with a causal rational stable allpass filter $F(z)$:*

$$G(z) = H(z)\Big|_{z^{-1}=F(z)}, \tag{14.70}$$

then the result is stable allpass. A proof was developed in Problem 7.11. Now observe that the first-order allpass filter in (14.67), and the first-order allpass filter in brackets in (14.68) can both be implemented using one of the lattice structures in Fig. 14.4 or Fig. 14.6(a), or even Fig. 14.6(b) if the k_i are real. The allpass filter (14.69) can therefore be implemented by the lattice structure shown in Fig. 14.7, that is, $Y(z)/U(z)$ in the figure will be equal to $G(z)$ in Eq. (14.69). This is an interconnection of two lattice sections, and is called a cascaded lattice structure. The quantities k_i are called lattice coefficients. ▽ ▽ ▽

The cascaded lattice structure has many useful properties, which we summarize next. All proofs can be found in Vaidyanathan [1993] and references therein.

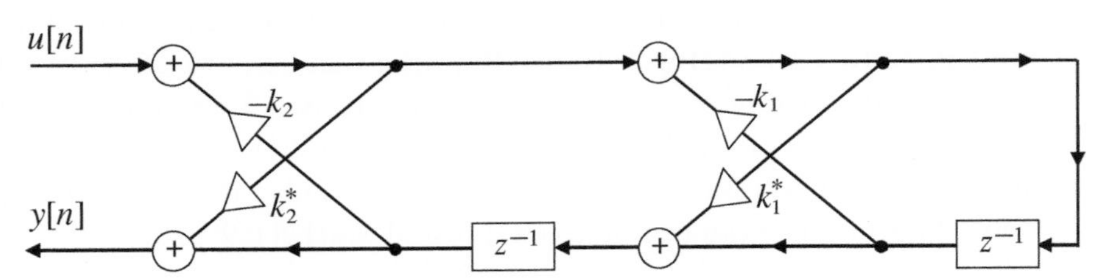

Figure 14.7 The cascaded lattice structure for implementing the second-order allpass filter in Eq. (14.69).

1. Any causal stable allpass filter of order N can be implemented as a cascade of N lattice sections (e.g., $N = 2$ in Fig. 14.7), and the lattice coefficients satisfy $|k_i| < 1$.
2. Conversely, given any cascaded lattice structure with all $|k_i| < 1$, the causal system $Y(z)/U(z)$ is guaranteed to be allpass and stable.
3. If the allpass filter has real coefficients, then the lattice coefficients k_i are real. In this case each lattice section can be replaced with the one-multiplier version in Fig. 14.6(b). *Thus, an Nth-order real-coefficient allpass filter can be implemented with N real multipliers and N delay elements.*
4. Finally, each lattice section can be independently chosen to be as in Fig. 14.4 or Fig. 14.6(a), or even Fig. 14.6(b) when the k_i are real.

14.6.2 Revisiting the Notch Filter

In Sec. 7.5 we showed that second-order notch filters can be implemented as

$$H(z) = \frac{1 + G(z)}{2}, \tag{14.71}$$

where $G(z)$ is the second-order real-coefficient allpass filter

$$G(z) = \frac{R^2 - 2R\cos\theta\, z^{-1} + z^{-2}}{1 - 2R\cos\theta\, z^{-1} + R^2 z^{-2}}, \tag{14.72}$$

with poles at $Re^{\pm j\theta}$. Here, $0 < R < 1$ is the pole radius and θ the pole angle. In Fig. 7.14(b) we showed a structure for this allpass filter with two multipliers and four delays. This is not a minimal structure, since a second-order filter should be realizable with only two delays. Such a minimal realization can be achieved if we use a cascaded lattice similar to Fig. 14.7. And since the coefficients are real, we can use one-multiplier lattice sections with real k_i. The final structure for the allpass filter $G(z)$ is shown in Fig. 14.8. The lattice coefficients k_1 and k_2 can be found by comparing Eq. (14.69) with Eq. (14.72). Thus

$$k_2 = R^2, \quad k_1 + k_1 k_2 = -2R\cos\theta, \tag{14.73}$$

so that

$$k_1 = -\frac{2R\cos\theta}{1 + k_2} = -\Big(\frac{2R}{1 + R^2}\Big)\cos\theta = -\cos\phi, \tag{14.74}$$

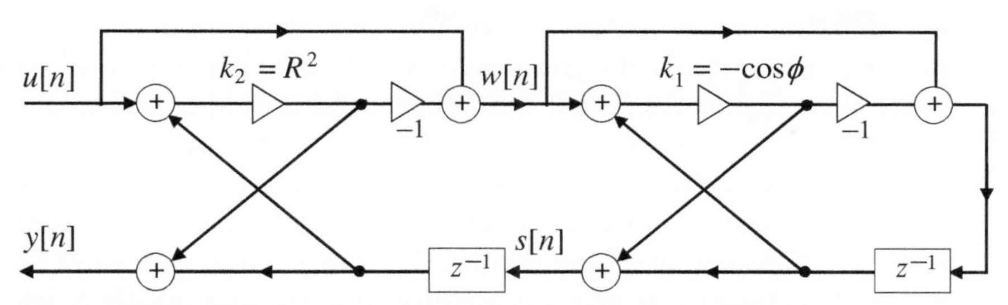

Figure 14.8 The lattice structure for the second-order allpass filter $G(z) = Y(z)/U(z)$ used in the notch filter. The notch quality and notch frequency can be controlled independently by tuning the multipliers k_2 and k_1, respectively.

where ϕ is the transmission zero of the notch filter (see Eq. (7.40)). The multipliers k_i in the lattice structures are therefore as shown in Fig. 14.8. Thus, the allpass filter (14.72) can be implemented with two multipliers and two delays using the cascaded lattice.

Decoupling the notch frequency and quality. There is another unique advantage of the lattice structure, when compared to Fig. 7.14(b) where the multipliers were R^2 and $-2R\cos\theta$. If we tune the multiplier R^2, it affects both the notch quality and the notch frequency ϕ, because $\cos\phi$ depends on R and θ as in Eq. (7.40). But the beauty of the lattice structure is that the multiplier k_1 controls the notch frequency ϕ only, and k_2 controls the notch quality only. The notch quality and notch frequency are therefore decoupled, and independently controllable in this stucture! ▽ ▽ ▽

14.7 Summary of Chapter 14

State-space descriptions (**A**,**B**,**C**,**D**) are very general and useful ways to describe computational graphs. In this chapter we studied state-space descriptions of a number of discrete-time LTI structures. We considered both SISO and MIMO systems. We derived general expressions for the transfer matrix $\mathbf{H}(z)$ and impulse response matrix $\mathbf{h}[n]$ in terms of (**A**,**B**,**C**,**D**). We also derived the homogeneous solution (output corresponding to zero input) in terms of the initial state vector and (**A**,**B**,**C**,**D**). Using the state-space framework it was shown that an LTI system in discrete time is realizable if and only if it is rational (i.e., $\mathbf{H}(z)$ has rational entries $H_{km}(z)$).

We introduced the concepts of reachability and observability for structures of MIMO LTI systems and showed that these properties can be checked by using the state-space descriptions (**A**,**B**,**C**,**D**). These are properties of the structure and not of the transfer matrix. We distinguished between the order and degree of a MIMO LTI system. A structure is minimal if the number of delay elements is equal to its degree. This property is satisfied if and only if the structure is reachable and observable. Given a transfer matrix, there exist an infinite number of structures to implement it, and we can get equivalent implementations by applying similarity transformations to an existing structure. In fact, all minimal realizations of an LTI system are related by similarity transformations. Given a minimal structure for $\mathbf{H}(z)$, the set of its poles is precisely the set of eigenvalues of $\mathbf{A}$ in the state-space description ($\mathbf{A}$,$\mathbf{B}$,$\mathbf{C}$,$\mathbf{D}$).

We also revisited IIR digital allpass filters and derived several equivalent structures for them using similarity transformations on state-space descriptions. Specifically, a number of lattice structures were presented for allpass filters. Digital notch filters, which can be expressed in terms of allpass filters, can therefore be implemented with lattice structures. For an Nth-order causal and stable allpass filter with real coefficients, there exists a lattice structure with exactly N real multipliers and N delays. This is therefore a minimal structure with minimal complexity. If such a structure is used to implement the second-order allpass filter in a notch filter, then the notch frequency and notch quality can be independently controlled by two separate multipliers.

State-space descriptions, and other ways to describe MIMO LTI systems, have been studied for many decades by system theorists. These have great importance in linear system theory, automatic controls, and circuit theory. Chapter 13 of Vaidyanathan [1993] gives a detailed discussion of MIMO LTI systems, including state-space descriptions. References for further reading include Anderson and Moore [1979], Anderson and Vongpanitlerd [1973], Chen [1984], and Kailath [1980]. An extensive tutorial on allpass filters can be found in Regalia et al. [1988].

PROBLEMS

14.1 For the structure shown in Fig. P14.1, write down the state-space description $(\mathbf{A}, \mathbf{B}, \mathbf{C}, \mathbf{D})$.

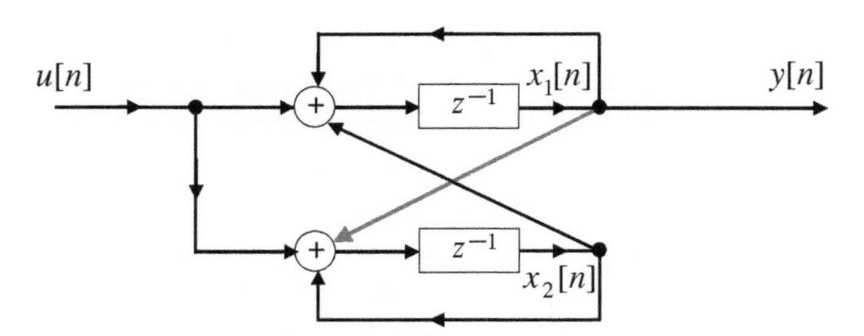

Figure P14.1 Structure for Problem 14.1.

14.2 Substituting the expressions for $\mathbf{A}, \mathbf{B}, \mathbf{C}, \mathbf{D}$ from (14.15) and (14.16) into Eq. (14.30), show that it reduces to the cascade-form transfer function given in Eq. (14.10).

14.3 The modified direct-form structure, which is different from the direct form shown in Fig. 14.1, was introduced in Sec. 5.9.2. This is reproduced in Fig. P14.3. Write down the state-space equations for this. Also find the state-space description $(\mathbf{A}, \mathbf{B}, \mathbf{C}, D)$. Assume $N = 3$ as in the figure.

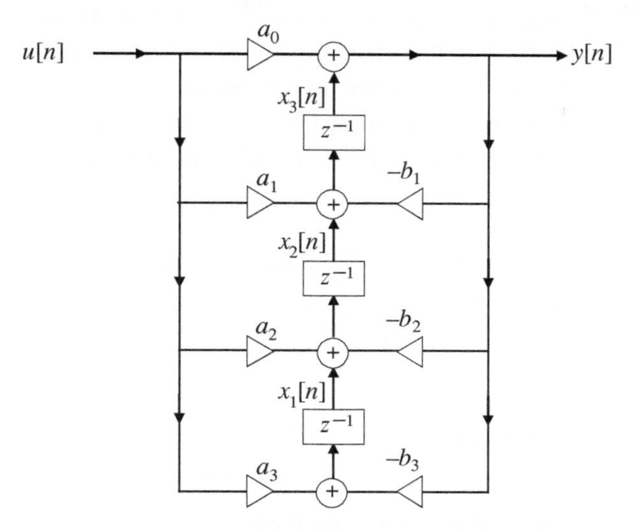

Figure P14.3 The modified direct-form structure.

14.4 Consider again the structure in Problem 14.1. (a) Is $\mathbf{A}$ stable? (b) Is $(\mathbf{A}, \mathbf{B})$ reachable? (c) Is $(\mathbf{C}, \mathbf{A})$ observable? (d) Is $(\mathbf{A}, \mathbf{B}, \mathbf{C}, \mathbf{D})$ minimal?

14.5 Consider the structure in Fig. P14.5. (a) Write down the state-space description $(\mathbf{A}, \mathbf{B}, \mathbf{C}, \mathbf{D})$. (b) For $m_1 = m_2 = 1$, is $\mathbf{A}$ stable? (c) For $m_1 = -1, m_2 = 2$, is $\mathbf{A}$ stable?

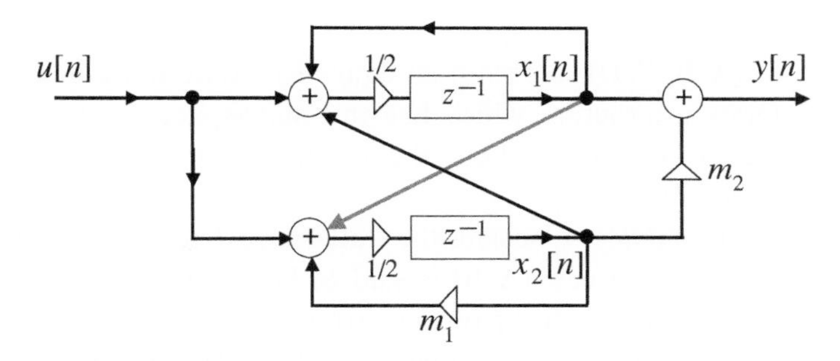

Figure P14.5 Structure for Problem 14.5.

14.6 Consider again the structure in Fig. P14.5. (a) For $m_1 = m_2 = 1$, is $(\mathbf{A}, \mathbf{B}, \mathbf{C}, \mathbf{D})$ minimal? (b) For $m_1 = -1, m_2 = 2$, is $(\mathbf{A}, \mathbf{B}, \mathbf{C}, \mathbf{D})$ minimal?

14.7 We know that for MIMO systems the degree and order are not necessarily equal. (a) Find a 2×2 example $\mathbf{H}(z)$ of a degree-two system with order one. (b) Find a 2×2 example $\mathbf{G}(z)$ of a degree-one system with order one. You should justify that the degree and order of your examples are as claimed. To avoid trivial answers, make sure $\mathbf{G}(z)$ is a nondiagonal matrix.

14.8 Consider the system $\mathbf{H}(z) = z^{-1}\mathbf{P}$, where $\mathbf{P}$ is $M \times M$ with rank ρ. Then we can write $\mathbf{H}(z) = \mathbf{U}\mathbf{V}z^{-1}$, where $\mathbf{U}$ and $\mathbf{V}$ are $M \times \rho$ and $\rho \times M$, respectively, with rank ρ. This leads to the implementation of Fig. P14.8.

(a) Find the state-space description $(\mathbf{A}, \mathbf{B}, \mathbf{C}, \mathbf{D})$ of this structure.
(b) Show that this structure for $\mathbf{H}(z)$ is minimal, that is, we cannot find a structure with fewer than ρ delays.

For the special case where $\mathbf{H}(z)$ is the diagonal matrix (14.51) with $a \neq 0$ and $b \neq 0$, it therefore follows that $\mathbf{H}(z)$ has degree two.

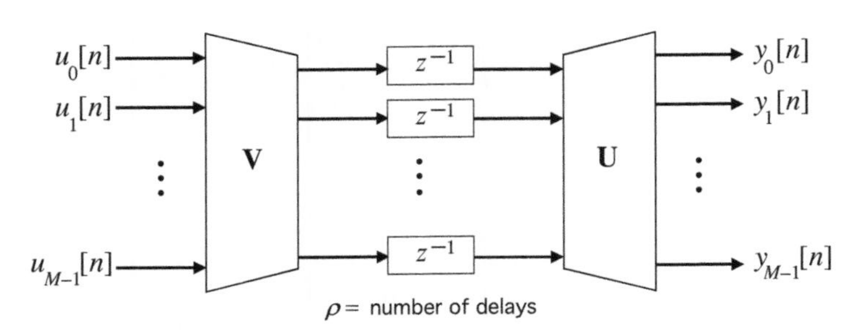

Figure P14.8 Structure for Problem 14.8.

14.9 Let $H(z) = (1 + z^{-1})/(1 - (5/6)z^{-1} + (1/6)z^{-2})$ and let $(\mathbf{A}, \mathbf{B}, \mathbf{C}, D)$ be the state-space description of the direct-form structure. Suppose we apply the similarity transformation

$$\mathbf{T} = \begin{bmatrix} 1 & 1 \\ -1 & 1 \end{bmatrix}$$

to $(\mathbf{A}, \mathbf{B}, \mathbf{C}, D)$ to obtain an equivalent state-space description $(\widehat{\mathbf{A}}, \widehat{\mathbf{B}}, \widehat{\mathbf{C}}, \widehat{D})$. Draw a structure which has this state-space description, using only two delays. Explicitly show the entries of $\widehat{\mathbf{A}}, \widehat{\mathbf{B}}, \widehat{\mathbf{C}}, \widehat{D}$ as multipliers in the structure.

14.10 *Controllability*. Consider the state-space description $(\mathbf{A}, \mathbf{B}, \mathbf{C}, \mathbf{D})$, where $\mathbf{A}$ is $N \times N$. The pair $(\mathbf{A}, \mathbf{B})$ is said to be controllable [Anderson and Moore, 1979] if we can start from any arbitrary initial state $\mathbf{x}[0]$ and force $\mathbf{x}[N] = \mathbf{0}$, by appropriate choice of $\mathbf{u}[0], \ldots, \mathbf{u}[N-1]$. Evidently if $(\mathbf{A}, \mathbf{B})$ is reachable, it is also controllable. However, the converse is not true.

(a) Show that the structure shown in Fig. P14.1 is controllable but not reachable.
(b) More generally, show that $(\mathbf{A}, \mathbf{B})$ is controllable as long as every column of $\mathbf{A}^N$ is in the column space of $\mathbf{R}$ defined in Eq. (14.56).

14.11 Recall the relation between the impulse response $\mathbf{h}[m]$ and state-space descriptions, given in Eq. (14.37). This was derived by first obtaining a general expression (14.35) for the output and then comparing it with the convolution sum (14.36). Assuming all eigenvalues of $\mathbf{A}$ satisfy $|\lambda_i| < 1$, a second procedure would be to start from the transfer function (14.29) and express $(z\mathbf{I} - \mathbf{A})^{-1}$ as a power series. Rederive (14.37) using this idea. *Hint:* $(\mathbf{I} - \mathbf{A})^{-1} = \mathbf{I} + \mathbf{A} + \mathbf{A}^2 + \mathbf{A}^3 + \cdots$ when $|\lambda_i| < 1$.

14.12 Consider the expression for the transfer function $\mathbf{H}(z)$ in Eq. (14.29). Here $z\mathbf{I} - \mathbf{A}$ is an $N \times N$ matrix whose entries are polynomials in z. Using the standard formula for matrix inversion (Sec. 18.5), we can write

$$(z\mathbf{I} - \mathbf{A})^{-1} = \frac{\mathbf{P}(z)}{\det(z\mathbf{I} - \mathbf{A})}, \tag{P14.12}$$

where the denominator is the determinant of $z\mathbf{I} - \mathbf{A}$, and $\mathbf{P}(z)$ is an $N \times N$ matrix whose entries contain cofactors of the entries of $z\mathbf{I} - \mathbf{A}$. Thus the entries of $\mathbf{P}(z)$ are polynomials in z.

(a) Based on Eq. (P14.12) show that if p is a pole of any entry $H_{km}(z)$ of the transfer matrix $\mathbf{H}(z)$, then p is an eigenvalue of the matrix $\mathbf{A}$.

(b) With no further assumptions, is the converse true? That is, if λ is an eigenvalue of $\mathbf{A}$, is it necessarily a pole of some element $H_{km}(z)$?

14.13 *The coupled-form structure.* Consider the structure shown in Fig. P14.13, which is a generalization of the coupled-form oscillator structure in Fig. 8.4. Write down the state-space description $(\mathbf{A}, \mathbf{B}, \mathbf{C}, \mathbf{D})$ for this structure. Using the expression (14.29), find the transfer function $H(z) = Y(z)/U(z)$, explicitly showing the coefficients of the numerator and denominator polynomials. What are the pole locations?

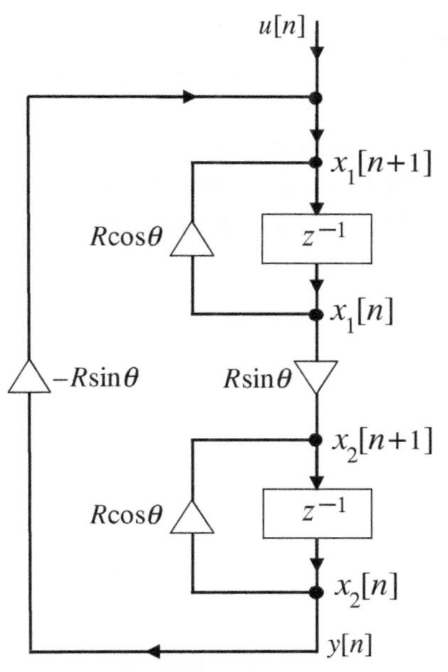

Figure P14.13 The coupled-form structure.

14.14 *An FIR lattice structure.* Figure P14.14 shows a two-input, two-output LTI system with no feedback loops. So this is an FIR system, and the transfer matrix is a polynomial in z^{-1}.

(a) Find the state-space description $(\mathbf{A}, \mathbf{B}, \mathbf{C}, \mathbf{D})$ for this structure.

(b) Find an expression for the transfer matrix $\mathbf{H}(z)$ using Eq. (14.29).

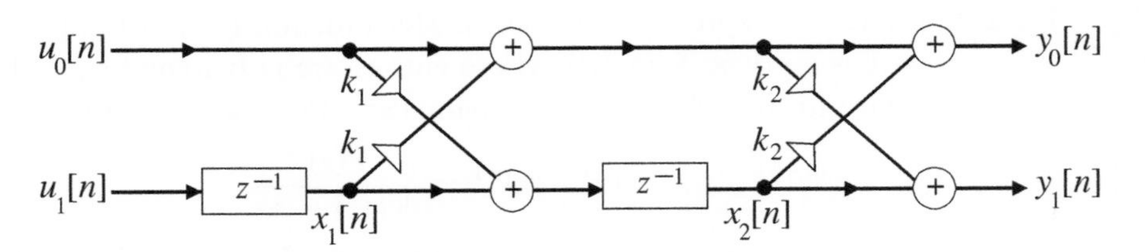

Figure P14.14 An FIR lattice structure.

14.15 Show that if the similarity transformation (14.54) is substituted into the right-hand side of Eq. (14.29) and simplified, the **T** matrix cancels off and leaves the transfer function unchanged.

14.16 Show that the two-multiplier lattice structure shown in Fig. 14.6(a) has the state-space description (14.62).

14.17 Show that the one-multiplier lattice structure shown in Fig. 14.6(b) has the state-space description (14.65).

14.18 (Computing assignment) Consider a structure with state-space description

$$\mathbf{A} = \begin{bmatrix} 1/2 & 0 & 0 \\ 2 & 1/3 & 0 \\ 3 & 2 & 1/2 \end{bmatrix}, \quad \mathbf{B} = \begin{bmatrix} 1 \\ 2 \\ 1 \end{bmatrix}, \quad \mathbf{C} = \begin{bmatrix} 1 & 1 & 2 \end{bmatrix}, \quad D = 1.$$

Let the input be of finite duration with only three nonzero samples, namely $u[0] = 2, u[1] = 1$, and $u[2] = 3$. Find an initial condition $\mathbf{x}[0]$ such that $y[n] = 0, n \geq 3$.

14.19 (Computing assignment) *State recursion.* Let $(\mathbf{A}, \mathbf{B}, \mathbf{C}, D)$ and $u[n]$ be as in Problem 14.18, and assume the initial condition is $\mathbf{x}[0] = \begin{bmatrix} 1 & 0 & 0 \end{bmatrix}^T$. Compute $\mathbf{x}[n]$ for $1 \leq n \leq 3$ and $y[n]$ for $0 \leq n \leq 3$.

15 More on Bandlimited Signals*

15.1 Introduction

Bandlimited signals are not only very important for engineers and signal processors, they are also appealing from a mathematical viewpoint. In this chapter we discuss a number of such interesting aspects of these signals. First we discuss the space of bandlimited L_2 signals. We will see that a set of uniformly shifted versions $\{h(t - nT)\}$ of an appropriately chosen sinc function $h(t)$ spans this space if T is small enough. In fact, this set is an orthogonal basis for this space if T is chosen right. A number of other interesting results about these spaces will also be discussed. We then show that the integral and the energy of a bandlimited signal can be obtained exactly from samples if the sampling rate is high enough. For non-bandlimited functions such a result is in general only approximately true, with the approximation getting better as the sampling rate increases. A number of less obvious consequences of these results are also presented. We also discuss a nice connection between bandlimited signals and analytic functions which arise in the theory of complex variables. The practical relevance of this will also be discussed.

Notations: Please see beginning of Chapter 10.

15.2 The Bandlimited Space of Signals

From the definition of a σ-BL signal (Sec. 10.3), it is clear that if $x_1(t)$ and $x_2(t)$ are σ-BL, then so is $a_1x_1(t) + a_2x_2(t)$. Repeated application shows that any linear combination

$$y(t) = \sum_{k=-\infty}^{\infty} a_k x_k(t) \tag{15.1}$$

of σ-BL signals $x_k(t)$ is σ-BL. Because of this property, we say that the set $\mathcal{B}_\sigma$ of all σ-BL signals is a **linear space**, called the **σ-BL space**. Note that if a signal is η-BL for some $\eta < \sigma$, then it is also σ-BL. So we can write

$$\mathcal{B}_\eta \subset \mathcal{B}_\sigma, \quad \text{for } \eta \leq \sigma. \tag{15.2}$$

That is, $\mathcal{B}_\sigma$ includes BL signals with smaller bandlimits as well.

15.2.1 A Set of Signals which Spans the σ-BL Space

Now, from the uniform sampling theorem we know that any σ-BL signal $x(t)$ can be represented as a linear combination

$$x(t) = \sum_{n=-\infty}^{\infty} x(nT)h(t - nT), \tag{15.3}$$

as long as $T \leq \pi/\sigma$ (i.e., $\omega_s \geq 2\sigma$). Here

$$h(t) = T\sin(\sigma t)/\pi t, \tag{15.4}$$

and its FT is

$$H(j\omega) = \begin{cases} T & -\sigma < \omega < \sigma, \\ 0 & \text{otherwise}, \end{cases} \tag{15.5}$$

as shown in Fig. 10.5(c). It is clear from Eq. (15.5) that $h(t)$ is itself σ-BL, that is, $h(t) \in \mathcal{B}_\sigma$. Furthermore, since

$$h(t - nT) \longleftrightarrow e^{-j\omega nT}H(j\omega), \tag{15.6}$$

it follows that $h(t - nT) \in \mathcal{B}_\sigma$ for all n. So, any linear combination of the functions $\{h(t - nT)\}$ belongs in $\mathcal{B}_\sigma$. Conversely, since any $x(t) \in \mathcal{B}_\sigma$ can be expressed as in Eq. (15.3) as long as $T \leq \pi/\sigma$, it follows that the *shifted sinc functions* $\{h(t - nT)\}$ *span the entire* σ*-BL space* $\mathcal{B}_\sigma$ in this case. That is,

$$\text{span}\Big(\{h(t - nT\}\Big) = \mathcal{B}_\sigma \quad \text{when } T \leq \pi/\sigma. \tag{15.7}$$

We can slightly generalize the above conclusions and summarize as follows:

1. Any linear combination of the set of shifted sinc functions $\{h(t - nT)\}$ belongs in $\mathcal{B}_\sigma$ *regardless* of the value of T.
2. As for the converse, any $x(t) \in \mathcal{B}_\sigma$ can be expressed as a linear combination of $\{h(t - nT)\}$ as long as $T \leq \pi/\sigma$ (i.e., $\omega_s \geq 2\sigma$).

15.2.2 Orthogonal Basis for the σ-BL Space

Now consider the case $T = \pi/\sigma$, which corresponds to the Nyquist rate. This is the minimum sampling rate (maximum sample spacing) allowed by the sampling theorem, and in this case

$$h(t) = \frac{\sin(\pi t/T)}{\pi t/T}, \tag{15.8}$$

which was shown in Fig. 10.8(d). We will show in this case that the following property holds:

$$\int_{-\infty}^{\infty} h(t - nT)h(t - mT)dt = T\delta[n - m] \quad (\text{when } T = \pi/\sigma). \tag{15.9}$$

The left-hand side is the inner product of $h(t-nT)$ and $h(t-mT)$. So the above says that these two shifted versions of the sinc function are orthogonal, that is, the set

$\{h(t-nT)\}$ not only spans $\mathcal{B}_\sigma$ (as discussed in Sec. 15.2.1), it is also an **orthogonal basis** for $\mathcal{B}_\sigma$ when $T=\pi/\sigma$.

Proof of Eq. (15.9). Since $h(t-\tau) \leftrightarrow e^{-j\omega\tau}H(j\omega)$, it follows from Parseval's theorem that

$$\begin{aligned}\int_{-\infty}^{\infty} h(t-nT)h(t-mT)dt &= \frac{1}{2\pi}\int_{-\infty}^{\infty} e^{-j\omega nT}H(j\omega)e^{j\omega mT}H^*(j\omega)d\omega \\ &= \frac{T^2}{2\pi}\int_{-\pi/T}^{\pi/T} e^{j(m-n)\omega T}d\omega \\ &= \frac{T^2}{2\pi}\frac{e^{j(m-n)\omega T}}{j(m-n)T}\Big|_{\omega=-\pi/T}^{\pi/T} \quad \text{when } n\neq m \\ &= 0. \end{aligned} \tag{15.10}$$

The second equality is because of Eq. (15.5) with $\sigma=\pi/T$. And when $n=m$, the second line simplifies to $(T^2/2\pi)\times(2\pi/T)=T$, proving Eq. (15.9). $\bigtriangledown\bigtriangledown\bigtriangledown$

When a set of functions $\{\eta_k(t)\}$ is orthogonal, it can be shown that they are in particular **linearly independent** (Problem 15.3). So, when $T=\pi/\sigma$ the set of shifted sinc functions $\{h(t-nT)\}$ is linearly independent, that is, it forms a **basis** for the space that it spans. Moreover, as shown in Sec. 15.2.1, it spans the σ-BL space $\mathcal{B}_\sigma$. In short, *when $T=\pi/\sigma$, the set $\{h(t-nT)\}$ forms an orthogonal basis for the σ-BL space $\mathcal{B}_\sigma$*. When $h(t)$ is as in Eq. (15.4) with $T\neq\pi/\sigma$, we can distinguish two cases:

1. When $T<\pi/\sigma$ (case of oversampling), it can be shown (Sec. 15.2.3) that the set $\{h(t-nT)\}$ is not linearly independent (hence not a basis). But since Eq. (15.7) is true, we have span $(\{h(t-nT)\})=\mathcal{B}_\sigma$. It is just that $\{h(t-nT)\}$ has redundancy due to linear dependence. So we say that $\{h(t-nT)\}$ is a **frame**, rather than a basis, for $\mathcal{B}_\sigma$.
2. When $T>\pi/\sigma$ (case of undersampling), it can be shown that the set $\{h(t-nT)\}$ is linearly independent (Sec. 15.2.3). But its span does not include all of $\mathcal{B}_\sigma$, it is a basis only for a proper subspace of $\mathcal{B}_\sigma$.

So we see that the set of shifted sinc functions $\{h(t-nT)\}$ forms a **basis** for the σ-BL space $\mathcal{B}_\sigma$ if and only if $T=\pi/\sigma$. When this happens, the basis is actually an orthogonal basis, satisfying Eq. (15.9).

15.2.2.1 Clarification on Orthogonality

The orthogonality property of sinc functions can be stated in its own right without reference to bandlimited functions, sampling, and so forth. Thus, suppose $H(j\omega)$ is the ideal lowpass filter with cutoff frequency ω_c, that is,

$$H(j\omega)=\begin{cases} T & -\omega_c<\omega<\omega_c, \\ 0 & \text{otherwise}, \end{cases} \tag{15.11}$$

so that

$$h(t) = T\frac{\sin(\omega_c t)}{\pi t}. \tag{15.12}$$

Then $\{h(t - nT)\}$ is an orthogonal set of functions, that is,

$$\int_{-\infty}^{\infty} h(t - nT)h(t - mT) = 0, \quad n \neq m, \tag{15.13}$$

if and only if the cutoff ω_c and the shift parameter T are related as

$$\omega_c = K\frac{\pi}{T} \quad \text{(orthogonality condition)}, \tag{15.14}$$

for some integer $K \geq 1$. This can be proved by a slight modification of the proof shown in Eq. (15.10) (Problem 15.4). In the specific context of sampling theory, the sampling rate is defined as $\omega_s = 2\pi/T$. So Eq. (15.14) translates to

$$\omega_c = K\frac{\omega_s}{2}. \tag{15.15}$$

For $K = 1$ we get the reconstruction filter shown in Fig. 10.11(c). If $K > 1$, then Eq. (15.15) yields $\omega_c \geq \omega_s$, so the filter passband extends to ω_s or beyond. This means the filter will not be able to filter out the copies $X(j(\omega - k\omega_s), k \neq 0$, in the sampled version, that is, the filter output is not a perfectly reconstructed version of $x(t)$. Thus, for sampling a σ-BL signal with $\omega_s \geq 2\sigma$, the **only perfect reconstruction filter with orthogonality property** corresponds to

$$\omega_c = \frac{\omega_s}{2} = \frac{\pi}{T}. \tag{15.16}$$

As we know, two cases are possible:

1. If $\omega_s > 2\sigma$, this filter has excess BW as shown in Fig. 10.11(c).
2. If $\omega_s = 2\sigma$ (Nyquist sampling), there is only one filter for reconstruction, and it satisfies the orthogonality requirement (15.14) automatically because filter cutoff $\omega_c = \sigma = \pi/T$ in this case.

15.2.3 Linear Independence, Subspace Spanned ...

Now consider signals of the form

$$x(t) = \sum_{n=-\infty}^{\infty} c[n]h(t - nT), \tag{15.17}$$

where $h(t)$ is the impulse response of a bandlimiting filter, and $c[n]$ are just the coefficients in the linear combination. Note that $c[n]$ are not necessarily the samples $x(nT)$ as in earlier sections (also see Problem 15.1). It will be insightful to understand how such linear combinations of uniformly shifted functions $\{h(t - nT)\}$ work. Let us begin by taking the Fourier transform on both sides of Eq. (15.17). Since $h(t - nT)$ has FT equal to $e^{-jnT\omega}H(j\omega)$, it follows that

$$X(j\omega) = \Big(\sum_{n=-\infty}^{\infty} c[n]e^{-j\omega Tn}\Big)H(j\omega) = C(e^{j\omega T})H(j\omega), \tag{15.18}$$

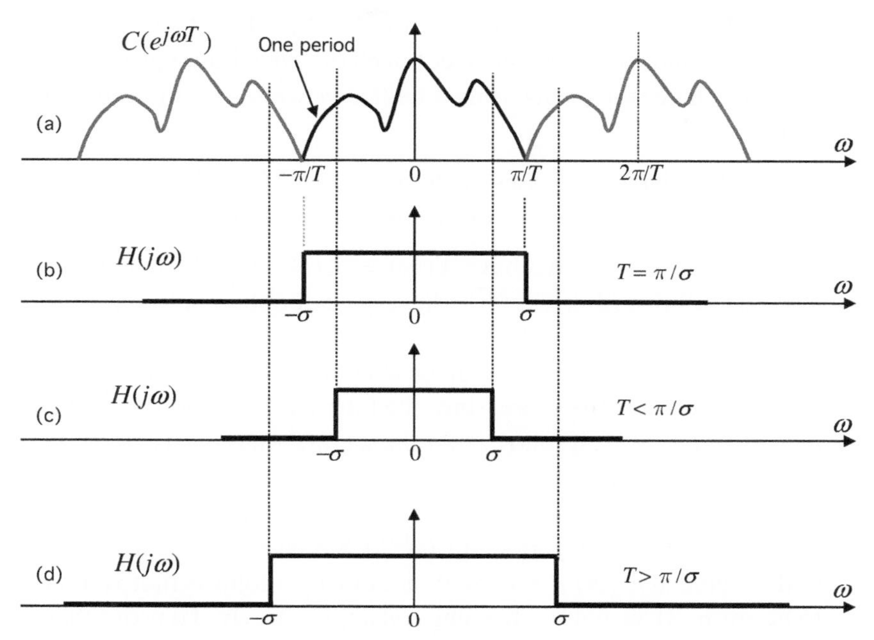

Figure 15.1 Generating bandlimited L_2 functions using $x(t) = \sum c[n]h(t - nT)$ which has FT $X(j\omega) = C(e^{j\omega T})H(j\omega)$. (a) The quantity $C(e^{j\omega T})$ and (b)–(d) three choices of $H(j\omega)$. See text.

where

$$C(e^{j\omega}) = \sum_{n=-\infty}^{\infty} c[n]e^{-j\omega n}. \tag{15.19}$$

Here $C(e^{j\omega})$ is the DTFT of $c[n]$, and $H(j\omega)$ is the CTFT of $h(t)$. Thus, both types of FT are mixed up in the FT of $x(t)$. Since $C(e^{j\theta})$ has period 2π in θ, the function $C(e^{j\omega T})$ is periodic in ω with period $2\pi/T$. Figure 15.1(a) shows a typical plot of $C(e^{j\omega T})$. We wish to understand the nature of the bandlimited space that can be generated by Eq. (15.17), or equivalently Eq. (15.18), as $c[n]$ is varied over all possible sequences for which a DTFT $C(e^{j\omega})$ exists.

Let us consider an example where $H(j\omega)$ is an ideal σ-BL lowpass filter with different choices for the cutoff σ, as shown in Fig. 15.1(b)–(d). Three cases are shown: $T = \pi/\sigma$, $T < \pi/\sigma$, and $T > \pi/\sigma$. The conclusions for these cases are as follows:

1. When $T = \pi/\sigma$, $\{h(t-nT)\}$ spans all of $\mathcal{B}_\sigma$ and in fact $\{h(t-nT)\}$ is an *orthogonal basis* for $\mathcal{B}_\sigma$.
2. When $T < \pi/\sigma$, $\{h(t - nT)\}$ spans all of $\mathcal{B}_\sigma$ but $\{h(t - nT)\}$ is not a linearly independent set. That is, it has redundancy. So, it is a *frame* (and not a basis) for $\mathcal{B}_\sigma$.
3. When $T > \pi/\sigma$, $\{h(t - nT)\}$ spans only a subspace of $\mathcal{B}_\sigma$, but $\{h(t - nT)\}$ is a *linearly independent* set.

The first conclusion was already proved in Sec. 15.2.2 based on sampling theory, and the second and third were mentioned therein without proof. In what follows we prove all three statements using a unified viewpoint, without recourse to sampling theory.

Proof. Crucial to all cases is the fact that $\sum_n c[n]h(t-nT)$ has FT $X(j\omega) = C(e^{j\omega T})H(j\omega)$, as we proved above. We consider each case separately:

Case 1. $T = \pi/\sigma$. We have $X(j\omega) = C(e^{j\omega T})$ in the region $|\omega| < \sigma$, which is also one full period of $C(e^{j\omega T})$. Since we can obtain any shape for $C(e^{j\omega T})$ in the region $|\omega| < \pi/T = \sigma$ by choice of $c[n]$, it follows that $X(j\omega)$ can achieve any shape in $|\omega| < \sigma$, that is, $x(t)$ can be chosen to be any σ-BL signal by choosing $c[n]$ appropriately. Furthermore, since the filter cutoff σ and the shift parameter T are related as $T = \pi/\sigma$, it follows that the set $\{h(t-nT)\}$ satisfies orthogonality (15.9). This proves Case 1.

Case 2. $T < \pi/\sigma$. We have $X(j\omega) = C(e^{j\omega T})$ in the region $|\omega| < \sigma < \pi/T$, which is a portion of one full period of $C(e^{j\omega T})$. Since we can obtain any shape for $C(e^{j\omega T})$ in the region $|\omega| < \pi/T > \sigma$ by choice of $c[n]$, it follows that $x(t)$ can still be chosen to be any σ-BL signal by choosing $c[n]$ appropriately. To prove that $\{h(t-nT)\}$ has linear dependence, we have to show that $\sum_n c[n]h(t-nT)$ can be identically zero even if $c[n]$ is not zero for all n. Indeed, it is clear from the figure that if $C(e^{j\omega T}) = 0$ in the region $|\omega| < \sigma$ and nonzero in

$$\sigma < |\omega| < \pi/T, \tag{15.20}$$

we get a nonzero $c[n]$ such that the right-hand side of (15.18) is identically zero, that is, $\sum_n c[n]h(t-nT) = 0$ for all t. This proves Case 2.

Case 3. $T > \pi/\sigma$. We still have $X(j\omega) = C(e^{j\omega T})$ in $|\omega| < \sigma$. But as seen from the figure, the value of $C(e^{j\omega T})$ in the region

$$\pi/T < |\omega| < \sigma \tag{15.21}$$

is a repetition of portions of $C(e^{j\omega T})$ in its first period $|\omega| < \pi/T$. This means that the shape of $X(j\omega)$ which can be achieved in $|\omega| < \sigma$ is restricted. Thus $\{h(t-nT)\}$ spans only a subspace of $\mathcal{B}_\sigma$ in this case. However, $\{h(t-nT)\}$ is a linearly independent set because, as seen from the figure, $\sum_n c[n]h(t-nT) \longleftrightarrow H(j\omega)C(e^{j\omega T})$ can be identically zero only if $C(e^{j\omega T})$ is identically zero, that is, $c[n]$ is zero for all n. This proves Case 3. ▽ ▽ ▽

15.3 The Bandlimited L_2 Space

Recall that the energy of a signal is defined as the integral of $|x(t)|^2$, and from Parseval's relation we have

$$\int_{-\infty}^{\infty} |x(t)|^2 dt = \frac{1}{2\pi} \int_{-\infty}^{\infty} |X(j\omega)|^2 d\omega. \tag{15.22}$$

Signals for which the energy is finite are said to be L_2 signals. The set of all finite-energy σ-BL signals is said to be the σ-BL L_2 space. These are bandlimited signals

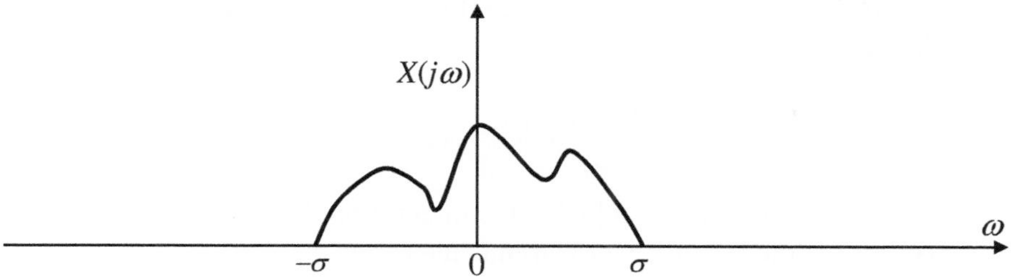

Figure 15.2 Fourier transform of a bandlimited L_2 signal.

with an arbitrary shape for $X(j\omega)$ in the region $-\sigma < \omega < \sigma$, as shown in Fig. 15.2, such that $|X(j\omega)|^2$ has finite integral. Thus, we can create examples of σ-BL L_2 signals by taking an arbitrary shape $X(j\omega)$ in this figure (such that its magnitude squared is integrable), and taking its inverse transform $x(t)$.

For example, consider the sinc function $h(t)$ in Eq. (15.4) whose FT is the pulse (15.5). This has finite energy, since

$$\int_{-\infty}^{\infty} |h(t)|^2 dt = \frac{1}{2\pi} \int_{-\infty}^{\infty} |H(j\omega)|^2 d\omega = \frac{T^2}{2\pi} \int_{-\sigma}^{\sigma} d\omega = \frac{T^2\sigma}{\pi}. \tag{15.23}$$

Since $h(t)$ is also σ-BL, it follows that it is a σ-BL L_2 signal. Another way to create more examples is to take linear combinations of the shifted sinc function

$$x(t) = \sum_{n=-\infty}^{\infty} c[n]h(t - nT), \tag{15.24}$$

where $h(t)$ is as in Eq. (15.4). We already proved that this is a σ-BL signal. This is also an L_2 signal if $c[n]$ is a finite-duration sequence (so that the above sum is finite). When $c[n]$ has infinite duration, Eq. (15.24) still represents an L_2 signal under some mild restrictions, as we shall see below. (Also see Sec. 15.7 in this regard.) Notice that the sinusoid

$$x(t) = \sin(\omega_0 t), \tag{15.25}$$

although bandlimited, is *not* an L_2 signal, because $\int_{-\infty}^{\infty} \sin^2(\omega_0 t)dt$ is not finite. Except for rare examples like this, most of the examples we encounter are L_2 signals. The L_2 assumption brings more structure into the mathematics, and makes it easy to prove some interesting properties, as we shall see.

15.3.1 Bandlimited L_2 Signals of the Form (15.24)

For signals of the form (15.24), we now restrict the coefficient sequence $c[n]$ to be a finite-energy sequence, that is,

$$\sum_{n=-\infty}^{\infty} |c[n]|^2 < \infty. \tag{15.26}$$

Such $c[n]$ is said to be an ℓ_2 **sequence**. In view of Parseval's relation for discrete-time FT, we have

$$\sum_{n=-\infty}^{\infty} |c[n]|^2 = \frac{1}{2\pi}\int_{-\pi}^{\pi} |C(e^{j\omega})|^2 d\omega, \tag{15.27}$$

which shows that when $c[n] \in \ell_2$, the integral on the right is also finite. As shown in Sec. 15.2.3, the FT of Eq. (15.24) has the form

$$X(j\omega) = C(e^{j\omega T})H(j\omega). \tag{15.28}$$

Assuming that $c[n]$ is an ℓ_2 signal and $h(t)$ is σ-BL, we now show that the signal

$$x(t) = \sum_{n=-\infty}^{\infty} c[n]h(t - nT) \tag{15.29}$$

is a σ-BL L_2 signal if

$$|H(j\omega)| \leq B < \infty \tag{15.30}$$

for some $B > 0$, that is, as long as the frequency response of $h(t)$ is bounded.

Proof. Since $h(t)$ is σ-BL, it is obvious from Eq. (15.29) that $x(t)$ is σ-BL. It only remains to prove that $x(t) \in L_2$. For this, observe that

$$\begin{aligned} \frac{1}{2\pi}\int_{-\sigma}^{\sigma} |X(j\omega)|^2 d\omega &= \frac{1}{2\pi}\int_{-\sigma}^{\sigma} \left|C(e^{j\omega T})H(j\omega)\right|^2 d\omega \\ &\leq \frac{B^2}{2\pi}\int_{-\sigma}^{\sigma} |C(e^{j\omega T})|^2 d\omega. \end{aligned} \tag{15.31}$$

We have used Eq. (15.28) in the first line and Eq. (15.30) in the second line. Since $c[n] \in \ell_2$, the integral in Eq. (15.27) is finite. So the last integral in Eq. (15.31) is finite for finite σ (see Fig. 15.1). $\nabla \nabla \nabla$

Equation (15.30) is satisfied by all the filters we consider in our discussions, and is readily satisfied in engineering practice, so there is little loss of generality in the assumption (15.30). Now let us impose one further restriction on the filter $h(t)$, namely the orthogonality property

$$\int_{-\infty}^{\infty} h(t - nT)h(t - mT)dt = T\delta[n - m]. \tag{15.32}$$

As shown in Sec. 15.2.2, this happens, for example, when $h(t)$ is the σ-BL filter with frequency response

$$H(e^{j\omega}) = \begin{cases} T & -\sigma < \omega < \sigma, \\ 0 & \text{otherwise}, \end{cases} \tag{15.33}$$

and T satisfies $T = \pi/\sigma$. Assuming Eq. (15.32) holds, we will show that

$$\sum_{n=-\infty}^{\infty} |c[n]|^2 = \frac{1}{T}\int_{-\infty}^{\infty} |x(t)|^2 dt, \tag{15.34}$$

where $x(t)$ is as in Eq. (15.29). Thus, the energy in the ℓ_2 sequence $c[n]$ and the energy in the L_2 signal $x(t) = \sum_{n=-\infty}^{\infty} c[n]h(t - nT)$ are equal (up to a scale factor $1/T$) when $h(t)$ satisfies orthogonality (15.32).

Proof of Eq. (15.34). We have

$$\begin{aligned}\int_{-\infty}^{\infty} |x(t)|^2 dt &= \int_{-\infty}^{\infty} \sum_{m=-\infty}^{\infty} \sum_{n=-\infty}^{\infty} c[m]c^*[n]h(t - mT)h^*(t - nT)dt \\ &= \sum_{m=-\infty}^{\infty} \sum_{n=-\infty}^{\infty} c[m]c^*[n] \int_{-\infty}^{\infty} h(t - mT)h^*(t - nT)dt \\ &= T \sum_{m=-\infty}^{\infty} \sum_{n=-\infty}^{\infty} c[m]c^*[n]\delta[m - n] \\ &= T \sum_{n=-\infty}^{\infty} |c[n]|^2, \end{aligned} \tag{15.35}$$

which proves the claim. ▽ ▽ ▽

15.4 Integrals Involving Bandlimited Functions

Sampling theory not only has an impact on digital signal processing and information sciences, it also has a number of mathematical implications. In this section we mention a few of these.[1] We first show that if $x(t)$ is σ-bandlimited and $T \leq \pi/\sigma$ (i.e., sampling rate $\omega_s \geq 2\sigma$), then

$$\int_{-\infty}^{\infty} x(t)dt = T \sum_{n=-\infty}^{\infty} x(nT). \tag{15.36}$$

The significance of Eq. (15.36) is that the integral of a bandlimited signal can be evaluated exactly just by adding up the samples taken at or above the Nyquist rate. This is unlike arbitrary signals, for which the integrals can only be approximated by summation of samples. For example, when we learn about integrals first, we are told that $\int x(t)dt$ is the limit of a summation of the form

$$\sum_n x(n\Delta t)\Delta t \tag{15.37}$$

as $\Delta t \to 0$, and that the approximation

$$\int x(t)dt \approx \sum_n x(n\Delta t)\Delta t \tag{15.38}$$

gets better as Δt gets smaller. The interesting thing about bandlimited signals is that the above **approximation becomes exact** as soon as the sample spacing Δt becomes

[1] The inspiration for this section comes from a lecture delivered by Dr. E. C. Posner at the California Institute of Technology many years ago. Some of these results are also summarized in the appendix of Vaidyanathan [2001].

small enough to satisfy the sampling theorem (i.e., $\Delta t \leq \pi/\sigma$ or $2\pi/\Delta t \geq 2\sigma$). Decreasing Δt further does not improve the approximation (15.38), as it has already become a perfect approximation!

Proof of Eq. (15.36). If $x(t)$ is σ-BL, we know that we can write

$$x(t) = \sum_{n=-\infty}^{\infty} x(nT)h(t - nT), \tag{15.39}$$

as long as $T \leq \pi/\sigma$ and $h(t)$ satisfies the perfect reconstruction condition (10.48). From Eq. (15.39) we have

$$\int_{-\infty}^{\infty} x(t)dt = \sum_{n=-\infty}^{\infty} x(nT) \int_{-\infty}^{\infty} h(t - nT)dt. \tag{15.40}$$

Observe now that

$$\int_{-\infty}^{\infty} h(t - nT)dt = \int_{-\infty}^{\infty} h(t)dt = H(0) = T, \tag{15.41}$$

where the first equality is trivial, the second equality follows from the definition of the Fourier transform, and the last equality follows from Eq. (10.48). This proves the claim (15.36). ▽ ▽ ▽

15.4.1 Energy of a Bandlimited Function

A relation similar to Eq. (15.36) is true for the energy of a bandlimited signal. To be precise, we will see that if $x(t)$ is σ-BL and $T \leq \pi/\sigma$ (i.e., $\omega_s \geq 2\sigma$), then

$$\int_{-\infty}^{\infty} |x(t)|^2 dt = T \sum_{n=-\infty}^{\infty} |x(nT)|^2. \tag{15.42}$$

Since $x(t - t_0)$ is also σ-BL for any shift t_0, and has the same energy as $x(t)$, it also follows from the above that

$$\int_{-\infty}^{\infty} |x(t)|^2 dt = T \sum_{n=-\infty}^{\infty} |x(nT - t_0)|^2 \qquad \text{(for any real } t_0\text{)}. \tag{15.43}$$

Observe carefully that Eq. (15.42) does not follow from Eq. (15.36). This is because, if $x(t)$ is σ-BL, then $|x(t)|^2$ is not necessarily σ-BL. We can only say that it is 2σ-BL, as its bandwidth can double (see Problem 10.11). So the samples $|x(nT)|^2$ are not at or above the Nyquist rate of $|x(t)|^2$.

Proof of Eq. (15.42). To prove Eq. (15.42) we start again from Eq. (15.39), which is valid for any $h(t)$ satisfying the perfect reconstruction condition (10.48). To facilitate the proof, we will choose $H(j\omega)$ to be

$$H(j\omega) = \begin{cases} T & -\dfrac{\pi}{T} < \omega < \dfrac{\pi}{T}, \\ 0 & \text{otherwise.} \end{cases} \tag{15.44}$$

This was plotted earlier in Fig. 10.11(c). The impulse response of this filter, $h(t) = \sin(\pi t/T)/(\pi t/T)$, satisfies the orthogonality property

$$\int_{-\infty}^{\infty} h(t-nT)h(t-mT) = T\delta[n-m], \tag{15.45}$$

as shown in Sec. 15.2.2. Thus, from Eq. (15.39) we get

$$\begin{aligned}\int_{-\infty}^{\infty} |x(t)|^2 dt &= \sum_{m=-\infty}^{\infty}\sum_{n=-\infty}^{\infty} x(mT)x^*(nT)\int_{-\infty}^{\infty} h(t-mT)h^*(t-nT)dt \\ &= T\sum_{m=-\infty}^{\infty}\sum_{n=-\infty}^{\infty} x(mT)x^*(nT)\delta[m-n] = T\sum_{n=-\infty}^{\infty} |x(nT)|^2,\end{aligned}$$

which completes the proof. ▽ ▽ ▽

Equation (15.42) says that the energy of $x(t)$ can be calculated exactly, by calculating the energy in the samples. Again, for arbitrary (non-BL) signals, Eq. (15.42) is only approximately true, the approximation getting better as T gets smaller. For bandlimited signals, the approximation becomes exact when T becomes small enough.

15.4.2 Applying Eq. (15.42) to the Sinc Function

In what follows we assume $T \leq \pi/\sigma$, so that Eq. (15.42) can be applied. First consider the lowpass filter

$$H(j\omega) = \begin{cases} T & -\sigma < \omega < \sigma, \\ 0 & \text{otherwise,} \end{cases} \tag{15.46}$$

which has $h(t) = T\sin\sigma t/\pi t$. From Parseval's relation

$$\int_{-\infty}^{\infty} |h(t)|^2 dt = \frac{1}{2\pi}\int_{-\infty}^{\infty} |H(j\omega)|^2 d\omega = \frac{T^2}{2\pi} * 2\sigma = \frac{\sigma T^2}{\pi}. \tag{15.47}$$

Since $h(t)$ is evidently σ-BL, it follows from Eq. (15.42) that

$$\sum_{n=-\infty}^{\infty} |h(nT)|^2 = \frac{1}{T}\int_{-\infty}^{\infty} |h(t)|^2 dt = \frac{\sigma T}{\pi}. \tag{15.48}$$

Similar to Eq. (15.43), the above readily generalizes to

$$\sum_{n=-\infty}^{\infty} |h(nT-t_0)|^2 = \frac{\sigma T}{\pi}, \quad \text{for any real } t_0. \tag{15.49}$$

Substituting $h(t) = T\sin\sigma t/\pi t$ and rearranging, it follows that

$$\sum_{n=-\infty}^{\infty}\left(\frac{\sin\sigma(nT-t_0)}{\sigma(nT-t_0)}\right)^2 = \frac{\pi}{\sigma T} \quad (\text{for } T \leq \pi/\sigma). \tag{15.50}$$

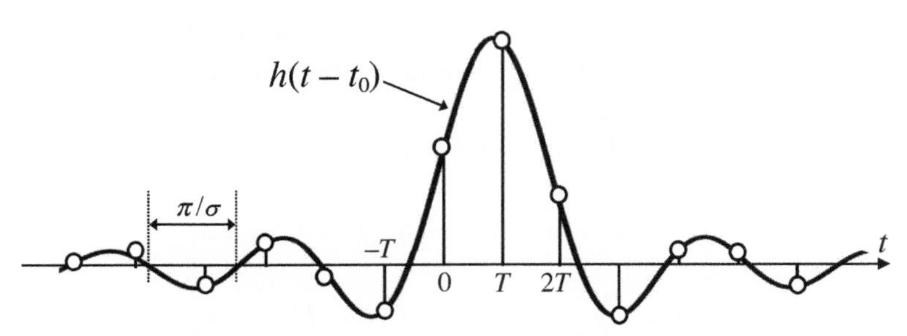

Figure 15.3 Plot of $h(t - t_0)$ for some constant $t_0 \neq 0$, where $h(t) = T \sin \sigma t / \pi t$. Samples $h(nT - t_0)$ are shown for $T < \pi/\sigma$, where π/σ is the distance between zero crossings.

For the special case where $t_0 = 0$ and $\sigma = \pi/T$ (filter cutoff = half the sampling rate), Eq. (15.49) reduces to

$$\sum_{n=-\infty}^{\infty} |h(nT)|^2 = 1. \tag{15.51}$$

This is admittedly obvious from the plot of $h(t)$, which satisfies $h(nT) = \delta[n]$ in this case (Fig. 10.8(d)). But for arbitrary $\sigma \leq \pi/T$ and arbitrary t_0, if we did not know sampling theory for bandlimited functions, then starting just from the expression $h(t) = T \sin \sigma t / \pi t$ it would be difficult to prove identities like (15.50). To give further insight, Fig. 15.3 shows the samples of $h(t - t_0)$ for some $T < \pi/\sigma$. The fact that the sum of squares of these samples is $\sigma T/\pi$ for *any* choice of t_0 is not an easy guess! But as we saw, the proof becomes easy with sampling theory.

15.4.3 More Fun with Sampling Theory ...

We can write Eq. (15.50) in normalized form by letting $\theta = \sigma T$ and $\theta_0 = \sigma t_0$ as follows:

$$\sum_{n=-\infty}^{\infty} \left(\frac{\sin(\theta n - \theta_0)}{\theta n - \theta_0} \right)^2 = \frac{\pi}{\theta} \qquad (\text{for } \theta \leq \pi). \tag{15.52}$$

This is indeed a beautiful mathematical identity that was obtained quite easily with the help of sampling theory. We will conclude this section by mentioning a few more "fun things" like this, that can be done with sampling theory. Consider the following well-known infinite series, which is useful for computing π:

$$1 - \frac{1}{3} + \frac{1}{5} - \frac{1}{7} + \frac{1}{9} + \cdots = \frac{\pi}{4}. \tag{15.53}$$

This is sometimes referred to as the Madhava–Leibniz formula, and is one of the earliest known ways to compute π up to arbitrary number of decimal places. The above series can be derived simply by using the expression (10.17) for the Fourier transform of the sampled version $x_d[n] = x(nT)$ of a signal $x(t)$:

$$X_d(e^{j\omega}) = \frac{1}{T} \sum_{k=-\infty}^{\infty} X\left(j\left(\frac{\omega - 2\pi k}{T}\right)\right). \tag{15.54}$$

Thus, if we take $x(t)$ to be the pulse signal (one of the simplest signals we know) and substitute the appropriate expressions for $X_d(e^{j\omega})$ and $X(j\omega)$ into Eq. (15.54), we can obtain Eq. (15.53). See Problem 15.8 for details. Similarly, the well-known infinite series for π^2

$$1+\frac{1}{3^2}+\frac{1}{5^2}+\frac{1}{7^2}+\frac{1}{9^2}+\cdots=\frac{\pi^2}{8} \tag{15.55}$$

can be obtained by applying Eq. (15.54) to the triangle signal $x(t)$ (Problem 15.9).

15.4.4 Noise Analysis

In Sec. 10.6 we mentioned that when the samples $x(nT)$ are contaminated with noise $e(nT)$, the reconstructed signal has a noise term $e(t)$ (see Eq. (10.54)). We now explain how the mean square value of this noise term can be computed. Some basic background on random processes [Peebles, 1987] is required in order to read this section. (Readers can skip this discussion without loss of continuity.) The noise term in Eq. (10.54) is

$$e(t)=\sum_{n=-\infty}^{\infty} e(nT)h(t-nT). \tag{15.56}$$

Since $h(t-nT)$ has the same passband as $h(t)$, the noise $e(t)$ is a random waveform bandlimited to the passband of the filter $h(t)$. It is typical to assume that $e(nT)$ is zero-mean white noise, that is, $E[e(nT)]=0$ and

$$E[e(nT)e^*(mT)]=\sigma_e^2\delta[m-n]. \tag{15.57}$$

Thus $e(nT)$ has a fixed variance σ_e^2 for all n, and it is uncorrelated from sample to sample. The mean square value of the noise in the reconstructed signal is

$$\begin{aligned}
E[|e(t)|^2] &= \sum_{n=-\infty}^{\infty}\sum_{m=-\infty}^{\infty} E\Big[e(nT)e^*(mT)\Big]h(t-nT)h^*(t-mT)\\
&=\sigma_e^2\sum_{n=-\infty}^{\infty}|h(t-nT)|^2 \qquad \text{(from Eq. (15.57))}\\
&=\frac{\sigma_e^2}{T}\int_{-\infty}^{\infty}|h(t)|^2dt \qquad \text{(see below)}\\
&=\frac{\sigma_e^2}{T}\int_{-\infty}^{\infty}|H(j\omega)|^2\frac{d\omega}{2\pi} \qquad \text{(Parseval's relation).}
\end{aligned} \tag{15.58}$$

To understand how the third line above is obtained, apply Eq. (15.43) to $h(t)$:

$$\sum_{n=-\infty}^{\infty}|h(nT-t_0)|^2=\frac{1}{T}\int_{-\infty}^{\infty}|h(t)|^2dt. \tag{15.59}$$

The $h(nT - t_0)$ in the summation can clearly be replaced by $h(-nT - t_0)$. Since Eq. (15.59) is valid for all real t_0, replacing $-t_0$ with t makes no difference, so we have

$$\sum_{n=-\infty}^{\infty} |h(t - nT)|^2 = \frac{1}{T}\int_{-\infty}^{\infty} |h(t)|^2 dt, \tag{15.60}$$

which gives the third line in Eq. (15.58). Now recall that the reconstruction filter has to satisfy $H(j\omega) = T$ in $-\sigma < \omega < \sigma$. Since the filter has excess bandwidth, let us say that it extends to the region $\sigma \leq |\omega| < \sigma + \Delta\omega$. For example, $\Delta\omega = \omega_s - 2\sigma$ in Fig. 10.11(b) and $\Delta\omega = \pi/T - \sigma$ in Fig. 10.11(c). The mean square reconstruction error can now be rewritten as

$$\begin{aligned} E[|e(t)|^2] &= \frac{\sigma_e^2}{T}\int_{-\sigma}^{\sigma} T^2 \frac{d\omega}{2\pi} \\ &+ \frac{\sigma_e^2}{T}\int_{\sigma}^{\sigma+\Delta\omega} |H(j\omega)|^2 \frac{d\omega}{2\pi} \;+\; \frac{\sigma_e^2}{T}\int_{-\sigma-\Delta\omega}^{-\sigma} |H(j\omega)|^2 \frac{d\omega}{2\pi}. \end{aligned}$$

The first line is the same as the noise at the ouput of the minimum bandwidth filter. The second line is due to excess bandwidth, and is positive unless $\Delta\omega = 0$. Assuming, for example, that $H(j\omega) = T$ throughout the passband, the above becomes

$$E[|e(t)|^2] = \left(\frac{T\sigma}{\pi}\right)\sigma_e^2 + \left(\frac{T\Delta\omega}{\pi}\right)\sigma_e^2. \tag{15.61}$$

This shows that the additional mean square error due to excess bandwidth grows linearly with the excess bandwidth.

15.5 Bandlimited Signals and Analytic Functions

The bandlimited property of a signal imposes several other restrictions on it. For example, we found in Sec. 10.9 that bandlimited functions cannot also be time limited, unless they are identically zero. We will now see that if $x(t)$ is bandlimited then it has to be a continuous function of t, in fact differentiable everywhere. It can even be shown that all higher derivatives of $x(t)$ exist for all t. In this sense a bandlimited signal is infinitely smooth. All these results follow from a single mathematical result that relates bandlimited functions to analytic functions. This is explained in this section.

15.5.1 Analytic Functions

The reader will recall from early math classes that a complex function $f(z)$ of the complex variable z is said to be analytic at the point z_0 if it is differentiable everywhere in a neighborhood of z_0. That is, there is a constant $\epsilon > 0$ such that $f(z)$ is differentiable for all z in the region

$$|z - z_0| < \epsilon. \tag{15.62}$$

It turns out that the existence of the first derivative in the neighborhood of z_0 guarantees that $f(z)$ has all the higher derivatives as well, that is, $f(z)$ is infinitely differentiable at z_0. A function $f(z)$ is said to be **entire** if it is analytic everywhere (except possibly at infinity). For example, the functions

$$e^z, \ \sin z, \quad 1 - 2z^2 \tag{15.63}$$

are entire. Any polynomial in z is an entire function. The function $1/(z-1)$ is not entire because of the pole at $z = 1$. A function $x(t)$ of the real argument t is said to be the **real-axis restriction** of the entire function $f(z)$, if

$$x(t) = f(z)\Big|_{z=t}. \tag{15.64}$$

In words, $f(z)$ evaluated on the real-axis of the z-plane equals $x(t)$. For example, the real-axis restriction of $f(z) = e^z$ is $x(t) = e^t$, the real-axis restriction of $f(z) = \sin z$ is $x(t) = \sin t$, and so forth. When $f(z)$ is an entire function, it is clear that its real-axis restriction $x(t)$ is differentiable with respect to t. In fact, it is differentiable any number of times, for all t. In particular, of course, it is a continuous function of t. If a signal $x(t)$ is not continuous and infinitely differentiable for all t, it cannot therefore be regarded as (the real-axis restriction of) an entire function $f(z)$. For example, rectangular pulses and triangular waveforms cannot be regarded as real-axis restrictions of entire functions, because the former is not continuous and the latter is not differentiable.

15.5.2 Bandlimited Signals are Analytic

The most important result of this section is the following:

Theorem 15.1 *Bandlimited functions and entire functions.* Let $x(t)$ be bandlimited and assume further that

$$\int_{-\sigma}^{\sigma} |X(j\omega)| d\omega < \infty. \tag{15.65}$$

That is, the Fourier transform $X(j\omega)$ is absolutely integrable. Then $x(t)$ can be regarded as the real-axis restriction of an entire function. That is, there exists an entire function $f(z)$ of the complex variable z such that $f(t) = x(t)$. ◇

For most bandlimited signals of interest, the condition (15.65) is easily satisfied. For example, if $X(j\omega)$ is bounded, that is, $|X(j\omega)| \leq B_X < \infty$ for all ω, then (15.65) holds. Similarly, if $x(t)$ is a bandlimited L_2 signal (Sec. 15.3), then (15.65) holds (Problem 15.10). Notice incidentally that there are plenty of examples of *non*-bandlimited signals for which $\int_{-\infty}^{\infty} |X(j\omega)| d\omega < \infty$ is *not* true. For example, if $x(t)$ is the rectangular pulse, then $X(j\omega)$ (sinc function) is not absolutely integrable.

Sketch of the proof of Theorem 15.1. Since $x(t)$ is bandlimited, we can write

$$x(t) = \frac{1}{2\pi} \int_{-\sigma}^{\sigma} X(j\omega) e^{j\omega t} d\omega. \tag{15.66}$$

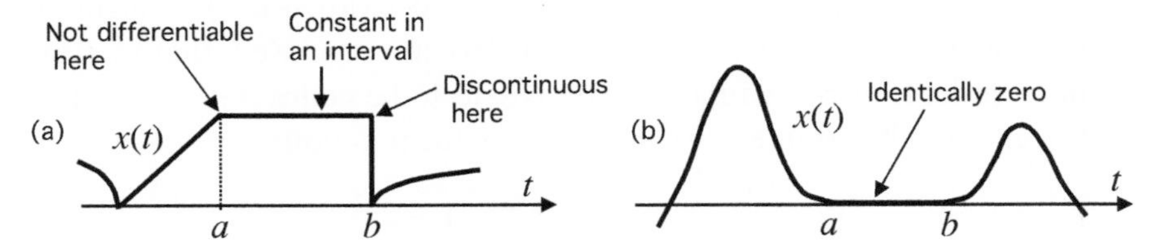

Figure 15.4 (a) A signal which violates a number of properties of bandlimited signals and (b) a signal which is identically zero in an interval; this cannot be bandlimited.

Now define a function $f(z)$ of the complex variable z as follows:

$$f(z) = \frac{1}{2\pi} \int_{-\sigma}^{\sigma} X(j\omega) e^{j\omega z} d\omega. \tag{15.67}$$

Obviously, $x(t)$ is the real-axis restriction of $f(z)$. If we prove that $f(z)$ is an entire function, then the proof is complete. For this we have to show that the derivative $df(z)/dz$ exists by proving that the limit

$$\lim_{\epsilon \to 0} \frac{f(z+\epsilon) - f(z)}{\epsilon} \tag{15.68}$$

exists. Under the assumption (15.65) this can indeed be shown to be the case. Notice carefully that the limiting operation has to be done as ϵ approaches 0 in all directions in the complex plane. See Sec. 15.8 for details. $\bigtriangledown \bigtriangledown \bigtriangledown$

The fact that a bandlimited function $x(t)$ can be regarded as an analytic function $f(z)$ evaluated for real argument z can be used to obtain a number of conclusions which would otherwise not be so easy to derive.

1. *Unlimited differentiability.* Since $f(z)$ is continuous and differentiable any number of times for all z, the bandlimited signal $x(t)$ is continuous for all t and furthermore differentiable any number of times for all t. Qualitatively this means that a bandlimited function is infinitely smooth everywhere. As mentioned earlier, this fact can be used to conclude by inspection that signals with discontinuities (e.g., rectangular pulses) and sharp corners (e.g., triangles) are not bandlimited. Figure 15.4(a) shows an example which violates some of the requirements for bandlimitedness.
2. *Taylor series.* If a function $f(z)$ is analytic at a point, we can express it in the form of a Taylor series around that point [Churchill and Brown, 1984]. Suppose we write $f(z)$ as a Taylor series around a real point t_0, then

$$f(z) = f(t_0) + f^{(1)}(t_0)(z - t_0) + f^{(2)}(t_0)\frac{(z - t_0)^2}{2!} + f^{(3)}(t_0)\frac{(z - t_0)^3}{3!} + \cdots,$$

where $f^{(k)}(t_0)$ is the kth derivative at $z = t_0$. Thus, a bandlimited function $x(t)$ can be expressed as a Taylor series around any point in time t_0 as

$$x(t) = x(t_0) + x^{(1)}(t_0)(t - t_0) + x^{(2)}(t_0)\frac{(t - t_0)^2}{2!} + x^{(3)}(t_0)\frac{(t - t_0)^3}{3!} + \cdots. \tag{15.69}$$

3. *Derivatives have all information.* The preceding equation shows that a bandlimited signal $x(t)$ is completely determined for *all time t* if we know the signal $x(t_0)$ and all its derivatives $x^{(k)}(t_0)$ at a *single point* t_0. Of course, in practice, this is not necessarily useful to reconstruct $x(t)$ for all t from this localized information, as any slight error can get severely amplified.
4. *BL signals cannot vanish in an interval.* A bandlimited signal $x(t)$ cannot be identically zero in a nontrivial time interval like $a < t < b$ (Fig. 15.4(b)) unless it is zero everywhere. To see this, consider the Taylor series (15.69) written around a point t_0 in this interval. If $x(t)$ is identically zero in the neighborhood of t_0, then all its derivatives are zero and from Eq. (15.69) we conclude that $x(t) \equiv 0$ for all t. Similarly, it can be argued that a bandlimited signal **cannot be a constant** in a nontrivial time interval like $a < t < b$, as demonstrated in Fig. 15.4(a), unless it is constant everywhere. A special case is the result that a nonzero bandlimited signal cannot be time limited. This was proved in Sec. 10.9 using more direct methods.
5. *Bandlimited extrapolation.* We cannot find two different bandlimited signals such that they are identical in an interval $a < t < b$, no matter how small the interval is. For if we could, then the difference between these signals would be a nonzero bandlimited signal which is identically zero in that interval, and that is not possible as explained above. This implies that a bandlimited signal $x(t)$ is completely determined for all t if it is known everywhere in an interval $a < t < b$, no matter how small that interval is. If $x(t)$ is given in such an interval, we can, in principle, **extrapolate** and compute it for points outside this interval. This is called bandlimited extrapolation [Papoulis, 1975, 1977a; Sauer and Allebach, 1987; Vaidyanathan, 1987c]. In practice, the accuracy of extrapolation depends on how large the given segment is, how much noise there is in the data, and the range of time for which the extrapolation needs to be done.

15.5.3 Are all Entire Functions Bandlimited?

Theorem 15.1 says that bandlimited functions are entire functions under some mild conditions. How about the converse? Suppose $f(z)$ is an entire function, and we define the signal $x(t)$ to be the real-axis restriction $f(t)$. Can we claim that $x(t)$ is a bandlimited signal? In general the answer is no. For example, the function $f(z) = e^{-z^2/2}$ is analytic everywhere in the z-plane and is therefore entire. However, we know that the real-axis restriction

$$x(t) = e^{-t^2/2} \tag{15.70}$$

is not bandlimited because it has the shape of the Gaussian, and its Fourier transform has similar shape (Sec. 9.7), namely,

$$X(j\omega) = \sqrt{2\pi}e^{-\omega^2/2}. \tag{15.71}$$

The additional conditions on an entire function which make it bandlimited on the real axis are given by the celebrated **Paley–Wiener theorem**. Thus, suppose the entire function $f(z)$ satisfies the following two additional properties called the Paley–Wiener conditions:

1. *Finite energy*. The real-axis restriction satisfies $\int |f(t)|^2 dt < \infty$.
2. *Exponential boundedness*. For any z, the magnitude of $f(z)$ satisfies

$$|f(z)| < Ae^{b|z|}, \tag{15.72}$$

   for some positive and finite constants A and b.

Then the real-axis restriction $x(t) = f(t)$ is indeed bandlimited. For a proof see Papoulis [1977a].

An example. Consider again the entire function $f(z) = e^{-z^2/2}$. Since the real-axis restriction is not bandlimited, we expect that the Paley–Wiener conditions are somehow violated by this function. Let us see how. Actually the first condition is satisfied by $f(z)$ because on the real axis $f^2(t) \leq f(0)f(t)$, so that

$$\int f^2(t)dt \leq f(0) \int f(t)dt < \infty, \tag{15.73}$$

since the Gaussian $f(t)$ has finite integral. However, the second condition, exponential boundedness, is violated. To see this, consider the imaginary axis of the z-plane where $z = jy$. Then $f(z) = e^{y^2/2}$ and we cannot find finite positive constants A and b such that $e^{y^2/2} < Ae^{b|y|}$ for all real y. ▽ ▽ ▽

15.5.3.1 Checking Exponential Boundedness

To demonstrate how one might check whether a function is exponentially bounded or not, consider the function

$$f(z) = \frac{\sin z}{z} \cdot \tag{15.74}$$

We will show that this is exponentially bounded. For this, recall [Churchill and Brown, 1984] that $\sin z$ can be expressed in a power series

$$\sin z = z - \frac{z^3}{3!} + \frac{z^5}{5!} - \frac{z^7}{7!} + \cdots, \tag{15.75}$$

which converges for all finite z. Thus

$$\frac{\sin z}{z} = 1 - \frac{z^2}{3!} + \frac{z^4}{5!} - \frac{z^6}{7!} + \cdots \tag{15.76}$$

for all z. This can be regarded as a Taylor series around $z = 0$, and since it is valid for all z, this shows that $\sin z/z$ is an entire function. Its magnitude can be bounded as follows:

$$\left|\frac{\sin z}{z}\right| \leq 1 + \frac{|z|^2}{3!} + \frac{|z|^4}{5!} + \cdots$$
$$\leq 1 + \frac{|z|^2}{2!} + \frac{|z|^4}{4!} + \cdots$$
$$\leq 1 + |z| + \frac{|z|^2}{2!} + \frac{|z|^3}{3!} + \frac{|z|^4}{4!} + \cdots$$
$$= e^{|z|}.$$

This shows that $f(z)$ satisfies the condition (15.72) for $b = 1$ and any $A > 1$. So $f(z)$ is exponentially bounded. The real-axis restriction of the function $f(z)$ in the preceding example is

$$x(t) = \frac{\sin t}{t}. \tag{15.77}$$

It can be shown that this has finite energy. Thus, $f(z)$ satisfies both of the Paley–Wiener conditions and the real-axis restriction $x(t)$ should therefore be bandlimited. This, of course, is not new to us. We know that the Fourier transform of $\sin t/t$ is a rectangular pulse, which shows that $\sin t/t$ is bandlimited.

15.6 Summary of Chapter 15

This chapter presented a detailed discussion of bandlimited signals. First, the σ-BL signal space $\mathcal{B}_\sigma$ was studied, and an orthogonal basis for this space introduced. This basis is nothing but the set of uniformly shifted versions $\{h(t-nT)\}$ of the sinc function $h(t)$, where $T = \pi/\sigma$. When $T \neq \pi/\sigma$, $\{h(t-nT)\}$ is not an orthogonal basis for σ-BL space, and the detailed behavior depends on whether T is smaller or larger than π/σ. These details were also explained. The σ-BL L_2 space was then studied. This is the collection of σ-BL signals with finite energy.

It was then shown that integrals involving a σ-BL signal $x(t)$ have some interesting properties. For example, the integral of $x(t)$ is equal to $T\sum x(nT)$ as long as the sample spacing T is small enough (i.e., $1/T \geq \sigma/\pi$). Similarly, the energy of $x(t)$ is equal to $T\sum |x(nT)|^2$. Based on these observations and other observations on bandlimited signals, many well-known mathematical identities can be derived just by using sampling theory. Some examples were given in Sec. 15.4.3. When the samples of a bandlimited signal are contaminated with noise, the reconstructed signal is also noisy. This noise depends on the reconstruction filter. Excess bandwidth in this filter increases the noise, and this was quantitatively analyzed.

A beautiful connection between bandlimited signals and analytic functions in complex variable theory was pointed out. Namely, a bandlimited signal $x(t)$ is the real-axis restriction of an entire function $f(z)$ (i.e., a function of complex z, analytic everywhere in $|z| < \infty$). That is, $x(t) = f(z)$ when $z = t$. Based on this, it was shown that a bandlimited signal is infinitely smooth. It is continuous and differentiable any number of times for all t, it cannot be identically constant in any interval (unless it is constant everywhere), and so on.

15.7 Appendix: Infinite Sum of L_2 Signals

Let $f_k(t)$ be L_2 functions, that is, $\int_{-\infty}^{\infty} |f_k(t)|^2 dt < \infty$. Consider a linear combination of the form

$$x(t) = \sum_{k=1}^{\infty} c_k f_k(t), \tag{15.78}$$

where $c_k \in \ell_2$, that is, $\sum_k |c_k|^2 < \infty$. Can we say that $x(t) \in L_2$? In general, the answer is no. We can construct a counterexample as follows. Let $\{g_k(t)\}, 0 \le k \le \infty$, be an orthonormal set, that is,

$$\int_{-\infty}^{\infty} g_k(t) g_m^*(t) dt = \delta[k-m]. \tag{15.79}$$

Define

$$f_k(t) = g_k(t) + \epsilon g_0(t), \quad 1 \le k \le \infty, \tag{15.80}$$

where $\epsilon > 0$. Then

$$\begin{aligned}
\int_{-\infty}^{\infty} f_k(t) f_m^*(t) dt &= \int_{-\infty}^{\infty} \Big(g_k(t) + \epsilon g_0(t)\Big)\Big(g_m^*(t) + \epsilon g_0^*(t)\Big) dt \\
&= \int_{-\infty}^{\infty} \Big(g_k(t) g_m^*(t) + \epsilon g_k(t) g_0^*(t) + \epsilon g_0(t) g_m^*(t) + |\epsilon g_0(t)|^2\Big) dt \\
&= \int_{-\infty}^{\infty} g_k(t) g_m^*(t) dt + \epsilon^2 \\
&= \delta[k-m] + \epsilon^2.
\end{aligned} \tag{15.81}$$

Now from Eq. (15.78) we have

$$\begin{aligned}
\int_{-\infty}^{\infty} |x(t)|^2 dt &= \sum_{k=1}^{\infty} \sum_{m=1}^{\infty} c_k c_m^* \int_{-\infty}^{\infty} f_k(t) f_m^*(t) dt \\
&= \sum_{k=1}^{\infty} \sum_{m=1}^{\infty} c_k c_m^* \Big(\delta[k-m] + \epsilon^2\Big) \\
&= \sum_{k=1}^{\infty} |c_k|^2 + \epsilon^2 \sum_{k=1}^{\infty} \sum_{m=1}^{\infty} c_k c_m^* \\
&= \sum_{k=1}^{\infty} |c_k|^2 + \epsilon^2 \sum_{k=1}^{\infty} c_k \sum_{m=1}^{\infty} c_m^*,
\end{aligned} \tag{15.82}$$

where we have used Eq. (15.81) in the second line. Consider the example where

$$c_k = \frac{1}{k}, \quad k \ge 1. \tag{15.83}$$

Then $c_k \in \ell_2$, although $c_k \notin \ell_1$. In this case Eq. (15.82) reduces to

$$\int_{-\infty}^{\infty} |x(t)|^2 dt = \sum_{k=1}^{\infty} |c_k|^2 + \epsilon^2 \left(\sum_{k=1}^{\infty} \frac{1}{k}\right)^2. \tag{15.84}$$

The first term converges to a finite value because $c_k \in \ell_2$. But in the second term, $\sum_{k=1}^{\infty} 1/k$ does not converge to a finite value. So $|x(t)|^2$ does not have a finite integral. Thus, even though $c_k \in \ell_2$ and $f_k(t) \in L_2$, it is not true that $x(t) \in L_2$. Notice incidentally that $f_k(t)$ in Eq. (15.80) is a linearly independent set because any linear combination can be written as

$$\sum_{k=1}^{\infty} \alpha_k f_k(t) = \sum_{k=1}^{\infty} \alpha_k g_k(t) + \epsilon \left(\sum_{k=1}^{\infty} \alpha_k \right) g_0(t). \tag{15.85}$$

Assume the α_k are not all zero. If the above sum is zero, then it contradicts the fact that $\{g_k(t)\}$ are orthonormal and hence linearly independent.

Now consider a different situation: suppose $f_k(t) \in L_2$ is itself an orthonormal set in Eq. (15.78). Then if $c_k \in \ell_2$, it is indeed true that $x(t) \in L_2$ because

$$\begin{aligned}\int_{-\infty}^{\infty} |x(t)|^2 dt &= \sum_{k=1}^{\infty}\sum_{m=1}^{\infty} c_k c_m^* \int_{-\infty}^{\infty} f_k(t) f_m^*(t) dt \\ &= \sum_{k=1}^{\infty}\sum_{m=1}^{\infty} c_k c_m^* \delta[k-m] \\ &= \sum_{k=1}^{\infty} |c_k|^2 < \infty. \end{aligned} \tag{15.86}$$

15.8 Appendix: Details in Proof of Theorem 15.1

We now show that the limit (15.68) exists under the condition (15.65), and

$$\lim_{\epsilon \to 0} \frac{f(z+\epsilon) - f(z)}{\epsilon} = \frac{1}{2\pi} \int_{-\sigma}^{\sigma} j\omega X(j\omega) e^{j\omega z} d\omega, \tag{15.87}$$

where z and ϵ are complex quantities. For this we will show that

$$\lim_{\epsilon \to 0} \underbrace{\left| \frac{f(z+\epsilon) - f(z)}{\epsilon} - \frac{1}{2\pi} \int_{-\sigma}^{\sigma} j\omega X(j\omega) e^{j\omega z} d\omega \right|}_{\text{call this } A(\epsilon)} = 0. \tag{15.88}$$

Details. Substituting for $f(z)$ from Eq. (15.67), we have

$$A(\epsilon) = \left| \frac{1}{2\pi} \int_{-\sigma}^{\sigma} X(j\omega) e^{j\omega z} \Big(\frac{e^{j\omega\epsilon} - 1 - j\omega\epsilon}{\epsilon} \Big) d\omega \right|. \tag{15.89}$$

Since $|\omega| \leq \sigma$, we can write $|e^{j\omega z}| \leq e^{\sigma|z|}$, so that

$$A(\epsilon) \leq \frac{e^{\sigma|z|}}{2\pi} \int_{-\sigma}^{\sigma} \left| X(j\omega) \Big(\frac{e^{j\omega\epsilon} - 1 - j\omega\epsilon}{\epsilon} \Big) \right| d\omega. \tag{15.90}$$

Now consider the quantity $e^{j\omega\epsilon} - 1 - j\omega\epsilon$ appearing above. This has the form $e^v - 1 - v$. The reader will recall that e^v can be expressed as a power series

$$e^v = 1 + v + \frac{v^2}{2!} + \frac{v^3}{3!} + \cdots \tag{15.91}$$

and that the series converges for all v. We therefore have

$$\begin{aligned} |e^v - 1 - v| &= \left|\frac{v^2}{2!} + \frac{v^3}{3!} + \frac{v^4}{4!} + \cdots\right| \\ &\leq |v|^2\Big(\frac{1}{2!} + \frac{|v|}{3!} + \frac{|v^2|}{4!} + \cdots\Big) \\ &\leq |v|^2\Big(1 + |v| + \frac{|v^2|}{2!} + \cdots\Big) = |v|^2 e^{|v|}. \end{aligned}$$

Using this in Eq. (15.90), we see that

$$\begin{aligned} A(\epsilon) &\leq \frac{e^{\sigma|z|}}{2\pi}\int_{-\sigma}^{\sigma} |\epsilon X(j\omega)|\omega^2 e^{|\omega\epsilon|}d\omega \\ &\leq \frac{e^{\sigma|z|}}{2\pi}\int_{-\sigma}^{\sigma} |\epsilon X(j\omega)|\sigma^2 e^{|\sigma\epsilon|}d\omega = \frac{|\epsilon|\sigma^2 e^{\sigma(|z|+|\epsilon|)}}{2\pi}\int_{-\sigma}^{\sigma}|X(j\omega)|d\omega. \end{aligned}$$

In view of Eq. (15.65), the integral on the right is finite. So when $\epsilon \to 0$, the right-hand side above goes to zero. This proves Eq. (15.88). $\triangledown\triangledown\triangledown$

One might wonder if proving the existence of $df(z)/dz$ might just be a matter of starting from Eq. (15.67) and writing

$$\frac{df(z)}{dz} = \frac{1}{2\pi}\frac{d}{dz}\int_{-\sigma}^{\sigma} X(j\omega)e^{j\omega z}d\omega = \frac{1}{2\pi}\int_{-\sigma}^{\sigma} X(j\omega)\frac{de^{j\omega z}}{dz}d\omega, \tag{15.92}$$

which indeed yields Eq. (15.87). However, this involves interchanging the derivative and integral operators, and assuming that the integral on the right of Eq. (15.92) exists. The more detailed proof given above makes all this rigorous, and makes it clear why the assumption (15.65) is needed.

PROBLEMS

15.1 Let $h(t) = T\sin\sigma t/\pi t$, and define a signal $x(t) = \sum_n c[n]h(t - nT)$ for some $c[n] \in \ell_2$.

(a) For $T = \pi/\sigma$, show that this $x(t)$ satisfies $x(nT) = c[n]$.

(b) Assume $T < \pi/\sigma$. Is it necessarily true that $x(nT) = c[n]$? Explain.

(c) Repeat (b) for $T > \pi/\sigma$.

15.2 Let $x(t)$ be σ-BL and let $x(nT)$ be the samples obtained at sampling rate $2\pi/T \geq 2\sigma$. Then we know $x(t) = \sum_n x(nT)h(t - nT)$, where $h(t)$ is the appropriate sinc function. Using the Cauchy–Schwarz inequality, show that

$$|x(t)|^2 \le \frac{\sigma T}{\pi} \sum_{n=-\infty}^{\infty} |x(nT)|^2 \le \sum_{n=-\infty}^{\infty} |x(nT)|^2.$$

This shows that for a bandlimited signal, the signal value *at every instant of time t* is bounded by the square root of the energy in the samples!

Remark: The usefulness of this result arises when there is noise in the samples and we have to reconstruct the signal. Thus, if the samples $x(nT)$ are replaced by $x(nT) + e(nT)$, where $e(nT)$ is error (say due to quantization), then the reconstructed signal $x(t) + e(t)$ has the error term $e(t)$ bounded as $|e(t)|^2 \le \sum_n |e(nT)|^2$. The interested reader should also see Sec. 15.4.4.

15.3 Suppose a set of functions $\{x_i(t)\}$ with nonzero energy is orthogonal, that is, they satisfy

$$\int_{-\infty}^{\infty} x_i(t)x_m^*(t)dt = 0 \quad \text{for} \quad i \ne m.$$

Show that $\{x_i(t)\}$ are also linearly independent. That is, if $\sum_i c_i x_i(t) = 0$ for all t, then it is necessarily true that $c_i = 0$ for all i.

15.4 Suppose $H(j\omega)$ is the following ideal lowpass filter:

$$H(j\omega) = \begin{cases} T & -\omega_c < \omega < \omega_c, \\ 0 & \text{otherwise.} \end{cases} \qquad \text{(P15.4a)}$$

Let $h(t)$ denote its impulse response. Show that the set $\{h(t - nT)\}$ (where n runs through all integers) is an orthogonal set, that is, Eq. (15.13) holds, if and only if $\omega_c = K\pi/T$ for some integer $K \ge 1$.

15.5 Let $h(t) = \sin(\pi t)/(\pi t)$, and consider the set $\mathcal{S} = \{h(t - kT)\}$ for $-\infty < k < \infty$.

(a) Find three examples of T, namely T_1, T_2, and T_3 such that (i) for $T = T_1$, $\mathcal{S}$ is not linearly independent; (ii) for $T = T_2$, $\mathcal{S}$ is linearly independent but not orthogonal; and (iii) for $T = T_3$, $\mathcal{S}$ is orthogonal (hence linearly independent).

(b) In which of these cases does $\mathcal{S}$ span π-BL L_2 signals?

(c) In which of these cases is $\mathcal{S}$ a basis for π-BL L_2 signals?

15.6 With $h(t) = \sin(\pi t/T)/(\pi t/T)$, we know that the set $\mathcal{S} = \{h(t - kT)\}$ (with $-\infty < k < \infty$) is an orthogonal basis for (π/T)-BL L_2 space.

(a) Find a new basis $\widehat{\mathcal{S}} = \{g(t - kT)\}$ for the same space such that $g(t) \ge 0$ for all t.

(b) Does the new basis you found satisfy orthogonality?

15.7 Find an example of a nonzero bandlimited signal $x(t)$ which decays to zero at least as fast as $1/t^6$, that is, $|x(t)| \le c/|t|^6$ for all t, for some finite $c > 0$. What is the bandwidth of the example you found?

15.8 *Deriving well-known mathematical identities.* In this problem we demonstrate another amusing aspect of sampling theory, namely that it allows us to derive some standard mathematical identities with very little effort. For example, we know that the pulse $p(t)$ in Fig. P15.8 has Fourier transform

$$P(j\omega) = \frac{\sin(\omega/2)}{(\omega/2)}. \tag{P15.8a}$$

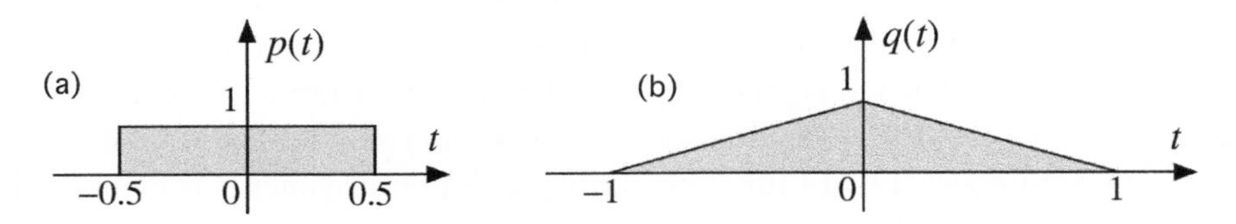

Figure P15.8 A rectangular pulse and a triangular pulse.

Let $x_d[n] = p(n)$ be the discrete-time signal obtained by sampling $p(t)$ with spacing $T = 1$. Clearly $x_d[n] = \delta[n]$, so that $X_d(e^{j\omega}) = 1$ for all ω. But $X_d(e^{j\omega})$ can also be written in terms of shifted copies of $P(j\omega)$, according to the sampling theorem. Using this, show that

$$\sum_{k=-\infty}^{\infty} \frac{(-1)^k}{\omega + 2\pi k} = \frac{1}{2\sin(\omega/2)}, \tag{P15.8b}$$

for $0 < \omega < 2\pi$. Setting $\omega = \pi$ and simplifying, show that

$$1 - \frac{1}{3} + \frac{1}{5} - \frac{1}{7} + \frac{1}{9} - \cdots = \frac{\pi}{4}. \tag{P15.8c}$$

This is the well-known Madhava–Leibniz formula for evaluating π to arbitrary digits of accuracy.

15.9 Continuing with the idea of Problem 15.8, we know the triangular pulse $q(t)$ shown in Fig. P15.8 has Fourier transform $[P(j\omega)]^2$, where $P(j\omega)$ is as in Eq. (P15.8a). Let $y_d[n] = q(nT)$ be the discrete-time signal obtained by sampling $q(t)$ with spacing $T = 1$. Clearly $y_d[n] = \delta[n]$. By using the fact that $Y_d(e^{j\omega})$ can be expressed as a sum of shifted copies of $[P(j\omega)]^2$, show that

$$\sum_{k=-\infty}^{\infty} \frac{1}{(\omega + 2\pi k)^2} = \frac{1}{4\sin^2(\omega/2)}, \tag{P15.9a}$$

for $0 < \omega < 2\pi$. In particular, with $\omega = \pi$, show that

$$1 + \frac{1}{3^2} + \frac{1}{5^2} + \frac{1}{7^2} + \frac{1}{9^2} + \cdots = \frac{\pi^2}{8}. \tag{P15.9b}$$

15.10 If $x(t)$ is a σ-BL L_2 signal, show that $X(j\omega)$ satisfies Eq. (15.65), that is, $\int_{-\sigma}^{\sigma} |X(j\omega)| d\omega < \infty$.

15.11 Let $x[n]$ be a nonzero discrete-time signal such that it is zero everywhere in the interval $0 \leq n \leq 9$. Can such a signal be σ-BL at all (with $\sigma < \pi$)? If so, construct an example. *Note:* For discrete-time signals, the σ-BL property was defined in Sec. 10.9.1.

15.12 Figure P15.12 shows portions of two continuous-time signals $x(t), y(t)$ and a discrete-time signal $s[n]$. For each signal, answer the following and give reasons: based only on the portions shown, can you conclude that the signal is not bandlimited? *Note:* For discrete-time signals, the bandlimited property was defined in Sec. 10.9.1.

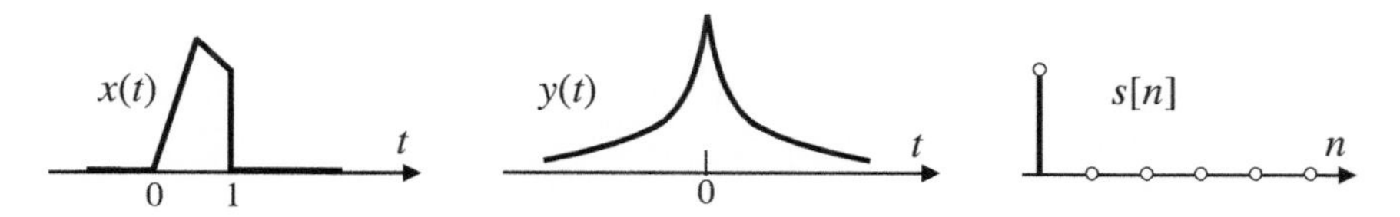

Figure P15.12 Signals for Problem 15.12.

16 Sampling Based on Sparsity*

16.1 Introduction

In Chapter 10 we showed that if we have the a priori knowledge that a signal is bandlimited, then it can be recovered from its samples as long as the sampling rate is large enough. A natural question is, if a signal is not necessarily bandlimited but has some other property $\mathcal{P}$, and if we know a priori that it has property $\mathcal{P}$, will it enable us to reconstruct the signal from an appropriate set of samples? This is indeed sometimes possible and leads to very novel methods to sample a signal. One case in point, which had great impact in recent years, is sampling based on the so-called sparsity property. Another example was given in Problem 10.16. Another ingeneous instance is sampling based on the so-called "finite rate of innovations," which was introduced by Vetterli et al. [2002].

In this chapter we give a flavor for sampling based on sparsity. While at first sight our discussion may look like a digression into linear algebra, it will connect nicely with some major results in sampling theory, which are very different from the results of Shannon and Nyquist. The presentation here is very introductory, the purpose being to inspire the reader to explore the details by studying some of the mentioned references.

At this point the reader may find it useful to review Chapter 18.

16.2 Sparse Sampling and Reconstruction

For a bandlimited signal $x(t)$, the energy is limited to a narrow band of frequencies. Because of this, we can sample the signal without loss of information, and sampling theorems for σ-BL signals and bandpass signals are based on this (Chapter 10). In Sec. 10.9.1 it was also shown how the concept of bandlimitedness can be extended to discrete time. We can even extend it to finite-duration sequences $g[n]$, which can be regarded as $M \times 1$ vectors $\mathbf{g}$. If the DFT $\mathbf{G} = \mathbf{W}\mathbf{g}$ (Chapter 12) is zero for all frequencies except some low frequencies, as in

$$\mathbf{G} = [\times \quad \times \quad \times \quad 0 \quad 0 \quad 0 \quad 0 \quad \cdots \quad 0 \quad 0 \quad 0 \quad 0 \quad \times \quad \times]^T, \qquad (16.1)$$

where $\times$ denotes nonzero entries, we can regard $\mathbf{g}$ as "bandlimited." (Note that the nonzero samples at the right end are close to $\omega = 2\pi$, which is equivalent to zero frequency.) More generally, if $\mathbf{G}$ is zero everywhere except on some arbitrary subset of frequencies, we can regard it as bandlimited in a generalized sense. Instead of calling it bandlimited, we say that the DFT is **sparse**, which just means that the DFT only has a small number $s << M$ of nonzero coefficients. We also sometimes say that such a signal $g[n]$ is "sparse in the DFT basis."

For such signals, is it possible to recover the vector $\mathbf{g}$ from a small subset of its samples, say the first K components of $\mathbf{g}$? The answer is yes, perhaps not surprisingly. More interestingly, even if we do *not* know the locations of the nonzero frequency samples in $\mathbf{G}$, this is often possible. In this chapter we explain the fundamental ideas behind such methods, which are often referred to as sparse-sampling and sparse-reconstruction methods. The mention of DFT above serves as a motivation, but the results are more general than that. Most of our discussions will be based on vectors (i.e., finite-length sequences) and matrices. Sparsity of a signal in some basis is also at the heart of being able to compress a signal, that is, represent it in terms of a small number of samples. Sampling techniques like this, which are not based on classical approaches such as alias suppression or cancellation, can be regarded as truly **sub-Nyquist** sampling methods (see discussion in Sec. 10.10.3).

16.3 Sparsity Constraints

Let us begin by examining an equation of the form

$$\mathbf{y} = \underbrace{\mathbf{A}}_{K\times M}\mathbf{x}, \tag{16.2}$$

where $\mathbf{A}$ is $K \times M$ and $K < M$. Then the vector $\mathbf{x}$ satisfying the above equation is not unique, for a given $\mathbf{y}$ (Sec. 18.9). But suppose we have the a priori information that there is a solution $\mathbf{x}$ with at most s nonzero elements for some integer $s < M$. If s is sufficiently small, then we can show that this solution $\mathbf{x}$ is unique (even though we may not know a priori where the nonzero elements of $\mathbf{x}$ are). Such a solution is called s-**sparse** because at most s components of $\mathbf{x}$ are nonzero. Here are some examples of s-sparse vectors:

$$\begin{aligned}
\text{1-sparse: } & \begin{bmatrix}0 & 0 & 0 & \times & 0\end{bmatrix}^T, \\
\text{2-sparse: } & \begin{bmatrix}0 & \times & 0 & 0 & \times\end{bmatrix}^T, \\
\text{3-sparse: } & \begin{bmatrix}0 & \times & \times & 0 & \times\end{bmatrix}^T.
\end{aligned} \tag{16.3}$$

The set $\mathcal{S}$ of nonzero locations in $\mathbf{x}$ is called the **support** of $\mathbf{x}$. In the above examples, assuming the locations are numbered $0, 1, 2, 3, 4$, the supports $\mathcal{S}$ are, respectively,

$$\{3\}, \quad \{1, 4\}, \quad \{1, 2, 4\}.$$

To be precise with the statement of the main result, let ρ_{Kr} be the Kruskal rank of $\mathbf{A}$ (defined in Sec. 18.6.2). Then the claim is that the s-sparse solution $\mathbf{x}$ is unique if and only if

$$\rho_{Kr} \geq 2s. \tag{16.4}$$

You can think of this as analogous to the condition $\omega_s \geq 2\sigma$ for sampling σ-BL signals (Theorem 10.1), but the two results are based on very different principles. The proof of the claim is given below.

Proof. To avoid trivialities, we assume no column of $\mathbf{A}$ is equal to $\mathbf{0}$. We first assume $\rho_{Kr} \geq 2s$ and show that there cannot be more than one s-sparse solution. Assume the contrary: let $\mathbf{x}_1 \neq \mathbf{x}_2$ be two s-sparse solutions. Then $\mathbf{y} = \mathbf{A}\mathbf{x}_1 = \mathbf{A}\mathbf{x}_2$, so that $\mathbf{A}\mathbf{v} = \mathbf{0}$, where $\mathbf{v} = \mathbf{x}_1 - \mathbf{x}_2 \neq \mathbf{0}$. Since $\mathbf{x}_i$ have at most s nonzero entries, $\mathbf{v}$ has at most $2s$ nonzero entries. So $\mathbf{A}\mathbf{v} = \mathbf{0}$ means that there are $2s$ or fewer columns of $\mathbf{A}$ which are linearly dependent. This contradicts the assumption that $\rho_{Kr} \geq 2s$.

To prove the converse, let us assume $\rho_{Kr} < 2s$. So there exists a set of $\mu < 2s$ columns of $\mathbf{A}$ that are linearly dependent. So $\mathbf{A}\mathbf{u} = \mathbf{0}$ for some $\mathbf{u} \neq \mathbf{0}$, where $\mathbf{u}$ has $\leq \mu < 2s$ nonzero elements. But $\mathbf{u}$ has at least two nonzero components (because of the assumption that no column of $\mathbf{A}$ is $\mathbf{0}$). So we can rewrite $\mathbf{A}\mathbf{u}_1 = \mathbf{A}\mathbf{u}_2$, where $\mathbf{u}_1$ and $\mathbf{u}_2$ are both nonzero with at most s nonzero elements each, and such that their nonzero components do not occur in the same place. In particular, this means $\mathbf{u}_1$ and $\mathbf{u}_2$ are s-sparse, $\mathbf{u}_1 \neq \mathbf{u}_2$, and $\mathbf{A}\mathbf{u}_1 = \mathbf{A}\mathbf{u}_2$. Labeling the vector $\mathbf{A}\mathbf{u}_1$ as $\mathbf{y}$, we see that indeed $\mathbf{y} = \mathbf{A}\mathbf{x}$ has two s-sparse solutions, $\mathbf{u}_1$ and $\mathbf{u}_2$. $\triangledown\triangledown\triangledown$

16.3.1 Remarks on Sparse Reconstruction

The problem of reconstructing an s-sparse signal $\mathbf{x}$ from a limited number of samples of a transformed version $\mathbf{y} = \mathbf{A}\mathbf{x}$ is called the *sparse reconstruction* problem. It has received much attention in the last two decades and has also led to the field of compressive sensing. These ideas have changed our view of sampling in profound ways. There is a great deal of mathematics behind this [Candès and Tao, 2005, 2006; Candès et al., 2006; Donoho, 2006; Donoho and Tanner, 2005; Khajehnejad et al., 2011; Tropp and Gilbert, 2007]. Our discussion here is admittedly quite simplified, the goal being to motivate the student at an introductory level. Some remarks are now in order.

1. *What we need to know a priori.* The claim made before Eq. (16.4) is useful only if it is known a priori that there is an s-sparse solution. Fortunately, many practical situations can be found where this is so. One example is the problem of identifying the directions of arrival (DOA) of s sources (plane electromagnetic waves) from space, using a spatial array of sensors [Malioutov et al., 2005]. Another is when we are trying to model a periodic signal as a linear combination of elementary periodic components [Vaidyanathan and Tenneti, 2019].
2. *Finding the sparse solution.* The above theory does not tell us *how* to solve for the unique s-sparse solution $\mathbf{x}$. Note that we do *not* even know the support of

the s-sparse solution $\mathbf{x}$. It turns out that the main challenge is to find the support $\mathcal{S}$ of $\mathbf{x}$, that is, the locations m where the components x_m are nonzero; we will remark on this later. Assuming for a moment that the support *is* known, we can find the s-sparse solution $\mathbf{x}$ as follows: write

$$\mathbf{y} = \mathbf{A}\mathbf{x} = \mathbf{A}_s\mathbf{x}_s, \tag{16.5}$$

where $\mathbf{x}_s$ is an $s \times 1$ vector containing the nonzero components of $\mathbf{x}$ and $\mathbf{A}_s$ is the $K \times s$ submatrix of $\mathbf{A}$ obtained by retaining the corresponding s columns. Under the condition (16.4), we have

$$s < \rho_{Kr} \leq K, \tag{16.6}$$

which implies that $\mathbf{A}_s$ is a tall matrix with rank s. So there exist s linearly independent rows. We can then write

$$\mathbf{y}_s = \mathbf{A}_{ss}\mathbf{x}_s, \tag{16.7}$$

where $\mathbf{A}_{ss}$ is an $s \times s$ submatrix of $\mathbf{A}_s$ with rank s, and $\mathbf{y}_s$ is the corresponding subvector of $\mathbf{y}$. So we can uniquely solve for

$$\mathbf{x}_s = \mathbf{A}_{ss}^{-1}\mathbf{y}_s, \tag{16.8}$$

and hence find $\mathbf{x}$ uniquely. It should be mentioned that this is not the best way to find $\mathbf{x}_s$, because it does not use all the K measurements (data) in $\mathbf{y}$. When there is measurement noise, there are better ways to estimate $\mathbf{x}$ with minimum error by using all the K measurements in $\mathbf{y}$. In any case, the main point here is that, once the support $\mathcal{S}$ of the s-sparse solution $\mathbf{x}$ is "somehow" known, we can solve for $\mathbf{x}$ uniquely.

3. *Finding the support* $\mathcal{S}$. The problem of identifying the support $\mathcal{S}$ is nontrivial. One of the major contributions in recent years [Candès and Tao, 2005, 2006] is to show that this can be achieved with the help of linear programming or other efficient methods. This has been a very active area of research in recent decades, and there exist many important papers, as mentioned above. We will make some brief remarks on this in Sec. 16.4.

The set of columns in $\mathbf{A}$ is sometimes called a **dictionary**, and each column called an **atom**. Saying that $\mathbf{x}$ is sparse is the same as saying that very few atoms from the dictionary are required to represent $\mathbf{y}$. Given $\mathbf{y}$, the nontrivial issue is to find which atoms from $\mathbf{A}$ give the sparsest representation. Another interesting problem is dictionary design. Suppose we are given a class of signals, say speech signals, periodic signals, ECG signals, and so forth. Given this class, how do we design the dictionary $\mathbf{A}$ such that, for any example $\mathbf{y}$ from this class, we get a sparse representation using this dictionary? This problem is usually called the **dictionary-learning** problem.

16.3.2 The Vandermonde Example

Vandermonde matrices were discussed in Secs. 5.16 and 18.6.2. A $K \times M$ Vandermonde matrix has the form

$$\mathbf{A} = \begin{bmatrix} 1 & 1 & \cdots & 1 \\ \alpha_1 & \alpha_2 & \cdots & \alpha_M \\ \alpha_1^2 & \alpha_2^2 & \cdots & \alpha_M^2 \\ \vdots & \vdots & \vdots & \vdots \\ \alpha_1^{K-1} & \alpha_2^{K-1} & \cdots & \alpha_M^{K-1} \end{bmatrix}. \tag{16.9}$$

It will be assumed that $K \leq M$, and furthermore that the α_n are distinct. In Sec. 18.6.2 it is shown that under these assumptions, the above matrix has Kruskal rank $\rho_{Kr} = K$. For example, consider the special case where the α_n are the Mth roots of unity:

$$\alpha_n = e^{j2\pi(n-1)/M}, \quad 1 \leq n \leq M. \tag{16.10}$$

These are M distinct quantities, and $\mathbf{A}$ takes the form

$$\mathbf{A} = \begin{bmatrix} 1 & 1 & \cdots & 1 \\ 1 & e^{j2\pi/M} & \cdots & e^{j2\pi(M-1)/M} \\ 1 & e^{j4\pi/M} & \cdots & e^{j4\pi(M-1)/M} \\ \vdots & \vdots & \vdots & \vdots \\ 1 & e^{j2\pi(K-1)/M} & \cdots & e^{j2\pi(K-1)(M-1)/M} \end{bmatrix}. \tag{16.11}$$

Since the α_n are distinct, it follows that $\mathbf{A}$ has Kruskal rank $\rho_{Kr} = K$. This shows that if we know that $\mathbf{y} = \mathbf{Ax}$ has an s-sparse solution where $s \leq K/2$, then we can solve for $\mathbf{x}$ uniquely.

The matrix (16.11) is closely related to the $M \times M$ DFT matrix $\mathbf{W}$ (Sec. 12.3). To be precise, recall that the inverse DFT (IDFT) matrix is $\mathbf{W}^{-1} = \mathbf{W}^*/M$, so that

$$[\mathbf{W}^{-1}]_{km} = \frac{e^{j2\pi km/M}}{M}, \quad 0 \leq k, m \leq M-1. \tag{16.12}$$

Thus, except for the scale factor $1/M$, the matrix (16.11) is a submatrix of the IDFT matrix, obtained by retaining the first K rows. In short, the $K \times M$ submatrix of the IDFT obtained by retaining the first $K < M$ rows has Kruskal rank K. It is readily verified that the same is true if any set of consecutive K rows is retained.

16.3.3 An Unusual Sampling Theorem

To understand the significance of the Kruskal rank property of the IDFT matrix, let us switch back to traditional notations so that there is better clarity. Let

$$g[n], \ 0 \leq n \leq M-1, \tag{16.13}$$

be an M-point sequence with DFT $G[k]$, $0 \leq k \leq M-1$ (Sec. 12.2). Defining the $M \times 1$ vectors

$$\mathbf{g} = \begin{bmatrix} g[0] & g[1] & \cdots & g[M-1] \end{bmatrix}^T, \quad \mathbf{G} = \begin{bmatrix} G[0] & G[1] & \cdots & G[M-1] \end{bmatrix}^T,$$

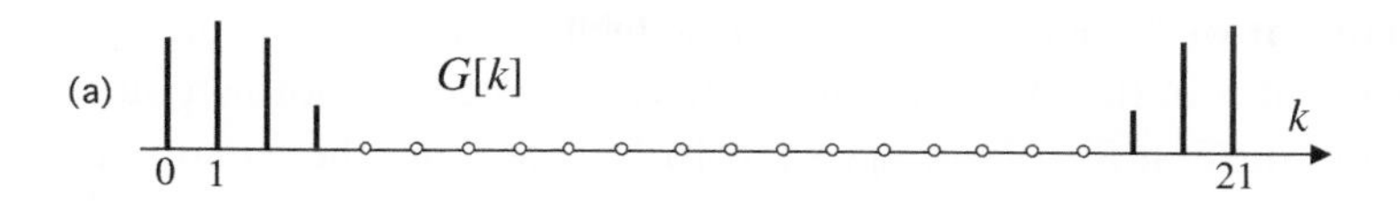

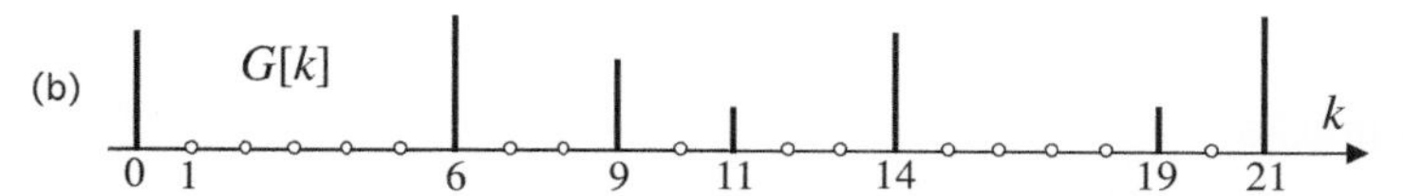

Figure 16.1 Two examples of the DFT of a 22-point sequence $g[n]$. (a) Bandlimited to the support $\mathcal{S} = \{0, 1, 2, 3, 19, 20, 21\}$ and (b) bandlimited to the support $\mathcal{S} = \{0, 6, 9, 11, 14, 19, 21\}$. The signal in (a) can be regarded as lowpass, since $k = 0$ and $k = 22$ correspond to $\omega = 0$ and $\omega = 2\pi \equiv 0$, respectively.

we then have $\mathbf{G} = \mathbf{Wg}$, where $\mathbf{W}$ is the $M \times M$ DFT matrix. Since $\mathbf{W}^{-1} = \mathbf{W}^*/M$ (Sec. 12.3), we have

$$\mathbf{g} = \frac{1}{M}\mathbf{W}^*\mathbf{G}. \tag{16.14}$$

Now let us assume that the DFT has only s nonzero coefficients

$$G[m_0], G[m_1], \ldots, G[m_{s-1}]. \tag{16.15}$$

Figure 16.1 shows some examples. We can think of this as a "bandlimited property" of the finite-duration sequence $g[n]$, in the sense that the DFT is restricted to some support

$$\mathcal{S} = \{m_0, m_1, \ldots, m_{s-1}\},$$

where $s < M$. We will say that $g[n]$ is **bandlimited** to $\mathcal{S}$. This is very different from our earlier definitions of a σ-BL signal for continuous-time or discrete-time cases (Chapter 10). The time-domain signal $g[n]$ has M samples. Assume now that we know only a subset of these samples, namely,

$$g[0], g[1], \ldots, g[K-1], \tag{16.16}$$

where $K \geq 2s$ and $K < M$. We now claim that *we can reconstruct all the M samples of $g[n]$ from these K samples* (16.16)!

This follows directly from the discussion of Sec. 16.3.2, and the idea is summarized below for clarity:

1. The matrix $\mathbf{W}^*$ is Vandermonde of the form (16.11), so the submatrix obtained from the first K rows has Kruskal rank K. Thus, given the K samples (16.16), we can first identify the support $\mathcal{S}$ and then identify the sparse coefficients (16.15) (see Sec. 16.3.1).
2. Once this is done, the entire $\mathbf{G}$ can be found by substituting $G[k] = 0$ for the remaining coefficients.
3. Then the entire vector $\mathbf{g}$ (i.e., all the samples $g[n]$) can be found using Eq. (16.14).

Thus, if we have the a priori knowledge that the M-point signal $g[n]$ is bandlimited to a set $\mathcal{S}$ in the DFT domain with size s, where s is known but $\mathcal{S}$ *is unknown*, then from the first $K \geq 2s$ samples (16.16)[1] we can find the remaining $M - K$ samples in $g[n]$. This is therefore a "sampling theorem," which can be regarded as a *sparse-DFT sampling theorem*!

Remarks

1. In the traditional sampling theorem, the frequency band where the signal is nonzero should be known, in order to reconstruct the signals from samples using LTI filters. But in the above sampling theorem, the support $\mathcal{S}$ itself need not be known. Finding it is part of the algorithm's job, and it is the nontrivial part.
2. The result can be generalized to matrices other than the DFT/IDFT type of matrices. Thus, suppose $\mathbf{g} = \mathbf{AG}$, where $\mathbf{A}$ is $M \times M$ and nonsingular. Suppose we have the information that $\mathbf{G}$ is s-sparse (with unknown support $\mathcal{S}$). Then all M components of $\mathbf{g}$ can be found just by knowing $K \geq 2s$ samples of $\mathbf{g}$, provided there is a $K \times M$ submatrix of $\mathbf{A}$ with Kruskal rank K.
3. Our proof of the above claims is not complete because we have not shown how to find the sparse support from the K measurements (16.16). The reader should study the references mentioned above for that. However, the discussion in the next section will give some insight.

16.4 Compressive Sensing

We now consider a different situation. Suppose the $M \times 1$ vector $\mathbf{x}$ is not sparse by itself but it is sparse in some orthogonal basis. That is, $\mathbf{x}$ can be written as $\mathbf{x} = \mathbf{Bs}$, where $\mathbf{B}$ is an $M \times M$ unitary matrix and $\mathbf{s}$ is s-sparse with $s << M$. So, even though $\mathbf{x}$ is M-dimensional for some large M, it has a sparse representation in some basis $\mathbf{B}$. There are many practical examples like this. Signals such as speech, music, and images are sparse in certain standard bases such as the *Fourier basis, filter bank basis,* or *wavelet basis* (Sec. 11.10); we do not go into the details here. Given that $\mathbf{x}$ has such a sparse representation, we wish to create a compressed version $\mathbf{y} = \mathbf{Ax}$, where $\mathbf{A}$ is $K \times M$ and $K << M$. The term "compressed" just means that $\mathbf{y}$ has fewer components than $\mathbf{x}$. The components of $\mathbf{y}$ can be regarded as "samples" of the components of $\mathbf{x}$ in a generalized sense. Thus, if the rows of $\mathbf{A}$ are denoted $\mathbf{r}_k^H$, we can write $\mathbf{y} = \mathbf{Ax}$ as

$$\begin{bmatrix} y_0 \\ y_1 \\ \vdots \\ y_{K-1} \end{bmatrix} = \begin{bmatrix} \mathbf{r}_0^H\mathbf{x} \\ \mathbf{r}_1^H\mathbf{x} \\ \vdots \\ \mathbf{r}_{K-1}^H\mathbf{x} \end{bmatrix}. \tag{16.17}$$

[1] In fact, from any consecutive set of $K \geq 2s$ samples of $g[n]$.

So each sample y_k is an inner product between a "measuring vector" $\mathbf{r}_k$ and $\mathbf{x}$. Summarizing, we have

$$\mathbf{y} = \underbrace{\mathbf{A}}_{K\times M}\mathbf{x} = \underbrace{\mathbf{A}}_{K\times M}\underbrace{\mathbf{B}}_{M\times M}\mathbf{s}. \tag{16.18}$$

Where K is the number of *samples* or *measurements* of $\mathbf{x}$, M is the "ambient dimension" of $\mathbf{x}$ and $\mathbf{s}$ (i.e., they belong in the M-dimensional space), s is the sparsity of $\mathbf{s}$, and the support $\mathcal{S}$ of $\mathbf{s}$ is not known a priori. The question now is this: *Assuming* $s << M$, *can we really reconstruct* $\mathbf{x}$ *from* $\mathbf{y}$, *using very few samples* $K << M$?

One of the important results in the last two decades is that this is indeed possible [Candès and Tao, 2005, 2006]. In fact, it is possible to choose the matrix $\mathbf{A}$ such that the basis $\mathbf{B}$ almost does not matter. To put it briefly and informally, the main result is this: even if $\mathbf{A}$ has its entries a_{km} *randomly* chosen (in particular, independent of $\mathbf{B}$), it is possible to make the number of measurements as small as

$$K = \beta s \log(M/s), \tag{16.19}$$

where β is a constant of proportionality [Baraniuk, 2007; Candès and Wakin, 2008]. Note that M (the size of the vectors $\mathbf{x}$ and $\mathbf{s}$) itself can be very large but only $\log(M/s)$ enters in Eq. (16.19). Thus, the number of measurements is essentially linear in the sparsity s.

The method to reconstruct $\mathbf{s}$ is based on an interesting algorithm, which works as follows: it looks for a vector $\widehat{\mathbf{s}}$ such that $\|\widehat{\mathbf{s}}\|_1$ (the ℓ_1 norm of $\widehat{\mathbf{s}}$) is minimized subject to the constraint $\mathbf{y} = \mathbf{AB}\widehat{\mathbf{s}}$:

$$\min_{\widehat{\mathbf{s}}} \|\widehat{\mathbf{s}}\|_1 \qquad \text{such that} \qquad \mathbf{y} = \mathbf{AB}\widehat{\mathbf{s}}. \tag{16.20}$$

Here, the ℓ_1 norm is defined as $\|\widehat{\mathbf{s}}\|_1 = \sum_m |\widehat{s}_m|$. If we solve (16.20), the solution $\widehat{\mathbf{s}}$ is equal to $\mathbf{s}$ "with high probability" under some mild conditions. To be more clear, there is a condition called the *restricted isometry property* (RIP) [Candès and Tao, 2005]. If it is satisfied by the matrix $\mathbf{AB}$, then (16.20) yields the right solution $\mathbf{s}$. And the beauty is that this property is satisfied "with high probability" by choosing $\mathbf{A}$ with random entries. Once $\widehat{\mathbf{s}}$ is found, the estimate of $\mathbf{x}$ can be found as $\widehat{\mathbf{x}} = \mathbf{B}\widehat{\mathbf{s}}$. In fact, randomness of $\mathbf{A}$ is an important ingredient for the success of the method. The reason for minimizing the ℓ_1 norm of $\widehat{\mathbf{s}}$ is that such an approach promotes sparsity, that is, it makes sure that the solution turns out to be sparse. See Baraniuk [2007] for geometrical insights as to why the ℓ_1 norm, and *not* the ℓ_2 norm, promotes sparsity.

Although we have kept it simple here, the actual theorems are very precise and elegant and the proofs involve sophisticated mathematics [Candès and Tao, 2005, 2006]. Another important observation is that the optimization problem (16.20) can be solved using efficient convex optimization tools, which can often be as simple as linear programming.

There are many variations of this algorithm that work in this generalized setting, such as the *Lasso* algorithm, *orthogonal matching pursuit* (OMP), and so forth.

These algorithms come with rigorous theoretical guarantees about the recoverability of the $M \times 1$ vector $\mathbf{s}$ from only about $K << M$ measurements. An excellent survey of the algorithms can be found in Tropp and Wright [2010]. In practice, of course, $\mathbf{s}$ may not be exactly sparse but approximately sparse (in the sense that the largest s magnitudes in $\mathbf{s}$ are much larger than the others). Furthermore there will be noise, that is, we can only have access to $\mathbf{x} + \mathbf{e}$ where $\mathbf{e}$ is some form of additive noise. Generalizations of the above results exist for these nonideal situations as well.

Summarizing, the above ideas extend the traditional sampling theorem in a major way. Instead of thinking about bandlimited sequences $x[n]$, we now simply think of our data as a finite (large)-dimensional vector $\mathbf{x}$. Instead of the bandlimited property we have the sparse representation property ($\mathbf{x} = \mathbf{Bs}$, where $\mathbf{s}$ is sparse). Then we can sample $\mathbf{x}$ to get the compressed version $\mathbf{y} = \mathbf{Ax}$, where the sampling matrix $\mathbf{A}$ should be such that $\mathbf{AB}$ satisfies the RIP property. The latter is almost surely satisfied for almost any $\mathbf{B}$ if $\mathbf{A}$ is randomly selected from certain distributions. And finally there are efficient algorithms for reconstructing $\mathbf{x}$ from the "samples" $\mathbf{y}$, and there are theoretical guarantees about the accuracy of these reconstructions.

Remarks

The careful student will notice an important gap in the above discussion. It was assumed that the $M \times 1$ vector $\mathbf{x}$ is available to us. From it we created a compressed version $\mathbf{y} = \mathbf{Ax}$ which could be stored or transmitted efficiently. When desired, we can then recover $\mathbf{x}$ from this compressed version $\mathbf{y}$ by first recovering $\mathbf{s}$, from which $\mathbf{x}$ can be reconstructed. While this may be very nice and elegant, in some situations it would be more sensible to take advantage of this compressibility at an earlier stage, namely the acquisition stage where the data $\mathbf{x}$ is acquired, possibly by sampling some continuous-time (or continuous-space) signal. This process of compressed sampling, while a continuous-time signal is being converted into discrete time, is called *compressive sensing*. It is a major topic [Baraniuk, 2007; Candès and Wakin, 2008], and we shall not go into details here.

16.5 DFT and Sparsity

We now discuss a very interesting property of the DFT. Let $x[n], 0 \leq n \leq N-1$, be an N-point sequence with DFT $X[k], 0 \leq k \leq N-1$. Suppose the number of nonzero samples in $x[n]$ is N_t, and let N_w be the number of nonzero samples in $X[k]$. Clearly $0 \leq N_t, N_w \leq N$. Now we claim that two things are true:

Claim A: No consecutive N_t samples of the DFT $X[k]$ can be all zero.
Claim B: $N_t N_w \geq N$, and there exists $x[n]$ to achieve equality if and only if N is a multiple of N_t.

The second claim says that a sequence and its DFT cannot both be "too sparse." This is analogous to the uncertainty principle in Fourier transform theory, which says that the product of the duration in time and duration in frequency cannot be arbitrarily small (Problem 9.15). Notice that the picket-fence sequence (Problem 12.10) is an example where $N_tN_w = N$.

Proofs of the Two Claims

To prove the first claim, assume the contrary. Suppose $X[k] = 0$ for $k_1 \leq k \leq k_1 + N_t - 1$. Then

$$\mathbf{0} = \begin{bmatrix} X[k_1] \\ X[k_1+1] \\ \vdots \\ X[k_1+N_t-1] \end{bmatrix}$$

$$= \begin{bmatrix} W^{k_1n_1} & W^{k_1n_2} & \dots & W^{k_1n_{N_t}} \\ W^{(k_1+1)n_1} & W^{(k_1+1)n_2} & \dots & W^{(k_1+1)n_{N_t}} \\ \vdots & \vdots & \vdots & \vdots \\ W^{(k_1+N_t-1)n_1} & W^{(k_1+N_t-1)n_2} & \dots & W^{(k_1+N_t-1)n_{N_t}} \end{bmatrix} \begin{bmatrix} x[n_1] \\ x[n_2] \\ \vdots \\ x[n_{N_t}] \end{bmatrix}, \tag{16.21}$$

where $x[n_1], x[n_2], \ldots, x[n_{N_t}]$ are the N_t nonzero samples, and $W = e^{-j2\pi/N}$. The $N_t \times N_t$ matrix above can be rewritten as the product

$$\begin{bmatrix} 1 & 1 & \dots & 1 \\ W^{n_1} & W^{n_2} & \dots & W^{n_{N_t}} \\ \vdots & \vdots & \vdots & \vdots \\ W^{n_1(N_t-1)} & W^{n_2(N_t-1)} & \dots & W^{n_{N_t}(N_t-1)} \end{bmatrix} \begin{bmatrix} W^{k_1n_1} & 0 & \dots & 0 \\ 0 & W^{k_1n_2} & \dots & 0 \\ \vdots & \vdots & \vdots & \vdots \\ 0 & 0 & \dots & W^{k_1n_{N_t}} \end{bmatrix}.$$

In the matrix product above, the matrix on the left is Vandermonde, and since the n_i are distinct with $0 \leq n_i \leq N-1$, it follows that the W^{n_i} are distinct. So this Vandermonde matrix is nonsingular. The diagonal matrix on the right is clearly nonsingular as well. So the $N_t \times N_t$ matrix in Eq. (16.21) is nonsingular. Since the $x[n_i]$ are nonzero, the matrix vector product on the right side of Eq. (16.21) is therefore nonzero, which establishes the contradiction. This proves Claim A.

Meaning of "consecutive samples." Now recall from Sec. 12.5 that $X[k]$ is periodic in k, that is,

$$X[k] = X[k+N],$$

so that $X[k]$ is defined for all integer k. Thus the statement "N_t consecutive samples" in Claim A should be read as "N_t consecutive samples modulo N." For example, if $N = 8$ and $N_t = 4$, we know that $X[0], X[1], X[2], X[3]$ are consecutive and cannot all be zero. Similarly, the samples $X[6], X[7], X[0], X[1]$ are also consecutive (modulo 8), and cannot all be zero. ▽ ▽ ▽

To prove the second claim, we use vector notation as in Sec. 12.3, so that $\mathbf{X} = \begin{bmatrix} X[0] & X[1] & \cdots & X[N-1] \end{bmatrix}^T$. Since $N_t \leq N$, we can always write

$$N = aN_t + b, \tag{16.22}$$

where a and b are non-negative integers with $0 \leq b \leq N_t - 1$. Then $\mathbf{X}$ can be partitioned into blocks as follows:

$$\mathbf{X} = \begin{bmatrix} \mathbf{X}_1 \\ \mathbf{X}_2 \\ \vdots \\ \mathbf{X}_a \\ \cdots\cdots \\ \mathbf{X}_{a+1} \end{bmatrix}, \tag{16.23}$$

where the first a blocks have N_t samples each, and the last block $\mathbf{X}_{a+1}$ has b samples. Since no N_t consecutive values of $X[k]$ can all be zero, it follows that each of the first a blocks has at least one nonzero element.

If N is a multiple of N_t (i.e., $b = 0$), then $\mathbf{X}_{a+1}$ is not there, and the number of nonzero elements N_w in $\mathbf{X}$ satisfies $N_w \geq a$, so that $N_t N_w \geq N_t a = N$, as stated in Claim B. Next assume N is not a multiple of N_t (i.e., $b \neq 0$). Consider the extreme case where each of the first a blocks in Eq. (16.23) has exactly one nonzero element. Then these elements are equispaced in $\mathbf{X}$, for otherwise, Claim A will be violated. Under this situation, if $\mathbf{X}_{a+1} = \mathbf{0}$, then the trailing set of zeros in $\mathbf{X}_a$, the set of zeros in $\mathbf{X}_{a+1}$, and the leading set of zeros in $\mathbf{X}_1$ together form a consecutive set of zeros (modulo N), and you can verify that there are more than N_t such zeros. This violates Claim A. Summarizing, in view of Claim A, if the first a blocks have only one nonzero sample each, then $\mathbf{X}_{a+1} \neq \mathbf{0}$. So the number of nonzero samples N_w in $\mathbf{X}$ satisfies

$$N_w \geq a + 1. \tag{16.24}$$

In the other situation, where at least one of the first a blocks has more than one nonzero sample, Eq. (16.24) is trivially true as well. So we have proved that when N is not a multiple of N_t (i.e., $b \neq 0$),

$$N_t N_w \geq N_t(a+1) = aN_t + N_t > aN_t + b = N, \tag{16.25}$$

which proves Claim B for the case $b \neq 0$. This completes the proof.

16.6 Summary of Chapter 16

In this chapter we gave a brief overview of sampling and reconstruction for signals which have certain sparsity properties. This can be regarded as a generalization of traditional sampling theory, which is based on the bandlimited property. Examples include finite-duration signals whose DFTs are sparse. Sparse reconstruction methods are also closely related to the theory of compressive sensing.

Compressive sensing and reconstruction based on sparsity are major topics. The reader interested in this should first read the tutorial articles written by Baraniuk [2007] and Candès and Wakin [2008]. These will explain the wisdom behind these unconventional sampling methods in greater detail. Then the student can read some of the key papers mentioned in the above two articles. Papers which make the connection from theory to practice include Duarte et al. [2008] and Mishali et al. [2011], among many others. Comprehensive books on sparse reconstruction and compressive sensing include Elad [2013] and Eldar [2015].

17 Mathematical Intricacies*

17.1 Introduction

In this chapter we present some mathematical details relating to the Fourier transform, Fourier series, and their inverses. These details were omitted in the preceding chapters in order to enable the reader to focus on the engineering aspects. But it is nice to have a good appreciation of some of these mathematical subtleties. Section 17.2 discusses the discrete-time case, wherein two types of Fourier transforms are distinguished, namely the ℓ_1-FT and the ℓ_2-FT. A similar distinction between L_1-FT and L_2-FT for the continuous-time case is made in Sec. 17.4, after some preliminary discussion of L_p spaces in Sec. 17.3. A discussion on the pointwise convergence of the Fourier series representation is then given in Sec. 17.5, where a number of sufficient conditions for such convergence are presented. This involves concepts such as bounded variation, one-sided derivatives, and so on. We take the time to explain the meanings of these concepts. Detailed discussions of these conditions, along with several illuminating examples, are presented in Sec. 17.6. In Sec. 17.7, the discussion is extended to the case of the Fourier integral.

17.2 ℓ_p Spaces and Discrete-Time FT

In this section we discuss two types of discrete-time Fourier transforms, namely the ℓ_1 and ℓ_2 transforms. A sequence $x[n]$ is said to be an ℓ_1 signal if it is absolutely summable, that is,

$$\sum_{n=-\infty}^{\infty} |x[n]| < \infty. \tag{17.1}$$

In Sec. 3.3 we showed that an LTI system is BIBO stable if and only if the impulse response $h[n]$ is absolutely summable. Thus, stability is equivalent to saying $h[n] \in \ell_1$. A sequence $x[n]$ is said to be an ℓ_2 signal if it has finite energy, that is,

$$\sum_{n=-\infty}^{\infty} |x[n]|^2 < \infty. \tag{17.2}$$

More generally, a signal $x[n]$ for which the quantity

$$\|x[n]\|_p \triangleq \Big(\sum_{n=-\infty}^{\infty} |x[n]|^p\Big)^{1/p} \tag{17.3}$$

is finite for some positive integer p is said to be an ℓ_p signal. For a given p, the set of all ℓ_p signals is a linear space in the sense that any finite linear combination of signals in this set continues to be an ℓ_p signal. The space of ℓ_p signals is said to be the ℓ_p **space**.

The quantity $\|x[n]\|_p$ is said to be the ℓ_p norm of $x[n]$. For example, the ℓ_2 **norm** is the square root of the energy of $x[n]$. Each element in the space ℓ_p is a sequence $x[n]$, where $-\infty < n < \infty$. This sequence can be regarded as a vector of infinite number of components. The space ℓ_p is therefore an **infinite-dimensional** space, that is, the number of basis functions required to span the space is infinite.

The space ℓ_∞ is defined to be the space of **bounded sequences** (i.e., signals satisfying $|x[n]| \leq B$ for all n for some finite B), and the ℓ_∞ norm is defined as

$$\|x[n]\|_\infty = \sup_n |x[n]|, \tag{17.4}$$

where "sup" denotes the supremum, which is similar to the maximum.[1] We now summarize some of the key properties of these spaces. More details can be found in a number of books [e.g., Naylor and Sell, 1982]. First we prove that ℓ_p is a subspace of ℓ_{p+1}, that is,[2]

$$\ell_p \subset \ell_{p+1}, \tag{17.5}$$

for any positive integer p. Writing Eq. (17.5) more explicitly, we have

$$\ell_1 \subset \ell_2 \subset \ell_3 \subset \cdots \subset \ell_\infty, \tag{17.6}$$

as demonstrated in Fig. 17.1.

Proof of Eq. (17.5). Notice first that if the quantity (17.3) is finite, then $x[n]$ is bounded because

$$|x[n]| = (|x[n]|^p)^{1/p} \leq \Big(\sum_{n=-\infty}^{\infty} |x[n]|^p\Big)^{1/p} = \|x[n]\|_p < \infty$$

for all n. So $x[n] \in \ell_\infty$. It is then obvious that $\ell_p \subset \ell_\infty$. We now have

$$\|x[n]\|_{p+1}^{p+1} = \sum_n |x[n]|^{p+1} \leq \sup_n |x[n]| \sum_n |x[n]|^p = \|x[n]\|_\infty \|x[n]\|_p^p.$$

Thus, if $\|x[n]\|_p$ is finite, then so is $\|x[n]\|_{p+1}$. This proves $\ell_p \subset \ell_{p+1}$. $\triangledown \triangledown \triangledown$

[1] In mathematics, the *supremum* is used whenever the existence of a maximum is not guaranteed. In Eq. (17.4) the supremum is the smallest number A_x such that $|x[n]| \leq A_x$. If a real number y belongs to the open interval $(0, 1)$, then the supremum of y is 1, whereas the maximum does not exist. If z belongs to the closed interval $[0, 1]$, then the supremum and the maximum are identical $(= 1)$.

[2] We shall use the standard set-theoretic notations; thus, $x[n] \in \ell_2$ means that $x[n]$ belongs to the space ℓ_2.

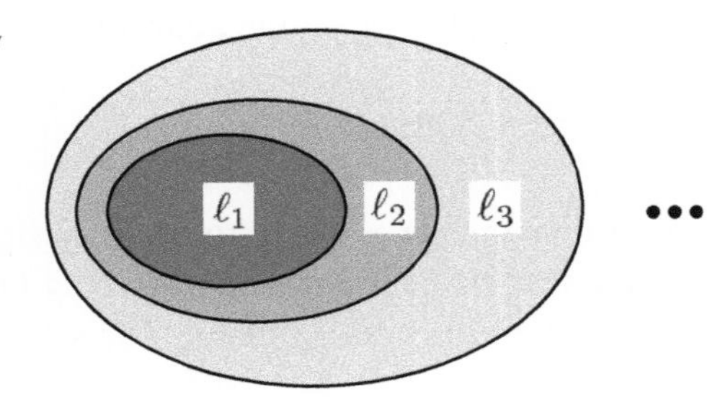

Figure 17.1 Inclusion property of the ℓ_p spaces.

17.2.1 Two Types of Discrete-Time Fourier Transforms

We now focus on Fourier transforms of ℓ_1 and ℓ_2 signals. Observe first that since $\ell_1 \subset \ell_2$, we can find examples of ℓ_2 signals which are not in ℓ_1. A familiar example is

$$x[n] = \frac{1}{n}\mathcal{U}[n-1]. \tag{17.7}$$

It is well known that $\sum_{n>0} 1/n$ does not converge, but $\sum_{n>0} 1/n^2$ does. So $x[n]$ is in ℓ_2 but not in ℓ_1. Using this, we can also show that

$$h[n] = \frac{\sin(\omega_c n)}{\pi n} \tag{17.8}$$

is an ℓ_2 signal but not an ℓ_1 signal. This $h[n]$ is nothing but the impulse response of an ideal lowpass filter which is unstable (Sec. 4.3.4).

17.2.1.1 The ℓ_1 Fourier Transform

Consider the Fourier transform summation

$$X(e^{j\omega}) = \sum_{n=-\infty}^{\infty} x[n]e^{-j\omega n}. \tag{17.9}$$

For an ℓ_1 signal $x[n]$, the summation (17.9) converges to a finite value, that is, the Fourier transform exists pointwise for all ω. This is called the ℓ_1-FT of $x[n]$. To prove convergence of Eq. (17.9), just observe that

$$\Big|\sum_{n=-\infty}^{\infty} x[n]e^{-j\omega n}\Big| \leq \sum_{n=-\infty}^{\infty} |x[n]e^{-j\omega n}| = \sum_{n=-\infty}^{\infty} |x[n]|, \tag{17.10}$$

and the sum on the right is indeed finite by definition of the ℓ_1 property. The ℓ_1-FT satisfies some interesting properties:

1. *ℓ_1-FT is bounded.* Equation (17.10) shows that

$$|X(e^{j\omega})| \leq \sum_{n=-\infty}^{\infty} |x[n]| = \|x[n]\|_1, \tag{17.11}$$

   that is, the ℓ_1-FT is upper bounded by the ℓ_1 norm $\|x[n]\|_1$.
2. *ℓ_1-FT is continuous.* The ℓ_1-FT $X(e^{j\omega})$ is a continuous function of ω. For proof, see Sec. 6.9.4 under "Proof of continuity."

17.2.1.2 The ℓ_2 Fourier Transform

Since an ℓ_2 signal need not necessarily be in ℓ_1, the sum $\sum_n |x[n]|$ is not necessarily finite for an ℓ_2 signal, and we cannot guarantee that the Fourier transform summation (17.9) converges for all ω. Consider, however, the partial summations

$$X_k(e^{j\omega}) = \sum_{n=-k}^{k} x[n]e^{-j\omega n}, \tag{17.12}$$

which obviously converge for all ω. For $x[n] \in \ell_2$ it can be shown (see below) that the sequence of functions

$$X_1(e^{j\omega}), X_2(e^{j\omega}), X_3(e^{j\omega}), \ldots \tag{17.13}$$

converges to a function $X(e^{j\omega})$ as $k \to \infty$, in the sense that

$$\lim_{k\to\infty} \int_0^{2\pi} \left| X_k(e^{j\omega}) - X(e^{j\omega}) \right|^2 d\omega = 0. \tag{17.14}$$

The limit $X(e^{j\omega})$ is called the ℓ_2 Fourier transform or ℓ_2-FT. Since $X_k(e^{j\omega})$ and $X(e^{j\omega})$ are periodic, the convergence of $X_k(e^{j\omega})$ to $X(e^{j\omega})$ is similar to the convergence of Fourier series for periodic functions (if $x[n]$ is interpreted as the Fourier series coefficients for $X(e^{j\omega})$; Sec. 12.7). While the ℓ_1-FT is continuous and bounded, the ℓ_2-FT in general is not continuous nor guaranteed to be bounded (as we shall see in the examples in this section).

Proof of convergence of partial sums. The function $X_k(e^{j\omega})$ defined in Eq. (17.12) has finite energy $\sum_{n=-k}^{k} |x[n]|^2$, that is,

$$\frac{1}{2\pi} \int_0^{2\pi} |X_k(e^{j\omega})|^2 d\omega < \infty. \tag{17.15}$$

Such functions are called $L_2[0, 2\pi]$ functions (Sec. 17.3), the argument $[0, 2\pi]$ indicating that the domain is $0 \le \omega \le 2\pi$. We now claim that the sequence $X_k(e^{j\omega})$ has the property that $X_{k_1}(e^{j\omega})$ and $X_{k_2}(e^{j\omega})$ get closer and closer together as k_1 and $k_2 > k_1$ get larger and larger. To be more precise, given an $\epsilon > 0$, we will show that we can always find an N such that

$$\frac{1}{2\pi} \int_0^{2\pi} |X_{k_2}(e^{j\omega}) - X_{k_1}(e^{j\omega})|^2 d\omega < \epsilon, \tag{17.16}$$

for all $k_1, k_2 \ge N$. Such a sequence $X_k(e^{j\omega})$ is called a *Cauchy sequence*. To see that the preceding claim is indeed true, first observe that

$$X_{k_2}(e^{j\omega}) - X_{k_1}(e^{j\omega}) = \sum_{n=k_1+1}^{k_2} x[n]e^{-j\omega n} + \sum_{n=-k_2}^{-(k_1+1)} x[n]e^{-j\omega n}. \tag{17.17}$$

By Parseval's relation, the left-hand side of Eq. (17.16) equals

$$\sum_{n=k_1+1}^{k_2} |x[n]|^2 + \sum_{n=-k_2}^{-(k_1+1)} |x[n]|^2. \tag{17.18}$$

Now, since $x[n]$ is in ℓ_2, the sum $\sum_{n=-\infty}^{\infty}|x[n]|^2$ converges, which means that the summation (17.18) can be made smaller than ϵ by making $k_1, k_2 > N$ for appropriate N, and this proves Eq. (17.16). Next, it is known [Naylor and Sell, 1982] that the space $L_2[0, 2\pi]$ is *complete*, that is, any Cauchy sequence $X_k(e^{j\omega})$ in $L_2[0, 2\pi]$ converges to a limit $X(e^{j\omega})$ in $L_2[0, 2\pi]$. This concludes the proof. ▽ ▽ ▽

Notice in particular that the ℓ_2-FT $X(e^{j\omega})$ is in $L_2[0, 2\pi]$.

17.2.1.3 Example: ℓ_1 and ℓ_2 Fourier Transforms

Consider the ideal lowpass response $H(e^{j\omega})$ shown in Fig. 17.2(a). The impulse response is

$$h[n] = \frac{\sin \omega_c n}{\pi n}. \tag{17.19}$$

We know this filter is not BIBO stable, that is, $h[n]$ is not in ℓ_1. But

$$\sum_{n=-\infty}^{\infty} h^2[n] = \left(\frac{\omega_c}{\pi}\right)^2 + 2\sum_{n=1}^{\infty}\left(\frac{\sin \omega_c n}{\pi n}\right)^2 < \left(\frac{\omega_c}{\pi}\right)^2 + \frac{2}{\pi^2}\sum_{n=1}^{\infty}\frac{1}{n^2}. \tag{17.20}$$

Since the series $\sum_{n=1}^{\infty} 1/n^2$ converges to a finite number ($= \pi^2/6$), we see that $\sum_n h^2[n]$ is finite, that is, $h[n] \in \ell_2$ though $h[n] \notin \ell_1$. So the Fourier transform shown in Fig. 17.2(a) should be regarded as an ℓ_2-FT and not an ℓ_1-FT.

Next consider Fig. 17.2(b), which is proportional to the convolution of Fig. 17.2(a) with itself. The impulse response of this filter is

$$g[n] = h^2[n] = \left(\frac{\sin \omega_c n}{\pi n}\right)^2, \tag{17.21}$$

when $c = \omega_c/\pi$. Since $h[n] \in \ell_2$ it is clear that $g[n] \in \ell_1$. So the Fourier transform in Fig. 17.2(b) can be regarded as an ℓ_1-FT. We proved earlier that the ℓ_1-FT is a continuous function of ω. Indeed, $G(e^{j\omega})$ in Fig. 17.2(b) is continuous (the sharp corner at $\omega = 0$ does not violate continuity, the plot is just not *differentiable* there). Notice, however, that the ℓ_2-FT in Fig. 17.2(a) is discontinuous at ω_c.

Finally, **stable rational** digital filters, such as FIR filters and causal IIR filters with poles inside the unit circle, have $h[n] \in \ell_1$. For example, suppose $h[n] = p^n\mathcal{U}[n]$, with the pole p satisfying $|p| < 1$. Then $h[n] \in \ell_1$ and

$$H(e^{j\omega}) = \frac{1}{1 - pe^{-j\omega}} \tag{17.22}$$

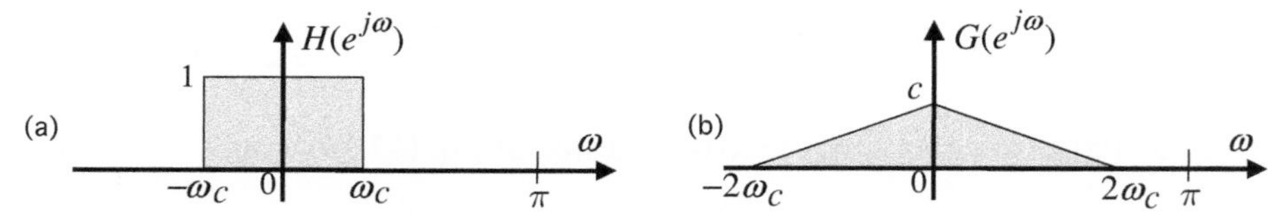

Figure 17.2 (a) Frequency response of the ideal lowpass filter ($h[n] \in \ell_2$); (b) a triangular lowpass filter ($g[n] \in \ell_1$).

is an ℓ_1-FT (Fig. 5.3). It is not only continuous and bounded but also infinitely differentiable in ω – this is true of any rational $H(e^{j\omega})$. Usually the distinction between ℓ_1-FT and ℓ_2-FT is clear from the context. Since an ℓ_1 signal is also in ℓ_2, there is no need to distinguish the two types in that case. Both types of transform satisfy standard Fourier transform properties such as *convolution theorems, Parseval relation*, and so forth. Notice that Parseval's relation can be written using the ℓ_2-norm notation as follows: $\|x[n]\|_2^2 = \int_0^{2\pi} |X(e^{j\omega})|^2 d\omega/2\pi$.

Fourier transforms with impulses. It is clear that signals such as $x[n] = e^{j\omega_1 n}$ are not ℓ_p functions for any finite p. Their Fourier transform is usually regarded as the delta function

$$X(e^{j\omega}) = 2\pi\delta_c(\omega - \omega_1), \quad 0 \leq \omega < 2\pi. \tag{17.23}$$

This is not an ℓ_1 or ℓ_2-FT but is considered a pointwise limit, for each $\omega \neq \omega_1$, of the function

$$X_N(e^{j\omega}) \stackrel{\Delta}{=} \sum_{n=-N}^{N} x[n]e^{-j\omega n} = \frac{\sin\left(L\left(\frac{\omega-\omega_1}{2}\right)\right)}{\sin\left(\frac{\omega-\omega_1}{2}\right)}, \tag{17.24}$$

as $N \to \infty$ (where $L = 2N + 1$). This was explained in detail when we derived Eq. (3.101) for the case where $\omega_1 = 0$. ▽ ▽ ▽

17.2.2 Fine Points About the ℓ_2-FT

We now present an example to show that the ℓ_2 Fourier transform need not be a bounded function. But first a discussion of the integral $\int dx/x^\alpha$ will be helpful. Thus consider the integral

$$\int_1^\infty \frac{dx}{x^\alpha}, \tag{17.25}$$

where $\alpha > 0$. The integrand is a bounded signal for all x in the range of integration. We know that the integral exists (i.e., has finite value) as long as $\alpha > 1$. For example, $\int_1^\infty dx/x^2$ exists, whereas $\int_1^\infty dx/x$ does not, nor does $\int_1^\infty dx/\sqrt{x}$. The condition $\alpha > 1$ ensures a sufficiently rapid decay as $x \to \infty$, enabling the integral to exist.

Next consider the integral $\int_0^1 dx/x^\alpha$, where $\alpha > 0$. The integrand is now *unbounded* at $x = 0$ (i.e., as x approaches 0 from the positive side). So it is not obvious that the integral exists. It can be shown, however (see p. 270 of Apostol [1974]), that

$$\int_0^1 \frac{dx}{x^\alpha} = \frac{1}{1-\alpha}, \tag{17.26}$$

as long as $\alpha < 1$. For example, $\int_0^1 dx/\sqrt{x}$ exists, but $\int_0^1 dx/x^2$ does not. In some sense the conditions for integrability of $\int_0^1 dx/x^\alpha$ and $\int_1^\infty dx/x^\alpha$ are *complementary*. The condition $\alpha < 1$ makes sure that the unboundedness at $x = 0$ is "not too severe," and integration is possible. Mathematicians are more careful when stating

Eq. (17.26) because the integrand is undefined at $x = 0$. What they do is to define a function

$$f(x) = \begin{cases} \frac{1}{x^{\alpha}} & x > 0, \\ 0 & x = 0. \end{cases} \tag{17.27}$$

This is discontinuous at $x = 0$, and unbounded as x approaches 0 from above, but it is finite for any fixed $x \geq 0$. The integral $\int_0^1 f(x)dx$ has the value shown in Eq. (17.26) for $\alpha < 1$.

17.2.2.1 An ℓ_2-FT which is Unbounded

As an example of the application of Eq. (17.26), consider the Fourier transform

$$X(e^{j\omega}) = \begin{cases} \frac{1}{\omega^{1/4}} & 0 < |\omega| \leq \pi, \\ 0 & \text{at } \omega = 0. \end{cases} \tag{17.28}$$

See Fig. 17.3, which shows this for $\omega > 0$. Using Eq. (17.26) with an appropriate change of variables, we get

$$\frac{1}{2\pi}\int_{-\pi}^{\pi} |X(e^{j\omega})|^2 d\omega = \frac{1}{\pi}\int_0^{\pi} \frac{1}{\omega^{1/2}} d\omega = \frac{2}{\sqrt{\pi}}. \tag{17.29}$$

The function $X(e^{j\omega})$ is therefore in the space $L_2[0, 2\pi]$, though unbounded. It is therefore a valid Fourier transform for an ℓ_2 signal $x[n]$. Such a signal *cannot be an ℓ_1 signal* because $X(e^{j\omega})$ violates boundedness.

17.2.2.2 The "Almost Everywhere" Jargon

Suppose $X_1(e^{j\omega})$ and $X_2(e^{j\omega})$ are two ℓ_2-FTs so that they are $L_2[0, 2\pi]$ functions. Suppose they differ at only one point, say ω_0. Then we say that $X_1(e^{j\omega}) = X_2(e^{j\omega})$ almost everywhere. If we take their difference and compute the integral

$$\frac{1}{2\pi}\int_0^{2\pi} [X_1(e^{j\omega}) - X_2(e^{j\omega})]e^{j\omega n} d\omega, \tag{17.30}$$

then the result is zero for any n. This shows that they have the same inverse transform. Thus, if the ℓ_2-FT of a signal $x[n]$ is altered at one point, the result continues

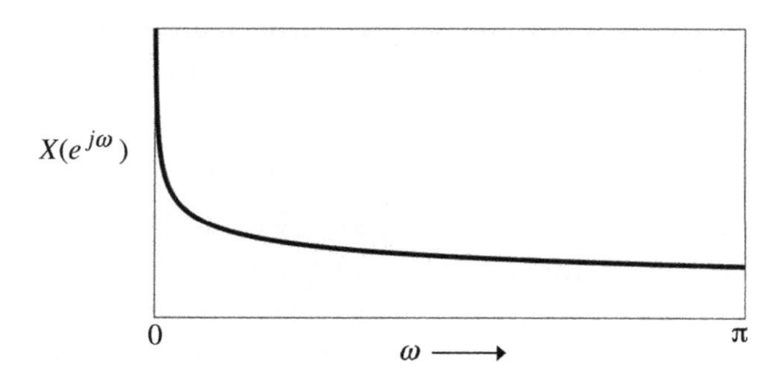

Figure 17.3 Example of an unbounded ℓ_2 Fourier transform.

to be a valid ℓ_2-FT for the same $x[n]$. More generally, if the ℓ_2-FT is altered at a countable number of points (say at the integers), the result is still an ℓ_2-FT for $x[n]$. In this sense the ℓ_2-FT is not unique. On the other hand, since the ℓ_1-FT of a signal is continuous, we cannot add arbitrary values to it at a countable number of frequencies without destroying continuity. So the ℓ_1-FT is unique.

Return now to ℓ_2-FTs which are not unique. Functions of a continuous variable ω, which differ from each other at a countable number of points, are said to be identical **almost everywhere** (abbreviated as a.e.). The idea is that their disagreement at a countable number of points is insignificant because the **measure** of a countable subset of the real line is zero. Whenever we compare two ℓ_2-FTs, we implicitly assume that the comparison is a.e. This is indicated by expressions such as

$$X_a(e^{j\omega}) = X_b(e^{j\omega}) \quad a.e., \qquad |X_a(e^{j\omega})| > |X_b(e^{j\omega})| \quad a.e., \tag{17.31}$$

and so forth.

Engineer versus mathematician. An ironical fact is that the set of rational numbers is a *countable* subset of real numbers, and therefore has measure zero on the real line. This is in spite of the fact that rational numbers can be used to approximate any irrational number arbitrarily closely. The latter is expressed by the statement that the set of rationals is *dense* on the real line [Naylor and Sell, 1982]. If we change the Fourier transform $X_1(e^{j\omega})$ of a ℓ_2-sequence $x_1[n]$ at every rational value of ω (say by adding unity at these points), then the result $X_2(e^{j\omega})$ is essentially the same function because

$$X_1(e^{j\omega}) = X_2(e^{j\omega}), \quad a.e., \tag{17.32}$$

but computer plots of the two functions (which necessarily are shown only at a countable number of points) can look completely different from each other! Two things that look essentially identical to the mathematician can look very different in the eyes of the engineer! ▽ ▽ ▽

17.2.3 Filtering ℓ_2 Signals

Consider Fig. 17.4, which shows a digital filter with impulse response $h[n]$. The output is given by the convolution $y[n] = \sum_m h[m]x[n-m]$. Assume that $x[n] \in \ell_2$ and $h[n] \in \ell_2$, so that the ℓ_2-FTs $X(e^{j\omega})$ and $H(e^{j\omega})$ exist. Define the product

$$Y_1(e^{j\omega}) \stackrel{\Delta}{=} H(e^{j\omega})X(e^{j\omega}). \tag{17.33}$$

In view of the convolution theorem (Sec. 3.4.3), it is tempting to think that $Y_1(e^{j\omega})$ is the Fourier transform of $y[n]$. But this statement is meaningful only if $y[n]$ is in ℓ_2.

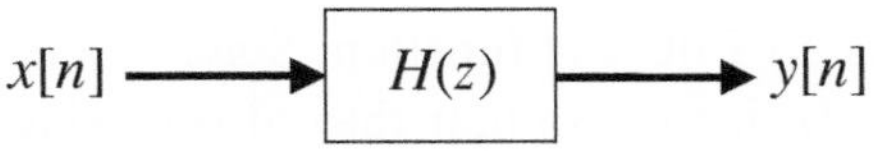

$x[n]$ → $H(z)$ → $y[n]$

Figure 17.4 A digital filter with ℓ_2 input $x[n]$.

Notice that even if $x[n]$ and $h[n]$ are in ℓ_2, the signal $y[n]$ may not be in ℓ_2. For example, suppose $X(e^{j\omega})$ and $H(e^{j\omega})$ are both taken to be the unbounded function in Eq. (17.28). Then the product $Y_1(e^{j\omega}) = H(e^{j\omega})X(e^{j\omega})$ remains unbounded. Since

$$|Y_1(e^{j\omega})|^2 = \begin{cases} \dfrac{1}{\omega} & 0 < |\omega| \leq \pi, \\ 0 & \text{at } \omega = 0, \end{cases} \tag{17.34}$$

this is not integrable, as explained earlier. Thus, $Y_1(e^{j\omega})$ is not in $L_2[0, 2\pi]$. So the product $Y_1(e^{j\omega}) = H(e^{j\omega})X(e^{j\omega})$ cannot be interpreted as the ℓ_2-FT of the output $y[n]$. This shows that the output $y[n]$ is not in ℓ_2. For if it were, then its ℓ_2-FT would be an L_2 function and would be equal to $H(e^{j\omega})X(e^{j\omega})$ by the convolution theorem, resulting in a contradiction.

17.2.4 Bounded ℓ_2 Filters

We now define a class of ℓ_2 filters for which ℓ_2 inputs necessarily produce ℓ_2 outputs. A filter for which $h[n] \in \ell_2$, and $H(e^{j\omega})$ is bounded, that is,

$$|H(e^{j\omega})| \leq B < \infty, \tag{17.35}$$

is called a **bounded-ℓ_2 filter**. Since the ℓ_1-FT is bounded (Sec. 17.2.1) and since $\ell_1 \subset \ell_2$, it follows that *a stable filter (ℓ_1 filter) is also a bounded ℓ_2 filter*. But bounded ℓ_2 filters are not necessarily stable. For example, the ideal lowpass filter is bounded-ℓ_2 but not stable. The reader should notice that all commonly encountered filters, including ideal ones, are bounded ℓ_2, though not necessarily ℓ_1. An important property of bounded-ℓ_2 filters is that ℓ_2 inputs always produce ℓ_2 outputs:

Proof. Just observe that

$$\frac{1}{2\pi}\int_0^{2\pi} |Y(e^{j\omega})|^2 d\omega = \frac{1}{2\pi}\int_0^{2\pi} |H(e^{j\omega})X(e^{j\omega})|^2 d\omega \leq \frac{B^2}{2\pi}\int_0^{2\pi} |X(e^{j\omega})|^2 d\omega.$$

Thus, $Y(e^{j\omega})$ is an $L_2[0, 2\pi]$ function, and its inverse FT $y[n]$ is therefore in ℓ_2. Using Parseval's relation, we see in fact that $\|y[n]\|^2 \leq B^2\|x[n]\|^2$. $\triangledown \triangledown \triangledown$

Thus, if an ℓ_2 sequence $x[n]$ is input to a bounded ℓ_2 filter $h[n]$, then the ouput $y[n]$ is an ℓ_2 signal and

$$Y(e^{j\omega}) = H(e^{j\omega})X(e^{j\omega}) \quad \text{(convolution theorem)}, \tag{17.36}$$

where $Y(e^{j\omega})$ is an $L_2[0, 2\pi]$ function. Since stable filters (ℓ_1 filters) are automatically bounded-ℓ_2, it follows that they also produce ℓ_2 outputs in response to ℓ_2 inputs. The preceding discussions can also be used to conclude that the *cascade of two bounded ℓ_2 filters is a bounded ℓ_2 filter.*

17.3 L_p Spaces

In Sec. 17.2.1 we discussed two types of discrete-time Fourier transform, namely ℓ_1-FT and ℓ_2-FT. Similarly, in the continuous-time case we will distinguish between L_1-FT and L_2-FT. First we give a preliminary discussion on L_p spaces.

1. A signal $x(t)$ is said to be an $L_p[a,b]$ signal if

$$\left(\int_a^b |x(t)|^p \, dt\right)^{1/p} < \infty. \tag{17.37}$$

That is, the integral exists and has finite value. As mentioned in Sec. 17.9, when integrals are interpreted as Lebesgue integrals, an $L_1[a,b]$ signal is the same thing as a Lebesgue integrable signal (when all signals are "measurable," which we always assume). As another example, an $L_2[a,b]$ signal is a signal with finite energy $\int_a^b |x(t)|^2 dt$.

2. $L_p(-\infty,\infty)$ denotes the space $L_p[a,b]$ with $a \to -\infty$ and $b \to \infty$. If the interval is clear from the context, or irrelevant for a particular discussion, we omit the interval and write "L_p-space."
3. For fixed p, the collection of all $L_p[a,b]$ signals is a linear vector space (i.e., any finite linear combination of signals in $L_p[a,b]$ is in $L_p[a,b]$). This is called the $L_p[a,b]$ space. The norm of an L_p signal is defined as

$$\|x\|_p = \left(\int_a^b |x(t)|^p dt\right)^{1/p}. \tag{17.38}$$

This is called the L_p norm. Note that $\|x\|_2^2$ is nothing but the energy of the signal. Given two signals $x(t)$ and $y(t)$, suppose they differ only at a countable set of points $t_1, t_2, \ldots$. Then

$$\|x(t) - y(t)\|^p = \int_a^b |x(t) - y(t)|^p dt = 0. \tag{17.39}$$

So we say that $x(t) = y(t)$ as far as the L_p norm is concerned. We also say that $x(t) = y(t)$ *almost everywhere* (a.e.), which signifies that they differ only at a countable set of points. Thus, each signal in an L_p space is in reality an infinite collection of signals which are identical a.e.

4. A signal $x(t)$ is said to be *bounded* if there is a constant $A_x < \infty$ such that

$$|x(t)| \leq A_x, \tag{17.40}$$

almost everywhere. The constant A_x might depend on $x(t)$ but not on time t. If we take a bounded signal $x(t)$ and modify it at isolated points $t_1, t_2, \ldots$, by changing the magnitude to ∞, this does not change the boundedness property because of the "almost everywhere" phrase. So the above boundedness is sometimes referred to as *essential boundedness*. We will leave out the phrase "essential" for simplicity.

5. The $L_\infty[a, b]$ space consists of all bounded signals. The L_∞ norm of $x(t)$ is defined as

$$\|x(t)\|_\infty = \sup_t |x(t)|, \tag{17.41}$$

where "sup" represents the supremum (see footnote at the beginning of Sec. 17.2).

17.3.1 Properties of L_p Functions

For discrete-time sequences we found in Sec. 17.2 that the spaces ℓ_p had the following inclusive property: $\ell_1 \subset \ell_2 \subset \ell_3 \cdots \subset \ell_\infty$. A similar property does not hold for the continuous-time case. However, we will show below that if an L_1 function $x(t)$ is bounded (i.e., it is also in L_∞), then it is in L_2. That is,

$$L_1 \cap L_\infty \subset L_2. \tag{17.42}$$

Furthermore, when discussing bounded intervals $[a, b]$, it is true that

$$L_2[a, b] \subset L_1[a, b]. \tag{17.43}$$

Equation (17.42) implies $L_1[a, b] \cap L_\infty[a, b] \subset L_2[a, b] \cap L_\infty[a, b]$, whereas Eq. (17.43) implies $L_2[a, b] \cap L_\infty[a, b] \subset L_1[a, b] \cap L_\infty[a, b]$. Combining these, it follows that

$$L_1[a, b] \cap L_\infty[a, b] = L_2[a, b] \cap L_\infty[a, b]. \tag{17.44}$$

That is, *for bounded functions on bounded intervals, L_1 and L_2 spaces are identical.* Finally, we will see that Eq. (17.42) can readily be generalized to

$$L_1 \cap L_\infty \subset L_p, \quad p \geq 1. \tag{17.45}$$

That is, a bounded L_1 function is also an L_p function for any $p > 1$.

Proof of Eqs. (17.42)–(17.45). Let $x(t) \in L_1 \cap L_\infty$. Then $|x(t)| \leq A_x < \infty$, and $\int |x(t)| dt < \infty$. We have

$$\int |x(t)|^p \, dt \leq A_x^{p-1} \int |x(t)| \, dt < \infty, \tag{17.46}$$

proving that $x(t) \in L_p$ indeed. This proves Eq. (17.45) and hence Eq. (17.42). Next, to prove Eq. (17.43), assume $x(t) \in L_2[a, b]$, so $\int_a^b |x(t)|^2 dt < \infty$. Using the Cauchy–Schwarz inequality (Sec. 18.3.3), we have

$$\left(\int_a^b |x(t)| dt\right)^2 \leq \int_a^b |x(t)|^2 \, dt \int_a^b 1^2 dt = (b-a) \int_a^b |x(t)|^2 \, dt < \infty, \tag{17.47}$$

because $(b - a)$ is finite. So $x(t) \in L_1[a, b]$ indeed, proving Eq. (17.43). $\triangledown \triangledown \triangledown$

17.3.2 Inner Products and Orthogonality in L_2 Spaces

Given two signals $x_1(t)$ and $x_2(t)$ in L_2, their **inner product** is defined as

$$< x_1(t), x_2(t) >= \int x_1(t)x_2^*(t)dt. \tag{17.48}$$

Using the Cauchy–Schwarz inequality (Sec. 18.3.3), it follows that

$$| < x_1(t), x_2(t) > |^2 \leq \|x_1(t)\|_2^2 \times \|x_2(t)\|_2^2 < \infty. \tag{17.49}$$

So the inner product is always finite. Notice in particular that the product of two L_2 functions is integrable (the integral, being an inner product, always exists). So the product is an L_1 function. We say that $x_1(t)$ and $x_2(t)$ are **orthogonal** if the inner product is zero. Given a set of L_2 signals $\{x_k(t)\}$, we say that it forms an **orthonormal** set if

$$\int x_k(t)x_m^*(t)dt = \delta[k-m], \tag{17.50}$$

that is, any pair of signals is orthogonal, and each signal has unit norm (unit energy). Similar comments hold for discrete-time signals in ℓ_2 spaces (Sec. 17.2):

1. The inner product is defined as $\sum_n x_1[n]x_2^*[n]$ and the signals are orthogonal if the inner product is zero.
2. The product $x_1[n]x_2[n]$ of two ℓ_2 sequences is in ℓ_1.

17.4 The L_1 and L_2 Fourier Transforms

We now distinguish between the L_1 and L_2 Fourier transforms. Our discussion will be brief because of similarities with the discrete-time case (Sec. 17.2).

17.4.1 The L_1 Fourier Transform

For a signal $x(t)$ in L_1, the Lebesgue integral

$$X(j\omega) = \int_{-\infty}^{\infty} x(t)e^{-j\omega t}dt \tag{17.51}$$

exists for all ω (Sec. 17.9), and is called the L_1 Fourier transform or L_1-FT of $x(t)$. It can be shown (see Sec. 11 in Chapter 11 of Haaser and Sullivan [1991]) that the L_1-FT satisfies the following properties:

1. $X(j\omega)$ is a *continuous function* of ω.
2. $X(j\omega)$ is *bounded*. In fact, $|X(j\omega)| \leq \int_{-\infty}^{\infty} |x(t)|dt$.
3. $X(j\omega) \to 0$ as $\omega \to \pm\infty$. This is called the *Riemann–Lebesgue* lemma.

If a Fourier transform is a discontinuous function, we immediately conclude that it is not an L_1-FT, although it can possibly be an L_2-FT. For example, the ideal lowpass frequency response defined by

$$H(j\omega) = \begin{cases} 1 & -\omega_c < \omega < \omega_c, \\ 0 & \text{otherwise,} \end{cases} \tag{17.52}$$

is discontinuous at $\omega = \omega_c$, and therefore does not represent an L_1-FT. Rather, it is the L_2-FT (defined below) of the sinc function. We will see later that the L_1-FT $X(j\omega)$ itself may not be an L_1 function (Sec. 17.7).

17.4.2 The L_2 Fourier Transform

Now consider a signal $x(t)$ in $L_2(-\infty, \infty)$, which may not necessarily be in $L_1(-\infty, \infty)$. The truncated version

$$x_k(t) = \begin{cases} x(t) & -k < t < k, \\ 0 & \text{otherwise,} \end{cases} \tag{17.53}$$

is an $L_2[-k, k]$ signal and is therefore also an $L_1[-k, k]$ signal because of Eq. (17.43). So we can define its L_1 Fourier transform

$$X_k(j\omega) = \int_{-k}^{k} x(t)e^{-j\omega t}dt. \tag{17.54}$$

Since the energy of $X_k(j\omega)$ is equal to that of $x(t)$ in the interval $-k < t < k$, and since $x(t)$ has finite energy ($x(t) \in L_2$), it follows that $X_k(j\omega)$ is an L_2 function. It can be shown (e.g., see p. 200 of Rudin [1974]), that there exists an L_2 function $X(j\omega)$ such that

$$\lim_{k\to\infty} \int_{-\infty}^{\infty} \left| X_k(j\omega) - X(j\omega) \right|^2 \frac{d\omega}{2\pi} = 0. \tag{17.55}$$

That is, there exists $X(j\omega)$ such that the L_2 norm (or energy) of the difference $X_k(j\omega) - X(j\omega)$ goes to zero as the length of the support of $x_k(t)$ increases. This limit $X(j\omega)$ is called the L_2 Fourier transform or L_2-FT of the L_2 signal $x(t)$. Thus, any L_2 signal has an L_2 Fourier transform.

Pointwise convergence. If we replace $X(j\omega)$ with a function which differs from it on a countable set (i.e., a set of measure zero), the result $X_{new}(j\omega)$ is also regarded as the L_2-FT of $x(t)$ because the L_2 norm $\|X(j\omega) - X_{new}(j\omega)\|$ is zero. Similarly, if $x(t)$ is changed on a set of measure zero, it continues to be the inverse L_2-FT of $X(j\omega)$. Thus, in the theory of L_2-FT we do not talk about pointwise convergence of integrals. All convergence is in the L_2 sense. For the L_1-FT we will talk about pointwise convergence of the inverse FT (Sec. 17.7), and present conditions under which the inverse Fourier transform formula represents $x(t)$ pointwise. ▽ ▽ ▽

The L_2-FT has a number of important properties. We mention these below, along with some connections to the L_1-FT.

1. *Time–frequency symmetry.* The L_2-FT $X(j\omega)$ is itself an L_2 function, and furthermore the L_2 norms of $x(t)$ and $X(j\omega)$ are related by Parseval's relation:

$$\int_{-\infty}^{\infty} |X(j\omega)|^2 \frac{d\omega}{2\pi} = \int_{-\infty}^{\infty} |x(t)|^2 dt. \tag{17.56}$$

Since $X(j\omega)$ itself is in L_2, we can define its inverse L_2-FT, and it is equal to $x(t)$. Thus, there is complete time–frequency symmetry in the properties of an L_2 function and its Fourier transform. If the L_2-FT $X(j\omega)$ happens to also be an L_1 function, the inverse FT $\int_{-\infty}^{\infty} X(j\omega)e^{j\omega t}d\omega/2\pi$ can also be regarded as a Lebesgue integral.

2. *L_1-FT lacks symmetry.* Notice by contrast that the L_1-FT of an L_1 function is not necessarily an L_1 function (the FT of the pulse, which is a sinc, being the familiar example). Thus, the L_1-FT lacks the time–frequency symmetry enjoyed by the L_2-FT. Another asymmetry of the L_1-FT is that $X(j\omega)$ is a continuous function, although $x(t)$ itself may not be continuous.
3. *Consistency.* If $x(t)$ is in L_1 and L_2, then the L_1-FT is also an L_2-FT.
4. *Case when L_1-FT is in L_1.* Suppose $x(t)$ is in L_1, and assume the L_1-FT $X(j\omega)$ happens to be in L_1 as well. Then *both $x(t)$ and $X(j\omega)$ are in L_2 as well.* This can be proved as follows: since the L_1-FT is bounded, it follows that $X(j\omega)$ is in L_1 and L_∞. Using Eq. (17.42) it then follows that $X(j\omega)$ is in L_2, so $\int |X(j\omega)|^2 d\omega$ is finite. From Parseval's relation we see that $\int |x(t)|^2 dt$ is finite as well, so $x(t)$ is in L_2.

17.4.3 Convolutions and L_p Space

Let $x_1(t)$ be an L_p function $(1 \leq p \leq \infty)$ and $x_2(t)$ an L_1 function. Then the convolution

$$y(t) = \int_{-\infty}^{\infty} x_1(\tau)x_2(t-\tau)d\tau \tag{17.57}$$

exists and is an L_p function (see p. 158 of Rudin [1974]). Recall that an LTI system is stable if and only its impulse response is an L_1 signal. The following conclusions can be drawn based on the above:

1. An L_p signal input to a stable LTI system yields an L_p output.
2. In particular, an L_∞ signal input to a stable LTI system yields an L_∞ output. That is, a bounded input produces a bounded output whenever the impulse response $h(t)$ is in L_1.
3. The convolution of two L_1 signals is an L_1 signal. Two familiar consequences are: (a) an L_1 signal input to a stable LTI system produces an L_1 output; (b) the cascade of two stable LTI systems is stable.

We often use L_2 signals $x(t)$ as inputs to L_2 filters. As long as we use **bounded L_2 filters**, that is, $h(t) \in L_2$ and $|H(j\omega)| \leq B$ for all ω for some finite B, it can be shown that the output $y(t)$ is an L_2 function. The proof is similar to our arguments in Sec. 17.2.4. It also follows readily that the cascade of two bounded L_2 filters is a bounded L_2 filter.

17.5 Convergence of the Fourier Series

In the preceding chapters we assumed that the Fourier series representation works for any periodic (or finite-duration) signal and proceeded to study several properties and examples. We now look more closely into the problem of convergence of the Fourier series, and present several sets of sufficient conditions for the same. Given a periodic function $x(t)$ with period T, recall that its Fourier series is given by

$$x(t) = \sum_{k=-\infty}^{\infty} a_k e^{jk\omega_0 t}, \tag{17.58}$$

where $\omega_0 = 2\pi/T$. The main point of discussion in this section is the convergence of the infinite series given by the right-hand side, which is

$$\sum_{k=-\infty}^{\infty} a_k e^{jk\omega_0 t}. \tag{17.59}$$

Here, the a_k are the Fourier coefficients of $x(t)$:

$$a_k = \frac{1}{T}\int_{-T/2}^{T/2} x(t)e^{-jk\omega_0 t}dt. \tag{17.60}$$

In all earlier chapters, the integral above has implictly been assumed to exist. This is indeed the case if $x(t)$ is integrable in $[-T/2, T/2]$ in the Lebesgue sense (see Sec. 17.9). As reviewed in Sec. 17.9, this implies in particuar that $x(t)$ is absolutely integrable, that is,

$$\int_{-T/2}^{T/2} |x(t)|dt \tag{17.61}$$

exists, and furthermore that the integral in Eq. (17.60) exists. In the following discussions, "integrability" stands for Lebesgue integrability.

Convergence in the L_2 sense. From the results of Sec. 11.6.2 we can see that when $x(t)$ is an $L_2[-T/2, T/2]$ function (i.e., has finite energy in the interval $[-T/2, T/2]$), the partial sums

$$x_M(t) = \sum_{k=-M}^{M} a_k e^{jk\omega_0 t} \tag{17.62}$$

converge to $x(t)$ in the L_2 sense, as $M \to \infty$. That is,

$$\int_{-T/2}^{T/2} |x(t) - x_M(t)|^2 dt \to 0, \tag{17.63}$$

as $M \to \infty$. In this case it is said that the Fourier series summation (17.59) converges to $x(t)$ in the L_2 sense. The fact that any $L_2[-T/2, T/2]$ function can be written as a Fourier series means in particular that the functions $\{e^{jk\omega_0 t}\}$ form a basis for $L_2[-T/2, T/2]$, namely, the Fourier basis. In fact, $\{e^{jk\omega_0 t}/\sqrt{T}\}$ is an **orthonormal basis** for $L_2[-T/2, T/2]$ because of Eq. (11.29). Note that if a series converges to $x(t)$ in the L_2 sense, it also converges to any other function $y(t)$ which differs from $x(t)$ at a countable number of points. For example, we can define

$$y(t) = \begin{cases} x(t) + 1 & \text{when } t = \text{integer}, \\ x(t) & \text{otherwise.} \end{cases} \quad (17.64)$$

From the L_2 viewpoint, $x(t)$ and $y(t)$ are identical, that is,

$$\int_{-T/2}^{T/2} |x(t) - y(t)|^2 dt = 0.$$

For this reason, convergence in the L_2 sense does not mean that Eq. (17.59) converges to $x(t)$ *pointwise* for all t. ▽ ▽ ▽

The main topic of this section is **pointwise convergence**, that is, convergence for each specific value of t in $[-T/2, T/2]$. While L_2 convergence does not imply pointwise convergence, different types of sufficient conditions for pointwise convergence have been developed over the decades. The rest of this section is devoted to explaining some of these conditions.

17.5.1 Overview of Sufficient Conditions

We now summarize several sets of sufficient conditions for pointwise convergence of the Fourier series. These conditions involve a number of mathematical concepts such as "bounded variation," "piecewise continuity," "left-hand derivative," "left-hand limit," and so forth. For readers not familiar with these concepts, we will explain the meanings of these terms in Sec. 17.6, along with several examples. For proofs of the results stated in this section, the reader should refer to appropriate texts in mathematics cited throughout this section. In particular, most of the results reviewed here can be found in one or other of the following: Chapter 2 in Vol. 1 of Zygmund [1968]; Chapters 3 and 7 of Churchill and Brown [1987], Chapter 11 of Apostol [1974] and Chapter 11 of Haaser and Sullivan [1991].

17.5.1.1 Two Sets of Sufficient Conditions for Pointwise Convergence

The first result, based on the existence of one-sided derivatives, is mostly the work of Dini:

Theorem 17.1 *Existence of one-sided derivatives is sufficient.* Suppose $x(t)$ is integrable in $[-T/2, T/2]$ and has finite left-hand derivative $x'_L(t_0)$ and right-hand derivative $x'_R(t_0)$ at a point $t_0 \in [-T/2, T/2]$. Then the Fourier series summation (17.59) converges to

$$\frac{x(t_0^+) + x(t_0^-)}{2}, \quad (17.65)$$

where $x(t_0^-)$ and $x(t_0^+)$ denote the left and right-hand limits at t_0. In particular, if $x(t)$ is continuous at t_0, the expression (17.65) equals $x(t_0)$, so the Fourier series converges to $x(t_0)$. ◇

It will be seen in Sec. 17.6.2 that the existence of $x'_L(t_0)$ and $x'_R(t_0)$ implicitly means the existence of $x(t_0^-)$ and $x(t_0^+)$. When the left and right-hand derivatives $x'_L(t_0)$ and $x'_R(t_0)$ exist, we say that $x(t)$ is **almost differentiable** at t_0 (Chapter 11 of Haaser and Sullivan [1991]). Thus, *almost-differentiability is sufficient for pointwise convergence!* Notice in particular the following special cases of the above result:

1. If the left and right-hand derivatives exist everywhere, then the Fourier series converges to $[x(t^+) + x(t^-)]/2$ for all t. It is not necessary for $x(t)$ to be differentiable (or even continuous) at t.
2. If $x(t)$ is *differentiable* everywhere, then the Fourier series converges pointwise to $x(t)$ everywhere. This is because differentiability implies continuity as well as the existence of left and right-hand derivatives [Churchill and Brown, 1987].

The next result is based on the work of mathematicians Dirichlet and Jordan:

Theorem 17.2 *Bounded variation is sufficient.* If $x(t)$ is integrable in the interval $[-T/2, T/2]$ and has bounded variation in a region $[t_0 - \epsilon, t_0 + \epsilon]$ within the interval $[-T/2, T/2]$ for some $\epsilon > 0$, then the Fourier series evaluated at $t = t_0$ converges to the value $[x(t_0^+) + x(t_0^-)]/2$. ◇

We will see later that bounded variation implicitly guarantees the existence of $x(t_0^-)$ and $x(t_0^+)$ (Sec. 17.6.6). The theorem implies, in particular, that if $x(t)$ is *continuous and has bounded variation* in the entire interval $[-T/2, T/2]$, then the Fourier series converges pointwise to $x(t)$ everywhere (because $x(t^-) = x(t^+) = x(t)$ for continuous functions).

From the preceding two results we see that the convergence of the Fourier series at a point t_0 depends only on the *local behavior* of $x(t)$ in a neighborhood of that point. Almost-differentiability at t_0 is sufficient. So is bounded variation in some neighborbood of t_0.

17.5.1.2 Cesàro Summability

A third type of convergence result was proved by Fejér in 1903 after the development of the Lebesgue integral in 1902. To describe this we first need some definitions. The Fourier series (17.59) is an infinite sum. Define the mth partial sum by

$$S_m(t) = \sum_{k=-m}^{m} a_k e^{jk\omega_0 t}, \tag{17.66}$$

and let $\mathcal{C}_N(t)$ denote the arithmetic mean of the first N partial sums:

$$\mathcal{C}_N(t) = \frac{1}{N} \sum_{m=0}^{N-1} S_m(t). \tag{17.67}$$

If $\mathcal{C}_N(t_0)$ has a limit $\mathcal{C}(t_0)$ as $N \to \infty$, we say that the series (17.59) is Cesàro summable to $\mathcal{C}(t_0)$ at t_0. Fejér proved that the Fourier series is always Cesàro summable:

Theorem 17.3 *Cesàro summability.* If $x(t)$ is integrable in $[-T/2, T/2]$, the arithmetic mean $C_N(t_0)$ of partial sums of the Fourier series converges to

$$\frac{x(t_0^+) + x(t_0^-)}{2}, \tag{17.68}$$

for each $t_0 \in [-T/2, T/2]$, as long as $x(t_0^+)$ and $x(t_0^-)$ exist. ◇

So the Fourier series is always Cesàro summable to $[x(t_0^+) + x(t_0^-)]/2$, which reduces to $x(t_0)$ when $x(t)$ is continuous at t_0. No sophisticated conditions such as bounded variation or one-sided differentiability are required. Integrability is necessary anyway, to ensure the existence of Fourier series coefficients (17.60). The existence of $x(t_0^+)$ and $x(t_0^-)$ is also a minimal expectation, for otherwise there will be nothing to compare the value of the Fourier series (17.59) with!

It should be noticed that the sum (17.67) is nothing but the truncated Fourier series with a_n multiplied by $(N-n)/N$. That is, we apply a triangular set of weights to the coefficients. As N gets larger, the triangle gets wider, which means that for small n, coefficients are nearly unaltered and for large n they are weighted down linearly.

17.5.2 Other Results on Convergence

In Sec. 17.6 we will explain Theorems 17.1 and 17.2 in greater detail, with examples. We conclude this section by mentioning two more results.

1. *Convergence almost everywhere.* A deep result due to Carleson (proved in 1966, quite late in the history of Fourier series) states that as long as $x(t)$ has finite energy (i.e., it is an $L_2[-T/2, T/2]$ function, which is a light requirement), the Fourier series converges to $x(t)$ pointwise "almost everywhere." This means that pointwise convergence can fail at most on a set of measure zero, such as a countable set. For example, the set of integers is countable, and so is the set of all *rational numbers* (ratios of integers such as $7/8, -11/21$, and so on). Convergence "almost everywhere," while mathematically beautiful, can be less interesting to the engineer for reasons explained at the end of Sec. 17.2.2.
2. *Uniform convergence.* For readers familiar with uniform convergence we mention two more results. Recall that $S_m(t)$ in Eq. (17.66) denotes the mth partial sum in the Fourier series. We say that the Fourier series converges to $x(t)$ *uniformly* in an interval $a \le t \le b$ if the following is true: given $\epsilon > 0$, we can find a large enough integer M such that for all $m \ge M$,

$$|S_m(t) - x(t)| < \epsilon, \quad \text{for all } t \text{ such that } a \le t \le b. \tag{17.69}$$

This is a stronger requirement than pointwise convergence. Two results will now be mentioned. Assume as usual that $x(t)$ is integrable in the interval $[-T/2, T/2]$.

(a) If $x(t)$ is continuous in $[-T/2, T/2]$ with $x(-T/2) = x(T/2)$, and its derivative is piecewise continuous in $(-T/2, T/2)$, then the Fourier series summation converges to $x(t)$ uniformly and absolutely everywhere in $[-T/2, T/2]$ (see Chapter 3 of Churchill and Brown [1987] for a proof).

(b) If $x(t)$ is continuous in $[-T/2, T/2]$ and has bounded variation in $(-T/2, T/2)$, then the Fourier series converges uniformly in the interval $[-T/2, T/2]$ (see p. 57, vol. 1 of Zygmund [1968]).

From the results of this section we see that, in practical applications, the convergence of the Fourier series is not an issue. Most engineering texts on Fourier series do not therefore get into a deep discussion of this topic. Mathematicians, however, have constructed many clever examples of **divergent Fourier series**. In fact, Kolmogorov has constructed examples of Fourier series which diverge almost everywhere! A detailed discussion of many such constructions can be found in Chapter 8 in Vol. 1 of Zygmund [1968]. Kolmogorov's construction does not contradict Carleson's result stated above, because Kolmogorov's example is for L_1 functions whereas Carleson's results are for L_2 functions.

17.6 Elaboration of Convergence Results

For completeness, we now elaborate on some of the mathematical terms that are used in Sec. 17.5, and provide a number of examples to demonstrate these.

17.6.1 Continuity

Given a function $x(t)$, define

$$x(t_0^-) \triangleq \lim_{t \to t_0^-} x(t). \tag{17.70}$$

This is the limit as t approaches t_0 from below (i.e., $t \to t_0$ with $t < t_0$), and is called the **left-hand limit** at t_0. Similarly,

$$x(t_0^+) \triangleq \lim_{t \to t_0^+} x(t) \tag{17.71}$$

is the *right-hand limit.* We say that $x(t)$ is **continuous** at t_0 if the above limits exist, $x(t_0)$ itself is defined, and furthermore

$$x(t_0^+) = x(t_0^-) = x(t_0). \tag{17.72}$$

The student must have seen many examples of continuous and discontinuous functions. Here are a few examples for completeness:

1. Consider the function $x(t) = \sin(1/t), t \neq 0$, with $x(0) = 0$. This function is plotted in Fig. 17.5 for $t > 0$. Note that it oscillates faster and faster as $t \to 0^+$, and the limit $x(0^+)$ does not exist. Similarly, $x(0^-)$ does not exist. So $x(t)$ is not continuous at $t = 0$, although it is continuous for other values of t.
2. Consider the function $x(t) = t\sin(1/t), t \neq 0$, with $x(0) = 0$. Even though $\sin(1/t)$ oscillates faster and faster as $t \to 0^+$, it satisfies $-1 \leq \sin(1/t) \leq 1$, so that $t\sin(1/t)$ gets smaller and smaller as $t \to 0^+$. So the limits $x(0^+)$ and $x(0^-)$ exist, and

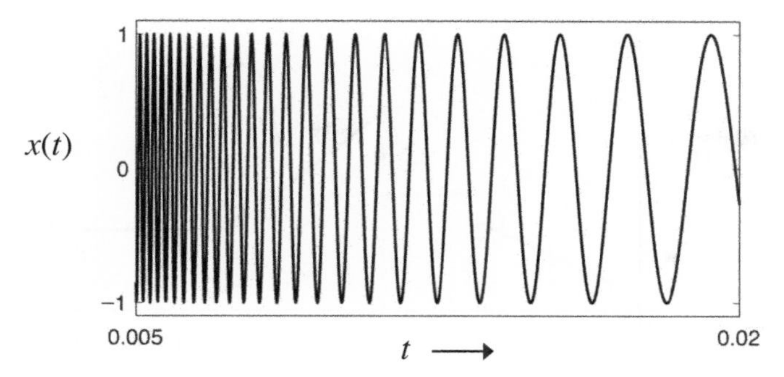

Figure 17.5 Plot of the function $x(t) = \sin(1/t)$, for $t > 0$. This function oscillates faster and faster as $t \to 0^+$, and the limit $x(0^+)$ does not exist. Neither does $x(0^-)$. So the function is not continuous at $t = 0$.

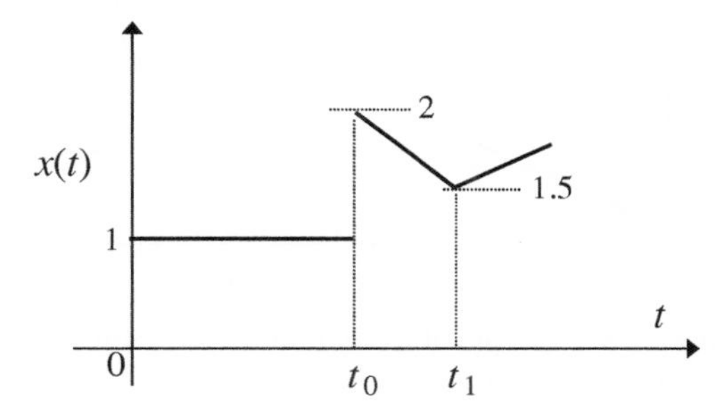

Figure 17.6 A function discontinuous at $t = t_0$. We have $x(t_0^-) = 1$ and $x(t_0^+) = 2$. This is an irremovable discontinuity. Note that $x(t_1^-) = x(t_1^+) = 1.5 = x(t_1)$, so $x(t)$ is continuous at t_1.

$$x(0^+) = x(0^-) = x(0) = 0. \tag{17.73}$$

So $x(t)$ is continuous at $t = 0$, and therefore continuous everywhere. If $x(0)$ had been defined to be something else, say $x(0) = 2$, then the function would be discontinuous at $t = 0$, but this is a **removable discontinuity** [Apostol, 1974]. That is, we can make $x(t)$ continuous at $t = 0$ simply by redefining $x(0)$.

3. Consider the function shown in Fig. 17.6. In this case, $x(t_0^-) = 1$ and $x(t_0^+) = 2$, and these are unequal. So the function is discontinuous at t_0. This is an irremovable discontinuity because we cannot achieve continuity just by redefining $x(t_0)$. At the point t_1 there is a sharp corner but the function is still continuous, since $x(t_1^+) = x(t_1^-) = x(t_1) = 1.5$.

A function $x(t)$ is said to be **piecewise continuous** in an interval $[a, b]$ if we can partition $[a, b]$ into a finite number of open subintervals

$$a < t < t_1, \quad t_1 < t < t_2, \quad \ldots, \quad t_{M-1} < t < b,$$

such that (a) $x(t)$ is continuous in each open subinterval and (b) $x(t)$ has finite limits as t approaches the boundary point of any of these intervals from the interior of

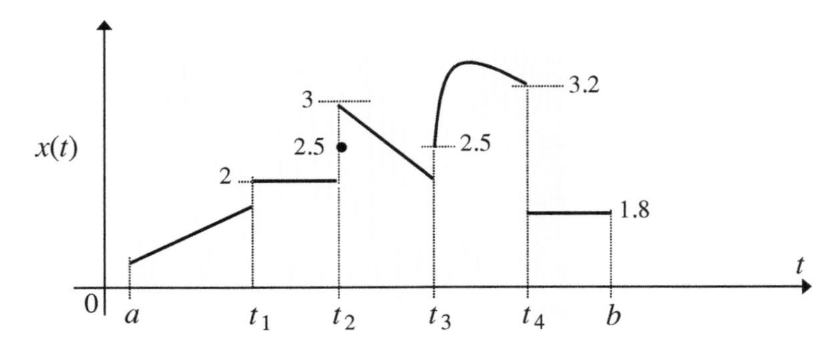

Figure 17.7 Example of a piecewise continuous function.

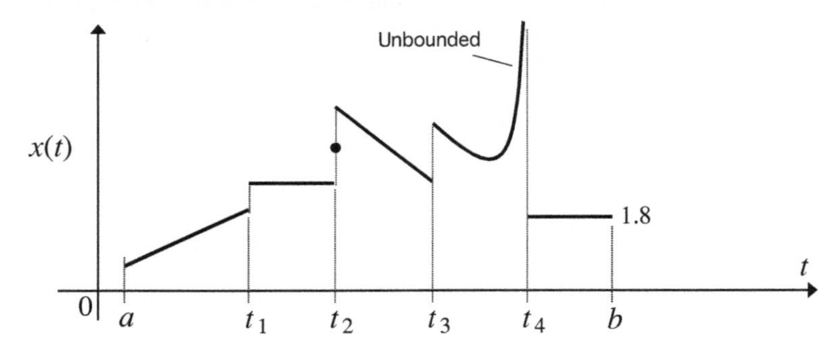

Figure 17.8 Example of a function which is not piecewise continuous because $x(t_4^-)$ is not finite.

the interval. At the boundary points $a, t_1, t_2, \ldots$, the function $x(t)$ can be discontinuous and may or may not even be defined. Figure 17.7 shows an example. In the example, the limits as t approaches some of the boundary points from the interior are indicated:

$$x(t_2^-) = 2, \; x(t_2^+) = 3, \; x(t_4^-) = 3.2, \; x(t_4^+) = 1.8, \tag{17.74}$$

and so on. All these limits exist and are finite. In the example, $x(t_2) = 2.5$ and $x(t_k)$ is undefined for other k. Figure 17.8 shows an example which is not piecewise continuous because the limit $x(t_4^-)$ does not exist, although $x(t_4^+) = 1.8$.

17.6.2 One-Sided Derivatives

Let $x(t)$ have the left-hand limit $x(t_0^-)$, and define

$$x'_L(t_0) = \lim_{t \to t_0^-} \frac{x(t) - x(t_0^-)}{t - t_0}. \tag{17.75}$$

Note that $t \to t_0^-$ means that t approaches t_0 from below, maintaining $t < t_0$ all the while. If the above limit exists, it is called the **left-hand derivative** at t_0. This definition presupposes the existence of the left-hand limit $x(t_0^-)$. So whenever this derivative exists, so does $x(t_0^-)$, implicitly. Similarly, the limit

$$x'_R(t_0) = \lim_{t \to t_0^+} \frac{x(t) - x(t_0^+)}{t - t_0} \tag{17.76}$$

is called the **right-hand derivative** at t_0 if it exists. Its existence implicitly also means that $x(t_0^+)$ exists. These one-sided derivatives may exist even when the ordinary (two-sided) derivative does not exist. Recall that the ordinary derivative of $x(t)$ at t_0 is defined as the limit

$$x'(t_0) = \lim_{t \to t_0} \frac{x(t) - x(t_0)}{t - t_0}. \tag{17.77}$$

Its existence presupposes that $x(t)$ is continuous at t_0, that is, $x(t_0^-)$ and $x(t_0^+)$ exist, $x(t_0)$ is defined, and $x(t_0^-) = x(t_0^+) = x(t_0)$. It can be shown [Churchill and Brown, 1987] that whenever the derivative $x'(t_0)$ exists, so do $x'_L(t_0)$ and $x'_R(t_0)$, and furthermore,

$$x'_L(t_0) = x'_R(t_0) = x'(t_0). \tag{17.78}$$

Referring to Fig. 17.7, the quantity $x'_L(t_1)$ is simply the slope of the plot as we approach t_1 from the left, so $x'_L(t_1) > 0$. Next, $x'_R(t_1)$ is the slope of the plot just to the right of t_1, so that $x'_R(t_1) = 0$. Note that the ordinary derivative $x'(t_1)$ does not exist because $x(t)$ is not even continuous at t_1. In this example, $x'_L(t_k)$ and $x'_R(t_k)$ exist and are finite at the four discontinuities. Furthermore, $x'_R(a) > 0$ and $x'_L(b) = 0$ at the boundaries. Refer next to Fig. 17.8. This has all the properties of the function in Fig. 17.7, with the exception that $x(t_4^-)$ and $x'_L(t_4)$ do not exist as finite limits.

17.6.3 Fourier Series of Piecewise Continuous Functions

It is clear that Fig. 17.7 shows a piecewise continuous function in the interval $[a, b]$ with finite $x'_L(t)$ and $x'_R(t)$ at discontinuities and at the boundaries. Furthermore, it is differentiable in the interior of the continuous regions, so in the interior regions $x'_L(t) = x'_R(t) = x'(t)$. Such a function, which is *piecewise continuous and differentiable within each region of continuity*, clearly has a pointwise convergent Fourier series, as seen from an application of Theorem 17.1. In the example where $x(t) = \sin(1/t)$, since the derivative is not bounded as $t \to 0^+$, the function does not satisfy the above-mentioned property.

This important result on piecewise continuous functions is formally presented as a theorem in some books [Churchill and Brown, 1987; Kreyszig, 1972]. In the example of Fig. 17.8, since $x'_L(t_4)$ does not exist, we cannot apply Theorem 17.1 at the point t_4. As long as the unbounded behavior at t_4 does not disallow integrability of $x(t)$ (see discussion at the beginning of Sec. 17.2.2), the Fourier series exists. Furthermore, Theorem 17.1 can still be applied to conclude that there is pointwise convergence at points other than t_4.

17.6.3.1 A Rather Subtle Example

Consider a period-T function defined in $[-T/2, T/2]$ by

$$x(t) = \begin{cases} t^2 \sin(1/t) & t \neq 0, \\ 0 & t = 0. \end{cases} \tag{17.79}$$

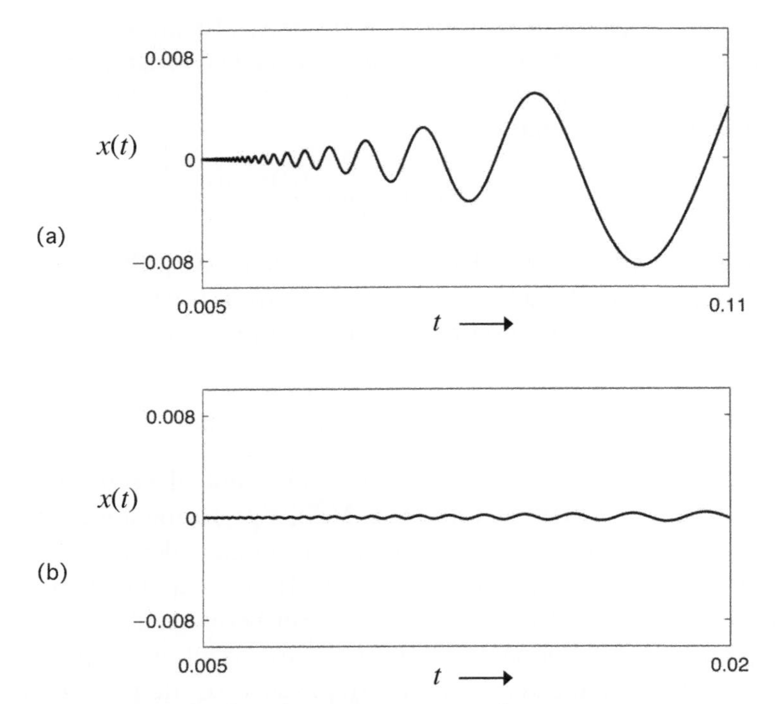

Figure 17.9 (a) A plot of the function $x(t) = t^2 \sin(1/t)$ for $t > 0$; (b) details closer to $t = 0$. With $x(0)$ defined to be 0, this is continuous and differentiable everywhere. In particular, $x(0^-) = x(0^+) = x(0) = 0$ and $x'_L(0) = x'_R(0) = x'(0) = 0$. But the limits $x'(0^-)$ and $x'(0^+)$ do not exist. See text.

The $\sin(1/t)$ factor oscillates faster and faster as $t \to 0^+$. But in this example, the factor t^2 attenuates the function to zero as $t \to 0^+$ or 0^- (Fig. 17.9). So $x(0^+) = x(0^-) = x(0) = 0$, and the function is continuous at $t = 0$.

Observe next that for $t \neq 0$ the ordinary (two-sided) derivative is

$$x'(t) = 2t \sin(1/t) - \cos(1/t). \tag{17.80}$$

For $t \to 0^+$ the first term vanishes, but the second term oscillates faster and faster, approaching no limiting value. That is, the limit $x'(0^+)$ (and similarly $x'(0^-)$) does not exist. This leads us to suspect that the Fourier series may not converge at $t = 0$. However, in this example, the *one-sided derivatives* exist! Thus, the right-hand derivative at $t = 0$ is

$$x'_R(0) = \lim_{t \to 0^+} \frac{x(t) - x(0^+)}{t} = \lim_{t \to 0^+} t \sin(1/t) = 0, \tag{17.81}$$

and similarly $x'_L(0) = 0$. Thus, the conditions of Theorem 17.1 for pointwise convergence are satisfied at $t = 0$ as well. Since $x'(t)$ has finite value for all $t \neq 0$, the same theorem shows that the Fourier series converges pointwise everywhere.

In this particular example it turns out that the ordinary (two-sided) derivative $x'(t)$ exists at $t = 0$, though the limits $x'(0^+)$ and $x'(0^-)$ do not! To see this, simply observe that

$$x'(0) = \lim_{t\to 0} \frac{x(t) - x(0)}{t} = \lim_{t\to 0} t\sin(1/t) = 0. \tag{17.82}$$

This example demonstrates the subtle fact (see p. 72 of Churchill and Brown [1987]) that the existence of $x'(t_0)$ implies that of $x'_R(t_0)$ and $x'_L(t_0)$, but does not imply the existence of the one-sided limits $x'(t_0^+)$ and $x'(t_0^-)$. And Theorem 17.1 only uses the existence of $x'_R(t_0)$ and $x'_L(t_0)$ (but not $x'(t_0^+)$ and $x'(t_0^-)$) for pointwise convergence.

17.6.4 Fourier Series of Unbounded Functions

An interesting example of an unbounded periodic function was presented in Tolstov [1962]. We now point out the main features of this example, referring the reader to Tolstov for detailed derivations. Thus, let

$$x(t) = -\ln|2\sin(t/2)|. \tag{17.83}$$

Whenever $t = 2k\pi$ for integer k, we have $\sin(t/2) = 0$ and the logarithm becomes unbounded. The plot of $x(t)$ is shown in Fig. 17.10, and it is a periodic function with period 2π. Even though $x(t)$ is unbounded, it can be shown that it is integrable in its period $[-\pi, \pi]$. So its Fourier series can be computed and the result is [Tolstov, 1962]

$$\cos t + \frac{\cos 2t}{2} + \frac{\cos 3t}{3} + \cdots. \tag{17.84}$$

Since $x(t)$ is differentiable everywhere except at $t = 2k\pi$, we use Theorem 17.1 to conclude that this Fourier series converges to $x(t)$ pointwise for all $t \neq 2\pi k$. Furthermore, at $t = 2\pi k$ the series reduces to

$$1 + \frac{1}{2} + \frac{1}{3} + \cdots.$$

The partial sums of this series get monotonically larger and larger in an unbounded manner, so the above sum can be regarded as ∞.[3] Thus we can write the Fourier representation

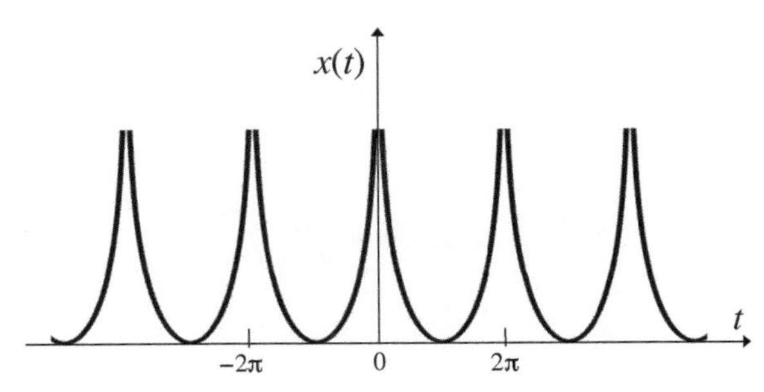

Figure 17.10 Plot of the function $x(t) = -\ln|2\sin(t/2)|$. This is unbounded at $t = 2\pi k$ for all integer k.

[3] This is unlike the infinite series $1 - 1 + 1 - 1\ldots$ which does not converge nor does it "blow up" to ∞.

$$-\ln|2\sin(t/2)| = \cos t + \frac{\cos 2t}{2} + \frac{\cos 3t}{3} + \cdots, \qquad -\infty < t < \infty, \quad (17.85)$$

with the understanding that both sides become ∞ at $t = 2k\pi$. This example shows that even if $x(t)$ is unbounded at a certain isolated point in $[-T/2, T/2]$, the Fourier series can sometimes exist, and furthermore exhibit pointwise convergence! The fact that some unbounded functions are integrable to a finite answer should not come as a surprise to the reader. For example, see Eq. (17.26) and discussions therein.

17.6.5 Functions with Bounded Variation

From Theorem 17.2 we know that if $x(t)$ has a property called "bounded variation," then the Fourier series converges pointwise. We now explain what this property means. The reader will realize after this discussion that bounded functions that do not have bounded variation are quite rare. Thus, practical signals encountered in science and engineering always have pointwise convergent Fourier series. Let us begin with an example. Thus, consider again the function

$$x_1(t) = \begin{cases} \sin(1/t) & t \neq 0, \\ 0 & t = 0, \end{cases} \quad (17.86)$$

reproduced in Fig. 17.11(a). Given a finite interval like $[-T/2, T/2]$ including the origin, it is clear that the function shows more and more "variation" (faster oscillations) as $t \to 0$. Compare this with the function

$$x_2(t) = \begin{cases} t^2\sin(1/t) & t \neq 0, \\ 0 & t = 0. \end{cases} \quad (17.87)$$

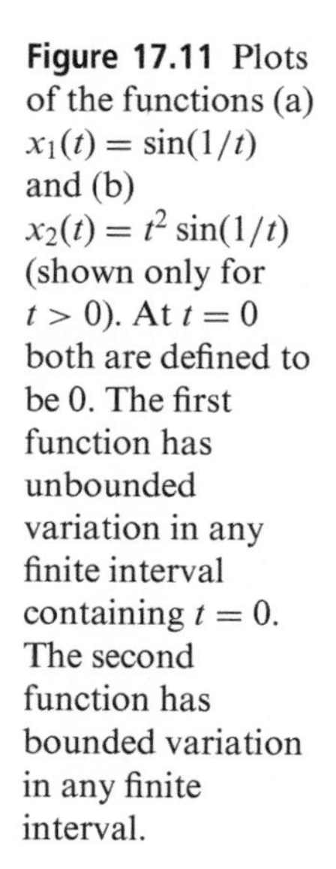

Figure 17.11 Plots of the functions (a) $x_1(t) = \sin(1/t)$ and (b) $x_2(t) = t^2\sin(1/t)$ (shown only for $t > 0$). At $t = 0$ both are defined to be 0. The first function has unbounded variation in any finite interval containing $t = 0$. The second function has bounded variation in any finite interval.

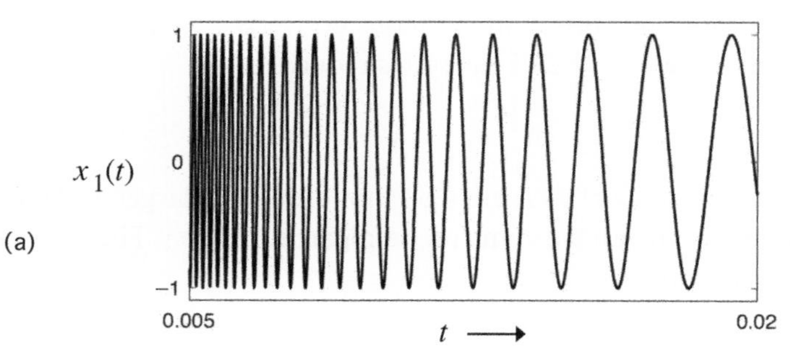

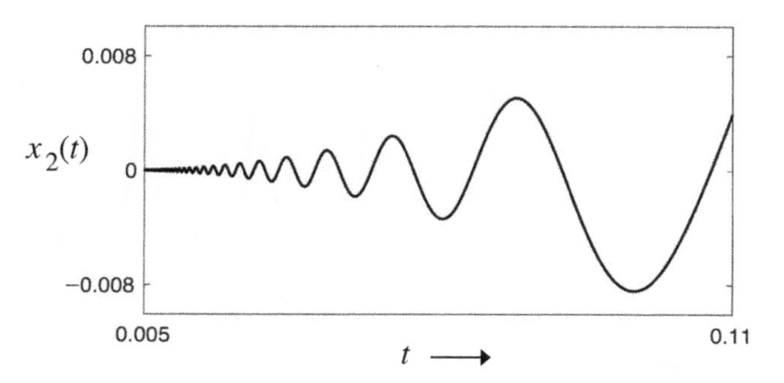

This is plotted in Fig. 17.11(b). Even though there is faster oscillation as $t \to 0$, the function t^2 damps the amplitude. So the "variation" of $x_2(t)$ in a finite interval $[-T/2, T/2]$ is quite small compared to that of $x_1(t)$. In mathematics there is a way to define variation formally, according to which $x_1(t)$ has infinite (unbounded) variation, whereas $x_2(t)$ has bounded variation.

Let $x(t)$ be defined everywhere in the interval $[a, b]$. Let $t_0, t_1, \ldots, t_n$ be such that

$$a = t_0 < t_1 < t_2 < \cdots < t_n = b. \tag{17.88}$$

Then the set of numbers $\{t_n\}$ defines a **partition** of the interval $[a, b]$ into n subintervals $[t_{i-1}, t_i]$. Given such a partition $\mathcal{P}$, the quantity

$$\mathcal{V}_{\mathcal{P}}(x) = \sum_{i=1}^{n} |x(t_{i-1}) - x(t_i)| \tag{17.89}$$

is called the variation of $x(t)$ with respect to the partition $\mathcal{P}$. The maximum possible value of this variation, as $\mathcal{P}$ is varied over all possible partitions, is called the **total variation** of $x(t)$ in $[a, b]$. Formally,

$$\mathcal{V}(x) = \sup_{\mathcal{P}} \mathcal{V}_{\mathcal{P}}(x) \qquad \text{(total variation)}, \tag{17.90}$$

where "sup" stands for *supremum* (see footnote at the beginning of Sec. 17.2).

Definition 17.1 *Bounded variation.* Let $x(t)$ be defined in the closed interval $[a, b]$. If its total variation $\mathcal{V}(x)$ in $[a, b]$ is finite, then we say that $x(t)$ has bounded variation in $[a, b]$. $\diamond$

It will be seen as part of the discussions below that $x_1(t)$ in Fig 17.11(a) has unbounded variation, whereas $x_2(t)$ (Fig 17.11(b)) has bounded variation. The acronym *BV* is used to denote bounded variation.

17.6.6 Properties of Functions with Bounded Variation

Functions with BV have several beautiful properties. Some are summarized next without proof. For proofs, the interested reader should consult the lucid expositions given in Apostol [1974] and Haaser and Sullivan [1991]. All discussions below are for a function $x(t)$ defined everywhere in $[a, b]$.

1. *Bounded variation implies boundedness.* If $x(t)$ has bounded variation in $[a, b]$, then it is bounded in $[a, b]$. The converse is not true. For example, $x_1(t)$ in Fig. 17.11 is bounded but does not have bounded variation.
2. *Relation to increasing functions.* $x(t)$ has bounded variation if and only if it is a difference of two monotone increasing functions. This is a strikingly beautiful property because increasing functions[4] are easy to understand. To a large extent this demystifies the concept of bounded variation. Notice in particular that *a monotone increasing or decreasing function is automatically of bounded variation.*

[4] $y(t)$ is said to be increasing if $y(t + \epsilon) \geq y(t)$ for all $\epsilon > 0$ and all t.

3. *Bounded variation implies one-sided limits exist.* If $x(t)$ has bounded variation in $[a, b]$, then $x(t_0^+)$ and $x(t_0^-)$ exist for all t_0 in (a, b). This follows from a similar property for monotone functions (see p. 95 of Apostol [1974]). In the example $x_1(t)$ of Fig. 17.11, since $x_1(0^+)$ and $x_1(0^-)$ do not exist, $x_1(t)$ has unbounded variation.
4. *Bounded variation implies continuity and differentiability a.e.* If $x(t)$ has bounded variation then the number of discontinuities is countable. So it has to be continuous almost everywhere. Similarly, it is differentiable almost everywhere (see Sec. 4 in Chapter 9 of Haaser and Sullivan [1991]).
5. *Bounded variation implies integrability.* If $x(t)$ has bounded variation in $[a, b]$ it is integrable (both in the Riemann and the Lebesgue sense) on $[a, b]$. In particular, the Fourier coefficients exist (see p. 232 of Haaser and Sullivan [1991]).
6. *Bounded derivatives imply bounded variation.* If $x(t)$ is continuous in $[a, b]$ and has bounded derivative in the interior ($|x'(t)| \leq B < \infty$ in (a, b)), then $x(t)$ has bounded variation in $[a, b]$. The function $x_2(t)$ in Fig. 17.11 has derivative

$$x_2'(t) = 2t \sin(1/t) - \cos(1/t), \tag{17.91}$$

which is bounded in any open interval of the form $(0, b)$ or $(a, 0)$. So $x_2(t)$ has bounded variation in the closed intervals $[a, 0]$ and $[0, b]$, and therefore in $[a, b]$.

In the preceding discussions we saw that

$$x_1(t) = \begin{cases} \sin(1/t) & t \neq 0, \\ 0 & t = 0, \end{cases} \tag{17.92}$$

does not have bounded variation in an interval which includes 0 because a necessary condition is violated ($x_1(0^+)$ and $x_1(0^-)$ do not exist). On the other hand, we found that

$$x_2(t) = \begin{cases} t^2 \sin(1/t) & t \neq 0, \\ 0 & t = 0, \end{cases} \tag{17.93}$$

is of bounded variation in any interval because it satisfies a sufficient condition (boundedness of derivatives). How about the function

$$x(t) = \begin{cases} t \sin(1/t) & t \neq 0, \\ 0 & t = 0, \end{cases} \tag{17.94}$$

which has t instead of t^2 in front of $\sin(1/t)$? Once again $\sin(1/t)$ has faster and faster oscillations as $t \to 0$, but $t \sin(1/t)$ does approach zero as $t \to 0+$ or $0-$. Thus, $x(0^+) = x(0^-) = x(0) = 0$, implying continuity. Though a necessary condition for bounded variation is satisfied, this function **does not** have bounded variation. This can be shown by explicitly constructing a partition with n subintervals in the fixed interval $[a, b]$ such that the variation increases unboundedly as n increases [Apostol, 1974].

Applying the definition of one-sided derivatives given in Sec. 17.6.2, we find that for the function $x(t)$ in Eq. (17.94), $x_R'(0) = \lim_{t \to 0^+} \sin(1/t)$, which does not exist.

Thus $x'_R(0)$, and similarly $x'_L(0)$, do not exist in this case. Notice also that since $x'(t) = \sin(1/t) - t^{-1}\cos(1/t)$ for $t \neq 0$, it follows that $x'(0^+)$ and $x'(0^-)$ do not exist either. Since $x'(0) = \lim_{t\to 0} x(t)/t = \lim_{t\to 0}\sin(1/t)$, it does not exist.

Bounded variation is not necessary. In Sec. 17.6.4 we presented the example of an unbounded function and its Fourier series. See Eq. (17.85). In this example there is pointwise convergence. At the points $t = 2\pi k$ where $x(t)$ is unbounded, the Fourier series (right-hand side of Eq. (17.85)) is also unbounded. This is an example where the Fourier series represents $x(t)$ pointwise everywhere, even though $x(t)$ does not have bounded variation (because unbounded functions cannot have bounded variation; see Property 1 above). In this sense, *bounded variation is also not necessary*, although it is *sufficient* for the pointwise convergence of the Fourier series. ▽ ▽ ▽

17.6.6.1 The Cube-Root Function

Given any number z, there are three cube roots. For example, the cube roots of -1 are $\{-1, e^{j\pi/3}, e^{-j\pi/3}\}$ and those of 1 are $\{1, e^{j2\pi/3}, e^{-j2\pi/3}\}$. For real time-argument t, let us define

$$x(t) = t^{1/3}$$

to be the unique real-valued cube root. This is sketched in Fig. 17.12. The right-hand derivative at $t = 0$ is

$$x'_R(0) = \lim_{t\to 0+} \frac{x(t) - x(0^+)}{t - 0} = \lim_{t\to 0+} \frac{t^{1/3} - 0}{t - 0} = \lim_{t\to 0+} \frac{1}{t^{2/3}}.$$

This limit does not exist (as a finite value). The same is true of $x'_L(0)$. Thus the sufficient condition required by Theorem 17.1 for pointwise convergence of Fourier series is not satisfied.

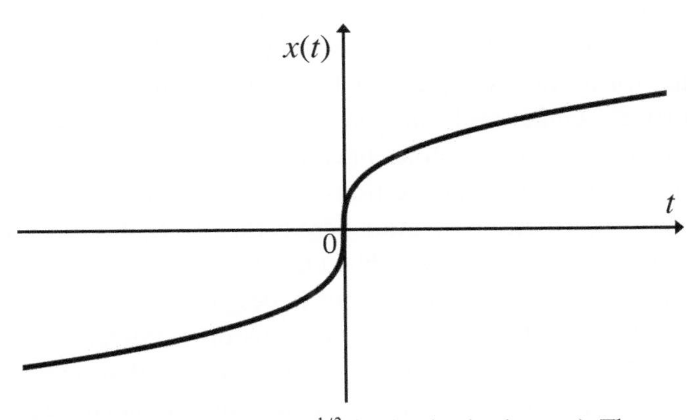

Figure 17.12 Plot of $x(t) = t^{1/3}$ (real-valued cube root). The one-sided derivatives do not exist at $t = 0$ but the function has bounded variation, so the sufficient conditions of Theorem 17.2 for pointwise convergence of the Fourier series are satisfied.

Observe, however, that since $x(t)$ is a monotone increasing function, it has bounded variation in any interval $[-T/2, T/2]$ (see second property at the beginning of this section). So it satisfies the sufficient conditions required by Theorem 17.2. The Fourier series of $x(t)$ (limited to $[-T/2, T/2]$, and repeating periodically) therefore converges pointwise everywhere, including at $t = 0$. This example therefore demonstrates a function which has bounded variation (as required by Theorem 17.2) but does not have one-sided derivatives (as required by Theorem 17.1).

17.7 The Fourier Integral

In this section we give an overview of the mathematical conditions under which a nonperiodic function $x(t)$ in $-\infty < t < \infty$ can be represented as a Fourier integral. Recall (Sec. 17.3) that an L_1 function $x(t)$ is any function for which $\int |x(t)|dt$ exists and is finite. If the integral is regarded as a Lebesgue integral (see Sec. 17.9), then an L_1 function and a Lebesgue-integrable function are essentially synonymous. If $x(t)$ is Lebesgue integrable in $-\infty < t < \infty$, then the L_1 Fourier transform (Sec. 17.4) exists and is given by

$$X(j\omega) = \int_{-\infty}^{\infty} x(t)e^{-j\omega t}dt. \tag{17.95}$$

The inverse Fourier transform was introduced in Chapter 9, and has the form

$$x(t) = \frac{1}{2\pi}\int_{-\infty}^{\infty} X(j\omega)e^{j\omega t}d\omega. \tag{17.96}$$

This is also called the Fourier integral representation of $x(t)$. For fixed t, say $t = t_0$, what are the conditions under which the integral on the right, namely,

$$\frac{1}{2\pi}\int_{-\infty}^{\infty} X(j\omega)e^{j\omega t}d\omega, \tag{17.97}$$

exists and represents $x(t_0)$? This is the pointwise convergence problem (more correctly, the pointwise Fourier representation problem). Convergence was taken for granted for all the examples in Chapter 9. For the case of the Fourier series, we considered a similar problem of pointwise convergence of the sum $\sum_k a_k e^{jk\omega_0 t}$, and summarized two sets of sufficient conditions in Sec. 17.5.1. In this section we shall present similar sufficient conditions for the Fourier integral.

A very important observation here is that an L_1 Fourier transform $X(j\omega)$ itself need not be a Lebesgue-integrable function. For example, if $x(t)$ is the pulse, its L_1-FT is the sinc function which is not Lebesgue integrable (Sec. 17.9). This implies, in particular, that the inverse Fourier transform integral (17.96) does not in general exist as a Lebesgue integral. It turns out, however, that since $x(t)$ is an L_1 function, its Fourier transform $X(j\omega)$ is continuous and bounded (Sec. 17.4.1). This implies that the integral

$$\int_{-a}^{a} X(j\omega)e^{j\omega t}d\omega \tag{17.98}$$

exists for any finite $a > 0$ as a Reimann integral, hence as a Lebesgue integral [Haaser and Sullivan, 1991]. If the signal $x(t)$ satisfies certain additional conditions, then the limit

$$\lim_{a\to\infty} \int_{-a}^{a} X(j\omega)e^{j\omega t}\,\frac{d\omega}{2\pi} \tag{17.99}$$

exists for any t, and furthemore this limit is exactly equal to $x(t)$ (or the average of $x(t^+)$ and $x(t^-)$ where there is discontinuity). After the following interlude we will present two examples of such sufficient conditions.

Improper integrals and Cauchy principal values. The limit in Eq. (17.99) is called the Cauchy principal value [Apostol, 1974; Kreyszig, 1972]. It should not be confused with the Lebesgue integral $\int_{-\infty}^{\infty} X(j\omega)e^{j\omega t}d\omega/2\pi$, which exists only when the Fourier transform $X(j\omega)$ is itself an L_1 function. Nor should it be confused with the limit of

$$\int_{-a}^{b} X(j\omega)e^{j\omega t}\frac{d\omega}{2\pi}, \tag{17.100}$$

as a and b separately go to ∞; such a limit would be called the *improper Riemann integral* [Apostol, 1974]. For example, consider the integral $\int_{-\infty}^{\infty} \omega d\omega$. Interpreted as an improper Riemann integral, this is

$$\int_{-\infty}^{\infty} \omega d\omega = \lim_{a\to\infty} \int_{-a}^{0} \omega d\omega \quad + \quad \lim_{b\to\infty} \int_{0}^{b} \omega d\omega, \tag{17.101}$$

which does not exist. But if it is interpreted as a Cauchy principal value, then

$$\int_{-\infty}^{\infty} \omega d\omega = \lim_{a\to\infty} \int_{-a}^{a} \omega d\omega = 0. \tag{17.102}$$

Thus the Cauchy principal value exists whereas the improper Riemann integral does not. $\nabla\ \nabla\ \nabla$

Here is the result on the inverse Fourier transform, which is the counterpart of Theorem 17.2:

Theorem 17.4 *Bounded variation is sufficient.* Suppose $x(t)$ is Lebesgue integrable in $(-\infty, \infty)$, so that the L_1 Fourier transform $X(j\omega)$ exists. If $x(t)$ has bounded variation in a region $[t_0 - \epsilon, t_0 + \epsilon]$ for some $\epsilon > 0$, then the limit in Eq. (17.99) exists at $t = t_0$, and furthermore

$$\frac{x(t_0^+) + x(t_0^-)}{2} = \lim_{a\to\infty} \int_{-a}^{a} X(j\omega)e^{j\omega t_0}\,\frac{d\omega}{2\pi}, \tag{17.103}$$

where $x(t_0^-)$ and $x(t_0^+)$ denote the left and right-hand limits at t_0. $\diamond$

Note that in Eq. (17.103), the right-hand side is the Cauchy principal value. For functions which do not necessarily have bounded variation, a result somewhat similar to Theorem 17.1 is true:

Theorem 17.5 Suppose $x(t)$ is Lebesgue integrable in $(-\infty, \infty)$, so that the L_1 Fourier transform $X(j\omega)$ exists. If $x(t_0^+)$ and $x(t_0^-)$ exist, and if the Lebesgue integrals

$$\int_0^\delta \frac{x(t_0+\tau)-x(t_0^+)}{\tau}d\tau \quad \text{and} \quad \int_0^\delta \frac{x(t_0-\tau)-x(t_0^-)}{\tau}d\tau \tag{17.104}$$

exist for some $\delta > 0$, then the limit in Eq. (17.99) exists at $t = t_0$, and Eq. (17.103) holds. ◇

Proofs of the above two theorems can be found in Apostol [1974]. Some remarks are now in order:

1. If $x(t)$ is continuous at t_0, then the left-hand side of Eq. (17.103) reduces to $x(t_0)$.
2. The one-sided limits $x(t_0^-)$ and $x(t_0^+)$ automatically exist because of bounded variation in Theorem 17.4 (see Sec. 17.6.6). But in Theorem 17.5 the existence of $x(t_0^-)$ and $x(t_0^+)$ has to be explicitly assumed.
3. In particular, if $x(t)$ has bounded variation in all finite intervals and is also continuous everywhere, then the familiar inverse transformation formula (17.96) holds for all t, provided the integral on the right is interpreted as the Cauchy principal value.

We conclude this section with one more result. If the L_1-FT $X(j\omega)$ itself is in L_1, then the integral (17.97) exists as a Lebesgue integral. Denote this as

$$x_1(t) = \frac{1}{2\pi}\int_{-\infty}^{\infty} X(j\omega)e^{j\omega t}d\omega. \tag{17.105}$$

Since $X(j\omega)e^{j\omega t}$ is in L_1, the function $x_1(t)$ has properties similar to an L_1-FT. For example, it is continuous and bounded (Sec. 17.4.1). The following result says that this continuous and bounded function $x_1(t)$ represents $x(t)$ almost everywhere (though the L_1 function $x(t)$ may not be continuous). See Theorem 9.11 in Rudin [1974].

Theorem 17.6 *Convergence almost everywhere.* Let $x(t)$ be Lebesgue integrable in $(-\infty, \infty)$, so that the L_1 Fourier transform $X(j\omega)$ exists. Assume further that $X(j\omega)$ is itself an L_1 function. Then the Fourier integral (17.105) exists as a Lebesgue integral, and $x(t) = x_1(t)$ almost everywhere (i.e., for almost all t). ◇

In particular, *if $x(t)$ is continuous for all t and both $x(t)$ and $X(j\omega)$ are Lebesgue integrable, then the inverse transform formula (17.96) holds for all t.*

17.8 Summary of Chapter 17

In this chapter we have reviewed some of the interesting mathematical aspects of Fourier representations which are usually not emphasized in introductory engineering classes. We explained the differences between the ℓ_1 and ℓ_2 Fourier transforms for the discrete-time case, and between the L_1 and L_2 Fourier transforms for the continuous-time case. When the L_1 and L_2 integrals do not exist, the Fourier transform or its inverse can still often exist if the integrals are interpreted as either the Cauchy principal value or an improper Riemann integral.

We also reviewed some sufficient conditions under which the Fourier series of $x(t)$ converges to $x(t)$ (or to $[x(t^+)+x(t^-)]/2$). One of these is bounded variation of $x(t)$, and the other is the existence of one-sided derivatives for $x(t)$. The meanings of terms such as "bounded variation" and "one-sided derivative" were also presented, along with a number of simple examples. Next, we explained how the Fourier series always converges to $[x(t^+) + x(t^-)]/2$ in the sense of a Cesàro sum. We also found that bounded variation of $x(t)$ plays a role in the convergence of the Fourier integral to the value $x(t)$ (or to $[x(t^+) + x(t^-)]/2$).

The material reviewed in this chapter is fundamental and of lasting value. It is good to be aware of these mathematical subtleties, even though from the engineer's viewpoint their importance may not manifest in day-to-day applications of the Fourier transform.

References for this chapter include Apostol [1974], Carslaw [1950], Churchill and Brown [1987], Haaser and Sullivan [1991], Kreyszig [1972] and [Rudin, 1974]. A recent tutorial can be found in Singh et al. [2022].

17.9 Appendix: Lebesgue and Riemann Integrals

In this appendix we make brief remarks on the Lebesgue integral. It will take us too far afield to get into the detailed definitions and properties, so we shall be content making brief remarks which might be useful to the reader. Detailed treatments of this topic can be found in many books [e.g., Apostol, 1974; Haaser and Sullivan, 1991].

When students are introduced to integrals for the first time, they learn the Riemann integral. All the standard theory, properties, and examples they learn in high schools and introductory college-level courses are based on this. But when it comes to certain subtler applications in mathematics, the Riemann integral has some limitations. The Lebesgue integral, introduced in 1902, is defined differently and is mathematically more convenient. For most applications the difference is not noticeable, but when we get into subtleties of certain proofs, the difference matters. For example, one often interchanges infinite sums with infinite integrals, infinite integrals with limits, and so forth. The formal theory which establishes conditions allowing such interchange becomes simpler with the use of the Lebesgue integral. The theory of inverse Fourier transformation, its existence, and so forth, becomes simpler with Lebesgue integration, and so does the question of pointwise convergence of the Fourier series. Some further points of difference between the Lebesgue and Riemann integrals are clear from the following.

1. *Case of finite intervals.* It can be shown that if $x(t)$ is Riemann integrable in a finite interval $[a, b]$, then it is also Lebesgue integrable (and the integrals are

equal). The converse is not true. A classic example of a function which is not Riemann integrable is the Dirichlet function, defined as

$$x(t) = \begin{cases} 1 & \text{if } t \text{ is rational and } 0 \leq t \leq 1, \\ 0 & \text{otherwise.} \end{cases} \tag{17.106}$$

For this function the Lebesgue integral exists, and is $\int_0^1 x(t)dt = 0$ (see pp. 107 and 121 of Haaser and Sullivan [1991]).

2. *Case of bounded continuous functions.* If $x(t)$ is bounded and continuous almost everywhere in a finite interval $[a, b]$, then it is Reimann integrable and hence Lebesgue integrable in $[a, b]$ (see pp. 119 of Haaser and Sullivan [1991]).
3. *Unounded intervals.* On unbounded intervals the situation is different. For example,

$$\int_{-\infty}^{\infty} \frac{\sin t}{t} dt = \pi, \tag{17.107}$$

if the integral is interpreted as the improper Riemann integral. But the function $\sin t/t$ is not Lebesgue integrable (see p. 285 of Apostol [1974]).

4. *If $x(t)$ is Lebesgue integrable, then $|x(t)|$ is also Lebesgue integrable* (see p. 140 of Haaser and Sullivan [1991]). It is well known that this is not true for the Riemann integral. For example, consider the Riemann integral (17.107). We know the absolute value $|\sin t/t|$ cannot be integrated – $\sin t/t$ represents the impulse response of an unstable, ideal, lowpass filter. Neither $\sin t/t$ nor $|\sin t/t|$ can be integrated in $(-\infty, \infty)$ in the Lebesgue sense.
5. *If $|x(t)|$ is Lebesgue integrable, then $x(t)$ is also Lebesgue integrable,* as long as $x(t)$ is a "measureable" function (see p. 141 of Haaser and Sullivan [1991]). All examples we encounter in practice are "measureable," so we will not worry about this restriction. The reader interested in the definition and mathematical subtleties of measurability should refer to Apostol [1974] or Haaser and Sullivan [1991].
6. *L_1 functions and L_1 Fourier transforms.* If $|x(t)|$ is Lebesgue integrable in an interval $[a, b]$, it is called an $L_1[a, b]$ function. The fact that this implies integrability of $x(t)$ also means that $x(t)e^{-j\omega t}$ can be integrated. Thus, for any $L_1(-\infty, \infty)$ function the Fourier transform exists pointwise for all ω. This is called an L_1 Fourier transform (to distinguish it from an L_2-FT; see Sec. 17.4 for a comparison).

18 Review of Matrices*

18.1 Introduction

Even though we make only light use of matrices in most chapters, a review of matrices might be beneficial to the student, especially to study some of the optional sections in some chapters. So we provide an overview of matrices here. We assume that the student would have taken a class on linear algebra, so this review is brief. After introducing basic matrix operations, we define determinants and matrix inverses. The rank and Kruskal rank of matrices are defined and explained. Eigenvalues and eigenvectors are then reviewed. Unitary matrices are also reviewed. Finally, linear equations and existence and uniqueness of their solutions are discussed.

18.2 Definition and Examples

A $K \times M$ matrix is an array of KM numbers. For example, here is a 2×3 matrix:

$$\mathbf{A} = \begin{bmatrix} a_{00} & a_{01} & a_{02} \\ a_{10} & a_{11} & a_{12} \end{bmatrix}. \tag{18.1}$$

The element in the kth row and mth column is called the (k, m)th element and is denoted $[\mathbf{A}]_{km}$. It is also denoted A_{km} or a_{km}, according to convenience. The row index k and column index m have the range $0 \leq k \leq K - 1$ and $0 \leq m \leq M - 1$. (In rare cases we also use the range $1 \leq k \leq K$ and $1 \leq m \leq M$ if it is more convenient.) A matrix with one row is called a row vector and a matrix with one column is a column vector:

$$\text{row vector } \mathbf{r} = \begin{bmatrix} 3 & 4 & 5 \end{bmatrix} \qquad \text{column vector } \mathbf{c} = \begin{bmatrix} 1 \\ 2 \\ 6 \end{bmatrix}. \tag{18.2}$$

Usually the term "vector" without qualifiers means a column vector. Clearly a vector is also a matrix, by definition. If $K = M$ we say that $\mathbf{A}$ is a **square** matrix, and if $K \neq M$ we say $\mathbf{A}$ is **rectangular**. $\mathbf{A}$ is **tall** if $K > M$ and **fat** if $K < M$. When $K = M = 1$ the matrix reduces to a *scalar*.

The elements a_{ii} are called the diagonal elements of $\mathbf{A}$. A square matrix is called a **diagonal matrix** if all nondiagonal elements are zero, as in

$$\mathbf{A} = \begin{bmatrix} 2 & 0 \\ 0 & 4 \end{bmatrix}, \quad \mathbf{B} = \begin{bmatrix} 2 & 0 & 0 \\ 0 & j & 0 \\ 0 & 0 & -1 \end{bmatrix}, \quad \mathbf{I}_3 = \begin{bmatrix} 1 & 0 & 0 \\ 0 & 1 & 0 \\ 0 & 0 & 1 \end{bmatrix}. \tag{18.3}$$

The matrix $\mathbf{I}_3$ is called an **identity matrix**. By definition, an $N \times N$ identity matrix $\mathbf{I}_N$ is a diagonal matrix, with all diagonal elements equal to 1. Thus

$$[\mathbf{I}_N]_{km} = \begin{cases} 0 & k \neq m, \\ 1 & k = m. \end{cases} \tag{18.4}$$

Sometimes we drop the subscript N and use $\mathbf{I}$ for the identity matrix. The zero matrix or **null matrix** has all elements equal to zero, and is denoted $\mathbf{0}$. Matrices of the form (demonstrated for size 3×3)

$$\boldsymbol{\Delta}_l = \begin{bmatrix} a_{00} & 0 & 0 \\ a_{10} & a_{11} & 0 \\ a_{20} & a_{21} & a_{22} \end{bmatrix}, \quad \boldsymbol{\Delta}_u = \begin{bmatrix} a_{00} & a_{01} & a_{02} \\ 0 & a_{11} & a_{12} \\ 0 & 0 & a_{22} \end{bmatrix} \tag{18.5}$$

are said to be triangular. $\boldsymbol{\Delta}_l$ is lower triangular and $\boldsymbol{\Delta}_u$ is upper triangular.

18.3 Working with Matrices and Vectors

We now describe some basic operations that are commonly done on matrices. The transpose $\mathbf{A}^T$ of a matrix $\mathbf{A}$ is defined such that $[\mathbf{A}^T]_{km} = [\mathbf{A}]_{mk}$. For $\mathbf{A}$ in (18.1) we have

$$\mathbf{A}^T = \begin{bmatrix} a_{00} & a_{10} \\ a_{01} & a_{11} \\ a_{02} & a_{12} \end{bmatrix}. \tag{18.6}$$

The notation $\mathbf{A}^*$ stands for the complex conjugate of $\mathbf{A}$, that is, every element a_{km} is conjugated. The transpose conjugate $\mathbf{A}^H$ is defined as $\mathbf{A}^H = [\mathbf{A}^T]^*$. That is, we transpose the matrix and conjugate the elements. For $\mathbf{A}$ in (18.1) we have

$$\mathbf{A}^H = \begin{bmatrix} a_{00}^* & a_{10}^* \\ a_{01}^* & a_{11}^* \\ a_{02}^* & a_{12}^* \end{bmatrix}, \tag{18.7}$$

where $\mathbf{A}^H$ is read as $\mathbf{A}$-Hermitian, and is also called the Hermitian transpose. A matrix satisfying $\mathbf{A}^H = \mathbf{A}$ is said to be **Hermitian**, and a matrix satisfying $\mathbf{A}^T = \mathbf{A}$ is said to be **symmetric**. We often write a column vector as the transpose of a row vector, to save space:

$$\begin{bmatrix} 1 \\ 2 \\ 0 \\ 6 \end{bmatrix} \quad \text{is written as} \quad \begin{bmatrix} 1 & 2 & 0 & 6 \end{bmatrix}^T. \tag{18.8}$$

Next, a submatrix of **A** is a matrix obtained from **A** by deleting some columns and/or rows. For example, here are some submatrices of (18.1):

$$\begin{bmatrix} a_{10} & a_{11} & a_{12} \end{bmatrix}, \begin{bmatrix} a_{11} & a_{12} \end{bmatrix}, [a_{11}], \begin{bmatrix} a_{00} & a_{01} \\ a_{10} & a_{11} \end{bmatrix}, \begin{bmatrix} a_{00} & a_{02} \\ a_{10} & a_{12} \end{bmatrix}, \begin{bmatrix} a_{01} \\ a_{11} \end{bmatrix}, \quad (18.9)$$

and so on. Denoting the columns of **A** as $\mathbf{a}_m$, we sometimes write the matrix as

$$\mathbf{A} = \begin{bmatrix} \mathbf{a}_0 & \mathbf{a}_1 & \cdots & \mathbf{a}_{M-1} \end{bmatrix}. \quad (18.10)$$

Similarly, if the rows of **A** are denoted as $\mathbf{r}_k^H$, we have

$$\mathbf{A} = \begin{bmatrix} \mathbf{r}_0^H \\ \mathbf{r}_1^H \\ \vdots \\ \mathbf{r}_{K-1}^H \end{bmatrix}. \quad (18.11)$$

18.3.1 Addition and Multiplication of Matrices

The sum of two matrices of the same size $K \times M$ is obtained by adding the individual elements. Thus, if $\mathbf{C} = \mathbf{A}+\mathbf{B}$, then $c_{km} = a_{km}+b_{km}$. Matrix multiplication is a more involved operation. Given two matrices **P** and **Q**, the product $\mathbf{R} = \mathbf{PQ}$ is defined only when the number of columns of **P** is equal to the number of rows of **Q**:

$$\mathbf{R} = \underbrace{\mathbf{P}}_{K\times L} \underbrace{\mathbf{Q}}_{L\times M}. \quad (18.12)$$

The product is defined by defining the elements of the matrix **R** as follows:

$$r_{km} = \sum_{l=0}^{L-1} p_{kl} q_{lm}. \quad (18.13)$$

For example,

$$\underbrace{\begin{bmatrix} 1 & 2 \\ 3 & 4 \end{bmatrix}}_{\mathbf{P}} \underbrace{\begin{bmatrix} 1 & 3 & -1 \\ 3 & 1 & 1 \end{bmatrix}}_{\mathbf{Q}} = \begin{bmatrix} 1*1+2*3 & 1*3+2*1 & 1*-1+2*1 \\ 3*1+4*3 & 3*3+4*1 & 3*-1+4*1 \end{bmatrix}$$

$$= \begin{bmatrix} 7 & 5 & 1 \\ 15 & 13 & 1 \end{bmatrix}. \quad (18.14)$$

Note that **QP** does not exist in this example because the number of columns of **Q** (= 3) is not equal to the number of rows of **P** (= 2). Even in cases where **PQ** and **QP** both exist, we do not in general have $\mathbf{PQ} = \mathbf{QP}$. That is, matrix multiplication is **not commutative**. For example, if

$$\mathbf{P} = \begin{bmatrix} 1 & 1 \\ 1 & 1 \end{bmatrix}, \quad \mathbf{Q} = \begin{bmatrix} 1 & 1 \\ 1 & -1 \end{bmatrix}, \quad (18.15)$$

you can verify that $\mathbf{PQ} \neq \mathbf{QP}$. In fact, $\mathbf{PQ}$ and $\mathbf{QP}$ may not even have the same size, as you can see in this example:

$$\mathbf{P} = \begin{bmatrix} 1 & 1 \end{bmatrix}, \ \mathbf{Q} = \begin{bmatrix} 1 \\ 1 \end{bmatrix} \quad \Longrightarrow \quad \mathbf{PQ} = 2, \quad \mathbf{QP} = \begin{bmatrix} 1 & 1 \\ 1 & 1 \end{bmatrix}. \tag{18.16}$$

In this example $\mathbf{PQ}$ is a scalar (1×1 matrix), whereas $\mathbf{QP}$ is a 2×2 matrix.

Matrix-vector products of the form $\mathbf{Pv}$ arise commonly. Here, $\mathbf{P}$ is a matrix and $\mathbf{v}$ is a column vector. The result is a column vector unless $\mathbf{P}$ itself is a row vector, in which case the result is a scalar. Here are some examples:

$$\underbrace{\begin{bmatrix} 1 & 1 \\ 1 & -1 \\ 1 & 2 \end{bmatrix}}_{\mathbf{P}} \underbrace{\begin{bmatrix} 1 \\ 1 \end{bmatrix}}_{\mathbf{v}} = \begin{bmatrix} 2 \\ 0 \\ 3 \end{bmatrix}, \quad \underbrace{\begin{bmatrix} 1 & 1 \end{bmatrix}}_{\mathbf{P}} \underbrace{\begin{bmatrix} 1 \\ 1 \end{bmatrix}}_{\mathbf{v}} = 2, \quad \underbrace{\begin{bmatrix} 1 & -1 \end{bmatrix}}_{\mathbf{P}} \underbrace{\begin{bmatrix} 1 \\ 1 \end{bmatrix}}_{\mathbf{v}} = 0. \tag{18.17}$$

Given a matrix $\mathbf{P}$ and a scalar c, the notation $c\mathbf{P}$ stands for a matrix whose elements are the scaled versions cp_{km}. We say $c\mathbf{P}$ is a *scaled version* of $\mathbf{P}$.

18.3.2 Inner Products, Norms, and Orthogonality

Given two $K \times 1$ column vectors

$$\mathbf{x} = \begin{bmatrix} x_0 \\ x_1 \\ \vdots \\ x_{K-1} \end{bmatrix}, \quad \mathbf{y} = \begin{bmatrix} y_0 \\ y_1 \\ \vdots \\ y_{K-1} \end{bmatrix}, \tag{18.18}$$

the quantity

$$\mathbf{x}^H \mathbf{y} = \sum_{k=0}^{K-1} x_k^* y_k \tag{18.19}$$

is called the inner product of $\mathbf{x}$ with $\mathbf{y}$. Similarly, the inner product of $\mathbf{y}$ with $\mathbf{x}$ is $\mathbf{y}^H\mathbf{x}$. Clearly $\mathbf{y}^H\mathbf{x} = (\mathbf{x}^H\mathbf{y})^*$. The inner product of $\mathbf{x}$ with itself is therefore

$$\mathbf{x}^H \mathbf{x} = \sum_{k=0}^{K-1} x_k^* x_k = \sum_{k=0}^{K-1} |x_k|^2 \geq 0, \tag{18.20}$$

and its square root is called the norm, or more precisely the ℓ_2 norm, of $\mathbf{x}$. The ℓ_2 norm is denoted $\|\mathbf{x}\|_2$, although the subscript is often dropped. Thus

$$\|\mathbf{x}\|^2 = \mathbf{x}^H \mathbf{x} = \text{norm-square of } \mathbf{x}. \tag{18.21}$$

This quantity is also called the **energy** of the vector $\mathbf{x}$. The norm of $\mathbf{x}$ can also be regarded as the "length" of the vector, although this is difficult to visualize geometrically when $\mathbf{x}$ is complex. Two vectors $\mathbf{x}$ and $\mathbf{y}$ are said to be **orthogonal** if their inner product is zero:

$$\mathbf{x}^H \mathbf{y} = 0 \quad \text{(orthogonality)}, \tag{18.22}$$

and a set of vectors $\{\mathbf{v}_0, \mathbf{v}_1, \ldots, \mathbf{v}_{N-1}\}$ is said to be an orthogonal set if every pair of vectors is orthogonal: $\mathbf{v}_k^H \mathbf{v}_m = 0$ for $k \neq m$.

18.3.3 Cauchy–Schwarz Inequality

For two $K \times 1$ vectors $\mathbf{x}$ and $\mathbf{y}$, the Cauchy–Schwarz inequality says that

$$|\mathbf{x}^H \mathbf{y}|^2 \;\leq\; \|\mathbf{x}\|^2 \|\mathbf{y}\|^2, \tag{18.23}$$

or more explicitly,

$$\Big| \sum_{k=0}^{K-1} x_k y_k^* \Big|^2 \;\leq\; \sum_{k=0}^{K-1} |x_k|^2 \sum_{k=0}^{K-1} |y_k|^2, \tag{18.24}$$

with equality if and only if $\mathbf{y} = c\mathbf{x}$ for some scalar c. This can be generalized to infinite-dimensional vectors (i.e., sequences defined over $-\infty < n < \infty$) as follows:

$$\Big| \sum_{n=-\infty}^{\infty} x[n] y^*[n] \Big|^2 \;\leq\; \sum_{n=-\infty}^{\infty} |x[n]|^2 \sum_{n=-\infty}^{\infty} |y[n]|^2, \tag{18.25}$$

with equality if and only if $y[n] = cx[n]$ for some constant c. These can also be generalized to functions of continuous time as follows:

$$\Big| \int_{-\infty}^{\infty} x(t) y^*(t) dt \Big|^2 \;\leq\; \int_{-\infty}^{\infty} |x(t)|^2 dt \int_{-\infty}^{\infty} |y(t)|^2 dt, \tag{18.26}$$

with equality if and only if $y(t) = cx(t)$ for some constant c. The same holds if all infinite integrals are replaced with finite integrals of the form $\int_a^b$.

18.4 Determinant of a Matrix

Given a square matrix $\mathbf{P}$ of size $N \times N$, we define the determinant recursively as follows:

$$\det \mathbf{P} = \sum_{k=0}^{N-1} (-1)^{k+m} p_{km} M_{km}, \tag{18.27}$$

where p_{km} is the (k, m)th element of $\mathbf{P}$ and M_{km} is the determinant of the smaller $(N-1) \times (N-1)$ matrix formed by deleting the kth row and mth column of $\mathbf{P}$. The numbers M_{km} and $(-1)^{k+m} M_{km}$ are, respectively, called the **minor** and **cofactor** of the element p_{km}. It should be clear from definition (18.27) that

$$\det \mathbf{I}_N = 1 \tag{18.28}$$

for any $N \times N$ identity matrix. More generally, the *determinant of a diagonal matrix is the product of its diagonal elements.* The **determinant of a triangular matrix** is also the product of its diagonal elements.

Note that in Eq. (18.27) the summation is over the row index k, with the column index fixed as m. It can be shown that the answer is the same no matter which m we choose. Thus, the determinant can be computed by using any column as a reference. Similarly, we can compute it by using any row as a reference:

$$\det \mathbf{P} = \sum_{m=0}^{N-1} (-1)^{k+m} p_{km} M_{km}, \tag{18.29}$$

and the result can be shown to be the same as Eq. (18.27). Given two square matrices **P** and **Q**, it can be shown that

$$\det(\mathbf{PQ}) = (\det \mathbf{P})(\det \mathbf{Q}). \tag{18.30}$$

It can also be verified that

$$\det \mathbf{P}^T = \det \mathbf{P} \quad \text{and} \quad \det \mathbf{P}^H = (\det \mathbf{P})^*. \tag{18.31}$$

Simple examples of determinants are shown below. Determinants are indicated by using vertical bars, as in the examples below. The vertical bars, should not be confused with absolute values, as determinants can be negative or even complex.

$$\det \begin{bmatrix} p & q \\ r & s \end{bmatrix} = \begin{vmatrix} p & q \\ r & s \end{vmatrix} = ps - qr, \tag{18.32}$$

$$\begin{vmatrix} p_{00} & p_{01} & p_{02} \\ p_{10} & p_{11} & p_{12} \\ p_{20} & p_{21} & p_{22} \end{vmatrix} = p_{00} \underbrace{\begin{vmatrix} p_{11} & p_{12} \\ p_{21} & p_{22} \end{vmatrix}}_{M_{00}} - p_{01} \underbrace{\begin{vmatrix} p_{10} & p_{12} \\ p_{20} & p_{22} \end{vmatrix}}_{M_{01}} + p_{02} \underbrace{\begin{vmatrix} p_{10} & p_{11} \\ p_{20} & p_{21} \end{vmatrix}}_{M_{02}}. \tag{18.33}$$

18.5 Inverse of a Matrix

Let **P** be a square matrix of size $N \times N$. We say **Q** is an inverse of **P** if **Q** is also an $N \times N$ matrix, such that $\mathbf{PQ} = \mathbf{I}_N$. The inverse may or may not exist. But when it exists for a square matrix, it is unique, and furthermore $\mathbf{PQ} = \mathbf{QP} = \mathbf{I}_N$. We denote the inverse as $\mathbf{P}^{-1}$, that is,

$$\mathbf{Q} = \mathbf{P}^{-1} = \text{inverse of } \mathbf{P}. \tag{18.34}$$

From $\mathbf{PQ} = \mathbf{I}_N$ it follows that

$$(\det \mathbf{P})(\det \mathbf{Q}) = 1, \tag{18.35}$$

where we have used Eqs. (18.30) and (18.28). This shows that when the inverse of **P** exists, det **P** is nonzero. There is an important closed-form expression for the inverse of a square matrix **P**, namely,

$$\mathbf{P}^{-1} = \frac{\text{Adj}\, \mathbf{P}}{\det \mathbf{P}}, \tag{18.36}$$

where Adj **P** is also an $N \times N$ matrix like **P**. It is called the **adjugate** of **P**. It is defined as follows:

$$[\text{Adj}\, \mathbf{P}]_{km} = \text{cofactor of } p_{mk}, \tag{18.37}$$

where p_{mk} is the (m, k)th element of $\mathbf{P}$. A square matrix for which $\det \mathbf{P} = 0$ is said to be a **singular** matrix. For such a matrix the inverse does not exist. So, a square matrix is invertible if and only if it is **nonsingular**. For 2×2 matrices, Eq. (18.36) simplifies as follows:

$$\begin{bmatrix} p & q \\ r & s \end{bmatrix}^{-1} = \frac{1}{ps - qr} \begin{bmatrix} s & -q \\ -r & p \end{bmatrix}, \tag{18.38}$$

where we have used Eq. (18.32). For a diagonal matrix with nonzero diagonal elements p_{mm}, the inverse is a diagonal matrix with diagonal elements $1/p_{mm}$.

The concept of an inverse can be generalized to rectangular matrices. Given a $K \times M$ matrix $\mathbf{A}$, if there exists $\mathbf{B}$ such that $\mathbf{BA} = \mathbf{I}_M$, we say $\mathbf{B}$ is a left-inverse of $\mathbf{A}$. Similarly, if there exists $\mathbf{C}$ such that $\mathbf{AC} = \mathbf{I}_K$, we say $\mathbf{C}$ is a right-inverse of $\mathbf{A}$. These inverses do not always exist – it depends on a property called the rank of the matrix $\mathbf{A}$. When such an inverse exists, it may not be unique. When $K \neq M$ it is not possible for *both* the left and right-inverses to exist. At most one can exist. When $K = M$ these two inverses exist if and only if $\det \mathbf{A} \neq 0$, and in this case the two inverses are equal.

18.6 Rank of a Matrix

A set of column vectors $\mathbf{v}_0, \mathbf{v}_1, \ldots, \mathbf{v}_{n-1}$ (with identical size) is said to be **linearly independent** if the equation

$$\sum_{m=0}^{n-1} c_m \mathbf{v}_m = \mathbf{0} \tag{18.39}$$

cannot be satisfied unless $c_m = 0$ for all m. The same definition holds for row vectors. The column-rank of a $K \times M$ matrix $\mathbf{A}$ is defined to be an integer $\rho > 0$ such that there exists a set of ρ linearly independent columns, but no set of n columns is linearly independent if $n > \rho$. The row-rank is similarly defined. It can be shown that *column-rank = row-rank,* so we just call it the **rank** of the matrix. Sometimes we use the notation $\rho(\mathbf{A})$ for the rank of $\mathbf{A}$. From the above discussion it is clear that

$$\rho(\mathbf{A}) = \rho(\mathbf{A}^T) = \rho(\mathbf{A}^H). \tag{18.40}$$

For a $K \times M$ matrix $\mathbf{A}$, it is clear that

$$\rho(\mathbf{A}) \leq \min(K, M). \tag{18.41}$$

If $\rho = M$ we say the matrix has full column-rank (all columns are linearly independent), and if $\rho = K$ we say the matrix has full row-rank (all rows are linearly independent). It should be clear to the reader that if any two columns of $\mathbf{A}$ are identical up to scale (e.g., $\mathbf{a}_1 = c\mathbf{a}_0$), then $\rho(\mathbf{A}) < M$. Similarly, if two rows are identical up to scale, $\rho(\mathbf{A}) < K$.

18.6.1 Basic Properties of Rank

We now state a number of well-known results, the proofs for which can be found in standard books on linear algebra.

1. An $N \times N$ matrix $\mathbf{P}$ has rank N if and only if it is nonsingular, that is, $\det \mathbf{P} \neq 0$. This is also the condition for $\mathbf{P}^{-1}$ to exist (by Eq. (18.36)).
2. A $K \times M$ matrix $\mathbf{A}$ has rank $\rho < M$ *if and only if* there exists a vector $\mathbf{v} \neq \mathbf{0}$ such that

$$\mathbf{Av} = \mathbf{0}. \tag{18.42}$$

   We say that $\mathbf{v}$ annihilates $\mathbf{A}$ from the right. Similarly, a $K \times M$ matrix $\mathbf{A}$ has rank $\rho < K$ if and only if there exists a vector $\mathbf{v} \neq \mathbf{0}$ such that $\mathbf{v}^H\mathbf{A} = \mathbf{0}$. We say that $\mathbf{v}^H$ annihilates $\mathbf{A}$ from the left.
3. *Sylvester's inequality*. Consider a product of two matrices

$$\mathbf{R} = \underbrace{\mathbf{P}}_{K\times L}\underbrace{\mathbf{Q}}_{L\times M}. \tag{18.43}$$

   Then the rank of $\mathbf{R}$ is bounded as follows:

$$\rho(\mathbf{P}) + \rho(\mathbf{Q}) - L \leq \rho(\mathbf{R}) \leq \min\Big(\rho(\mathbf{P}), \rho(\mathbf{Q})\Big). \tag{18.44}$$

18.6.2 The Kruskal Rank

The Kruskal rank of a $K \times M$ matrix $\mathbf{A}$ is the largest integer $\rho_{Kr} > 0$ such that any set of ρ_{Kr} columns is linearly independent. If ρ is the rank of $\mathbf{A}$, it is clear that $\rho \geq \rho_{Kr}$. Let us consider some examples: let

$$\mathbf{A} = \begin{bmatrix} 1 & 2 & 1 \\ 1 & 2 & -1 \end{bmatrix} \quad \text{and} \quad \mathbf{B} = \begin{bmatrix} 1 & 1 & 1 & 0 \\ 1 & -1 & 0 & 1 \\ 1 & 0 & 0 & 0 \end{bmatrix}. \tag{18.45}$$

For $\mathbf{A}$, the rank $\rho = 2$ because the last two columns are linearly independent. But $\rho_{Kr} = 1$ because the first two columns are linearly dependent. For $\mathbf{B}$, the rank $\rho = 3$ because the first three columns are linearly independent. And what is ρ_{Kr}? You can verify that any pair of columns is linearly independent, but the last three columns are not. So $\rho_{Kr} = 2$ only. A slight variation of the Kruskal rank, called "**spark**," is sometimes used. The spark of $\mathbf{A}$ is nothing but $\rho_{Kr} + 1$ (unless $\rho_{Kr} = M$, in which case the spark is determined through some conventions).

Vandermonde matrix. We now consider an illuminating example. In Sec. 5.16 we discussed square Vandermonde matrices. These have the form

$$\mathbf{V} = \begin{bmatrix} 1 & 1 & \cdots & 1 \\ \alpha_1 & \alpha_2 & \cdots & \alpha_N \\ \alpha_1^2 & \alpha_2^2 & \cdots & \alpha_N^2 \\ \vdots & \vdots & \vdots & \vdots \\ \alpha_1^{N-1} & \alpha_2^{N-1} & \cdots & \alpha_N^{N-1} \end{bmatrix}. \tag{18.46}$$

We showed in Sec. 5.16 that such a matrix is nonsingular if and only if the α_i are distinct. Now consider a $K \times M$ Vandermonde matrix with $K < M$, that is, a *fat Vandermonde matrix*, demonstrated below for $K = 3$ and $M = 4$:

$$\mathbf{V}_f = \begin{bmatrix} 1 & 1 & 1 & 1 \\ \alpha_1 & \alpha_2 & \alpha_3 & \alpha_4 \\ \alpha_1^2 & \alpha_2^2 & \alpha_3^2 & \alpha_4^2 \end{bmatrix}. \tag{18.47}$$

It is clear that any set of K columns defines a square Vandermonde matrix which is nonsingular when the α_i are distinct. But any set of $K + 1$ columns is linearly dependent because the rank of the matrix cannot exceed K. This shows that $\rho_{Kr} = K$. Summarizing, a $K \times M$ Vandermonde matrix with $K \leq M$ and distinct α_i has Kruskal rank K.

$\bigtriangledown \bigtriangledown \bigtriangledown$

18.7 Eigenvalues and Eigenvectors

Given an $N \times N$ matrix $\mathbf{A}$ and an $N \times 1$ vector $\mathbf{v}$, we know that $\mathbf{Av}$ is also an $N \times 1$ vector. Suppose $\mathbf{v}$ is such that $\mathbf{v} \neq \mathbf{0}$ and

$$\mathbf{Av} = \lambda \mathbf{v}. \tag{18.48}$$

That is, $\mathbf{Av}$ is nothing but $\mathbf{v}$ itself, except that it is scaled by some scalar λ. Then we say that $\mathbf{v}$ is an eigenvector of $\mathbf{A}$, and λ is the eigenvalue of $\mathbf{A}$ corresponding to $\mathbf{v}$. For example, if

$$\mathbf{A} = \frac{1}{\sqrt{2}} \begin{bmatrix} 1 & 1 \\ 1 & -1 \end{bmatrix}, \tag{18.49}$$

then the eigenvalues are $\lambda_1 = 1$ and $\lambda_2 = -1$, and the corresponding eigenvectors are

$$\mathbf{v}_1 = \begin{bmatrix} 1 \\ \sqrt{2} - 1 \end{bmatrix}, \quad \mathbf{v}_2 = \begin{bmatrix} \sqrt{2} - 1 \\ -1 \end{bmatrix}. \tag{18.50}$$

As we can see from Eq. (18.48), if $\mathbf{v}$ is an eigenvector, then so is $c\mathbf{v}$ for any nonzero scalar c. The scaled version $c\mathbf{v}$ is not considered a different eigenvector from $\mathbf{v}$. Note that by definition, an eigenvector has to be nonzero. Notice also that we can always scale an eigenvector such that it has unit norm, that is, $\mathbf{v}^H\mathbf{v} = \|\mathbf{v}\|^2 = 1$. Finally, note that the eigenvalue can turn out to be zero in some examples. For example,

$$\begin{bmatrix} 1 & 1 \\ 1 & 1 \end{bmatrix} \begin{bmatrix} 1 \\ -1 \end{bmatrix} = \begin{bmatrix} 0 \\ 0 \end{bmatrix} = 0 * \begin{bmatrix} 1 \\ -1 \end{bmatrix}, \tag{18.51}$$

which proves that, for this 2×2 matrix, $\begin{bmatrix} 1 & -1 \end{bmatrix}^T$ is an eigenvector with eigenvalue 0.

18.7.1 Connection to Determinants

Next, for an $N \times N$ matrix $\mathbf{A}$, it can be shown that the quantity

$$D(z) = \det(z\mathbf{I} - \mathbf{A}) \tag{18.52}$$

is a polynomial in z with degree N. So it has N roots $\lambda_k,\ 0 \leq k \leq N-1$. It turns out that these roots are eigenvalues of $\mathbf{A}$ and conversely any eigenvalue of $\mathbf{A}$ is a root of $D(z)$. The polynomial $D(z)$ is called the **characteristic polynomial** of $\mathbf{A}$. Another important relation between determinants and eigenvalues is that

$$\det(\mathbf{A}) = \prod_{k=0}^{N-1} \lambda_k. \tag{18.53}$$

And it turns out that the sum of eigenvalues is equal to the sum of diagonal elements of $\mathbf{A}$. This sum is called the **trace** of $\mathbf{A}$, denoted $\text{Tr}(\mathbf{A})$. Thus

$$\text{Tr}(\mathbf{A}) \triangleq \sum_{k=0}^{N-1} a_{kk} = \sum_{k=0}^{N-1} \lambda_k. \tag{18.54}$$

18.7.2 Further Properties Relating to Eigen Analysis

A number of points should be noted:

1. An $N \times N$ matrix $\mathbf{A}$ always has N eigenvalues (N roots of $D(z)$) but they may not all be distinct. For example, the identity matrix has N eigenvalues, all of them equal to unity. So we say that the eigenvalue is unity with multiplicity N. If a 6×6 matrix has eigenvalues $\{1, 1, 2, 2, 2, 3\}$, then eigenvalue 1 has multiplicity two; eigenvalue 2 has multiplicity three; and eigenvalue 3 has multiplicity one.
2. For an eigenvector $\mathbf{v}$ of $\mathbf{A}$, the corresponding eigenvalue is obviously unique. (Otherwise we would have $\lambda_1\mathbf{v} = \lambda_2\mathbf{v}$ with $\lambda_1 \neq \lambda_2$ and $\mathbf{v} \neq \mathbf{0}$, which is absurd.) But given an eigenvalue, the eigenvector is not necessarily unique. For example, consider the identity matrix $\mathbf{I}_N$: the only eigenvalue is $\lambda = 1$, but any vector is an eigenvector!
3. Let $\mathbf{A}_1 = \mathbf{T}^{-1}\mathbf{A}\mathbf{T}$ for some nonsingular $\mathbf{T}$. This is called a similarity transformation. We say that $\mathbf{A}_1$ is a similarity-transformed version of $\mathbf{A}$. Then $\mathbf{A}_1$ has the same set of N eigenvalues as $\mathbf{A}$.
4. For a triangular matrix, the eigenvalues are equal to the diagonal elements. So a triangular matrix is nonsingular if and only if all the diagonal elements are nonzero.
5. For an $N \times N$ matrix $\mathbf{A}$ with N distinct eigenvalues (i.e., $\lambda_k \neq \lambda_m$ for $k \neq m$), the corresponding set of N eigenvectors is unique (up to scale), and linearly independent. (If the N eigenvalues are not distinct, there may or may not be N linearly independent eigenvectors.)

Here is a proof of the last item above. First we will prove linear independence of the N eigenvectors $\mathbf{v}_i$ corresponding to N distinct λ_i. Suppose they are not linearly independent. Let us say the dependent set of smallest size is $L \leq N$ and without loss of generality assume $\mathbf{v}_0, \mathbf{v}_1, \ldots, \mathbf{v}_{L-1}$ are linearly dependent. Then

$$\sum_{i=0}^{L-1} \alpha_i \mathbf{v}_i = \mathbf{0}, \tag{18.55}$$

for some set of nonzero constants α_i. Premultiplying the above equation by $\mathbf{A}$ and using $\mathbf{A}\mathbf{v}_i = \lambda_i \mathbf{v}_i$, we get

$$\sum_{i=0}^{L-1} \alpha_i \lambda_i \mathbf{v}_i = \mathbf{0}. \tag{18.56}$$

Multiplying Eq. (18.55) by λ_0 and subtracting from Eq. (18.56), we get

$$\sum_{i=1}^{L-1} \alpha_i (\lambda_i - \lambda_0) \mathbf{v}_i = \mathbf{0}. \tag{18.57}$$

Since the eigenvalues are distinct, $\lambda_i - \lambda_0 \neq 0$ in Eq. (18.57). Furthermore, since $\alpha_i \neq 0$, Eq. (18.57) means that the set of $L - 1$ vectors $\mathbf{v}_1, \mathbf{v}_2, \ldots, \mathbf{v}_{L-1}$ is linearly dependent, violating our starting assumption that the dependent set of smallest size is L. This proves that the N eigenvectors are linearly independent.

We next prove that the eigenvector $\mathbf{v}_i$ corresponding to each eigenvalue λ_i is unique up to scale. Assume the contrary. Say there are two eigenvectors $\mathbf{v}_0$ and $\mathbf{u}_0$ for λ_0 such that $\mathbf{u}_0 \neq c\mathbf{v}_0$ for any c. Since we already proved that $\mathbf{v}_0, \mathbf{v}_1, \ldots, \mathbf{v}_{N-1}$ are linearly independent, it follows that $\mathbf{u}_0$ (which has size N) can be expressed as

$$\mathbf{u}_0 = \sum_{i=0}^{N-1} c_i \mathbf{v}_i. \tag{18.58}$$

Premultiplying by $\mathbf{A}$ and using $\mathbf{A}\mathbf{v}_i = \lambda_i \mathbf{v}_i$, we get

$$\lambda_0 \mathbf{u}_0 = \sum_{i=0}^{N-1} \lambda_i c_i \mathbf{v}_i. \tag{18.59}$$

Multiplying Eq. (18.58) by λ_0 and subtracting from Eq. (18.59), we get

$$\mathbf{0} = \sum_{i=1}^{N-1} (\lambda_i - \lambda_0) c_i \mathbf{v}_i. \tag{18.60}$$

Since the $\mathbf{v}_i$ are linearly independent, this shows that $(\lambda_i - \lambda_0)c_i = 0$ for all $i \geq 1$. But since $\lambda_i \neq \lambda_0$ for $i \geq 1$, this means $c_1 = c_2 = \cdots = c_{N-1} = 0$, so $\mathbf{u}_0 = c_0 \mathbf{v}_0$ (from Eq. (18.58)), contradicting the starting assumption that $\mathbf{u}_0$ is not a scaled version of $\mathbf{v}_0$. This proves that each eigenvector is unique up to scale.

18.7.3 Singular Matrix: A Summary

We conclude by summarizing the different equivalent ways of saying that an $N \times N$ matrix $\mathbf{A}$ is singular:

1. $\mathbf{A}$ is singular.
2. $\det \mathbf{A} = 0$.
3. $\mathbf{A}$ does not have an inverse.
4. There is an eigenvalue equal to zero (by Eq. (18.53)).
5. There exists a vector $\mathbf{v} \neq \mathbf{0}$ such that $\mathbf{Av} = \mathbf{0}$.
6. There exists a vector $\mathbf{v} \neq \mathbf{0}$ such that $\mathbf{v}^H\mathbf{A} = \mathbf{0}$.
7. The N columns of $\mathbf{A}$ (equivalently, N rows of $\mathbf{A}$) are linearly dependent.
8. Rank of $\mathbf{A} < N$. We also say $\mathbf{A}$ *is rank-deficient*.

18.8 Unitary Matrices

An $N \times N$ matrix $\mathbf{U}$ is said to be unitary if

$$\mathbf{U}^H\mathbf{U} = \mathbf{I}_N. \tag{18.61}$$

Denoting the columns of $\mathbf{U}$ as $\mathbf{u}_k$, that is,

$$\mathbf{U} = \begin{bmatrix}\mathbf{u}_0 & \mathbf{u}_1 & \cdots & \mathbf{u}_{N-1}\end{bmatrix}, \tag{18.62}$$

we see that the unitary property is equivalent to $\mathbf{u}_k^H\mathbf{u}_m = \delta[k-m]$. That is, the columns are pairwise orthogonal, and each column has unit norm. Equation (18.61) implies that the inverse of a square unitary matrix is

$$\mathbf{U}^{-1} = \mathbf{U}^H. \tag{18.63}$$

Premultiplying both sides by $\mathbf{U}$, we have $\mathbf{I}_N = \mathbf{U}\mathbf{U}^{\mathbf{H}}$. So, for a square unitary matrix

$$\mathbf{U}^H\mathbf{U} = \mathbf{U}\mathbf{U}^{\mathbf{H}} = \mathbf{I}_N. \tag{18.64}$$

Thus, the rows of $\mathbf{U}$ are pairwise orthogonal as well. When a unitary matrix is real, we have $\mathbf{U}^T\mathbf{U} = \mathbf{I}_N$, and $\mathbf{U}$ is also referred to as an **orthogonal** matrix. The matrix $\boldsymbol{\Theta}$ in Eq. (8.53) (also reproduced below in Eq. (18.67)) is an example of a real unitary matrix. An example of an $N \times N$ complex unitary matrix is

$$\mathbf{U} = \mathbf{I} - 2\mathbf{u}\mathbf{u}^H, \tag{18.65}$$

where $\mathbf{u}$ is any column vector with unit norm (i.e., $\mathbf{u}^H\mathbf{u} = 1$). The unitary property is easily verified as follows:

$$\begin{aligned}\mathbf{U}^H\mathbf{U} &= \left(\mathbf{I} - 2\mathbf{u}\mathbf{u}^H\right)\left(\mathbf{I} - 2\mathbf{u}\mathbf{u}^H\right)\\ &= \mathbf{I} - 2\mathbf{u}\mathbf{u}^H - 2\mathbf{u}\mathbf{u}^H + 4\mathbf{u}\mathbf{u}^H\mathbf{u}\mathbf{u}^H\\ &= \mathbf{I} - 2\mathbf{u}\mathbf{u}^H - 2\mathbf{u}\mathbf{u}^H + 4\mathbf{u}\mathbf{u}^H = \mathbf{I}_N.\end{aligned} \tag{18.66}$$

The unitary matrix (18.65) is called the **Householder** matrix. Here are some important properties of any square unitary matrix $\mathbf{U}$:

1. If $\mathbf{y} = \mathbf{Ux}$ then $\mathbf{y}^H\mathbf{y} = \mathbf{x}^H\mathbf{x}$, that is, the norms are equal: $\|\mathbf{y}\| = \|\mathbf{x}\|$. Thus, a unitary matrix preserves the length of a vector. Equivalently, it preserves the energy $\|\mathbf{x}\|^2$ in the vector.

2. Conversely, if a square matrix $\mathbf{A}$ is such that $\|\mathbf{A}\mathbf{x}\| = \|\mathbf{x}\|$ for all $\mathbf{x}$, then it can be shown that $\mathbf{A}$ is unitary.
3. Geometrically, the property $\|\mathbf{y}\| = \|\mathbf{x}\|$ says that a unitary matrix $\mathbf{U}$ merely **rotates** $\mathbf{x}$ in the N-dimensional space to produce $\mathbf{y} = \mathbf{U}\mathbf{x}$. When $\mathbf{U}$ is a real matrix, this is easy to visualize. For example, consider the $N = 2$ case. The real matrix

$$\mathbf{U} = \begin{bmatrix} \cos\theta & -\sin\theta \\ \sin\theta & \cos\theta \end{bmatrix} \tag{18.67}$$

is unitary as it satisfies $\mathbf{U}^T\mathbf{U} = \mathbf{I}$. It can be shown that if $\mathbf{y} = \mathbf{U}\mathbf{x}$ then $\mathbf{y}$ is the rotated version of $\mathbf{x}$, where the rotation is counterclockwise by an angle θ (Problem 8.9).
4. Taking determinants on both sides of $\mathbf{U}^H\mathbf{U} = \mathbf{I}$ and using Eq. (18.31), it follows that $|\det \mathbf{U}| = 1$ for a unitary matrix $\mathbf{U}$.
5. If λ is an eigenvalue of a unitary matrix, then $|\lambda| = 1$. *Proof:* $\mathbf{U}\mathbf{v} = \lambda\mathbf{v}$ implies $\mathbf{v}^H\mathbf{U}^H\mathbf{U}\mathbf{v} = |\lambda|^2\mathbf{v}^H\mathbf{v}$. Using $\mathbf{U}^H\mathbf{U} = \mathbf{I}$ and $\mathbf{v} \neq \mathbf{0}$, it follows that $|\lambda| = 1$ indeed.

For matrices that are not square it is possible to define a restricted form of unitary property, but we shall not discuss it here.

18.9 Solutions to Linear Equations

Consider an equation of the form

$$\mathbf{y} = \mathbf{A}\mathbf{x}, \tag{18.68}$$

where $\mathbf{A}$ is $K \times M$. Suppose we are given $\mathbf{y}$, and want to solve for $\mathbf{x}$ such that Eq. (18.68) is satisfied. The size and rank of $\mathbf{A}$ determine the nature of the solutions:

1. If $K = M$ and $\mathbf{A}$ is nonsingular we can find a unique solution, namely, $\mathbf{x} = \mathbf{A}^{-1}\mathbf{y}$.
2. If $K > M$ then there are more equations (K) than the number of free variables (M). Such equations are said to be *overdetermined*, and we may not be able to get a solution. For example, consider

$$\underbrace{\begin{bmatrix} 1 \\ 1 \\ 1 \end{bmatrix}}_{\mathbf{y}} = \underbrace{\begin{bmatrix} 1 & 0 \\ 0 & 1 \\ 1 & 1 \end{bmatrix}}_{\mathbf{A}} \underbrace{\begin{bmatrix} x_0 \\ x_1 \end{bmatrix}}_{\mathbf{x}}. \tag{18.69}$$

There is no solution $\mathbf{x}$ for this because the last equation $1 = x_0 + x_1$ cannot be satisfied if the first two equations ($x_0 = 1$ and $x_1 = 1$) are satisfied. (More generally, the right-hand side of Eq. (18.68) is a linear combination of columns of $\mathbf{A}$. So, unless $\mathbf{y}$ belongs to the so-called column span of $\mathbf{A}$, there does not exist a solution $\mathbf{x}$.)
3. If $K < M$, there are fewer equations (K) than the number of free variables (M). Such equations are called *underdetermined equations*. In this case, if there exists

one solution $\mathbf{x}$ then there will exist infinitely many solutions. The reason for this is as follows: since $K < M$, the rank of $\mathbf{A}$ is at most K. So the M columns of $\mathbf{A}$ are linearly dependent, so there is a vector $\mathbf{v} \neq \mathbf{0}$ such that $\mathbf{Av} = \mathbf{0}$. For example, if $\mathbf{A}$ is the fat matrix in Eq. (18.45), then

$$\mathbf{A}\underbrace{\begin{bmatrix} 2 \\ -1 \\ 0 \end{bmatrix}}_{\mathbf{v}} = \begin{bmatrix} 1 & 2 & 1 \\ 1 & 2 & -1 \end{bmatrix}\begin{bmatrix} 2 \\ -1 \\ 0 \end{bmatrix} = \begin{bmatrix} 0 \\ 0 \end{bmatrix}. \tag{18.70}$$

So if $\mathbf{x}$ is a solution to $\mathbf{y} = \mathbf{Ax}$, then so is $\mathbf{x} + c\mathbf{v}$ for any c. Summarizing, this is what we can say when $K \neq M$:

$K > M$ (tall $\mathbf{A}$): overdetermined; there may not exist a solution.

$K < M$ (fat $\mathbf{A}$): underdetermined; if $\exists$ a solution, $\exists$ infinitely many.

The above results are important in many signal processing scenarios, although we shall not require them a great deal in this introductory book. However, Chapter 16 shows that these are useful in the context of discussing some unconventional sampling theorems.

18.10 Summary of Chapter 18

We presented a brief overview of matrix algebra. We started with basic definitions and simple matrix operations such as addition, multiplication, transposition, and so forth. Then we defined the determinant of a square matrix and discussed matrix inversion in some detail. The rank of a matrix was then discussed. The connection between rank, determinant, and invertibility was also explained. Then the Kruskal rank of a matrix was defined, along with some examples. Eigenvalues and eigenvectors were defined and a number of properties discussed. Many equivalent meanings of singularity (non-invertibility) of matrices were also summarized. Finally, linear equations were discussed. The conditions under which a solution exists and the condition for the solution to be unique were also explained and demonstrated with examples.

Even though matrices are not used extensively in this book, there are some parts where this review can be helpful. There are many excellent books on matrices [e.g., Horn and Johnson, 1985; Strang, 2016]. A detailed review can also be found in Appendix A of Vaidyanathan [1993].

References

1. Abramowitz, M., and Stegun, I. A. *Handbook of Mathematical Functions*, Dover, New York, 1965.
2. Agarwal, R. C., and Burrus, C. S. "Number theoretic transforms to implement fast digital convolution," *Proceedings of the IEEE*, vol. 63, no. 4, pp. 550–560, 1975.
3. Akansu, A. N., and Haddad, R. A. *Multiresolution Signal Decomposition: Transforms, Subbands, and Wavelets*, Academic Press, New York, 1992.
4. Anderson, B. D. O., and Moore, J. B. *Optimal Filtering*, Prentice Hall, Englewood Cliffs, NJ, 1979.
5. Anderson, B. D. O., and Vongpanitlerd, S. *Network Analysis and Synthesis*, Prentice Hall, Englewood Cliffs, NJ, 1973.
6. Antoniou, A. *Digital Filters: Analysis, Design, and Applications*, McGraw-Hill, New York, 1993.
7. Antoniou, A. *Digital Signal Processing: Signals, Systems, and Filters*, McGraw-Hill, New York, 2006.
8. Apostol, T. M. *Mathematical Analysis*, Addison-Wesley, Reading, MA, 1974.
9. Backus, J. *The Acoustical Foundations of Music*, W. W. Norton & Co., New York, 1977.
10. Baraniuk, R. G. "Compressive sensing: Lecture notes," *IEEE Signal Processing Magazine*, vol. 24, no. 4, pp. 118–124, 2007.
11. Beevers, C. A., and Lipson, H. "A brief history of Fourier methods in crystal-structure determination," *Australian Journal of Physics*, vol. 38, pp. 263–272, 1985.
12. Blahut, R. E. *Fast Algorithms for Digital Signal Processing*, Addison-Wesley, Reading, MA, 1985.
13. Bracewell, R. N. *The Fourier Transform and its Applications*, McGraw-Hill, New York, 1986.
14. Brigham, E. O. *The Fast Fourier Transform and its Applications*, Prentice Hall, Englewood Cliffs, NJ, 1988.
15. Burrus, C. S., Gopinath, R. A., and Guo, H. *Introduction to Wavelets and Wavelet Transforms*, Prentice Hall, Upper Saddle River, NJ, 1998.
16. Candès, E. J., Romberg, J., and Tao, T. "Robust uncertainty principles: Exact signal reconstruction from highly incomplete frequency information," *IEEE Transactions on Information Theory*, vol. 52, no. 2, pp. 489–509, 2006.
17. Candès, E. J., and Tao, T. "Decoding by linear programming," *IEEE Transactions on Information Theory*, vol. 51, no. 12, pp. 4203–4215, 2005.
18. Candès, E. J., and Tao, T. "Near-optimal signal recovery from random projections: Universal encoding strategies?," *IEEE Transactions on Information Theory*, vol. 52, no. 12, pp. 5406–5425, 2006.
19. Candès, E. J., and Wakin, M. B. "An introduction to compressive sampling," *IEEE Signal Processing Magazine*, vol. 25, no. 2, pp. 21–30, 2008.
20. Carslaw, H. S. *Introduction to the Theory of Fourier's Series and Integrals*, Dover, New York, 1950.
21. Castleman, K. R. *Digital Image Processing*, Prentice Hall, Upper Saddle River, NJ, 1996.
22. Chen, C.-T. *Linear System Theory and Design*, Holt, Rinehart & Winston, New York, 1984.
23. Chui, C. K. *An Introduction to Wavelets*, Academic Press, New York, 1992.
24. Churchill, R. V., and Brown, J. W. *Introduction to Complex Variables and Applications*, McGraw-Hill, New York, 1984.

25. Churchill, R. V., and Brown, J. W. *Fourier Series and Boundary Value Problems*, McGraw-Hill, New York, 1987.
26. Constantinides, A. G. "Spectral transformations for digital filters," *Proceedings of the IEE*, vol. 117, no. 8, pp. 1585–1590, 1970.
27. Cooley, J. W., Lewis, P. A. W., and Welch, P. D. "The finite Fourier transform," *IEEE Transactions on Audio Electroacoustics*, vol. AU-17, pp. 77–85, 1969.
28. Cooley, J. W., and Tukey J. W. "An algorithm for the machine computation of complex Fourier series," *Mathematics of Computation*, vol. 19, no. 90, pp. 297–301, 1965.
29. Coppens, P. *X-Ray Charge Densities and Chemical Bonding*, Oxford University Press, Oxford, 1997.
30. Cybenko, G. "Approximation by superpositions of a sigmoidal function," *Mathematics of Control, Signals, and Systems*, vol. 2, pp. 303–314, 1989.
31. Daubechies, I. "Orthonormal bases of compactly supported wavelets," *Communications on Pure and Applied Mathematics,* vol. 4, pp. 909–996, 1988.
32. Daubechies, I. *Ten Lectures on Wavelets*, CBMS Series, SIAM, Philadelphia, PA, 1992.
33. Desoer, C. A., and Schulman, J. D. "Zeros and poles of matrix transfer functions and their dynamic interpretation," *IEEE Transactions on Circuits and Systems*, vol. CAS-21, pp. 1–8, 1974.
34. Deutsch, D. (ed.). *The Psychology of Music*, Academic Press, New York, 1999.
35. Diniz, P. S. R., da Silva, E. A. B., and Netto, S. L. *Digital Signal Processing: System Analysis and Design*, Cambridge University Press, Cambridge, 2010.
36. Donoho, D. L. "Compressed sensing," *IEEE Transactions on Information Theory*, vol. 52, no. 4, pp. 1289–1306, 2006.
37. Donoho, D. L., and Tanner, J. "Sparse nonnegative solution of underdetermined linear equations by linear programming," *Proceedings of the National Academy of Sciences*, vol. 102, no. 27, pp. 9446–9451, 2005.
38. Duarte, M. F., Davenport, M. A., Takhar, D., Laska, J. N., Sun, T., Kelly, K. F., and Baraniuk, R. G. "Single-pixel imaging via compressive sampling," *IEEE Signal Processing Magazine*, vol. 25, no. 2, pp. 83–91, 2008.
39. Elad, M. *Sparse and Redundant Representations*, Springer, New York, 2013.
40. Eldar, Y. C. *Sampling Theory: Beyond Bandlimited Systems*, Cambridge University Press, Cambridge, 2015.
41. Feynman, R. P., Leighton, R. B., and Sands, M. *The Feynman Lectures on Physics*, Vols. 1–3, Addison-Wesley, Reading, MA, 1963.
42. Fletcher, N. H., and Rossing, T. D. *The Physics of Musical Instruments*, Springer, New York, 1998.
43. Gamow, G. *The Great Physicists from Galileo to Einstein*, Dover, New York, 1961.
44. Gold, B., and Rader, C. M. *Digital Processing of Signals*, McGraw-Hill, New York, 1969.
45. Gonzalez, R. C., and Woods, R. E. *Digital Image Processing*, Addison-Wesley, Reading, MA, 1993.
46. Goodfellow, I., Bengio, Y., and Courville, A. *Machine Learning*, MIT Press, Cambridge, MA, 2016.
47. Goodman, J. W. *Introduction to Fourier Optics*, Macmillan, New York, 2017.
48. Graves, A., Mohamed, A., and Hinton, G. "Speech recognition with deep recurrent neural networks," *Proceedings of the IEEE International Conference on Acoustics, Speech, and Signal Processing*, Vancouver, Canada, pp. 6645–6649, May 2013.
49. Gray Jr., A. H. "Passive cascaded lattice digital filters," *IEEE Transactions on Circuits and Systems*, vol. CAS-27, pp. 337–344, 1980.
50. Gray Jr., A. H., and Markel, J. D. "Digital lattice and ladder filter synthesis," *IEEE Transactions on Audio Electroacoustics*, vol. AU-21, pp. 491–500, 1973.
51. Gray Jr., A. H., and Markel, J. D. "A normalized digital filter structure," *Transactions on Acoustics, Speech and Signal Processing,* vol. ASSP-23, pp. 268–277, 1975.

52. Guenther, R. D. *Modern Optics*, Wiley, New York, 1990.
53. Haar, A. "Zur theorie der orthogonalen funktionen-systeme," *Mathematische Annalen*, vol. 69, pp. 331–371, 1910.
54. Haaser, N. B., and Sullivan, J. A. *Real Analysis*, Dover, New York, 1991.
55. Halliday, D., and Resnick, R. *Physics*, Wiley, New York, 1978.
56. Hamming, R. W. *Digital Filters*, Prentice Hall, Englewood Cliffs, NJ, 1989.
57. Harris, F. J. "On the use of windows for harmonic analysis with the discrete Fourier transform," *Proceedings of the IEEE*, vol. 66, no. 1, pp. 51–83, 1978.
58. Haykin, S. *Digital Communications*, Wiley, New York, 1988.
59. Heideman, M. T., Johnson, D. H., and Burrus, C. S. "Gauss and the history of the fast Fourier transform," *IEEE Acoustics, Speech, and Signal Processing Magazine*, vol. 1, no. 4, pp. 14–21, 1984.
60. Hertz, J., Krogh, A., and Palmer, R. G. *Introduction to the Theory of Neural Computation*, Addison-Wesley, Reading, MA, 1991.
61. Hirano, K., Nishimura, S., and Mitra, S. K. "Design of digital notch filters," *IEEE Transactions on Communications*, vol. 22, no. 7, pp. 964–970, 1974.
62. Horn, R. A., and Johnson, C. R. *Matrix Analysis*, Cambridge University Press, Cambridge, 1985.
63. Hornik, K. "Approximation capabilities of multilayer feedforward networks," *Neural Networks*, vol. 4, pp. 251–257, 1991.
64. Hornik, K., Stinchcombe, M., and White, H. "Multilayer feedforward networks are universal approximators," *Neural Networks*, vol. 2, pp. 359–366, 1989.
65. Jackson, L. B. *Signals, Systems, and Transforms*, Addison-Wesley, Reading, MA, 1991.
66. Jackson, L. B. *Digital Filters and Signal Processing*, Kluwer Academic, Boston, MA, 1996.
67. Jury, E. L. "Contribution to the modified z-transform theory," *Journal of the Franklin Institute*, vol. 270, no. 2, pp. 114–129, 1960.
68. Kailath, T. *Linear Systems*, Prentice Hall, Englewood Cliffs, NJ, 1980.
69. Kaiser, J. F. "Nonrecursive digital filter design using the I_0-sinh window function," *Proceedings of the IEEE International Symposium on Circuits and Systems*, pp. 20–23, San Francisco, CA, April 1974.
70. Khajehnejad, M. A., Xu, W., Avestimehr, A. S., and Hassibi, B. "Analyzing weighted ℓ_1 minimization for sparse recovery with nonuniform sparse models," *IEEE Transactions on Signal Processing*, vol. 59, no. 5, pp. 1985–2001, 2011.
71. Kolba, D. P., and Parks, T. W. "A prime factor FFT algorithm using high-speed convolution," *IEEE Transactions on Acoustics, Speech, and Signal Processing*, vol. ASSP-25, no. 4, pp. 281–294, 1977.
72. Kreyszig, E. *Advanced Engineering Mathematics*, Wiley, New York, 1972.
73. Krim, H., Tucker, D., Mallat, S., and Donoho, D. "On denoising and best signal representation," *IEEE Transactions on Information Theory*, vol. 45, no. 7, pp. 2225–2238, 1999.
74. Krishnaswamy, A. "Inflexions and microtonality in South Indian classical music," in *Proceedings of Frontiers of Research on Speech and Music*, Annamalainagar, India, 2004.
75. Lathi, B. P. *Modern Digital and Analog Communication Systems*, Holt, Rinehart & Winston, New York, 1989.
76. Lathi, B. P. *Linear Systems and Signals*, Berkeley-Cambridge Press, Carmichael, CA, 1992.
77. Lee, E. A., and Varaiya, P. *Structure and Interpretation of Signals and Systems*, Addison-Wesley, Boston, MA, 2003.
78. Lewis, A. S., and Knowles, G. "Image compression using the 2-D wavelet transform," *IEEE Transactions on Image Processing*, vol. 1, no. 2, 244–250, 1992.
79. Liu, Z.-M., Zhang, C., and Yu, P. S. "Direction-of-arrival estimation based on deep neural networks with robustness to array imperfections," *IEEE Transactions on Antennas and Propagation*, vol. 66, no. 12, pp. 7315–7327, 2018.

80. Malioutov, D., Cetin, M., and Willsky, A. S. "A sparse signal reconstruction perspective for source localization with sensor arrays," *IEEE Transactions on Signal Processing*, vol. 53, no. 8, 3010–3022, 2005.
81. Mallat, S. "Multiresolution approximations and wavelet orthonormal bases of $L^2(R)$," *Transactions of the American Mathematical Society,* vol. 315, pp. 69–87, 1989.
82. Mallat, S. *A Wavelet Tour of Signal Processing*, Academic Press, San Diego, CA, 1998.
83. Mallat, S. "Understanding deep convolutional networks," *Philosophical Transactions of the Royal Society, Series A*, 2016. https://doi.org/10.1098/rsta.2015.0203
84. Malvar, H. S. *Signal Processing with Lapped Transforms*, Artech House, Norwood, MA, 1992.
85. McClellan, J. H., and Parks, T. W. "Eigenvalue and eigenvector decomposition of the discrete Fourier transform," *IEEE Transactions on Audio and Electroacoustics*, vol. 20, no. 1, pp. 66–74, 1972.
86. McClellan, J. H., and Parks, T. W. "A unified approach to the design of optimum FIR linear-phase digital filters," *IEEE Transactions on Circuit Theory*, vol. CT-20, pp. 697–701, 1973.
87. McClellan, J. H., and Rader, C. M. *Number Theory in Digital Signal Processing*, Prentice Hall, Englewood Cliffs, NJ, 1979.
88. McClellan, J. H., Schafer, R. W., and Yoder, M. A. *Signal Processing First*, Prentice Hall, Upper Saddle River, NJ, 2003.
89. Mishali, M., Eldar, Y. C., and Elron, A. J. "Xampling: Signal acquisition and processing in union of subspaces," *IEEE Transactions on Signal Processing*, vol. 59, no. 10, pp. 4719–4734, 2011.
90. Mitra, S. K. *Digital Signal Processing: A Computer-Based Approach*, McGraw-Hill, New York, 2011.
91. Mitra, S. K. *Signals and Systems*, Oxford University Press, New York, 2015.
92. Mitra, S. K., and Hirano, K. "Digital all-pass networks," *IEEE Transactions on Circuits and Systems*, vol. 21, no. 5, pp. 688–700, 1974.
93. Naylor, A. W., and Sell, G. R. *Linear Operator Theory in Engineering and Science*, Springer-Verlag, New York, 1982.
94. Neuvo, Y., Dong, C.-Y., and Mitra, S. K. "Interpolated finite impulse response filters," *IEEE Transactions on Acoustics, Speech and Signal Processing*, vol. ASSP-32, pp. 563–570, 1984.
95. Nyquist, H. "Certain topics in telegraph transmission theory," *Transactions of the AIEE*, vol. 47, pp. 617–644, 1928.
96. Olson, H. F. *Music, Physics, and Engineering*, Dover, New York, 1967.
97. Oppenheim, A. V., and Schafer, R. W. *Digital Signal Processing*, Prentice Hall, Englewood Cliffs, NJ, 1975.
98. Oppenheim, A. V., and Schafer, R. W. *Discrete-Time Signal Processing*, Prentice Hall, Upper Saddle River, NJ, 2010.
99. Oppenheim, A. V., and Willsky, A. S., with Nawab, S. H. *Signals and Systems*, Prentice Hall, Upper Saddle River, NJ, 1997.
100. Oster, G. "Auditory beats in the brain," *Scientific American*, pp. 94–102, 1973.
101. Papoulis, A. *The Fourier Integral and its Applications*, McGraw-Hill, New York, 1962.
102. Papoulis, A. *Systems and Transforms with Applications in Optics*, McGraw-Hill, New York, 1968.
103. Papoulis, A. "A new algorithm in spectral analysis and band-limited extrapolation," *IEEE Transactions on Circuits and Systems*, vol. 22, no. 9, pp. 735–742, 1975.
104. Papoulis, A. *Signal Analysis*, McGraw-Hill, New York, 1977a.
105. Papoulis, A. "Generalized sampling expansion," *IEEE Transactions on Circuits and Systems*, vol. 24, no. 11, pp. 652–654, 1977b.
106. Papoulis, A. *Circuits and Systems: A Modern Approach*, Holt, Rinehart & Winston, New York, 1980.
107. Peebles Jr., P. Z. *Probability, Random Variables, and Random Signal Principles*, McGraw-Hill, New York, 1987.

108. Pei, S.-C., Guo, B.-Y., and Lu, W.-Y. "Narrowband notch filter using feedback structure," *IEEE Signal Processing Magazine*, vol. 33, no. 3, pp. 115–118, 2016.
109. Pickles, J. O. *An Introduction to the Physiology of Hearing*, 2nd ed., Academic Press, New York, 1988.
110. Poulakis, M. "Metamaterials could solve one of 6G's big problems," *Proceedings of the IEEE*, vol. 110, no. 9, pp. 1151–1158, 2022.
111. Proakis, J. G., and Manolakis, D. G. *Digital Signal Processing: Principles, Algorithms, and Applications*, Prentice Hall, Upper Saddle River, NJ, 2007.
112. Proakis, J. G., and Salehi, M. *Digital Communications*, McGraw-Hill, New York, 2008.
113. Rabiner, L. R., and Gold, B. *Theory and Application of Digital Signal Processing*, Prentice Hall, Englewood Cliffs, NJ, 1975.
114. Rader, C. M. "Discrete Fourier transforms when the number of data samples is prime," *Proceedings of the IEEE*, vol. 56, no. 6, pp. 1107–1108, 1968.
115. Ragazzini, J. R., and Zadeh, L. A. "The analysis of sampled-data systems," *Transactions of the American Institute of Electrical Engineers, Part II: Applications and Industry*, vol. 71, no. 5, pp. 225–234, 1952.
116. Raman, C. V. "The Indian musical drum," *Proceedings of the Indian Academy of Sciences*, vol. 1, 1934.
117. Rasch, R., and Plomp, R. "The perception of musical tones," in D. Deutsch (ed.), *The Psychology of Music*, Academic Press, New York, 1999.
118. Regalia, P. A., Mitra, S. K., and Vaidyanathan, P. P. "The digital all-pass filter: A versatile signal processing building block," *Proceedings of the IEEE*, vol. 76, no. 1, pp. 19–37, 1988.
119. Rioul, O., and Vetterli, M. "Wavelets and signal processing," *IEEE Signal Processing Magazine*, pp. 14–38, 1991.
120. Risset, J-C., and Wessel, D. L. "Exploration of timbre by analysis and synthesis," in D. Deutsch (ed.), *The Psychology of Music*, Academic Press, New York, 1999.
121. Rudin, W. *Real and Complex Analysis*, McGraw-Hill, New York, 1974.
122. Sambamurthy, P. *South Indian Music*, The Indian Music Publishing House, Madras, 1963.
123. Saramäki, T. "On the design of digital filters as a sum of two allpass filters," *IEEE Transactions on Circuits and Systems*, vol. CAS-32, pp. 1191–1193, 1985.
124. Sauer, K. D., and Allebach, J. P. "Iterative reconstruction of band-limited images from nonuniformly spaced samples," *IEEE Transactions on Circuits and Systems*, vol. CAS-34, pp. 1497–1506, 1987.
125. Sayed, A. H. *Inference and Learning from Data*, vols. 1–3. Cambridge University Press, Cambridge, 2023.
126. Sayood, K. *Introduction to Data Compression*, Morgan Kaufmann, San Francisco, CA, 2000.
127. Schoenberg, I. J. *Cardinal Spline Interpolation*, SIAM, Philadelphia, PA, 1973.
128. Shalev-Shwartz, S., and Ben-David, S. *Understanding Machine Learning: From Theory to Algorithms*, Cambridge University Press, Cambridge, 2014.
129. Shannon, C. E. "A mathematical theory of communication," *The Bell System Technical Journal*, vol. 27, pp. 379–423, 1948.
130. Shannon, C. E. "Communications in the presence of noise," *Proceedings of the IRE*, vol. 37, pp. 10–21, 1949.
131. Shiung, D. "A trick for designing composite filters with sharp transition bands and highly suppressed stopbands," *IEEE Signal Processing Magazine*, vol. 39, no. 5, pp. 70–76, 2022.
132. Singh, P., Singhal, A., Fatimah, B., Gupta, A., and Joshi, S. D. "Proper definitions of Dirichlet conditions and convergence of Fourier representations," *IEEE Signal Processing Magazine*, vol. 39, no. 5, pp. 77–84, 2022.
133. Stout, G. H., and Jensen, L. H. *X-Ray Structure Determination: A Practical Guide*, Wiley, New York, 1989.

134. Strang, G. *Introduction to Linear Algebra*, Wellesley-Cambridge Press, Wellesley, MA, 2016.
135. Strang, G., and Nguyen, T. *Wavelets and Filter Banks*, Wellesley-Cambridge Press, Wellesley, MA, 1996.
136. Tolstov, G. P. *Fourier Series*, Dover, New York, 1962.
137. Treichler, J. R., Fijalkow, I., and Johnson, C. R. "Fractionally spaced equalizers," *IEEE Signal Processing Magazine*, vol. 13, no. 3, pp. 65–81, 1996.
138. Tropp, J. A., and Gilbert, A. C. "Signal recovery from random measurements via orthogonal matching pursuit," *IEEE Transactions on Information Theory*, vol. 53, no. 12, pp. 4655–4666, 2007.
139. Tropp, J. A., and Wright, S. J. "Computational methods for sparse solution of linear inverse problems," *Proceedings of the IEEE*, vol. 98, no. 6, 948–958, 2010.
140. Unser, M. "Sampling – 50 years after Shannon," *Proceedings of the IEEE*, vol. 88, no. 4, pp. 569–587, 2000.
141. Unser, M., Aldroubi, A., and Eden, M. "A family of polynomial spline wavelet transforms," *Signal Processing*, vol. 30, no. 2, pp. 141–162, 1993.
142. Vaidyanathan, P. P. "Theory and design of M-channel maximally decimated quadrature mirror filters with arbitrary M, having perfect reconstruction property," *IEEE Transactions on Acoustics, Speech and Signal Processing*, vol. ASSP-35, pp. 476–492, 1987a.
143. Vaidyanathan, P. P. "Quadrature mirror filter banks, M-band extensions and perfect-reconstruction techniques," *IEEE ASSP Magazine*, vol. 4, pp. 4–20, 1987b.
144. Vaidyanathan, P. P. "On predicting a band-limited signal based on past sample values," *Proceedings of the IEEE*, vol. 75, pp. 1125–1127, 1987c.
145. Vaidyanathan, P. P. *Multirate Systems and Filter Banks*, Prentice Hall, Englewood Cliffs, NJ, 1993. Available at www.systems.caltech.edu/dsp/ or https://authors.library.caltech.edu/records/rmhds-22q28.
146. Vaidyanathan, P. P. "Homogeneous time-invariant systems," *IEEE Signal Processing Letters*, vol. 6, no. 4, 76–77, 1999.
147. Vaidyanathan, P. P. "Generalizations of the sampling theorem: Seven decades after Nyquist," *IEEE Transactions on Circuits and Systems I: Fundamental Theory and Applications*, vol. 48, no. 9, 1094–1109, 2001.
148. Vaidyanathan, P. P., and Chen, T. "Role of anticausal inverses in multirate filter-banks – Part I: System-theoretic fundamentals," *IEEE Transactions on Signal Processing*, vol. 43, pp. 1090–1102, 1995a.
149. Vaidyanathan, P. P., and Chen, T. "Role of anticausal inverses in multirate filter-banks – Part II: The FIR case, factorizations, and biorthogonal lapped transforms," *IEEE Transactions on Signal Processing*, vol. 43, pp. 1103–1115, 1995b.
150. Vaidyanathan, P. P., and Chen, T. "Structures for anticausal inverses and application in multirate filter banks," *IEEE Transactions on Signal Processing*, vol. 46, pp. 507–514, 1998.
151. Vaidyanathan, P. P., and Liu, V. C. "Classical sampling theorems in the context of multirate and polyphase digital filter bank structures," *IEEE Transactions on Acoustics, Speech and Signal Processing*, vol. ASSP-36, pp. 1480–1495, 1988.
152. Vaidyanathan, P. P., and Liu, V. C. "Efficient reconstruction of bandlimited sequences from nonuniformly decimated versions by use of polyphase filter banks," *IEEE Transactions on Acoustics, Speech and Signal Processing*, vol. ASSP-38, pp. 1927–1936, 1990.
153. Vaidyanathan, P. P., and Mitra, S. K. "On the construction of a digital transfer function from its real part on unit circle," *Proceedings of the IEEE*, vol. 70, pp. 198–199, 1982.
154. Vaidyanathan, P. P., Mitra, S. K., and Neuvo, Y. "A new approach to the realization of low sensitivity IIR digital filters," *IEEE Transactions on Acoustics,*

Speech and Signal Processing, vol. ASSP-34, pp. 350–361, 1986.
155. Vaidyanathan, P. P., Phoong, S.-M., and Lin, Y.-P. *Signal Processing and Optimization for Transceiver Systems*, Cambridge University Press, Cambridge, 2010.
156. Vaidyanathan, P. P., Regalia, P., and Mitra, S. K. “Design of doubly complementary IIR digital filters using a single complex allpass filter, with multirate applications,” *IEEE Transactions on Circuits and Systems*, vol. CAS-34, pp. 378–389, 1987.
157. Vaidyanathan, P. P., and Tenneti, S. “Srinivasa Ramanujan and signal-processing problems,” *Philosophical Transactions of the Royal Society, Series A*, vol. 378, no. 2163, 2019.
158. Vetterli, M., and Herley, C. “Wavelets and filter banks,” *IEEE Transactions on Signal Processing*, vol. SP-40, 1992.
159. Vetterli, M., and Kovačević, J. *Wavelets and Subband Coding*, Prentice Hall, Englewood Cliffs, NJ, 1995.
160. Vetterli, M., Kovačević, J., and Goyal, V. K. *Foundations of Signal Processing*, Cambridge University Press, Cambridge, 2014.
161. Vetterli, M., Marziliano, P., and Blu, T. “Sampling signals with finite rate of innovation,” *IEEE Transactions on Signal Processing*, vol. 50, no. 6, pp. 1417–1428, 2002.
162. Watson, J. D. *The Double Helix*, Simon & Schuster, New York, 1968.
163. Whittaker, J. M. “The Fourier theory of the cardinal functions,” *Proceedings of the Mathematical Society Edinburgh*, vol. 1, pp. 169–176, 1929.
164. Winograd, S. “On computing the discrete Fourier transform,” *Mathematics of Computation*, vol. 32, no. 141, pp. 175–199, 1978.
165. Zwicker, E., and Fastl, H. *Psychoacoustics: Facts and Models*, Springer, New York, 1990.
166. Zygmund, A. *Trigonometric Series*, vols. 1, 2, Cambridge University Press, London, 1968.

Index